Intermediate Mechanics of Materials

SOLID MECHANICS AND ITS APPLICATIONS

Volume 175



*Aims and Scope of the Series*

The fundamental questions arising in mechanics are: *Why?*, *How?*, and *How much?* The aim of this series is to provide lucid accounts written by authoritative researchers giving vision and insight in answering these questions on the subject of mechanics as it relates to solids.

The scope of the series covers the entire spectrum of solid mechanics. Thus it includes the foundation of mechanics; variational formulations; computational mechanics; statics, kinematics and dynamics of rigid and elastic bodies: vibrations of solids and structures; dynamical systems and chaos; the theories of elasticity, plasticity and viscoelasticity; composite materials; rods, beams, shells and membranes; structural control and stability; soils, rocks and geomechanics; fracture; tribology; experimental mechanics; biomechanics and machine design.

The median level of presentation is the first year graduate student. Some texts are monographs defining the current state of the field; others are accessible to final year undergraduates; but essentially the emphasis is on readability and clarity.

For other titles published in this series, go to
www.springer.com/series/6557

J.R. Barber

# Intermediate Mechanics of Materials

J.R. Barber
University of Michigan
Department of Mechanical Engineering
2350 Hayward Street
48109-2125 Ann Arbor
Michigan
USA
jbarber@umich.edu

This is a revised and updated second edition of Intermediate Mechanics of Materials, McGraw-Hill, 2000.

ISSN 0925-0042
ISBN 978-94-007-3416-6 ISBN 978-94-007-0295-0 (eBook)
DOI 10.1007/978-94-007-0295-0
Springer Dordrecht Heidelberg London New York



Printed on acid-free paper

Springer is part of Springer Science+Business Media (www.springer.com)

# Contents

# Preface

Most engineering students first encounter the subject of mechanics of materials in a course covering the concepts of stress and strain and the elementary theories of axial loading, torsion, bending and shear. There is broad agreement as to the content of such courses, there are many excellent textbooks and it is easy to motivate the students by using simple examples with obvious engineering relevance.

The second course in the subject presents considerably more challenge to the instructor. There is a very wide range of possible topics and different selections will be made (for example) by civil engineers and mechanical engineers. The concepts tend to be more subtle and the examples more complex making it harder to motivate the students, to whom the subject may appear merely as an intellectual excercise. Existing second level texts are frequently pitched at too high an intellectual level for students, many of whom will still have a rather imperfect grasp of the fundamental concepts.

Most undergraduate students are looking ahead to a career in industry, where they will use the methods of mechanics of materials in design. Many will get a foretaste of this process in a capstone design project and this provides an excellent vehicle for motivating the subject. In mechanical or aerospace engineering, the second course in mechanics of materials will often be an elective, taken predominantly by students with a design concentration. It is therefore essential to place emphasis on the way the material is used in design.

Mechanical design typically involves an initial conceptual stage during which many options are considered. During this phase, quick approximate analytical methods are crucial in determining which of the initial proposals are feasible. The ideal would be to get within $\pm 30\%$ with a few lines of calculation. The designer also needs to develop experience as to the kinds of features in the geometry or the loading that are most likely to lead to critical conditions. With this in mind, I try wherever possible to give a physical and even an intuitive interpretation to the problems under investigation. For example, students are encouraged to estimate the location of weak and strong bending axes and the resulting neutral axis of bending by eye and methods are discussed for getting good accuracy with a simple one degree of freedom

Rayleigh-Ritz approximation. Students are also encouraged to develop a feeling for the mode of deformation of engineering components by performing simple experiments in their outside environment, for example, estimating the radius to which an initially straight bar can be bent without producing permanent deformation, or convincing themselves of the dramatic difference between torsional and bending stiffness for a thin-walled open beam section by trying to bend and then twist a structural steel beam by hand-applied loads at the ends.

In choosing dimensions for mechanical components, designers will expect to be guided by criteria of minimum weight, which with elementary calculations, often leads to a thin-walled structure as the optimal solution. This demands that students be introduced to the limits imposed by elastic instability. Some emphasis is also placed on the effect of manufacturing errors on such highly-designed structures — for example, the effect of load misalignment on a beam with a large ratio between principal stiffnesses and the large magnification of initial alignment or loading errors in a column below, but not too far below the buckling load.

No modern text of mechanics on materials would be complete without a discussion of the finite element method. However, students and even some instructors are often confused as to the respective rôles played by analytical and numerical methods in engineering practice. Numerical methods provide accurate solutions for complex practical problems, but the results are specific to the geometry and loading modelled and the solution involves a significant amount of programming effort. By contrast, analytical methods may be very idealized and hence approximate, but they are often quick to apply and they provide generality, permitting a whole family of designs to be compared or even optimized.

The traditional approach to mechanics is to define the basic concepts, derive a general theory and then illustrate its application in a variety of examples. As a student and later as a practising engineer, I have never felt comfortable with this approach, because it is impossible to understand the nuances of the definitions or the general treatment until after they are seen in examples which are simple enough for the mathematics and physics to be transparent. Over the years, I have therefore developed rather untraditional ways of proving and explaining things, relying heavily on simple examples during the derivation process and using only the bare minimum of specialist terminology. I try to avoid presenting to the student anything which he or she cannot reasonably be expected to understand fully *now*.

The problems provided at the end of each chapter range from routine applications of standard methods to more challenging problems. Particularly lengthy or challenging problems are identified by an asterisk. The solution manual to accompany this book is prepared to the same level of detail as the example problems in the text and in many cases introduces additional discussion. It is available to *bona fide* instructors on application to the author at jbarber@umich.edu. Answers to even-numbered problems are provided in Appendix D.

This book evolved out of a set of notes that I wrote for a second-level course at the University of Michigan and the resulting interaction with my students and colleagues has played a crucial rôle in the development of my thinking about the subject. Special thanks go to Przemislaw Zagrodzki of Warsaw University of Technology and Raytech Composites Inc. for his invaluable help with the appendix on finite element methods. I also wish to thank the many people who have made suggestions for improvements and corrections to the first edition which I have incorporated wherever possible.

J.R.Barber
Ann Arbor
2010

[illegible] has played [illegible] in [illegible] of my thinking about the subject. Special thanks [illegible] Polytech Company [illegible] methods. I also [illegible] improvements and corrections to the first edition [illegible] possible.

[illegible]

# 1

# Introduction

Mechanics of materials is principally concerned with the determination of the stresses and deformation of engineering devices and structures. Stresses are important because, in combination with fundamental tests on engineering materials, they enable us to determine the loading conditions under which material failure will occur. Deformations may also be important insofar as they affect the kinematic function of a device.

As engineers, the principal use we make of such calculations is to provide guidance in the design process. Typically, we may perform a mechanics of materials calculation to determine the minimum dimension for a component if it is not to fail under the expected loading or, more generally, to choose between several competing designs on the basis of strength or weight. To ask the right questions in mechanics of materials, it is therefore important to have some understanding of the process of engineering design.

## 1.1 The engineering design process

The engineering designer typically starts with a well-defined specification that the proposed device is required to meet, but no very clear idea of the form it is to take. The first step is therefore to generate a large number of potential solutions to the problem in an initial conceptual phase. This is a creative process and it is advantageous to give the imagination free rein by refraining deliberately from eliminating possible solutions based on any criteria other than the fundamental function of the device.

Once a collection of potential concepts has been established, the choice can be narrowed down by evaluative arguments involving estimates of cost, weight, maximum stress, manufacturability etc. For a complex device, these arguments will initially only concern major characteristics of the design, details being left for determination at a later stage. Towards the end of the process, the choices will have been narrowed down to one or two only, and calculations will be performed to determine appropriate dimensions for the components.

J.R. Barber, *Intermediate Mechanics of Materials*, Solid Mechanics and Its Applications 175, 2nd ed., DOI 10.1007/978-94-007-0295-0_1, © Springer Science+Business Media B.V. 2011

**Economics of design calculations**

It is important to recognize that, even when a design is more or less finalized, it is seldom possible to perform detailed stress calculations for all the components because of time and cost constraints. For example, a finite element calculation of the stresses in a fairly simple three-dimensional component might involve two days work for an experienced analyst and cost $2000, including salary, computing services and overhead (indirect) costs. At this rate, the stress analysis alone for a fairly modest device might cost around $50,000. In addition, we shall probably need to perform other design calculations associated with dynamics, hydraulic or electrical control, lubrication, selection of bearings, thermal effects, materials selection, etc. If the device is to market for $100, these costs would be prohibitive unless we expect to sell at least 100,000 units. Even when the costs are acceptable, the time required to perform the analysis may place an unreasonable delay between the initial concept stage and the device coming into production.

Clearly, the size of the production run is a critical parameter in deciding how much time and money can be invested in design calculations. For this reason, some of the simplest and most common artefacts are the subject of intensive and sophisticated design and manufacturing studies. An example is the formed aluminium beverage can, for which modest cost reductions (typically by a reduction in the amount of aluminium required) have led to significant savings because of the millions of items purchased each year.

## 1.2 Design optimization

Design can be seen as a complex optimization process. In the broadest terms we wish to find the design which meets a specified function at minimum cost.[1] The functional requirements will often be defined as *constraints* which take the form of inequalities.[2] For example, we want a design that will not fail in service, so the stresses must be below some failure threshold, but we don't mind it being too strong, provided the cost is not thereby increased. However, reducing the size and hence the weight of a component will typically reduce the cost and increase the maximum stress, so the optimal design will often be one that is limited by a maximum stress criterion.

It is tempting to generalize this argument and to seek a design in which every point in the body is at its maximum permissible stress under the most severe service conditions.[3] However, the diversity of function in mechanical engineering components generally makes it impossible to do this. Consider, for example, the simple gearbox shaft shown in Figure 1.1 *(a)*. The shaft is supported in bearings at $A, C$ and it carries a gear at $B$ and a coupling to transmit torque to another component at $D$.

[1] Notice that, as remarked above, the costs of developing the design must themselves be figured into this calculation, which threatens to make the argument circular!

[2] Constraints involving inequalities are known to as *unilateral constraints*.

[3] Some structural optimization methods use an algorithm not too different from this.

The gear force will be equally shared between the two bearings due to symmetry and the bending moment will be a maximum at $B$, falling to zero at $A$ and $C$. The transmitted torque is constant between $B$ and $D$, but zero in $AB$. Because the shaft transmits only a shear force at $A$, a strictly optimal design would require only a very small local diameter there, but bearing design considerations will probably require a larger diameter. It is superficially tempting to save material by choosing a design like that of Figure 1.1 *(b)*, but the changes in section will themselves introduce locally high stresses due to stress concentration effects and the forming or machining costs will probably more than outweigh any saving in material costs due to the reduced weight. Also, the reduced stiffness of the design 1.1 *(b)* will lower the natural frequency of vibration of the shaft and if this falls in the operating speed range there is a danger of excessive noise and accelerated fatigue failure.

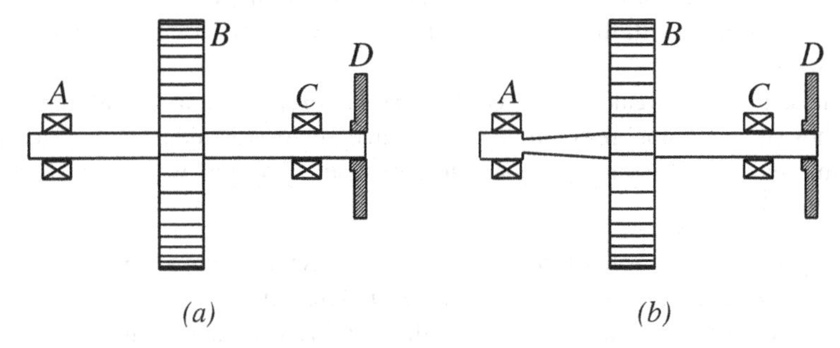


*Figure 1.1: (a) Geometry of gearbox shaft; (b) 'optimal' design based on elementary bending and torsion calculations*

More generally, the maximum stresses reached are only one of many criteria to be satisfied by a successful design and this fact plus considerations of simplicity in manufacture and assembly will usually result in a design for which the maximum acceptable stress level is reached only at certain critical locations. If we can identify these locations, we can then also save time and money in stress analysis by concentrating attention on the critical points. Details of the geometry and loading distant from these points can be safely approximated by idealized shapes and distributions. Thus, the two crucial questions to ask at the beginning of a mechanics of materials problem are:-

(i) How and where is the component or device most likely to fail?
(ii) Can we estimate the stresses and/or displacements in critical regions by a simple calculation (in particular, without also calculating the stresses and displacements everywhere else in the component)?

### 1.2.1 Predicting the behaviour of the component

The experienced engineering designer can look at a drawing or a conceptual sketch of a proposed device and predict with high reliability the problems which are likely to be encountered in realizing the design without doing any calculations at all. What is less obvious is that we all have this ability in some degree, because we have been performing 'experiments' on engineering components continuously in our everyday 'non-engineering' life. Every time we sit in a chair, lean against a tree, break a cup accidentally or blow up a balloon we are gaining experience about the strength and the stiffness of mechanical devices.

Suppose you were confronted with a set of chairs made of diverse materials and in different, perhaps rather bizarre, designs (You might include for example a cardboard carton standing on end, a piece of wood supported by pieces of string or thin strips of rubber, or a bed sheet stretched between two supports). Without touching any of the chairs, ask yourself the following questions:-

- Which ones are strong enough to support your weight without breaking?
- Which will deflect or deform the most when you sit on them?
- About how much will they deform — i.e. what will they look like when you are sitting on them?
- Of those that will break, where will the fracture or other mode of failure occur?
- Of those that will not break, about how much extra load would be needed to make them break (e.g. if someone is standing on your shoulders when you sit down, will they break)?

Most people will be able to answer these questions with quite a high degree of reliability with no engineering training whatsoever. After all, the ability to answer them is almost a condition of making one's way through life without continually breaking things or falling over! Engineering knowledge can even impose a cultural block between this experiential knowledge and the problem. If you ask a non-engineer how much weight could be supported by a cardboard carton before it collapses, he/she, having no other source of information, will give an experience-based estimate. By contrast, the engineer is likely to spend a few hours searching in the library for calculation methods for rectangular plates and may even, as a result of inappropriate assumptions or erroneous calculations, give a totally implausible answer.

The successful design engineer is one who is able to use intuition and experience as complementary to calculation in predicting performance. This skill can be cultivated by practice. Draw sketches of the proposed device from various perspectives and then try to visualize how it will behave under loading. Try to develop the ability to run an imaginary video of the loading process in the mind's eye and note in particular which points have excessive deflection and where fracture or permanent deformation occurs. You can improve this skill by performing simple experiments on (disposable!) objects like packaging materials, disposable cups, beverage cans etc. More complex 'components' can be made out of cardboard and paper.

Non-destructive tests can be performed on a wider range of objects to determine the mode of deformation and the stiffness (i.e. the relation between force and displacement). For example, push against a flat plate structure like a window pane or the side of a filing cabinet, or sit on the hood of your car to see how far it deflects downwards as a result of the suspension spring compression.

In each case, guess what will happen before you carry out the test. The purpose of the excercise is not to find the answer to the problem, but rather to improve your ability to guess reasonably accurately. You will be surprised how often your guess is close to the truth.

The moral of this rather laboured discussion is 'Make a conscious effort to unlock the door between your engineering thinking and your everyday experience'. You may already have several years of engineering training, but you have a lifetime of hands-on experience of the physical world.

### 1.2.2 Approximate solutions

The techniques described above can be used with some reliability to identify the modes of deformation and failure of a loaded structure and hence determine a limited number of critical locations for which stress calculations are needed. The accuracy of numerical estimates of failure loads etc. are less certain, unless one has a wide experience of structures similar to that under consideration. At this stage, we therefore generally wish to perform some mechanics of materials calculations. However, probably we are still considering a number of competing designs, so economic considerations favour the use of idealized or approximate calculations wherever possible. In this context, a method that gives an accuracy of $\pm 30\%$ with 4 lines of calculation is very much more useful than one that takes 4 pages of calculation to reach an accuracy of $\pm 5\%$. You should keep this in mind throughout the rest of this book. Often the motivation for performing a more accurate (and hence extended) calculation is to establish the accuracy of the simpler procedure for future reference, or to determine the range of parameters for which a simpler theory is sufficiently accurate. For example, how 'curved' does a curved beam need to be before we need to take the curvature into account, rather than assuming that the elementary straight beam theory gives an adequate approximation to the stress distribution.[4]

It is worth remarking that calculations of high accuracy are needed only when the component in question is expected to operate fairly close to the maximum permissible load. An approximate calculation is often sufficient to show either (i) that even with a large overload, the component is far from the failure condition and hence the design is not critical, or at the other extreme (ii) that the design condition is so far above the failure load that the concept is not feasible without a major redesign. For cases that fall between these extremes, we may be able to achieve an adequate factor of safety by minor changes of geometry or material, but these are considerations that will come into play only at a later stage of detail design, when the number of options has been reduced.

## 1.3 Relative magnitude of different effects

In many engineering components, stresses and displacements may result from several different effects. For example, the automotive brake disc of Figure 1.2 may experience

[4] This question is addressed in Chapter 11, §11.4.

(i) surface loading (normal pressure and tangential (frictional) tractions) at the brake pads,
(ii) bending due to unequal loading on the two sides of the disc balanced by support loads at the disc centre,
(iii) self-weight of the disc,
(iv) loading associated with the centripetal acceleration during rotation,
(v) surface tractions due to air resistance (drag),
(vi) stresses associated with non-uniform temperatures due to frictional heating.

*Figure 1.2: Automotive disc brake*

A full analysis of all these effects threatens to make the design of even the simplest component a major project. However, fortunately the stresses from one source will usually dominate all the others, permitting them to be neglected. This important principle is one of the few known exceptions to Murphy's law.[5] It is used continually by any experienced design engineer, but as far as I am aware has never been formally enunciated. I therefore stake a tongue-in-cheek claim to immortality by defining

**Barber's exception**

*If a component is subjected to several different kinds of loading process, the relative magnitude of the resulting stresses will be such that only one of the processes needs to be taken into account. The stresses from the remaining processes will usually be within the range of uncertainty associated with this dominant process.*

Furthermore, the difference in magnitudes is often so large that the simplest of estimates is sufficient to establish which loading process is dominant. If $A = 1000 \times B$, estimates of $A$ and $B$ which might be in error by as much as a factor of 5 are still sufficient to establish that $A \gg B$.

[5] In the formulation that everything will generally happen for the worst.

**Example 1.1**

To illustrate this procedure, consider the various kinds of loading experienced by the disc brake of Figure 1.2 and enumerated above. The disc will probably be made of steel or cast iron, for which representative material properties are yield stress $S_Y \approx 30-60\times 10^3$ psi, Young's modulus $E=30\times 10^6$ psi, coefficient of thermal expansion $\alpha=7\times 10^{-6}$ per °F, density $\rho=0.284$ lbs per in$^3$.

(i) The hydraulic system will provide some magnification of the force applied at the brake pedal, depending on the relative sizes of the hydraulic cylinders at the pedal and the brake. However, with a brake pedal force of about 20 lbs, it is unlikely that the force on a given brake pad could exceed 100 lbs (say) and with a pad area of around 5 in$^2$, this gives a nominal pressure of 20 psi which is negligible.

(ii) Caliper brakes are supposed to exert equal forces on each side of the disc, so that bending loads do not arise. If there is inequality, it should be substantially less than the maximum brake force. To avoid calculations, imagine loading a typical brake disc (say 10 in outside diameter, 5 in inside diameter and $\frac{1}{2}$ in thick) by a 10 lb force and supporting it at the inner radius. It is inconceivable that this could cause failure. Placing a disc between two supports and standing on it would exert a force of 150 lbs (depending on your weight!) and would not be sufficient to cause failure. Thus, bending is not important.

(iii) Imagine the brake disc resting on the floor. Would you expect it to be weak enough to fail under self weight? Obviously not. If you want to be more scientific about this, you can estimate the stresses due to self weight by recalling that the pressure in a fluid under gravity is $\rho gh$ if $h$ is the depth below the free surface. The same equilibrium arguments show that the pressure under (say) a rectangular block of height $h$ resting on the ground is $\rho gh$. For the disc, the biggest dimension is the diameter and we calculate[6] $0.284\times 10\approx 3$ psi. This is clearly negligible.

(iv) Think in the same way as in (iii), but replace $g$ by the maximum centripetal acceleration. At a road speed of 60 mph (88 ft/s), with a tyre diameter of 30 inches (2.5 ft), the rotational speed is $(88/1.25)=70$ rad/s. The acceleration is then $\Omega^2 r=70^2\times(5/12)\approx 2000$ ft/s$^2$, compared with the gravitational acceleration of 32 ft/s$^2$. Thus, stresses associated with acceleration are in the order $3\times 2000/32\approx 200$ psi, which is still small.

(v) Stresses due to air resistance must be small. If they were even comparable with those in (i) above due to the brake pad, the brakes would essentially be on all the time!

(vi) Experience suggests that brake discs would be burning hot to the touch after a severe stop (say $\approx$ 300°F). However, only temperature *difference* is important in determining stresses here, since an unrestrained disc heated to a uniform temperature will simply expand freely without stress. There will be differences in the

[6] Notice that the gravitation constant $g$ is already included in the weight when lbs are used as the units, but would need to be included explicitly when working in SI units.

heating and cooling rates at different radii, due to the higher speed at the outer radius (linear speed is proportional to radius for a given rotational speed), but the resulting temperature difference will probably not exceed $\approx 50^{o}$F, since steel and cast iron are quite good conductors.
The order of magnitude of thermal stresses can be estimated as $\sigma \approx E\alpha\Delta T = 30\times 10^6\times 7\times 10^{-6}\times 50 \approx 10\times 10^3$ psi.

These 'back of the envelope' calculations tell us that the stresses due to temperature difference are significantly greater than the others, though still probably not as high as the yield stress of the material. Since the loading is cyclic (stresses are developed and then relaxed each time the brake is applied), there is a possibility of fatigue failure resulting from thermal stresses and the above estimate is large enough to justify a more exact calculation at a later stage in the design process.[7] All the other loading processes can be neglected.

## 1.4 Formulating and solving problems

Once we have decided to perform some calculations to evaluate or specify a design, we shall have to write governing equations defining the appropriate physical laws and other equations describing the geometry and the boundary conditions specific to the system under consideration. It is always useful to examine first where these equations will come from, how many equations there are, how many unknowns, and what sequence of mathematical steps will lead to the answer. In other words, plan a strategy for the solution before you start performing any calculations. Don't memorize a set of solution procedures and then try to force every question into one or other of them. There are always ways in which the question might be posed that will elude these categories and engineering teachers are adept at finding them.

There is a significant psychological dimension to solving problems accurately. If you have total confidence that the method you are using is going to work, you will be much less likely to make errors during the solution. By contrast, if you have a nagging feeling that you may have overlooked something in the formulation, you will be distracted during the process and will probably make mistakes. This applies in the real world as well as in the examination room.

### 1.4.1 Use of procedures

One of the most effective ways of improving your problem solving abilities is to break the problem down into sub-tasks that can be treated as applications of standard procedures. The word 'procedure' is used in a technical sense in computer programming to denote a set of instructions that can be called into play when needed to perform certain operations and generate given results from a possible set of input data.

[7] Methods of determining the thermal stresses in a heated disc will be discussed in Chapter 10. See Example 10.2 and Problem 10.3.

Programs organized with the use of procedures are much easier to understand and trouble-shoot than those that are written 'linearly' — i.e. simply as a set of instructions in sequence. The same benefits are obtained from treating mechanics problems in a similar way.

For example, suppose we are given the cross section of a beam and the applied bending moment $M$, and are asked to determine the maximum bending stress $\sigma_{max}$. Elementary bending theory tells us that this is given by

$$\sigma_{max} = \frac{Mc}{I} , \tag{1.1}$$

where $c$ is the maximum distance from the neutral axis and $I$ is the appropriate second moment of area.

However, in order to use this relation, we must first find $c$ and $I$ for the section. We know that the neutral axis passes through the centroid of the section and the location of the centroid is also needed in the determination of $I$. We might therefore define the following simple procedures:-

$P_1$: Given $M, c, I$, determine $\sigma_{max}$ from equation (1.1).
$P_2$: Given the geometrical description of the section, determine the location of the centroid.
$P_3$: Given the geometrical description of the section and the location of the centroid, determine $c$.
$P_4$: Given the geometrical description of the section and the location of the centroid, determine $I$.

When the problem is defined in this way, it is clear that the *output* of $P_3$ and $P_4$ is required as *input* to $P_1$ and the output of $P_2$ is required as input to both $P_3$ and $P_4$. Thus, our strategy must be to perform $P_2$ first, followed by $P_3$ and $P_4$ and then $P_1$, as shown schematically in Figure 1.3.

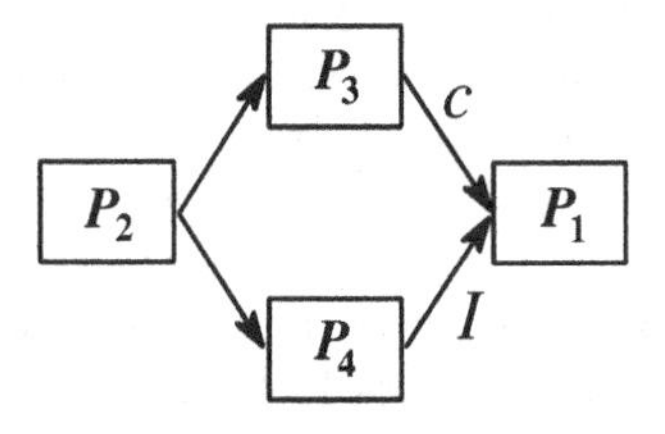


*Figure 1.3: Strategy for determining maximum bending stress*

The major benefits of thinking this way are:-

(i) When we are planning the strategy for the solution, we can think of the procedures as single steps or building blocks in the development of the solution, without having to concern ourselves with the details.
(ii) Conversely, when we are actually performing the operations inside one of the procedures, we can temporarily erase from the mind the larger solution strategy

and concentrate on the smaller task,[8] which is generally something that we are completely confident we know how to do.

### 1.4.2 Inverse problems

Mechanics teachers are well aware of the fact that many different problems can be generated from the same set of calculations by interchanging the order of the dependent and independent variables. For example, in a given bending problem we might:-

(i) Give the dimensions of the cross section, the material and the loading and ask for the factor of safety against yielding.
(ii) Give the loading and the maximum permissible tensile stress and ask for one of the dimensions of the cross section to be determined.
(iii) Give the dimensions of the cross section and the maximum permissible tensile stress and ask for the maximum bending moment that can be transmitted.

In more complicated problems, the number of permutations is enormous.

It is tempting in each case to try to devise a sequence of procedures that will start from the given numerical data and lead by successive arithmetic calculations to the desired result. However, *algebra was invented to deal with precisely this problem.* The thinking here is 'I don't know how to solve this problem. If instead I were given $x, y, z$ and asked to find $a, b, c$, I *would* be able to do it. Therefore, if I write $x, y, z$ as symbols, pretend that they are given and perform the calculations I know until I reach expressions for $a, b, c$, I can then use the given values of the latter to define equations for the unknowns'. Solving the equations may be a mathematical challenge, but (i) we know that we need as many equations as there are unknowns, which provides a check on the viability of the method and (ii) it is easier to lose confidence in the formulation of the mechanics problem than it is in the mathematical manipulations.[9]

As a simple example, in the bending problem (ii) above we might denote the unknown dimension of the cross section by $d$, find the maximum stress in the beam in terms of $d$ and the given loading, and then equate it to the maximum permissible tensile stress to obtain an equation that can be solved for $d$. Of course, this is a very simple example, but the technique becomes more important when the overall complexity of the problem is greater. A major advantage of using algebra to transform the sequence in which the problem is solved is that we are thereby able to choose a sequence with which we are familiar, or for which we have a high degree of confidence of success.

[8] It is interesting to note that procedures in computer programs generally adopt this same 'restricted vision'. Variables defined inside procedures have no definition outside the procedure and *vice versa*, unless an explicit 'common' definition is invoked. The more general parallel between problem solving and programming is also instructive. Computer programs have to be written so as to be completely 'foolproof' and it also makes sense to devise problem solving techniques that are as immune as possible to our occasional periods of intellectual weakness.

[9] These days, many of the more routine mathematical operations can be performed in symbolic languages such as Mathcad, Mathematica, MatLab and Maple.

### 1.4.3 Physical uniqueness and existence arguments

The problems which are easiest to formulate are those which mimic a conceivable experimental situation. For example, a beam of given dimensions and material is placed on simple supports and a given set of external loads is applied to it. The information given is sufficient to enable someone to manufacture the beam and load it in the laboratory. Suppose we imagine performing the experiment. In an ideal world, the beam will behave in the same way each time, so we must presume that the given information is sufficient to define unique values for the moments, stresses and displacements throughout the beam. More generally, whenever the information given in the problem is sufficient for the component to be manufactured (i.e. the dimensions and material are completely specified) and loaded in the laboratory (the loading conditions and history are completely specified), the mechanics problem must necessarily be well-posed, since the resulting experiment can only logically have one outcome.[10]

'Thought experiments' of this kind give us confidence that the given information is indeed sufficient for the problem to be solved.[11] In cases where the problem is not posed in a manner that could form the subject of an experiment, it is helpful to compare it with a related 'experimental' problem obtained by interchanging some of the dependent and independent variables. In particular, make sure that the number of unknowns in the 'experimental' formulation is equal to the number of additional conditions given in the problem statement.

## 1.5 Review of elementary mechanics of materials

We shall end this introductory chapter by briefly reviewing some topics which we assume the reader has already encountered in a first course in mechanics of materials. This material is covered in depth in the numerous excellent elementary texts, a selection of which is listed under 'Further reading' at the end of this chapter. Many of these topics will be revisited in the following chapters — indeed, one of the purposes of more advanced study of the subject is to explore the limitations of the elementary theories and to extend their range of application. This review will also serve to introduce the notation that will be used for some of the more commonly occurring quantities.

### 1.5.1 Definition of stress components

Stress is the force per unit area transmitted across an internal surface in the body. It can be resolved into components, one normal to the surface (*the normal stress*)

[10] There are some situations in mechanics where the loaded state is non-unique, notably those involving elastic instability or buckling (see Chapter 12). However, this kind of behaviour is generally unacceptable in service, so our interest in such problems is generally restricted to determining the unique load at which buckling first occurs.

[11] As we explained above, having confidence that the problem is well-posed is an important factor in successfully solving it.

and (in three dimensions) two *shear stresses* in the plane of the surface. If we set up a local Cartesian coordinate system $x,y,z$ such that the $x$-direction is the outward normal from the surface in question, the normal stress component is usually denoted by $\sigma_x$ and the two shear stresses by $\tau_{xy}, \tau_{xz}$, where the second suffix denotes the direction of the shear stress component. In more advanced work, it is usual to use the symbol $\sigma$ for all stress components and a double suffix notation, such that the first suffix denotes the direction of the outward normal to the surface and the second suffix denotes the direction of the stress component. The quantities $\sigma_x, \tau_{xy}, \tau_{xz}$ would therefore be written $\sigma_{xx}, \sigma_{xy}, \sigma_{xz}$. Notice that the $x$-plane (i.e. the plane normal to the $x$-direction) can also be identified by an equation of the form $x=$constant.

With this notation, the stress components acting on a small rectangular brick element of material are labelled as shown in Figure 1.4. Notice that the notation implies that tensile normal stresses are positive and compressive stresses are negative.

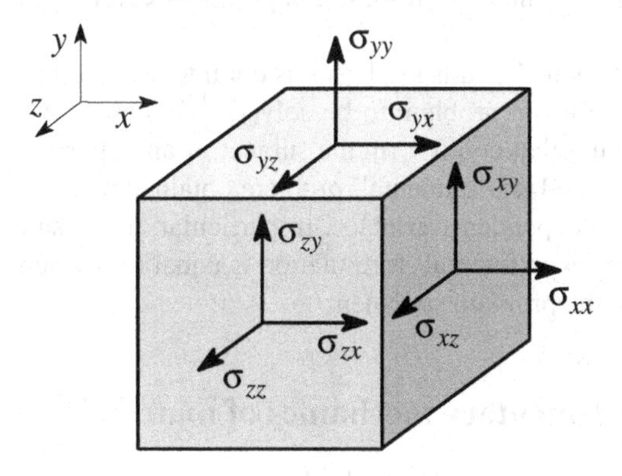


*Figure 1.4: Notation for stress components*

It is often convenient to consider the components in the form of a *stress matrix*

$$\sigma \equiv \begin{bmatrix} \sigma_{xx} & \sigma_{yx} & \sigma_{zx} \\ \sigma_{xy} & \sigma_{yy} & \sigma_{zy} \\ \sigma_{xz} & \sigma_{yz} & \sigma_{zz} \end{bmatrix} . \tag{1.2}$$

Equilibrium considerations demand that the shear stresses be equal in pairs (for example $\sigma_{xy} = \sigma_{yx}$). This can be proved by taking moments about the three axes for the element of Figure 1.4. It follows that the stress matrix (1.2) is symmetric. The equal shear stress components $\sigma_{xy}, \sigma_{yx}$ are referred to as *complementary shear stresses.* One advantage of the double suffix notation is that it lends itself to the use of matrix methods in manipulating stress components in general problems.

In the following text, we shall generally use the double suffix notation, but we shall continue to use the symbol $\tau$ without suffices to denote shear stresses in situations where the direction of the stress is either obvious or not important (for example, in Chapter 6).

### 1.5.2 Transformation of stress components

If all the stress components — i.e. all the elements of the stress matrix — are known in a given Cartesian coordinate system, the corresponding components in rotated coordinate systems at the same point can be determined from equilibrium considerations. In two dimensions, the resulting equations permit a convenient graphical representation known as Mohr's circle. An important result is that there exists a special Cartesian coordinate system — the *principal axes* — in which the corresponding shear stress components are zero. We shall review these results in §2.1 and extend them to the three-dimensional case.

### 1.5.3 Displacement and strain

The displacement of a point is a vector representing the distance moved by the point during a process of deformation. Displacement will be denoted by the symbol $u$, with suffices to indicate direction where required. Thus, $u_x$ is the component of displacement in the positive $x$-direction.

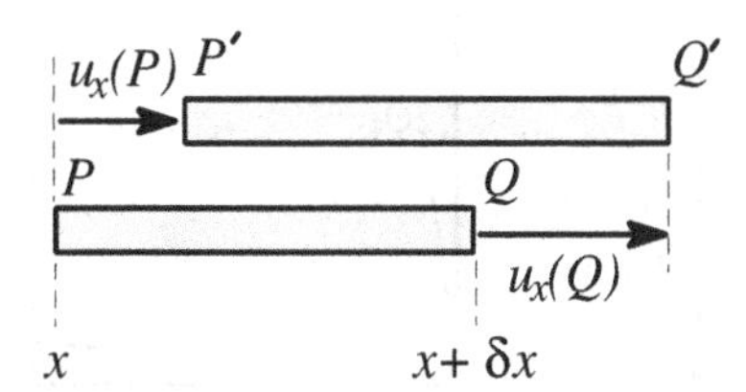


*Figure 1.5: Relation between normal strain and displacement*

Normal strain is defined as the ratio extension/original length for an element of material. It can also be written in terms of displacement. Consider the small element $PQ$ shown in Figure 1.5, which is subjected to tensile loading. The element may extend, but it may also move in the $x$-direction as shown. If the two ends of the element move the same distance, there will be no extension. More generally, the extension is $\Delta = u_x(Q) - u_x(P)$. If the ends, $P, Q$ are originally at the points $x$, $x+\delta x$, respectively, the original length of the element is $\delta x$ and the normal strain can then be written

$$\varepsilon_x = \frac{\Delta}{\delta x} = \frac{u_x(x+\delta x) - u_x(x)}{\delta x} .$$

Taking the limit as $\delta x \to 0$, we obtain[12]

$$\varepsilon_x = \frac{\partial u_x}{\partial x} . \tag{1.3}$$

[12] Recall that the mathematical definition of the derivative is

$$f'(x) = \lim_{\delta x \to 0} \frac{f(x+\delta x) - f(x)}{\delta x} .$$

Similar expressions can of course be written for $\varepsilon_y, \varepsilon_z$.

The shear strain $\gamma_{xy}$ is defined as the change (measured in radians) in the angle at the corner of a rectangular element, as shown in Figure 1.6*(a)*. Notice that in general, the element may both deform and rotate, as shown in Figure 1.6*(b)*, in which case $\gamma_{xy} = \phi_1 + \phi_2$. It can be shown by geometric arguments that for sufficiently small angles

$$\phi_1 = \frac{\partial u_y}{\partial x} \; ; \; \phi_2 = \frac{\partial u_x}{\partial y} \tag{1.4}$$

and hence

$$\gamma_{xy} = \frac{\partial u_y}{\partial x} + \frac{\partial u_x}{\partial y} \; . \tag{1.5}$$

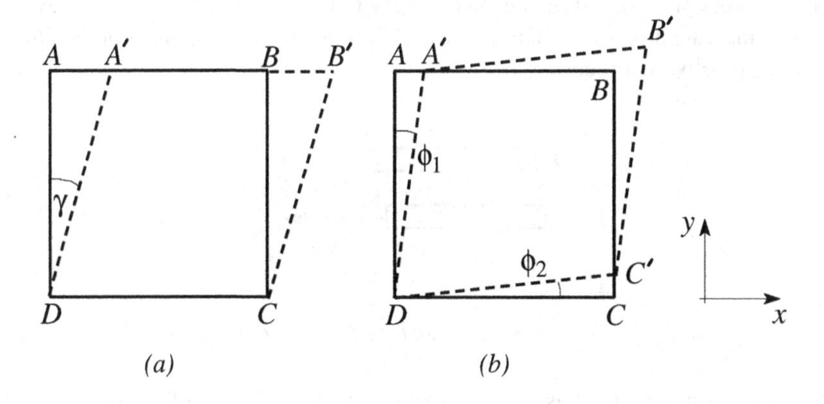


*Figure 1.6: Geometrical definition of shear strain*

As with stress, there is a double suffix notation for strain which is helpful in the matrix manipulation of general results. We use the symbol $e$ for all strain components and define $e_{xx} = \varepsilon_x$ and $2e_{xy} = \gamma_{xy}$ etc. Notice the factor of 2 in the definition of shear strain.[13] The quantity $e_{xy}$ is referred to as the *mathematical shear strain* in contrast to the *engineering shear strain* $\gamma_{xy}$.

The factor of 2 is not introduced just to make the subject more confusing! It turns out that many of the relationships between strains are simplified with the mathematical definition — notably those associated with strain transformation. The reader may have noticed that the equations used to obtain strain components in a rotated coordinate system (those associated with Mohr's circle of strain) are almost the same as the corresponding equations for stress transformation except for a mysterious factor of 2

[13] The double suffix notation is widely used in more advanced texts, particularly those concerned with theoretical elasticity. Confusion is avoided by the universal use of $\gamma$ for engineering shear strain and $e$ or $\varepsilon$ for mathematical shear strain.

in certain expressions. This difference disappears when the mathematical definition of shear strain is used.

In this book, we shall generally use the double suffix mathematical definitions

$$e_{xx} = \varepsilon_x = \frac{\partial u_x}{\partial x} \quad ; \quad e_{xy} = \frac{1}{2}\gamma_{xy} = \frac{1}{2}\left(\frac{\partial u_y}{\partial x} + \frac{\partial u_x}{\partial y}\right) . \tag{1.6}$$

Notice that, as with stress, a repeated suffix indicates a normal strain and differing suffices a shear strain. In other words, the diagonal elements of the strain matrix are normal strains, whilst the off-diagonal elements are shear strains. It is clear from the definition (1.6) that the strain matrix is symmetric — i.e. $e_{xy} = e_{yx}$.

### 1.5.4 Hooke's law

Most engineering components are designed to operate in the elastic régime (i.e. to return to the undeformed configuration on unloading) and deformations are generally small. Even in components where the deflections are visible to the eye, such as a coil spring, the *strains* are usually small compared with unity. Under these conditions, most materials obey Hooke's law — i.e. they exhibit a linear relation between stress and strain. The most general linear relation would permit each strain component to be a linear function of all six independent stress components $\sigma_{xx}, \sigma_{yy}, \sigma_{zz}, \sigma_{xy}, \sigma_{yz}, \sigma_{zx}$, but if the material is *isotropic* (i.e. if it has no directional properties, so that a specimen cut with any orientation from a block of material will have the same properties), it can be argued from symmetry considerations that normal stresses will produce only normal strains and shear stresses only shear strains.

The constants of proportionality between stress and strain have to be determined by experiment and the simplest experiment is the uniaxial tensile test leading to the definitions of Young's modulus $E$ and Poisson's ratio $\nu$. In terms of these quantities, we have

$$e_{xx} = \frac{\sigma_{xx}}{E} - \frac{\nu\sigma_{yy}}{E} - \frac{\nu\sigma_{zz}}{E} , \tag{1.7}$$

$$e_{yy} = \frac{\sigma_{yy}}{E} - \frac{\nu\sigma_{zz}}{E} - \frac{\nu\sigma_{xx}}{E} , \tag{1.8}$$

$$e_{zz} = \frac{\sigma_{zz}}{E} - \frac{\nu\sigma_{xx}}{E} - \frac{\nu\sigma_{yy}}{E} . \tag{1.9}$$

If there is only one normal stress $\sigma_{xx}$ (i.e. if $\sigma_{yy} = \sigma_{zz} = 0$), these relations predict the occurrence of a corresponding normal strain $e_{xx} = \sigma_{xx}/E$ and transverse strains $e_{yy} = e_{zz} = -\nu\sigma_{xx}/E$. The transverse strains are negative indicating that the cross section of the tensile test specimen will get smaller as the specimen is stretched.[14] Equations (1.7–1.9) are essentially a superposition of the simultaneous effects of the three orthogonal normal stresses $\sigma_{xx}, \sigma_{yy}, \sigma_{zz}$.

[14] This conclusion depends upon $\nu$ being positive. There is no logical objection to negative values of $\nu$, but no practical materials are known which exhibit this behaviour.

Shear stresses and shear strains are also linearly related through the equations

$$e_{xy} = \frac{\sigma_{xy}}{2G} \;\; ; \;\; e_{yz} = \frac{\sigma_{yz}}{2G} \;\; ; \;\; e_{zx} = \frac{\sigma_{zx}}{2G} \,, \tag{1.10}$$

where $G$ is a new elastic constant known as the *shear modulus* or the *modulus of rigidity*.[15] The shear modulus can be found experimentally by performing a simple torsion test on a circular cylindrical specimen and using the elementary theory of torsion to intepret the results. However, the three constants $E, \nu, G$ are not independent of each other. To explain this we note that, even in a bar in uniaxial tension, shear stresses and strains occur in coordinate systems inclined to the axis. These stresses and strains can be found from the transformation equations and their ratio leads to the relation[16]

$$G = \frac{E}{2(1+\nu)} \,. \tag{1.11}$$

To maintain consistency, it is therefore advantageous to rewrite relations (1.10) in the form

$$e_{xy} = \frac{\sigma_{xy}(1+\nu)}{E} \;\; ; \;\; e_{yz} = \frac{\sigma_{yz}(1+\nu)}{E} \;\; ; \;\; e_{zx} = \frac{\sigma_{zx}(1+\nu)}{E} \tag{1.12}$$

in applications where both tensile and shear stresses are involved.

**Thermal expansion**

If the temperature of a piece of isotropic material is increased uniformly by $\Delta T$, it expands in such a way that all the dimensions increase in the same proportion. In other words, the deformed body looks like a photographic enlargement of the original. It retains the same shape and hence no angles change, so there are no shear strains. If we assume that the thermal expansion is linearly proportional to the temperature change, we can define this effect in terms of a thermal strain

$$e_{xx} = e_{yy} = e_{zz} = \alpha \Delta T \;\; ; \;\; e_{xy} = e_{yz} = e_{zx} = 0 \,, \tag{1.13}$$

where $\alpha$ is a material constant known as the *coefficient of thermal expansion.*

Strains due to thermal expansion can be superposed on those due to stresses [equations (1.7–1.9)], so that

$$e_{xx} = \frac{\sigma_{xx}}{E} - \frac{\nu\sigma_{yy}}{E} - \frac{\nu\sigma_{zz}}{E} + \alpha\Delta T \,, \tag{1.14}$$

$$e_{yy} = \frac{\sigma_{yy}}{E} - \frac{\nu\sigma_{zz}}{E} - \frac{\nu\sigma_{xx}}{E} + \alpha\Delta T \,, \tag{1.15}$$

$$e_{zz} = \frac{\sigma_{zz}}{E} - \frac{\nu\sigma_{xx}}{E} - \frac{\nu\sigma_{yy}}{E} + \alpha\Delta T \,, \tag{1.16}$$

whilst (1.10, 1.12) remain unchanged.

---

[15] Notice that the definition of the mathematical shear strain introduces a factor of 2 into these relations.

[16] See, for example, E.P. Popov and T.A. Balan (1999), *Engineering Mechanics of Solids*, Prentice Hall, Upper Saddle River, NJ, 2nd edn., §5.7.

### 1.5.5 Bending of beams

In defining the equation for the elementary bending theory, we choose a coordinate system such that the $z$-axis passes through the centroid of the beam cross section, as shown in Figure 1.7. If a moment $M_x$ is applied about the $x$-axis, the *neutral plane* (where the bending stress is zero) will be defined by $y=0$ and the resulting stress and deformation is defined by the equation

$$\frac{M_x}{I_x} = \frac{\sigma_{zz}}{y} = \frac{E}{R} = -E\frac{d^2u_y}{dz^2} , \tag{1.17}$$

where $I_x$ is the second moment of area of the beam cross section about the $x$-axis, $R$ is the radius of curvature of the initially straight beam due to bending and $u_y$ is the beam deflection. Notice that following §§1.5.1, 1.5.3, the bending stress is denoted by $\sigma_{zz}$ because it acts in the $z$-direction on planes $z=$ constant and the beam deflection by $u_y$, since it constitutes a displacement of the particles of the beam in the positive $y$-direction.

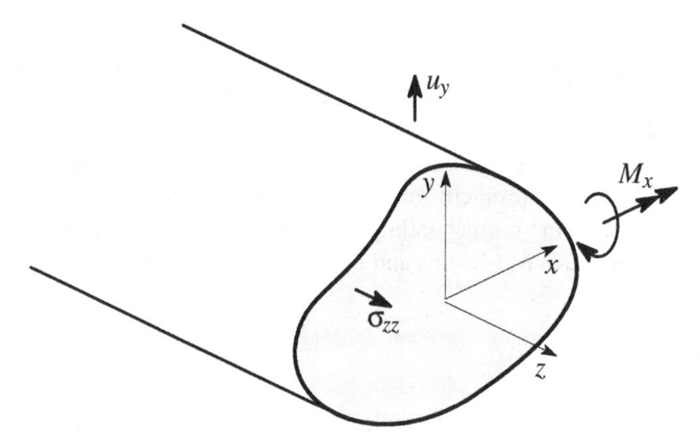


*Figure 1.7: Coordinate system for elementary bending*

If a bending problem is determinate (i.e. if the support reactions can be determined from equilibrium considerations alone), the bending moment for a given loading can be found as a function of $z$ and substitution into (1.17) leads to a second order ordinary differential equation for $u_y$. The solution of this equation is easily obtained and contains two arbitrary constants which are determined by imposing two kinematic boundary conditions derived from the way in which the beam is supported.

Indeterminate problems, in which there are more supports than are needed to prevent motion, are most conveniently treated by using the inverse method described in §1.4.2. Suppose there are $N$ unknown support reactions, $R_1, R_2, ..., R_N$. Since there are only 2 equilibrium equations for two-dimensional beam problems, the problem is statically indeterminate if $N>2$. We invert the problem by 'replacing' $N-2$ of the reactions by external forces which we treat as given. It helps here to use a different symbol for these forces — for example, we can leave $R_1, R_2$ as reactions and write the remaining reactions as the external forces $S_3, S_4, ..., S_N$. Now solve this new *equivalent determinate problem.* Notice that since it is determinate, we can see this phase of the process as the use of a procedure for the solution of determinate problems, which we denote by $\boldsymbol{P_D}$, and gain the advantages listed in §1.4.1. The output from $\boldsymbol{P_D}$ will be an expression for the displacement $u_y$ of the beam which will contain the unknown forces $S_3, S_4, ..., S_N$ and two arbitrary constants of integration — i.e. $N$ unknowns in all. Since there are $N$ reactions, there must be $N$ kinematic conditions at the supports, which will provide $N$ equations for the $N$ unknowns.

This strategy can also be applied to other indeterminate problems, such as those involving trusses or curved beams. We shall give examples in §3.10.5 where a new energy-based method is used for the determinate procedure $\boldsymbol{P_D}$.

### 1.5.6 Torsion of circular bars

The elementary theory of torsion leads to the formula

$$\frac{T}{J} = \frac{\tau}{r} = G\phi \;, \tag{1.18}$$

where $T$ is the applied torque, $J$ is the polar second moment for the cross section, $\tau$ is the shear stress,[17] $r$ is the radial distance from the axis of the bar and $\phi$ is the twist per unit length. This theory is only applicable to solid or hollow circular cylinders and can give very large errors if it is used for bars of non-axisymmetric cross section. The torsion problem for more general cross sections is considerably more challenging than the corresponding bending problem. The important case of thin-walled sections, such as I-beams and box sections is discussed in Chapter 6.

## 1.6 Summary

In this chapter, we have discussed the rôle of calculations in the engineering design process and introduced some ideas and definitions from elementary mechanics of materials. During the early stages of an engineering design, many competing concepts may be under review and only a limited time is available to judge their feasibility. The effective designer needs to posess insight into the qualitative behaviour of elastic systems and make extensive use of simple approximate techniques that provide estimates for the stresses and displacements with the absolute minimum of calculation. More time consuming rigorous analytical and numerical methods are appropriate at a later stage in the design process, when the geometry and loading are largely defined.

This theme will recur throughout the remaining chapters of this book. Emphasis will be placed on qualitative aspects of the behaviour revealed by the analytical methods discussed. Simple approximate techniques will be introduced wherever possible

[17] Notice that the shear stress $\tau$ acts in the circumferential direction on the cross-sectional plane $z =$ constant. It would therefore be denoted by $\sigma_{z\theta}$ in the double suffix notation in cylindrical polar coordinates $(r, \theta, z)$.

and their predictions compared with more exact treatments. Each chapter concludes with a summary in which the important messages for engineering design are emphasised.

## Further reading

### The design process

D.G. Ullman (1992), *The Mechanical Design Process*, McGraw-Hill. New York.

### Elementary mechanics of materials

F.P. Beer and E.R. Johnston (1992), *Mechanics of Materials*,McGraw-Hill, New York, 2nd edn.
J.M. Gere and S.P. Timoshenko (1997), *Mechanics of Materials*, PWS Publishing Co., Boston, 4th edn.
R.C. Hibbeler (1997), *Mechanics of Materials*, Prentice Hall, Upper Saddle River, NJ, 3rd edn.
T.J. Lardner and R.R. Archer (1994), *Mechanics of Solids: An Introduction*, McGraw-Hill, New York.
E.P. Popov and T.A. Balan (1998), *Engineering Mechanics of Solids*, Prentice Hall, Upper Saddle River, NJ, 2nd edn.
W.F. Riley and L.W. Zachary (1989), *Introduction to Mechanics of Materials*, John Wiley, New York.

## Problems

### Section 1.1

**1.1** You are project manager for the gearbox for a manual transmission car expected to sell a total of 200,000 units over a five year life. The gearbox comprises a two-part housing, four shafts, 14 gears, two couplings, eight bearings, a shift mechanism and assorted smaller items and is estimated as constituting about $500 of the estimated retail vehicle price of $18,000.

You have a bright young University of Michigan graduate on your staff who has an idea for a redesign of the input shaft that he claims will save 30% of the manufacturing costs for that component. He asks you for two weeks full time to evaluate it analytically. You think his chance of success is 50%. His annual salary is $50,000 and carries 30% fringe benefit and 70% infrastructure overhead.

Make a few rough calculations to see whether you think you should approve his request.

**1.2.** A manufacturer of specialized test equipment sells a machine for testing automotive friction materials at $150,000. They are approached by a customer who needs a custom-made machine to operate at twice the maximum operating speed of the normal production model. From informal discussions, it appears that the customer might be willing to pay up to $250,000, but the specialized nature of the operating conditions makes it unlikely that more than one or two such machines would ever be sold.

The design challenges of meeting the specification are significant. You have a few ideas, such as using two counter-rotating systems to achieve the required relative speed, or improving the level of machining and balancing to permit higher speed operation. However, you do not know yet whether these ideas are even feasible, let alone whether you can meet the specification within a realistic price range.

How many hours design work at $180 per hour would you do before deciding whether to quote *at all*. Assuming the results of this work appear promising, how many hours can be committed to developing the quote (which would then commit the company to supplying a machine to the stated specification at the quoted price)?

**1.3.** A well-known manufacturer of lawnmower engines is developing a new generation of quieter engines which it is hoped will be capable of being manufactured for about the same price as the present standard, which retails for $100. The manufacturer presently has 20% market share in the US and hopes to increase this to 25% as a result of the improved design. Assume that stress and failure analysis amounts to 5% of the expected design costs and that there are eight significantly stressed components in a typical engine (block, crankcase, piston, connecting rod, crankshaft etc.) each of which will require about $20,000 in finite element analysis and interpretation of results to provide confidence that the design is adequate against fracture. Perceived reliability is a major factor in consumer choice so this level of design rigour is deemed necessary if the new engine is to be introduced.

Make some elementary estimates based on plausible assumptions to decide whether (in the area of stress analysis at least) the project is financially viable.

## Section 1.2

*Instructors are encouraged to use some of the problems in this section as the basis for a classroom discussion. Ask for a few alternative answers from the class and then take a vote on the results. In some cases, a simple experiment can be performed as well.*

**No bending calculations are to be performed in answering problems 1.4–1.8.**

**1.4.** A 1.5 m wooden beam of 25 mm square cross section is built-in at one end. If you put all your weight on the other end of the beam will it break? What is the longest beam of this cross section that would be strong enough to support your weight?

**1.5.** A 2 ft steel bar of 1/4 inch diameter is built-in at one end. What lateral load at the other end would be just sufficient to cause permanent deformation of the bar?

**1.6.** A wooden plank is 200 mm wide, 25 mm thick and 1.5 m long. It is simply supported at the two ends. How far will it deflect if you stand on it at the centre?

**1.7.** A cylindrical log of diameter 12 inches and length 10 feet is floating in water so that the highest point on the cylinder is 2 inches above the water level. If you stand on the log, will you get your feet wet (i) if you stand at the mid point, or (ii) if you stand at one end?

**1.8.** We wish to support a textbook weighing 2.5 kg by two strips of paper, one attached to each cover by adhesive tape. What is the minimum width of the strips if they are not to fail in tension?

**1.9.** What is the maximum tensile force you can apply to a strip of paper by gripping one end in each hand and pulling?

**1.10.** Check your answer to Problem 1.8 by (i) testing strips of various widths until you determine the maximum width of strip that you can break and then comparing the answer with that from Problem 1.8. Be careful not to fail the strips in a tearing mode where you attach them to the book or grip them.

**1.11.** Estimate the maximum bending moment you can apply to a one inch diameter broom handle using only your hands. (Suppose the broom is lying on the ground and you want to pick it up with one hand from the end of the handle, keeping the handle horizontal all the time. Would you be able to do so and if so, how much weight at the other (broom) end could be added before the moment required from your grip would be too large?)

**1.12.** Find a way to support a 5 lb textbook 8 inches above a flat surface using a structure made only of one sheet of paper (it can be cut into several pieces if desired) and 2 inches of adhesive tape. Learn from your failures as well as your success.

**1.13.** What is the maximum twisting moment that you can apply by hand to the broom handle of Problem 1.11?

**1.14.** Estimate the coefficient of friction between steel and rubber by placing an eraser on a steel plane surface (e.g. the bottom of a biscuit tin) and tilting it until slip occurs. [This is just as good an estimate as you will find in (for example) an Engineer's Handbook, which would be the usual source of information chosen by an engineering student.]

**1.15.** Repeat problem 1.14 with a range of different materials.

**1.16.** Roughly how far does the rear end of your car descend when you fill the tank with gas? Use this result to estimate the stiffness of the car rear springs in N/mm.

**1.17.** A tree is 60 feet high and the trunk is 2 feet in diameter at the base. Estimate the total weight of wood in the tree.

**1.18.** A rubber bushing consists of an annulus of rubber, 60 mm outside diameter, 25 mm inside diameter and 25 mm thick. The outside diameter is perfectly bonded to a rigid support and a 25 mm diameter rigid bar is perfectly bonded to the inner diameter. The bar projects 300 mm axially from the bushing.

(i) How much axial deflection would be produced by a 10 N axial force applied to the bar?
(ii) If a 10 N lateral force is applied to the end of the bar, through what angle will the bar rotate away from the axis?
(iii) If you apply the largest twisting moment to the bar that you can using the grip of one hand, how much axial rotation of the bar will occur?

**1.19.** A steel beam is of L-shaped cross section, each leg of the L being of length 1 inch and thickness $1/16$ inch. It is 2 feet long. If you grip it with one hand at each end and try to twist it (i.e. load it in torsion), how much relative rotation (in degrees) between the ends will you be able to produce?

Now answer the same question for a beam consisting of a hollow tube of 1 inch diameter and $1/16$ inch wall thickness.

Try these experiments in your local hardware store. L-shaped beams (angle irons) are sold for the construction of (e.g.) shelving units. Comment on the results and speculate on possible explanations for the difference between the behaviour of the two sections.

**1.20.** Repeat Problem 1.19 (comparing an angle iron and a circular tube) in bending instead of torsion. Which is the stiffer and by how much? Bending stiffness should be judged by the relative rotation between the two ends under a given moment.

**1.21.** Construct a thin-walled paper tube by rolling up a piece of paper into a cylinder (say 50 mm diameter) and taping it up along the side.

Pinch the tube between your fingers at one end. What happens to the initially circular cross section of the tube at the other end? Can you find a simple explanation of this result?

Try the same experiment on a cardboard tube such as the centre from a roll of toilet paper. Does it behave differently and if so, what is different here?

**1.22.** What is the maximum compressive force that can be transmitted by a rolled up paper tube of diameter 0.5 in, length 8 in and one layer of paper wall thickness? Guess first and then make a tube as in Problem 1.21 and test it. You can use books of various sizes as weights.

**1.23.** An initially straight piece of steel wire of diameter 1 mm is wrapped around a rigid drum of diameter 100 mm. The wire is then released. Will it return to the straight shape or will it be permanently deformed? If it is deformed, what radius would you expect it to have after being released?

**Section 1.3**

**1.24.** An office building is five stories high. Which of the following kinds of loading is likely to be most significant and by roughly what ratios:-

(i) self weight of the building structure,
(ii) wind load during a severe storm,
(iii) loading due to furniture and employees.

Would the relative magnitude of the first two effects be different if the building were 50 stories high?

**1.25.** A steam boiler is 5 feet in diameter, 20 feet long and has a wall thickness of 1 inch. It experiences

(i) an internal pressure of 250 psi,
(ii) weight of contained water (during normal service it is half full of water),
(iii) a temperature difference between the outside and the inside wall in the heated area of 100°F,
(iv) bending stresses due to self weight, if the boiler is supported on conforming structures at each end over a rectangular area 2 ft square.

Which of these effects must be taken into account in the design of the boiler?

**1.26.** What determines the minimum wall thickness usable for an aluminium beverage can:-

(i) resistance to damage during shipping,
(ii) retention of shape when standing on a surface,
(iii) possible failure due to internal pressure of fermented beverages,
(iv) tearing during the manufacturing process,
(v) thermal stresses during refrigeration.

**1.27.** One of the shafts in an automotive gearbox is 200 mm long and it carries two gears, one of 50 mm diameter and one of 100 mm diameter. Which two of the following factors are most likely to determine the minimum diameter that can be used for the shaft:-

(i) bending stresses in the shaft due to gear tooth forces,
(ii) torsional stresses in the shaft due to gear tooth forces,
(iii) shaft deflection affecting gear meshing action,
(iv) internal diameter needed to accommodate adequate roller bearings,
(v) natural frequency of shaft vibrations.

**1.28.** Which two of the following factors are most likely to be important in determining the maximum stresses in an automotive cylinder block:-

(i) gas pressure in the cylinders during the power stroke,
(ii) fluid pressure in the coolant passages,
(iii) self weight of the block,
(iv) thermal stresses due to temperature variation during warm-up,
(v) thermal stresses due to temperature gradients during normal operation,
(vi) normal forces between piston rings and cylinder walls,
(vii) frictional forces between piston rings and cylinder walls,
(viii) localized loading at the bolts which attach the cylinder head.

# 2

# Material Behaviour and Failure

Most calculations in mechanics of materials are performed in order to predict whether a given component will fail in service or to determine dimensions or other design specifications to ensure that the probability of failure is acceptably low.

An engineering component may be said to *fail* when it does not perform its specified function. For example, if a bridge over a river deflects so much under load that we get our feet wet when crossing it, the design is a failure, even though the bridge returns to its original configuration after the load is removed. In this chapter, we shall use the more restrictive concept of *material failure* corresponding to situations in which an irreversible change occurs in the material in response to the applied stress field. Major categories of material failure include *ductile failure* or *yielding*, where the component does not return to its original state on unloading, *brittle failure* or *fracture*, where an originally coherent structure breaks apart catastrophically, and *fatigue*, where a progressive failure develops as a result of many cycles of loading and unloading. We shall see below that yielding is largely determined by the maximum shear stress and fracture by the maximum tensile stress in the body.

Brittle failure and fatigue failure are characterized by the separation of the body into two or more pieces and are therefore always catastrophic, but yielding simply implies that the body is permanently deformed and this may or may not be acceptable in a given design application, depending on the function of the component. For example, if the crankshaft of an internal combustion engine is permanently deformed, it will cause additional dynamic loads due to out of balance mass, misalignment of the pistons in the cylinders and misalgnment of the bearings and hence rapidly precipitate more serious damage to the engine. By contrast, a limited amount of permanent deformation of the members of a bridge structure may be acceptable. In the latter case, a more efficient design can often be achieved by analyzing the behaviour of the structure in the plastic régime and applying appropriate safety factors.[1]

In order to predict whether a component will fail under a given load, we must first estimate the stresses at potentially critical areas and then compare them with the results of standardized material tests, such as the uniaxial tensile test. This procedure

[1] We shall see applications of this procedure in Chapters 5 and 10 below.

J.R. Barber, *Intermediate Mechanics of Materials*, Solid Mechanics and Its Applications 175, 2nd ed., DOI 10.1007/978-94-007-0295-0_2, © Springer Science+Business Media B.V. 2011

is based on the argument that failure is a localized process involving irreversible behaviour at a particular point in the component. The material in the immediate vicinity of the failure site that participates in the process can only be influenced by local conditions. A particle of material embedded in a beam has no way of knowing whether the tensile stresses it experiences are the result of bending or tensile loading and therefore the tensile stress at failure will be the same for both modes of loading. Thus, *differences in global loading conditions can only influence the material behaviour insofar as they influence local conditions.* Stresses are arguably the most important determinant of material failure, but other local quantities can be significant. For example, the yield stress may be temperature-dependent, in which case local temperature will feature in the yield criterion.

For a simple loading geometry, the failure load can be determined by comparison with experimental results that produce a similar state of stress. For example, if a bar is loaded in bending, the stress state is one of uniaxial tension and the behaviour of the component can therefore be predicted using the results of uniaxial tensile tests. In particular, plastic deformation will be predicted when the tensile stress reaches the yield stress $S_Y$ for the material.

However, many engineering components are subjected to more complex states of stress involving both normal and shear components. The most general state of stress involves six independent stress components — three normal stresses $\sigma_{xx}, \sigma_{yy}, \sigma_{zz}$ and three[2] shear stresses $\sigma_{xy}, \sigma_{yz}, \sigma_{zx}$. In principle, we might carry out tests to determine the conditions at failure under all possible combinations of these components, but in practice the experimental effort required would be enormous and the presentation of the data would require volumes of tables or graphs. To put this in perspective, note that, if failure depended on a single parameter (such as the maximum tensile stress), one experiment would suffice to determine it and the results could be presented as a single numerical value. If there were *two* parameters, the results could be presented as a graph or as a column of values. *Three* parameters requires a family of graphs or a table. Imagine what we would need to present results with six independent parameters! In this section, we shall examine ways of reducing the complexity of the problem, using a variety of arguments, assumptions and approximations. However, we must first review the relations between stress components in different coordinate systems, in order to determine the most efficient way of characterizing a given state of stress.

## 2.1 Transformation of stresses

If the six stress components are known in a given Cartesian coordinate system $x, y, z$, we can determine the corresponding components at the same point in a new system $x', y', z'$ by writing equilibrium equations for an appropriate infinitesimal element of material.

[2] Strictly there are six shear stress components, but they are equal in complementary pairs — i.e. $\sigma_{xy} = \sigma_{yx}$ etc., as discussed in §1.5.1.

### 2.1.1 Review of two-dimensional results

The reader should be familiar with this process for the simpler, two-dimensional case in which the only non-zero stress components are $\sigma_{xx}$, $\sigma_{yy}$, $\sigma_{xy}$. However, this has important consequences in failure analysis and it will therefore be useful to review the more important results.

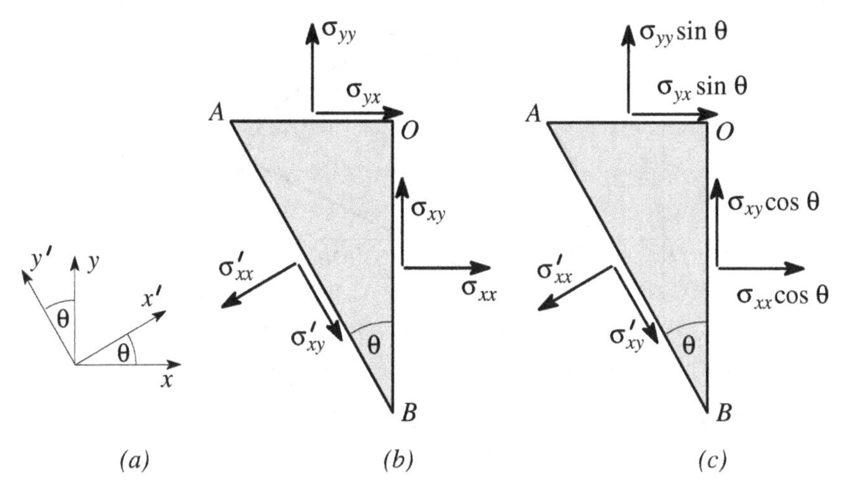


*Figure 2.1: (a) Rotated coordinate system Ox′y′; (b) stress components; (c) force components*

Figure 2.1 *(b)* shows the side view of a triangular prismatic element which is of constant thickness in the direction into the page. If the inclined face *AB* of the element has unit area, the areas of the faces *OB*, *OA* will be $\cos\theta$, $\sin\theta$, respectively and, for example, the force on *OB* associated with the stress component $\sigma_{xx}$ will be $\sigma_{xx}\cos\theta$. This and the forces associated with the other stress components are shown in Figure 2.1 *(c)*.

We can now sum the resolved forces in the directions $x', y'$ respectively in Figure 2.1 *(c)* to obtain the equilibrium relations

$$\sigma'_{xx} = \sigma_{xx}\cos^2\theta + \sigma_{yy}\sin^2\theta + 2\sigma_{xy}\sin\theta\cos\theta \tag{2.1}$$

$$\sigma'_{xy} = \sigma_{xy}(\cos^2\theta - \sin^2\theta) + (\sigma_{yy} - \sigma_{xx})\sin\theta\cos\theta\ . \tag{2.2}$$

These are the two-dimensional *stress transformation relations*. They can also be written in the double-angle form

$$\sigma'_{xx} = \frac{\sigma_{xx}+\sigma_{yy}}{2} + \frac{\sigma_{xx}-\sigma_{yy}}{2}\cos 2\theta + \sigma_{xy}\sin 2\theta \tag{2.3}$$

$$\sigma'_{xy} = \sigma_{xy}\cos 2\theta - \frac{\sigma_{xx}-\sigma_{yy}}{2}\sin 2\theta\ , \tag{2.4}$$

making use of well-known trigonometrical identities.

### Mohr's circle

If a graph is plotted comprising all points whose coordinates are $\sigma'_{xx}, \sigma'_{xy}$, with $\theta$ as a parameter, the resulting figure is a circle known as *Mohr's circle*. Each point on the circle corresponds to a particular value of $\theta$ and hence to a particular inclination for the plane $AB$ in Figure 2.1 *(b)*.

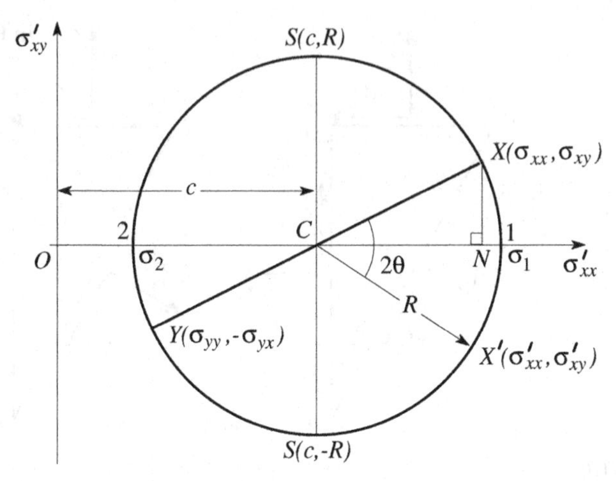


*Figure 2.2: Mohr's stress circle*

The special case $\theta = 0$ corresponds to the face $OB$ (the $x$-plane) and has coordinates $(\sigma_{xx}, \sigma_{xy})$. This point is labelled $X$ in Figure 2.2. As we increase $\theta$ from zero, the plane $AB$ will rotate anticlockwise in Figure 2.1 *(b)*, but the corresponding point moves around the circle *clockwise* (i.e. in the opposite direction) in Figure 2.2, such that the angle subtended at the centre of the circle is $2\theta$. Thus, the point $X'$ in Figure 2.2 corresponds to the plane $AB$ in Figure 2.1 *(b)*. If we were to make $\theta = \pi/2$, $AB$ would be horizontal in Figure 2.1 *(b)*, but would face vertically downwards. It follows that in this case, we have $\sigma'_{xx} = \sigma_{yy}$ and $\sigma'_{xy} = -\sigma_{yx}$. Notice the minus sign in the expression for the shear stress on this plane, which is a consequence of the sign convention used in Figure 2.1 *(b)* for $x', y',$. Thus, the point $Y$ corresponding to the $y$-plane $OA$ has coordinates $(\sigma_{yy}, -\sigma_{yx})$ in Figure 2.2.

The strategy for finding the centre of the circle and its radius is therefore as follows:-

(i) Identify the points $X(\sigma_{xx}, \sigma_{xy})$ and $Y(\sigma_{yy}, -\sigma_{yx})$.
(ii) Draw the diametral line $XY$, which crosses the horizontal axis at the centre $C$ of the circle $(c, 0)$, where

$$c = \frac{(\sigma_{xx} + \sigma_{yy})}{2} \; . \tag{2.5}$$

(iii) The radius $R$ of the circle can now be found using Pythagoras' theorem on the triangle $CNX$, with the result

$$R = \sqrt{\left(\frac{\sigma_{xx} - \sigma_{yy}}{2}\right)^2 + \sigma_{xy}^2} \,. \tag{2.6}$$

### Principal planes and principal stresses

Figure 2.2 enables us to draw the important conclusion that there exist two orthogonal *principal planes*, labelled 1,2, on which the shear stress $\sigma'_{xy} = 0$. The normal stresses $\sigma'_{xx}$ on these planes (the *principal stresses* $\sigma_1, \sigma_2$) are respectively the maximum and minimum values of normal stress. From Figure 2.2 we conclude that

$$\sigma_1, \sigma_2 = c \pm R = \frac{\sigma_{xx} + \sigma_{yy}}{2} \pm \sqrt{\left(\frac{\sigma_{xx} - \sigma_{yy}}{2}\right)^2 + \sigma_{xy}^2} \,. \tag{2.7}$$

### Maximum shear stress

The maximum shear stress $\tau_{\max}$ corresponds to the points $S$ in Figure 2.2 and hence occurs on planes at 45° to the principal planes. Clearly

$$\tau_{\max} = R = \sqrt{\left(\frac{\sigma_{xx} - \sigma_{yy}}{2}\right)^2 + \sigma_{xy}^2} \,. \tag{2.8}$$

### Example 2.1

*The stress components at a point are $\sigma_{xx} = 100$ MPa, $\sigma_{yy} = -20$ MPa, $\sigma_{xy} = -50$ MPa. Find the principal stresses, the maximum in-plane shear stress and the angles between the principal directions and the x-direction. Illustrate the inclination of the principal planes with a sketch.*

The points $X(\sigma_{xx}, \sigma_{xy})$ and $Y(\sigma_{yy}, -\sigma_{xy})$ are located as (100,−50) and (−20,50) respectively, as shown in Figure 2.3*(a)*. The centre of the Mohr's circle is therefore at the point (40,0), since

$$c = \frac{(\sigma_{xx} + \sigma_{yy})}{2} = \frac{100 - 20}{2} = 40 \text{ MPa.}$$

From the triangle *CNX*, we determine the radius of the circle as

$$R = \sqrt{60^2 + 50^2} = 78.1 \text{ MPa}$$

and hence

$$\sigma_1 = 40 + 78.1 = 118.1 \text{ MPa}, \;\; \sigma_2 = 40 - 78.1 = -38.1 \text{ MPa} \,.$$

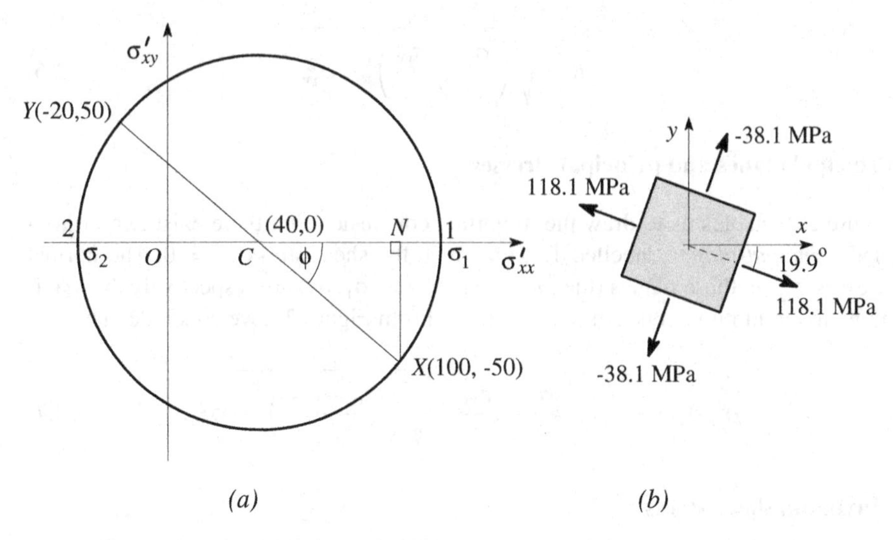


*Figure 2.3: (a) Mohr's circle; (b) orientation of the principal planes*

To determine the inclination of the principal planes, we first calculate the angle $\phi$ in Figure 2.3 *(a)*. The tangent of this angle is 50/60=0.833 and hence

$$\phi = \arctan(0.833) = 39.8^\circ .$$

To get from $X$ to 1 on the circle, we have to move $39.8^\circ$ anticlockwise around the circle. It follows that to get to the principal plane 1 from the $x$-plane, we must rotate *clockwise* through half this angle — i.e.

$$\theta_1 = 19.9^\circ .$$

Notice that we rotate the plane in the opposite direction to the rotation around the circle. This defines the plane on which the larger principal stress $\sigma_1$ acts, as shown in Figure 2.3 *(b)*. Plane 2 is at right angles to plane 1 and hence corresponds to $\theta_2 = 109.9^\circ$. Figure 2.3 *(b)* shows both principal planes in their correct orientation and the principal stresses acting upon them.

### 2.1.2 Principal stresses in three dimensions

For the more general three-dimensional case there are six stress components $\sigma_{xx}$, $\sigma_{yy}$, $\sigma_{zz}$, $\sigma_{xy}$, $\sigma_{yz}$, $\sigma_{zx}$. The corresponding components in a new coordinate system can be obtained by considering the equilibrium of the tetrahedron *OABC* of Figure 2.4, three of whose faces lie in the $x, y, z$ coordinate system and one normal to direction $x'$ in a new Cartesian coordinate system $x', y', z'$.

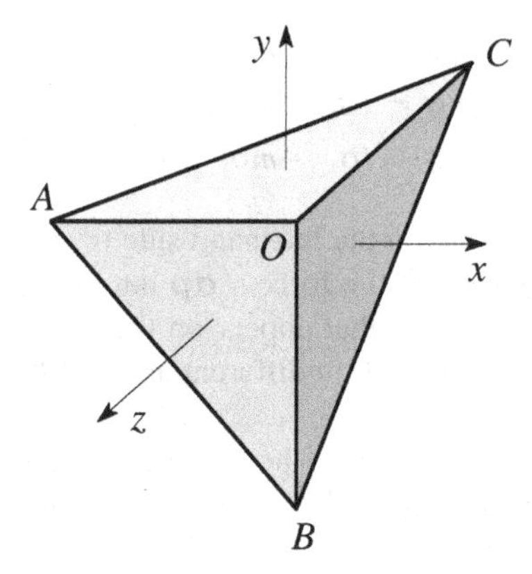


*Figure 2.4: Transformation of stress components in three dimensions*

As in the two-dimensional case, there is a special choice of the coordinate system $x', y', z'$ — the *principal directions* — such that the *three* shear stresses $\sigma'_{xy}, \sigma'_{yz}, \sigma'_{zx}$ are all zero. The corresponding normal stresses are the principal stresses, $\sigma_1, \sigma_2, \sigma_3$. In this section, we shall show how the principal stresses and directions can be found when the stress state is defined in terms of the components $\sigma_{xx}, \sigma_{yy}, \sigma_{zz}, \sigma_{xy}, \sigma_{yz}, \sigma_{zx}$.

We define a set of unit vectors $\boldsymbol{i}, \boldsymbol{j}, \boldsymbol{k}$ corresponding to the Cartesian coordinate system $x, y, z$. The tetrahedron of Figure 2.4 is drawn such that $\boldsymbol{i}, \boldsymbol{j}, \boldsymbol{k}$ are the outward normals to the faces *OBC, OCA, OAB* respectively and the *inward*[3] unit vector normal to the remaining inclined face *ABC* is denoted by $\boldsymbol{p}$. We also define the three *direction cosines* of $\boldsymbol{p}$,

$$l = \boldsymbol{i} \cdot \boldsymbol{p} \; ; \; m = \boldsymbol{j} \cdot \boldsymbol{p} \; ; \; n = \boldsymbol{k} \cdot \boldsymbol{p} \,, \tag{2.9}$$

so named because they are the cosines of the angles between $\boldsymbol{p}$ and $\boldsymbol{i},\boldsymbol{j},\boldsymbol{k}$ respectively. They are also the components of $\boldsymbol{p}$ in the three directions $x, y, z$ and hence

$$l^2 + m^2 + n^2 = 1 \,, \tag{2.10}$$

since $\boldsymbol{p}$ is a unit vector.

If the face *ABC* is taken to have unit area, the area of the face *OBC* will be

$$OBC = l \,, \tag{2.11}$$

since *OBC* is the projection of *ABC* on the $x$-plane. Similarly, the areas $OCA = m$ and $OAB = n$.

Consider now the special case in which $\boldsymbol{p}$ is a principal direction and hence *ABC* is a principal plane. The only stress component on this face is then a normal stress $\sigma$ in direction $-\boldsymbol{p}$ and it corresponds to a force $-\sigma\boldsymbol{p}$, since *ABC* has unit area. Equilibrium of the element in the three directions $x, y, z$ then yields the three equations

[3] The reader should compare with the two-dimensional Figure 2.1, where the positive $x'$-direction is also the *inward* normal to the inclined face *AB*.

$$l\sigma = l\sigma_{xx} + m\sigma_{yx} + n\sigma_{zx} \quad (2.12)$$
$$m\sigma = l\sigma_{xy} + m\sigma_{yy} + n\sigma_{zy} \quad (2.13)$$
$$n\sigma = l\sigma_{xz} + m\sigma_{yz} + n\sigma_{zz} \quad (2.14)$$

Notice that the direction cosine on the left-hand side of equation (2.12) (for example) comes from the resolution of the force $-\sigma \boldsymbol{p}$ into the $x$-direction. By contrast, the stress components $\sigma_{xx}, \sigma_{xy}, \sigma_{xz}$ that appear on the right-hand side of this equation all act in the $x$-direction, but act on different areas $OBC = l, OCA = m, OAB = n$ respectively.[4]

Considering (2.12–2.14) as a set of homogeneous equations for the direction cosines $l, m, n$, we note that there will be a non-trivial solution if and only if the determinant of coefficients is zero — i.e.

$$\begin{vmatrix} (\sigma_{xx}-\sigma) & \sigma_{yx} & \sigma_{zx} \\ \sigma_{xy} & (\sigma_{yy}-\sigma) & \sigma_{zy} \\ \sigma_{xz} & \sigma_{yz} & (\sigma_{zz}-\sigma) \end{vmatrix} = 0 . \quad (2.15)$$

Expanding the determinant, we obtain the cubic equation

$$\sigma^3 - I_1\sigma^2 + I_2\sigma - I_3 = 0 , \quad (2.16)$$

where

$$I_1 = \sigma_{xx} + \sigma_{yy} + \sigma_{zz} \quad (2.17)$$
$$I_2 = \sigma_{xx}\sigma_{yy} + \sigma_{yy}\sigma_{zz} + \sigma_{zz}\sigma_{xx} - \sigma_{xy}^2 - \sigma_{yz}^2 - \sigma_{zx}^2 \quad (2.18)$$
$$I_3 = \sigma_{xx}\sigma_{yy}\sigma_{zz} - \sigma_{xx}\sigma_{yz}^2 - \sigma_{yy}\sigma_{zx}^2 - \sigma_{zz}\sigma_{xy}^2 + 2\sigma_{xy}\sigma_{yz}\sigma_{zx} \quad (2.19)$$

are known as *stress invariants*.

Equation (2.16) has three solutions, which are the three principal stresses $\sigma_1$, $\sigma_2$, $\sigma_3$. When $\sigma$ takes any one of these three values, the three equations (2.12–2.14) are not linearly independent — i.e. any one of them can be derived from the other two. We can then determine the direction cosines defining the corresponding principal direction from (2.10) and any two of (2.12–2.14).

The set of equations (2.12–2.14) defines a linear eigenvalue problem, in which $\sigma$ is the eigenvalue and $\{l, m, n\}$ is the eigenvector. Furthermore, the matrix of coefficients for this problem is the matrix of stress components in $x, y, z$ and is therefore symmetric. It follows from a well-known theorem in linear algebra[5] that the three eigenvalues are all real and the three eigenfunctions mutually orthogonal. This latter result is equivalent to the geometric statement that the three principal directions corresponding to $\sigma_1, \sigma_2, \sigma_3$ are orthogonal to each other.

[4] Remember that equilibrium is a statement about *forces*, not stress components. Another advantage of the double suffix notation is that it reminds us that stress components are associated with specific planes as well as directions.

[5] See for example M. O'Nan (1976), *Linear Algebra*, Harcourt Brace Jovanovich, New York, 2nd edn., §7.2.

**Example 2.2**

*The stress components at a point are* $\sigma_{xx} = -2$ *ksi,* $\sigma_{yy} = 10$ *ksi,* $\sigma_{zz} = -5$ *ksi,* $\sigma_{xy} = 7$ *ksi,* $\sigma_{yz} = 5$ *ksi,* $\sigma_{zx} = -3$ *ksi.*

*Find the three principal stresses and the direction cosines of the plane on which the maximum tensile stress acts.*

We first calculate the three stress invariants, which are

$$\begin{aligned}
I_1 &= \sigma_{xx} + \sigma_{yy} + \sigma_{zz} = 3 \text{ ksi} \\
I_2 &= \sigma_{xx}\sigma_{yy} + \sigma_{yy}\sigma_{zz} + \sigma_{zz}\sigma_{xx} - \sigma_{xy}^2 - \sigma_{yz}^2 - \sigma_{zx}^2 = -143 \text{ (ksi)}^2 \\
I_3 &= \sigma_{xx}\sigma_{yy}\sigma_{zz} - \sigma_{xx}\sigma_{yz}^2 - \sigma_{yy}\sigma_{zx}^2 - \sigma_{zz}\sigma_{xy}^2 + 2\sigma_{xy}\sigma_{yz}\sigma_{zx} = 95 \text{ (ksi)}^3 .
\end{aligned}$$

Substitution into (2.16) then yields the cubic equation

$$\sigma^3 - 3\sigma^2 - 143\sigma - 95 = 0$$

for the principal stresses.

Some calculators and mathematical packages will be able to solve this equation directly. Alternatively, we can use the closed form solution for cubic equations with three real roots, which in the present case yields

$$\sigma_1 = \frac{I_1}{3} + \frac{2}{3}\sqrt{I_1^2 - 3I_2}\cos(\phi) \tag{2.20}$$

$$\sigma_2 = \frac{I_1}{3} + \frac{2}{3}\sqrt{I_1^2 - 3I_2}\cos\left(\phi + \frac{2\pi}{3}\right) \tag{2.21}$$

$$\sigma_3 = \frac{I_1}{3} + \frac{2}{3}\sqrt{I_1^2 - 3I_2}\cos\left(\phi + \frac{4\pi}{3}\right) , \tag{2.22}$$

where

$$\phi = \frac{1}{3}\arccos\left(\frac{2I_1^3 - 9I_1I_2 + 27I_3}{2(I_1^2 - 3I_2)^{3/2}}\right) . \tag{2.23}$$

If the principal value of $\theta = \arccos(x)$ is defined such that $0 \le \theta \le \pi$, the principal stresses defined by equations (2.20–2.22) will always satisfy the inequality

$$\sigma_1 \ge \sigma_3 \ge \sigma_2 \tag{2.24}$$

and, in particular, $\sigma_1$ will be the maximum tensile stress.

Substituting the above numerical values for $I_1, I_2, I_3$, we obtain the three principal stresses

$$\sigma_1 = 13.83 \text{ ksi}, \ \sigma_2 = -10.16 \text{ ksi}, \ \sigma_3 = -0.68 \text{ ksi} .$$

The maximum tensile stress is therefore $\sigma_1 = 13.83$ ksi. To find the direction cosines of the corresponding plane, we substitute this value for $\sigma$ into any two of the equations (2.12–2.14) and solve for the ratios $l/n, m/n$. For example, dividing (2.12, 2.13) by $n$, we have

$$(\sigma_{xx}-\sigma_1)\frac{l}{n}+\sigma_{yx}\frac{m}{n}+\sigma_{zx}=0$$
$$\sigma_{xy}\frac{l}{n}+(\sigma_{yy}-\sigma_1)\frac{m}{n}+\sigma_{zy}=0\,,$$

which constitute two simultaneous linear equations for the two unknowns $l/n, m/n$. Substituting the numerical values for the stresses, we have

$$-15.83\frac{l}{n}+7\frac{m}{n}-3=0$$
$$7\frac{l}{n}-3.83\frac{m}{n}+5=0$$

with solution

$$\frac{l}{n}=2.01\ ;\ \ \frac{m}{n}=4.97\,.$$

We now write equation (2.10) in the form

$$\frac{1}{n^2}=\frac{l^2}{n^2}+\frac{m^2}{n^2}+1=29.75\,,$$

from which $n=0.183$. The values of $l,m$ are then recovered as

$$l=\frac{l}{n}\times n=2.01\times 0.183=0.368\ ;\ \ m=\frac{m}{n}\times n=4.97\times 0.183=0.910\,.$$

To summmarize, the principal stresses are

$$\sigma_1=13.83\text{ ksi},\ \sigma_2=-10.16\text{ ksi},\ \sigma_3=-0.68\text{ ksi}$$

and the maximum tensile stress $\sigma_1$ acts on the plane defined by[6]

$$\{l,m,n\}=\{0.368,0.910,0.183\}\,.$$

**Stress invariants**

Equation (2.16) defines the principal stresses in terms of the stress components in any convenient Cartesian coordinate system $x,y,z$. The principal stresses cannot depend on the choice of $x,y,z$ and hence, for a given state of stress, the same equation (2.16) must be obtained regardless of the original choice of coordinate system. It follows that the coefficients $I_1, I_2, I_3$ in this equation must have the same value in all Cartesian coordinate systems — in other words, that equations (2.17–2.19) define quantities that are *invariant* under arbitrary rotation of the coordinate system (hence the name). In particular, if we choose $x,y,z$ to be aligned with the three principal directions, we have

[6] If we had taken the negative root $n=-0.183$, we would have obtained the eigenvector $\{-0.368,-0.910,-0.183\}$ which defines the same principal plane.

$$I_1 = \sigma_1 + \sigma_2 + \sigma_3 \tag{2.25}$$
$$I_2 = \sigma_1\sigma_2 + \sigma_2\sigma_3 + \sigma_3\sigma_1 \tag{2.26}$$
$$I_3 = \sigma_1\sigma_2\sigma_3\ , \tag{2.27}$$

which could also be deduced directly by noting that equation (2.16) must be capable of factorization in the form

$$(\sigma - \sigma_1)(\sigma - \sigma_2)(\sigma - \sigma_3) = 0\ . \tag{2.28}$$

The importance of stress invariants stems from the fact that the failure of an isotropic material cannot depend upon the orientation of the stress field and hence must depend on a stress measure that is invariant under coordinate transformation. This question is discussed in §2.2 below.

### Cases where one principal direction is known

The largest stresses in engineering components almost always occur at the surface, rather than at interior points. For example, in bending, the tensile stress is linearly proportional to the distance from the neutral axis and hence the maximum stress occurs at the greatest distance from this axis, which must necessarily be adjacent to an exposed surface. Most exposed surfaces are traction-free, so if we choose a local coordinate system such that the $z$-direction is the outward normal from the surface, the components $\sigma_{zx} = \sigma_{zy} = \sigma_{zz} = 0$. In some cases, there may be a non-zero normal traction at the surface (i.e. $\sigma_{zz} \neq 0$) due to contact with another body or pressure of a gas, but it is really quite difficult to apply shear tractions to an exposed surface.[7]

If the shear stresses $\sigma_{zx} = \sigma_{zy}$ are zero at the point of interest, it follows immediately that the $z$-direction is one of the principal directions and the other two must therefore lie in the orthogonal $x, y$ plane. These directions and the corresponding principal stresses can then be found from the two-dimensional analysis of §2.1.1, since, although there may now be non-zero tractions $\sigma_{zz}$ on the triangular faces of the three-dimensional element equivalent to Figure 2.1 *(b)*, the corresponding forces will not enter into the in-plane equilibrium equations (2.1, 2.2).

Thus, although there are strictly no two-dimensional problems (all bodies are three-dimensional), most of the stress transformation questions that arise in design reduce to an application of the two-dimensional equations of §2.1.1.

### Three-dimensional Mohr's circles

Once we have determined the three principal stresses $\sigma_1, \sigma_2, \sigma_3$, we can use a Mohr's circle to find the stress components on any plane that can be reached by rotating any of the principal planes about one of the principal coordinate axes. Three such circles can be drawn, as shown in Figure 2.5 *(a)*.

[7] Tangential tractions may result from frictional contact or from viscous drag of a fluid, but the resulting tractions are generally small compared with failure strengths for the material.

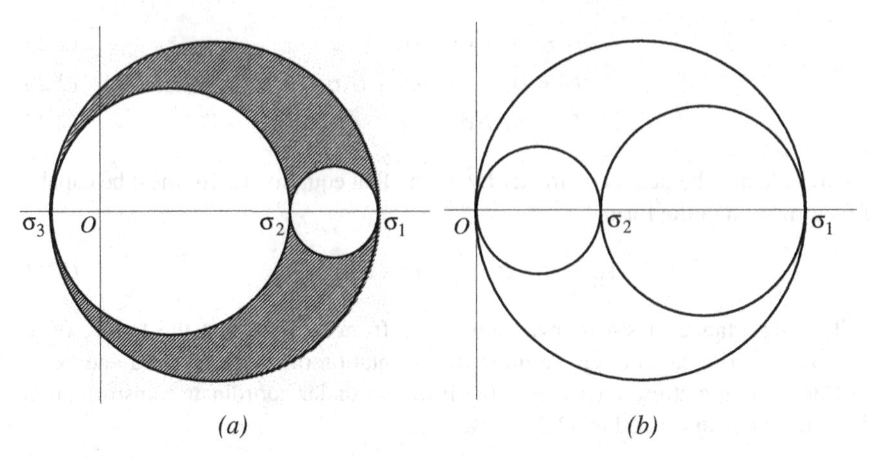


*Figure 2.5: Three dimensional Mohr's circles: (a) the general case, (b) the case where* $\sigma_3 = 0$ *and* $\sigma_1 > \sigma_2 > 0$.

Of course, there exist a whole range of planes that cannot be reached by such a rotation, but it can be shown that the normal and shear stresses on all these planes correspond to points in the area bounded by the three circles and shown shaded in Figure 2.5 *(a)* (Boresi *et al.* (1993), pp.48–51). The maximum tensile stress is therefore the largest of $\sigma_1, \sigma_2, \sigma_3$ and the maximum shear stress is the largest of the three radii — i.e.

$$\sigma_{\max} = \max(\sigma_1, \sigma_2, \sigma_3) \tag{2.29}$$

$$\tau_{\max} = \max\left(\frac{|\sigma_1 - \sigma_2|}{2}, \frac{|\sigma_2 - \sigma_3|}{2}, \frac{|\sigma_3 - \sigma_1|}{2}\right) . \tag{2.30}$$

It is important to realize that all real world problems are three-dimensional, even if the three stress components $\sigma_{zz}, \sigma_{yz}, \sigma_{zx}$ are zero (and hence one of the three principal stresses is zero). Suppose we analyze a system and determine that both the non-zero principal stresses $\sigma_1, \sigma_2$ are tensile (positive) — for example $\sigma_1 > \sigma_2 > 0$. A naïve analysis, neglecting the three-dimensionality of the problem would identify the maximum shear stress as $(\sigma_1 - \sigma_2)/2$, whereas it is clear from the three-dimensional Mohr's circles of Figure 2.5 *(b)* (using $\sigma_3 = 0$) that it is $\sigma_1/2$.

## 2.2 Failure theories for isotropic materials

If we take a block of material and machine a test specimen from it, the behaviour of the specimen (e.g. the tensile strength or the elastic properties) might depend on the orientation of the specimen relative to the original block. Such a material is referred to as *anisotropic* — i.e. it does not have the same properties in all directions. A simple example of anisotropy is a fibre-reinforced composite in which all the fibres are parallel. In this case, a tensile specimen cut with the fibres along the test axis will generally be stronger and stiffer than one cut across the fibre axis. Other sources of anisotropy include the orientation of rows of atoms in a single crystal.

For most design calculations (with the important exception of fibre-reinforced composites and wood), we make the simplifying assumption that the material is *isotropic*, meaning that it has the same properties in all directions and hence that all test specimens will have the same properties regardless of orientation. In many cases, the material may be quite inhomogeneous and anisotropic on the microscopic scale. For example, it may have a granular structure and each grain may have its atoms aligned in different directions. Also there may be distributions of impurities, microcracks and voids. However, if these irregularities are sufficiently small in size and are randomly oriented and distributed, a larger specimen of the material will appear isotropic.

Local failure of an isotropic material must depend only on measures of the severity of the local stress state without reference to direction, and hence must be capable of description in terms of quantities that are invariant under coordinate transformation. These quantities include the set of three principal stresses $\sigma_1, \sigma_2, \sigma_3$ or the three stress invariants $I_1, I_2, I_3$ of equations (2.17–2.19). Any one of these sets can be determined from the other, so a comprehensive set of failure tests for an isotropic material involves the variation of only three parameters.

### 2.2.1 The failure surface

Suppose we could devise a machine that is capable of applying arbitrarily chosen values of the three principal stresses $\sigma_1, \sigma_2, \sigma_3$ to a suitable test specimen of a given material. Testing would involve varying one or more of the stresses until failure occurs.

#### Failure under plane stress

Consider first the simple case of *plane stress*, where one of the three principal stresses, $\sigma_3$ is always zero. Possible stress states can then be presented as points on a graph with coordinates $(\sigma_1, \sigma_2)$ and loading scenarios will be directional lines on this plot as shown in Figure 2.6.

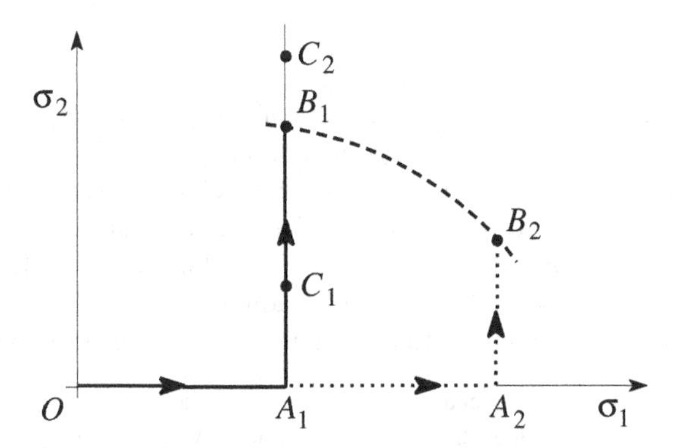


*Figure 2.6: Plane stress loading scenarios leading to failure*

For example, the line $OA_1B_1$ describes a test in which all three stresses are originally zero; $\sigma_1$ is then increased slowly until it reaches the value $OA_1$, after which it is held constant whilst $\sigma_2$ is increased, until failure occurs at $B_1$. A second test could be performed using a similar scenario, but with a different value $OA_2$ for $\sigma_1$, in which case the critical value of $\sigma_2$ ($A_2B_2$) would also generally be different, as shown.

If we connect the failure points $B_1, B_2$ etc. in Figure 2.6 by a line, this line will define a boundary separating *safe* states of stress from *unsafe* states. Thus, the point $C_1$ is safe because it is passed without failure during the test $OA_1B_1$, but the point $C_2$ is unsafe, because failure occurs at $B_1$ before $C_2$ is reached.

Failure can occur in compression as well as tension, so the complete failure envelope is likely to be a closed curve as shown in Figure 2.7. Everywhere inside the curve, including the origin, where all three stresses are zero, is safe, but failure will occur if we try to cross the failure envelope into the surrounding unsafe region.

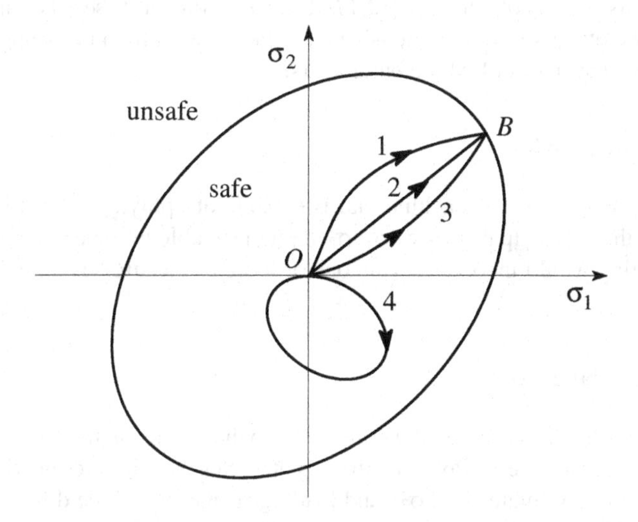


*Figure 2.7: Failure envelope for plane stress*

It is important to recognize that this argument depends on the assumption that failure is determined by the instantaneous state of stress and is not influenced by the history of the loading before the failure point. For most materials, this is a reasonable assumption if the deformation inside the failure envelope is elastic and the loading scenario does not involve stress reversal — for example, failure will occur at the point $B$ in Figure 2.7 for any of the 'monotonic'[8] loading scenarios 1,2,3. However, scenario 4, involving a number of cycles of loading and unloading within the safe region, is likely to lead to failure before the monotonic failure envelope is reached.

[8] We shall refer to a loading scenario as *monotonic* if the none of the six stress rates $\dot{\sigma}_{xx}, \dot{\sigma}_{yy}, \dot{\sigma}_{zz}, \dot{\sigma}_{xy}, \dot{\sigma}_{yz}, \dot{\sigma}_{zx}$ change sign during the process, where $\dot{\sigma} \equiv \partial\sigma/\partial t$ and $t$ is time.

This process is known as *fatigue* and will be discussed in more detail in §2.3 below. For the rest of this section, we shall restrict attention to failure under monotonic loading.

#### The three-dimensional failure envelope

The ideas of the previous section are easily extended to the case where all three principal stresses can be non-zero. For this purpose, we need to imagine a three-dimensional graph in which $\sigma_1, \sigma_2, \sigma_3$ define the three axes. Loading scenarios will again correspond to lines in this three-dimensional space and failure points can be joined to define a closed *surface* such that all points inside the surface correspond to safe states of stress and all points outside the surface are unsafe.

### 2.2.2 The shape of the failure envelope

So far we have discussed ways of presenting data about the failure of materials under general states of stress, but we have not given any indication of the form the failure criterion is likely to take. In other words, what shape would we expect the failure envelope to have? Many authors have proposed theories of how materials might be expected to behave, but ultimately the question can only be answered by performing appropriate experiments. Notice however that we can draw one further conclusion from the isotropy of the material, which is that failure cannot depend on the *order* of the three stresses $\sigma_1, \sigma_2, \sigma_3$. For example, if failure occurs at the point $(a, b, c)$ it must also occur at $(b, a, c)$. It follows that the curve in Figure 2.7 must be symmetrical about the $45^\circ$ line $\sigma_1 = \sigma_2$ and additional symmetries of the same kind apply in the three-dimensional envelope.

### 2.2.3 Ductile failure (yielding)

To proceed further, we need to consider separately the cases of ductile and brittle materials, since they differ qualitatively in mechanism and behaviour.

Ductile materials suffer irreversible plastic deformation before fracture occurs. Whether this constitutes failure depends on the specific design application, so to avoid confusion, we shall refer to ductile failure as *yielding* and the corresponding failure surface as the *yield surface*. Thus, if a component is loaded such that the stress in some region passes outside the yield surface, it will not return to its original state on unloading. A simplistic model of the distinction between elastic and plastic deformation can be developed from atomic theory.

Consider first a sphere resting on an undulating surface as in Figure 2.8 *(a)*. When unloaded, the sphere will rest at the bottom of the groove. Application of a horizontal force will cause the sphere to move as shown in Figure 2.8 *(b)*, but if the force is removed slowly, the sphere will return to the bottom of the groove. This is an analogue of elastic deformation. However, if a sufficiently large force is applied, the sphere will be pushed over into the next groove and it will remain there even if the force is removed. Once the critical force is reached, the sphere could be pushed over many grooves, producing much larger displacements than are possible in the elastic range. Notice also that when the sphere is being pushed down into the next groove, the system is unstable. We could remove the force at this point but the sphere would keep moving and would be left oscillating under gravity about the equilibrium position in the next groove.

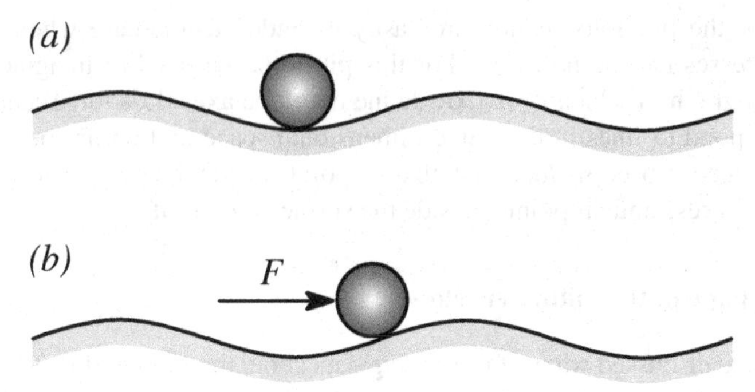


*Figure 2.8: The sphere and groove model — (a) unloaded equilibrium position, (b) displacement due to a horizontal force*

Now consider a piece of material as a large assembly of atoms or molecules held together under the influence of interatomic forces. When there is no external load, the assembly will adopt an equilibrium position.[9] Application of external forces will cause the atoms to make small relative motions to a new equilibrium configuration. However, sufficiently large forces might cause an atom or a block of atoms to jump into a new configuration, analogously with the movement from one groove to the next in Figure 2.8. In this case, removal of the load will not cause the system to return to its original configuration and when the forces are removed, some of the atoms will be left vibrating about their new equilibrium positions. The kinetic energy associated with this vibration cannot be recovered through the external forces during unloading and in fact has already been converted to heat.[10]

Calculations of the force needed to move one block of atoms relative to another grossly overestimate the theoretical yield stress in shear. In reality, crystal structures are generally not geometrically perfect, but contain local defects known as *dislocations*. Near the dislocations, the interatomic distances differ from the theoretical value for a regular structure and a considerably smaller force is needed to make a pivotal atom move to a new equilibrium position.[11]

---

[9] We shall argue in Chapter 3 that this can also be seen as a minimum energy configuration, as it clearly is for the sphere at the bottom of the groove in Figure 2.8 *(a)*.

[10] Recall that the distinction between a hot body and a cold one is that the atoms or molecules in the hot body are vibrating with larger amplitude about their mean position.

[11] J.E. Gordon (1968), *The New Science of Strong Materials*, Princeton University Press, Princeton NJ, pp.91–98.

**Change in volume**

Elastic deformation usually involves a change in volume or *dilatation*. For example, the increase in length of a bar in uniaxial tension will more than compensate for the reduction in diameter due to Poisson's ratio, leading to an increase in volume except in the special case of a material for which $\nu = 0.5$. In the atomic model, this corresponds to a stretching of the interatomic bonds, so that the average interatomic distance is increased.

If we load a body into the plastic range and then remove the load, the final shape will differ from the original shape, but there is no reason to expect the volume of the body to have changed, since the average interatomic distance is governed by similar equilibrium arguments in both original and final states. This is confirmed by experiments which show that plastic deformation does not generally cause any change in volume.[12]

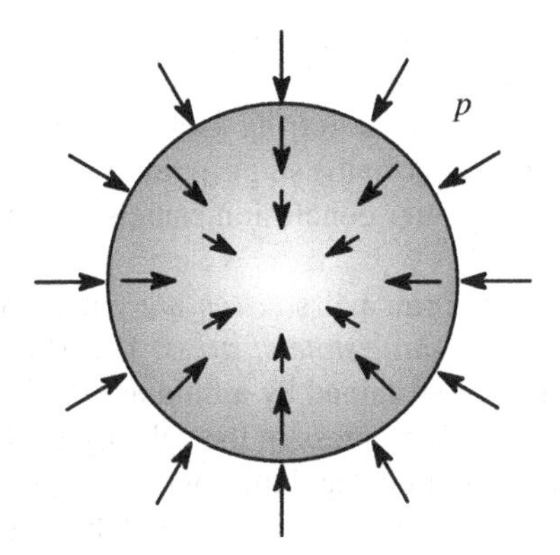


*Figure 2.9: Sphere loaded by uniform compressive tractions*

A corollary of this result is that stress states causing no change of shape cannot result in plastic deformation regardless of their magnitude. Consider for example a solid sphere of material loaded by a uniform external pressure $p$ as shown in Figure 2.9. Because of symmetry, we would expect the sphere to remain spherical,[13] but to

[12] It must be emphasised that this conclusion relates to a comparison of the volume before loading and after unloading. Consider a uniaxial tensile test in which a specimen is loaded into the plastic range and then unloaded. The unloading process is elastic and will involve a reduction in volume. If the material yields at constant stress, the increase in volume during the elastic phase of the loading process will be exactly equal to the decrease during unloading and there will be no change in volume during the plastic phase of the process. However, if the material work hardens (i.e. the yield stress increases with plastic strain), the unloading process will occur over a larger stress range than the elastic phase of the loading and will involve correspondingly more change in volume. The final volume will still be equal to the initial volume and therefore some increase of volume is associated with work hardening during the plastic phase of the process.

[13] Strictly, we should allow the possibility that the body deviates from sphericity under the load. However, it can be shown that the work done by the uniform applied tractions must be

get smaller under the influence of the pressure. Thus, there is no change in shape — only a change in volume — and so no plastic deformation can be involved. In principle, we should be able to increase $p$ as much as we like without causing yielding to occur. Perhaps more surprising is that the same argument would apply if the tractions were tensile instead of compressive — i.e. if we could exert a uniform tensile traction on the surface of the sphere. Engineers have been quite ingenious[14] in trying to verify this conclusion by experiment. Such results as are obtained suggest that tensile failure eventually occurs, but in an essentially brittle manner (i.e. by propagation of microcracks) even for usually ductile materials.

### Hydrostatic stress

If we argue from symmetry that the stresses in the sphere of Figure 2.9 depend only on the distance $R$ from the centre, equilibrium arguments can then be used to show that the stress state everywhere is defined by $\sigma_1 = \sigma_2 = \sigma_3 = -p$ — i.e. the three principal stresses are equal to the applied uniform tractions (remember that compressive stresses are negative by definition). Referring back to Figure 2.5, we conclude that the three Mohr's circles will reduce to the single point $\sigma = -p$ and that the maximum shear stress will be zero. This latter conclusion can also be obtained from equation (2.30).

The case where the three principal stresses are equal and there are no shear stresses is referred to as a *hydrostatic state of stress*. The name arises because this is the *only* stress state that can be sustained by a fluid at rest. Recall that the constitutive law for a fluid relates the shear stress to the velocity gradient (for a Newtonian fluid, these quantities are proportional and the constant of proportionality is defined as the viscosity). If the fluid is at rest, the velocity and the velocity gradient are both zero and hence there must be no shear stress. Notice also that when the three Mohr's circles condense to the point $\sigma = -p$, the normal stress on all planes must be equal to $-p$. This constitutes a proof of the well-known result in hydrostatics that the pressure at a point in a fluid at rest is equal in all directions.

It is no coincidence that an example chosen to exhibit no change of shape under load involves no shear stress. Shear stress causes shear strain, which in turn is defined in terms of the change in angles between faces of an element of material.[15] Changes of angles necessarily involve change of shape, so the only states in which there is no change in shape during deformation are those in which the stress is everywhere hydrostatic.

---

zero under a constant volume plastic deformation. Since plasticity is a dissipative process, it can only occur when the body deforms in such a way as to allow the external loads to do work, so we can still conclude that plastic deformation is impossible in the sphere loaded by a uniform pressure.

[14] It is far from easy to apply a sufficiently large uniform tensile traction to the surface of a body. Generally, the load application mechanism introduces areas of non-uniform stress which serve as yield sites.

[15] See §1.5.3 and Figure 1.6.

The arguments of the preceding sections lead to the conclusion that yielding should never occur under hydrostatic stress, regardless of its magnitude, and hence that the complete straight line $\sigma_1 = \sigma_2 = \sigma_3$ should lie within the safe region for the three-dimensional yield envelope. The two most important yield theories used for ductile materials both meet this condition.

### Tresca's maximum shear stress criterion

Since plastic deformation involves irreversible shear strain and cannot occur if there are no shear stresses, an obvious hypothesis for yielding is that it will occur when the maximum shear stress reaches a critical value $\tau_Y$ — i.e.

$$\tau_{\max} \equiv \max\left(\frac{|\sigma_1 - \sigma_2|}{2}, \frac{|\sigma_2 - \sigma_3|}{2}, \frac{|\sigma_3 - \sigma_1|}{2}\right) = \tau_Y \,. \tag{2.31}$$

The elastic region is then defined by the inequality $\tau_{\max} < \tau_Y$. This is known as the *maximum shear stress theory* or *Tresca's yield theory.*

Tresca's theory has one major advantage and one disadvantage. The governing equation at the yield envelope is linear in the stresses and hence leads to linear equations which are algebraically easy to solve. However, the fact that $\tau_{\max}$ is the largest of three quantities can be inconvenient. Generally we shall know enough about the probable stress state to know which Mohr's circle will have the largest radius, but this choice makes it necessary to think about the problem carefully rather than applying the criterion mindlessly. This opens up the possibility of error and it also makes for complication when the theory is used as part of a numerical algorithm.

### Von Mises' deviatoric strain energy criterion

An alternative theory due to *von Mises* states that yielding will occur when the strain energy density associated with the shear stresses reaches a critical value. This concept requires some preliminary discussion before the criterion can be formally enunciated.

When an element of material is loaded elastically, the applied tractions do work which is recovered on unloading. In the loaded state, it is natural to think of this energy as being stored in the material as *strain energy*.

Consider an infinitesimal element of material of volume $\delta V = \delta x \delta y \delta z$, over which the stresses $\sigma_{xx}, \sigma_{yy}, \sigma_{zz}, \sigma_{yz}, \sigma_{zx}, \sigma_{xy}$ can be considered uniform. The stress $\sigma_{xx}$ acts over a surface $\delta y \delta z$ and hence corresponds to a force

$$F = \sigma_{xx} \delta y \delta z \,.$$

The relative displacement of the two opposite faces experiencing this force is

$$\delta u_x = e_{xx} \delta x$$

and hence the contribution to the strain energy is

$$\delta U = \frac{1}{2} F \delta u_x = \frac{1}{2} \sigma_{xx} e_{xx} \delta x \delta y \delta z = \frac{1}{2} \sigma_{xx} e_{xx} \delta V \; .$$

Similar results are obtained for the other stress components and lead to the expression

$$\delta U = \frac{\delta V}{2} (\sigma_{xx} e_{xx} + \sigma_{yy} e_{yy} + \sigma_{zz} e_{zz} + 2\sigma_{yz} e_{yz} + 2\sigma_{zx} e_{zx} + 2\sigma_{xy} e_{xy})$$

for the strain energy contained in the volume $\delta V$. The *strain energy density* $U_0$ is defined as the strain energy per unit volume and is therefore

$$U_0 = \frac{1}{2} (\sigma_{xx} e_{xx} + \sigma_{yy} e_{yy} + \sigma_{zz} e_{zz} + 2\sigma_{yz} e_{yz} + 2\sigma_{zx} e_{zx} + 2\sigma_{xy} e_{xy}) \; .$$

Hooke's law [equations (1.7–1.9, 1.12)] can be used to express this result in terms of stresses only in the form

$$U_0 = \frac{1}{2E} \Big[ \sigma_{xx}^2 + \sigma_{yy}^2 + \sigma_{zz}^2 - 2\nu(\sigma_{xx}\sigma_{yy} + \sigma_{yy}\sigma_{zz} + \sigma_{zz}\sigma_{xx}) + 2(1+\nu)\left(\sigma_{yz}^2 + \sigma_{zx}^2 + \sigma_{xy}^2\right) \Big] \; . \tag{2.32}$$

If the Cartesian axes $x, y, z$ are chosen to coincide with the principal directions, this expression reduces to the simpler form

$$U_0 = \frac{1}{2E} \left[ (\sigma_1^2 + \sigma_2^2 + \sigma_3^2) - 2\nu(\sigma_1\sigma_2 + \sigma_2\sigma_3 + \sigma_3\sigma_1) \right] \; . \tag{2.33}$$

Early researchers argued that a material could only store a limited amount of strain energy per unit volume before yield occurred, but this leads to a criterion that conflicts with the conclusion that yield cannot occur under hydrostatic stress. This difficulty can be overcome by decomposing the actual stress state into a hydrostatic *mean stress* $\bar{\sigma}$ and a *deviatoric stress* $\hat{\sigma}$, defined by

$$\sigma = \bar{\sigma} + \hat{\sigma} \; , \tag{2.34}$$

where

$$\bar{\sigma}_1 = \bar{\sigma}_2 = \bar{\sigma}_3 = \bar{\sigma} \equiv \frac{\sigma_1 + \sigma_2 + \sigma_3}{3} \tag{2.35}$$

$$\hat{\sigma}_1 = \sigma_1 - \bar{\sigma} = \frac{2\sigma_1 - \sigma_2 - \sigma_3}{3} \tag{2.36}$$

$$\hat{\sigma}_2 = \sigma_2 - \bar{\sigma} = \frac{2\sigma_2 - \sigma_3 - \sigma_1}{3} \tag{2.37}$$

$$\hat{\sigma}_3 = \sigma_3 - \bar{\sigma} = \frac{2\sigma_3 - \sigma_1 - \sigma_2}{3} \; . \tag{2.38}$$

Substituting these expressions into (2.33) and using the result $\hat{\sigma}_1 + \hat{\sigma}_2 + \hat{\sigma}_3 = 0$ which can be verified by adding (2.36–2.38), we can write $U_0$ in the form[16]

[16] We cannot simply *assume* that the total strain energy density will be the sum of the values obtained on substituting $\bar{\sigma}$ and $\hat{\sigma}$ respectively into equation (2.33), since the equation is

$$U_0 = \bar{U}_0 + \hat{U}_0 , \tag{2.39}$$

where

$$\bar{U}_0 = \frac{3\bar{\sigma}^2(1-2\nu)}{2E} = \frac{(\sigma_1+\sigma_2+\sigma_3)^2(1-2\nu)}{6E} = \frac{I_1^2(1-2\nu)}{6E} \tag{2.40}$$

is the *dilatational strain energy density* and

$$\hat{U}_0 = \frac{\left[(\hat{\sigma}_1-\hat{\sigma}_2)^2+(\hat{\sigma}_2-\hat{\sigma}_3)^2+(\hat{\sigma}_3-\hat{\sigma}_1)^2\right](1+\nu)}{6E}$$
$$= \frac{\left[(\sigma_1-\sigma_2)^2+(\sigma_2-\sigma_3)^2+(\sigma_3-\sigma_1)^2\right](1+\nu)}{6E} \tag{2.41}$$
$$= \frac{(I_1^2-3I_2)(1+\nu)}{3E} \tag{2.42}$$

is the *deviatoric strain energy density*. Deviatoric strain energy is sometimes referred to as the *strain energy of distortion* or *shear strain energy*.

Von Mises' theory states that *yielding will occur when the deviatoric strain energy density reaches a critical value for the material* — i.e.

$$\hat{U}_0 \equiv \frac{\left[(\sigma_1-\sigma_2)^2+(\sigma_2-\sigma_3)^2+(\sigma_3-\sigma_1)^2\right](1+\nu)}{6E} = \hat{U}_Y . \tag{2.43}$$

The elastic region is then defined by the inequality $\hat{U}_0 < \hat{U}_Y$.

Notice that using von Mises' criterion, we don't have to decide which of the three stress differences (Mohr's circle diameters) is the greatest. Instead, we just square them (thereby ensuring a positive result) and add them. Von Mises' criterion is therefore easier to apply than Tresca's, but it leads to non-linear equations.

### Octahedral shear stress

There exist eight *octahedral planes* which make equal angles with the three principal directions. If we define the $x, y, z$ axes to coincide with the principal directions, the three direction cosines of the normals to the octahedral planes must be of equal magnitude and hence must be $1/\sqrt{3}$, in view of (2.10). Thus, the octahedral planes are defined by the set $\{l, m, n\} = \{\pm 1/\sqrt{3}, \pm 1/\sqrt{3}, \pm 1/\sqrt{3}\}$. One of the octahedral planes is illustrated in Figure 2.10. There are eight such planes in all and they define a regular octahedron.

The normal stress and resultant shear stress on all these planes can be shown[17] to be given by

---

non-linear (quadratic) in the stresses. However, the algebra shows that the strain energy density does in fact decompose in this way, indicating that the processes of dilatation and distortion are orthogonal to each other. This implies that no work is done by a purely deviatoric stress field during a change in volume without change in shape and conversely that no work is done by a hydrostatic stress field during a distortion without dilatation.

[17] See for example Boresi *et al.* (1993), equation (2.22).

$$\sigma_{\text{oct}} = \bar{\sigma} = \frac{I_1}{3} \tag{2.44}$$

$$\tau_{\text{oct}} = \frac{1}{3}\sqrt{(\sigma_1-\sigma_2)^2+(\sigma_2-\sigma_3)^2+(\sigma_3-\sigma_1)^2} = \frac{1}{3}\sqrt{2(I_1^2-3I_2)} \tag{2.45}$$

and are known as the *octahedral normal stress* and *octahedral shear stress*, respectively.

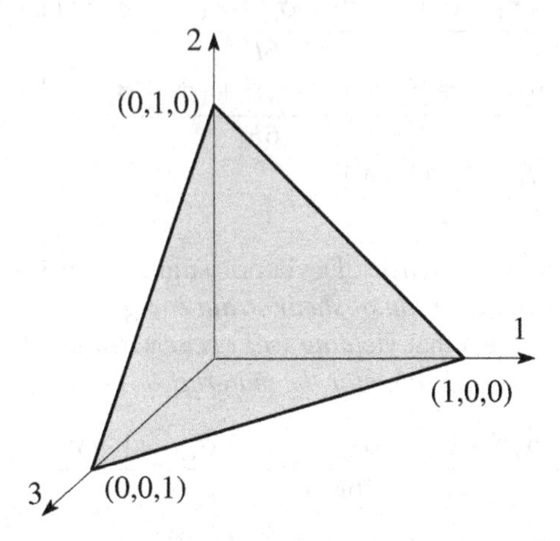


*Figure 2.10: The octahedral plane* $\{l,m,n\} = \{1/\sqrt{3}, 1/\sqrt{3}, 1/\sqrt{3}\}$

Comparing these results with equation (2.43), we see that the von Mises theory is mathematically equivalent to the statement that *yielding will occur when the octahedral shear stress reaches a critical value for the material.* The von Mises theory is therefore also sometimes referred to as the *octahedral shear stress theory*.

It is instructive to note that two essentially different measures of the severity of a state of stress can lead to the same expression for the yield envelope. There is no way that a set of experiments could determine which (if indeed either) of the octahedral shear stress or the deviatoric strain energy density is most influential in causing yielding.[18] All we can do is to observe the experimental shape of the yield envelope (which is usually closely approximated by equation (2.43) for an isotropic ductile material) and conclude that either of the two physical quantities provide plausible physical 'explanations' of this observation. For design purposes, this is also all we *need* to do. A mathematical or graphical description of the experimentally determined yield surface is sufficient to tell us whether a proposed state of stress is safe or not.

[18] The reader might like to ponder whether some different kind of experiment could be devised that *would* enable the balance to be tilted in favour of one or other of these theories. For example, would the theories predict the same yield envelope for an anisotropic material and if not, what would the results of the corresponding test tell us?

### Equivalent tensile stress

The theories of Tresca and von Mises define the shape of the yield surface, so the result of a single experiment is sufficient to define the surface completely. For example, suppose we perform a uniaxial tensile test and determine the uniaxial yield stress $S_Y$ for the material. It follows that the point $(S_Y, 0, 0)$ must lie on the yield surface and this is sufficient to determine

$$\tau_Y = \frac{S_Y}{2} \tag{2.46}$$

from (2.31) or

$$\hat{U}_Y = \frac{S_Y^2(1+\nu)}{3E} \tag{2.47}$$

from (2.43).

It is convenient to define the *equivalent tensile stress* $\sigma_E$ as that function of the stress components that is equal to $S_Y$ on the yield surface. For example, we can substitute (2.47) into (2.43) to restate von Mises' criterion as

$$\begin{aligned}\frac{\left[(\sigma_1-\sigma_2)^2+(\sigma_2-\sigma_3)^2+(\sigma_3-\sigma_1)^2\right](1+\nu)}{6E} &= \frac{(I_1^2-3I_2)(1+\nu)}{3E} \\ &= \frac{S_Y^2(1+\nu)}{3E}\end{aligned} \tag{2.48}$$

and hence

$$\sigma_E \equiv \sqrt{\frac{(\sigma_1-\sigma_2)^2+(\sigma_2-\sigma_3)^2+(\sigma_3-\sigma_1)^2}{2}} = \sqrt{I_1^2-3I_2} = S_Y \,. \tag{2.49}$$

A similar procedure with equations (2.46, 2.31) gives

$$\sigma_E \equiv \max(|\sigma_1-\sigma_2|, |\sigma_2-\sigma_3|, |\sigma_3-\sigma_1|) = S_Y \tag{2.50}$$

for the Tresca criterion.

The formulations (2.49, 2.50) are very convenient for design purposes, since all we need to do is to evaluate $\sigma_E$ and compare it with tabulated values of the uniaxial yield stress $S_Y$. The concept of equivalent tensile stress also enables us to define a safety factor against yield as

$$SF = \frac{S_Y}{\sigma_E} \,. \tag{2.51}$$

If the stress components were increased proportionally, the safety factor is the ratio by which they would need to be increased for yield to occur.

### Graphical representation

In the case of plane stress ($\sigma_3 = 0$), the Tresca theory (2.50) reduces to

$$\sigma_E \equiv \max(|\sigma_1-\sigma_2|, |\sigma_2|, |\sigma_1|) = S_Y \tag{2.52}$$

and the von Mises theory (2.49) to

$$\sigma_E \equiv \sqrt{\sigma_1^2 + \sigma_2^2 - \sigma_1\sigma_2} = S_Y \,. \tag{2.53}$$

These expressions plot out as the hexagon and ellipse respectively in Figure 2.11.

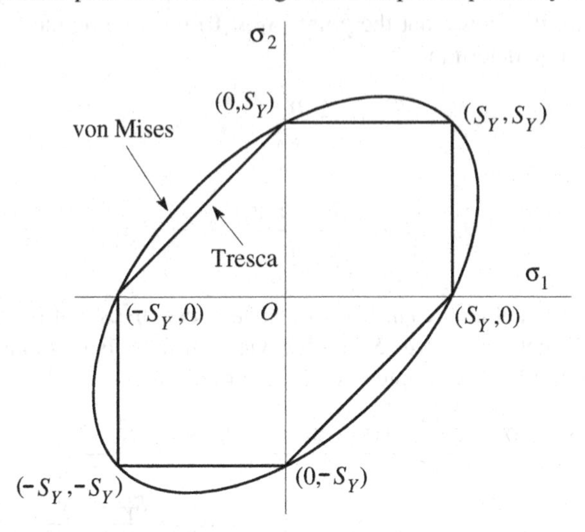


*Figure 2.11: Tresca and von Mises yield envelopes for plane stress*

The corresponding three-dimensional yield envelope is a regular hexagonal cylinder for the Tresca criterion and a circumscribed circular cylinder for von Mises' theory, the axis in each case being the inclined line $\sigma_1 = \sigma_2 = \sigma_3$ (which we know is always in the safe region. These envelopes are illustrated in Figure 2.12.

The plane stress results of Figure 2.11 are simply the intersection of these regular three-dimensional surfaces with the plane $\sigma_3 = 0$. Notice that in the three-dimensional case, the von Mises criterion has the very simple graphical interpretation that *yielding occurs when the shortest distance in principal stress space to the hydrostatic line* $\sigma_1 = \sigma_2 = \sigma_3$ *exceeds a critical value*. This value (the radius of the cylinder in Figure 2.12) can be shown to be $\sqrt{2/3}\,S_Y$.

Careful experiments for ductile materials such as metals tend to favour von Mises' theory over Tresca's, but the difference is comparatively small and mostly lies within the range of variability of material between one sample and another. Thus, in practice, we use whichever theory is more convenient for design calculations. As we explained above, Tresca's criterion has the advantage of leading to linear equations and hence to a simpler formulation in problems involving plasticity.[19] However, if we simply wish to calculate a safety factor in design using equation (2.51), the von Mises theory is more convenient since it involves the computation of a single algebraic expression.

[19] See Chapters 5,9.

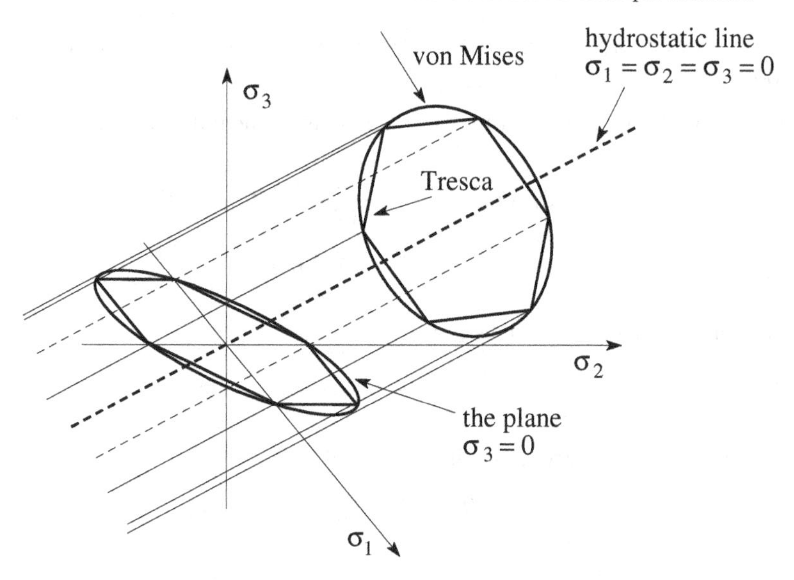


*Figure 2.12: Tresca and von Mises yield envelopes for three-dimensional states of stress*

An additional advantage is that the von Mises equivalent tensile stress is stated in terms of the stress invariants $I_1, I_2$ and hence can be evaluated directly from the stress components in a general coordinate system, without first finding the principal stresses. Substituting for $I_1, I_2$ from (2.17, 2.18) into (2.49) and simplifying, we find

$$\sigma_E = \sqrt{\sigma_{xx}^2 + \sigma_{yy}^2 + \sigma_{zz}^2 - \sigma_{xx}\sigma_{yy} - \sigma_{yy}\sigma_{zz} - \sigma_{zz}\sigma_{xx} + 3\sigma_{xy}^2 + 3\sigma_{yz}^2 + 3\sigma_{zx}^2} \,. \quad (2.54)$$

For the special case of plane stress ($\sigma_{zx} = \sigma_{zy} = \sigma_{zz} = 0$), this reduces to

$$\sigma_E = \sqrt{\sigma_{xx}^2 + \sigma_{yy}^2 - \sigma_{xx}\sigma_{yy} + 3\sigma_{xy}^2} \,. \quad (2.55)$$

Finally, if $\sigma_{yy}$ is also zero — i.e. there is only one tensile stress $\sigma_{xx}$ and one shear stress $\sigma_{xy}$ acting on the same plane — we have the simple expression

$$\sigma_E = \sqrt{\sigma_{xx}^2 + 3\sigma_{xy}^2} \,. \quad (2.56)$$

This situation occurs quite frequently — for example, when we have a bar subjected to combined bending and torsion. Design handbooks (and many experienced designers) tend to quote and use equation (2.56) without mentioning that it is a special case of (2.54).

**Example 2.3**

*A cylindrical steel shaft of diameter* 30 *mm is subjected to a compressive axial force* $F = 10$ *kN, a bending moment* $M = 170$ *Nm and a torque* $T = 220$ *Nm. Estimate the* *factor of safety against yielding using von Mises' theory, if the steel has a yield stress $S_Y = 300$ MPa.*

The area $A$, second moment of area $I$, and polar moment of area $J$ for the shaft are

$$A = \frac{\pi \times 30^2}{4} = 707 \text{ mm}^2 \quad ; \quad I = \frac{\pi \times 30^4}{64} = 39.8 \times 10^3 \text{ mm}^4 \, ;$$

$$J = \frac{\pi \times 30^4}{32} = 79.5 \times 10^3 \text{ mm}^4 \, .$$

The axial force produces a normal stress

$$\sigma_{zz} = -\frac{F}{A} = -\frac{10 \times 10^3}{707} = -14.1 \text{ N/mm}^2 \text{ (MPa)} \, ,$$

the bending moment produces a maximum normal stress

$$\sigma_{zz} = \frac{Mc}{I} = \pm \frac{170 \times 10^3 \times 15}{39.8 \times 10^3} = \pm 64.1 \text{ MPa} \, ,$$

where $c$ is the maximum distance from the neutral axis, and the torque produces a maximum shear stress

$$\sigma_{z\theta} = \frac{TR}{J} = \frac{220 \times 10^3 \times 15}{79.5 \times 10^3} = 41.5 \text{ MPa} \, ,$$

where $R$ is the radius of the shaft.

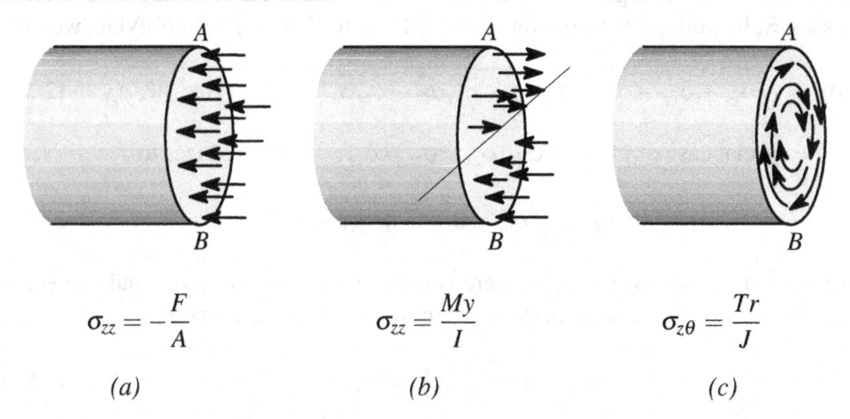


*Figure 2.13: Stress distributions due to (a) axial force, (b) bending and (c) torsion*

The complete stress distributions due to axial force, bending moment and torque are illustrated in Figure 2.13 *(a,b,c)* respectively. The most likely points for yielding to occur are $A$ and $B$, where the maximum bending stress and the maximum shear stress due to torsion coincide. At $A$ we have

$$\sigma_{zz} = -14.1 + 64.1 = 50 \text{ MPa} \quad ; \quad \sigma_{z\theta} = 41.5 \text{ MPa} \, ,$$

whereas at $B$

$$\sigma_{zz} = -14.1 - 64.1 = -78.2 \text{ MPa} \; ; \quad \sigma_{z\theta} = 41.5 \text{ MPa} .$$

The equivalent tensile stress $\sigma_E$ by von Mises' theory is given by the polar coordinate version of equation (2.54), which here reduces to

$$\sigma_E = \sqrt{\sigma_{zz^2} + 3\sigma_{z\theta}^2} .$$

Clearly the maximum value will occur at $B$ and is

$$\sigma_E = \sqrt{78.2^2 + 3 \times 41.5^2} = 106 \text{ MPa} ,$$

so the safety factor against yield is

$$SF = \frac{S_Y}{\sigma_E} = \frac{300}{106} = 2.83 .$$

### 2.2.4 Brittle failure

We now turn our attention to the failure of brittle materials by catastrophic fracture. Look around you at some common objects. Those that can be caused to change their shape permanently by the application of a sufficiently large load are ductile, whereas those that will always return to their original shape on unloading until the load reaches a value sufficient for catastrophic failure are brittle. Typical brittle materials are ceramics, glass and most kinds of rock. Metals are usually ductile, but some can be made relatively brittle by appropriate heat treatment and most become brittle at extremely low temperatures.

Brittle failure typically occurs by the propagation of cracks in the material. Most materials contain a random array of microscopic defects such as cracks, voids, inclusions of foreign materials, grain boundaries etc. and these serve as stress concentrations when loads are applied to the body. When the material is ductile, the result is some local plastic deformation that establishes a more favourable local microstructure and relieves the stress. Plastic deformation also causes some of the work done by the applied loads to be dissipated as heat. By contrast, if the material is brittle, all the work done is stored as elastic strain energy and is recoverable on unloading. In many cases, crack growth is unstable once it starts because more energy is released than is needed to sustain crack propagation, and the component fractures catastrophically. The excess of strain energy converts to kinetic energy of the particles of the body, so the body is left in a state of vibration, which is why fracture is noisy. Try (i) bending a thin metal rod until it deforms plastically and (ii) breaking a similar glass rod in bending and notice that the first process is silent, whereas the second is not. Of course, you know this will be the case before you perform the experiment. It is part of your experiential knowledge that we are trying to unlock.

#### Stress intensity factor

The stress concentration due to a crack is caused by the fact that the crack is unable to transmit tensile tractions, so the load has to be transmitted around the the crack tips, as shown in Figure 2.14. Stresses are highest near the crack tips, where the load lines are closest together. If the crack has zero thickness — i.e. the two faces just touch when the body is unloaded — it can be shown that the theoretical elastic solution of the problem involves stresses that tend to infinity with the inverse square root of the distance from the crack tip.[20]

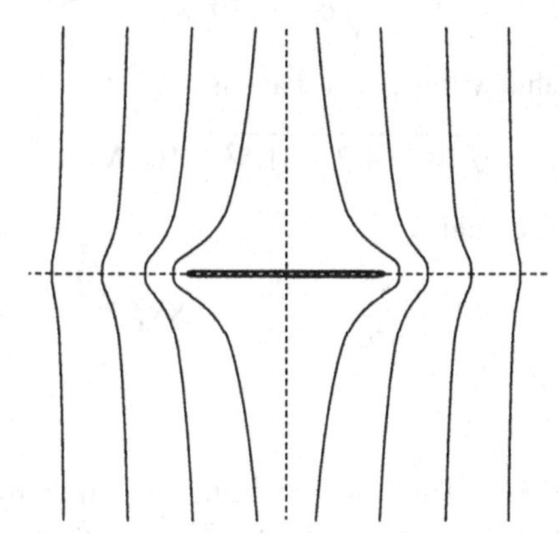

*Figure 2.14: Transmission of load around a crack*

The simplest and most prevalent theory of brittle fracture — that due to Griffith — states in essence that crack propagation will occur when the scalar multiplier on this singular stress field exceeds a certain critical value. More precisely, Griffith proposed the thesis that a crack would propagate when propagation caused a reduction in the total energy of the system. Crack propagation causes a reduction in strain energy in the body, but also generates new surfaces which have *surface energy*. Surface energy is related to the force known as surface tension in fluids and follows from the fact that to cleave a solid body along a plane involves doing work against the interatomic forces across the plane. When this criterion is applied to the stress field in particular cases, it turns out that for a small change in crack length, propagation is predicted when the multiplier on the singular term, known as the *stress intensity factor*, exceeds a certain critical value, which is a constant for the material known as the *fracture toughness*. This is the central thesis of *Linear Elastic Fracture Mechanics (LEFM)*.

Stress intensity factors are conveniently defined in terms of the stress components on the crack plane. If the crack is taken to occupy the region $x<0$ on the plane $y=0$, as shown in Figure 2.15, there are three possible stress components on the rest of the plane $(x>0, y=0)$, comprising a tensile stress $\sigma_{yy}$, an in-plane shear stress $\sigma_{yx}$ and an out-of-plane (or *antiplane*) shear stress $\sigma_{yz}$. Unless they are zero, they will all tend to infinity with $x^{-1/2}$ and we can define[21] stress intensity factors $K_I, K_{II}, K_{III}$ as

[20] See, for example, J.R. Barber (2010), *Elasticity*, Springer, Dordrecht, 3rd edn., §11.2.3.

[21] Notice that the factor $(2\pi)$ in these definitions is conventional, but essentially arbitrary. In applying fracture mechanics arguments, it is important to make sure that the results used for the stress intensity factors (a theoretical or numerical calculation) and the fracture toughness (from experimental data) are based on the same definition of $K$.

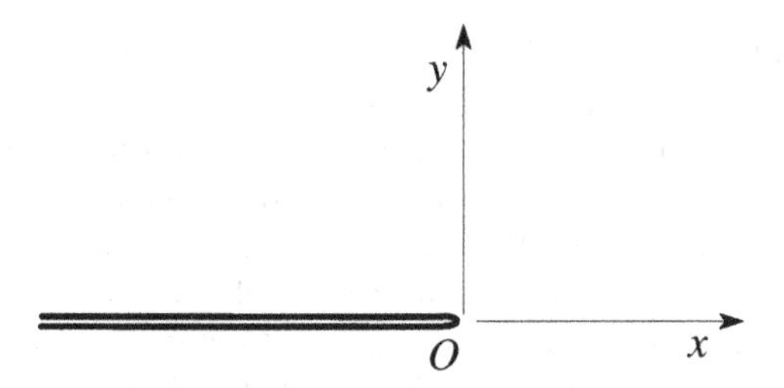


*Figure 2.15 Coordinate system for the semi-infinite crack*

$$K_I = \lim_{x \to 0}(2\pi x)^{\frac{1}{2}} \sigma_{yy} \; ; \; K_{II} = \lim_{x \to 0}(2\pi x)^{\frac{1}{2}} \sigma_{yx} \; ; \; K_{III} = \lim_{x \to 0}(2\pi x)^{\frac{1}{2}} \sigma_{yz} \, . \tag{2.57}$$

These are referred to as Mode I, II and III stress intensity factors respectively.

The deformations associated with the three modes are illustrated in Figure 2.16. Notice how in mode I (opening mode) the crack faces are pulled apart giving rise to a finite gap known as the *crack opening displacement (COD)*. By contrast, the other two modes involve tangential relative displacements at the crack faces, but there is no crack opening displacement.

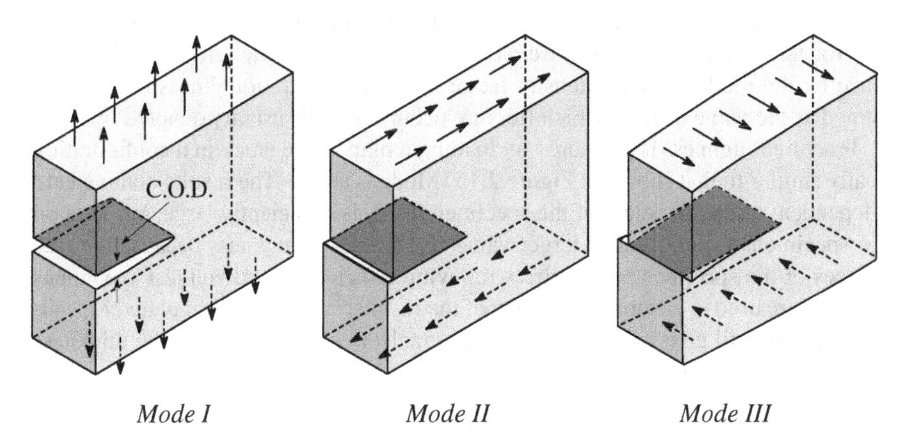


*Mode I* *Mode II* *Mode III*

*Figure 2.16: Loading of a cracked body in modes I,II,III*

**Reasons for the success of LEFM**

It may seem paradoxical to found a theory of real material behaviour on properties of a singular elastic field, which clearly cannot accurately represent conditions in the precise region where the failure is actually to occur. After all, the theory of elasticity is based on the assumption that the strains are small and this will clearly be violated at sufficiently small distances $r$ from the crack tip. Also, at very small $r$, we would need to recognize that the material is not really a continuum, but has atomic structure and the crack cannot really be a geometric line with zero thickness, since if it were, the interatomic forces would act across it and it would cease to be a crack. However, if the material is brittle, the region in which the elastic solution is inapplicable may be restricted to a relatively small *process zone* surrounding the crack tip. Furthermore, the certainly very complicated conditions in this process zone can only be influenced by the surrounding elastic material and hence the conditions for failure must be expressible in terms of the characteristics of the much simpler surrounding elastic field. As long as the process zone is small compared with the other linear dimensions of the body (notably the crack length), it will have only a very localized effect on the surrounding elastic field, which will therefore be adequately characterized by the dominant singular term in the linear elastic solution, whose multiplier (the stress intensity factor) then determines the conditions for crack propagation.

It is notable that this argument requires no assumption about or knowledge of the actual mechanism of failure in the process zone and, by the same token, the success of linear elastic fracture mechanics as a predictor of the strength of brittle components provides no evidence for or against any particular failure theory.[22]

### Design based on fracture mechanics

Fracture mechanics methods enable us to predict the load at which a crack in a component will start to propagate. The procedure is to calculate the stress intensity factor at the crack tip and compare it with measured and tabulated values of fracture toughness for the material. In most cases, fracture is dominated by tensile loading and hence by the mode I stress intensity factor $K_I$. The fracture toughness is therefore defined as the value of $K_I$ at which fracture occurs and it is usually denoted by $K_{Ic}$.

Fracture toughness is measured by loading a plane edge crack in a configuration locally similar to that shown in Figure 2.15 (Mode I) above. The results obtained are independent of the thickness of the specimen if this is sufficiently large, but tests on thin specimens generally give larger values for fracture toughness because the free surfaces of the specimen have a stress relieving effect. Standardized test specimens are therefore used to ensure consistency of the results.[23] The use of tabulated $K_{Ic}$ values in design will give good predictions for cracks in plates as long as the thickness of the plate $t$ satisfies the condition

$$t \geq 2.5 \left( \frac{K_{Ic}}{S_Y} \right)^2 .$$

For thinner plates, the condition $K_I < K_{Ic}$ is conservative and a more efficient design may be possible using more advanced concepts from elastic-plastic fracture mechanics.[24]

[22] Indeed, experimental values of fracture toughness are not compatible with theoretical predictions based on Griffith's theory and estimates of surface energy. For more details of the extensive development of the field of fracture mechanics, the reader is referred to the many excellent texts on the subject, such as Kanninen and Popelar (1985), Leibowitz (1971).

[23] See, for example, Boresi *et al.* (1993), Figure 15.2.

[24] See, for example, S.T. Rolfe and J.M. Barsom (1977), *Fracture and Fatigue Control in Structures*, Prentice-Hall, Englewood Cliffs, NJ, Chapter 19.

Stress intensity factors for a wide range of geometries are tabulated in standard reference works.[25] The simplest configuration is that in which a large plate contains a single through-thickness crack of length $2a$ and is loaded by an otherwise uniform tensile stress $\sigma$ normal to the crack plane, as shown in Figure 2.17.

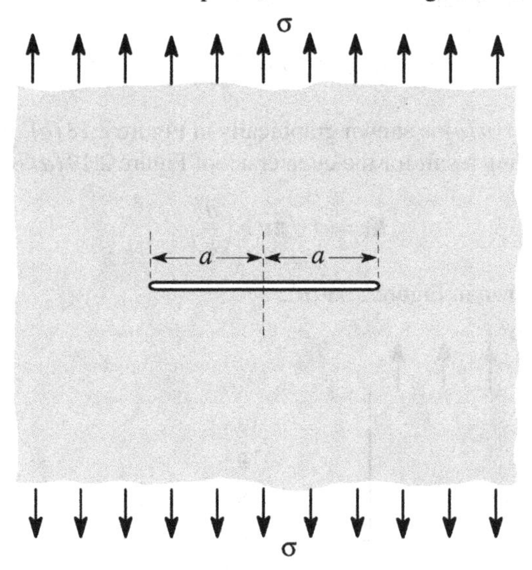


*Figure 2.17: Plane crack in an infinite plate*

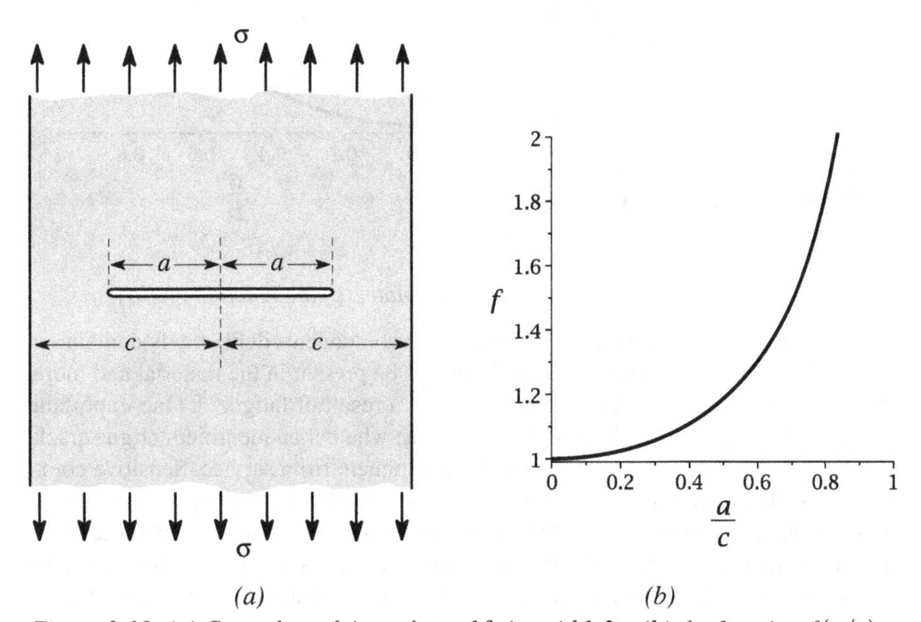


*Figure 2.18: (a) Central crack in a plate of finite width $2c$, (b) the function $f(a/c)$*

[25] See for example, Sih (1973), Pilkey (1994).

In this case, $K_I$ is given by

$$K_I = \sigma\sqrt{\pi a}\,. \tag{2.58}$$

If the crack is at the centre of a plate of finite width $2c$ as shown in Figure 2.18 *(a)*, $K_I$ is increased, being given by

$$K_I = \sigma\sqrt{\pi a}\, f\left(\frac{a}{c}\right)\,, \tag{2.59}$$

where the function $f(a/c)$ is shown graphically in Figure 2.18 *(b)*.

The corresponding result for the edge crack of Figure 2.19 *(a)* is

$$K_I = \sigma\sqrt{\pi a}\, g\left(\frac{a}{b}\right)\,, \tag{2.60}$$

where $g(a/b)$ is shown in Figure 2.19 *(b)*.

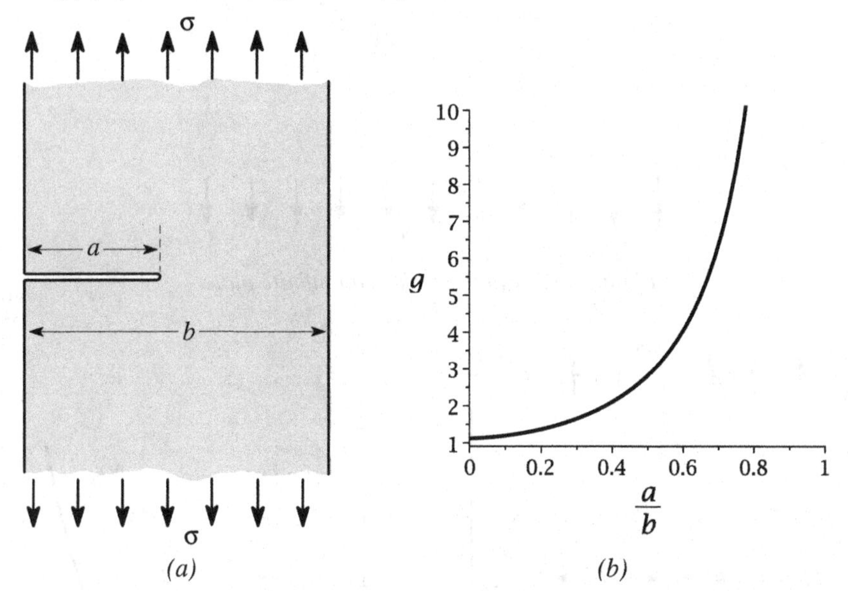


*Figure 2.19: (a) Edge crack in a finite plate, (b) the function $g(a/b)$*

Of course, we do not generally manufacture components deliberately containing identifiable cracks, but microcracks will always be present in the material and more extended cracks may develop during service as a result of fatigue.[26] One important application of fracture mechanics is to determine whether an identified fatigue crack is serious enough to warrant removing the component from service. Sensitive components such as aircraft engines, mountings and structural components and reactor pressure vessels are routinely tested by non-destructive methods such as ultrasonics in order to detect potentially dangerous cracks. The stress intensity factor generally increases with crack length, as in equations (2.58–2.60) above, so it is possible to identify a maximum permissible crack length for safe performance.

[26] See §2.3 below.

**Example 2.4**

*Figure 2.20 shows part of an aircraft engine mounting bracket that is subject to a maximum tensile force of 20 kN. It is routinely examined by an ultrasonic NDE (non-destructive evaluation) method that is capable of detecting cracks of length greater than 2 mm. The material has a fracture toughness $K_{Ic} = 30$ MPa$\sqrt{m}$. Is this testing method adequate? If so, what is the safety factor against fracture when the plate has an edge crack of length 2 mm?*

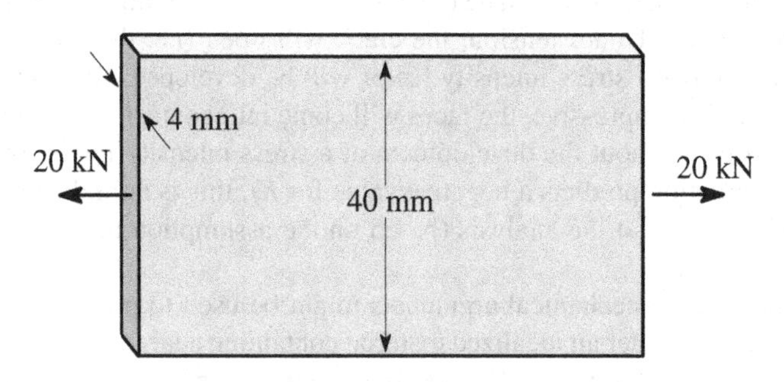


*Figure 2.20*

The uncracked plate experiences a maximum tensile stress

$$\sigma = \frac{20,000}{4 \times 40} = 125 \text{ MPa}.$$

If there is a 2 mm edge crack, we have

$$\frac{a}{b} = \frac{2}{40} = 0.05$$

and hence

$$g\left(\frac{a}{b}\right) = 1.15\,,$$

from Figure 2.19 *(b)*.

Substituting into equation (2.60), we have

$$K_I = 125\sqrt{\pi \times 0.002} \times 1.15 = 11.39 \text{ MPa}\sqrt{\text{m}}\,.$$

This is less than the fracture toughness value, so the NDE method is adequate. The safety factor is

$$SF = \frac{30}{11.39} = 2.63\,.$$

Notice however that the mounting bracket must be examined with sufficient frequency to ensure that cracks of a dangerous length do not develop between examinations.

### Consequences for the failure envelope

In principle, we also ought to be able to use fracture mechanics to estimate the effect of a distribution of microcracks in a brittle material on the resulting failure envelope. In practice, it is not that easy, since to obtain appropriate fracture toughness data would require the manufacture of a specimen with a solitary microcrack. However, arguments from micromechanics do provide some guidance as to the *shape* of failure envelope to be expected for a brittle material.

We must first remark that a plane crack will behave differently under tensile and compressive loading. Under tension, the crack will open (i.e. the crack faces will separate) and a mode I stress intensity factor will be developed. By contrast, when the applied load is compressive, the faces will come into contact and will be able to transmit the stress without the development of a stress intensity factor. In fact, if a theoretical calculation predicts a negative value for $K_I$, this is an indication that the crack will close and that the analysis (based on the assumption that it transmits no tractions) is invalid.

To see how micromechanical arguments might be used to predict the resulting failure envelope, consider an idealized material containing a large number of circular cracks all of radius $a$, but with different orientations. Suppose a block of this material is subjected to a nominally uniform stress field defined by the principal stresses $\sigma_1, \sigma_2, \sigma_3$. If the cracks are sufficiently far apart, the perturbation in the stress field associated with a particular crack will depend only on the normal stress $\sigma$ and the resultant shear stress $\tau$ that would have been transmitted across the crack plane if the crack had not been present. The stress intensity factors associated with this perturbation can be found in reference works[27] and are given by

$$K_I = 2\sigma\sqrt{\frac{a}{\pi}} \tag{2.61}$$

$$K_{II} = \frac{4\tau}{(2-\nu)}\sqrt{\frac{a}{\pi}}\cos\theta \tag{2.62}$$

$$K_{III} = \frac{4(1-\nu)\tau}{(2-\nu)}\sqrt{\frac{a}{\pi}}\sin\theta\ , \tag{2.63}$$

where $\theta$ describes the position around the circumference of the crack in a suitable polar coordinate system.

Since all three modes are present, fracture will be governed by a combination of the three stress intensity factors. We recall Griffith's argument that crack will propagate when the resulting release of strain energy in the body exceeds the energy required for crack extension. The *strain energy release rate*, defined as the strain energy released per unit area of crack extension, is related to the stress intensity factors by the equation[28]

[27] See, for example, Sih (1973), pp. 3.2.1-1 and 3.2.1-5. For the analytical methods used to obtain these results, see M.K. Kassir and G.C. Sih (1975), Three dimensional crack problems, Noordhoff, Leyden, Netherlands.

[28] M.F. Kanninen and C.H. Popelar (1985), §6.2.4.

$$G = \frac{(1-\nu^2)}{E}\left[K_I^2 + K_{II}^2 + (1+\nu)K_{III}^2\right] . \tag{2.64}$$

It follows that an equivalent statement of the fracture criterion is that failure will occur when $G$ reaches the *critical strain energy release rate* $G_c$. We note incidentally that this definition implies the relation

$$G_c = \frac{(1-\nu^2)\,K_{Ic}^2}{E} , \tag{2.65}$$

in order that the criteria in terms of energy release rate and fracture toughness should be consistent in the special case of mode I loading.

Substituting for $K_I, K_{II}, K_{III}$ from (2.61–2.63) into (2.64), we find that the maximum value of $G$ occurs when $\theta = 0$ and we conclude that fracture will occur when

$$\sigma^2 + \frac{4\tau^2}{(2-\nu)^2} = \frac{\pi E G_c}{4(1-\nu^2)a} \equiv S_t^2 , \tag{2.66}$$

where we have defined the new material property $S_t$.

Since the present analysis is for illustrative purposes only, we shall restrict attention to the simplest case where $\nu = 0$ and hence (2.66) reduces to

$$S^2 \equiv \sigma^2 + \tau^2 = S_t^2 . \tag{2.67}$$

Now $S$ defined by equation (2.67) is the distance from the origin to a point on the Mohr's circle and it is easy to show by reference to Figures 2.21 *(a,b)* that the maximum will occur at $\max\{|\sigma_1|, |\sigma_2|, |\sigma_3|\}$. If this maximum is tensile, we then conclude that the failure criterion is

$$\max\{\sigma_1, \sigma_2, \sigma_3\} = S_t \tag{2.68}$$

which is the criterion of *maximum tensile stress*.

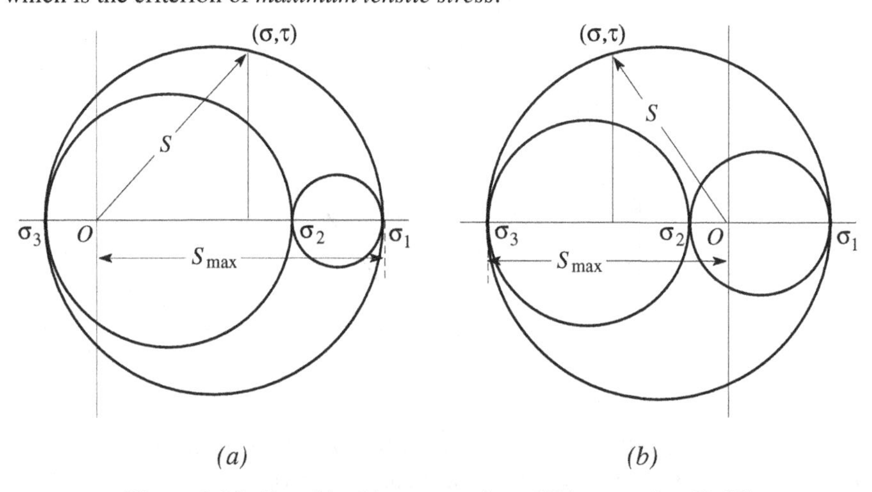


*Figure 2.21: Graphical interpretation of S in equation (2.67)*

However, if the largest value of $S$ corresponds to a compressive principal stress as shown in Figure 2.21 *(b)*, (2.67) does not apply, since we argued above that $K_I = 0$ when a crack is loaded in compression. In this case, the compressive traction will be transmitted by the crack faces, but stress intensity factors in shear will still be obtained, since relative tangential motion between the crack faces can occur. The simplest case is where we assume that this tangential motion is unresisted (i.e. there is no friction at the crack faces), in which case the contribution from $\tau$ in (2.67) is unaffected, leaving the criterion $\tau^2 = S_t^2$ or $|\tau| = S_t$.

For the biaxial case, where $\sigma_3 = 0$, this leads to the failure envelope *(a)* in Figure 2.22, where we see a maximum tensile stress envelope for $\sigma_1 + \sigma_2 > 0$ and a maximum shear stress (Tresca) envelope for $\sigma_1 + \sigma_2 < 0$.

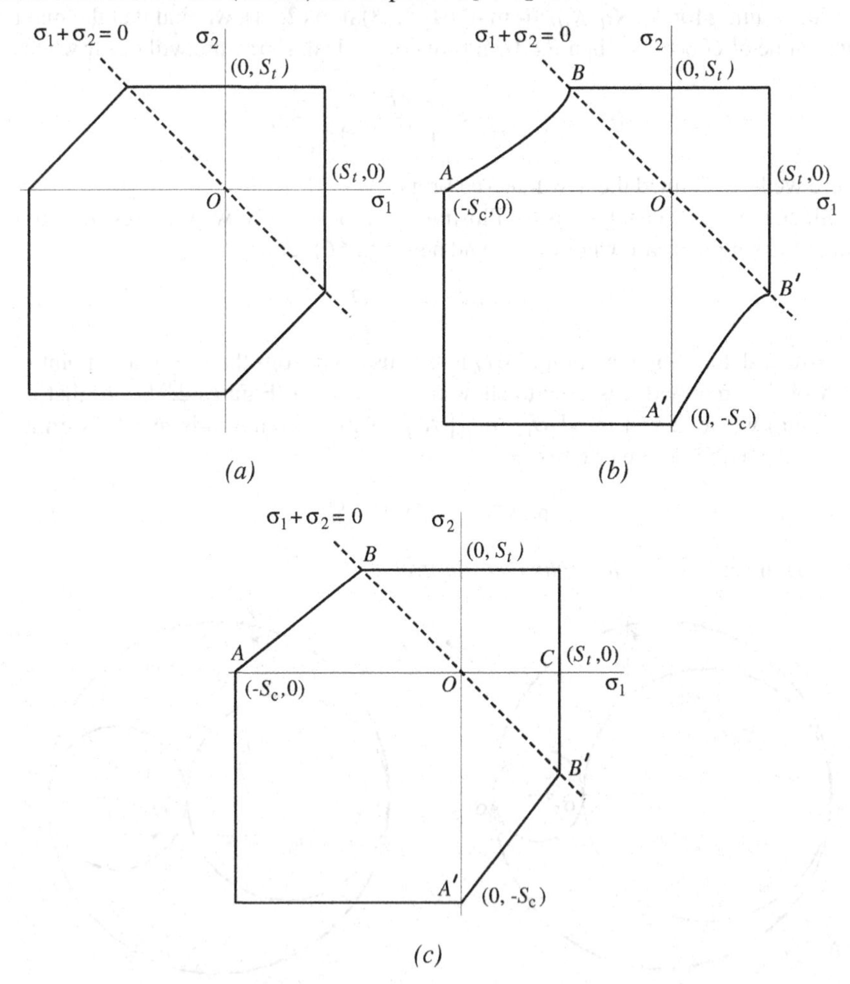


*Figure 2.22: Biaxial failure envelopes for brittle materials: (a) No friction at the crack faces, (b) finite friction, (c) the modified Mohr criterion*

More realistically, we must anticipate some friction at the crack faces. If we assume Coulomb friction with a coefficient of friction $\mu$, slip (and hence stress intensity factors in shear) will now only occur if $|\tau| > \mu(-\sigma)$ (remember that $\sigma$ is compressive and hence negative). The resulting stress intensity factor depends on the *difference* between the actual shear stress and this limiting value and we obtain the criterion

$$\max\{|\tau| - \mu(-\sigma)\} = S_0 \quad ; \quad \sigma < 0 \,. \tag{2.69}$$

Once again, we have to determine the points on the Mohr's circle at which this expression is a maximum. After some algebra, it can be shown to define the envelope of Figure 2.22*(b)*, where the point $A$, corresponding to uniaxial compression is defined by $\sigma_1 \equiv -S_c = -2S_t / \left(\sqrt{1+\mu^2} - \mu\right)$ and $S_c$ is the failure strength in uniaxial compression.

A reasonable approximation to curve *(b)* is obtained by joining the points $A, B$ by a straight line, giving the envelope *(c)*. This criterion is known as the *modified Mohr criterion* and it is found to give quite a good description of the shape of the biaxial failure surface for brittle materials. Notice however that experimental agreement between the modified Mohr criterion and the behaviour of a real material emphatically does *not* prove that the material contains a distribution of identical circular cracks! In fact any assumed distribution of initially closed microdefects will predict a similar failure envelope.

To define the envelope completely, we need to perform two mechanical tests, since the predicted behaviour in tension and compression are qualitatively different. For example, we could perform uniaxial tests to failure in tension and compression, defining the points $C, A$ respectively in Figure 2.22*(c)*. Once these points are determined, the rest of the envelope *(c)* is completely defined.

We have developed the above results for a biaxial state of stress with $\sigma_3 = 0$, but the more general triaxial state follows the same pattern, provided we focus attention on the largest of the three-dimensional Mohr's circles. If we label the principal directions such that $\sigma_1 > \sigma_2 > \sigma_3$, the modified Mohr criterion can be stated formally as predicting failure when

$$\sigma_1 = S_t \quad ; \quad (\sigma_1 + \sigma_3) > 0 \tag{2.70}$$

$$\left(\frac{S_c}{S_t} - 1\right)\sigma_1 - \sigma_3 = S_c \quad ; \quad (\sigma_1 + \sigma_3) < 0 \,. \tag{2.71}$$

In three-dimensional stress space, this defines a regular hexagonal pyramid centred on the isotropic line, but truncated by the three planes $\sigma_1 = S_t, \sigma_2 = S_t, \sigma_3 = S_t$, as shown in Figure 2.23. Thus, when the stresses are predominantly compressive, even a brittle material follows an essentially ductile failure criterion and this parallel extends to the fact that energy can be dissipated in compressive deformation through the work done against friction at the crack faces. However, notice that with this internal friction mechanism, the shear stress needed to cause failure increases with increasingly compressive mean stress $\bar{\sigma}$.

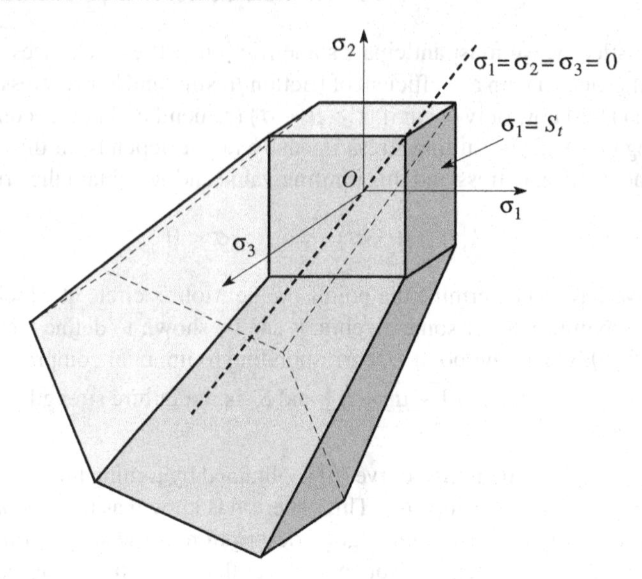


*Figure 2.23: Three-dimensional failure surface for the modified Mohr theory*

For most brittle materials, the tensile strength $S_t$ is significantly lower than the compressive strength $S_c$ and potential failure sites are generally those associated with the maximum tensile stress. Indeed, the challenge of designing with brittle materials is to develop structures that will transmit loads without the development of substantial tensile stresses.

**Materials exhibiting both ductile and brittle behaviour**

Many materials can exhibit either brittle or ductile failure patterns depending on the state of stress and the average size of pre-existing cracks. Also, the fracture toughness is a strong function of temperature and hence behaviour may change from ductile to brittle under tensile stress at low temperatures.

For such materials, both fracture toughness and yield stress values are tabulated and it is necessary to perform design calculations on both criteria to determine which mode of failure is limiting for the particular stress state.[29]

Two consequences of this dual failure behaviour are worth noting:-

(i) A ductile material can always be made to fail in a brittle mode if it is subjected to a sufficiently large hydrostatic tensile stress $(\sigma_1 = \sigma_2 = \sigma_3)$.

(ii) Certain brittle materials can be made to deform plastically and hence formed, if the required shear stresses are superposed on a sufficiently large hydrostatic compressive stress. Notice however that if the material has significant porosity (i.e. if it contains microscopic voids), the first failure under hydrostatic compression may be by crushing and void collapse.

[29] See for example Problem 2.33.

**Example 2.5**

*A brittle material is tested in uniaxial tension and compression and found to fracture at stresses of* 50 *MPa and* 350 *MPa respectively. It is to be used in an application where the design stresses are* $\sigma_1 = 20$ *MPa,* $\sigma_2 = 0$ *MPa,* $\sigma_3 = -80$ *MPa. Determine the factor of safety using the modified Mohr criterion of failure.*

We have $\sigma_1 > \sigma_2 > \sigma_3$ and $\sigma_1 + \sigma_3 < 0$, so equation (2.71) defines the appropriate section of the failure surface.

If the factor of safety is $SF = \lambda$, the stresses

$$\sigma_1 = 20\lambda \text{ MPa} \quad ; \quad \sigma_2 = 0 \text{ MPa} \quad ; \quad \sigma_3 = -80\lambda \text{ MPa}$$

must lie on the failure surface — i.e.

$$\left(\frac{350}{50} - 1\right) \times 20\lambda + 80\lambda = 350$$

$$200\lambda = 350$$

and the safety factor is

$$SF = \lambda = 1.75 \, .$$

## 2.3 Cyclic loading and fatigue

Take a piece of wire — a paper clip will do — and bend it. You will be able to produce a substantial permanent deformation without fracture occurring. However, if you straighten the wire out and then bend it again a few times, fracture will eventually occur. Count the number of cycles of bending and straightening needed to cause fracture. Then take a new piece of wire and stop a few cycles before the final fracture. You should be able to see that the wire is cracked at the point where failure is about to occur. This process, in which the ductility of the material is exhausted by reversed or cyclic loading is known as *fatigue failure*.

Of course, we do not usually design engineering components to experience reversed stresses in the plastic range, but even if the stresses are nominally well inside the 'safe' region of Figure 2.12, there will be inhomogeneities in the material such as microcracks, irregular grain boundaries and impurity inclusions that might be sufficient to raise the *local* stresses to the level at which damage occurs. Sometimes the microcracks so developed will grow into a less heavily stresses region and be 'arrested', but more often the stress concentration associated with the tip of the newly extended crack will itself cause further local failure (crack extension) and the damage proceeds until the remaining material is insufficient to support the load statically and a final fracture occurs.

The process appears to take place discontinuously — the crack does not extend a fixed small amount each time the stress is reversed. Instead, a number of cycles are needed to weaken the material ahead of the crack tip sufficiently for extension to occur after which the crack extends a finite amount. This can be seen by examining the failure surfaces of fatigued components, which show a series of lines representing the intermediate positions of the crack tip during the process. These lines are sometimes referred to as *beach marks*.

Failure by yielding or brittle fracture will occur as soon as a sufficiently large load is applied, but fatigue failure requires that the load be applied and removed a large number of times and hence generally only occurs after several months or even years in service. During this time, large numbers of components may have been manufactured and sold, so design errors resulting in fatigue failures can be very expensive in terms of warranty claims, recalls and loss of customer confidence.

Many of the characteristics of fatigue failure can be traced to the fact that the process starts from a microscopic initial defect in the material or the surface. These defects are randomly distributed and oriented and hence apparently identical components may exhibit significantly different fatigue behaviour. Even under idealized laboratory conditions, where the specimen characteristics are controlled within tight limits, the number of cycles of loading at a given stress level required to produce failure may differ between specimens by as much as a factor of 100.

If the company for which you work has sold 1,000,000 components to your design and 5 of them failed in fatigue after 6 months, you should immediately look for another job before it is too late! The reason is that the first few failures lie in the tail of the statistical distribution curve which will usually be approximately Gaussian when plotted against the logarithm of the number of cycles to failure. In this region, the curve climbs steeply, so 5 failures after 6 months will probably translate into 200,000 failures in the first 3 years.

In view of these considerations, it is surprising that design against fatigue is a fairly recently developed science. In fact, widespread interest in fatigue can be traced to the 1950s, following a series of dramatic aircraft crashes featuring the DeHaviland Comet — one of the earliest passenger jets. The debris from these crashes was collected and meticulously reassembled to reveal that the fracture started in each case from the same small hole towards the rear of the fuselage. A full scale aircraft was then tested on the ground to simulate the cyclic effects of cabin pressurization and depressurization and a similar fatigue failure was thereby produced under experimental conditions. Previous to this, design against structural failure was largely based on static yield stress or ultimate strength, using a large factor of safety. Of course, if the safety factor is large enough, this procedure gives a design that is safe in fatigue as well, but it is then a matter of experience or guesswork just how large is large enough. It is characteristic that difficulties with this procedure first appeared in the aircraft industry, where the payoff from weight reduction makes it advantageous to pare safety factors as far as current knowledge permits.

### 2.3.1 Experimental data

As with static yield and brittle fracture, design against fatigue requires (i) some basic experimental data obtained under idealized conditions and (ii) some kind of 'failure theory' to reduce the vast range of possible service stress fields and histories to a set of conditions of manageable proportions, which can be tested and the results tabulated.

The most basic fatigue test is one in which a specimen is subjected to cyclic tension and compression, such that the axial force $F$ varies according to the equation

$$F = F_0 \sin(\omega t) , \tag{2.72}$$

where $t$ is time. For obvious reasons, this is referred to as completely reversed loading.

In practice, it is a great deal easier to achieve this stress history by loading the specimen in bending and then rotating it, so that a given portion of the specimen near the surface passes through the zones of maximum tensile and maximum compressive stress once per revolution. A typical experimental system is shown schematically in Figure 2.24. A force $F$ is applied to the carrier $ABCD$, causing a constant bending moment $M = Fd/2$ in the central section of the specimen.

Notice how the specimen is gently rounded to a minimum diameter, so as to ensure that the failure occurs in a region where the nominal stresses are well-defined, rather than at a change of section or other stress concentration. We shall see below that fatigue strength is significantly influenced by the size of the specimen, so a standard specimen of 0.3 inch (7.6 mm) diameter is generally used.

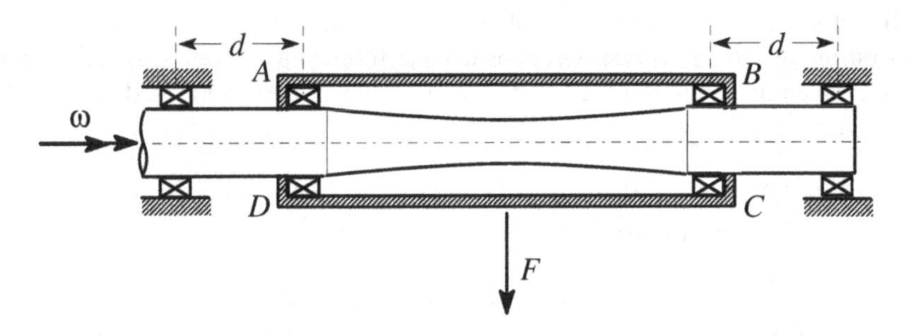


*Figure 2.24: Rotating bending test for fatigue failure*

The test procedure is to apply a load below that required to produce immediate static yield, set the machine rotating and count the number of rotations until failure occurs. The results are then plotted on an *'S–N curve'*, which shows the stress level $S$ as a function of the number of cycles to failure $N$. Curves for a typical steel and aluminium alloy are shown in Figure 2.25. A logarithmic scale is generally used, since the lives of different specimens can vary by several orders of magnitude. It is not practical to conduct tests for much more than $10^8$ cycles, because they are very time consuming. Even so, for these long running tests, the test machine has to be left running continuously for several days and therefore has to be instrumented to detect failure and switch itself off.

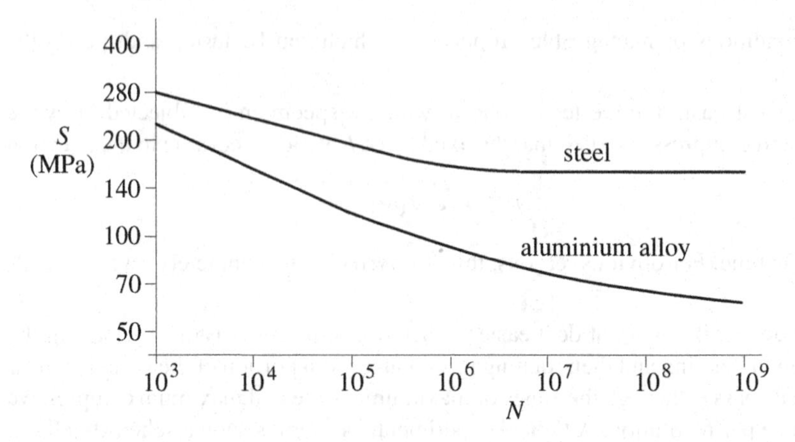


*Figure 2.25: S–N curves for a typical steel and an aluminium alloy*

For steels, the fatigue strength decays with number of cycles up to about $N=10^6$ cycles, after which the curve levels off. There is therefore a *fatigue limit* denoted by $S_n$ (also called the *endurance limit*), below which it seems[30] that failure will not occur, no matter how long the test is run. This is a very useful result in design, since it means that we do not have to design for a specific number of cycles of loading if that number is greater than $10^6$.

To get some idea of the number of cycles accumulated in some typical components during a normal service life, consider the following examples. At the upper extreme, an engine crankshaft running 24 hours a day at 5000 rpm for 10 years will accumulate

$$N = 5000\text{ rpm} \times 60\text{ min/hr} \times 24\text{ hr/day} \times 365\text{ days/year} \times 10\text{ years}$$
$$= 26 \times 10^9\text{ cycles.}$$

This is about the largest value of $N$ you are likely to encounter in rotating machinery, since most machines don't run all the time and 5000 rpm is quite a high speed. Larger numbers of cycles can be accumulated in devices subjected to high frequency vibration. A machine running at 200 rpm for 2 hours a day and 5 days per week would accumulate

$$N=200\text{ rpm}\times 60\text{ min/hr}\times 2\text{ hr/day}\times 5\text{ days/week}\times 52\text{ weeks/year}\times 10\text{ years}$$
$$= 62 \times 10^6\text{ cycles.}$$

However, there are some applications involving much smaller numbers of cycles. For example, every time a machine is started, it 'warms up', and there will be thermal stresses associated with the differential thermal expansion of the parts. These stresses may be constant during operation, so one complete reversal occurs only every start

[30] We qualify this statement because no-one knows what would happen in such cases if the number of cycles were several orders of magnitude greater than those reached in tests or service conditions.

up, which may be once or twice a day. Fatigue under these conditions is referred to as *low cycle fatigue* (for obvious reasons) and in such applications a more efficient design can be obtained by making use of the sloping portion of the $S$–$N$ curve. The Comet disasters referred to above were eventually attributed to fatigue associated with cabin pressurization, which typically cycles only once per take off and landing. However, the above examples show that in many applications, $N$ is large enough to justify using the fatigue limit only.

The concept of a fatigue limit is so useful that we invent one even when it is not discovered experimentally. Thus, with aluminium and its alloys, the $S$–$N$ curves continue to fall with $N$ throughout the range, although there is a levelling trend as shown in Figure 2.25. Under these conditions, we typically design to a fatigue limit $S_n$ equal to the stress level $S$ for a life of $10^8$ or $10^9$ cycles. In other words, we approximate the experimental curve by a horizontal straight line in the practical range of $N$.

### Estimates for the fatigue limit

Ideally, we should use values for $S_n$ for the same material as in the component being designed. However, in practice there are so many different alloys, with a corresponding number of different heat treatments and production methods (all of which affect the fatigue strength) that we are often unable to obtain appropriate data. More significantly, the exact choice of material and heat treatment will often not be made until a fairly late stage in the design process, so it is useful to have a rough idea of what values of $S_n$ to expect for a range of common materials.

A good rule of thumb is that the fatigue limit for a polished specimen will be between a quarter and a half of the ultimate tensile strength for the same material. The ultimate strength $S_u$ is the stress required to cause rupture of a tensile test specimen. It can be significantly greater than the stress at first yield $S_Y$, due to the phenomenon of work hardening. There is much more data available for tensile strength than for fatigue limit and in the worst case we can always find $S_u$ by testing a specimen ourselves — it is much easier to perform a single monotonic tensile test than to conduct a series of fatigue tests.

For steels, Juvinall (1983) §8.3 suggests the relation

$$S'_n \approx 0.5 S_u \approx 0.25 \times \text{ BHN (in ksi)} \approx 1.73 \times \text{ BHN (in MPa)}, \tag{2.73}$$

where BHN is the hardness value obtained in the Brinell hardness test and the prime ($'$) in $S'_n$ is used to indicate that this is the value obtained in a rotating bending test on a standard polished specimen. Note however that $S'_n$ for extremely hard steels levels off at a value of about 100 ksi (700 MPa) and this value should be used if equation (2.73) predicts a *larger* value.

A corresponding result for aluminium alloys is

$$S'_n \approx 0.4 S_u \text{ if } S_u < 48 \text{ ksi (330 MPa)} \tag{2.74}$$

$$\approx 19 \text{ ksi (130 MPa) if } S_u > 48 \text{ ksi (330 MPa)}. \tag{2.75}$$

However, it should be emphasised that these results are only very approximate. They may be used in the spirit of §1.2.2 to obtain estimates of the fatigue strength of a component, but there is no substitute for a fatigue test on the actual material in applications where large safety factors cannot be used.

### 2.3.2 Statistics and the size effect

The stress concentration due to a fatigue crack increases with its length, so the crack grows more rapidly during the later stages of the process. Indeed, by the time the crack is big enough to be detected, the fatigue life of the component is usually almost exhausted. You can test this crudely by subjecting a piece of wire to half the number of loading cycles needed to cause failure and then looking (probably unsuccessfully) for signs of cracking.

The greater part of the fatigue life is therefore associated with initiation and development of cracks on the microscopic scale. Cracks are believed to start from pre-existing defects in the material or the component and it is a matter of luck (or to be more scientific, of statistics) whether a suitably-oriented defect exists in the most heavily stressed region. This has the unfortunate effect of making the process of designing against fatigue very dependent on the exact condition of the material and on minor details of the geometry. Most of these effects can only be assessed experimentally, so we tend to allow for them by 'correction factors' multiplying the fatigue limit. Such factors will be denoted by the symbol $C$ with appropriate suffices. It also follows that fatigue data is subject to considerable statistical scatter and, of two otherwise identical specimens, one may fail at a significantly lower stress (or smaller number of cycles) than the other. In this section we shall examine some of these questions using the simplest appropriate statistical hypothesis. For a more sophisticated approach to the question of statistical design against fatigue, the reader is referred to Collins (1981).

#### The Weibull distribution

Suppose that a given volume of material $V$ is subjected to uniform alternating tensile stress $S$ and contains a random distribution of microdefects of various dimensions and orientations. Fatigue failure will eventually occur if at least one of these defects weakens the material sufficiently for crack initiation to occur at stress $S$.

The probability of finding a suitably oriented defect clearly increases if the volume $V$ increases. Suppose we define $R(S,V)$ as the *reliability* — i.e. the probability that a specimen of volume $V$ survives (i.e. does not fail in fatigue) either indefinitely or for a given number of cycles under an alternating stress $S$. The probability of failure $P(S,V)$ of the same specimen will then be

$$P(S,V) = 1 - R(S,V) . \tag{2.76}$$

If we now test a specimen of volume $nV$, it is essentially the same as if we were to test $n$ specimens each of volume $V$ simultaneously *and count the system as having failed if any one of the n were to fail.* The probability of survival in this case is $[R(S,V)]^n$ and we therefore conclude that

$$R(S,nV) = [R(S,V)]^n . \tag{2.77}$$

The simplest statistical distribution that satisfies[31] this condition is the two-parameter Weibull distribution[32], defined by

$$R(S,V) = \exp\left[-\left(\frac{S}{S_0}\right)^b \left(\frac{V}{V_0}\right)\right] , \tag{2.78}$$

where $V_0$ is a reference volume (e.g. the volume of stressed material in the standard 0.3 in diameter rotating bending test) and $S_0$ is the alternating stress at which the reliability of the reference volume is $R(S_0,V_0) = \exp(-1) = 0.368$.

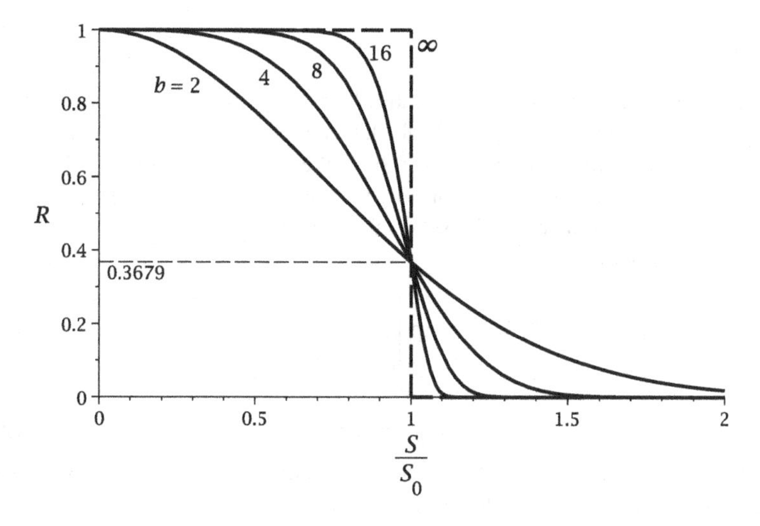


*Figure 2.26: Effect of b on the cumulative Weibull distribution*

The parameter $b$ in equation (2.78) determines the degree of scatter or variability in the distribution. Figure 2.26 shows the reliability as a function of normalized alternating stress $S/S_0$ for $V = V_0$ and various values of $b$. At very large $b$, the curve approaches the Heaviside step function $H(-S/S_0)$, corresponding to a material of

[31] The reader can verify by substitution that the definition (2.78) satisfies the condition (2.77).
[32] There is also a three-parameter Weibull distribution defined through the equation

$$R(S,V) = \exp\left[-\left(\frac{S - S_A}{S_0 - S_A}\right)^b \left(\frac{V}{V_0}\right)\right] ,$$

where $S_A$ is an additional parameter that can be used to improve the fit of the distribution to the statistical data.

which all specimens will exhibit a fatigue strength exactly equal to $S_0$. As $b$ is reduced, the variability in specimen strengths increases. For example, at $b=50$ the standard deviation in the strengths is about 2.5% of $S_0$, whereas at $b=10$ it is about 11%.

In an ideal world, we would have statistical information on the distribution of fatigue strengths for all the materials we might wish to use, but this would be a very large amount of experimental data. In practice, sufficiently accurate estimates of fatigue strength can be made using representative values of $b$ for broad classes of materials. For example, experimental data for steels and other common structural metals suggest that an appropriate standard deviation for the distribution is about 7% of the mean strength at any given life corresponding to a Weibull modulus $b \approx 16$. We shall therefore adopt this value in subsequent examples. More accurate data for the Weibull modulus for various materials can be found in the reference works at the end of this chapter.

### Reliability

Equation (2.78) essentially defines $S_0$ as the alternating stress at which a specimen of volume $V_0$ has a reliability of 0.368 (36.8%). Generally we shall wish to design for a higher value of reliability such as 90% or 99%. We first note that standard data from the rotating bending test usually report median values of the fatigue limit — i.e., $S_n'$ is the alternating stress at which the survival rate is 50%.

Under these conditions,

$$R(S_n', V_0) = \exp\left[-\left(\frac{S_n'}{S_0}\right)^b\right] = 0.5 \tag{2.79}$$

and hence

$$S_0 = S_n'[-\ln(0.5)]^{1/b} . \tag{2.80}$$

Suppose now that we want to design to a reliability of 99% — i.e., we wish to choose $S$ such that $R(S, V_0) = 0.99$. We have

$$R(S, V_0) = \exp\left[-\left(\frac{S}{S_0}\right)^b\right] = 0.99 \tag{2.81}$$

and hence

$$S = S_0[-\ln(0.99)]^{-1/b} = S_n'\left[\frac{\ln(0.99)}{\ln(0.5)}\right]^{1/b} = 0.77S_n' , \tag{2.82}$$

using $b = 16$. More generally, for a reliability $R$, we design to a strength

$$S = C_R S_n' , \tag{2.83}$$

where $C_R$ is a *reliability factor* defined by

$$C_R = \left[\frac{\ln(R)}{\ln(0.5)}\right]^{1/b} . \tag{2.84}$$

### The size effect

Equation (2.78) also shows that two components will have the same reliability if they have the same value of $S^b V$, where $V$ is the volume of material subjected to the maximum level of alternating stress. Thus to achieve 50% reliability in a specimen of stressed volume $V$, we must design to a strength

$$S = C_G S'_n \,, \tag{2.85}$$

where the *size factor* or *gradient factor* $C_G$ is defined by

$$C_G = \left(\frac{V_0}{V}\right)^{1/b} . \tag{2.86}$$

The name of the gradient factor derives from the fact that the size of the stressed region depends upon the rate at which the stress falls with distance from the point of maximum stress (usually at the surface). In particular, note that in a reversed axial loading (push-pull) test, *all* the specimen experiences the maximum stress cycle, whereas in a rotating bending test, only regions near to the surface experience it. To get an idea of the magnitude of $C_G$, notice that if $V/V_0 = 10$ (i.e. if the stressed volume is ten times greater than that in the standard specimen), $C_G = 0.87$.

Similar calculations can be performed for components of other sizes, but since the fatigue limit is not very sensitive to size (in the above calculation changing the size by a factor of 10 only reduced the strength by 13%), it is sufficient to use a rough estimate for $C_G$. A good rule of thumb is to reduce $C_G$ by 0.1 for each order of magnitude that the stressed volume exceeds that in the standard specimen. Also, if the component is comparable in size to the standard specimen but is loaded in reversed tension rather than in bending, use $C_G = 0.9$ on the grounds that only about 10% of the specimen is at the highest stress level in the bending test.

### Surface finish

It is well known that components with rough surfaces are more prone to fatigue failure than those which are polished. It is tempting to explain this by saying that rough surfaces are rough simply because they contain microcracks which act as initial defects from which fatigue cracks develop. However, more rigorous investigations show that the kind of surfaces which are most prone to fatigue failure (such as rough machined surfaces) have other properties which may be more important in fatigue than the roughness. They typically have a surface layer of heavily worked material where the ductility is nearly exhausted and hence where cyclic loading will produce failure earlier than in virgin material. Also in many cases the heat generated during the machining process causes yielding in the surface layers which leaves residual tensile stresses. Tensile residual stresses are undesirable, since if any microcracks exist normal to the surface, they will tend to open. We saw in §2.2.4 that crack propagation is inhibited when the loading tends to close the crack and hence any pre-existing stress tending to hold cracks open will increase the probability of propagation.

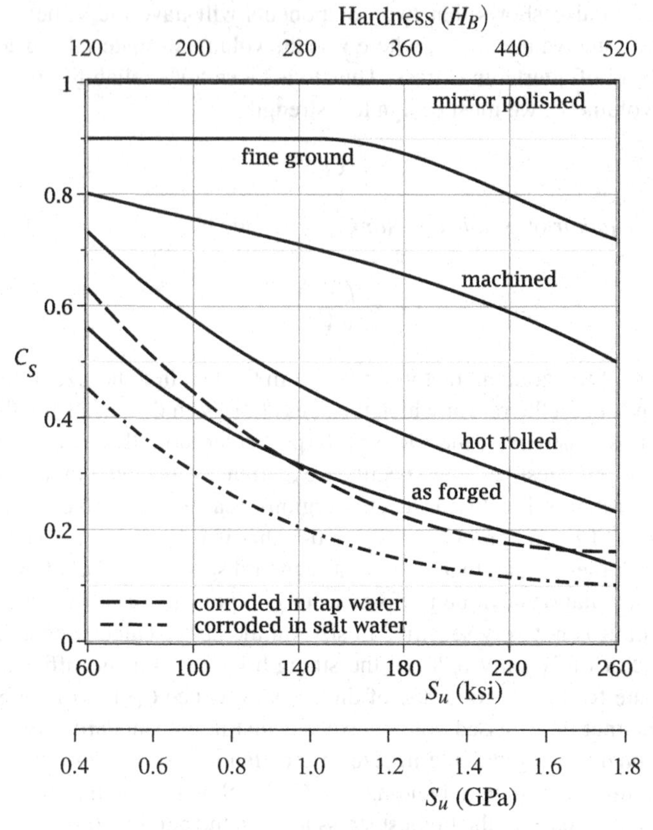


*Figure 2.27: Effect of surface condition on fatigue limit for steels with various manufacturing methods and hardnesses. Adapted from R.C.Juvinall, Fundamentals of Machine Component Design, John Wiley, New York, ©1983 by permission of John Wiley & Sons, Inc.*

These factors are almost impossible to quantify with any degree of confidence, so we tend to present data in terms of the measurable surface roughness and the method of manufacture. For example, typical values for steels are shown in Figure 2.27. Notice that the effect of the finishing technique on fatigue limit depends on the hardness of the material — the stronger steels generally have lower ductility and therefore are more sensitive to surface condition. This means that if we try to increase fatigue strength by using a stronger steel, there is a law of diminishing returns. Also, if we want to make full use of the stronger steel, we have to pay the price of making the part with a more expensive finishing operation such as grinding or polishing. Notice also that it is particularly important to ensure that this high quality finish is retained through any stress concentrations at section changes or notches and this can be quite hard to achieve.

Figure 2.27 is plotted in terms of a *surface factor* $C_S$, which acts to reduce the effective fatigue limit of the material. The standard rotating bending specimens are mirror polished, because this reduces the scatter in the experimental results, and they therefore have by definition a surface factor of unity.

### Summary

All of the above factors act together to modify the fatigue limit of a material. In design, we therefore use the value

$$S_n = S_n' C_R C_G C_S \,, \tag{2.87}$$

where $S_n'$ is the fatigue limit obtained in the standard rotating bending test.

### Example 2.6

*Figure 2.28 shows an automotive engine connecting rod which is forged from AISI 1040 steel, for which $S_u = 570$ MPa. Determine an appropriate design value for $S_n$ if a reliability of 99.9% is required.*

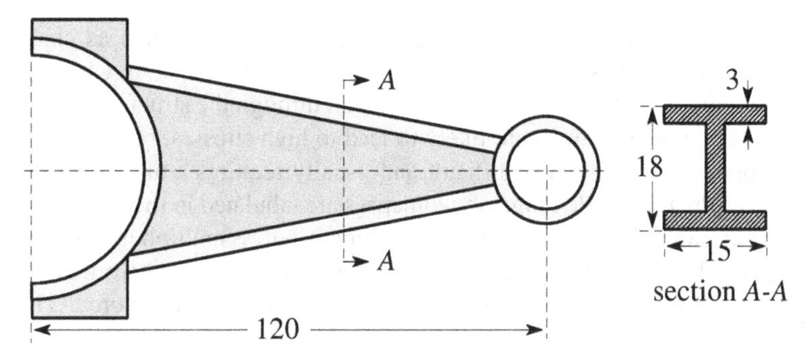


*Figure 2.28: Automotive engine connecting rod*

We first estimate $S_n'$ from equation (2.73) as

$$S_n' = 0.5 \times 570 = 285 \text{ MPa} \,.$$

The connecting rod is forged, so an appropriate surface factor is obtained from Figure 2.27 as $C_S = 0.45$.

The reliability factor

$$C_R = \left[ \frac{\ln(0.999)}{\ln(0.5)} \right]^{1/16} = 0.665 \,.$$

Finally, we need to estimate a size factor $C_G$. For this purpose it is sufficient to note that the linear dimensions of the connecting rod are about three times larger than those of the standard specimen, so $V \approx 27V_0$, giving

$$C_G = \left(\frac{1}{27}\right)^{1/16} = 0.81 \, .$$

Hence, an appropriate design value for $S_n$ is

$$S_n = 0.45 \times 0.665 \times 0.81 \times 285 = 69 \text{ MPa} \, .$$

Notice that a substantial improvement in fatigue strength could be achieved in this case by machining the surfaces of the connecting rod.

### 2.3.3 Factors influencing the design stress

Real engineering components will generally be a great deal more complex than the standard rotating bending fatigue specimen and they will also be subjected to more complex loading. In such cases, we need to calculate a design stress which can be compared with the results of the standardized tests discussed above.

#### Stress concentrations

Engineering components typically contain geometric features such as changes of section, holes or notches that cause local *stress concentrations*. As with the crack shown in Figure 2.13, the idea of the flow of load through the structure can often be used to predict which features are likely to lead to high stresses, but the calculation of the theoretical stress field is difficult and usually requires a numerical solution. However, results for a wide range of geometries are tabulated in the literature.[33]

The results are presented in the form of a dimensionless multiplier on the *nominal stress* $\sigma_{\text{nom}}$, which is usually the stress which would be obtained from elementary mechanics of materials theory if the stress concentration were ignored. Thus, the actual stress $\sigma$ is given by

$$\sigma = K_t \sigma_{\text{nom}} \, , \tag{2.88}$$

where $K_t$ is the *theoretical stress concentration factor*. In reference works, the mathematical expression for $\sigma_{\text{nom}}$ will be given explicitly either in the figure or in the accompanying text.

Figure 2.29 shows stress concentration factors for bending of a cylindrical shaft with a change in section from diameter $D$ to $d$ through a *fillet radius* $r$. The nominal stress is here based on the elementary theory using the smaller diameter $d$ and is therefore

$$\sigma_{\text{nom}} = \frac{32M}{\pi d^3} \, , \tag{2.89}$$

where $M$ is the bending moment.

Stress concentration charts for some other common geometries and loadings are given in Appendix C.

[33] See for example Peterson, (1974), Pilkey, (1994).

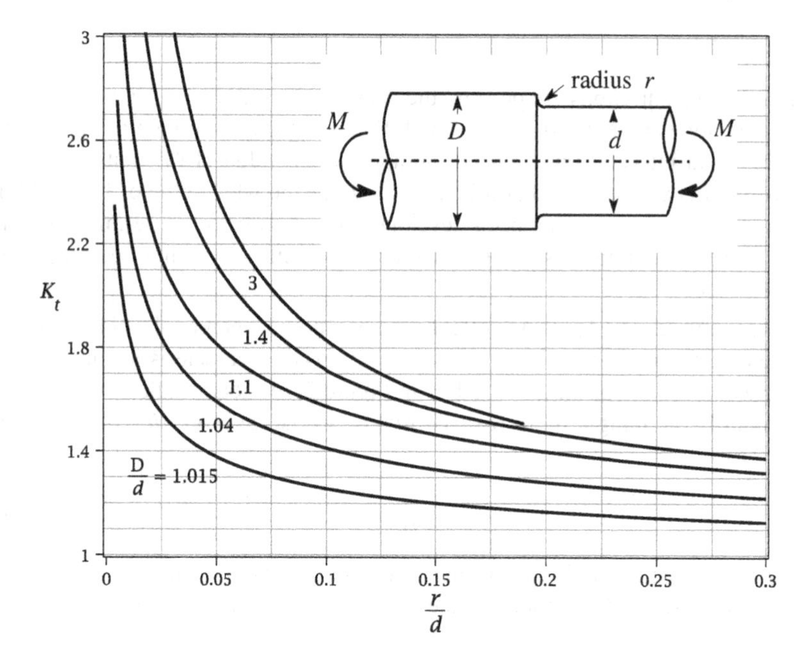


*Figure 2.29: Stress concentration factors for a cylindrical shaft with a change of section subjected to a bending moment M*

## Example 2.7

*A cylindrical shaft has a change of section from diameter* 30 *mm to diameter* 25 *mm through a fillet radius of* 1.2 *mm. The local bending moment is* 200 *Nm. Find the maximum stress at the root of the fillet.*

The nominal bending stress is

$$\sigma_{\text{nom}} = \frac{32 \times 200 \times 10^3}{\pi \times 25^3} = 130 \text{ MPa}.$$

The ratios

$$\frac{D}{d} = \frac{30}{25} = 1.2 \ ; \ \frac{r}{d} = \frac{1.2}{25} = 0.048$$

and hence interpolating on Figure 2.28, we have

$$K_t = 1.98 .$$

The maximum tensile stress in the root is therefore

$$\sigma_{\text{max}} = 1.98 \times 130 = 257.4 \text{ MPa}.$$

**Notch sensitivity**

When stress concentrations are present, the heavily stressed region for which the fatigue calculation is performed is very small. Following the discussion of §2.3.2, this should be taken into account by using an appropriate value of $C_G$, but in practice the stress field is so complex that it is difficult to decide what value of $V$ should be used. Instead, a procedure has been developed based on direct experimental comparison between the fatigue strengths of specimens with and without certain stress concentrations.

As we would expect from size effect considerations, theoretical estimates of fatigue life become progressively more conservative as the characteristic length of the stress concentration (e.g. the fillet radius in Figure 2.29) gets smaller. We take this into account by defining a more realistic *fatigue stress concentration factor* $K_f$ by the relation

$$K_f = 1 + q(K_t - 1) , \tag{2.90}$$

where $q$ is an experimentally determined parameter known as the *notch sensitivity factor*. Peterson suggests an approximate expression for $q$ in the form

$$q = \frac{1}{1+\sqrt{a/r}} , \tag{2.91}$$

where $r$ is the radius of the notch and $a$ is a length scale defined by the grain structure of the material. Figure 2.30 gives values for the notch sensitivity of some low-alloy steels, based on equation (2.91), with values of $a$ taken from Figure 13.20 of Juvinall (1967).

Notice that $q$ is always less than unity, showing that the effect of a stress concentration in fatigue is always less than the theoretical effect ($K_f < K_t$). Also, $q$ falls as the dimension $r$ of the dominant feature is reduced, as we would expect from size effect considerations. In principle, a perfectly sharp right angle corner (if such a thing could be manufactured) would have an infintely large $K_t$, but the volume of material so stressed would be zero, giving presumably a zero value of $q$. In practice, the increased tendency to fatigue failure with increasing sharpness tends to level out, albeit at too high a value for sharp corners to be desirable in design applications. One advantageous consequence of this is that the inevitable microscopic scratches in the surface of a component are less damaging to fatigue life than might be imagined.

Notice also that $q$ depends very much on the material. An interesting point is that cast iron has a very low notch sensitivity, so that geometric stress concentrations have almost no effect on its fatigue behaviour. This is believed to be due to the metallurgical structure of cast iron which is so inhomogeneous and permeated with internal material defects that any additional geometric discontinuities have comparatively little effect.

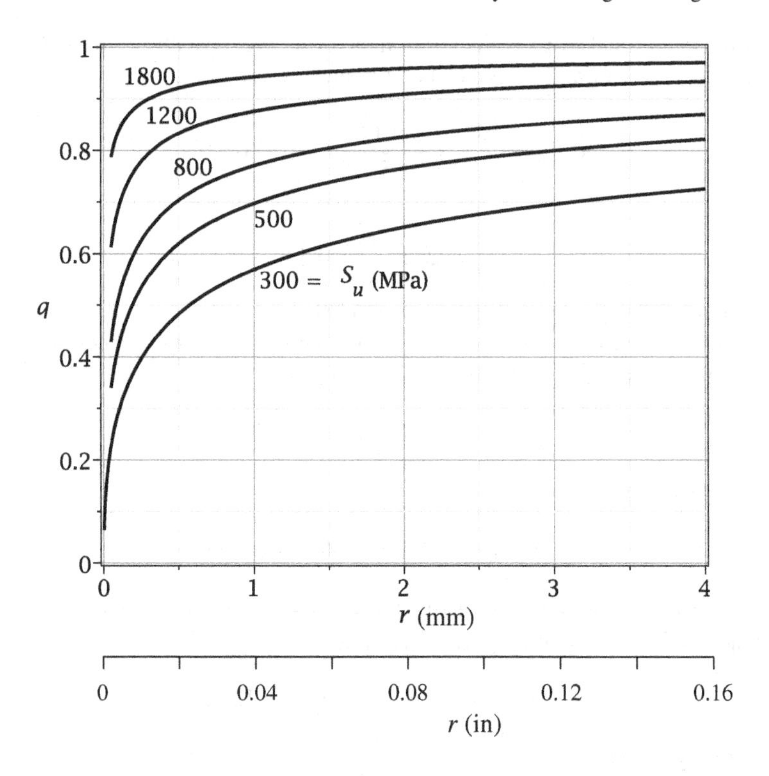


*Figure 2.30: Notch sensitivity for some low-alloy steels*

**Example 2.8**

*Figure 2.31 shows a bar with a change in section which is subjected to an alternating bending moment $M = M_0 \cos(\omega t)$, where $M_0 = 1.2$ Nm. The bar is machined from AISI 1010 steel for which $S_u = 324$ MPa.*

*Find the safety factor against fatigue failure using the values $C_R = 1$ (i.e. 50% reliability) and $C_G = 0.9$.*

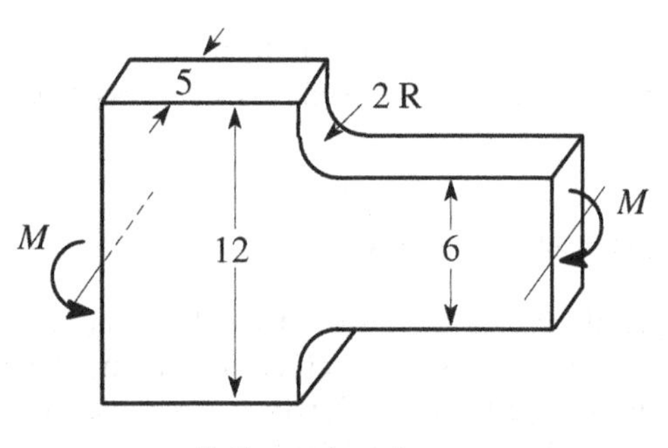


*Figure 2.31: Bar with a section change loaded in bending*

For the theoretical stress concentration factor $K_t$, we have $D=12$ mm, $d=6$ mm, $h=5$ mm and $r=2$ mm in Figure C.4 of Appendix C, so

$$\sigma_{\text{nom}} = \frac{6M}{hd^2} = \frac{6 \times 12 \times 10^3}{5 \times 6^2} = 40 \text{ MPa}$$

and

$$\frac{D}{d} = 2 \; ; \; \frac{r}{d} = 0.33 \, .$$

Interpolating in the figure, we estimate $K_t = 1.38$.

The notch sensitivity associated with a radius $r=2$ mm for a steel of $S_u = 324$ MPa is approximately $q=0.7$, from Figure 2.30, so

$$K_f = 1 + 0.7(0.38) = 1.266 \, .$$

Thus, the design stress is

$$\sigma = K_f \sigma_{\text{nom}} = 1.266 \times 40 = 50.6 \text{ MPa} \, .$$

For the material strength, we estimate $S'_n \approx 0.5 \times 324 = 162$ MPa and $C_S = 0.84$ for a machined surface from Figure 2.27. Thus,

$$S_n = 0.84 \times 0.9 \times 1.0 \times 162 = 122 \text{ MPa}$$

and the safety factor is

$$SF = \frac{122}{50.6} = 2.4 \, .$$

### 2.3.4 Effect of combined stresses

The discussion has so far been concerned with designing for a situation in which the cyclic stress state consists of uniaxial tension/compression, as in the rotating bending or push/pull tests. If a more complex cyclic stress system is present, we need something akin to a failure theory to reduce it to an equivalent tensile stress which can be compared with $S_n$. Fortunately it is found that the von Mises deviatoric strain energy theory correlates quite well with the results of biaxial fatigue tests and hence design can be based on the equivalent tensile stress as defined by equation (2.54).

### 2.3.5 Effect of a superposed mean stress

In many engineering applications, the stresses do not fluctuate between equal positive and negative values — there is also a superposed constant mean stress. A typical example is the rotating shaft of Figure 2.32, which experiences cyclic loading due to bending (as in the rotating bending test), but also transmits a torque $T$, leading to a constant torsional shear stress.

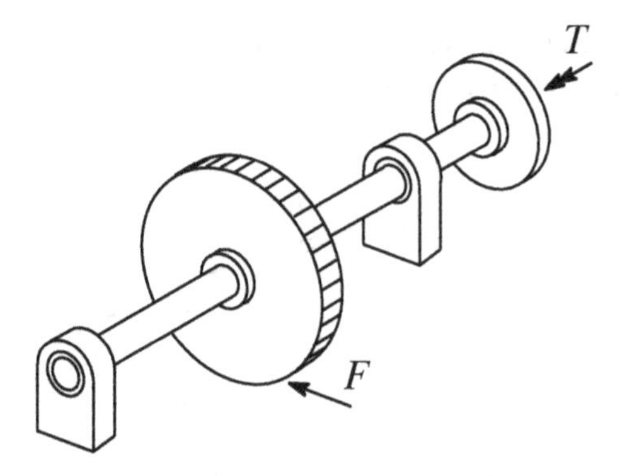


*Figure 2.32: Rotating shaft subjected to bending and torsion*

The stress state in such cases can be characterized in either of two ways. We can specify the maximum and minimum stresses $(\sigma_{max}, \sigma_{min})$, *or* the mean and alternating stresses $(\sigma_m, \sigma_a)$. Which is more appropriate depends on the application. In the example of Figure 2.32, the mean stress is caused by the torque and the alternating stress by the bending moment, so it is natural to choose this method of description, for which

$$\sigma = \sigma_m + \sigma_a \sin(\omega t) , \tag{2.92}$$

where $\omega$ is the rotational speed of the shaft.

However, a chair experiences one state of stress when you are sitting on it (presumably $\sigma_{max}$) and another ($\sigma_{min}$) when you are not, so these quantities will be the natural starting point for the calculation. It is easy to convert from one description to the other using the relations

$$\sigma_m = \frac{\sigma_{max} + \sigma_{min}}{2} \quad ; \quad \sigma_a = \frac{\sigma_{max} - \sigma_{min}}{2} \tag{2.93}$$

and

$$\sigma_{max} = \sigma_m + \sigma_a \quad ; \quad \sigma_{min} = \sigma_m - \sigma_a . \tag{2.94}$$

A special case that occurs quite frequently is that where the minimum stress state is one in which the component is unloaded ($\sigma_{min} = 0$), in which case $\sigma_m = \sigma_a = \sigma_{max}/2$.

We recall from §2.2.4 that cracks can transmit compressive tractions by crack face contact, but are opened by tensile stresses. Thus, we should anticipate that the superposition of a tensile mean stress will encourage fatigue failure, whilst a compressive mean stress will tend to inhibit it. Experimental results confirm this behaviour. Results are usually presented as a plot of $\sigma_a$ against $\sigma_m$ at the fatigue limit — i.e. as a failure surface in $\sigma_a, \sigma_m$ space. Results for a typical steel are shown in Figure 2.33. Points under the solid curve correspond to stress states that will not cause fatigue failure, regardless of the number of cycles of loading. Notice that two points on this curve can be located immediately from our previous discussion. If the mean stress $\sigma_m = 0$, the loading is completely reversed and the fatigue limit is defined by $\sigma_a = S_n$ as defined in §2.3.1. This corresponds to point $A$ on Figure 2.33. On the other hand, if $\sigma_a = 0$, there is no alternating stress and failure will occur when the mean stress reaches the ultimate tensile strength of the material ($\sigma_m = S_u$). This corresponds to point $B$ in Figure 2.33.

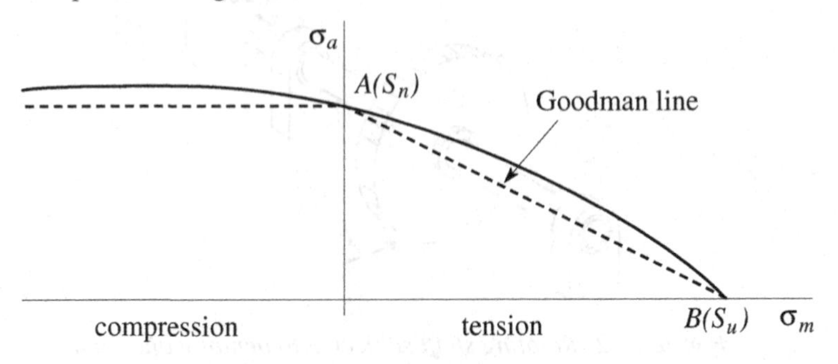


*Figure 2.33: Fatigue limit under combined mean and alternating stress*

Experimental curves such as Figure 2.33 are generally found to be convex upwards, so a conservative failure surface can be defined by locating the limiting points $A, B$ and joining them by a straight line. This line is known as the *Goodman line* and the corresponding diagram as the *Goodman diagram.* If the mean stress is compressive ($\sigma_m < 0$), the fatigue limit is generally increased, but it is usual to make the conservative approximation of neglecting this effect — i.e. to treat the failure surface as horizontal in the range $\sigma_m < 0$. These approximations are represented by the dotted straight lines in Figure 2.33. They are quite conservative and various methods have been proposed for making more accurate approximations to the experimental data.[34]

**Example 2.9**

*The critical point in an engineering component experiences a tensile stress that fluctuates between the maximum and minimum values $\sigma_{\max} = 140$ MPa, $\sigma_{\min} = 80$ MPa. The material has an ultimate strength $S_u = 560$ MPa and an endurance limit (after taking account of surface and gradient factors) $S_n = 180$ MPa. Estimate the factor of safety against fatigue failure.*

We first note that

$$\sigma_m = \frac{140 + 80}{2} = 110 \text{ MPa} \;\; ; \;\; \sigma_a = \frac{140 - 80}{2} = 30 \text{ MPa} \,.$$

The Goodman diagram has the form shown in Figure 2.34 and the operating point is at $P(110, 30)$. It is clearly on the safe side of the Goodman line and the safety factor is

$$SF = \frac{OQ}{OP} \,,$$

which can be estimated graphically to acceptable accuracy as

$$SF = 2.8 \,.$$

[34] See for example Shigley and Mitchell (1983), §7.14.

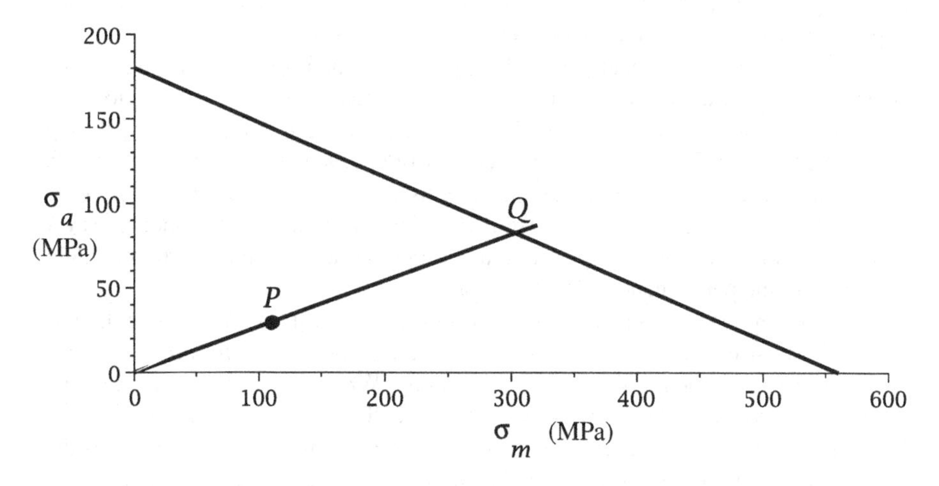


*Figure 2.34: Goodman diagram for Example 2.9.*

Alternatively, we can argue that the Goodman line is described by the equation

$$\frac{\sigma_a}{180} + \frac{\sigma_m}{560} = 1$$

and the load line $OPQ$ by

$$\frac{\sigma_m}{110} = \frac{\sigma_a}{30} .$$

Both equations are satisfied at $Q$, so eliminating $\sigma_m$, we obtain

$$\frac{\sigma_a^Q}{180} + \frac{11\sigma_a^Q}{3 \times 560} = 1 ,$$

with solution $\sigma_a^Q = 82.6$ MPa.

The safety factor then follows from

$$SF = \frac{\sigma_a^Q}{\sigma_a^P} = \frac{82.6}{30} = 2.75 .$$

However, we should emphasise that the accuracy of design calculations doesn't generally warrant such sophistication. The visual estimate from the sketch is as good as we are likely to need.

### Effect of contained plasticity

For ductile materials, the yield strength $S_Y$ is significantly lower than the ultimate strength $S_u$ and hence there are regions in the lower right-hand corner of the Goodman diagram where static yield will occur during the first application of the load. If in a design calculation we obtain high mean stresses in this range, their significance depends on whether or not a stress concentration is present.

If there is no stress concentration (for example, if the mean stress results from a constant torque or axial force on a uniform bar), extensive plastic deformation will occur, with strains that are an order of magnitude larger than those occurring during elastic deformation. The material will become work hardened — i.e. it will develop a higher yield stress in response to the high strains and this is why it can support maximum stresses in excess of those causing yield in a virgin material. Notice however that we would be unlikely to operate in this range because the plastic strains might compromise the performance of the component.

If the component contains a stress concentration and if the design calculations show a mean stress exceeding the yield stress *after the stress concentration factor has been included*, the predictions are misleading. When the loads are first applied, there will come a point when yield will occur locally near the stress concentration. Up to this point, the stresses are correctly given by the elastic calculation, but as the load is increased further, the stress in the stress concentration levels off at the yield stress $S_Y$. In order to produce a local stress in excess of $S_Y$ — i.e. to produce significant work hardening — we would need to have large plastic strains and these cannot occur in a local region at a stress concentration, since the strains are dominated by the elastic behaviour of the surrounding material.

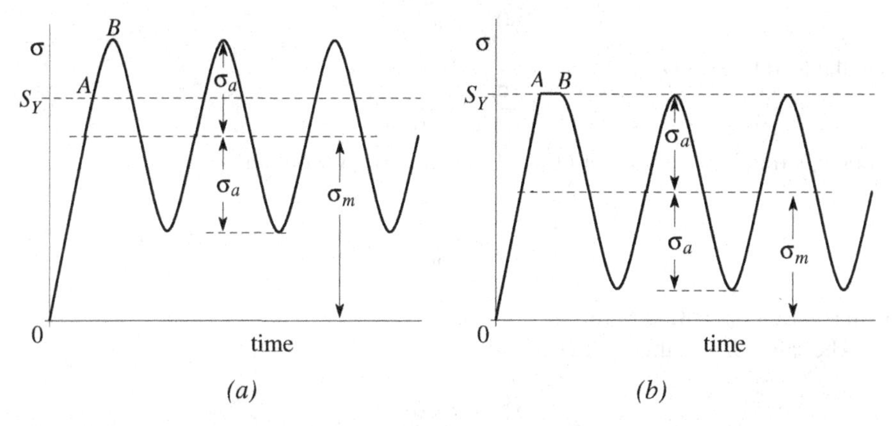


*Figure 2.35: Effect of contained plasticity at a stress concentration: (a) Theoretical stress history exceeding the yield stress, (b) Actual stress history taking into account local yielding*

Now suppose that a component with a stress concentration is loaded so as to produce the maximum theoretical elastic tensile stresses shown in Figure 2.35 *(a)*. In practice, yielding will start at $A$ and continue as long as the applied loads increase — i.e. to point $B$. The stress will remain at $S_Y$ during this yielding process, but after point $B$, unloading will occur elastically, so that the actual stress history will be as shown in Figure 2.35 *(b)*. In particular, we notice that the stress history is modified by yield in such a way that the alternating stress $\sigma_a$ remains unchanged, but the mean stress is reduced so as to reduce $\sigma_{\max}$ to $S_Y$.

We then have $\sigma_{\max} \equiv \sigma_m + \sigma_a = S_Y$ and hence

$$\sigma_m = S_Y - \sigma_a \, . \tag{2.95}$$

This equation corresponds to the line *CD* on the Goodman diagram of Figure 2.36. If the calculated point on the Goodman diagram lies to the right of *CD*, local yielding will reduce $\sigma_m$ until the effective operating point lies *on CD*. For example, the calculated point $P_1$ is moved to $P_1^*$, which is below *AB* and therefore safe, whilst $P_2$ is moved to $P_2^*$, which is above *AB* and therefore unsafe.

We conclude that if the calculated mean stress $\sigma_m > S_m^D$, fatigue failure will occur if and only if $\sigma_a \geq S_a^D$, where $S_m^D, S_a^D$ are the coordinates of the point *D*.

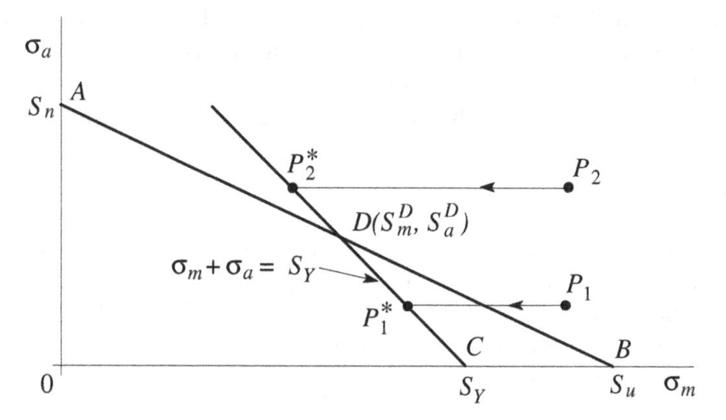


*Figure 2.36: Interpretation of local yielding on the Goodman diagram*

### 2.3.6 Summary of the design process

As we have seen, many factors are involved in fatigue failure and the design process can get quite complicated. We therefore give a concise summary of the steps as follows:-

(i) *Material properties*
  (a) Estimate $S_u$ either from tabulated values or by experiment.
  (b) Estimate $S_n'$, either directly from tabulated experimental data or by using equations (2.73–2.75).
  (c) Estimate $C_G, C_R, C_S$ (§2.3.2 above) and hence obtain the design fatigue limit $S_n = S_n' C_G C_R C_S$.
  (d) Sketch the Goodman diagram.
  (e) Find the notch sensitivity factor from Figure 2.30,

(ii) The nominal loading may involve (i) bending, (ii) torsion, (iii) axial force, (iv) shear force, *each of which* may have mean and alternating components. Separate the loading into these components (e.g. mean torque, alternating torque, mean axial force etc.).

(iii) *For each loading component,*
  (a) calculate the nominal stresses,
  (b) find $K_t$ by interpolation in the appropriate figure in Appendix C or other reference works,
  (c) calculate $K_f = 1 + q(K_t - 1)$,
  (d) calculate the actual stress $\sigma = K_f \sigma_{\text{nom}}$.

(iv) *Take all the stresses resulting from alternating loads* and combine them to obtain an alternating equivalent tensile stress $\sigma_{Ea}$ using equation (2.54) or (2.55, 2.56). Repeat using the mean stress components to find the mean equivalent tensile stress[35] $\sigma_{Em}$.

(v) Plot the operating point $P(\sigma_{Em}, \sigma_{Ea})$ on the Goodman diagram. The design is safe if $P$ lies under the Goodman line and the safety factor can be estimated by determining the ratio between the distances from the origin to $P$ and the distance along the same line to the failure surface.

(vi) If $\sigma_{Em} > S_m^D$ in Figure 2.36, the safety factor is given by $S_a^D / \sigma_{Ea}$.

**Example 2.10**

*Figure 2.37 shows part of the shaft for a concrete mixer that is subjected to a compressive axial force of 40 kN, a transverse bearing force of 50 kN and a torque of 2 kNm. The material of the shaft is 1018A carbon steel with $S_u = 341$ MPa, $S_Y = 220$ MPa and it has a machined finish. Estimate the safety factor against fatigue failure, based on a reliability of 50%.*

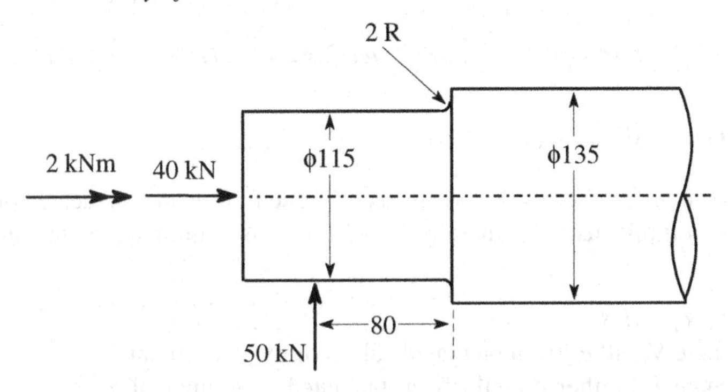


*Figure 2.37: Concrete mixer shaft. Notice the commonly used notation $\phi 115$ for a shaft of diameter 115 mm and 2 R for a fillet radius of 2 mm.*

[35] Note that since the principal effect of a mean stress is to hold the cracks open during fatigue, there is some justification for using the maximum principal stress

$$\sigma_{Em} = \sigma + \sqrt{\sigma^2 + \tau^2}$$

in place of the von Mises equivalent stress, particularly when the stresses are predominantly compressive.

(i) *Material properties*

(a) $S_u = 341$ MPa.

(b) $S'_n \approx 0.5 \times 341 = 170$ MPa.

(c) From Figure 2.27, $C_S = 0.8$. The reliability is 50%, so $C_R = 1$. Finally, to estimate $C_G$, note that the shaft diameter is 115 mm compared with 7 mm in the standard test, so the stressed volume is larger by a ratio of about

$$\left(\frac{115}{7}\right)^3 = 4500\,,$$

suggesting a factor

$$C_G = \left(\frac{1}{4500}\right)^{1/16} \approx 0.6\,.$$

Thus,

$$S_n = 1 \times 0.8 \times 0.6 \times 170 = 82 \text{ MPa}.$$

(d) The Goodman line is then constructed by joining the points (341,0) and (0,82) as shown in Figure 2.38.

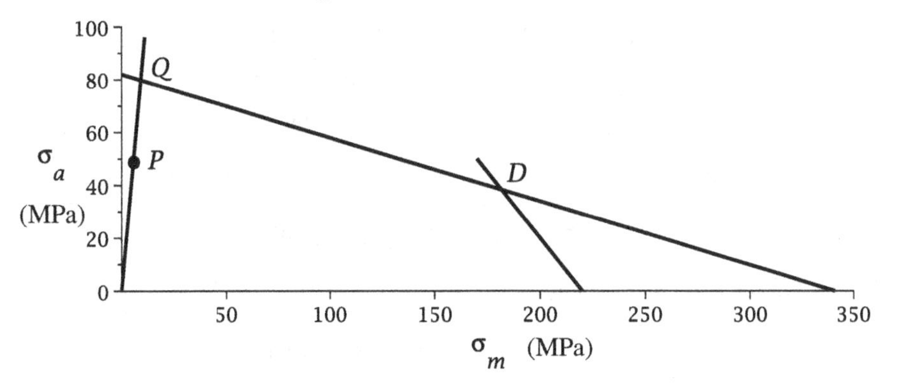


*Figure 2.38: Goodman diagram for Example 2.10.*

(e) From Figure 2.30, we have $q \approx 0.7$ for a steel with $S_u = 341$ MPa.

(ii) The 50 kN force generates a bending moment

$$M = 50 \text{ kN } \times 80 \text{ mm} = 4 \text{ kNm}.$$

The bending moment causes alternating stresses only, whilst the torque and axial force cause mean stresses only.

(iii) (a) The nominal stresses are:-

**Axial Force**

$$\sigma = \frac{F}{A} = \frac{-40 \times 10^3 \times 4}{\pi \times 115^2} = -3.85 \text{ MPa}$$

**Bending**

$$\sigma = \frac{32M}{\pi d^3} = \frac{32 \times 4 \times 10^6}{\pi \times 115^3} = 26.8 \text{ MPa}$$

**Torsion**

$$\tau = \frac{16T}{\pi d^3} = \frac{16 \times 2 \times 10^6}{\pi \times 115^3} = 6.7 \text{ MPa.}$$

(b) For $K_t$, we have

$$\frac{D}{d} = \frac{135}{115} = 1.174 \;\; ; \;\; \frac{r}{d} = \frac{4}{115} = 0.035\,,$$

and hence, using Figures C.5, C.6, C.7 from Appendix C,

$K_t = 2.24$ for axial loading, $K_t = 2.1$ in bending and $K_t = 1.74$ in torsion.

(c) It then follows that

$$\begin{aligned} K_f = 1 + (K_t - 1)q &= 1.87 \text{ for axial loading} \\ &= 1.77 \text{ in bending} \\ &= 1.52 \text{ in torsion.} \end{aligned}$$

(d) For the axial force

$$\sigma = -3.85 \times 1.87 = -7.2 \text{ MPa.}$$

For the bending moment

$$\sigma = 26.8 \times 1.77 = 47.4 \text{ MPa.}$$

For the torque

$$\tau_m = 6.7 \times 1.52 = 10.2 \text{ MPa.}$$

(iv) Alternating stresses arise only from bending and are given by

$$\sigma_{aE} = 47.4 \text{ MPa.}$$

Mean stresses arise from both the axial force and the torque. We have

$$\sigma_m = -7.2 \text{ MPa}$$

$$\tau_m = 10.2 \text{ MPa}$$

and hence

$$\sigma_{mE} = \sigma_m + \sqrt{\sigma_m^2 + \tau_m^2} = 5.3 \text{ MPa.}$$

(v) The operating point $P$ and the corresponding failure point $Q$ are shown in the Goodman diagram of Figure 2.38 above. The failure point can be determined by solving the equations

$$\frac{\sigma_a}{82} + \frac{\sigma_m}{341} = 1$$

representing the Goodman line and

$$\frac{\sigma_a}{47.4} = \frac{\sigma_m}{5.3}$$

representing the load line. We have

$$\frac{\sigma_a^Q}{82} + \frac{\sigma_a^Q}{341}\frac{5.3}{47.4} = 1$$

and hence

$$\sigma_a^Q = 79.8 \text{ MPa}$$

giving a safety factor

$$SF = \frac{79.8}{47.4} = 1.68.$$

(vi) We note that $Q$ is well to the left of the point $D$, so no yielding will occur in the stress concentration.

## 2.4 Summary

In this chapter, we have discussed the states of stress under which material failure may be expected. In particular,

(i) A general state of stress can be characterized by the three principal stresses or equivalently by the three stress invariants of equations (2.17–2.19). We have developed methods of determining the principal stresses and the maximum shear stresses from a general set of stress components.

(ii) Ductile failure or yield is associated with shear deformation without change of volume and is best predicted by the von Mises criterion (2.43), according to which yielding occurs when the distance in principal stress space of the point $(\sigma_1, \sigma_2, \sigma_3)$ from the hydrostatic line $(\sigma_1 = \sigma_2 = \sigma_3)$ exceeds a critical value. In design calculations, it is most conveniently implemented by comparing the equivalent tensile stress (2.54) with the uniaxial yield stress for the material.

(iii) Brittle materials fail by crack propagation which in turn is driven by the existence of a critical stress intensity factor at the crack tip. This implies a maximum tensile stress theory of failure. If all the applied stresses are compressive, considerations of crack face friction lead to the modified Mohr criterion. Full characterization of a brittle material requires experimental measurements of the strength in tension and in compression.

(iv) Under cyclic loading, fatigue failure can occur. The fatigue strength of a component is significantly affected by its size, the surface finish and the presence of geometric stress concentrations. The superposition of a tensile mean stress reduces the alternating stresses required to produce fatigue failure.

## Further reading

### Stress transformation and theories of failure

A.P. Boresi, R.J. Schmidt, and O.M. Sidebottom (1993), *Advanced Mechanics of Materials,* John Wiley, New York, 5th edn., Chapters 2,4.
J.E. Gordon (1968), *The New Science of Strong Materials*, Princeton University Press, Princeton NJ.
F.A. McClintock and A.S. Argon (1966), *Mechanical Behavior of Materials*, Addison-Wesley, Reading MA.

### Fracture mechanics

M.F. Kanninen and C.H. Popelar (1985), *Advanced Fracture Mechanics*, Clarendon Press, Oxford.
H. Leibowitz, **ed.** (1971), *Fracture, An Advanced Treatise*, 7 Vols., Academic Press, New York.

### Stress intensity factors and stress concentration factors

R.E. Peterson (1974), *Stress Concentration Factors*, John Wiley, New York.
W.D. Pilkey (1994), *Formulas for Stress, Strain and Structural Matrices*, John Wiley, New York.
R.J. Roark and W.C. Young (1975), *Formulas for Stress and Strain*, McGraw-Hill, New York, 5th edn. (Stress concentration factors only).
G.C. Sih (1973), *Handbook of Stess Intensity Factors*, Inst. of Fracture and Solid Mechanics, Lehigh University, Bethlehem, PA.

### Fatigue failure

J.A. Collins (1981), *Failure of Materials in Mechanical Design*, Wiley, New York.
R.C. Juvinall (1967), *Stress, Strain and Strength*, McGraw-Hill, New York.
R.C. Juvinall (1983), *Fundamentals of Machine Component Design*, Wiley, New York, Chapter 8.
J.E. Shigley and L.D. Mitchell (1983), *Mechanical Engineering Design*, McGraw-Hill, New York, 4th edn., Chapter 7.

## Problems

### Section 2.1

**2.1.** The stress at a point is defined by the components $\sigma_{xx} = -80$ MPa, $\sigma_{yy} = 10$ MPa, $\sigma_{xy} = \sigma_{yx} = 20$ MPa. Sketch the corresponding Mohr's circle and hence find the maximum in-plane shear stress and the orientation of the planes on which it acts.

Illustrate your answer with a sketch of an appropriately rotated rectangular element, taking care to indicate the correct direction for the shear stress on these planes.

**2.2.** The stress at a point is defined by the components $\sigma_{xx}=7$ ksi, $\sigma_{yy}=10$ ksi, $\sigma_{xy}=\sigma_{yx}=-20$ ksi. Sketch the corresponding Mohr's circle and hence find the maximum in-plane shear stress and the orientation of the planes on which it acts. Illustrate your answer with a sketch of an appropriately rotated rectangular element, taking care to indicate the correct direction for the shear stress on these planes.

**2.3.** The stress at a point is defined by the components $\sigma_{xx}=8$ ksi, $\sigma_{yy}=-10$ ksi, $\sigma_{xy}=\sigma_{yx}=-4$ ksi. Find the principal stresses $\sigma_1,\sigma_2$ and the inclination of the plane on which the maximum principal stress acts to the $x$-plane.

**2.4.** The stress at a point is defined by the components $\sigma_{xx}=0$ MPa, $\sigma_{yy}=100$ MPa, $\sigma_{xy}=\sigma_{yx}=-40$ MPa. Find the principal stresses $\sigma_1,\sigma_2$ and the inclination of the plane on which the maximum principal stress acts to the $x$-plane.

**2.5.** The principal stresses at a given point are $\sigma_1=-50$ MPa, $\sigma_2=-50$ MPa. Sketch Mohr's circle for this state of stress and hence determine the maximum in-plane shear stress. Comment on your results.

**2.6.** The principal stresses at a given point are $\sigma_1=10$ ksi, $\sigma_2=5$ ksi. Sketch Mohr's circle for this state of stress and hence determine the maximum in-plane shear stress.

**2.7.** The principal stresses at a given point are $\sigma_1=10$ MPa, $\sigma_2=-100$ MPa. Sketch Mohr's circle for this state of stress and determine the normal stress on a plane inclined at an angle $\theta$ to the principal plane 1. Hence find the range of values of $\theta$ for which the normal stress is tensile.

**2.8.** The stress at a point is defined by the components $\sigma_{xx}=220$ MPa, $\sigma_{yy}=220$ MPa, $\sigma_{zz}=0$ MPa, $\sigma_{xy}=-80$ MPa, $\sigma_{yz}=40$ MPa, $\sigma_{zx}=0$ MPa. Find the three principal stresses and the direction cosines of the plane on which the maximum tensile stress acts.

**2.9.** The stress at a point is defined by the components $\sigma_{xx}=120$ MPa, $\sigma_{yy}=-20$ MPa, $\sigma_{zz}=20$ MPa, $\sigma_{xy}=60$ MPa, $\sigma_{yz}=0$ MPa, $\sigma_{zx}=0$ MPa. Find the three principal stresses and the direction cosines of the plane on which the maximum tensile stress acts.

**2.10.** The stress at a point is defined by the components $\sigma_{xx}=0$ ksi, $\sigma_{yy}=0$ ksi, $\sigma_{zz}=0$ ksi, $\sigma_{xy}=-3$ksi, $\sigma_{yz}=-3$ ksi, $\sigma_{zx}=-3$ ksi. Find the three principal stresses and the direction cosines of the plane on which the maximum tensile stress acts.

**2.11.** The stress at a point is defined by the components $\sigma_{xx}=20$ ksi, $\sigma_{yy}=8$ ksi, $\sigma_{zz}=-15$ ksi, $\sigma_{xy}=0$ ksi, $\sigma_{yz}=4$ ksi, $\sigma_{zx}=16$ ksi. Find the three principal stresses and the direction cosines of the plane on which the maximum tensile stress acts.

**2.12.** The strain energy stored per unit volume in a body subjected to a uniform state of stress defined by the principal stresses $\sigma_1,\sigma_2,\sigma_3$ is

$$U_0 = \frac{1}{2}(\sigma_1 e_1 + \sigma_2 e_2 + \sigma_3 e_3) \,,$$

where $e_1, e_2, e_3$ are the corresponding principal strains. Use the stress-strain relations (1.7–1.9) to obtain an expression for $U_0$ in terms of the stress invariants and the material properties. Hence show that if $I_1 = 0$,

$$U_0 = -\frac{I_2}{2G} \,.$$

**2.13*.** Show that

$$I_2 < \frac{1}{3} I_1^2$$

for all states of stress.

**2.14*.** Establish a condition that must be satisfied by the six stress components $\sigma_{xx}, \sigma_{yy}, \sigma_{zz}, \sigma_{xy}, \sigma_{yz}, \sigma_{zx}$ if one of the principal stresses is to be zero and the other two negative — i.e. $\sigma_1 = 0, \sigma_2 < 0, \sigma_3 < 0$.

**Hint:** Solve the problem first in terms of the stress invariants using equations (2.25–2.27) and then substitute for the invariants in the final conditions using (2.17–2.19).

**2.15.** If the radii of the three Mohr's circles in Figure 2.5 *(a)* are denoted by $\tau_1, \tau_2, \tau_3$ respectively, show that

$$\tau_1^2 + \tau_2^2 + \tau_3^2 = \frac{1}{2}\left(I_1^2 - 3I_2\right) \,.$$

**2.16.** The stress at a point is defined by the components $\sigma_{xx} = 120$ MPa, $\sigma_{yy} = 300$ MPa, $\sigma_{zz} = 70$ MPa, $\sigma_{xy} = -80$ MPa, $\sigma_{yz} = \sigma_{zx} = 0$ MPa, so that $\sigma_{zz}$ is a principal stress. Find the other two principal stresses and sketch the three Mohr's circles. Hence determine the magnitude of the maximum shear stress.

**2.17.** The stress at a point is defined by the components $\sigma_{xx} = -20$ ksi, $\sigma_{yy} = 5$ ksi, $\sigma_{zz} = 2$ ksi, $\sigma_{xy} = 5$ ksi, $\sigma_{yz} = \sigma_{zx} = 0$ ksi, so that $\sigma_{zz}$ is a principal stress. Find the other two principal stresses, sketch the three Mohr's circles and hence determine the magnitude of the maximum shear stress.

**2.18.** The stress at a point is defined by the components $\sigma_{xx} = 85$ MPa, $\sigma_{yy} = 320$ MPa, $\sigma_{zz} = 100$ MPa, $\sigma_{xy} = -40$ MPa, $\sigma_{yz} = 50$ MPa, $\sigma_{zx} = -20$ MPa. Find the three principal stresses, sketch the three Mohr's circles and hence determine the magnitude of the maximum shear stress.

**2.19.** The stress at a point is defined by the components $\sigma_{xx} = 5$ ksi, $\sigma_{yy} = 0$ ksi, $\sigma_{zz} = -8$ ksi, $\sigma_{xy} = -4$ ksi, $\sigma_{yz} = 2$ ksi, $\sigma_{zx} = -2$ ksi. Find the three principal stresses, sketch the three Mohr's circles and hence determine the magnitude of the maximum shear stress.

## Section 2.2.3

**2.20.** The uniaxial yield strength of annealed AISI 1015 steel is 386 MPa. Estimate the shear stress at first yield in the torsion test for this material using (i) Tresca's theory and (ii) von Mises' theory.

**2.21.** Show that the radius of the cylinder in Figure 2.12 defining the von Mises yield surface is $\sqrt{2/3}\,S_Y$.

**2.22.** Show that the perpendicular distance $d$ of a general stress point $(\sigma_1, \sigma_2, \sigma_3)$ from the hydrostatic line in Figure 2.12 is given by $d = \sqrt{3}\,\tau_{oct}$.

**2.23.** A steel is found to yield in uniaxial tension at a stress $S_Y = 205$ MPa and in torsion at a shear stress $\tau_Y = 116$ MPa. Which of von Mises' and Tresca's theories is the most consistent with this experimental data.

**2.24.** The torsion of a circular cylinder produces a state of pure shear $\tau$, for which the principal stresses are $(\tau, -\tau, 0)$. If first yield occurs in the torsion test at a shear stress $\tau_Y$, find an expression for the *equivalent shear stress* $\tau_E$, defined as that function of the stress components that is equal to $\tau_Y$ on the von Mises yield surface.

**2.25.** A solid cylindrical bar of diameter $D$ is subjected to a bending moment $M$ and a torque $T$. Use equations (1.17, 1.18) to find the corresponding stresses and hence obtain an expression for the equivalent tensile stress $\sigma_E$ according to von Mises' theory.

**2.26.** A cylindrical tube of mean radius 1 in and wall thickness 0.1 in is subjected to an internal pressure $p = 300$ psi and a torque $T = 2000$ lb.in. The ends of the tube are closed. Find the safety factor against yielding if the material yields in uniaxial tension at $S_Y = 8$ ksi and von Mises' theory is used.

**2.27.** The stress components at a given point in an engineering component are estimated to be $\sigma_{xx} = 4.1$ ksi, $\sigma_{yy} = 0$ ksi, $\sigma_{zz} = 0.9$ ksi, $\sigma_{xy} = -3.1$ksi, $\sigma_{yz} = 1.2$ ksi, $\sigma_{zx} = 0$ ksi. Estimate the factor of safety against yielding using von Mises' theory if the uniaxial yield stress of the material is $S_Y = 17.2$ ksi.

**2.28.** A solid cylindrical shaft is found to yield when subjected to a torque of 5 kNm. An identical shaft is now subjected to a bending moment $M$. Find the value of $M$ that will cause yield assuming that (i) Tresca's theory and (ii) von Mises' theory of yielding applies.

**2.29.** A cylindrical steel shaft is required to transmit a bending moment of 1000 Nm and a torque of 2200 Nm with a safety factor against yielding of 3. If the steel has a yield stress of 250 MPa, find the minimum permissible diameter for the shaft. Use von Mises' theory of yielding.

**2.30.** A mass of 10 kg is suspended from the ceiling on a steel wire of 1 mm diameter and 2 m length. The mass is now rotated through an angle $\theta$ causing the wire to twist.

If the steel has a yield stress of 300 MPa and $G = 80$ GPa, estimate the maximum value of $\theta$ if there is to be no plastic deformation. Use von Mises' theory of yielding.

**2.31.** The three principal stresses at a given point are $\sigma_1 = 60$ MPa, $\sigma_2 = -100$ MPa, $\sigma_3 = -10$ MPa. If the material has a yield stress of 250 MPa, estimate the factor of safety against yielding using (i) the maximum shear stress theory and (ii) von Mises' theory.

**2.32*.** A series of experiments is conducted in which a thin plate is subjected to biaxial tension/compression $\sigma_1, \sigma_2$, the plane surfaces of the plate being traction-free (i.e. $\sigma_3 = 0$). Unbeknown to the experimenter, the material contains microscopic defects which can be idealized as a sparse distribution of small circular holes through the thickness of the plate. The hoop stress around the circumference of one of these holes when the plate is loaded in *uniaxial* tension $\sigma$ is known to be

$$\sigma_{\theta\theta} = \sigma(1 - 2\cos 2\theta) ,$$

where the angle $\theta$ is measured from the direction of the applied stress. Show graphically the relation that will hold at yield between the stresses $\sigma_1, \sigma_2$ applied to the defective plate if the Tresca criterion applies for the undamaged material.

**Hint:** The hoop stress due to biaxial stress can be constructed by superposition. The maximum must occur at either $\theta = 0$ or $\theta = \pi/2$, depending on the relative magnitude of $\sigma_1, \sigma_2$.

**Section 2.2.4**

**2.33.** AISI 403 stainless steel has a fracture toughness $K_{Ic} = 77$ MPa$\sqrt{\text{m}}$ and a uniaxial yield stress $S_Y = 690$ MPa. If a component contains a random distribution of through cracks up to 1 mm in length, determine whether it will exhibit ductile or brittle behaviour in tension.

**2.34.** A machine component can be idealized as a strip of width 20 mm and thickness 2 mm. It contains a central crack of length 2 mm. Determine the maximum load that the component can carry in tension without fracture if $K_{Ic} = 50$ MPa$\sqrt{\text{m}}$.

**2.35*.** An NDE test method can detect through cracks of lengths greater than 2 mm. A given component consists of a strip of width 30 mm and thickness 4 mm and is subjected to an alternating load in service of $20 \pm 10$ kN. The material has a fracture toughness $K_{Ic} = 30$ MPa$\sqrt{\text{m}}$. Laboratory experiments suggest that under these conditions fatigue cracks grow exponentially according to the equation

$$a = a_0 \exp(\lambda t) ,$$

where $t$ is the length of time in service in seconds, $a_0$ is the original crack half-length at $t = 0$ and $\lambda = 10^{-8}$ s$^{-1}$. Determine the minimum frequency of NDE testing if brittle failure during the intervening period is to be avoided.

**2.36.** A cylindrical beam of length $L$ and diameter $D$ is simply supported at its ends and loaded by a central force $F$. If the material is brittle and fails at maximum tensile stress $S_t$, find the force $F$ at fracture. Use your results to estimate the tensile strength of chalk by performing a simple test.

**2.37.** A ceramic rod of diameter 8 mm is loaded by an axial compressive force of 80 N and a torque $T$. Find the value of $T$ at failure if the material satisfies the modified Mohr theory of failure with $S_t = 0.8$ MPa, $S_c = 2.4$ MPa.

**2.38*.** A material exhibits brittle failure in uniaxial compression at a stress $\sigma_1 = -S_c$, $\sigma_2 = \sigma_3 = 0$. Sketch the corresponding Mohr's circle and obtain a general expression for the quantity

$$S = \tau - \mu(-\sigma)$$

as a function of the orientation $\theta$ of the plane to the principal direction 1. Find the value of $\theta$ for which $S$ is a maximum and hence show that if failure occurs when $S_{max} = S_t$,

$$S_c = \frac{2S_t}{\left(\sqrt{1+\mu^2} - \mu\right)} .$$

Comment on the implications of your results for the orientation of the failure plane in a uniaxial compression test.

**2.39.** A solid cylindrical shaft is found to fracture when subjected to a torque of 5 kNm. An identical shaft is now subjected to a bending moment $M$. Find the value of $M$ that will cause fracture assuming that the maximum tensile stress theory of failure applies.

**2.40.** A cylindrical pressure vessel of diameter 2 m and wall thickness 20 mm is subjected to an internal pressure of 20 bar (2 MPa). What is the maximum length of longitudinal through crack that can be permitted with a safety factor of 3, if the material has a fracture toughness $K_{Ic} = 50$ MPa$\sqrt{\text{m}}$?

**2.41.** The stress components at a given point in an engineering component are estimated to be $\sigma_{xx} = 4.1$ ksi, $\sigma_{yy} = 0$ ksi, $\sigma_{zz} = 0.9$ ksi, $\sigma_{xy} = -3.1$ksi, $\sigma_{yz} = 1.2$ ksi, $\sigma_{zx} = 0$ ksi. Estimate the factor of safety against fracture using the modified Mohr theory if the uniaxial tensile and compressive strengths are $S_t = 10$ ksi, $S_c = 30$ ksi, respectively.

**2.42.** A brittle material is found to obey the modified Mohr theory of failure with $S_t = 10$ ksi, $S_c = 30$ ksi. What value of the coefficient of internal friction $\mu$ is most consistent with these values?

## Sections 2.3.1, 2.3.2

**2.43.** A 1 in. diameter circular shaft is machined from AISI 1010 steel, for which $S_u = 47$ ksi. It is loaded in rotating bending and is required to resist fatigue failure with a reliability of 99.9%. Estimate the maximum bending moment that can be transmitted.

**2.44.** A high speed steel used for manufacturing twist drill bits is found experimentally to have a hardness of 480 BHN. The drills have a ground surface finish. What is an appropriate fatigue limit $S_n$ for a 2 mm diameter drill for 50% reliability?

**2.45*.** Figure P2.45 shows the geometry and dimensions of the standard rotating bending test specimen. Find the volume of material subjected to 95% or more of the maximum alternating tensile stress. Perform a similar calculation for a bar of 1 in. square cross section and length 2 in. subjected to a reversed bending moment $M = M_0 \cos(\omega t)$ and hence determine an appropriate value for $C_G$ for the bar.

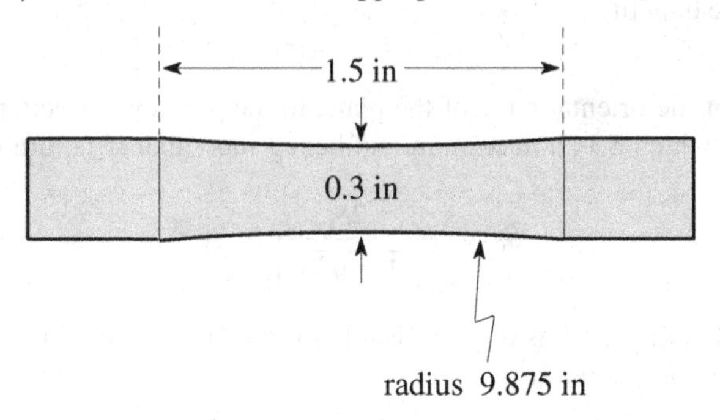


*Figure P2.45*

**2.46.** Figure P2.46 shows an I-beam of length 2m that is subjected to cyclic bending about the horizontal axis. What is an appropriate value of $C_G$ to use in this case?

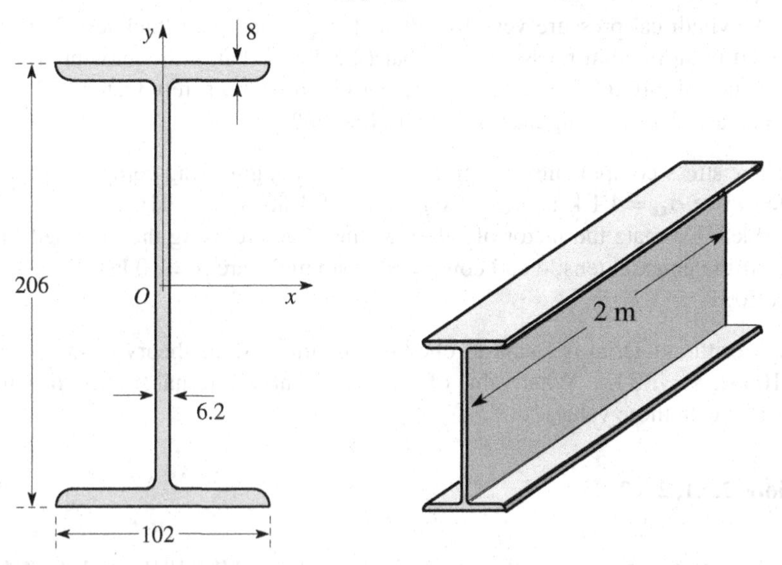


*Figure P2.46*

**2.47.** A designer chooses the dimensions of a component so as to give a safety factor of 1.5 against fatigue failure based on a reliability of 50%. What will be the probability of fatigue failure if the fatigue data for the material follow the Weibull distribution with modulus $b = 20$?

**2.48*.** Twelve standard rotating bending tests on nominally identical specimens of a certain steel yielded the following results for $S'_n$: 392, 401, 372, 386, 425, 417, 398, 407, 381, 400, 411, 391 MPa. Sketch the cumulative reliability distribution $R(S)$ and estimate $S_0$ and the Weibull parameter $b$ in equation (2.78).

## Section 2.3.3

**2.49.** A bar of rectangular cross section 20 mm × 3 mm has a central hole of diameter 6 mm. Find the maximum tensile stress if the bar is subjected to a tensile force of 5 kN.

**2.50.** The bar of Problem 2.49 is subjected to a bending moment $M = 1.8$ Nm about the more flexible bending axis. Find the maximum tensile stress near the hole.

**2.51.** A bar of thickness 1 in. has a symmetric step change in width from 2 in. to 1.5 in. through a fillet radius of 1/8 in., as shown in Figure P2.51. It is subjected to a bending moment of 4000 lb in. Find the maximum tensile stress at the change of section.

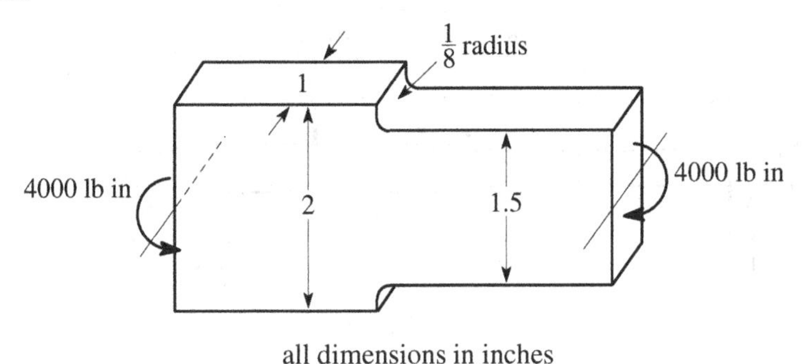


all dimensions in inches

*Figure P2.51*

**2.52.** The bar of Problem 2.49 is subjected to an alternating tensile load $F = F_0 \cos(\omega t)$, where $F_0 = 2$ kN. If the bar is machined from AISI 1010 steel ($S_u = 324$ MPa), find the safety factor against fatigue failure with a reliability of 50%.

**2.53.** The bar of Problem 2.51 is machined from AISI 1040 steel for which $S_u = 83$ ksi. It is subjected to an alternating bending moment $M = M_0 \cos(\omega t)$. Find the maximum value of $M_0$ if the reliability against fatigue failure is to be no less than 99%. An appropriate value of $C_G$ is 0.9.

**2.54.** Figure P2.54 shows a cylindrical shaft with a change in section through a fillet radius of 0.8 mm. The shaft transmits an alternating torque $T = T_0 \cos(\omega t)$, where $T_0 = 3$ Nm, it is to be manufactured from a steel with $S_u = 800$ MPa and is required to have a reliability of 99.99% against fatigue failure. Determine whether a machined shaft will meet this requirement, or whether it is necessary to grind the shaft.

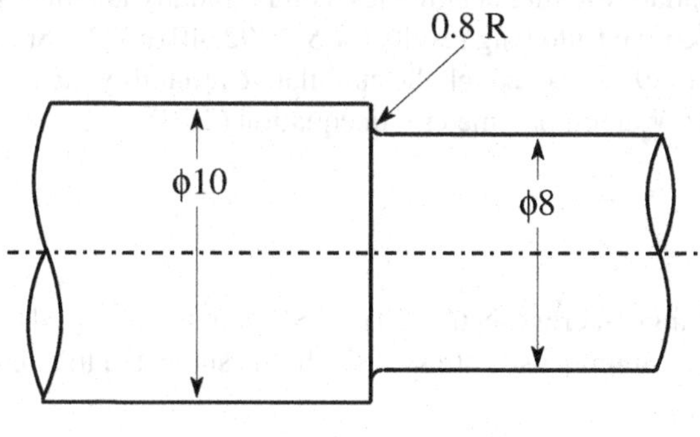


*Figure P2.54*

**2.55*.** The shaft shown in Figure P2.55 is loaded in rotating bending by the 400 lb force. The shaft is made from steel hardened to 350 BHN and the surface is ground. Find the factor of safety against fatigue failure, based on a reliability of 50%.

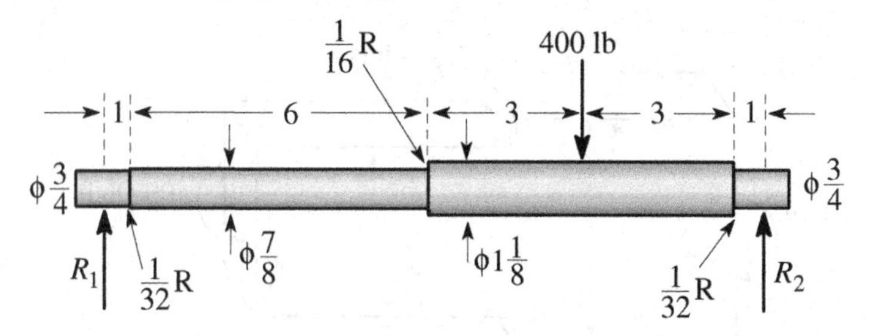


*Figure P2.55*

**Hint:** After finding the reactions $R_1, R_2$, sketch the bending moment diagram. It should then be obvious which point will be most likely to fail in fatigue.

**Section 2.3.4**

**2.56.** Figure P2.56 shows a bar of cross section 20 × 5 mm with a central hole of diameter 4 mm. It is subjected to a tensile force $F$ which fluctuates between maximum and minimum values of 10 kN and 8 kN respectively. Determine the safety factor against eventual fatigue failure with a reliability of 50% if the bar is machined from steel with $S_u = 324$ MPa and $S_Y = 210$ MPa.

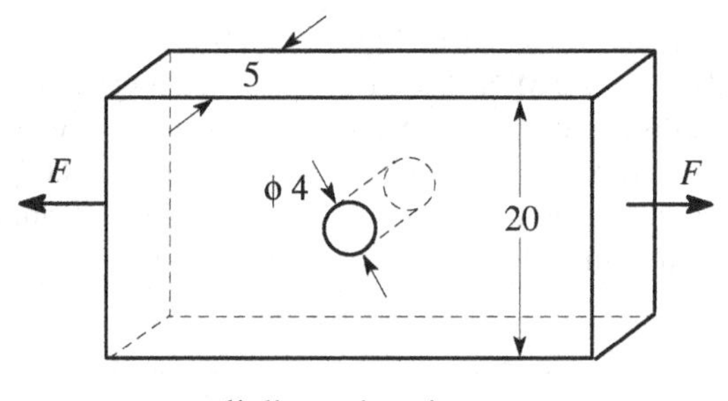


all dimensions in mm

*Figure P2.56*

**2.57.** A cold drawn rectangular bar is 1.6 in wide, 0.2 in thick and has a machined central hole of diameter 0.2 in. Estimate the maximum cyclic tensile force (loading zero to maximum) that can be applied to the bar if the steel has a hardness of 180 BHN and a yield stress of 60 ksi.

**2.58.** Figure P2.58 shows a spring which is formed from 20 mm × 2 mm cold drawn steel plate and then hardened to 510 BHN. The bends have an inner radius of 2 mm for which $K_t$ is estimated as 1.9. The force $F$ fluctuates between 40 N and 200 N during normal service. Estimate the probability of eventual fatigue failure and indicate where it is most likely to occur. Assume a surface factor intermediate between 'machined' and 'hot rolled' in Figure 2.27.

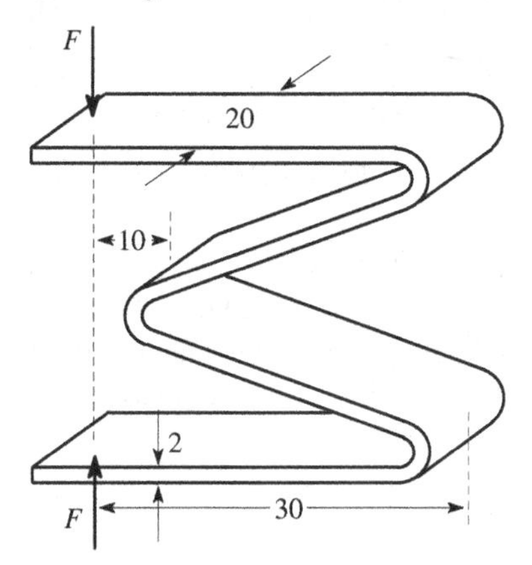


all dimensions in mm

*Figure P2.58*

**2.59.** The fatigue stress concentration factor in an engineering component is estimated to be $K_f = 1.72$ and the calculated nominal tensile stresses are $\sigma_m = 340$ MPa, $\sigma_a = 50$ MPa. If the material has an endurance limit $S_n = 185$ MPa, yield stress $S_Y = 350$ MPa and ultimate strength $S_u = 525$ MPa, estimate the safety factor against eventual fatigue failure.

**2.60*.** The shaft of the disk sander of Figure P2.60 is made of steel of $S_u = 900$ MPa, $S_Y = 750$ MPa, $S'_n = 480$ MPa. The most severe loading condition occurs when pressing the sanding disk against an object near the outer radius with sufficient force almost to stall the motor. Assuming a coefficient of friction of 0.6 between the disk and the object and a motor stall torque of 12 Nm, determine the safety factor against eventual fatigue failure if the relevant surfaces are machined.

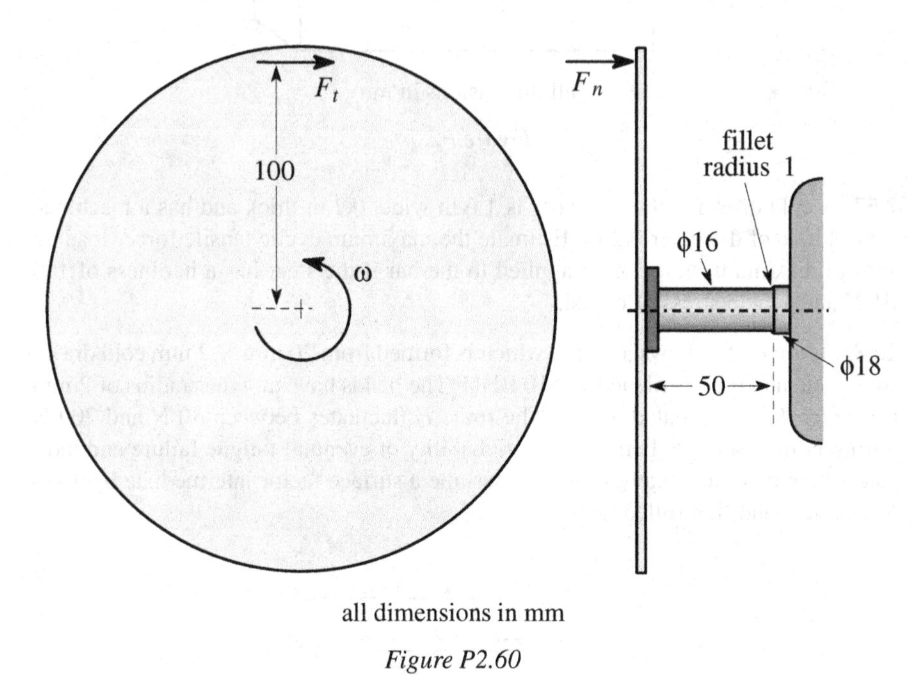


*Figure P2.60*

**2.61.** A cylindrical pressure vessel of 1 m diameter is made from hot rolled steel plate of thickness 4 mm, with properties $S_u = 800$ MPa, $S_Y = 560$ MPa, $S'_n = 425$ MPa. It is loaded by an internal pressure $p$, which fluctuates between zero and a maximum value. Find the maximum permissible working pressure if the vessel is to have an infinite life with a safety factor of 1.8.

# 3

# Energy Methods

The 'intuitive' or direct way to formulate problems in mechanics of materials is to analyze the structure into simple components, with internal forces acting between them. We then use the equations of equilibrium, geometrical conditions and stress-strain laws to develop a system of governing equations. In some cases, the equilibrium equations alone are sufficient to determine the internal forces and the problem is described as *statically determinate*. By contrast, in *statically indeterminate* problems, the full system of equations must generally be solved, even if only the internal forces are required.

Energy methods are an *alternative* to the direct approach — they do not provide additional information about the system, but instead generally replace one of the steps in the direct formulation. It is important to know which step is being replaced by a particular energy formulation, since otherwise we might write what is essentially the same equation in two different forms and end up with a redundant system of equations.

The advantages of energy methods will become apparent through the examples treated in this chapter. Notably, they can often be used to obtain the required answer without solving for a set of auxiliary quantities that are of no particular interest, and they also lend themselves to approximate solutions. In particular, they are central to the development of the finite element method, which is discussed in more detail in Appendix A.

Energy methods are a subset of a broader class of methods based on the variational calculus and known as *variational methods*. In fact it is possible to develop all the equations used in this chapter from purely mathematical arguments without ever making reference to the concept of energy. However, many of the following arguments will be easier for the reader to grasp in an energy formulation, simply because the physical ideas of conservation of energy and what is involved in doing work on a system can be called into play to aid our mathematical reasoning.

J.R. Barber, *Intermediate Mechanics of Materials*, Solid Mechanics and Its Applications 175, 2nd ed., DOI 10.1007/978-94-007-0295-0_3, © Springer Science+Business Media B.V. 2011

## 3.1 Work done on loading and unloading

When a deformable structure is loaded, the applied forces do work. We describe the structure as *elastic* if this work can be recovered on unloading, which in turn requires that the relation between force and deformation is the same on loading and unloading.

We can illustrate this for the case of the simplest system — a spring subject to a tensile force $F$, as shown in Figure 3.1 *(a)*. The displacement $u$ at the end of the spring increases with $F$ as shown in Figure 3.1 *(b)*.

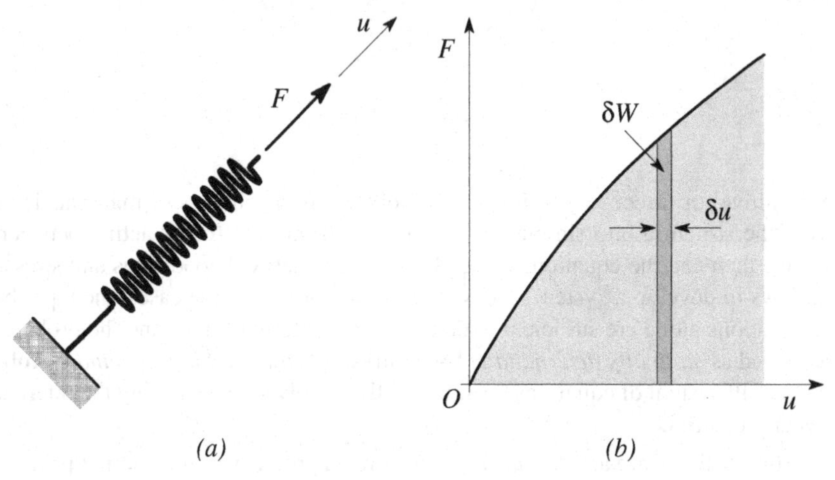


*Figure 3.1: Work done in extending a spring*

The *work done* during loading $W$ is given by the product of the force and the distance through which it acts. However, the force $F$ varies during the process because $F$ is a function of $u$, so we have to develop an expression for $W$ in integral form. We first note that the *increment* of work done $\delta W$ in increasing the displacement from $u$ to $u+\delta u$ is $F(u)\delta u$ (the product of the instantaneous force and the increment of displacement). This increment of work done is equal to the area of the darkly shaded strip in Figure 3.1 *(b)*. The total work done $W$ can be seen as the sum of a set of similar strips and is therefore equal to the area under the force-displacement curve, shown with lighter shading in Figure 3.1 *(b)*. It is obtained by integration as

$$W = \int_0^u F(u)du \,. \tag{3.1}$$

If the structure is elastic, the same relation will hold between $F$ and $u$ during both loading and unloading and the work done by the force on loading will be exactly equal to that done against the force (and hence recovered) during unloading. For *inelastic* structures, the unloading curve lies below the loading curve, as in Figure 3.2, and not all the work will be recovered. There will be a loss corresponding to the shaded area between the two curves. This is called *hysteresis loss* and the lost work is generally released in the form of heat.

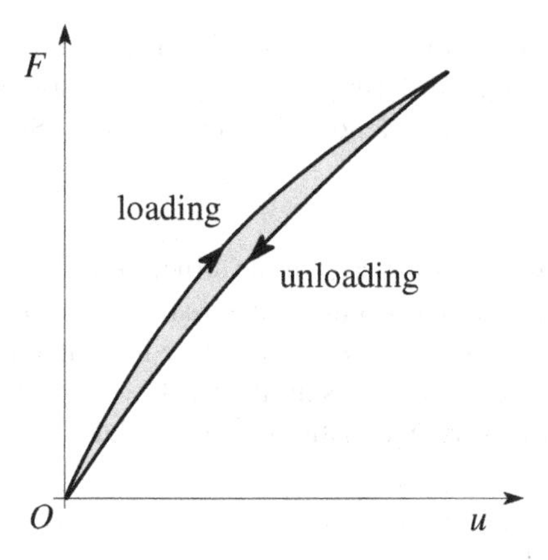


*Figure 3.2: Hysteresis losses on loading and unloading*

An important class of elastic structures is that for which the load-displacement relation is *linear* as shown in Figure 3.3. In this case, we can define a *stiffness* $k$ such that

$$F = ku\,. \tag{3.2}$$

In other words, $k$ is the slope of the straight line in Figure 3.3. It follows from equations (3.1,3.2) that the work done in extending a linear spring of stiffness $k$ is

$$W = \int_0^u ku du = \frac{1}{2}ku^2\,. \tag{3.3}$$

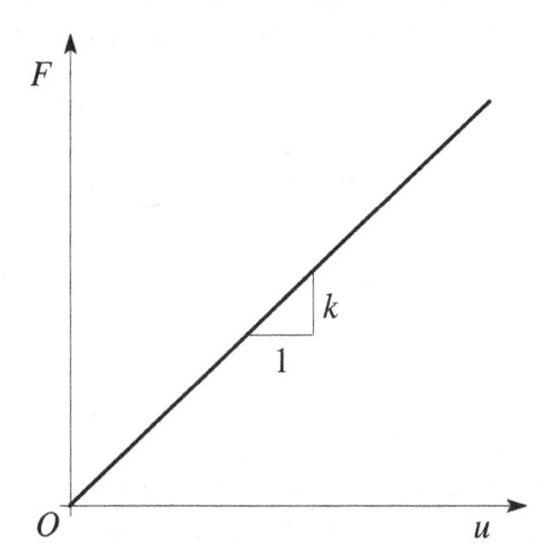


*Figure 3.3: Linear load-displacement relation*

## 3.2 Strain energy

Since the work we do on loading an elastic structure is recovered on unloading, it is natural to think of the energy as being 'stored' in the deformed structure. In other words, we think of the structure as a kind of reservoir for energy that is filled up during loading and emptied during unloading. The energy stored in this way is referred to as *strain energy* and is denoted by the symbol $U$. Since there is no energy loss,

$$U = W . \tag{3.4}$$

In other words, the strain energy stored in the structure in a deformed state is equal to the total net work we had to do on the structure to deform it.

The easiest way to find $U$ is to devise a simple loading history or scenario that will lead to the desired state and then sum the work we have to do to get there. For the simple case of a linear spring, equations (3.3, 3.4) lead to the expression

$$U = \frac{1}{2}ku^2 . \tag{3.5}$$

Two alternative forms of this equation can be obtained using equation (3.2). We have

$$U = \frac{1}{2}Fu \tag{3.6}$$

and

$$U = \frac{1}{2}\frac{F^2}{k} . \tag{3.7}$$

**Example 3.1**

*A uniform elastic beam of length L, and flexural rigidity EI, is subjected to equal and opposite moments M at each end. Find the strain energy stored in the beam.*

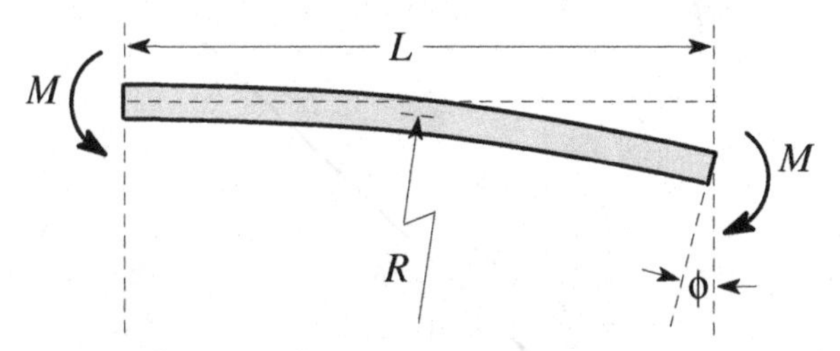


*Figure 3.4: Initially straight beam in bending*

If the beam is initially straight, it will bend into a circlular arc, as shown in Figure 3.4. The radius of curvature $R$ is given by the classical bending equation (1.17) as

$$\frac{1}{R} = \frac{M}{EI} \tag{3.8}$$

and hence, if the left end of the beam is fixed,[1] the right end will rotate through an angle

[1] It is not necessary to impose this restriction. If both ends of the beam rotate, both moments will do work, but the net work done depends only on the *difference* between the rotations of

$$\phi = \frac{L}{R} = \frac{ML}{EI} \; . \tag{3.9}$$

The moment at the right end of the beam therefore does work during its application, leading to the result

$$U = W = \int_0^\phi M d\phi = \frac{EI}{L}\int_0^\phi \phi d\phi = \frac{EI\phi^2}{2L} \; , \tag{3.10}$$

where we have used equation (3.9) to express $M$ in terms of $\phi$.

By analogy with equations (3.6,3.7), we can use (3.9) to obtain the alternative expressions

$$U = \frac{1}{2}M\phi \tag{3.11}$$

and

$$U = \frac{M^2 L}{2EI} \tag{3.12}$$

for the strain energy stored in a beam subject to pure bending.

Notice that the work done by a moment is the product of the moment and the angle (in radians) through which it acts. All the energy arguments in this chapter can be restated to apply to moments or torques by replacing forces by moments and displacements by rotations.

## 3.3 Load-displacement relations

We have suggested that an elastic component can be considered as a kind of reservoir for strain energy. It follows that, for a structure made up of several connected components, the strain energy stored is simply the sum of that stored separately in each of the components. This can be proved formally by drawing free-body diagrams for the separate components and noting that the work done by the internal forces thereby exposed must sum to zero, because internal forces will appear in equal and opposite pairs that move through the same displacements.

This result can be used to obtain a very efficient solution for the local displacement of a determinate elastic structure due to a single external force $F$. We first draw free-body diagrams for the various components of the structure, use the arguments of the previous section to find the strain energy stored in each component and add up the various contributions. We then equate this expression to the work done in applying the force, given by equation (3.1).

---

the two ends, which is still given by (3.9). Another way of stating this is to decompose the motion into (i) that in which the left end is fixed and (ii) an arbitrary rigid body rotation. During the rigid body rotation, the two end moments do equal and opposite amounts of work that are therefore self-cancelling. In fact, a more general theorem of this kind can be proved — *viz.* that a self-equilibrated system of forces acting on a body does no work if the body executes an arbitrary rigid mody motion.

If the structure is linear — i.e. if all the components obey Hooke's law and the deformations are small — we can then use (3.6) to write

$$U = \frac{1}{2}Fu \;\; ; \;\; u = \frac{2U}{F} \tag{3.13}$$

to get an immediate expression for the displacement $u$.

However, the method can also be used for non-linear problems. If we write $u(F)$ as the displacement due to the force $F$, the work done $W$ and hence the strain energy $U$ can be written in terms of $u(F)$ as

$$U = W = \int_0^F F\frac{du}{dF}dF \; . \tag{3.14}$$

Differentiating with respect to $F$, we then have

$$\frac{du}{dF} = \frac{1}{F}\frac{\partial U}{\partial F} \; . \tag{3.15}$$

Thus, if the strain energy is a known (non-linear) function of $F$, we can determine $du/dF$ and hence integrate it to obtain $u(F)$. In the special case of linearity, $U$ will be quadratic in $F$ [see for example equation (3.7)], the right hand side of (3.15) will be constant and the result reduces to (3.13).

**Example 3.2**

*Find the vertical displacement $u$ of the point $A$ for the structure of Figure 3.5(a), loaded by a vertical force, $F$. The linear springs are assumed to be capable of transmitting loads in either tension or compression, without buckling.*

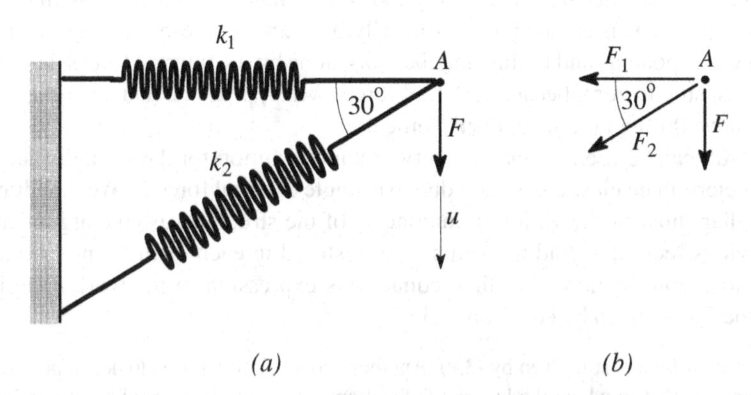


*Figure 3.5: A structure loaded by a single external force F*

First, we use the equilibrium equations to find the tensile forces $F_1, F_2$ in the two springs of stiffness $k_1, k_2$ respectively. Equilibrium of the pin at $A$ (Figure 3.5*(b)*) demands that

$$F_1 + F_2 \cos 30^\circ = 0 \tag{3.16}$$

$$F_2 \sin 30^\circ + F = 0 \tag{3.17}$$

and hence

$$F_1 = \sqrt{3}F \;\; ; \;\; F_2 = -2F \,. \tag{3.18}$$

As we might expect, spring 2 is in compression ($F_2 < 0$) and spring 1 is in tension.[2]

The energy stored in each spring $U_1, U_2$ is given by equation (3.7) with appropriate values of $F, k$, and hence the total strain energy is

$$U = U_1 + U_2 = \frac{1}{2}\frac{F_1^2}{k_1} + \frac{1}{2}\frac{F_2^2}{k_2}$$

$$= \frac{1}{2}\frac{\left(\sqrt{3}F\right)^2}{k_1} + \frac{1}{2}\frac{(-2F)^2}{k_2} = \left(\frac{3}{2k_1} + \frac{2}{k_2}\right)F^2 \,. \tag{3.19}$$

Now the total strain energy $U$ must also be equal to the work done by the force $F$ during its application. The structure as a whole is linear elastic and obeys Hooke's law, so the work done by $F$ is

$$W = \frac{1}{2}Fu \,, \tag{3.20}$$

by analogy with equation (3.6), where $u$ is the displacement component at $A$ *in the direction of* $F$ — i.e. the vertical component of the displacement at $A$.

Equating $U$ and $W$, we obtain

$$\frac{1}{2}Fu = \left(\frac{3}{2k_1} + \frac{2}{k_2}\right)F^2$$

and hence

$$u = \left(\frac{3}{k_1} + \frac{4}{k_2}\right)F \,, \tag{3.21}$$

which defines the vertical displacement of the point $A$ due to the force $F$.

Equation (3.21) can be regarded as defining a stiffness $k$ for the structure as a whole, such that

$$\frac{1}{k} = \left(\frac{3}{k_1} + \frac{4}{k_2}\right) \,. \tag{3.22}$$

In this context, it is worth noting that any linear elastic structure can be thought of as an equivalent single spring insofar as an applied force produces a proportional displacement at its point of application.

The procedure illustrated in this example is considerably easier than a direct solution, which would involve:-

[2] It is a good idea to adopt the convention that the forces in all axial members and springs are denoted by positive symbols when they are tensile, even when our intuition enables us to judge immediately that one or more of the forces in the system will be compressive.

(i) Finding the forces in the springs as in equations (3.16–3.18) above,
(ii) Using Hooke's law to determine the deformed lengths of the two springs,
(iii) Using kinematic (geometric) arguments to deduce how the deformed springs will fit together and hence determine the displacement of the point $A$.

The first two stages of the direct solution would be quite straightforward, but the kinematic argument (iii) can be quite complicated, particularly for a more complex interconnected structure such as that of Problem 3.12. The energy method used here replaces the kinematic argument to which it is therefore mathematically equivalent, though this fact is far from intuitively obvious.

The energy method developed in this section only works if there is a single (external) force $F$, since if there were more than one force, we would not be able to separate their individual contributions to the work done on the system. Furthermore, the method only gives the displacement in the direction of the force and at its point of application. In Example 3.2, the point $A$ will also have a horizontal component of displacement, which does not feature in the work done because it is orthogonal to the applied force, and which therefore cannot be found by the present method. However, later in this chapter, we shall develop ways of overcoming these limitations.

### 3.3.1 Beams with continuously varying bending moments

Equations (3.10–3.12) define the strain energy in a beam subjected to uniform bending, but in most beam problems the bending moment $M(z)$ varies with distance $z$ along the length. We can extend the argument to this case by regarding the beam as made up of a set of beam segments of infinitesimal length $\delta z$.

The strain energy in the segment between $z$ and $z+\delta z$ is

$$\delta U = \frac{M^2 \delta z}{2EI} , \tag{3.23}$$

from equation (3.12) and hence the total strain energy in the beam $0<z<L$ is

$$U = \frac{1}{2} \int_0^L \frac{M^2 dz}{EI} , \tag{3.24}$$

where we have left the flexural rigidity $EI$ under the integral sign, since this also may vary with $z$ in some cases. For example, the cross section of the beam (and hence the second moment of area $I$) may vary along the length.

**Example 3.3**

*Find the central deflection of the simply-supported beam of Figure 3.6(a) due to a central force $F$.*

By symmetry, the two end reactions are each $F/2$ and the bending moment at a distance $z$ from the left support in the segment $0<z<L/2$ can be found from the free-body diagram of Figure 3.6*(b)* as

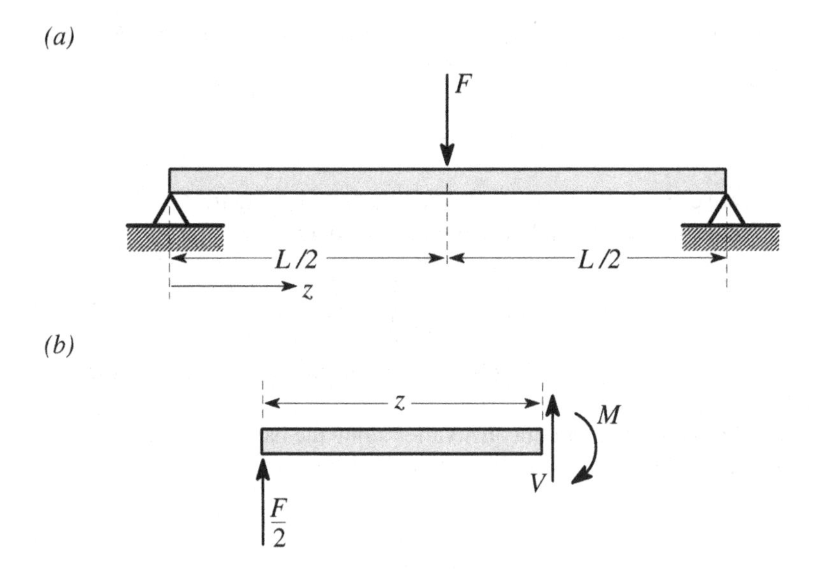


*Figure 3.6: A simply-supported beam with a central force*

$$M = -\frac{Fz}{2} \; . \tag{3.25}$$

The strain energy stored in the beam segment $0 < z < L/2$ is therefore

$$U_1 = \frac{1}{2}\int_0^{L/2} \frac{M^2 dz}{EI} = \frac{1}{2}\int_0^{L/2} \frac{F^2 z^2 dz}{4EI} = \frac{F^2 L^3}{192EI} \; ,$$

from equation (3.25). By symmetry, there must be an equal amount $U_2$ of energy stored in the other half of the beam $L/2 < z < L$ and hence the total strain energy,

$$U = U_1 + U_2 = \frac{F^2 L^3}{96EI} \; .$$

As in Example 3.2, we can deduce that the work done in (slowly) applying the force is $W = Fu/2$, where $u$ is the downward deflection under the load. The work done $W$ and the energy stored $U$ must be equal and hence the central deflection of the beam due to the force $F$ is

$$u = \frac{2W}{F} = \frac{2U}{F} = \frac{FL^3}{48EI} \; . \tag{3.26}$$

### 3.3.2 Axial loading and torsion

Essentially similar results can readily be obtained for bars loaded by axial forces and torques. We first note that a bar of cross-sectional area $A$ and length $L$, subjected to a uniform axial tensile force $F$ will experience a uniform tensile stress $\sigma = F/A$ and a strain $\varepsilon = F/EA$. The extension will therefore be

$$u = \frac{FL}{EA} . \tag{3.27}$$

The relation between $F$ and $u$ is similar to that illustrated in Figure 3.3 and it follows that the work done on loading and the stored strain energy are

$$U = W = \frac{1}{2}Fu = \frac{F^2L}{2EA} = \frac{EAu^2}{2L} , \tag{3.28}$$

where the alternative expressions are obtained by substituting for $u$ or $F$ respectively from (3.27).

If the axial force $F$ or the product $EA$ varies along the bar, we consider the bar as a set of segments of length $\delta z$ as in §3.3.1 obtaining the integral expression

$$U = \frac{1}{2}\int_0^L \frac{F^2 dz}{EA} . \tag{3.29}$$

Similarly, if a uniform circular bar[3] of length $L$ is loaded by a uniform torque $T$, the strain energy is

$$U = \frac{1}{2}T\theta = \frac{T^2L}{2GJ} = \frac{GJ\theta^2}{2L} \tag{3.30}$$

and if either the torque or the polar moment of area varies along the bar

$$U = \frac{1}{2}\int_0^L \frac{T^2 dz}{GJ} . \tag{3.31}$$

### 3.3.3 Combined loading

If a beam is subjected to simultaneous bending, torsion and axial load, the energy stored is the sum of that associated with the three loads separately. This is not a trivial result, since the energy expressions are quadratic (not linear) in the loads and the principle of superposition does not generally apply. However, it does apply in the present case because the deformations of each mode of loading are orthogonal to the other loads. For example, torsion does not cause any extension of the bar, so

[3] **Important Note** about the torsion equation. The elementary torsion formula (1.18) only applies for circular sections. However, the relation between torque and twist can be generalized to non-circular sections if the polar moment of area $J$ is replaced by a different quantity $K$ which reduces to $J$ in the case of the circle. For sections that deviate significantly from the circle (in particular for thin-walled open sections), $K$ can be several orders of magnitude smaller than $J$ so it can be catastrophic to use the elementary torsion formula out of context. The problem of finding $K$ for thin-walled sections will be addressed in Chapter 6 below. Results have been obtained either by analytical or numerical methods for a wide range of important sections and a table of such values is given by R.J. Roark and W.C. Young (1975), *Formulas for Stress and Strain*, McGraw-Hill, Table 20.

if a torque is applied to a bar in tension, the resulting deformation does not cause the axial force to do additional work. In order to preserve orthogonality between axial forces and bending moments, it is essential that the axial force be defined to act through the centroid of the beam cross section. An eccentric force must therefore be decomposed into an equal force acting through the centroid and an appropriate bending moment.

Beams are much stiffer in extension than they are in bending or torsion. For example, an ordinary wooden or plastic ruler can be visibly twisted and bent about its weaker axis by hand, whereas no visible extension can be produced. Since the energy stored is equal to the work done, which in turn is the product of the force and the distance through which it moves, it follows that more energy is generally stored in modes that involve significant deformations. Thus, if a beam is loaded simultaneously by a bending moment, a torque and an axial force, the strain energy associated with the axial force can generally be neglected.

### 3.3.4 More general expressions for strain energy

If the stress field in a body varies in a more general way, the strain energy can be computed by considering the body as made up of a set of infinitesimal elements of material of volume $\delta V$ over which the stresses can be considered uniform, and then summing the resulting expressions over the total volume of the body $V$ by integration.

We showed in §2.2.3 that the strain energy stored per unit volume is

$$U_0 = \frac{1}{2E}\Big[\sigma_{xx}^2 + \sigma_{yy}^2 + \sigma_{zz}^2 - 2\nu(\sigma_{xx}\sigma_{yy} + \sigma_{yy}\sigma_{zz} + \sigma_{zz}\sigma_{xx}) + 2(1+\nu)(\sigma_{yz}^2 + \sigma_{zx}^2 + \sigma_{xy}^2)\Big] ,$$

from equation (2.32). The total strain energy can therefore be written in the integral form

$$U = \iiint_V U_0 dV \tag{3.32}$$

$$= \frac{1}{2E}\iiint_V \Big[\sigma_{xx}^2 + \sigma_{yy}^2 + \sigma_{zz}^2 - 2\nu(\sigma_{xx}\sigma_{yy} + \sigma_{yy}\sigma_{zz} + \sigma_{zz}\sigma_{xx}) + 2(1+\nu)(\sigma_{yz}^2 + \sigma_{zx}^2 + \sigma_{xy}^2)\Big] dV . \tag{3.33}$$

### 3.3.5 Strain energy associated with shear forces in beams

If equation (3.33) is applied to a beam transmitting both bending moments and shear forces, it is clear that the shear stresses associated with the shear force will make a contribution to the total energy. However, for slender beams it is always small compared with the energy associated with the bending stresses. Inclusion of the energy due to shear is equivalent to other methods of taking account of additional deflections due to shear. It typically leads to an increase in the predicted displacements of 5% or less[4] and there are very few engineering applications where this level of accuracy is required.

## 3.4 Potential energy

Most readers will have encountered the concept of *potential energy* in physics or mechanics courses concerned with the motion of masses in a gravitational field. If we lift a body of mass $M$ through a height $h$, we have to do work against the gravitational force $Mg$. The work done is equal to the product of the force and the distance through which it acts — i.e.

$$W = Mgh\,. \tag{3.34}$$

Notice that there is no factor of $1/2$ in this equation, in contrast to equation (3.6). This is because in lifting the mass, the full gravitational force $Mg$ has to be opposed throughout the motion, whereas in extending the spring, the force increases gradually during the extension from zero to its maximum value, as shown in Figure 3.3.

The work done in lifting the mass can be recovered by lowering it again in a controlled manner, so that the gravitational force does work on the system. We therefore describe the mass as having *potential energy* in its raised state, equal[5] to the work done in lifting it $Mgh$.

Suppose a ball of mass $M$ is constrained to move in a frictionless groove, as shown in Figure 3.7. Intuitively, it is clear that the only positions where the ball may remain at rest are the points $A, B, C$, where the groove is locally horizontal. Furthermore, if the ball were slightly displaced from $A$ and then released, it would tend to roll back to $A$, whereas if it were displaced from $B$ it would tend to roll away

[4] It might be argued that we could find a counter example to this assertion by choosing a loading that involves a large shear force and a small bending moment. However, the two quantities are related through the equilibrium condition $V = dM/dz$, where $V$ is the shear force. A large shear force will therefore always be associated with a large bending moment, unless the beam is very short, in which case the theory of slender beams is arguably not applicable.

[5] Strictly speaking, only the *change* in potential energy can be defined this way, since there is some ambiguity about the datum — i.e. the location at which the potential energy is taken to be zero. For example, if the mass is resting on the ground, the gravitational force could be caused to do additional work by transporting the mass horizontally to a mine shaft and dropping it down. In Newtonian gravitational theory, all the mass in the universe was assumed to be located in a bounded region surrounded by an infinite empty space. Potential energy could then be defined unambiguously as the work that would have to be done to transport a mass from the 'point at infinity' (where there would be no gravitational force in view of the inverse-square law) to its actual location. However, we do not need to appeal to such abstruse notions, since we shall only be concerned with changes in potential energy and the choice of datum is therefore arbitrary. This will be apparent in problems in that any constants in the energy expressions will either be eliminated as a result of differentiation or will cancel in the final equations.

towards either $A$ or $C$. We therefore describe $A,C$ as *stable* equilibrium positions, whilst $B$ is an *unstable* equilibrium position.

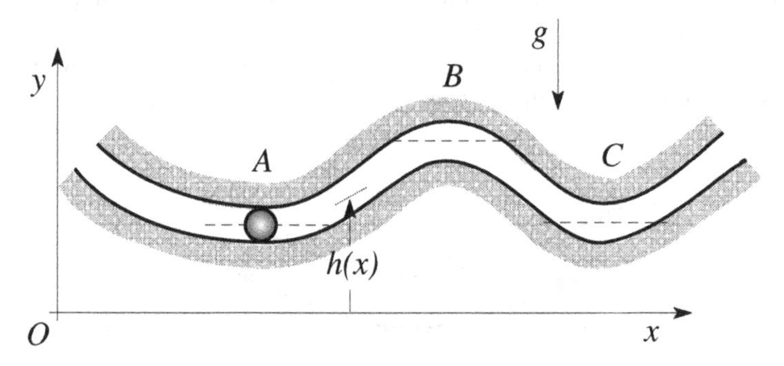


*Figure 3.7: The ball in a frictionless groove*

We can establish these claims by energy arguments. We first note that the ball has to remain within the groove, which we express in technical terms by saying that only positions within the groove are *kinematically permissible*. In the absence of this constraint, the ball would have two degrees of translational freedom,[6] which can be expressed by stating that we would need to know the horizontal and vertical coordinates $x,y$ of its centre in order to define its position. However, as long as the ball is in the groove, knowledge of the $x$ coordinate alone is sufficient to define its position, since there is only one position in the groove corresponding to any given value of $x$. Thus, the kinematically constrained ball has only one degree of translational freedom.

The potential energy $\Pi$ of the ball depends on its position in the groove and hence on $x$. In fact we can write

$$\Pi(x) = Mgh(x) , \tag{3.35}$$

where $y = h(x)$ is the equation defining the centreline of the groove in Figure 3.7. Thus, the equilibrium positions $A,B,C$ are all points where $h'(x) = 0$ and hence

$$\frac{\partial \Pi}{\partial x} = 0 . \tag{3.36}$$

Furthermore, the *stable* positions $A,C$ correspond to local minima in the potential energy — i.e. points where

$$\frac{\partial \Pi}{\partial x} = 0 \text{ and } \frac{\partial^2 \Pi}{\partial x^2} > 0 . \tag{3.37}$$

Now if the ball is instantaneously at rest at either $A$ or $C$, it can only move by passing initially through positions of higher potential energy. This in turn would

[6] We shall not consider rotations here, since rotation of the ball about its axis does not affect potential energy and, in a frictionless groove, no forces can be exerted on the ball tending to rotate it.

require that work would have to be done on the ball to increase its potential energy. In the absence of external forces other than gravity to do this work, the ball cannot move and we conclude that a minimum potential energy position must also be a position of equilibrium.

Suppose we now apply an external force to displace the ball slightly from $A$ to the point $D$ in Figure 3.8 and then release it.

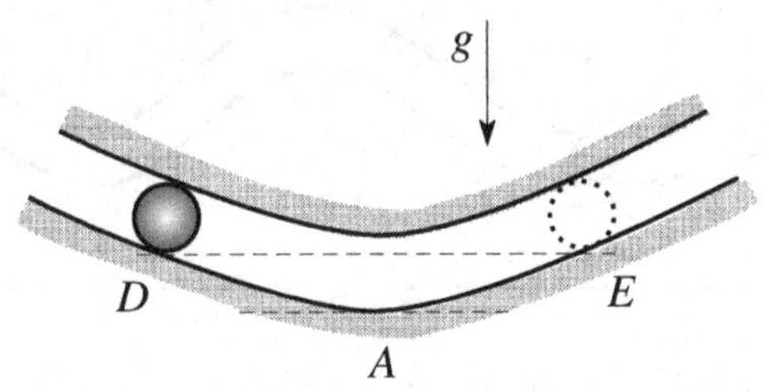


*Figure 3.8: If the ball is released at D, the subsequent motion will be oscillatory between the extremes D, E which have the same height h*

The potential energy at $D$ will exceed that at $A$, but again we can argue that the ball cannot pass through any points of higher energy than it posessed at the moment of release and hence it must remain in the region $DAE$. What will happen, of course, is that the ball will accelerate back towards $A$, at which point the surplus of potential energy will have been converted to kinetic energy. The ball will therefore oscillate about the equilibrium position, interchanging potential and kinetic energy, but the extremes of the motion will be circumscribed by the potential energy at the point of initial displacement — i.e. they will be defined by the point $D$ and the point $E$ on the other side of $A$ that has the same value of $h$ as has $D$.

By contrast, if we displace the ball slightly from the equilibrium point $B$ in Figure 3.7 where potential energy is a maximum, further motion away from $B$ will lower the potential energy. Since energy is conserved in the system (there is no friction), the lost potential energy is converted to kinetic energy and the ball accelerates away from $B$, which is therefore a position of unstable equilibrium.

All of these results could have been obtained using Newton's second law, without reference to potential energy. Since there is no friction, the reaction force at the groove must be normal to the surface and hence pass through the centre of the ball as shown in Figure 3.9.

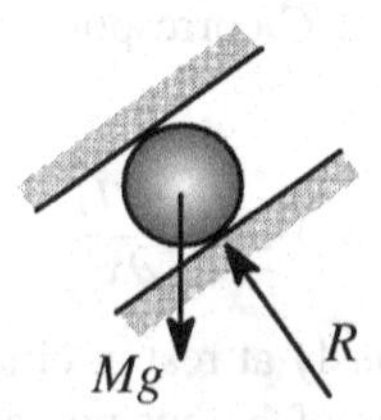


*Figure 3.9: Forces acting on the ball*

Since there are only two forces acting — the reaction force and the gravitational force — they can only be in equilibrium if they are equal and opposite, which will occur if and only if the groove is horizontal. Furthermore, if the ball is instantaneously at rest at a point where the groove is not horizontal, the acceleration will be in the direction of the resultant force, leading to the conclusion that the ball will accelerate along the groove in the direction of lower $h$. Thus, if the ball is displaced slightly from $A$ and released, it will tend to move back towards $A$, confirming that $A$ is a position of stable equilibrium, whereas a similar displacement from $B$ leads to forces tending to accelerate the ball further away from $B$, indicating instability as before.

## 3.5 The principle of stationary potential energy

The results established above are true for more general systems and can be expressed in the following statement:-

*Out of all kinematically permissible configurations of an elastic system, those at which the total potential energy is stationary with respect to any permissible motion are equilibrium configurations. Those that represent local minima of the total potential energy are stable equilibrium configurations, whilst those representing local maxima of the total potential energy are unstable equilibrium configurations.*

Total potential energy $\Pi$ is to be interpreted here as the total recoverable mechanical energy in the system and includes both energy stored as strain energy in elastic members and the potential energy of the external forces. In the following derivations, we shall generally use the symbol $U$ for strain energy and $\Omega$ for the potential energy of external forces such as gravity. Thus, the total potential energy can be written

$$\Pi = U + \Omega\,. \tag{3.38}$$

The intuitive reasoning behind the principle of stationary potential energy is the same as that used in discussing the ball sliding in the frictionless groove of Figure 3.7. If all kinematically possible small perturbations of the system from an equilibrium position result in an increase in total potential energy, such motion is precluded by the principle of conservation of energy. Furthermore, if the total potential energy is a differentiable function of the degree(s) of freedom and if any derivative at a given point is non-zero, it follows that there will be some direction of possible motion in which total potential energy can decrease and hence the system can (and will) accelerate.

**Example 3.4**

*Figure 3.10 (a) shows a rigid uniform beam ABC of length L and mass M, pinned at A and supported by springs, each of stiffness k, at B,C. Find the downward vertical displacement, $u_C$ of the end C. The forces in both the springs are zero when the bar is in the horizontal position.*

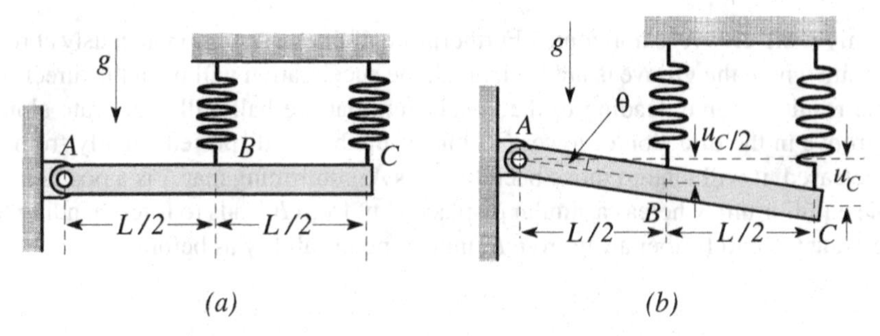


*Figure 3.10: A rigid beam supported by two springs*

Since the beam is rigid, the only permissible configuration consists of rotation about the end $A$ as shown in Figure 3.10*(b)*, in which case, the extension of spring $B$ will be $u_C/2$.

In the deformed configuration, there will be gravitational potential energy and elastic strain energy in the springs. Suppose that the gravitational potential energy $\Omega$ is given by $\Omega_0$ when the bar is horizontal. In the deformed configuration, we will then have

$$\Omega = \Omega_0 - \frac{1}{2}Mgu_C ,$$

since the centre of mass of the bar falls during the deformation through a distance $u_C/2$.

The strain energy in the springs is the sum of that stored in each of the two springs — i.e.

$$U = \frac{1}{2}k\left(\frac{1}{2}u_C\right)^2 + \frac{1}{2}ku_C^2 = \frac{5}{8}ku_C^2$$

and hence the total potential energy (strain energy + gravitational potential energy) is

$$\Pi = U + \Omega = \frac{5}{8}ku_C^2 + \Omega_0 - \frac{1}{2}Mgu_C . \tag{3.39}$$

The principle of stationary potential energy now states that the equilibrium position of the bar is that at which $\Pi$ is stationary with respect to permissible deformations — i.e.

$$\frac{\partial \Pi}{\partial u_C} = 0 .$$

Substituting for $\Pi$ from equation (3.39) and performing the differentiation, we find that

$$\frac{5}{4}ku_C - \frac{1}{2}Mg = 0$$

and hence the vertical displacement of the end $C$ is

$$u_C = \frac{2Mg}{5k} . \tag{3.40}$$

Notice that the constant $\Omega_0$ representing the potential energy in the horizontal position, disappears on differentiation. The value of $\Omega_0$ is therefore of no importance to the calculation and the datum for potential energy can be chosen arbitrarily.[7]

We can also determine the stability of the deformed configuration by performing a second differentation. We obtain

$$\frac{\partial^2 \Pi}{\partial u_C{}^2} = \frac{5k}{4} .$$

Since springs must have positive stiffnesses, this derivative is positive and it follows that equation (3.40) defines a minimum of total potential energy and hence a stable equilibrium configuration. We shall return to the question of stability of elastic systems in Chapter 12. The remainder of this chapter will be limited to the use of energy methods in the determination of equilibrium configurations.

### 3.5.1 Potential energy due to an external force

Example 3.4 was formulated in terms of gravity, since the concept of gravitational potential energy will be familiar to many readers from elementary physics and dynamics. However, the problem would be essentially unchanged if the bar, instead of having mass $M$, were subjected to a vertical external force $F$ of magnitude $Mg$ at its mid-point $B$. The displacement could then be written in terms of $F$ as

$$u_C = \frac{2F}{5k} .$$

We can generalize the idea of the potential energy to describe the effect of an external force $\boldsymbol{F}$ in any direction. The potential energy of such a force is

$$\Omega = -\boldsymbol{F} \cdot \boldsymbol{u} , \tag{3.41}$$

where $\boldsymbol{u}$ is the displacement of its point of application from some fixed datum. Thus, $\Omega$ is the work done against the force during the deformation. The force is assumed to be constant[8] and apply throughout the motion, so if the deformation were reversed, this work would be recovered and is therefore legitimately defined as potential energy.

### 3.5.2 Problems with several degrees of freedom

In Example 3.4, the bar has only one degree of freedom. Once $u_C$ is specified, the position of the bar and hence the extensions of the two springs are defined. The position of the bar could alternatively have been characterized using the angle $\theta$

[7] See footnote 5 on page 110.

[8] Notice once again that there is no factor of $1/2$ in this expression, since the force remains constant throughout the deformation, in contrast to that in a spring which increases linearly from zero during extension.

from the horizontal (see Figure 3.10*(b)*). We would then have to determine the strain energy and potential energy in terms of $\theta$ instead of $u_C$ and the statement of the principle of stationary potential energy would be written $\partial\Pi/\partial\theta=0$. The same final result would be obtained.

A good strategy is to sketch the system in a representative deformed configuration and then determine the minimum number $N$ of independent displacements and/or angles needed to define the configuration completely. We shall refer to these displacements and/or angles as the *degrees of freedom* of the system and refer to them as $u_i$, where $i=(1,N)$. In view of the above discussion, it is clear that the choice of the $u_i$ is not unique, but the *number* of degrees of freedom $N$ must be the same whatever choice is made.

The degrees of freedom $u_i$ completely define the deformed configuration and hence they also define the extended length of all springs and the distance moved by all external forces. We can therefore write the strain energy and potential energy in terms of the $u_i$. The principle of stationary potential energy is then written in the form that the partial derivative of $\Pi$ with respect to each degree of freedom must be zero — i.e.

$$\frac{\partial\Pi}{\partial u_i}=0 \tag{3.42}$$

for $i=(1,N)$. This provides $N$ equations to determine the $N$ unknowns $u_i$, and hence determine the deformed configuration of the system.

**Example 3.5**

*Determine the* horizontal *displacement of the point A for the structure of Figure 3.5(a), loaded by a vertical force F.*

This system has two degrees of freedom, since (i) we need two coordinates to define the position of the point $A$ and (ii) the springs are extensible so there are no kinematic constraints on this position within the plane. The choice of degrees of freedom is arbitrary, but a convenient choice is to take the horizontal and vertical displacements of $A$ as $u_1, u_2$ respectively, as shown in Figure 3.11.

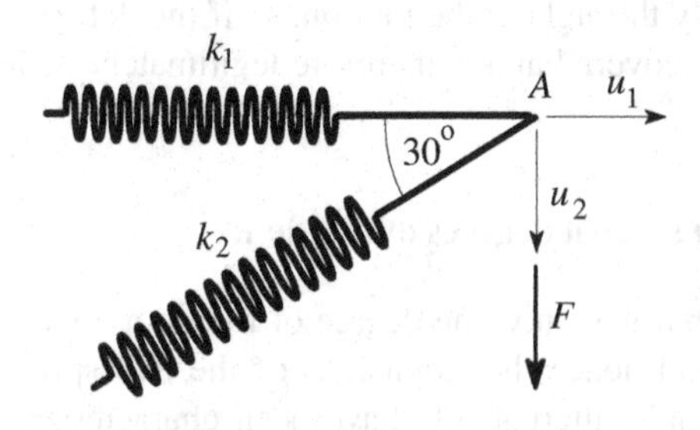


*Figure 3.11: Displacements of the point A*

With these displacements, the potential energy of the force $F$ is

$$\Omega=-Fu_2\,.$$

To find the strain energy stored in the two springs *in terms of the two degrees of freedom* $u_1, u_2$, we need to use kinematic arguments to find the extensions of the springs.

If the displacements $u_1, u_2$ are small in comparison with the original lengths of the springs, the change in length of each spring depends only on that component of displacement parallel to the spring. For example, the horizontal displacement $u_1$ causes an increase in length $\delta_1 = u_1$ in spring 1, but the vertical displacement $u_2$ causes no increase in length. Instead, it simply causes spring 1 to rotate about its fixed support through a small clockwise angle, as shown in Figure 3.12*(a)*.

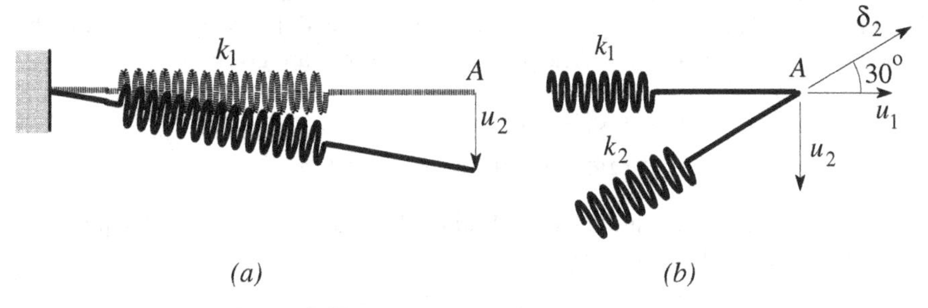


*Figure 3.12: Kinematics of spring extension*

Similarly, the extension $\delta_2$ of spring 2 depends only on the component of displacement parallel to spring 2. We can find this by resolving the components $u_1, u_2$ into this direction, as shown in Figure 3.12*(b)*, with the result

$$\delta_2 = u_1 \cos 30^\circ - u_2 \sin 30^\circ \, .$$

We can now write the strain energy in the two springs as

$$U_1 = \frac{1}{2} k_1 \delta_1^2 = \frac{1}{2} k_1 u_1^2$$

$$U_2 = \frac{1}{2} k_2 \delta_2^2 = \frac{1}{2} k_2 \left( \frac{\sqrt{3} u_1}{2} - \frac{u_2}{2} \right)^2 \, .$$

Collecting results, we have

$$\Pi = \Omega + U_1 + U_2 = -F u_2 + \frac{1}{2} k_1 u_1^2 + \frac{3}{8} k_2 u_1^2 - \frac{\sqrt{3}}{4} k_2 u_1 u_2 + \frac{1}{8} k_2 u_2^2 \, .$$

The principle of stationary potential energy now gives the two equations

$$\frac{\partial \Pi}{\partial u_1} = 0 \; ; \; \frac{\partial \Pi}{\partial u_2} = 0 \, ,$$

from which

$$k_1 u_1 + \frac{3}{4} k_2 u_1 - \frac{\sqrt{3}}{4} k_2 u_2 = 0 \tag{3.43}$$

$$-\frac{\sqrt{3}}{4} k_2 u_1 + \frac{1}{4} k_2 u_2 - F = 0 \, . \tag{3.44}$$

The solution of these equations is

$$u_2 = \left(\frac{3}{k_1} + \frac{4}{k_2}\right)F \; ; \; u_1 = \frac{\sqrt{3}F}{k_1} \, . \tag{3.45}$$

Notice that the vertical displacement $u_2$ is the same as we obtained in §3.3 [equation (3.21)], as we should expect, but the earlier method (equating the work done by the force to the total strain energy stored) was unable to give a value for the horizontal displacement $u_1$.

Notice also that the same equations (3.43, 3.44) could have been obtained by finding the forces in the springs from their extensions and then writing two equilibrium equations for the pin at $A$. The forces in the springs are given by

$$F_1 = k_1\delta_1 = k_1 u_2$$
$$F_2 = k_2\delta_2 = k_2(u_1\cos 30^\circ - u_2 \sin 30^\circ) \, .$$

Substituting these results into the equilibrium relations (3.16, 3.17) for the pin at $A$, we obtain (3.43, 3.44) as before.

### 3.5.3 Non-linear problems

In developing the principle of stationary potential energy, we appealed to the idea that a system cannot escape from a minimum energy state unless some external energy source is provided. This argument makes no assumptions about linearity and hence the principle can be applied to non-linear systems, provided the components are all elastic.[9]

Non-linear effects can arise in mechanics problems from two sources. *Material non-linearities* occur if the components are constructed from non-linear elastic materials — i.e. materials for which stress and strain are not linearly related, but which do follow the same curve on loading and unloading. *Geometric non-linearities* arise when the displacements are no longer small in comparison with the dimensions of the structure. In both cases, the principle of superposition fails — i.e. when two loads are applied to a structure, the deformation is not generally equal to the sum of that produced by each load acting separately.

The application of the principle of stationary potential energy involves the same steps as in linear problems, but of course the non-linearity of the resulting equations generally leads to more complicated algebra.

### Example 3.6

*Figure 3.13 (a) shows a mass M which is free to slide in a frictionless vertical groove and attached to a spring. The spring is unstretched in the horizontal configuration*

[9] This condition is important since for inelastic components, some of the work done during loading is dissipated as heat and is not recoverable on unloading.

*of Figure 3.13 (a) and has a non-linear relation between tensile force $F$ and the extension $u$ given by*

$$F = Cu^2 ,$$

*as shown in Figure 3.14. Find the equilibrium configuration of the system.*

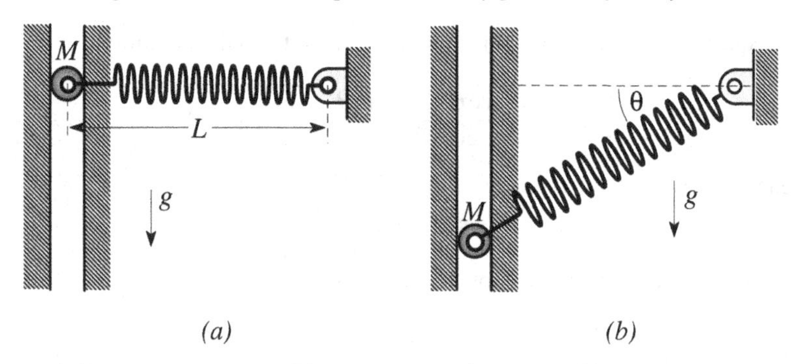


*Figure 3.13: Mass constrained by a groove and supported by a non-linear spring*

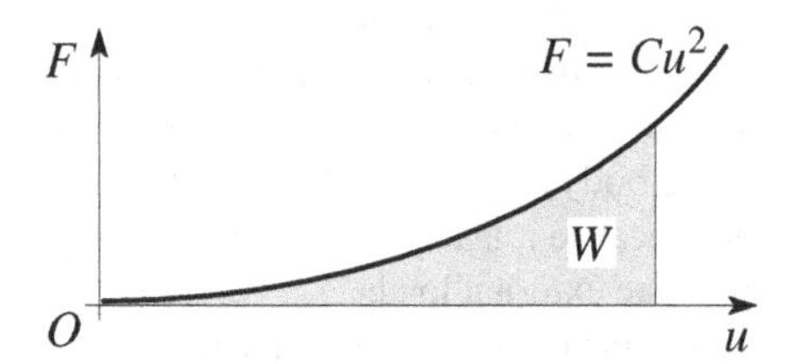


*Figure 3.14: Force extension curve for the non-linear spring*

As in equation (3.1), we can compute the strain energy stored in the spring by equating it to the work done during extension, with the result

$$U = \int_0^u Cu^2 du = \frac{1}{3} Cu^3 . \tag{3.46}$$

The mass has one degree of freedom which we shall characterize by the angle $\theta$ between the spring and the horizontal in the deformed position, as shown in Figure 3.13 *(b)*. We shall make no assumptions about smallness of the deformations and hence the problem exhibits both material and geometric non-linearity. In the deformed position, the potential energy of the mass is

$$\Omega = -MgL \tan \theta$$

and the deformed length of the spring is $L \sec \theta$, corresponding to an extension

$$u = L(\sec \theta - 1) .$$

Substituting into equation (3.46), we obtain

$$U = \frac{1}{3}CL^3(\sec\theta - 1)^3$$

and hence

$$\Pi = U + \Omega = \frac{1}{3}CL^3(\sec\theta - 1)^3 - MgL\tan\theta\,.$$

The equilibrium configuration is then determined from the condition

$$\frac{\partial \Pi}{\partial \theta} = \frac{CL^3(\sec\theta - 1)^2\sin\theta}{\cos^2\theta} - \frac{MgL}{\cos^2\theta} = 0$$

and hence

$$\sin\theta(\sec\theta - 1)^2 = \frac{Mg}{CL^2}\,,$$

which can be solved numerically for $\theta$.

## 3.6 The Rayleigh-Ritz method

To use the principle of stationary potential energy, we have to characterize the kinematics of the deformation in terms of a finite number of degrees of freedom $u_i$ and this can only be done for systems consisting of rigid bodies connected by springs. By contrast, a continuous elastic body (e.g. a laterally-loaded beam) has an infinite number of degrees of freedom, because it is capable of being deformed into an infinity of different shapes by appropriate external loads.

However, we can use the principle of stationary potential energy to develop an *approximate* solution to the problem. We first assume some approximate form for the solution, containing a number of arbitrary constants, which constitute the degrees of freedom in the solution. We then use the stationary potential energy theorem to determine 'optimal' values for these constants. This is known as the *Rayleigh-Ritz method*. It is similar to the process of fitting the best curve to a set of data points, where 'best' is here interpreted in the sense of minimizing some measure of error, such as the sum of the squares of the shortest distances from the points to the line (the so-called 'least squares' fit). The principle of stationary potential energy provides a natural criterion for choosing the best approximation in a given class.

By approximating the deformation by an expression with a finite number of degrees of freedom, we are constraining the solution to lie in a subset of the actually infinite number of kinematically possible states of deformation. The accuracy of the resulting approximation will clearly depend on the choice of this subset. There are no deterministic rules for choosing an appropriate approximate form, other than the requirement that it should satisfy the kinematic boundary conditions of the problem. The best strategy is to rely on our 'non-engineering' intuition (see Chapter 1) to sketch the expected deformed shape of the body and only then seek a mathematical expression that can describe this shape. We shall give various examples of this process and its effect on the accuracy of the resulting approximation in the following pages.

**Example 3.7**

*A cantilever beam is built-in at $z=0$ and subjected to a uniformly distributed load $w_0$ per unit length, as shown in Figure 3.15. Use the Rayleigh-Ritz method to find an approximate expression for the deflection of the beam.*

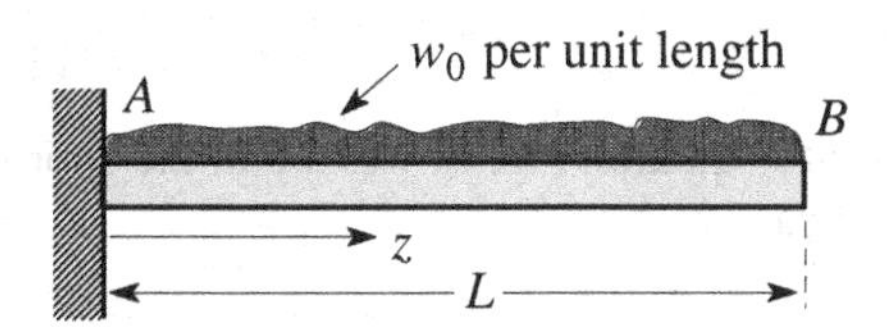


*Figure 3.15: Cantilever subjected to a uniform lateral load*

Clearly, the beam will bend downwards, but for consistency with other beam problems in Chapters 4,7,12, we shall use the sign convention that positive displacements $u(z)$ are upwards. The slope and deflection must be zero at $z=0$ — i.e. $u(0)=u'(0)=0$ — so we assume the simplest polynomial approximation that satisfies these conditions, which is

$$u = Cz^2 \; . \tag{3.47}$$

This is a one term approximation — i.e. there is only one degree of freedom, represented by the constant $C$ which must be determined by applying the principle of stationary potential energy.

In order to determine the strain energy, we need to express equation (3.24) in terms of the displacement, rather than the bending moment. Noting that

$$M = -EI\frac{d^2u}{dz^2}$$

from the elementary bending theory, and substituting for $M$ into (3.24), we obtain

$$U = \frac{1}{2}\int_0^L EI\left(\frac{d^2u}{dz^2}\right)^2 dz \; . \tag{3.48}$$

Substituting for $u$ from equation (3.47) and evaluating the integral, we find

$$U = \frac{EI}{2}\int_0^L (2C)^2 dz = 2EIC^2L \; .$$

The potential energy of the external load $\Omega$ can be found by thinking of the load as a set of concentrated forces, $w_0 dz$, each of which moves downwards through a distance $(-u)$, since the sign convention for $u$ is positive upwards. Thus,

$$\Omega = \int_0^L w_0 u(z)dz = w_0\int_0^L Cz^2 dz = \frac{w_0 CL^3}{3} \; .$$

The total potential energy is therefore

$$\Pi = U + \Omega = 2EIC^2L + \frac{w_0CL^3}{3}$$

and the principle of stationary potential energy $\partial\Pi/\partial C = 0$ then gives

$$4EICL + \frac{w_0L^3}{3} = 0 \quad \text{and hence} \quad C = -\frac{w_0L^2}{12EI} .$$

Substituting this result into equation (3.47), we obtain the approximate solution for the displacement of the beam as

$$u(z) = -\frac{w_0L^2z^2}{12EI} . \tag{3.49}$$

Of course, this problem can very easily be solved exactly by using equilibrium arguments to find the bending moment and then integrating the bending moment-curvature relation. The exact expression for the displacement is

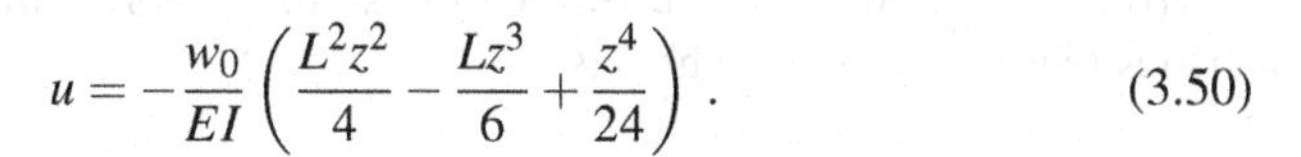

$$u = -\frac{w_0}{EI}\left(\frac{L^2z^2}{4} - \frac{Lz^3}{6} + \frac{z^4}{24}\right) . \tag{3.50}$$

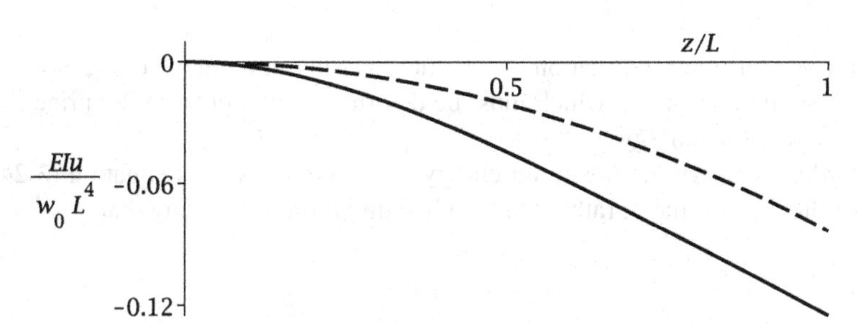


*Figure 3.16: Comparison of exact and approximate solutions for $u(z)$*

The exact and approximate deflection curves are compared in Figure 3.16. The approximate expression significantly underestimates the actual deflection. For example, it predicts only two thirds of the correct displacement at the free end $B$. The approximation is even poorer in other more detailed respects. For example, it is based on a constant curvature along the beam and hence implies a constant bending moment

$$M = -EI\frac{d^2u}{dz^2} = \frac{w_0L}{6} ,$$

whereas the actual bending moment is a function of $z$ given by

$$M = \frac{w_0(L-z)^2}{2} .$$

Thus, the approximation underestimates the *maximum* bending moment (at $z = 0$) by a factor of 3.

These results are fairly typical for a Rayleigh-Ritz approximation with only one degree of freedom. The approximation will generally be found to underestimate maximum deflections by anything up to 50% and significantly larger errors will be found in estimates for force resultants or stresses. The underestimate is associated with the fact that the assumed approximate form *constrains* the permitted shape of the body and hence tends to produce a stiffening effect. The true equilibrium state must have a lower total potential energy than the approximate one and this implies that if the constraint is relaxed, the external loads will lose potential energy by moving in their own directions. This trend will also be apparent if the constraint is partially removed by the addition of one or more additional degrees of freedom. Thus, the addition of an extra degree of freedom always moves the total energy nearer to the true equilibrium state and hence is associated with an improvement in accuracy.[10]

There are ways to improve the accuracy of the Rayleigh-Ritz approximation without making the calculation significantly more complicated, but we should not underestimate the value of a method that can produce an estimate even within 50% of the correct result using such a simple procedure. In this context, we note

(i) In the early stages of the design process, we often want just an order of magnitude estimate of a physical quantity, to determine which components, quantities or locations are critical in the proposed design and hence whether the concept is feasible (see Chapter 1). When these strategic questions have been answered, there are many techniques available (all of course much more time consuming) for getting more accurate predictions for the limited number of quantities that have been identified as critical.

(ii) Although the numerical values obtained from the method are grossly approximate, the parametric form is correct. For example, equation (3.49) correctly predicts that the maximum deflection will be a numerical multiple of $wL^4/EI$. This is very useful in the design process, since it gives us a good indication of what changes in the design parameters will be beneficial and how much effect they will have. By contrast, similar information can be obtained from purely numerical methods only by performing multiple solutions for different parameter values and analyzing the results.

(iii) In Example 3.7, the exact solution is not much more difficult than the approximate one and we have deliberately made this choice to enable simple comparisons to be made. To get a more realistic idea of the relative simplicity of the method, the reader should consider a more complex problem such as Problem 3.20. The exact solution for this indeterminate problem would involve the use of singularity functions and the imposition of boundary conditions would then lead to a set of simultaneous algebraic equations. By contrast, the Rayleigh-Ritz solution would be only marginally more complicated than Example 3.7. In fact, the reader is recommended to think through the procedure involved in a conventional solution to any of the problems he/she solves using any of the energy methods

[10] This is one of the major advantages of the use of total potential energy as a criterion for the 'best' approximation. It ensures that the addition of extra degrees of freedom always improves the approximation, regardless of the choice of form.

treated in this chapter, in order to develop a sense of the relative advantages of the various methods in different contexts.

### 3.6.1 Improving the accuracy

The obvious way to improve the accuracy of a Rayleigh-Ritz approximation is to use a representation with more degrees of freedom. From an algebraic point of view, there are practical limits to this process. We saw in Example 3.5 that even the use of two degrees of freedom significantly increases the complexity of the solution and it largely defeats the object of the excercise to use more than two. The only exceptions to this rule are those in which the terms corresponding to additional degrees of freedom can be described in a suitably condensed notation, such as a series.

#### Series solutions

Consider, for example, the simply-supported beam of Figure 3.17, subjected to a distributed load $w(z)$. A fairly general approximation to the deformed shape, satisfying the end conditions $u(0)=u(L)=0$ can be written in terms of the Fourier series

$$u(z) = \sum_{i=1}^{N} C_i \sin\left(\frac{i\pi z}{L}\right) . \tag{3.51}$$

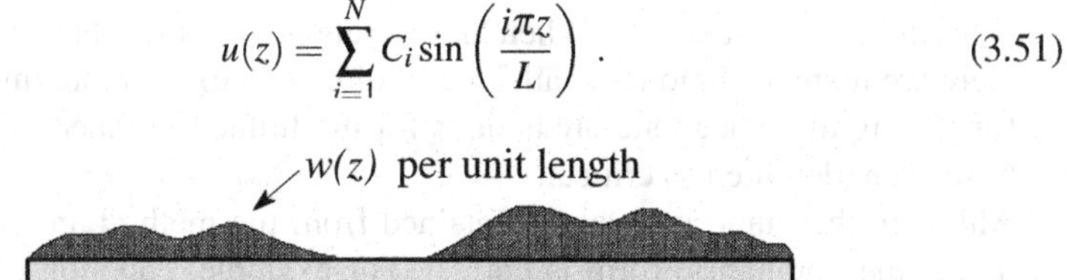
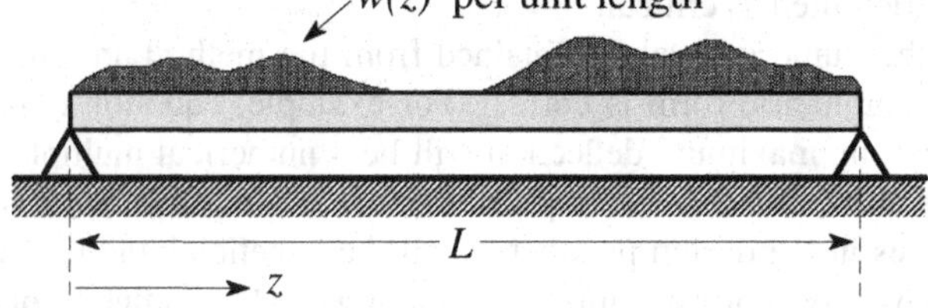


*Figure 3.17: Simply-supported beam subjected to arbitrary loading $w(z)$*

The strain energy is then found by substituting (3.51) into (3.48), with the result

$$U = \frac{EI\pi^4}{2L^4} \sum_{i=1}^{N} \sum_{j=1}^{N} i^2 j^2 C_i C_j \int_0^L \sin\left(\frac{i\pi z}{L}\right) \sin\left(\frac{j\pi z}{L}\right) dz . \tag{3.52}$$

To evaluate the integral, we use the results

$$\sin\left(\frac{i\pi z}{L}\right) \sin\left(\frac{j\pi z}{L}\right) = \frac{1}{2}\cos\left[\frac{(i-j)\pi z}{L}\right] - \frac{1}{2}\cos\left[\frac{(i+j)\pi z}{L}\right]$$

$$\begin{aligned} \int_0^L \cos\left(\frac{n\pi z}{L}\right) dz &= 0 \;\; ; \;\; n \neq 0 \\ &= L \;\; ; \;\; n = 0 , \end{aligned} \tag{3.53}$$

where $n$ is an integer, obtaining

$$U = \frac{EI\pi^4}{4L} \sum_{i=1}^{N} i^4 C_i^2 \,.$$

The potential energy of the load is

$$\Omega = \int_0^L w(z) u(z) dz \,, \tag{3.54}$$

where $u(z)$ is assumed to be positive upwards. The total potential energy is

$$\Pi = U + \Omega = \frac{EI\pi^4}{4L} \sum_{i=1}^{N} i^4 C_i^2 + \sum_{i=1}^{N} C_i \int_0^L w(z) \sin\left(\frac{i\pi z}{L}\right) dz \,.$$

Finally, we can find the coefficients $C_i$ from the conditions

$$\frac{\partial \Pi}{\partial C_i} = 0 = \frac{EIi^4\pi^4 C_i}{2L} + \int_0^L w(z) \sin\left(\frac{i\pi z}{L}\right) dz \quad ; \quad i = (1, N)$$

and hence

$$C_i = -\frac{2L}{EIi^4\pi^4} \int_0^L w(z) \sin\left(\frac{i\pi z}{L}\right) dz \,. \tag{3.55}$$

This result is remarkable in that the coefficients $C_i$ are obtained explicitly, without the need to solve a system of $N$ simultaneous equations. This is a consequence of the orthogonality of the trigonometric functions (3.53), which causes all the integrals in $U$ to be zero except those where $i = j$. For this reason, it is possible to allow the number of terms $N$ to increase without limit and hence obtain an exact solution for the problem of the simply-supported beam with arbitrary loading $w(z)$. This technique can also be used for other categories of problem, such as a rectangular plate simply-supported around its edges[11] and a rectangular plate subjected to in-plane loading.[12]

### Finite element solutions

A series approximation consists of a denumerable set of linearly independent functions, each of which is defined over the entire length of the beam. An alternative approach is to use a discrete approximation — i.e. to split the beam into a number of segments and use an independent low order approximation to the deflection in each segment. The combination of a discrete approximation to the deflection and the use of a variational principle (such as the stationary potential energy theorem) to choose

[11] S.P. Timoshenko and S. Woinowsky-Krieger (1959), *Theory of Plates and Shells*, McGraw-Hill, New York, §80.

[12] S.P. Timoshenko and J.N. Goodier (1970), *Theory of Elasticity*, Mc-Graw-Hill, New York, §94.

the 'best' values of the resulting degrees of freedom is known as the *finite element method*.

For relatively small numbers of degrees of freedom (say less than 10), there is little to choose between the two methods, but series methods become progressively more inaccurate due to rounding errors when large numbers of terms are used and the discrete method is therefore to be preferred for solutions of high accuracy.

The finite element method is an extremely important tool in modern mechanics of materials. It is therefore discussed in more depth in Appendix A.

### 3.6.2 Improving the back of the envelope approximation

Series and discrete approximations are useful ways of developing a relatively accurate approximation to a problem, but they generally involve susbtantial analytical or numerical work. We have already seen how a simple one term approximation can give estimates for deflection that are within 50% of the exact value with only a few lines of calculation. Fortunately, this 'back of the envelope' estimate can often be improved without adding extra degrees of freedom and hence with only a limited amount of additional calculation.

The solution is approximate only because we are seeking it within a restricted class of trial functions. The accuracy can therefore be improved if we can make use of additional information about the structure to choose a better approximating function. This information may be based on our knowledge of mechanics, or it may be more or less intuitive. We shall give examples of both kinds in this section.

#### Example 3.7 revisited

As an example of the first approach, we note that the bending moment is zero at $z=L$ in Example 3.7 so that the curvature there $(d^2u/dz^2)$ must be zero. The form assumed in equation (3.47) clearly does not satisfy this condition, since its second derivative is constant — in other words the curvature is constant along the beam.

We would therefore expect to get a better result if we chose a form that defined zero curvature at $z=L$ for all values of the constant $C$. We could develop such a form by starting with a polynomial of two-degrees of freedom — for example

$$u = C_1 z^2 + C_2 z^3 \tag{3.56}$$

and then choosing the second constant $C_2$ to satisfy the condition. Notice that (3.56) still satisfies the *essential* kinematic support conditions $u = du/dz = 0$ at $z = 0$. The curvature at a general point is

$$\frac{d^2u}{dz^2} = 2C_1 + 6C_2 z$$

and hence it will be zero at $z=L$ if $C_2 = -C_1/3L$. Substituting this result into (3.56), we obtain

$$u = C_1\left(z^2 - \frac{z^3}{3L}\right) ,$$

which is a one degree of freedom approximation (there is only one constant $C_1$ to determine from the principle of stationary potential energy) that satisfies the condition of zero curvature at $z=L$.

However, we are not restricted to polynomial forms. Trigonometric functions also give relatively straightforward algebraic solutions and in the interests of variety, we shall use the form

$$u = C\left[1 - \cos\left(\frac{\pi z}{2L}\right)\right] , \tag{3.57}$$

which also has only one degree of freedom ($C$), satisfies the kinematic support conditions $u=du/dz=0$ at $z=0$, and the additional condition that curvature be zero at $z=L$, since

$$\frac{d^2u}{dz^2} = \frac{\pi^2 C}{4L^2}\cos\left(\frac{\pi z}{2L}\right)$$

and the argument of the cosine is $\pi/2$ when $z=L$.

Repeating the calculation with this approximate shape, we find

$$\begin{aligned} U &= \frac{EIC^2}{2}\left(\frac{\pi}{2L}\right)^4 \int_0^L \cos^2\left(\frac{\pi z}{2L}\right) dx = \frac{EIC^2}{2}\left(\frac{\pi}{2L}\right)^3 \int_0^{\pi/2} \cos^2\theta d\theta \\ &= \frac{EI\pi^4 C^2}{64L^3} , \end{aligned} \tag{3.58}$$

from (3.48, 3.57) where we have used the change of variable $\theta=\pi z/2L$ to evaluate the integral.

Also,

$$\begin{aligned} \Omega &= w_0 C \int_0^L \left[1 - \cos\left(\frac{\pi z}{2L}\right)\right] dz = \frac{2Lw_0 C}{\pi} \int_0^{\pi/2} (1-\cos\theta) d\theta \\ &= w_0 CL\left(1 - \frac{2}{\pi}\right) , \end{aligned} \tag{3.59}$$

from equations (3.54, 3.57).

We then have

$$\Pi = U + \Omega = \frac{EI\pi^4 C^2}{64L^3} + w_0 CL\left(1 - \frac{2}{\pi}\right)$$

and the principle of stationary potential energy gives

$$\frac{\partial \Pi}{\partial C} = 0 = \frac{EI\pi^4 C}{32L^3} + w_0 L\left(1 - \frac{2}{\pi}\right)$$

— i.e.

$$C = -\frac{32w_0 L^4}{\pi^4 EI}\left(1 - \frac{2}{\pi}\right) = -0.119\frac{w_0 L^4}{EI} .$$

In particular, the end deflection is

$$u_B = u(L) = C = -0.119\frac{w_0 L^4}{EI} . \tag{3.60}$$

This improved approximate result is only 5% lower than the exact expression, which is $-0.125w_0L^4/EI$, from equation (3.50) with $z = L$. By contrast, the 'unthinking' trial function (3.47) underestimated the end deflection by 33%. Thus, a careful choice of approximate form results in this case in a quite respectable approximate solution for the deflection even though it uses only a single degree of freedom.

Physical intuition, as discussed in §1.2.1, can also play an important part in developing a more appropriate Rayleigh-Ritz approximation. We illustrate this process with the following example.

**Example 3.8**

*A semi-circular curved beam is pinned at its ends and subjected to a concentrated force F at the mid-point, as shown in Figure 3.18. Estimate the deflected shape of the beam and, in particular, the deflection under the load at B.*

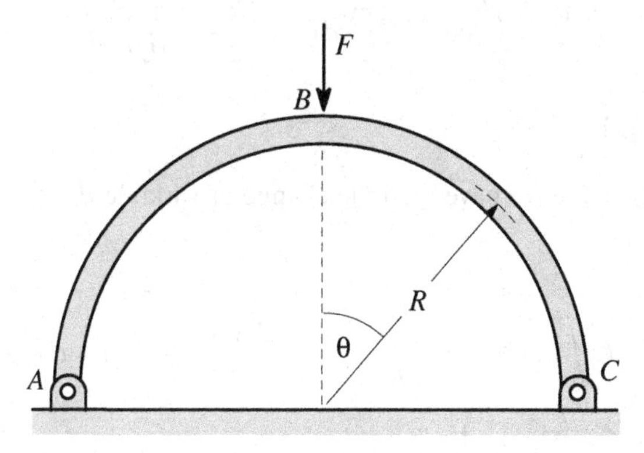


*Figure 3.18: A semi-circular beam subjected to a central load*

Defining the outward radial displacement $u$ as a function of angular position $\theta$, it is clear that the kinematic support conditions require

$$u\left(-\frac{\pi}{2}\right) = u\left(\frac{\pi}{2}\right) = 0 \tag{3.61}$$

and a simple one degree of freedom function satisfying these conditions is

$$u = C\cos\theta . \tag{3.62}$$

However, this defines a displacement which has the same sign throughout $-\pi/2 < \theta < \pi/2$ and hence corresponds to a mode of deformation like that shown in Figure

3.19 *(a)*, but our intuition tells us that a slender beam will deform instead into the shape shown in Figure 3.19 *(b)* — in other words, pushing down in the middle and restraining the ends will cause the beam to bulge outwards at each side.

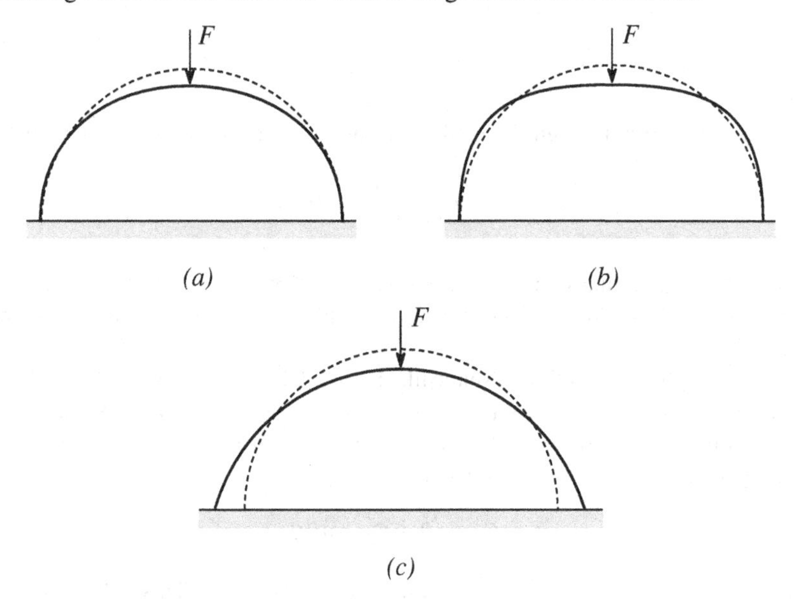


*Figure 3.19: Possible deformation modes for the curved beam*

It should be emphasised that people with no training in mechanics have sufficient intuitive or unconsciously learnt knowledge of the physical world to expect the deformation pattern 3.19 *(b)* rather than 3.19 *(a)*. The reader might like to verify this by asking their non-technical friends to sketch the deformed shape, without prompting them with any suggestions. It is of course easily tested by bending a piece of wire into a semicircle and loading it whilst restraining the ends from horizontal motion (If the ends are not restrained, the deformation pattern 3.19 *(c)* will be produced). By contrast, many engineers will approach the problem in a more formal manner by stating the boundary conditions (3.61) and will propose the inappropriate form (3.62), resulting in completely unrealistic estimates for the displacements. The moral here is "Always sketch the shape you expect the body to adopt, *before* seeking an appropriate approximating function."

An equally simple form that *does* look like Figure 3.19 *(b)* is

$$u = C\cos(3\theta) \tag{3.63}$$

since $\cos(3\theta)$ changes sign at the two points $\theta = \pm\pi/6$.

This function will give a much better approximation to the problem, but we can do even better, by combining our knowledge of mechanics with our intuition to ask *why* the beam deforms into a shape like Figure 3.19 *(b)*. The reason is to be found in the fact (discussed in §3.3.2 above) that beams are much stiffer in tension than in bending and hence tend to deform into configurations that substantially preserve their original length, if any such configuration is kinematically possible. The configuration 3.19 *(a)* implies that the beam is shorter in the deformed than in the undeformed state.

To a first approximation, the deformed length will be

$$L = \int_{-\pi/2}^{\pi/2} (R+u)d\theta \tag{3.64}$$

compared with an initial length $L_0 = \pi R$. Thus, there will be a change in length unless

$$\delta L = L - L_0 = \int_{-\pi/2}^{\pi/2} u d\theta = 0 \tag{3.65}$$

This in turn requires that there be some regions in $-\pi/2 < \theta < \pi/2$ in which $u$ is positive and some in which it is negative, the simplest such configuration being that of Figure 3.19 *(b)*.

Now although (3.63) defines a shape like Figure 3.19 *(b)*, it does not satisfy equation (3.65) and we should anticipate a significantly better approximation if we could choose a form that does so. As in the previous example, we can develop such a form by writing the two degree of freedom expression

$$u = C_1 \cos\theta + C_2 \cos(3\theta)$$

and then choosing $C_2$ to satisfy (3.65) with the result $C_2 = 3C_1$. This defines the one degree of freedom approximation

$$u = C_1 [\cos\theta + 3\cos(3\theta)] \tag{3.66}$$

We shall develop a Rayleigh-Ritz solution to the problem of Figure 3.18, using this approximating function.

We first need to adapt equation (3.48) to the curved beam. The elemental length of beam $dx$ is replaced by the arc length $Rd\theta$, and we might expect the change in curvature of the beam to be the second derivative of $u$ with respect to the arc length — i.e. $(1/R^2)d^2u/d\theta^2$. However, even if $u$ were constant (and hence had no derivative with respect to $\theta$), there would be a change of curvature, since the constant radial displacement $u$ would change the radius of the beam from $R$ to $R+u$, resulting in a reduction in curvature of magnitude $u/R$ to the first order. Using these results, the expression equivalent to (3.48) for the curved beam of Figure 3.18 is

$$U = \frac{EI}{2} \int_{-\pi/2}^{\pi/2} \left( \frac{1}{R^2} \frac{d^2u}{d\theta^2} + \frac{u}{R^2} \right)^2 Rd\theta \, , \tag{3.67}$$

which, after substituting for $u$ from equation (3.66) and evaluating the integral, yields

$$U = \frac{144\pi EIC_1^2}{R^3} \, .$$

The potential energy of the force $F$ is simply

$$\Omega = Fu(0) = 4FC_1$$

and hence

$$\Pi = U + \Omega = \frac{144\pi EIC_1^2}{R^3} + 4FC_1 \ .$$

Applying the principle of stationary potential energy, $\partial\Pi/\partial C_1 = 0$, we obtain

$$C_1 = -\frac{FR^3}{72\pi EI}$$

and hence the downward displacement of the force $F$ is

$$u_F = -u(0) = -4C_1 = \frac{FR^3}{18\pi EI} = 0.0177\frac{FR^3}{EI} \ . \tag{3.68}$$

This problem can be solved exactly using Castigliano's second theorem (see below §3.10 and Problem 3.51), the exact solution for $u_F$ being

$$u_F = 0.0189\frac{FR^3}{EI} \ , \tag{3.69}$$

so our one degree of freedom Rayleigh-Ritz approximation underestimates the displacement by 7%. By contrast, the result obtained using the superficially plausible expression (3.63) is $0.00995FR^3/EI$ which is only just over half of the correct answer.

## 3.7 Castigliano's first theorem

In this section, we shall introduce a method which is essentially equivalent to the principle of stationary total potential energy, but which looks at the problem in a slightly different way.

Suppose we have a complicated elastic system, supported in some way (it can be determinate or indeterminate) and subjected to $N$ external forces,$F_i$, $i = (1,N)$. The structure will generally deform under the action of the forces, causing them to do work which is stored as strain energy. We denote the distance through which $F_i$ moves in its own direction (i.e. the component of displacement of the point of application of $F_i$ in the direction parallel to $F_i$) by $u_i$.

Now consider the following scenario:-

(i) The system is first allowed to reach its equilibrium configuration under the influence of the forces $F_i$. Suppose that the strain energy in this state is $U$.

(ii) All the displacement components $u_i$ are then fixed at their equilibrium values, *except one*, which we denote by $u_k$.

(iii) A small additional force $dF_k$ is then applied, such that $u_k$ increases by $du_k$. There is no corresponding increase in $u_i, i \neq k$, since these displacement components are fixed, but additional forces will be induced at the fixed points. After $dF_k$ is applied the strain energy will generally be changed to $U + dU$.

The system is elastic, so the work done during the incremental loading [stage (iii) above] must be stored in the structure and is therefore equal to the *increase* in strain energy $dU$. However, the only force that moves during this stage (and hence does work) is $F_k$, since all the other displacement components are fixed.[13] The relation between $F_k$ and $u_k$ during stage (iii) may be non-linear, as shown in Figure 3.20, but for sufficiently small $dF_k$, the work done (equal to the shaded area under the curve) can be written

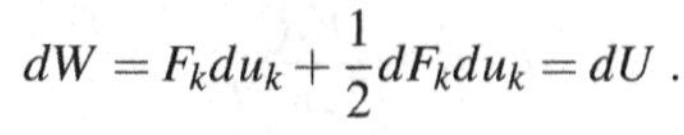

$$dW = F_k du_k + \frac{1}{2} dF_k du_k = dU \ .$$

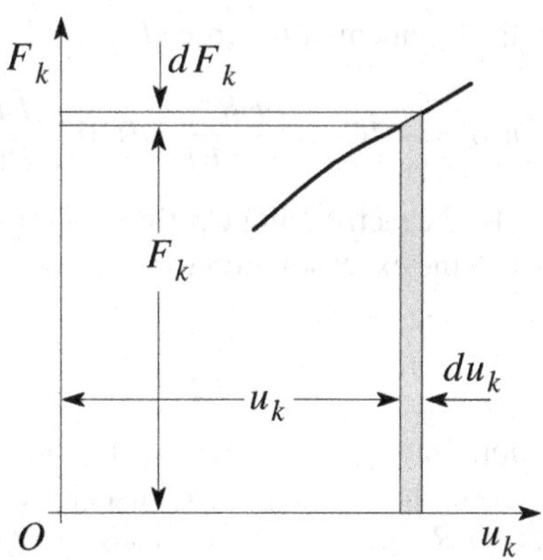


*Figure 3.20: Work done during the application of* $dF_k$

In fact, if $dF_k$ is sufficiently small, we can also drop the second term in this equation, giving in the limit

$$dU = F_k du_k$$

or

$$F_k = \frac{\partial U}{\partial u_k} \ . \tag{3.70}$$

This result was first obtained by Castigliano[14] and is known as *Castigliano's first theorem.* Strictly, following the conventions of thermodynamics, we should write it in the form

$$F_k = \left( \frac{\partial U}{\partial u_k} \right)_{u_i(i \neq k)}$$

to indicate that the partial derivative — like the incremental process it describes — is taking place under the constraint that the remaining points $i \neq k$ have zero incremental displacement.

Castigliano's first theorem is really a restatement of the principle of stationary potential energy. If we take the undeformed position of the system as the datum for potential energy of the each of the forces, we have

[13] By the same token, the support reactions do no work either.

[14] A. Castigliano (1879), *Théory de l'équilibre des systèmes élastiques et ses applications,* Turin.

$$\Omega_i = -F_i u_i \,,$$

from equation (3.41), for the potential energy of the force $F_i$ and hence

$$\Pi = U + \Omega = U - \sum_{i=1}^{N} F_i u_i \,. \tag{3.71}$$

Now the principle of stationary potential energy states that

$$\frac{\partial \Pi}{\partial u_k} = 0$$

for all $k$, for small perturbations about the equilibrium position. Substituting for $\Pi$ from (3.71), we recover

$$\frac{\partial U}{\partial u_k} - F_k = 0 \,,$$

which is identical with (3.70). In effect, the form (3.70) simply eliminates the stage of writing the total potential energy of the applied forces.

### Example 3.9 — The torque wrench

*Figure 3.21 shows an idealization of a torque wrench, designed to permit a nut to be tightened to a specific value of torque. The dimensions are defined such that the angle $\theta = \theta_0$ when the spring is relaxed. If the torque $T$ is increased from zero, $\theta$ decreases, but there is a maximum torque at some angle $\theta_1 < \theta_0$, beyond which the wrench would 'snap through' in an unstable manner to the other side of the horizontal — i.e. to negative values of $\theta$. Determine (i) the relation between the applied torque $T$ and the angle $\theta$ and (ii) the maximum value of torque that can be applied without the mechanism snapping through.*

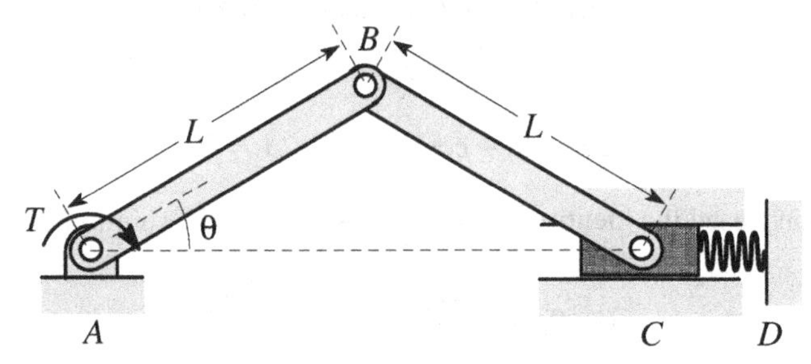


*Figure 3.21: The torque wrench*

The distance $AD$ is constant and hence the compression $u$ of the spring must be

$$u = 2L\cos\theta - 2L\cos\theta_0 \,.$$

The strain energy in the spring is therefore

$$U = \frac{1}{2}ku^2 = \frac{k}{2}(2L\cos\theta - 2L\cos\theta_0)^2$$

and Castigliano's first theorem immediately yields

$$T = -\frac{\partial U}{\partial \theta} = 4kL^2 \sin\theta(\cos\theta - \cos\theta_0)\,, \tag{3.72}$$

which is the required relation between torque and angle of rotation.[15] Notice that the only external forces acting are the torque $T$ and the reactions at $A, C$ and $D$ and, of these, only the torque does any work during the motion, thus meeting the requirements of the theorem.

Equation (3.72) is shown graphically in Figure 3.22 and exhibits a maximum torque in the range $0 < \theta < \theta_0$. If we increase $T$ monotonically, $\theta$ will fall steadily until this maximum is reached, when the mechanism will snap through dynamically to a new equilibrium point beyond $\theta = -\theta_0$.

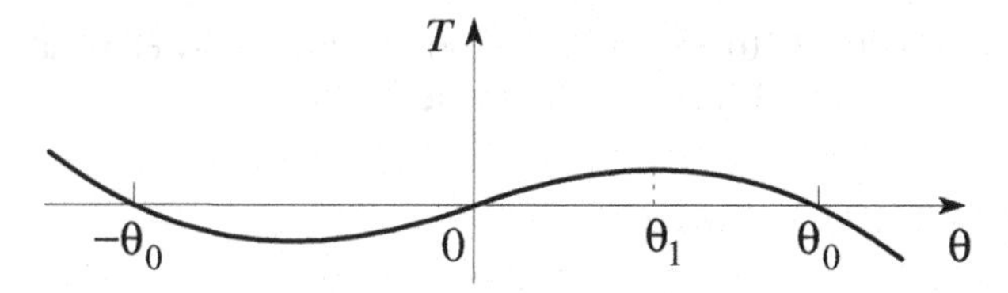


*Figure 3.22: Relation between torque and rotation for the torque wrench*

To find the maximum torque, we set

$$\frac{\partial T}{\partial \theta} = 0$$

obtaining

$$\cos\theta(\cos\theta - \cos\theta_0) + \sin\theta(-\sin\theta) = 0\,.$$

Thus, the critical value $\theta = \theta_1$ is defined by

$$2\cos^2\theta_1 - \cos\theta_0\cos\theta_1 - 1 = 0\,,$$

where we have used the identity $-\sin^2\theta_1 = \cos^2\theta_1 - 1$. The solution of this equation is

$$\cos\theta_1 = \frac{\cos\theta_0 \pm \sqrt{8 + \cos^2\theta_0}}{4}\,. \tag{3.73}$$

[15] The 'direct' method of solving this problem would be to draw a free-body diagram of the system, write equilibrium equations for the separate components to determine the reactions and internal forces and finally relate the force in the spring to its deformed length and hence to the angle $\theta$ through the kinematics of the mechanism. The reader might like to begin the solution of the problem by this method. It will rapidly become apparent that this is a case where the energy method is considerably more efficient.

The maximum torque corresponds to the positive sign in this equation.[16]

From a design point of view, the mechanism of Figure 3.21 is not very satisfactory for a torque wrench, since the wrench would feel 'springy' — i.e. significant motion of the handle would have to occur before the preset maximum torque was achieved and this might make the wrench unusable in some situations. A simple way of avoiding this problem would be to introduce a stop permitting the spring to be precompressed to a value slightly below that at snap-through, as shown in Figure 3.23.

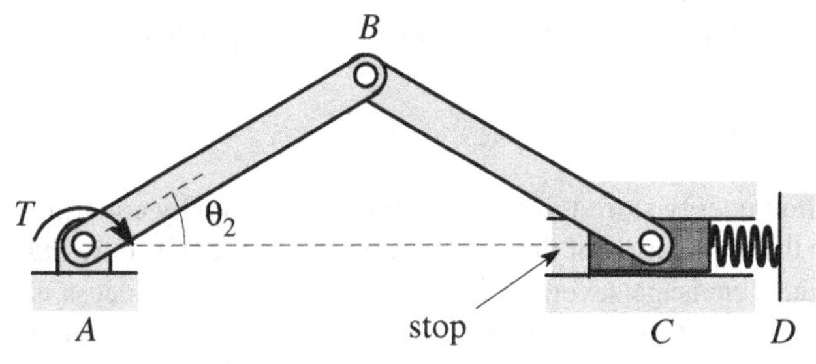


*Figure 3.23: Modification to the torque wrench design to reduce spring motion before snap through*

If we continue to define $\theta_0$ as the configuration at which the spring would be relaxed (although that position can now no longer be reached because of the stop), the stop would have to be located such that the angle $\theta_2$ with the slider touching the stop is slightly larger than $\theta_1$ as defined by (3.73). This defines a suitable position for the stop if the wrench is to be preset for a single value of torque. The wrench can be made adjustable by allowing the position of the fixed end of the spring $D$ to be changed, thus changing the rest position $\theta_0$. In this case, a mechanism would have to be devised to make a simultaneous change in the position of the stop.

## 3.8 Linear elastic systems

For the remainder of this chapter, we shall restrict attention to systems in which the displacements are small and linearly proportional to the applied loads, in which case the principle of linear superposition applies. In other words, the displacement at any point due to a system of loads is simply the sum of the displacements that would be produced by each of the loads acting separately.

Suppose the elastic system is kinematically (not necessarily determinately) supported and subjected to a system of $N$ external forces $F_i$. As in §3.7, we shall denote the displacement component in the direction of the force $F_i$ at its point of application by $u_i$.

[16] The reader might like to investigate the significance of the negative sign in equation (3.73), by sketching the complete graph of $T$ as a function of $\theta$ in $-\pi < \theta < \pi$. It will be found that there are two more stationary points — one maximum and one minimum — involving configurations in which the spring is in tension.

The principle of linear superposition then permits us to write

$$\begin{aligned} u_i &= C_{i1}F_1 + C_{i2}F_2 + \ldots C_{ij}F_j + \ldots \\ &= \sum_{j=1}^{N} C_{ij}F_j \,, \end{aligned} \tag{3.74}$$

where $C_{ij}$ is the displacement component $u_i$ due to a unit force $F_j$ acting alone.

We shall call the coefficients $C_{ij}$ *influence coefficients*. They can also be regarded as the elements of a matrix $\boldsymbol{C}$, which is called the *compliance matrix*.

### 3.8.1 Strain energy

The total strain energy stored in the system can be found by equating it to the work done during the application of the forces. However, if there is more than one external force, we could envisage several scenarios for this loading process, since the *order* of application of the forces is arbitrary. If the structure is elastic, the final state (and hence the total strain energy) must be independent of the loading history, and we can therefore conclude that the total work done by the external forces must be the same for all of these loading scenarios. This argument can be exploited to deduce relationships that must exist between the influence coefficients $C_{ij}$.

To fix ideas, suppose there are only two external forces $F_1, F_2$. We consider three scenarios:-

**(i) $F_1$ applied first, then $F_2$**

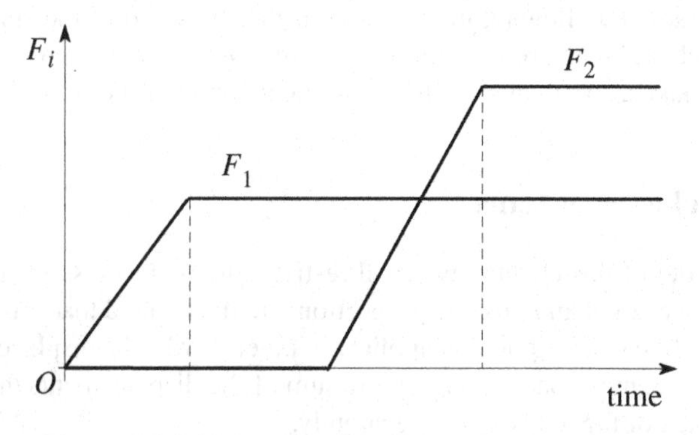


*Figure 3.24: Variation of the forces $F_1, F_2$ with time*

This scenario is depicted in Figure 3.24. During the first phase of the loading, $F_2=0$ and $F_1$ increases from zero to its maximum value. The work done is therefore

$$W_1 = \frac{1}{2}F_1 u_1 = \frac{1}{2}C_{11}F_1^2 \,,$$

from equation (3.6).

We now hold $F_1$ constant, whilst gradually applying $F_2$. The additional work done during this phase of the process is

$$W_2 = \frac{1}{2}F_2\Delta u_2 + F_1\Delta u_1$$
$$= \frac{1}{2}C_{22}F_2^2 + C_{12}F_1F_2 ,$$

where $\Delta u_1, \Delta u_2$ are the *changes* in $u_1, u_2$ respectively during (and due to) the application of $F_2$. Notice how $F_1$ does an additional amount of work 'involuntarily' as it moves due to the action of $F_2$. Notice also that there is no factor of $1/2$ in the term $F_1\Delta u_1$, since the force $F_1$ is at its maximum value throughout the period during which displacement $\Delta u_1$ occurs. You can check this by looking at Figure 3.24. By contrast, $F_2$ is increasing linearly from zero with $\Delta u_2$ during this process.

Since the total work done during both phases of the loading process must be stored as strain energy, we have

$$U = W_1 + W_2 = \frac{1}{2}C_{11}F_1^2 + \frac{1}{2}C_{22}F_2^2 + C_{12}F_1F_2 . \tag{3.75}$$

**(ii) $F_2$ applied first, then $F_1$**

Suppose we now consider a similar scenario, except that the load $F_2$ is applied first, followed by $F_1$. The argument is unchanged, except that the suffices 1,2 are reversed, with the result

$$U = \frac{1}{2}C_{22}F_2^2 + \frac{1}{2}C_{11}F_1^2 + C_{21}F_2F_1 \tag{3.76}$$

for the final strain energy.

**(iii) $F_1, F_2$ applied simultaneously and in proportion**

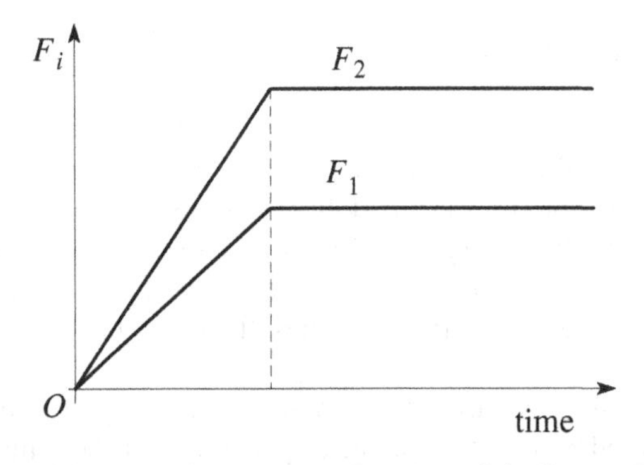


*Figure 3.25: Simultaneous application of the forces $F_1, F_2$*

Finally, we consider the scenario illustrated in Figure 3.25, in which both loads increase slowly from zero together, such that the ratio $F_1/F_2$ is constant and equal to its final value. In the time interval $\Delta t$,

$$\Delta u_1 = C_{11}\Delta F_1 + C_{12}\Delta F_2 \;;\; \Delta u_2 = C_{21}\Delta F_1 + C_{22}\Delta F_2$$

and the incremental work done (neglecting second order terms) is

$$\Delta U = F_1 \Delta u_1 + F_2 \Delta u_2 \, .$$

Integrating over the period of application of the forces, we find the total work done and hence the final strain energy is

$$U = W = \frac{1}{2}C_{22}F_2^2 + \frac{1}{2}C_{11}F_1^2 + \frac{1}{2}C_{12}F_1F_2 + \frac{1}{2}C_{21}F_2F_1 \, . \tag{3.77}$$

Each of the three scenarios considered lead to different expressions, (3.75, 3.76, 3.77) for the final strain energy, but the final state is independent of the order of application of the loads and hence these expressions must all be equal. Examination of the equations shows that this will be the case if and only if $C_{12}=C_{21}$. More generally, if there are $N$ forces and displacement components, as in equation (3.74), we must have[17]

$$C_{ij} = C_{ji} \, . \tag{3.78}$$

In other words, the compliance matrix $\boldsymbol{C}$ must be symmetric. An incidental advantage of this result is that it is not necessary to be too scrupulous about defining the order of the suffices $i, j$ in equations like (3.74). This is a statement of *Maxwell's reciprocal theorem* for linear elastic structures.

The expression for $U$ when there are $N$ forces is conveniently written in the form analogous to (3.77) as

$$U = \frac{1}{2}\sum_{i=1}^{N}\sum_{j=1}^{N} C_{ij}F_iF_j \, . \tag{3.79}$$

This result can be established using a scenario similar to (iii) above, in which all the forces $F_i$ are increased slowly and in proportion from zero to their final values.

### 3.8.2 Bounds on the coefficients

If we apply an external force or a system of external forces to an initially unloaded linear elastic structure, the work done must be positive and the strain energy must increase.[18] If this were not the case (i.e. if there were some system of loads that could be applied that would allow energy to be recovered from the system), a theoretically unbounded amount of energy could be recovered by applying a sufficiently large load, violating the first law of thermodynamics. It follows that $U \geq 0$ for all possible $F_i$ in equation (3.79).

This inequality imposes various constraints on the properties of the matrix $\boldsymbol{C}$, which can be explored by letting the $F_i$ take special values. The simplest example is

[17] Since (3.74) is true for arbitrary $F_i$, we can prove this more general result for any particular $i, j$ by replacing $1, 2$ in the above argument by $i, j$, respectively.

[18] Strictly, the work done may also be zero, if motion in that direction is regidly restrained, but it cannot be negative.

that in which all the forces are zero except one, which can be of arbitrary magnitude $F$. For example, writing $F_k=F$, $F_i=0\ (i\neq k)$, we obtain[19]

$$U=\frac{1}{2}C_{kk}F^2\geq 0\,,$$

from (3.79), which implies that

$$C_{kk}\geq 0\,. \tag{3.80}$$

In other words, all the diagonal elements of $\boldsymbol{C}$ must be non-negative. In physical terms, this means that when a force is applied to the structure, the displacement at the point of application must have a non-negative component in the direction of the force.

Intuitively, we might expect that a force will produce a greater displacement at its point of application than at other points, which would imply that the diagonal elements of $\boldsymbol{C}$ are generally greater than the off-diagonal elements. However, it is easy to find counter examples, such as the cantilever beam shown in Figure 3.26.

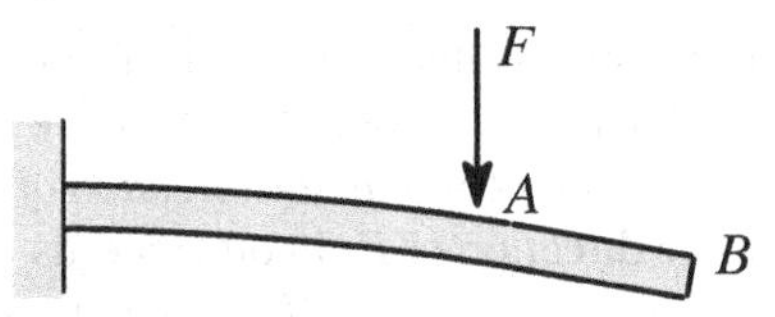


*Figure 3.26: The beam subjected to a force F*

If a force $F$ is applied at the point $A$, the beam will deform as shown and, in particular, the displacement at the end $B$ will be greater than that at the point of application of the load.

Nonetheless, inequalities can be established showing that the diagonal elements tend on average to be greater than the off-diagonal elements. For example, writing $F_j=1, F_k=-1, F_i=0\ (i\neq j,k)$, we obtain

$$U=\frac{1}{2}C_{jj}+\frac{1}{2}C_{kk}-C_{jk}\geq 0$$

and hence

$$C_{jj}+C_{kk}\geq 2C_{jk}\,. \tag{3.81}$$

The alternative choice $F_j=1, F_k=1, F_i=0\ (i\neq j,k)$ — i.e. changing the sign of $F_k$ — leads to the related condition

$$C_{jj}+C_{kk}\geq -2C_{jk}\,, \tag{3.82}$$

which would be a stronger constraint if $C_{jk}<0$. Notice that the off-diagonal elements are not constrained to be non-negative. In fact, their signs depend on the convention

[19] For those readers familiar with the summation convention in continuum mechanics and tensor notation, we note that, in this and subsequent expressions, the notation $C_{kk}$ etc does *not* imply any summation over permissible values of $k$.

adopted for the directions of the applied forces. Conditions (3.81, 3.82) can be summarized as

$$C_{jj}+C_{kk} \geq |2C_{jk}| \,. \tag{3.83}$$

Another inequality governing the relative magnitudes of diagonal and off-diagonal elements can be obtained by choosing $F_j, F_k$ such that $u_k=0$, which is achieved if $F_k=-C_{jk}F_j/C_{kk}$, $F_i=0$ $(i\neq j,k)$. We then obtain

$$U=\frac{1}{2}C_{jj}F_j^2+\frac{1}{2}\frac{C_{jk}^2}{C_{kk}}F_j^2-\frac{C_{jk}^2}{C_{kk}}F_j^2$$

and this expression is non-negative if and only if

$$C_{jj}C_{kk} \geq C_{jk}^2 \,. \tag{3.84}$$

### 3.8.3 Use of the reciprocal theorem

Maxwell's reciprocal theorem tells us that if we apply a force $F$ at point $A$ and measure a component of displacement $u$ at point $B$, the result we shall obtain is the same as if the force had instead been applied at $B$ *in the direction of* $u$ and the displacement had been measured at $A$ in the direction of $F$. In other words, two different problems for the same structure have the same solution. If we want to solve one of these problems, we therefore have a choice — we can solve the original problem or we can solve the *auxiliary problem* related to it by the reciprocal theorem, since they both have the same solution. The reciprocal theorem is useful in a particular case if and only if the auxiliary problem is easier to solve than the original problem.

**Example 3.10**

*Determine the displacement $u_B$ at the end of the beam in Figure 3.27(a) due to the force $F$ applied at $A$.*

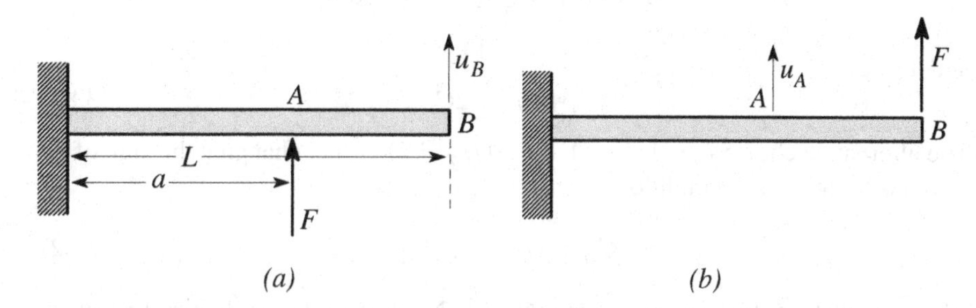


*Figure 3.27: Cantilever (a) loaded by a force at an intermediate point and (b) the auxiliary problem*

The reciprocal theorem tells us that the required displacement $u_B$ is the same as the displacement $u_A$ in the auxiliary problem of Figure 3.27*(b)*, in which the load is applied at the end of the beam. This is arguably an easier problem than that of Figure 3.27*(a)*, since using direct methods, we should need to use a singularity function description of the bending moment. By contrast, the bending moment in the auxiliary problem is

$$M = -F(L-z) = -EI\frac{d^2u}{dz^2}$$

and hence, after integration and imposing the built-in end conditions at $z=0$, we obtain

$$u = \frac{FLz^2}{2EI} - \frac{Fz^3}{6EI} .$$

Maxwell's theorem shows that this is also the solution for the end deflection $u_B$ in the problem of Figure 3.27*(a)*, provided we replace $z$ by $a$ — i.e.

$$u_B = \frac{FLa^2}{2EI} - \frac{Fa^3}{6EI} . \tag{3.85}$$

## 3.9 The stiffness matrix

The system of linear algebraic equations (3.74)

$$u_i = \sum_{j=1}^{N} C_{ij}F_j \quad \text{or} \quad \boldsymbol{u} = \boldsymbol{CF} \tag{3.86}$$

can be solved to give $F_j$ in terms of $u_i$ in the form

$$F_j = \sum_{i=1}^{N} K_{ji}u_i \quad \text{or} \quad \boldsymbol{F} = \boldsymbol{Ku} , \tag{3.87}$$

where

$$\boldsymbol{K} = \boldsymbol{C}^{-1} \tag{3.88}$$

is the inverse of $\boldsymbol{C}$. It follows that the matrix $\boldsymbol{K}$ is also symmetric and it is called the *stiffness matrix*.

The coefficient $K_{ij}$ is the force $F_i$ induced if $u_j=1$ and all other nodal displacements are maintained at zero (i.e. $u_k=0$ for all $k \neq j$). These coefficients are generally easier to calculate that the influence coefficients $C_{ij}$, since the deformation is localized near the node that is allowed to move. It also follows that many of the coefficients $K_{ij}$ will be zero. To understand this, consider the elastic truss shown in Figure 3.28, with nodes $A,B,C,...$, nodal displacements $u_1,u_2,u_3,...$ and interconnecting elastic bars.

To find a given coefficient $K_{ij}$ we fix all the nodes except one, at which the displacement component $u_j=1$. For example, the coefficients $K_{i6}$ correspond to the case where the node $C$ moves vertically upwards. The only bars that will change in length are $CE,CF$ and hence the only nodes at which forces will be induced are $E,F$ and of course $C$. This means that the elements $K_{16},K_{26},K_{36},K_{46},K_{76},K_{86}$ will all be zero, since no forces are induced at $A,B,D$ when $u_6$ is the only non-zero displacement. More generally, in a structure like that of Figure 3.28, an off-diagonal element $K_{ij}$ will be non-zero only if $u_i, u_j$ are two orthogonal displacement components for the same node or if there is a bar directly[20] connecting the nodes at which $u_i, u_j$ occur. For this reason, with a suitable labelling scheme, $\boldsymbol{K}$ is generally a *banded matrix* — i.e. the non-zero elements are clustered in a band near the diagonal.[21]

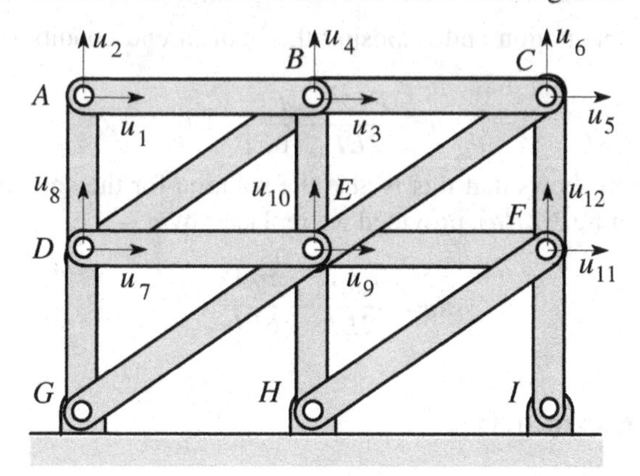


*Figure 3.28: A system of interconnected elastic bars*

We can now see why it is generally easier to determine the stiffness matrix than the compliance matrix. To determine $\boldsymbol{K}$ for the truss of Figure 3.28, we simply displace each node in turn; determine the new lengths of the bars connecting that node to the remaining fixed nodes; determine the bar forces from Hooke's law and hence use equilibrium to determine the non-zero nodal forces. For each node, this involves an analysis only of part of the structure — for example the section $BCEF$ when we are dealing with node $C$.

By contrast, if we wish to determine the compliance matrix directly, we have to find the displacements of all nodes due to the application of a force at each node in turn. Every calculation involves an equilibrium analysis and a kinematic analysis of the whole structure, which can be very lengthy if there are many nodes. It is therefore generally much simpler to calculate $\boldsymbol{K}$ and then use (3.88) to determine $\boldsymbol{C}$ if it is required.

### 3.9.1 Structures consisting of beams

Similar methods can be used to analyse systems of bars in bending and torsion. In this case, deformation can only be localized in a part of the structure if each non-active node is restrained against both translation and rotation. In the most general three-dimensional case, we therefore identify 6 degrees of freedom (3 translations

[20] In other words, not passing through some other node.

[21] Similar matrices arise in the formulation of the finite element method and special schemes have been developed for their efficient manipulation (see Appendix A, §A.6).

and 3 rotations) at each node and a corresponding set of 3 nodal forces and 3 nodal moments. If just one displacement component (degree of freedom) is non-zero, the deformation mode for each bar is then equivalent to that for a bar built in at one end and either translated or rotated (but not both) at the other.

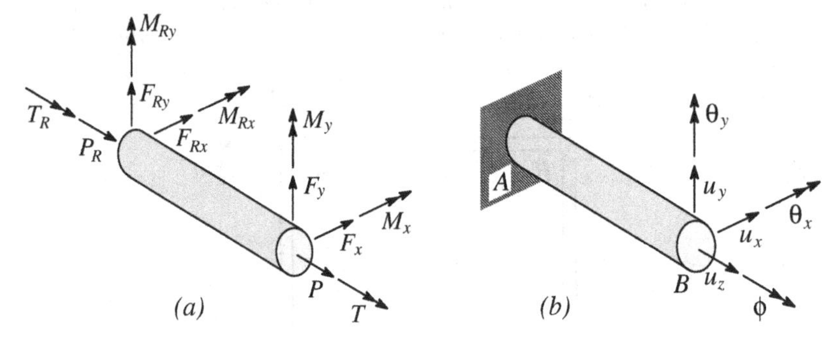


*Figure 3.29: (a) Load and (b) displacement components for a built-in elastic bar*

The load and displacement components for a typical bar are shown in Figure 3.29, where $A$ is the built-in end. The forces and moments at the free end $B$ can be found by elementary beam theory in terms of the degrees of freedom as

$$\begin{aligned}
P &= \frac{AE}{L}u_z \\
F_x &= \frac{12EI_y}{L^3}u_x - \frac{6EI_y}{L^2}\theta_y \\
F_y &= \frac{12EI_x}{L^3}u_y + \frac{6EI_x}{L^2}\theta_x \\
T &= \frac{GK}{L}\phi \\
M_x &= \frac{6EI_x}{L^2}u_y + \frac{4EI_x}{L}\theta_x \\
M_y &= -\frac{6EI_y}{L^2}u_x + \frac{4EI_y}{L}\theta_y \, .
\end{aligned} \tag{3.89}$$

We also record the reaction forces and moments

$$\begin{aligned}
P_R &= -\frac{AE}{L}u_z \\
F_{Rx} &= -\frac{12EI_y}{L^3}u_x + \frac{6EI_y}{L^2}\theta_y \\
F_{Ry} &= -\frac{12EI_x}{L^3}u_y - \frac{6EI_x}{L^2}\theta_x \\
T_R &= -\frac{GK}{L}\phi \\
M_{Rx} &= \frac{6EI_x}{L^2}u_y + \frac{2EI_x}{L}\theta_x \\
M_{Ry} &= -\frac{6EI_y}{L^2}u_x + \frac{2EI_y}{L}\theta_y \, .
\end{aligned} \tag{3.90}$$

**Example 3.11**

*Find the coefficients of the stiffness matrix for the two-dimensional rectangular frame of Figure 3.30(a), relating the force F and moments $M_B, M_C$ to the corresponding displacement u and rotations $\theta_B, \theta_C$. Assume that the beam segments are much stiffer in tension than in bending and that each has length L and flexural rigidity EI.*

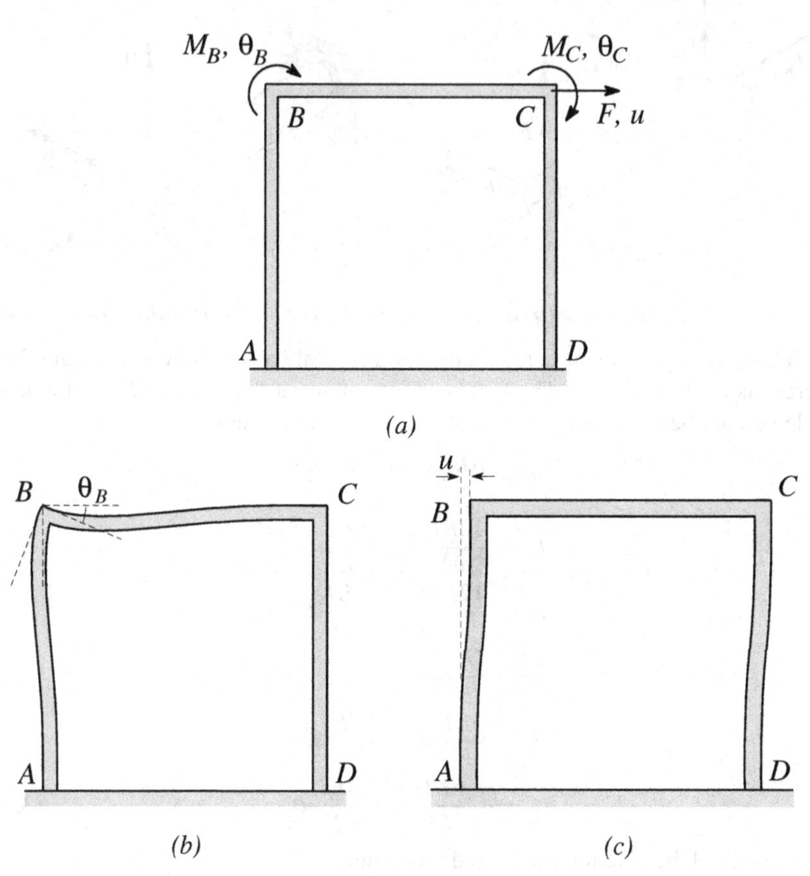


*Figure 3.30: A rectangular frame with three degrees of freedom: (a) Force and displacement components, (b) rotation of node B, (c) translation of bar BC*

Since the beam segments are much stiffer in tension than in bending, the only degrees of freedom comprise rotation of the joint $B$, rotation of the joint $C$ and horizontal translation of the member $BC$ without rotation. Figure 3.30*(b)* shows the configuration with $\theta_c = 0$ and $u = 0$. The beams $AB$ and $BC$ are both rotated through a clockwise angle $\theta_B$ at the end. To hold $AB$ in this configuration, we require a horizontal force

$$F = -\frac{6EI}{L^2}\theta_B$$

and a clockwise moment

$$M_B = \frac{4EI}{L}\theta_B \; ,$$

from equations (3.89). There will also be reactions induced at the support $A$ and given by equations (3.90), but these do not feature in the stiffness matrix because there are no degrees of freedom at $A$.

To hold the beam segment $BC$ in the deformed configuration of Figure 3.30*(b)*, we shall require a *vertical* force at $B$ and a clockwise moment

$$M_B = \frac{4EI}{L}\theta_B \; ,$$

from equations (3.89). In addition an equal and opposite vertical reaction force will be induced at $C$, along with a clockwise reaction moment

$$M_C = \frac{2EI}{L}\theta_B \; ,$$

from equations (3.90). The vertical forces do not appear in the stiffness matrix, since there can be no corresponding vertical displacement at $B, C$.

The forces and moments required to sustain the configuration 3.30*(b)* are the sum of those required for the beam segments $AB$, $BC$ considered separately and are therefore

$$F = -\frac{6EI}{L^2}\theta_B \;\; ; \;\; M_B = \frac{8EI}{L}\theta_B \;\; ; \;\; M_C = \frac{2EI}{L}\theta_B \; .$$

An exactly similar argument applied to the rotation $\theta_C$ shows that the required forces and moments will be

$$F = -\frac{6EI}{L^2}\theta_C \;\; ; \;\; M_B = \frac{2EI}{L}\theta_C \;\; ; \;\; M_C = \frac{8EI}{L}\theta_C \; .$$

Figure 3.30*(c)* shows the configuration obtained when $u$ is the only non-zero displacement. The beam segment $BC$ translates without deformation and the required forces and moments are readily found from equations (3.89) as

$$F = \frac{24EI}{L^3}u \;\; ; \;\; M_B = -\frac{6EI}{L^2}u \;\; ; \;\; M_C = -\frac{6EI}{L^2}u \; .$$

Thus, the stiffness formulation for the problem can be written in the matrix form

$$\left\{ \begin{array}{c} F \\ M_B \\ M_C \end{array} \right\} = \frac{EI}{L^3} \left[ \begin{array}{ccc} 24 & -6L & -6L \\ -6L & 8L^2 & 2L^2 \\ -6L & 2L^2 & 8L^2 \end{array} \right] \left\{ \begin{array}{c} u \\ \theta_B \\ \theta_C \end{array} \right\} . \tag{3.91}$$

If we wish to obtain the displacements and rotations at $B, C$ due to a specified set of loads, all we have to do is to invert this matrix to obtain the compliance matrix $\boldsymbol{C}$ and then use equation (3.86).

### 3.9.2 Assembly of the stiffness matrix

The forces and moments required to cause a single node to displace or rotate will contain contributions from each of the components connected to that node. These contributions can be summed, as in Example 3.11, using tabulated results for standardized components such as those given in equations (3.89, 3.90). Since the different components will generally be in different orientations, it is usually necessary to perform some coordinate transformation on these tabulated results to refer them to the coordinate system used for the system as a whole.

The process of summing the contributions from the separate components is referred to as the *assembly of the stiffness matrix*. This method is widely used, particularly in the elastic analysis of civil engineering structures,[22] which are often made up of beam segments. Generally, the number of degrees of freedom is large and numerical solution is required.

Essentially the same assembly procedure is used to obtain the global stiffness matrix in the finite element method, using the load-displacement characteristics of the individual elements in place of equations (3.89, 3.90). This will be further discussed in Appendix A (§A.2.2).

## 3.10 Castigliano's second theorem

In section §3.8.1, we showed that the strain energy $U$ can be written in terms of the applied forces and the influence coefficients as

$$U = \frac{1}{2}\sum_{i=1}^{N}\sum_{j=1}^{N} C_{ij}F_iF_j \, , \tag{3.92}$$

from equation (3.79). An alternative form of this expression can be obtained by using (3.74) to substitute for the sum on $j$, with the result

$$U = \frac{1}{2}\sum_{i=1}^{N} u_iF_i \, . \tag{3.93}$$

We can then substitute for $F_i$ from (3.87) to obtain

$$U = \frac{1}{2}\sum_{i=1}^{N}\sum_{j=1}^{N} K_{ij}u_iu_j \, , \tag{3.94}$$

which defines the strain energy in terms of the displacement components $u_i$ and the stiffness coefficients $K_{ij}$.

Suppose we differentiate (3.94) with respect to $u_i$. The only non-zero terms that will remain after differentiation will be those containing $u_i$ and these will occur in the

[22] See for example, J.B. Kennedy and M.K.S. Madugula (1990), *Elastic Analysis of Structures: Classical and Matrix methods*, Harper & Row, New York, Chapter 12.

$i$th row and the $i$th column of the matrix $\boldsymbol{K}$, as shown in Figure 3.31. In view of the reciprocal theorem, $K_{ij}u_iu_j = K_{ji}u_ju_i$ and hence each term in the double summation (3.94) occurs twice *except* the term on the diagonal $K_{ii}u_i^2$, whose derivative however also acquires a factor of 2 because $u_i$ is squared.[23] This factor cancels the 1/2 in (3.94) and we obtain

$$\frac{\partial U}{\partial u_i} = \sum_{j=1}^{N} K_{ij}u_j \,. \tag{3.95}$$

$$\begin{array}{cccc|c|cc}
K_{11}u_1^2 & K_{12}u_1u_2 & K_{13}u_1u_3 & \cdots & K_{1i}u_1u_i & \cdots & K_{1N}u_1u_N \\
K_{21}u_2u_1 & K_{22}u_2^2 & K_{23}u_2u_3 & \cdots & K_{2i}u_2u_i & \cdots & K_{2N}u_2u_N \\
K_{31}u_3u_1 & K_{32}u_3u_2 & K_{33}u_3^2 & \cdots & K_{3i}u_3u_i & \cdots & K_{3N}u_3u_N \\
\cdot & \cdot & \cdot & \cdots & \cdot & \cdots & \cdot \\
\cdot & \cdot & \cdot & \cdots & \cdot & \cdots & \cdot \\
\cdot & \cdot & \cdot & \cdots & \cdot & \cdots & \cdot \\
\cdot & \cdot & \cdot & \cdots & \cdot & \cdots & \cdot \\
\hline
K_{i1}u_iu_1 & K_{i2}u_iu_2 & K_{i3}u_iu_3 & \cdots & K_{ii}u_i^2 & \cdots & K_{iN}u_iu_N \\
\hline
\cdot & \cdot & \cdot & \cdots & \cdot & \cdots & \cdot \\
\cdot & \cdot & \cdot & \cdots & \cdot & \cdots & \cdot \\
\cdot & \cdot & \cdot & \cdots & \cdot & \cdots & \cdot \\
\cdot & \cdot & \cdot & \cdots & \cdot & \cdots & \cdot \\
K_{N1}u_Nu_1 & K_{N2}u_Nu_2 & K_{N3}u_Nu_3 & \cdots & K_{Ni}u_Nu_i & \cdots & K_{NN}u_Nu_N
\end{array}$$

*Figure 3.31: Terms in the double summation (3.94) involving a specific displacement component $u_i$*

[23] This argument is perhaps easier to follow by considering a particular example. Suppose there are only three external forces, in which case, (3.94) can be written

$$U = \frac{1}{2}\Big(K_{11}u_1^2 + K_{12}u_1u_2 + K_{13}u_1u_3 + K_{21}u_2u_1 + K_{22}u_2^2 + K_{23}u_2u_3$$
$$+K_{31}u_3u_1 + K_{32}u_3u_2 + K_{33}u_3^2\Big)$$

Differentiating with respect to $u_3$ (say), gives

$$\frac{\partial U}{\partial u_3} = \frac{1}{2}(K_{13}u_1 + K_{23}u_2 + K_{31}u_1 + K_{32}u_2 + 2K_{33}u_3)$$

which, using the reciprocal theorem, reduces to

$$\frac{\partial U}{\partial u_3} = K_{13}u_1 + K_{23}u_2 + K_{33}u_3 \,,$$

agreeing with (3.95).

Furthermore, we can use (3.87) to substitute for the right hand side of (3.95), obtaining

$$\frac{\partial U}{\partial u_i} = F_i \,. \tag{3.96}$$

This constitutes another proof of Castigliano's first theorem [equation (3.70)]. However, we should note that the present proof, being based on the stiffness matrix formulation, implies that the structure is linear, whereas no such limitation applies to the proof given in §3.7.

An exactly parallel argument can be used with equations (3.92, 3.74) to show that

$$\frac{\partial U}{\partial F_i} = \sum_{j=1}^{N} C_{ij} F_j = u_i \,. \tag{3.97}$$

This is a simplified version[24] of *Castigliano's second theorem.*

As in §3.7, we should note that the partial derivative implies that all the forces other than $F_i$ are being held constant.[25] This might be indicated, as before, by writing

$$u_i = \left( \frac{\partial U}{\partial F_i} \right)_{F_j (j \neq i)} \,.$$

### 3.10.1 Use of the theorem

Castigliano's second theorem permits us to find any displacement component, provided the strain energy can be written in terms of the external forces $F_i$. If the structure is determinate,[26] we can do this using equilibrium arguments only. Thus, the theorem is generally used in combination with an equilibrium analysis and replaces the kinematic arguments in the direct method. This contrasts with the principle of stationary potential energy (§3.5) and Castigliano's first theorem (§3.7), which are used in combination with kinematic analysis and replace the equilibrium arguments.

### Example 3.12

*The inverted U structure of Figure 3.32 is subjected to two horizontal forces $F_1, F_2$ as shown. Determine the horizontal displacement of the free end D. All sections of the structure have the same flexural rigidity EI and are of length L.*

---

[24] The more general version of the theorem refers to a quantity known as *complementary energy*, $C = \sum_{i=1}^{N} F_i u_i - U$. Complementary energy and strain energy are equal for elastic systems and small displacements, as can be seen from (3.93) and the definition of $C$. When the system is non-linear, $C$ and $U$ will generally differ, but Castigliano's second theorem is still applicable, provided $U$ is replaced by $C$. For more details on such applications, the student is referred to more advanced texts on energy and variational methods.

[25] Another way of saying this is that the $F_i$ are treated as an orthogonal set of *generalized coordinates*.

[26] We shall consider extensions to indeterminate structures in §3.10.5 below.

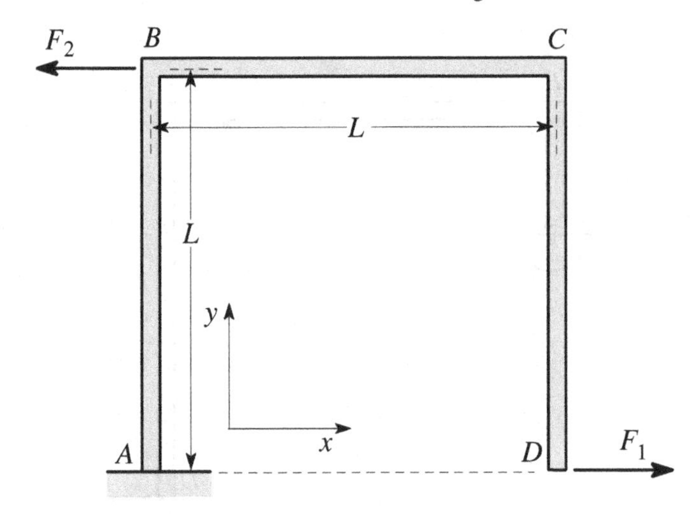


*Figure 3.32: A frame structure, built-in at A and loaded at B and D*

Following the discussion in §3.3 above, we shall assume that only the strain energy stored in bending is significant, which is equivalent to neglecting the extension of the horizontal section of the frame (for example) due to the axial force in comparison with the deflections due to bending.

Each part of the frame will store strain energy and the total energy $U$ will be the sum of that in each part. The procedure is to draw free-body diagrams for each part in turn, determine the bending moment *as a function of position along the beam* and substitute in expressions similar to equation (3.24). Finally, we sum the energy expressions for the three parts and apply Castigliano's second theorem (3.97).

Figure 3.33 shows the three free-body diagrams appropriate to the three beam segments $AB, BC, CD$. Notice that to find the bending moment in a particular segment (say in $AB$), we make an imaginary cut *at an arbitrary location* a distance $y$ from $A$ and then draw a free-body diagram of one of the two pieces separated by the cut, showing all the forces that act on that piece.[27] The dimensions $x, y$ defining the cuts in Figures 3.33 *(a,b,c)* are suggested by an implied Cartesian coordinate system centred on $A$, but any convenient definition can be used — for example, we could have measured $x$ from $C$ in the segment $BC$, in which case $(L-x)$ would be replaced by $x$, with some slight simplification in the evaluation of $U_2$ below. Notice also that we could have replaced Figure 3.33 *(a)* by the simpler segment $AC_1$, but we would then need to know the reactions at the support $A$, which would require a preliminary statical analysis of the whole structure.

[27] A common error is to make the cut correctly, but then only to show that part of the beam going from the cut to the point $B$ (say), neglecting the fact that there are internal forces and moments at $B$. Imagine making the cut with a hacksaw and see in your mind's eye the piece that drops off.

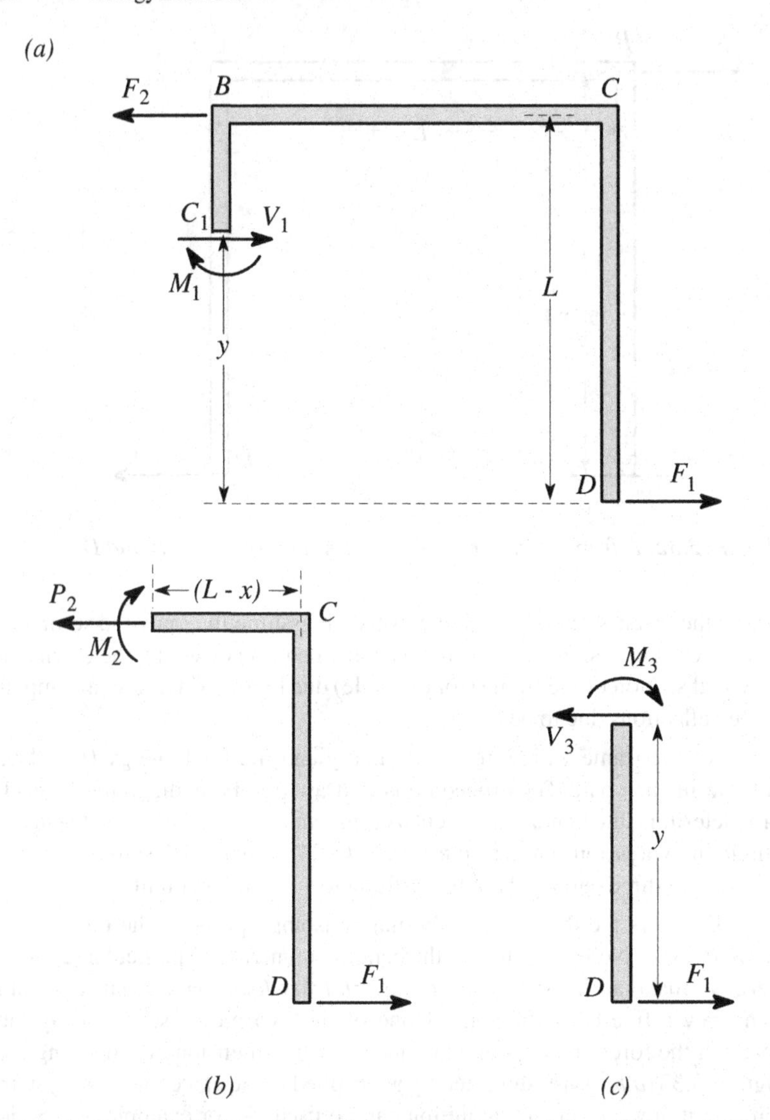


*Figure 3.33: Free-body diagrams for the problem of Figure 3.32*

Taking moments about the point $C_1$ for the segment $C_1BCD$ in Figure 3.33*(a)*, we find

$$M_1 = F_1y + F_2(L-y)$$

and hence, using equation (3.24), the strain energy stored in the segment $AB$ is

$$\begin{aligned} U_{AB} &= \frac{1}{2}\int_0^L \frac{M_1^2 dy}{EI} = \frac{1}{2EI}\int_0^L \left[F_1^2y^2 + 2F_1F_2y(L-y) + F_2^2(L-y)^2\right] dy \\ &= \frac{L^3}{6EI}\left(F_1^2 + F_1F_2 + F_2^2\right) , \end{aligned}$$

after evaluating the integrals. Similarly, from Figures 3.33 *(b,c)*, we have

$$M_2 = F_1 L$$
$$M_3 = F_1 y$$

in *BC*, *CD* respectively and the corresponding strain energy expressions are

$$U_{BC} = \frac{1}{2}\int_0^L \frac{M_2^2 dx}{EI} = \frac{F_1^2 L^2}{2EI}\int_0^L dy = \frac{F_1^2 L^3}{2EI}$$
$$U_{CD} = \frac{1}{2}\int_0^L \frac{M_3^2 dy}{EI} = \frac{F_1^2}{2EI}\int_0^L y^2 dy = \frac{F_1^2 L^3}{6EI} .$$

The total strain energy in the structure is therefore

$$U = U_{AB} + U_{BC} + U_{CD} = \frac{L^3}{6EI}\left(F_1^2 + F_1 F_2 + F_2^2\right) + \frac{F_1^2 L^3}{2EI} + \frac{F_1^2 L^3}{6EI}$$
$$= \frac{L^3}{6EI}\left(5F_1^2 + F_1 F_2 + F_2^2\right) .$$

To determine the horizontal displacement at *D*, we now apply Castigliano's second theorem (3.97) to obtain

$$u_D = \frac{\partial U}{\partial F_1} = \frac{L^3}{6EI}(10F_1 + F_2) , \tag{3.98}$$

which is the required result.

### 3.10.2 Dummy loads

Castigliano's second theorem enables us to determine the displacement of the point of application of any external force *in the direction of that force*, no matter how many external forces there are, in contrast to the method of §3.3 which only worked if there was only one external force. Thus, in Example 3.12, we could also find the horizontal displacement of the point *B* by taking the derivative with respect to $F_2$.

However, we often want to find displacements in a direction or at a point where there are no external forces. We can do this by the simple device of putting a dummy force $Q$ at the point where the displacement is to be found. We then proceed with the calculation of the required displacement and afterwards, set the dummy force $Q$ to zero.

**Example 3.13**

*Figure 3.34(a) shows a uniform cantilever beam of length L and flexural rigidity EI, subjected to a uniformly distributed load* $w_0$ *per unit length in the segment* $a<z<L$*. Find the downward vertical displacement at the end A of the cantilever.*

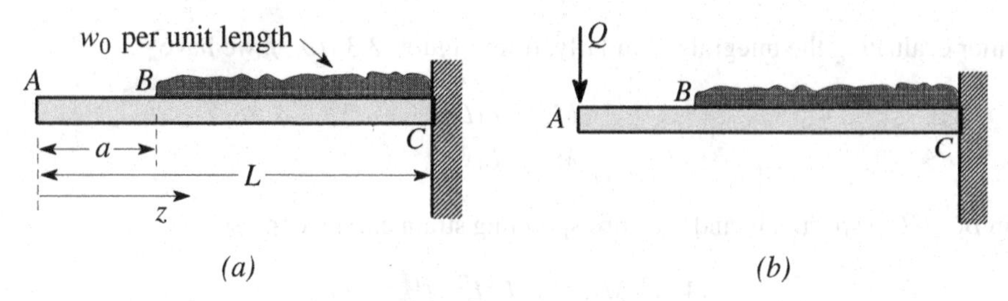


*Figure 3.34: Cantilever beam subjected to a distributed load $w_0$*

We first place a downward dummy force $Q$ at the end of the beam, as shown in Figure 3.34 *(b)*. With the dummy force in place, the bending moment in the beam is

$$\begin{aligned} M &= Qz & ; \; 0<z<a \\ &= Qz + \frac{w_0(z-a)^2}{2} & ; \; a<z<L\,. \end{aligned} \tag{3.99}$$

As in previous examples, we calculate the strain energy in the two regions $AB$ and $BC$ separately and add the result to get the total strain energy. In $AB$, we have

$$U_{AB} = \frac{1}{2EI}\int_0^a Q^2 z^2 dz = \frac{Q^2 a^3}{6EI}\,,$$

whilst, in $BC$,

$$\begin{aligned} U_{BC} &= \frac{1}{2EI}\int_a^L \left[Qz + \frac{w_0(z-a)^2}{2}\right]^2 dz \\ &= \frac{Q^2\left(L^3-a^3\right)}{6EI} + \frac{Qw_0}{24EI}\left(3L^4 - 8aL^3 + 6a^2L^2 - a^4\right) + \frac{w_0^2(L-a)^5}{40EI}\,. \end{aligned}$$

The total strain energy is therefore

$$U = U_{AB} + U_{BC} = \frac{Q^2L^3}{6EI} + \frac{Qw_0}{24EI}\left(3L^4 - 8aL^3 + 6a^2L^2 - a^4\right) + \frac{w_0^2(L-a)^5}{40EI}\,.$$

We now apply Castigliano's second theorem to obtain the displacement at $A$, which is

$$u_A = \frac{\partial U}{\partial Q} = \frac{QL^3}{3EI} + \frac{w_0}{24EI}\left(3L^4 - 8aL^3 + 6a^2L^2 - a^4\right)$$

and finally set the dummy force $Q$ to zero, obtaining

$$u_A = \frac{w_0}{24EI}\left(3L^4 - 8aL^3 + 6a^2L^2 - a^4\right)\,, \tag{3.100}$$

which is the required result.

It is important to note that the dummy force $Q$ cannot be set to zero until after the differentiation $\partial U/\partial Q$ has been performed, since otherwise the differentiation would yield a trivial result.

**Example 3.14**

*Use Castigliano's second theorem to find the horizontal displacement $u_H$ of the point A, in Example 3.2*

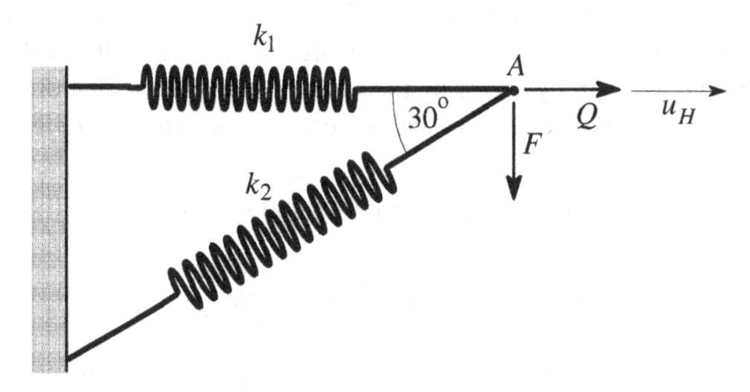


*Figure 3.35: Force supported by two springs*

In order to find the horizontal displacement, we need to introduce a horizontal dummy force $Q$ in addition to the applied vertical force $F$, as shown in Figure 3.35. The equilibrium equations (3.16, 3.17) are therefore modified to

$$\begin{aligned} F_1 + F_2 \cos 30^\circ - Q &= 0 \\ F_2 \sin 30^\circ + F &= 0 \end{aligned}$$

with solution

$$F_1 = Q + \sqrt{3}F \;\; ; \;\; F_2 = -2F \; .$$

As before, the strain energy in the two springs is calculated as

$$\begin{aligned} U &= \frac{1}{2}\frac{F_1^2}{k_1} + \frac{1}{2}\frac{F_2^2}{k_2} \\ &= \frac{1}{2}\frac{(Q+\sqrt{3}F)^2}{k_1} + \frac{1}{2}\frac{(-2F)^2}{k_2} \end{aligned}$$

and the horizontal displacement is then given by

$$u_H = \frac{\partial U}{\partial Q} = \frac{(Q+\sqrt{3}F)}{k_1} \; .$$

Finally, we set $Q=0$, with the result

$$u_H = \frac{\sqrt{3}F}{k_1} \; , \tag{3.101}$$

agreeing with (3.45) which was obtained rather more laboriously by the stationary potential energy method.

### 3.10.3 Unit load method

In the above examples, the algebra is complicated by the fact that the forces or moments are squared in the expressions for strain energy, but the expressions simplify considerably when the differentiation is performed and when the dummy load (if any) is set to zero. We can avoid some of this algebraic complexity, at the cost of a little clarity in the method, by changing the order in which these operations are performed. To illustrate this, we shall repeat the calculations for Example 3.13, using the alternative procedure.

We first note that the strain energy can be written symbolically in the form

$$U = U_{AB} + U_{BC} = \frac{1}{2EI}\int_0^a M^2 dz + \frac{1}{2EI}\int_a^L M^2 dz\,, \tag{3.102}$$

where $M$ is given by (3.99). We can perform the differentiation *before* the integration, with the result

$$u_A = \frac{\partial U}{\partial Q} = \frac{1}{EI}\int_0^a M\frac{\partial M}{\partial Q}dz + \frac{1}{EI}\int_a^L M\frac{\partial M}{\partial Q}dz\,. \tag{3.103}$$

Now that the differentiation has been performed, we can set $Q$ to zero. Thus, the displacement $u_A$ can be evaluated from equation (3.103) using the simpler equations

$$\begin{aligned} M &= 0 && ;\ 0 < z < a \\ &= \frac{w_0(z-a)^2}{2} && ;\ a < z < L \end{aligned} \tag{3.104}$$

[obtained by setting $Q=0$ in (3.99)] and

$$\begin{aligned} \frac{\partial M}{\partial Q} &= z\ ;\ 0 < z < a \\ &= z\ ;\ a < z < L\,, \end{aligned}$$

obtained by differentiation of (3.99).

Substituting these results into (3.103), we obtain

$$u_A = \frac{1}{EI}\int_a^L \frac{w_0(z-a)^2 z dz}{2} = \frac{w_0}{24EI}\left(3L^4 - 8aL^3 + 6a^2L^2 - a^4\right)\,, \tag{3.105}$$

as before. Notice that the first of the two integrals in (3.103) degenerates to zero, since $M$ is zero in this range from equation (3.104).

This procedure is known as the *unit load method*, since the expressions for $\partial M/\partial Q$ can be interpreted as the moments induced by a dummy load $Q$ of unit magnitude.

### 3.10.4 Formal procedure for using Castigliano's second theorem

Castigliano's second theorem is so useful in structural problems that we here list the steps in a formal way to make it easier to apply:-

(i) Put a dummy force (or moment) $Q$ at the point and in the direction that the displacement (rotation) component is required.
(ii) Use equilibrium arguments to determine the bending moments, torques etc. throughout the structure.
(iii) Differentiate these expressions with respect to $Q$. Notice that the results are identical to the bending moments, torques etc. due to a force (moment) $Q$ of unit magnitude acting alone.
(iv) Set $Q=0$ in the bending moments etc. obtained at step (ii).
(v) Write down the total strain energy as a sum or integral in terms of the bending moments, torques, axial forces etc, using equations (3.7, 3.24, 3.29, 3.31) as required. Do not substitute for these moments and forces or attempt to evaluate this sum or integral at this stage. Leave any integrals in a symbolic form analogous to (3.102).
(vi) Differentiate the expression for total strain energy with respect to $Q$.
(vii) Now substitute the results of steps (iii,iv) above (with $Q=0$) into this expression and perform any summations or integrations that result.

### 3.10.5 Statically indeterminate problems

A statically indeterminate problem is one having one or more redundant supports with corresponding unknown reactions that cannot be determined from equilibrium considerations alone. Castigliano's second theorem still applies to indeterminate structures, but step (ii) of the procedure in §3.10.4 cannot be completed, because the internal forces and moments in the structure cannot be found from equilibrium arguments alone. However, the procedure can be easily adapted to indeterminate problems and leads to a very efficient solution.

The indeterminacy is associated with the fact that there are more unknown reactions than can be found from the equilibrium conditions. The first step is to replace a sufficient number of these reactions by supposedly known external forces that we identify by a special symbol (say $S_1, S_2$ etc), in order to define an *equivalent determinate problem*.[28]

We can now use Castigliano's second theorem to find the displacements in this modified problem and in particular to determine the displacements $u_i$ at the points of application of $S_i$.

The configuration of the equivalent determinate problem and the original indeterminate problem will be the same, provided we choose the $S_i$ so as to make the $u_i$

[28] There may be more than one way of doing this and hence the equivalent determinate problem is not unique. For example, if a beam is supported on three simple supports, any one of the three may be replaced by an external force $S$ to develop a determinate problem. However, the final solution will be independent of the choice made.

zero, since then the points of application of the $S_i$ will not move and will be indistinguishable from immovable supports. Once the $S_i$ are known, other features of the solution can be obtained by further analysis of the equivalent determinate problem.

This, of course, is a general strategy for solving indeterminate problems. The particular advantage of Castigliano's second theorem is that it enables us to write the condition $u_i=0$ immediately in the form

$$u_i = \frac{\partial U}{\partial S_i} = 0\,, \tag{3.106}$$

where $U$ is the strain energy in the equivalent determinate problem.

**Example 3.15 — The propped cantilever**

*The cantilever shown in Figure 3.36 (a), loaded by a uniform load $w_0$ per unit length, is built-in at $z=L$ and also simply-supported at the end $z=0$. Find the reaction at the simple support.*

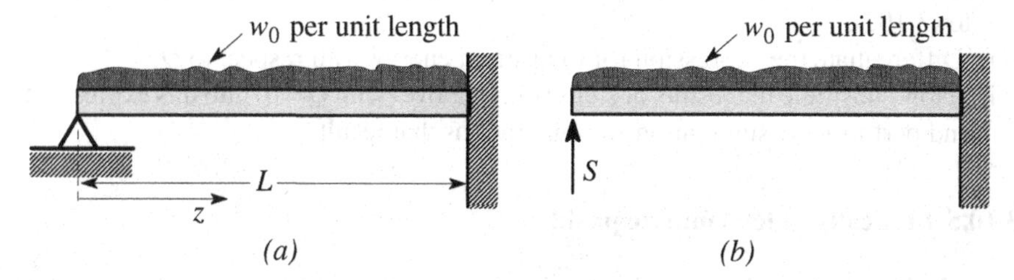


*Figure 3.36: (a) The propped cantilever, (b) the equivalent determinate problem*

There are three unknown reactions — a force and a moment at the built-in end and a force at $z=0$. We take the latter to be the redundant reaction $S$, which for the purposes of the theorem is then considered to be a known external force. The equivalent determinate problem is therefore defined by Figure 3.36 *(b)*

The bending moment can now be written as

$$M = \frac{w_0 z^2}{2} - Sz \tag{3.107}$$

and hence

$$u(0) = \frac{\partial U}{\partial S} = \frac{1}{EI}\int_0^L M\frac{\partial M}{\partial S}dz = 0\,.$$

Substituting for $M$ from (3.107), we obtain

$$\frac{1}{EI}\int_0^L \left(\frac{w_0 z^2}{2} - Sz\right)(-z)dz = \frac{1}{EI}\left(-\frac{w_0 L^4}{8} + \frac{SL^3}{3}\right) = 0$$

and hence the unknown reaction is

$$S = \frac{3w_0 L}{8}\,. \tag{3.108}$$

**Example 3.16: The curved bar**

*The complete circular bar of Figure 3.37 is compressed by equal and opposite forces $F$, as shown. Determine the change in the diameter AC due to the two forces.*

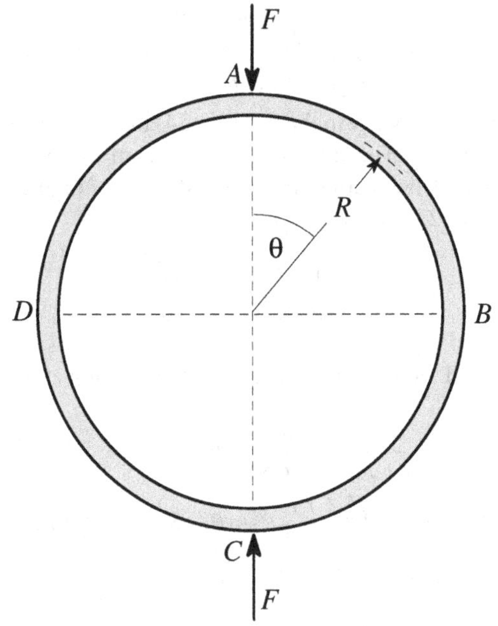


*Figure 3.37: The circular ring compressed by two forces*

The bar is symmetrical about both horizontal and vertical axes, so it is sufficient to consider the quadrant $0<\theta<\pi/2$.

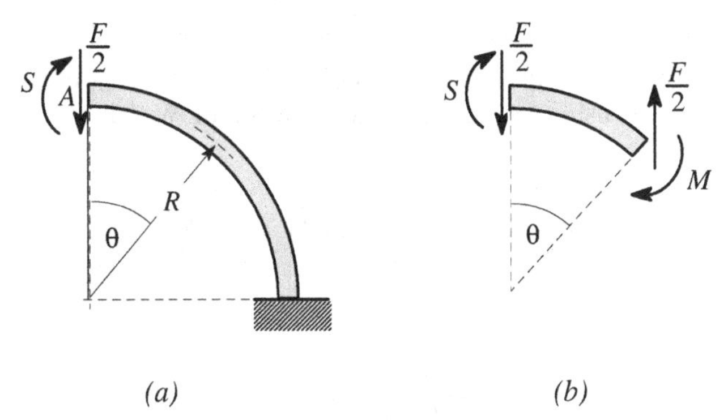


*Figure 3.38: One quarter of the circular ring*

The symmetry boundary conditions ensure that there is no rotation at either end of this quadrant, so we can treat the beam as built in at $\theta=\pi/2$ as shown in Figure 3.38 *(a)*. By symmetry, we can also say that the shear force at $\theta=0$ must be $F/2$, but the moment $S$ at that point is unknown and must be found from the symmetry condition that the slope of the beam must be zero at $\theta = 0$. This is what makes the problem indeterminate.

Treating $S$ as an external moment, we write the moment at a general point as

$$M = \frac{F}{2} R \sin\theta - S\,, \tag{3.109}$$

using the free-body diagram of Figure 3.38 *(b)*.

The equation for strain energy in the beam (3.24) has to be modified slightly here because the beam is curved. The small element of beam length $dL$ will be replaced by the arc length $Rd\theta$ and the quarter of the curved beam of Figure 3.38 *(a)* is defined by the range $0 < \theta < \pi/2$. Thus, we have

$$U = \frac{1}{2} \int_0^{\pi/2} \frac{M^2 R d\theta}{EI}\,. \tag{3.110}$$

Castigliano's second theorem then states that the rotation at $\theta = 0$ is[29]

$$\begin{aligned}\frac{\partial U}{\partial S} &= \frac{1}{EI} \int_0^{\pi/2} M \frac{\partial M}{\partial S} R d\theta \\ &= \frac{1}{EI} \int_0^{\pi/2} \left( \frac{F}{2} R \sin\theta - S \right) (-1) R d\theta\,,\end{aligned}$$

using (3.109, 3.110).

This rotation must be zero by symmetry and hence, after evaluating the integral, we obtain

$$S = \frac{FR}{\pi}\,. \tag{3.111}$$

We can now proceed to determine the vertical displacement $u_A$ of the point $A$ in Figure 3.38 *(a)*, again using Castigliano's second theorem. We have

$$u_A = \frac{\partial U}{\partial (F/2)} = 2\frac{\partial U}{\partial F} = \frac{2}{EI} \int_0^{\pi/2} M \frac{\partial M}{\partial F} R d\theta\,.$$

Differentiating (3.109), we have

$$\frac{\partial M}{\partial F} = \frac{R \sin\theta}{2}$$

and hence

[29] Notice that, as in other energy theorems, equation (3.97) may be modified by replacing the force $F_i$ by a moment and the displacement $u_i$ by the corresponding rotation.

$$
\begin{aligned}
u_A &= \frac{2}{EI}\int_0^{\pi/2}\left(\frac{FR\sin\theta}{2}-S\right)\frac{R^2\sin\theta d\theta}{2} \\
&= \frac{2}{EI}\left(\frac{\pi FR^3}{16}-\frac{SR^2}{2}\right) .
\end{aligned}
$$

We now substitute for $S$ from (3.111), to obtain

$$
u_A = \frac{FR^3}{EI}\left(\frac{\pi}{8}-\frac{1}{\pi}\right) = 0.0744\frac{FR^3}{EI} . \tag{3.112}
$$

If we assemble four quadrants like Figure 3.38(a) into the complete ring of Figure 3.37, it is clear that the diameter will change by $2u_A$, since the lower point $B$ will also move upwards by the same distance. Thus, the required change in diameter is $0.1488FR^3/EI$.

It is important to note that the Castigliano procedure is not applied to the indeterminate problem as such. The first stage is to identify redundant reactions and then treat them as if they were external forces, thus defining an equivalent determinate problem. The Castigliano procedure is then applied to this determinate problem throughout and the required results are only extracted at the end by substituting the values of the artificial external forces $S_i$ that cause the corresponding support displacements to be zero.

Most errors in the use of Castigliano's second theorem for indeterminate problems occur in this initial process of defining an equivalent determinate problem to which the theorem can be applied. It is therefore important to take great care at this stage. It might be helpful to try rephrasing the original problem, to make the relation between the indeterminate and determinate problems clearer. For example, we could rephrase Example 3.15 as follows:-

*A cantilever beam of length $L$ is built-in at $z=L$ and subjected to a uniformly distributed load $w_0$ and an end load $S$. Find the displacement at $z=0$ and hence find the value of $S$ if this displacement is to be zero.*

### 3.10.6 Three-dimensional problems

So far, we have restricted attention to problems of plane structures that remain within the plane during deformation. Three-dimensional structures, or plane structures that deform out of the plane due to applied loads, can be analyzed by the same methods, but beam segments will then generally experience torques as well as bending moments. The statical analysis of three-dimensional problems is also rather more challenging, if only because of the need to sketch three-dimensional figures.

### Example 3.17

*The L-shaped beam of Figure 3.39 is subjected to an out-of-plane load $F$. Find the displacement $u_C$ of the end in the direction of the force.*

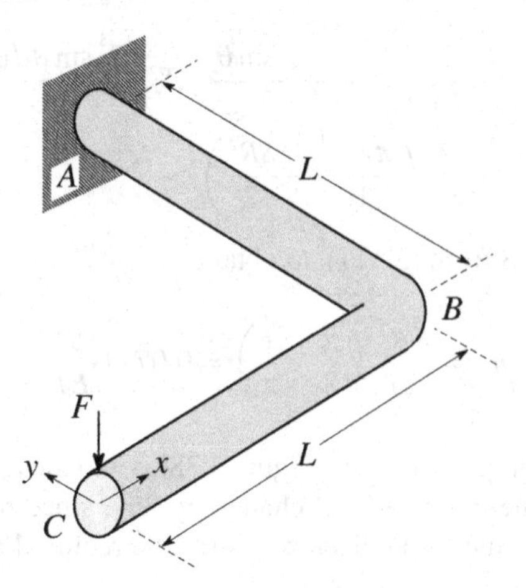


*Figure 3.39: The L-shaped beam subjected to an out-of-plane force*

The corresponding free-body diagrams for the segments $AB, BC$ are shown in Figure 3.40.

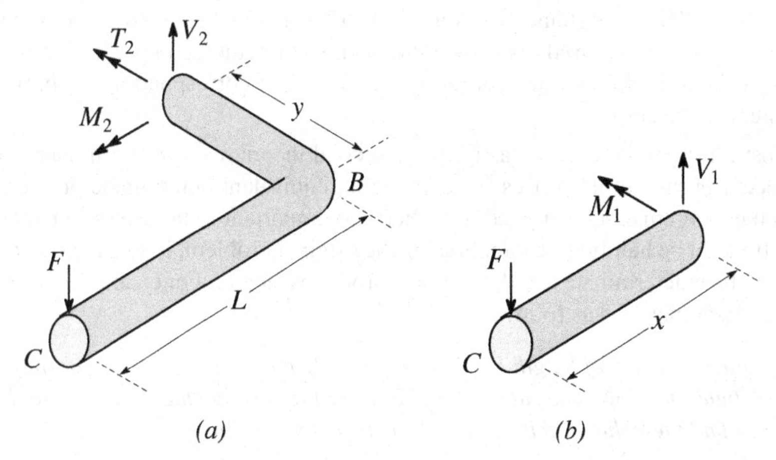


*Figure 3.40: Free-body diagrams for the problem of Figure 3.39*

Notice that the segment $BC$ experiences a bending moment $M_1$ only, but the segment $AB$ is in combined torsion and bending. Both segments also transmit a shear force, but following the arguments of §3.3.5, we neglect the shear contribution to the strain energy.

From equilibrium considerations, we have

$$\begin{aligned} M_1 &= Fx & &\text{in } BC \\ M_2 &= Fy \; ; \; T_2 = FL & &\text{in } AB. \end{aligned} \tag{3.113}$$

Thus, the strain energy in $BC$ is

$$U_1 = \frac{1}{2EI}\int_0^L M_1^2 dx$$

and that in $AB$ is

$$U_2 = \frac{1}{2EI}\int_0^L M_2^2 dy + \frac{1}{2GJ}\int_0^L T_2^2 dy \,.$$

Notice that the contributions to $U_2$ from the bending moment and the torque are simply added together, as explained in §3.3.3.

We could use either Castigliano's second theorem or the method of §3.3 for this problem, since there is only one external force. Using Castigliano's second theorem, we obtain

$$\begin{aligned} u_C &= \frac{\partial U}{\partial F} = \frac{\partial U_1}{\partial F} + \frac{\partial U_2}{\partial F} \\ &= \frac{1}{EI}\int_0^L M_1 \frac{\partial M_1}{\partial F} dx + \frac{1}{EI}\int_0^L M_2 \frac{\partial M_2}{\partial F} dy + \frac{1}{GJ}\int_0^L T_2 \frac{\partial T_2}{\partial F} dy \\ &= \frac{1}{EI}\int_0^L F x^2 dx + \frac{1}{EI}\int_0^L F y^2 dy + \frac{1}{GJ}\int_0^L F L^2 dy \\ &= \frac{2FL^3}{3EI} + \frac{FL^3}{GJ} \,, \end{aligned} \qquad (3.114)$$

using equations (3.113).

## 3.11 Summary

As we discussed in Chapter 1, the stresses and displacements in an engineering component are seldom critical in more than a few selected locations and success in engineering design depends on being able (i) to identify these critical locations and (ii) to estimate local stresses and/or displacements without an inordinate amount of calculation (in particular, without calculating the stresses and displacements everywhere in the component). Energy methods are extremely versatile tools for this purpose.

In this chapter, we have introduced several variational methods based on the concepts of strain energy and potential energy. The strain energy in an elastic structure can be determined by devising a scenario in which the loads are slowly applied and computing the work done on the structure. We have also introduced the idea of a stiffness matrix relating the external forces and the corresponding displacements and have shown by energy arguments that it must be symmetric.

The equilibrium configuration of an elastic system is that which minimizes the total potential energy (elastic strain energy + potential energy of external forces). This result can be used to replace equilibrium arguments in the solution of problems. It is particularly useful in engineering design in the sense of a Rayleigh-Ritz approximation, to select the best values for parameters in an approximate description of the kinematics of the deformation. The Rayleigh-Ritz method permits us to get a quick estimate for the deformation and hence the internal forces and moments in a structure, particularly in cases where an exact analytical solution is either impractical or outside the expertise of the designer.[30] Notice that the method depends on the choice of a good displacement function, which in turn requires us to practise the kind of intuitive thinking discussed in §1.2.1. Rayleigh-Ritz arguments are also fundamental to the finite element method, which is arguably the most used numerical method in engineering design (see Appendix A).

Castigliano's second theorem states that the derivative of the strain energy with respect to an external force is equal to the local displacement in the direction of that force. It is used in combination with an equilibrium analysis to determine the displacement of one or more specific points without solving for the displacement everywhere. It is particularly useful for indeterminate problems, where we need to enforce the condition that one or more displacements at redundant supports are zero.

## Further reading

T.M. Charlton (1973), *Energy Principles in Theory of Structures*, Oxford University Press, London.

H.L. Langhaar (1989), *Energy Methods in Applied Mechanics*, Krieger, Malabar, Florida.

K. Washizu (1975), *Variational Methods in Elasticity and Plasticity*, Pergamon, New York, 2nd edn.

## Problems

### Sections 3.1, 3.2

**3.1.** Find the strain energy stored in a straight bar of rectangular cross section $a \times b$ and length $L$, subjected to a tensile force $F$, if the material has Young's modulus $E$.

**3.2.** Find the strain energy stored in a solid circular cylindrical bar of diameter $d$ and length $L$, subjected to equal and opposite end torques $T$. The material has modulus of rigidity $G$.

**3.3** The ends $A, B$ of the bar of Figure P3.3 are subjected to the displacements $u_x^A, u_y^A, u_x^B, u_y^B$ as shown. Obtain an expression for the extension of the bar and hence

[30] In the latter case, the reader's first reaction may be to seek out someone who *does* know how to perform the calculation, such as a specialist stress analyst, but remember that we must figure the time and money spent on this in our optimization process, as discussed in §1.1.

show that the stored strain energy $U$ is

$$U = \frac{AE}{2L}[(u_x^B - u_x^A)\cos\theta + (u_y^B - u_y^A)\sin\theta]^2 ,$$

where $L,A$ are the length and cross-sectional area of the bar respectively and $E$ is Young's modulus.

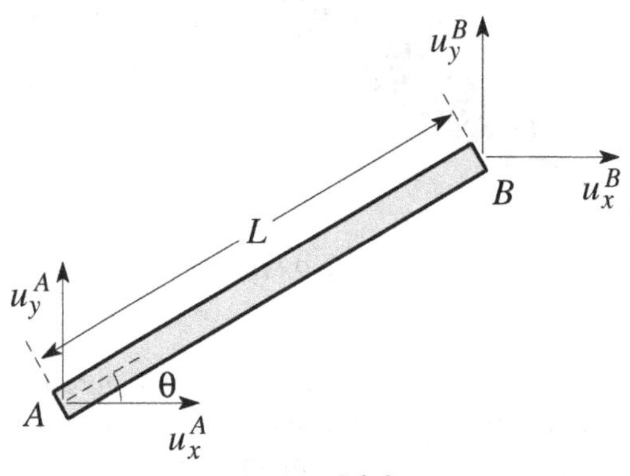


*Figure P3.3*

**3.4** Two springs of stiffness $k_1, k_2$ respectively are connected in series. Find the stiffness $k$ of the two-spring system and hence find the strain energy $U$ stored when the springs transmit a tensile force $F$. Then use equation (3.7) to find the strain energy stored in each spring separately ($U_1, U_2$) and verify that the sum $U_1 + U_2 = U$.

**3.5.** A spring is made of a non-linear material, such that the tensile force $F$ and the extension $u$ are related by the equation $F = F_0 \tan(u/u_0)$, where $F_0, u_0$ are constants. Find the strain energy stored $U$ (i) as a function of $u$ and (ii) as a function of $F$. Sketch the load-displacement curve and comment on the results when $|u/u_0| \ll 1$.

**3.6.** A cube of side $a$ is subjected to tensile stresses $\sigma_{xx}, \sigma_{yy}, \sigma_{zz}$ acting on the three orthogonal faces. Use the elastic constitutive relations (1.7–1.9) to find the relationship between the forces on the cube and the corresponding displacements, and hence find the strain energy $U$ stored in the cube.

**3.7.** In the system of Problem 3.4, suppose that one of the two springs is 100 times stiffer than the other (i.e. $k_1 = k$, $k_2 = 100k$). Show that a good approximation to the total energy stored $U$ can be obtained by neglecting $U_2$ — the energy stored in the stiffer spring.

**Section 3.3**

**3.8.** Find the displacement component $u$ due to the inclined force $F$ in Figure P3.8, by equating the work done during loading to the strain energy stored in the springs. Hence show that it reaches a minimum when $\alpha = 105^\circ$.

Find also the angle $\alpha$ for which $u$ is a maximum and the ratio $u_{max}/u_{min}$.

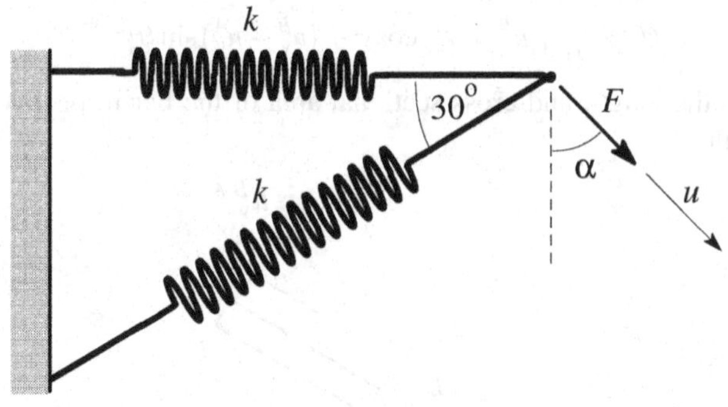


*Figure P3.8*

**3.9.** Find the strain energy stored in the cantilever of Figure P3.9, subjected to a concentrated force $F$ at the free end. The beam is of solid square cross section, of side $c$ and is made of a material of Young's modulus $E$.

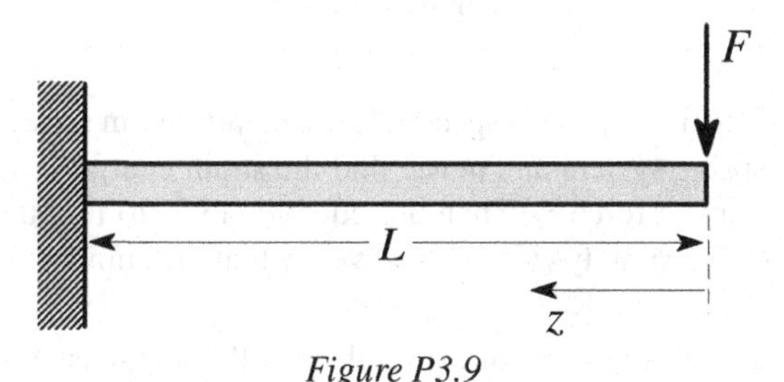


*Figure P3.9*

Hence, determine the downward vertical deflection $u$ under the load.

**3.10.** Suppose the cantilever of Figure P3.9 has a uniform width $c$, but its depth $d$ increases with distance $z$ from the free end according to the equation

$$d = c\left(\frac{1}{2} + \frac{z}{L}\right) .$$

Find the second moment of area $I$ as a function of $z$, the strain energy stored in the beam and the deflection under the load.

**Hint:** You will find it easier to perform the integration for the strain energy if you make the change of variable

$$y = \left(\frac{1}{2} + \frac{z}{L}\right) .$$

**3.11.** Figure P3.11 shows a beam of length $L$ and flexural rigidity $EI$ simply-supported on two springs each of stiffness $k$. Use an energy method to find the vertical displacement $u$ under the load. Hence show that if the dimensionless ratio

$$\lambda \equiv \frac{kL^3}{EI} < 1 \, ,$$

an acceptable approximation can be obtained for the displacement by treating the beam as a rigid body (i.e. by including only the strain energy in the springs). What would be the equivalent simplifying approximation for the case $\lambda \gg 1$ and how large would $\lambda$ need to be for this approximation to be within 5% of the exact answer?

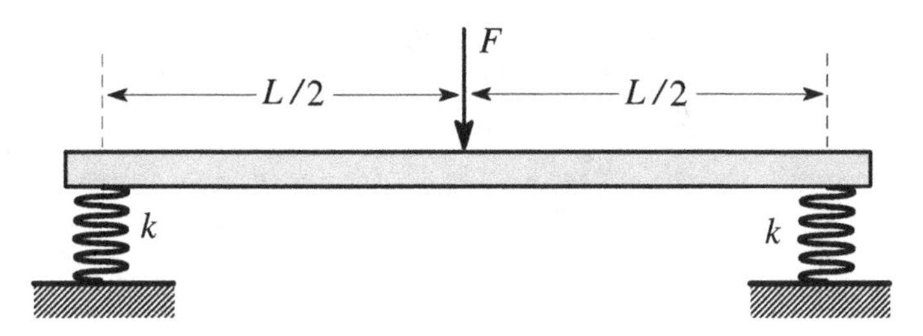


*Figure P3.11*

**3.12.** The determinate truss of Figure P3.12 is simply-supported at its ends $A, D$ and loaded by a vertical force $F$ at $C$. All the members have cross-sectional area $A$ and are of a material with Young's modulus $E$. Find the vertical displacement $u$ under the load at $C$.

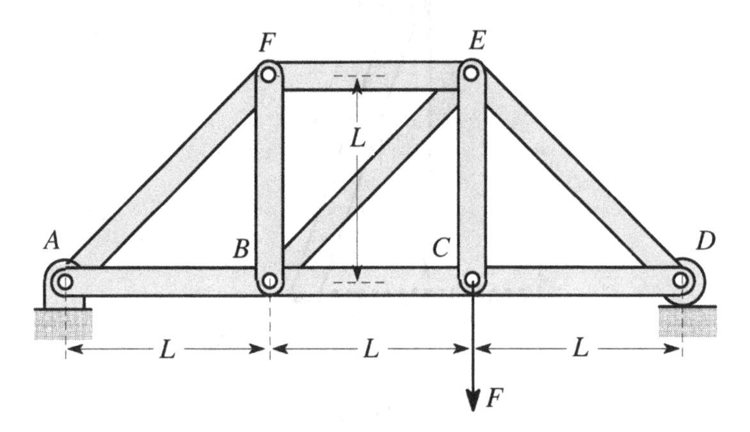


*Figure P3.12*

**3.13.** The cranked bar of Figure P3.13 is built in at $A$ and loaded by an out-of-plane force $F$ at $C$, so that the segment $BC$ is loaded in bending and the segment $AB$ in both bending and torsion. If the bar has a solid circular cross section with diameter $d$ and the elastic properties are $E, \nu$, find the vertical displacement $u$ at $C$. Note that $G = E/2(1+\nu)$.

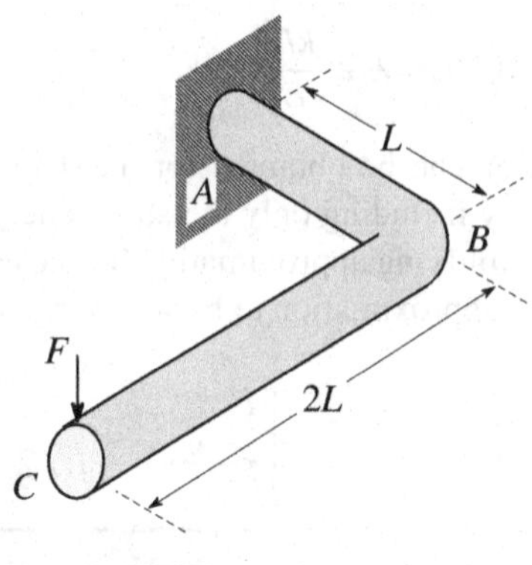


*Figure P3.13*

**Sections 3.4, 3.5**

**3.14.** The uniform rigid bar in Figure P 3.14 is of weight $W$ and slides without friction. The linear spring is unstretched when the bar is vertical. Use the principle of stationary potential energy to find equilibrium values of $\theta$ in terms of $W, k$ and $L$.

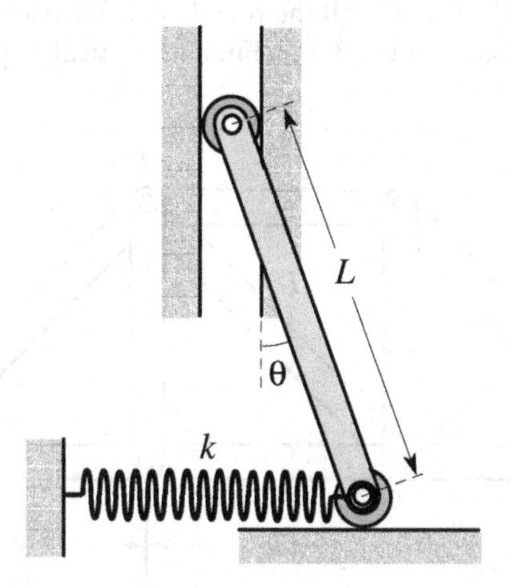


*Figure P3.14*

What condition must be satisfied in order that the position $\theta = 0$ should be one of stable equilibrium.

**3.15.** Figure P3.15 shows a mechanism consisting of a set of pin-jointed rigid bars, constrained by a spring of stiffness $k$ connecting the points $BC$. A force $F$ is applied at $E$. Find a kinematic relation between the vertical displacement of the force $u_E$ and the extension of the spring. This relation should not be restricted to the case of small displacements — i.e. it must also describe the case where the angle of rotation of $CE$ is not small. Use the result and the principle of stationary potential energy to deduce the (non-linear) relationship between $u_E$ and $F$.

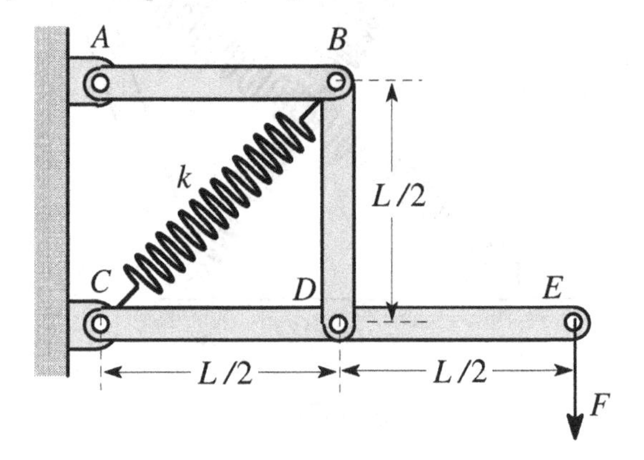


*Figure P3.15*

**3.16.** Use an energy argument to prove that the extension of the spring in Figure P3.16 is independent of the position of the mass $M$ on the platform $AB$.

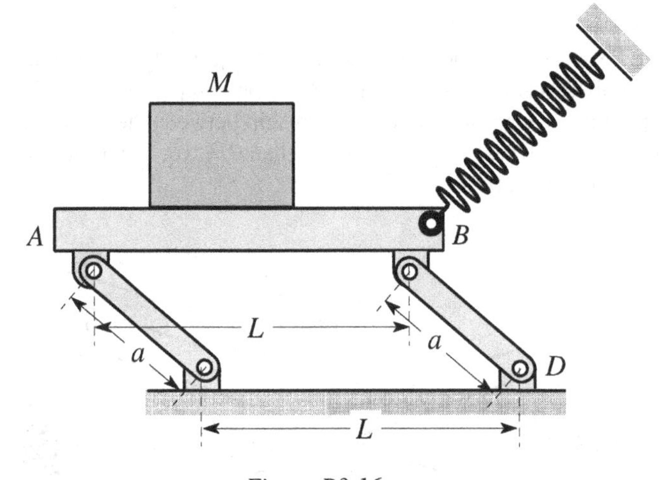


*Figure P3.16*

This type of mechanism is used in the design of weighing machines, in order that the measured weight be not dependent on the location of the mass.

**3.17.** The pin-jointed structure of Figure P3.17 consists of three rigid links each of length $L$ constrained by two springs on the diagonals, whose stiffnesses and original lengths are $k_1, L_1$ and $k_2, L_2$ respectively.

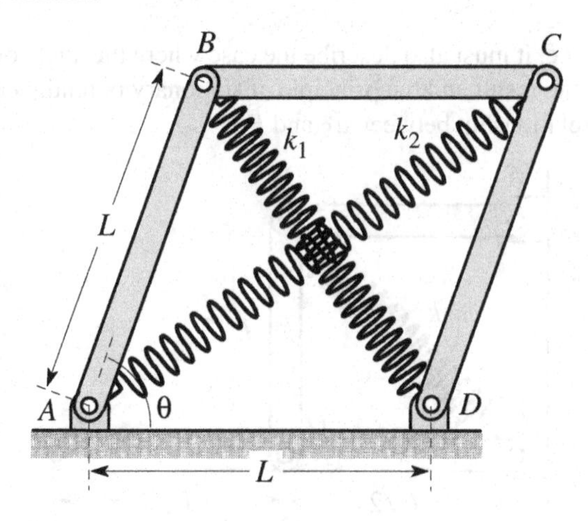


*Figure P3.17*

If no external forces are applied, the structure is found to adopt an equilibrium configuration in which $\theta = 60^{\circ}$. Show that the ratio of spring stiffnesses must satisfy the relation

$$\frac{k_1}{k_2} = \frac{(L_2/\sqrt{3} - L)}{(L_1 - L)} \; .$$

**3.18.** Figure P3.18 shows a mechanism of rigid pin-jointed links such that a horizontal force $F$ applied at $A$ causes the spring $BC$ to be compressed. Use the principle of stationary potential energy to find the relationship between the force $F$ and the angle $\theta$, if the spring has stiffness $k$ and is relaxed when $\theta = 30^{\circ}$. You may restrict attention to solutions in the range $0 < \alpha < 90^{\circ}$.

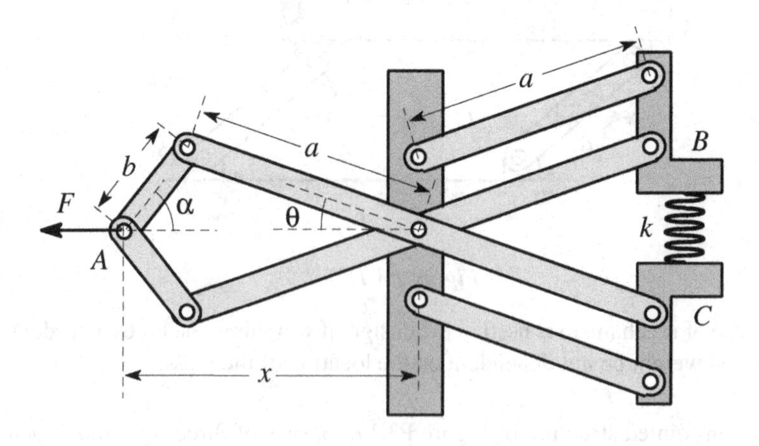


*Figure P3.18*

**3.19.** Figure P3.19 shows an aircraft door-closing mechanism. The door $DE$ and the links $ABC, EBF, CD, FG$ can be treated as rigid and all the connections are frictionless pin joints. The quadrilaterals $ABFG$ and $BCDE$ are parallelograms. Motion at $A$ is resisted by a torsion spring that exerts a moment $M$ in the direction shown of magnitude

$$M = K(\theta + \theta_0) ,$$

where $K, \theta_0$ are constants. A moment $T$ is applied to the door $DE$ as shown. Find the relationship between $T$ and $\alpha$.

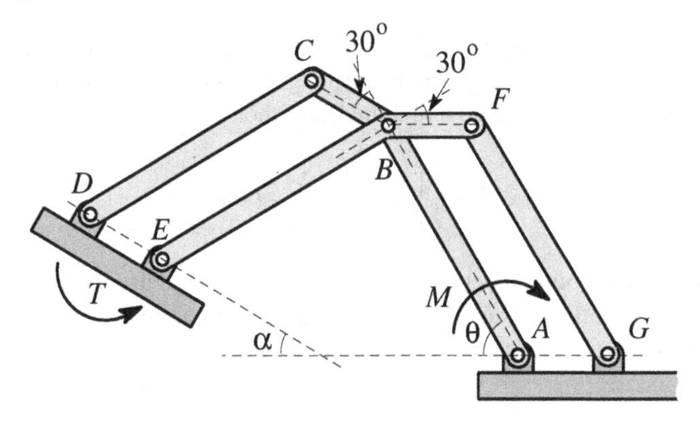


*Figure P3.19*

**Section 3.6**

**3.20.** Figure P3.20 shows a beam of length $L$ and flexural rigidity $EI$, which is built-in at $z=0$ and simply supported at the intermediate point $z=3L/4$. Use the Rayleigh-Ritz method with one degree of freedom to estimate the deflection at the free end.

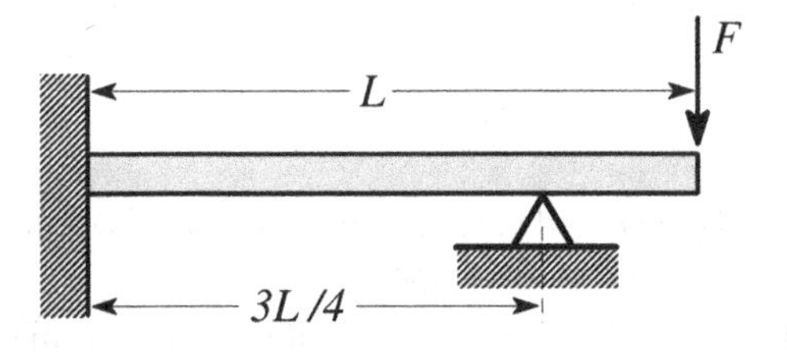


*Figure P3.20*

**3.21.** Figure P3.21 shows a beam of length $L$ and flexural rigidity $EI$, which is built-in at $z=0$, simply supported at $z=L$, and subjected to a uniformly distributed load $w_0$ per unit length. Use the Rayleigh-Ritz method with one degree of freedom to estimate the magnitude and location of the maximum deflection.

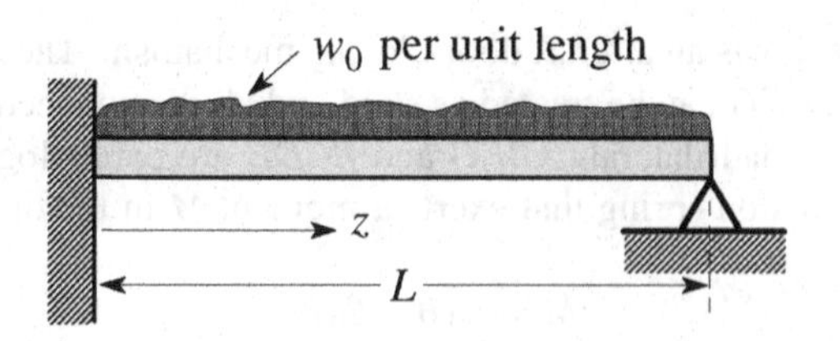


*Figure P3.21*

**3.22.** A beam of length $2L$ and flexural rigidity $EI$ is built-in at both ends $z=\pm L$ and subjected to a distributed load $w(z)$ per unit length, which is symmetric about the mid-point [i.e. $w(-z)=w(z)$]. Show that the trial function

$$u(z)=\sum_{i=1}^{N} C_i\left[\cos\left(\frac{i\pi z}{L}\right)-(-1)^i\right]$$

satisfies the kinematic boundary conditions of the problem and use the Rayleigh-Ritz method to find the coefficients $C_i$ in terms of integrals involving $w(z)$. Evaluate the resulting expressions for the special case where $w(z)=w_0$.

**3.23*.** Figure P3.23 shows a beam of length $L$ subjected to a uniformly distributed load $w_0$ per unit length. The ends of the beam are simply supported, but equal and opposite moments are applied at the ends whose magnitude is just sufficient to ensure that the deflection of the beam is exactly zero at the mid-point.

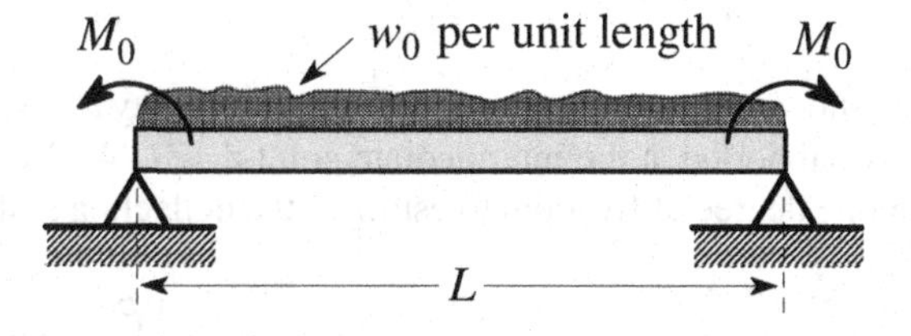


*Figure P3.23*

Use the Rayleigh Ritz method (with a two degree of freedom approximation) to estimate the slope of the beam at the supports.

**Hint:** Sketch the deformed shape of the beam before you make your choice of approximating function and make sure that your choice is capable of describing the shape of your sketch.

**3.24.** Use the Rayleigh-Ritz method with a one degree of freedom approximation to estimate the displacement under the load for the beam on three supports shown in Figure P3.24.

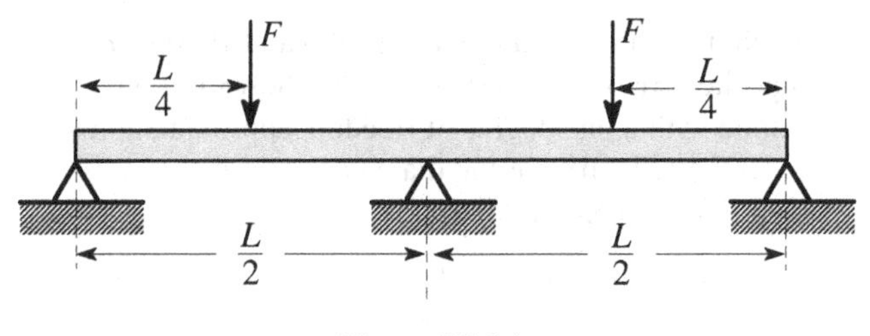


Figure P3.24

**3.25.** Figure P3.25 shows a cantilever beam of flexural rigidity $EI$ and length $L$, subjected to the distributed load

$$w(z) = w_0 \left( \frac{x}{L} - \frac{x^2}{L^2} \right) .$$

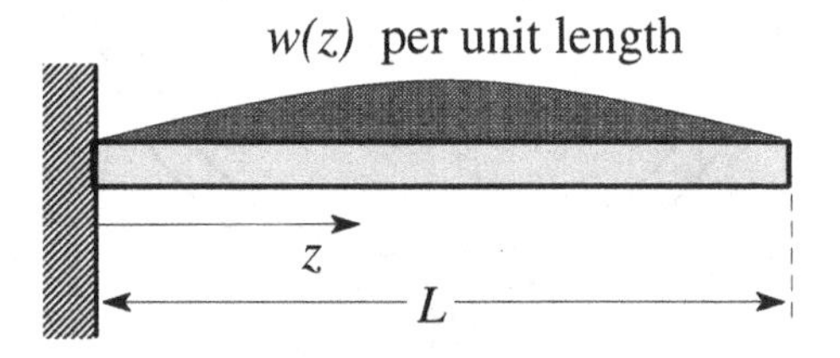


Figure P3.25

Use the Rayleigh-Ritz method to estimate the deflection at the free end. Use a polynomial approximating function with one degree of freedom, but which satisfies the condition of zero bending moment at the free end.

**3.26.** Figure P3.26 shows a simply-supported beam of length $L$ and flexural rigidity $EI$, loaded by a moment $M_0$ at one end. Use the Rayleigh-Ritz method to approximate the deformed shape of the beam and hence estimate the maximum deflection.

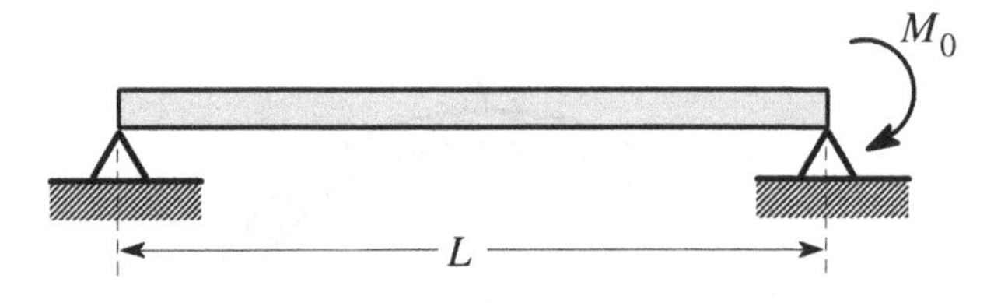


Figure P3.26

Use a one degree of freedom polynomial. Greater accuracy will be obtained if you choose a polynomial that gives zero bending moment at the unloaded end, in addition to satisfying the kinematic boundary conditions (see §3.6.2).

**3.27.** A complete circular ring of radius $R$ and flexural rigidity $EI$ is subjected to two equal and opposite forces $F$ on a diameter, as shown in Figure P3.27. Use the Rayleigh-Ritz method with a one degree of freedom approximation to the deformed shape to estimate (i) the reduction in the diameter $AC$, (ii) the *increase* in diameter $BD$ and (iii) the bending moment in the beam at $A$.

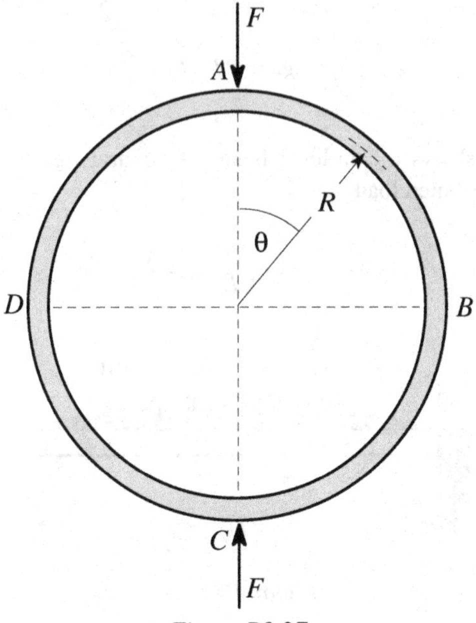


*Figure P3.27*

**3.28.** A semicircular curved bar of radius $R$ and flexural rigidity $EI$ is pressed against a rigid frictionless plane by a concentrated force $F$ at $B$, as shown in Figure P3.28. Use the Rayleigh-Ritz method with a one degree of freedom approximation to the deformed shape to estimate the change in the diameter $AC$ due to the load.

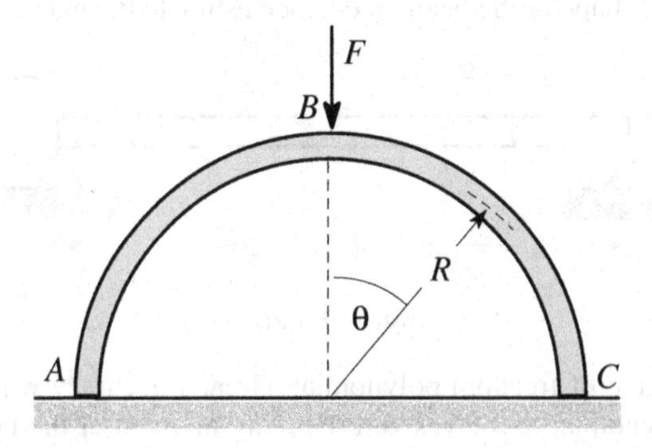


*Figure P3.28*

**Sections 3.7, 3.8, 3.9**

**3.29.** Use Castigliano's first theorem to solve Problem 3.15.

**3.30.** Use Castigliano's first theorem to solve Problem 3.18.

**3.31.** Use Castigliano's first theorem to solve Problem 3.19.

**3.32.** Find the displacements at $A, B$ for the problem of Figure P3.32 *(a)*, where $AB$ is a rigid bar and both springs have stiffness $k$. Use your results and Maxwell's reciprocal theorem to determine the displacement at $B$ due to the concentrated moment $M_0$ in Figure P3.32 *(b)*.

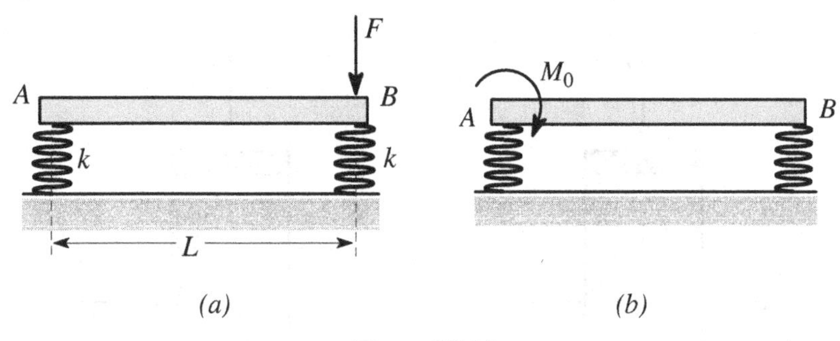


*Figure P3.32*

**3.33.** The displacement $u(z)$ for the simply supported beam of Figure P3.33 *(a)* loaded by an end moment $M_0$ is given by

$$u(z) = \frac{M_0 L^2}{6EI}\left(\frac{z}{L} - \frac{z^3}{L^3}\right) .$$

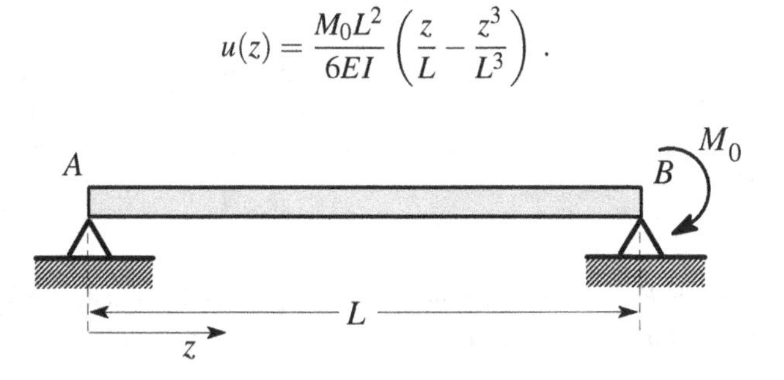


*Figure P3.33 (a)*

Use this result and Maxwell's reciprocal theorem to find where the force $F$ should be placed on the beam of Figure P3.33 *(b)* if the *slope* of the beam at $B$ is to be a maximum. What is the resulting maximum value?

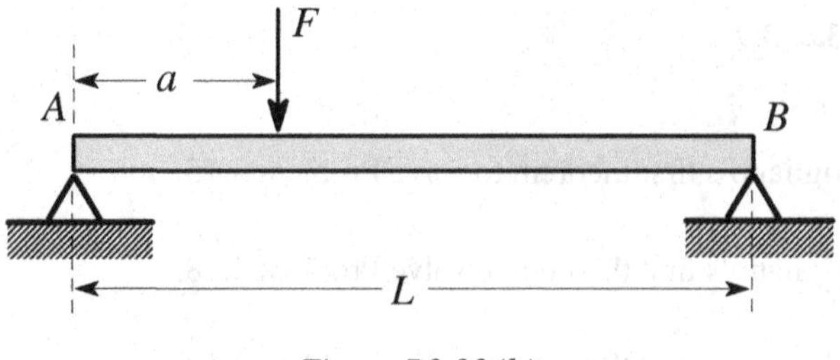


*Figure P3.33 (b)*

**3.34.** A cylinder of rubber of diameter $D$ and length $L$ rests on a flat surface and is loaded by a force $F_1$ applied through a flat rigid plate. Both surfaces are frictionless and the rubber has Young's modulus $E$ and Poisson's ratio $\nu = 0.5$.

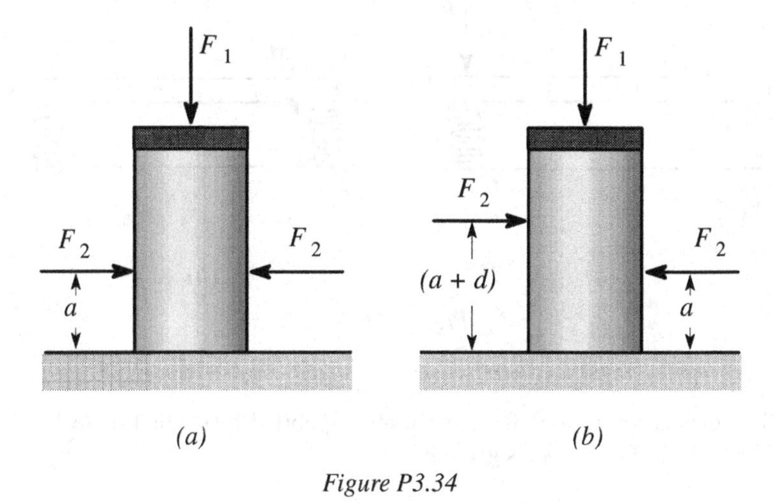


*Figure P3.34*

The cylinder is now pinched by two colinear forces $F_2$ as shown in Figure P3.34 *(a)*. Use Maxwell's reciprocal theorem with a suitable elementary state of stress as auxiliary solution to find the vertical displacement of the rigid plate due to the forces $F_2$.

**3.35.** The two forces $F_2$ in Problem 3.34 are now separated by a distance $d$ vertically as shown in Figure P3.34 *(b)*. How much does the rigid plate rotate as a result?

**3.36.** Find the coefficients of the $4 \times 4$ stiffness matrix $\boldsymbol{K}$ for the system of four masses connected by springs shown in Figure P3.36.

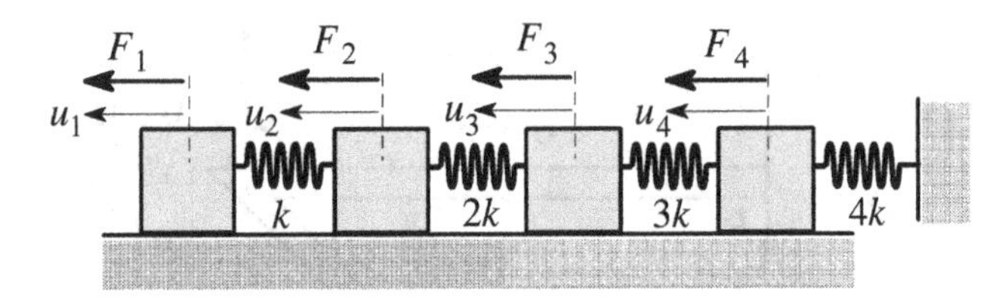


*Figure P3.36*

Invert the matrix to find the corresponding compliance matrix $\boldsymbol{C}$. Notice that $\boldsymbol{C}$ is a full matrix, whereas $\boldsymbol{K}$ is banded. Verify that both matrices are symmetric and that the inequalities (3.80, 3.83, 3.84) are satisfied for this case.

**3.37.** The pin-jointed bars in Figure P3.37 have a circular cross section of diameter 10 mm and are made of steel with Young's modulus $E = 210$ GPa. Find the $2 \times 2$ stiffness matrix for the system relating the forces $F_1, F_2$ and the corresponding displacements $u_1, u_2$. You might find it helpful to review the kinematics of Example 3.5 and Figure 3.12.

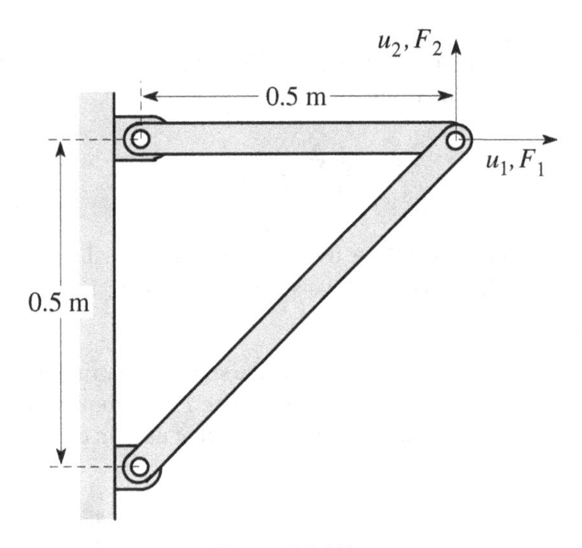


*Figure P3.37*

**3.38*.** The structure shown in Figure P3.38 consists of two circular steel bars, each of diameter 10 mm, built in at $B, C$ and welded together at $A$. Use equations (3.89, 3.90) to construct the $3 \times 3$ stiffness matrix relating the forces and moments $F_1, F_2, M$ and the corresponding displacements $u_1, u_2, \theta$. For steel, $E = 210$ GPa.

We recall from §3.3.2 that bars are much stiffer in extension than in bending. Discuss the implications of this result in the present problem.

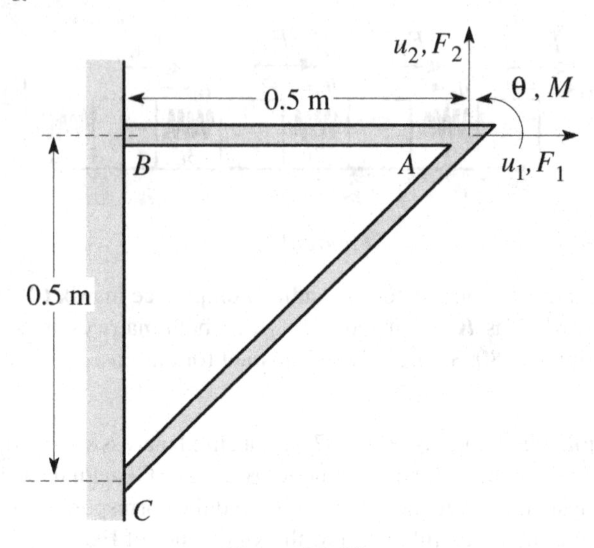


Figure P3.38

**3.39.** The strain energy in a structure can be written in terms of the stiffness matrix and the displacements in the form

$$U = \frac{1}{2}\sum_{i=1}^{N}\sum_{j=1}^{N} K_{ij}u_i u_j \; .$$

Use scenarios analogous to those used in §3.8.2 to establish inequalities that must be satisfied by the coefficients $K_{ij}$ of the stiffness matrix.

**3.40.** Use equations (3.89, 3.90) to construct the stiffness matrix for the built-in angle of Figure P3.40 subjected to out-of-plane loading $F$ and moments $M_1$, $M_2$. Notice that the coordinate system used here differs from that used in equations (3.89, 3.90).

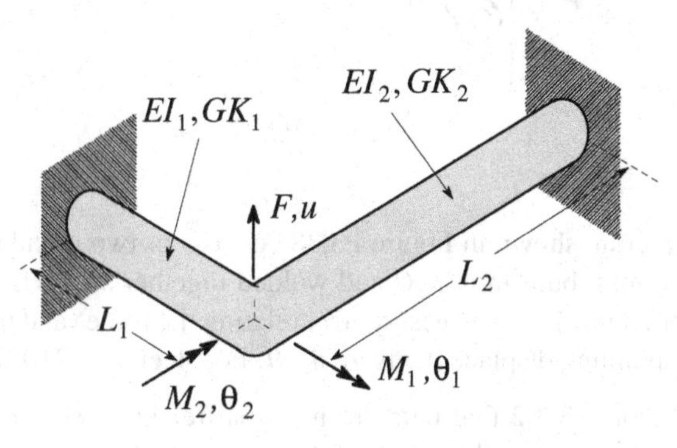


Figure P3.40

**Section 3.10**

**3.41.** The bars in Figure P3.41 are each made of steel of cross-sectional area $A = 1$ in$^2$. Use the result given in Problem 3.3 to express the total strain energy stored in the structure as a function of the displacement components, $u_i$, $i = (1,6)$. Hence find the $6 \times 6$ stiffness matrix relating the forces $F_i$ and the corresponding displacements $u_i$. Remember that any given product term (e.g. $u_3u_4$) occurs twice in the sum and that $K_{ij} = K_{ji}$. For steel, $E = 30 \times 10^6$ psi.

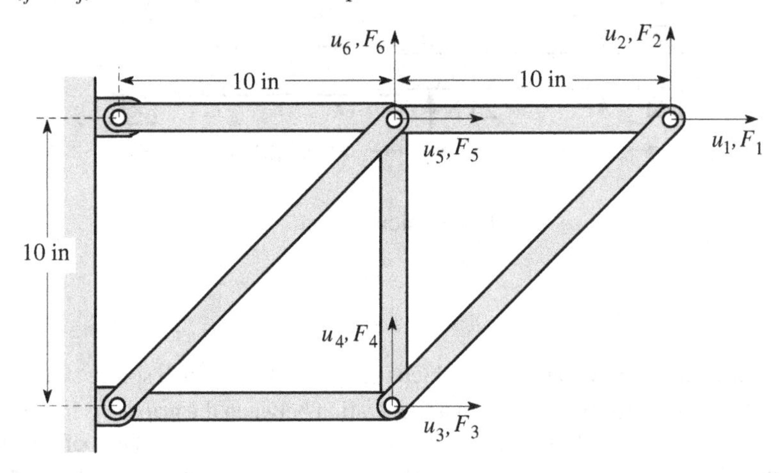


*Figure P3.41*

**3.42.** Both segments of the angled beam in Figure P3.42 have the same flexural rigidity $EI$ and length $L$.

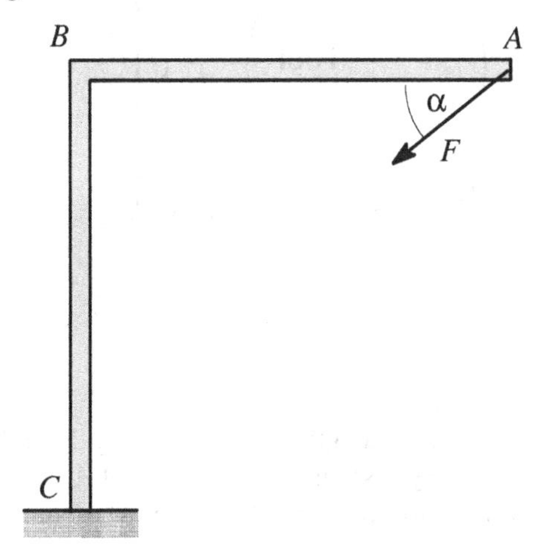


*Figure P3.42*

Find the angle $\alpha$ if the displacement of the point $A$ is to be in the same direction as the force $F$.

**3.43.** Figure P3.43 shows a stepped steel shaft supported in bearings at $A, C$ and loaded by a 2 kN force at $B$. Use Castigliano's second theorem to find the angular misalignment at the bearings (i.e. the local slope of the beam) due to the shaft deflection. For steel, $E = 210$ GPa.

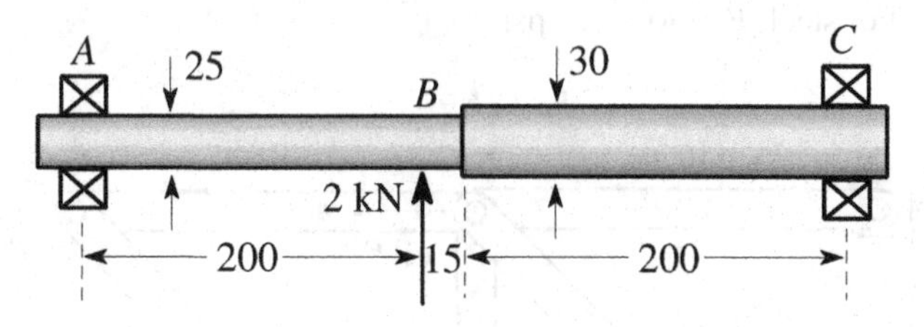


all dimensions in mm

*Figure P3.43*

**3.44.** Figure P3.44 shows a tool clip which is made of $1 \times \frac{1}{32}$ inch spring steel plate, for which $E = 30 \times 10^6$ psi, $\nu = 0.3$. The separation between the grips is $\frac{7}{8}$ inch in the undeformed position and diameter of the tool handle is 1 inch. If the coefficient of friction is 0.3, what is the maximum weight of tool that can be supported in a vertical position.

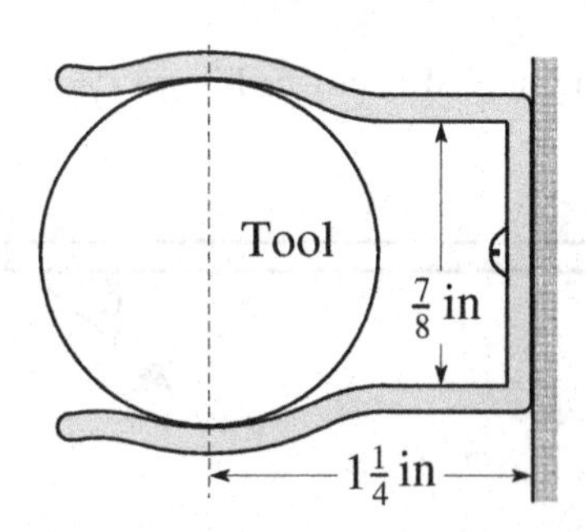


*Figure P3.44*

**3.45.** Figure P3.45 shows a beam of length $L$ and flexural rigidity $EI$, which is simply supported at $B$ and supported by a spring of stiffness $k$ at $A$. Find the slope of the beam at $A$ when the beam is subjected to a uniformly distributed load $w_0$ per unit length.

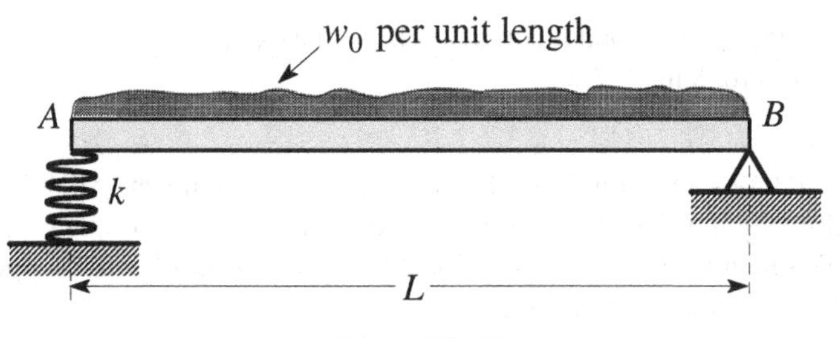


Figure P3.45

**3.46.** A student, asked to compute the downward vertical displacement $u_C$ at $C$ for the beam in Figure P3.46, equates it to the partial derivative $\partial U/\partial F$, where $U$ is the strain energy of the beam. Why is this an incorrect application of Castigliano's second theorem? Use the theorem to calculate the correct value of $u_C$, if the beam has flexural rigidity $EI$ and length $2L$. What is the physical interpretation of the incorrect result $\partial U/\partial F$?

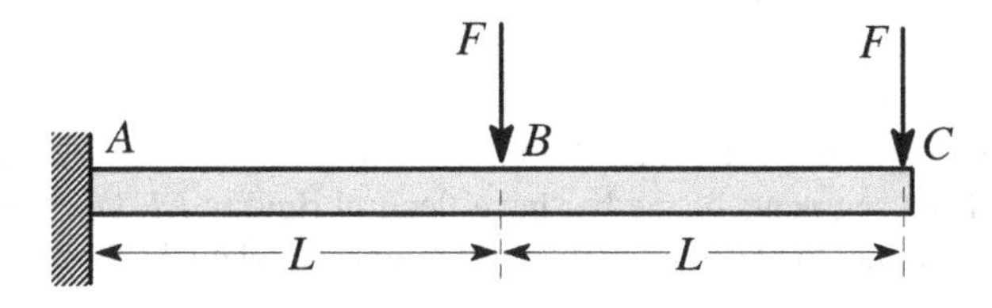


Figure P3.46

**3.47.** A thin semi-circular beam of radius $R$ and flexural rigidity $EI$ is subjected to a vertical force $F$ at the end, as shown in Figure P3.47. Find the horizontal and vertical components ($u_A, v_A$ respectively) of the displacement at the point $A$.

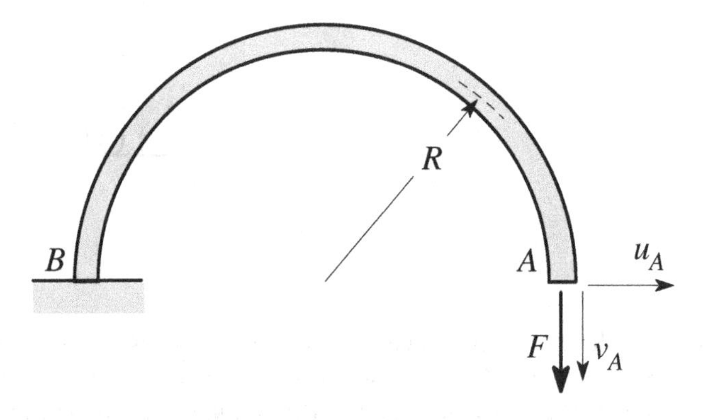


Figure P3.47

**3.48.** Use Castigliano's second theorem to obtain the exact solution for the change in diameter $BD$ in Problem 3.27.

**3.49.** Figure P3.49 shows a uniform beam subjected to a uniformly distributed load over its entire length, $L$. The beam rests on three equally spaced simple supports. Use Castigliano's second theorem to determine the distance $a$ between the supports if the load is to be shared equally between them.

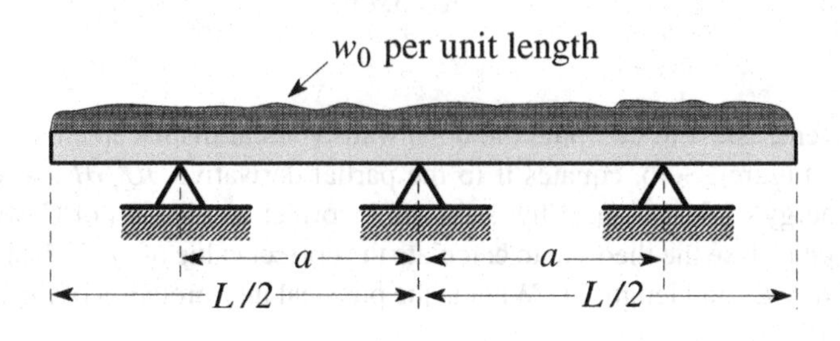


*Figure P3.49*

**3.50.** The right-angle bar of Figure P3.50 is built in at $A$ and rests on a frictionless support at $C$. Both segments of the bar have flexural rigidity $EI$. Use Castigliano's second theorem to find the horizontal displacement at $C$, when the bar is loaded by a moment $M_0$ at $B$.

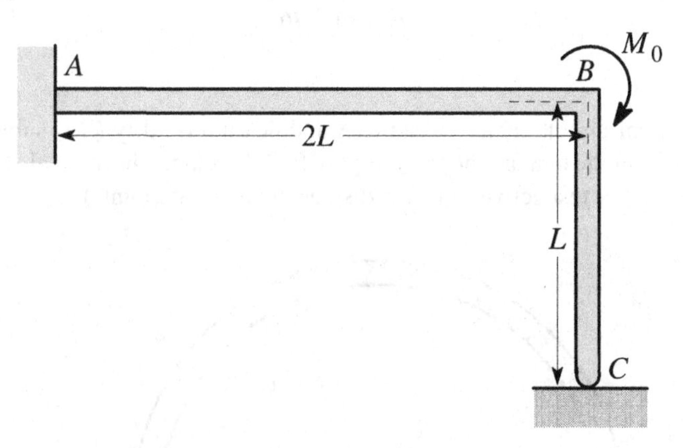


*Figure P3.50*

**3.51.** Figure P3.51 shows a semi-circular curved beam of radius $R$ and flexural rigidity $EI$, which is pinned at $A,C$ and subjected to a vertical load $F$ at $B$. Use Castigliano's second theorem to find the reactions at the supports and the displacement under the load.

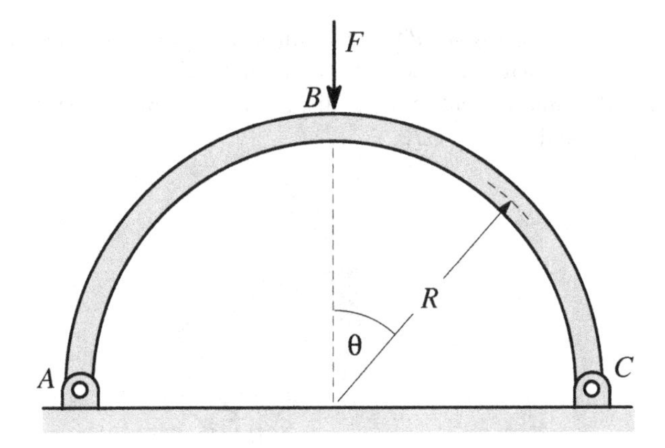


*Figure P3.51*

**3.52.** Use Castigliano's second theorem to find the reaction at the simple support in Problem 3.20 and hence to find the deflection at the free end of the beam.

**3.53.** The $L \times L$ square frame of Figure P3.53 is made from solid circular section bar of diameter $d$. It is suspended from the mid-point of the upper segment and loaded only by self-weight. Find the magnitude and location of the maximum tensile stress if the material has Young's modulus $E$ and density $\rho$.

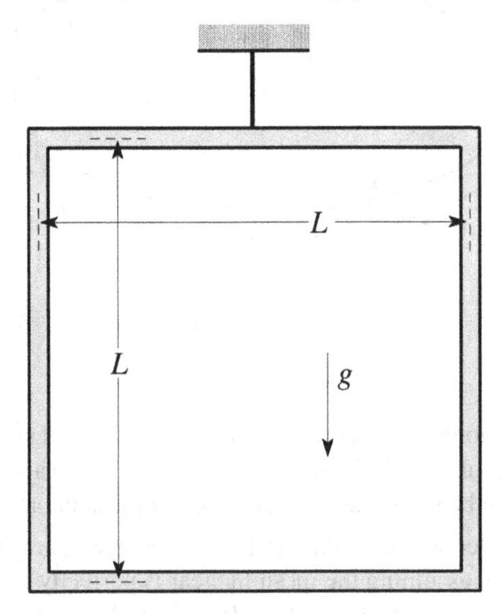


*Figure P3.53*

**3.54.** The curved beam of Figure P3.54 is built-in at $A$ and subjected to an out of plane load $F$ at $B$ as shown. Find the bending moment $M$ and torque $T$ as functions of $\theta$ and hence determine the out of plane displacement at $B$. The beam has flexural rigidity $EI$ and torsional rigidity $GJ$.

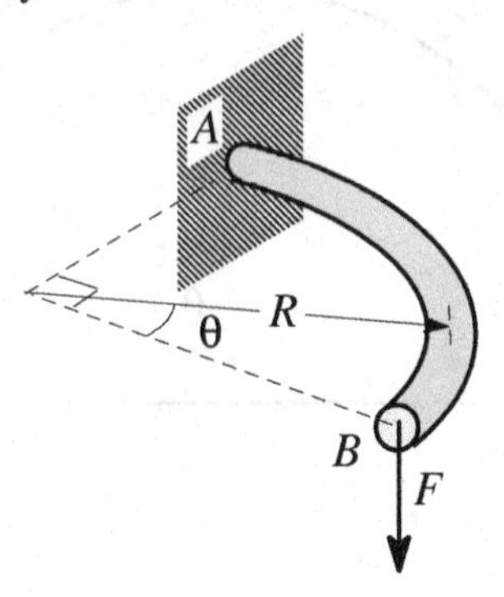


*Figure P3.54*

**3.55.** Figure P3.55 shows an open circular ring of mean radius 30 mm, whose cross section is a circle of diameter 2 mm. The two ends of the ring are subjected to equal and opposite 4 N forces out of the plane of the ring. Find the relative displacement of the two ends, if the ring is made of steel ($E = 210$ GPa, $\nu = 0.3$).

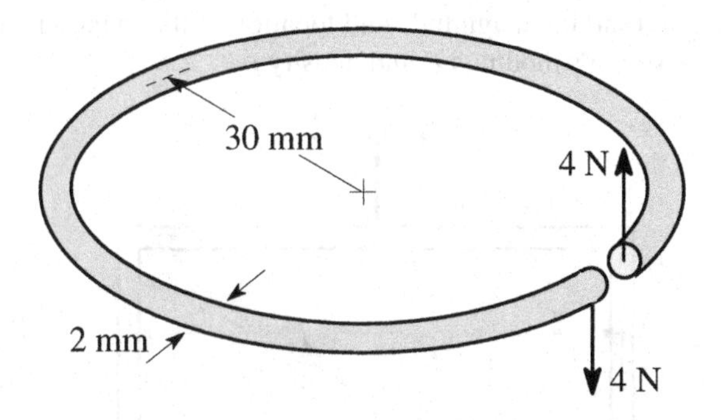


*Figure P3.55*

**3.56*.** Figure P3.56 shows two views of an outboard mounted spur gear that is subjected to a maximum force of 5 kN, which can be assumed to act half way along the length of the gear. The bearings can be treated as simple supports at their mid-point.

Use Castigliano's second theorem to estimate the misalignment of the gear teeth due to shaft deflection, assuming the meshing gear is perfectly aligned and does not deflect. The shaft is made of steel for which $E = 210$ GPa, $\nu = 0.3$. Which end of the gear will carry the most load as a result of this deformation?

Do you think your answer is sufficiently large to cause concern if the gear design were based on the assumption of uniform contact along the length?

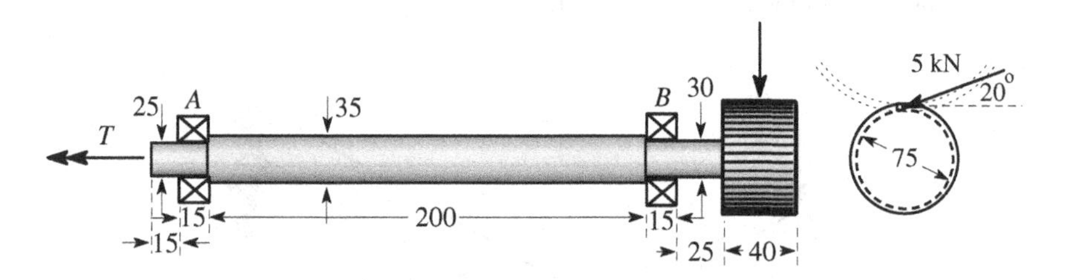


all dimensions in mm

*Figure P3.56*

**3.57*.** A long coil spring of mean coil diameter 20 mm is fabricated from steel wire of diameter 2 mm. The spring is close-coiled, which means that the adjacent coils touch in the unloaded state. The ends of the coil are now subjected to a bending moment $M$, as shown in Figure P3.57.

Find the effective flexural rigidity of the spring — i.e. the constant $K_B$ in the relation

$$M = \frac{K_B}{R} ,$$

where $R$ is the radius of curvature of the coil axis due to $M$. ($E = 210$ GPa, $\nu = 0.3$).

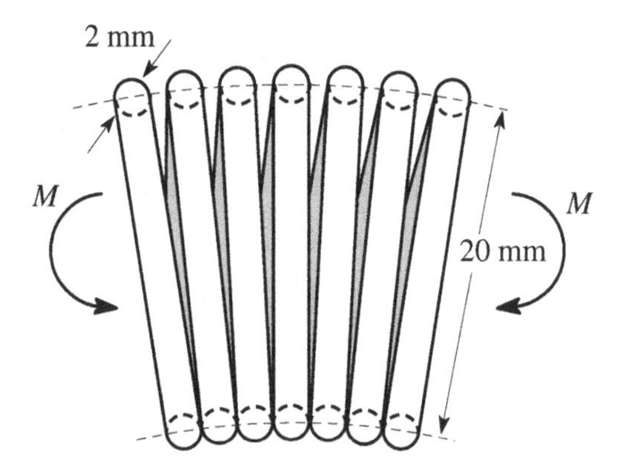


*Figure P3.57*

**3.58.** The right angle crank *ABC* of Figure P3.58 is made of steel bar of diameter 15 mm and is loaded by an out of plane force of 300 N at the end *C*.

Find the slope of the bar at *C*. ($E = 210$ GPa, $\nu = 0.3$).

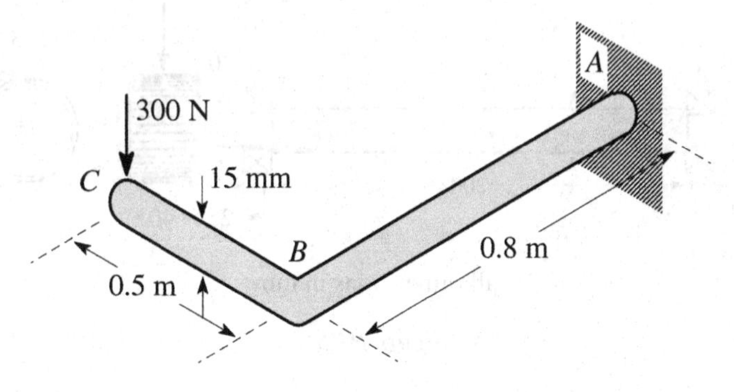


*Figure P3.58*

# 4

# Unsymmetrical Bending

The elementary theory of the bending of beams is restricted to the case where the beam has a cross section with at least one axis of symmetry. In the present chapter, we shall generalize the theory to beams of arbitrary cross section. All the remaining assumptions of the beam theory will remain unchanged.

## 4.1 Stress distribution in bending

Figure 4.1 shows a beam of arbitrary cross section $A$, whose axis is in the $z$-direction and which is loaded by an axial force $F$ and bending moments $M_x, M_y$. This figure also serves to define the sign convention for moments, which are regarded as positive when the moment vector acting on the positive $z$-surface[1] is directed along the positive $x$ or $y$ axes.

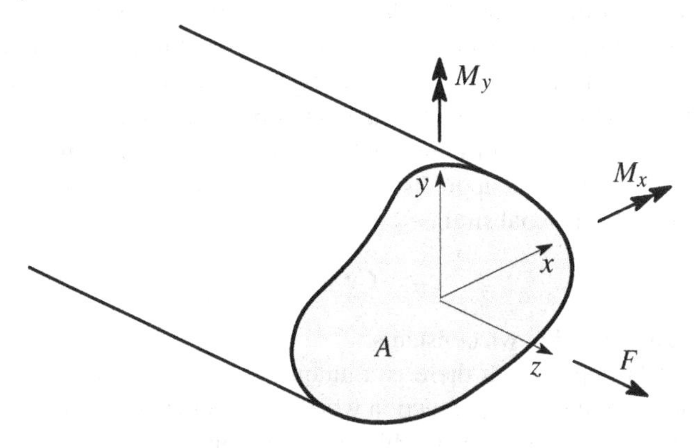


*Figure 4.1: Coordinate directions and sign convention for axial force and bending moments*

[1] i.e. the surface from which the $z$-direction is the outward normal.

J.R. Barber, *Intermediate Mechanics of Materials*, Solid Mechanics and Its Applications 175,
2nd ed., DOI 10.1007/978-94-007-0295-0_4, 

### 4.1.1 Bending about the $x$-axis only

The moments $M_x, M_y$ will generally cause the beam to bend about both the $x$ and $y$-axes, resulting in radii of curvature $R_x$, $R_y$ respectively. However, we first restrict attention to the simpler case in which there is no bending about $y$ — i.e. $R_y = \infty$, so that two orthogonal views of the bent beam will be as shown in Figure 4.2.

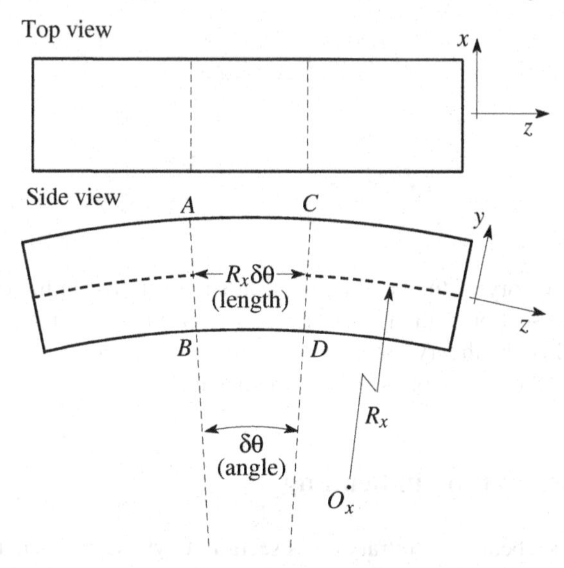


*Figure 4.2: Orthogonal views of the beam bent only about the x-axis*

We make the usual assumption of the beam theory that plane sections remain plane.[2] Thus, a segment of beam bounded by two parallel cross-sectional planes $AB$ and $CD$ in the undeformed state will deform as shown, such that there is a relative rotation between $AB$ and $CD$ of angle $\delta\theta$. If we extend the lines $AB, CD$ in Figure 4.2, they must therefore meet at a point $O_x$, defining the centre of curvature of the beam in the $yz$-plane. In the segment $ABCD$, imaginary fibres aligned in the $z$-direction that were initially straight and all of the same length will become deformed into arcs of circles whose lengths depend upon distance from this centre of curvature.

This implies that the axial strain $e_{zz}$ is a linear function of $y$ — i.e.

$$e_{zz} = Cy + D\,, \tag{4.1}$$

where $C, D$ are as yet unknown constants.

Equation (4.1) implies that there is a unique value of $y$ at which $e_{zz} = 0$. This defines a plane in the undeformed section which we call the *neutral plane*. Since the origin of coordinates can be chosen arbitrarily, it is convenient to choose it to lie in the neutral plane, which is then defined by the equation $y = 0$. Equation (4.1) then simplifies to

[2] We shall re-examine this assumption in §5.1 below in the more general context of an arbitrary non-linear constitutive law.

$$e_{zz} = Cy\,. \tag{4.2}$$

We also choose to define the radius of curvature $R_x$ as the distance from the centre of curvature $O_x$ to the neutral plane.

With these conventions, we can now determine the strain at any other point in the section. We first note that, by definition of the coordinate system, the original length of the fibre at $y=0$ is the same as its final length $R_x\delta\theta$ and this must be the original length of *all* the fibres in the segment $ABCD$, since $AB, CD$ were originally parallel. A fibre which is a distance $y$ from the neutral axis must deform into a circular segment of radius $(R_x+y)$ and hence have a deformed length $(R_x+y)\delta\theta$. The axial strain of such a fibre is therefore

$$e_{zz} = \frac{(R_x+y)\delta\theta - R_x\delta\theta}{R_x\delta\theta} = \frac{y}{R_x}\,, \tag{4.3}$$

showing that the constant $C$ in (4.2) is equal to $1/R_x$.

If we assume that the normal stress components in the plane of the section are zero ($\sigma_{xx}=\sigma_{yy}=0$), we then have

$$\sigma_{zz} = \frac{Ey}{R_x}\,, \tag{4.4}$$

where $E$ is Young's modulus. In other words, the axial stress is linearly proportional to the distance from the neutral axis, as in the elementary theory.

### 4.1.2 Bending about the $y$-axis only

Now suppose that instead of bending about the $x$-axis, we bend about the $y$-axis, with radius of curvature $R_y$. It is convenient to rotate the original view (Figure 4.1), so that the axis of bending points in the same direction as in the previous section (see Figure 4.3). Notice, however, how this now makes the $x$-axis point downwards in the side view as shown in Figure 4.4.

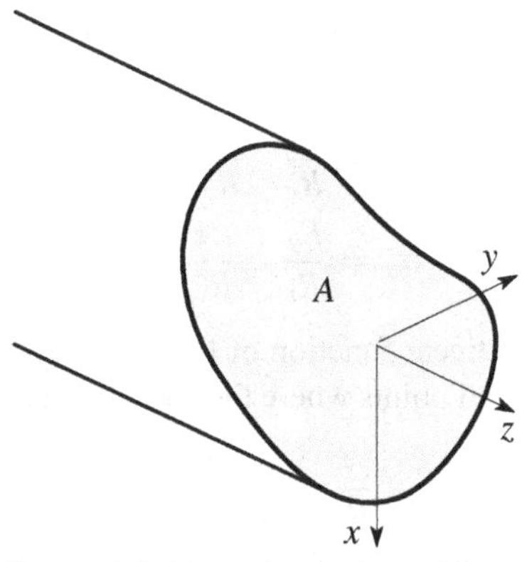


*Figure 4.3: Beam bent about the y-axis*

The convention used here is that in both cases, the curvature is such as would be produced by a positive bending moment as defined in Figure 4.1.

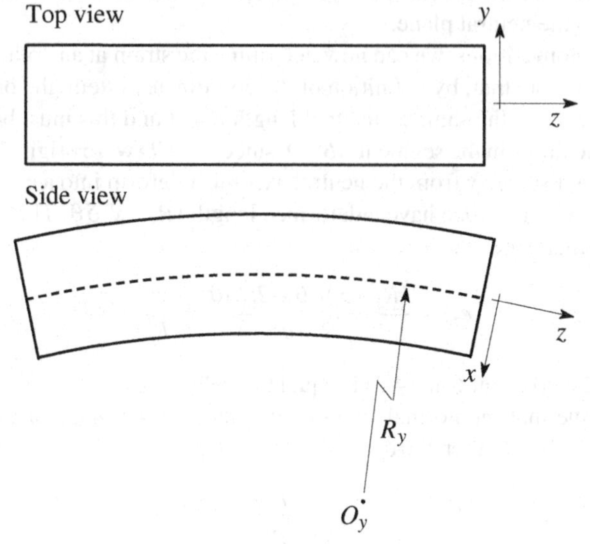


*Figure 4.4: Beam bent about the y-axis*

In other respects, the analysis is exactly similar to that for bending about the $x$-axis and leads to the results

$$e_{zz} = -\frac{x}{R_y} \; ; \; \sigma_{zz} = -\frac{Ex}{R_y} \; . \tag{4.5}$$

The negative sign in these equations arises because the $x$-axis now points downwards (i.e. towards the centre of curvature, rather than away from it), so that points with positive $x$ are in compression rather than tension.

### 4.1.3 Generalized bending

The most general kind of bending will involve curvature about both axes and can be obtained by superposing these results — i.e.

$$e_{zz} = \frac{y}{R_x} - \frac{x}{R_y} \tag{4.6}$$

$$\sigma_{zz} = \frac{Ey}{R_x} - \frac{Ex}{R_y} \; . \tag{4.7}$$

Notice that $\sigma_{zz}$ is now a linear function of both $x$ and $y$. We can again define a neutral plane as the locus of all points where the strain $e_{zz}$ and hence the stress $\sigma_{zz}$ is zero. This occurs when

$$\frac{y}{R_x} = \frac{x}{R_y} \; , \tag{4.8}$$

from (4.6) or (4.7). This equation is satisfied at the origin $x = y = 0$ and defines a plane surface which is generally inclined to both $x$ and $y$-axes.

### 4.1.4 Force resultants

Consider now the forces and moments on the beam cross section, due to the stress $\sigma_{zz}$ defined by equation (4.7). There will in general be an axial force $F$ as well as bending moments $M_x, M_y$ about both $x$ and $y$-axes.

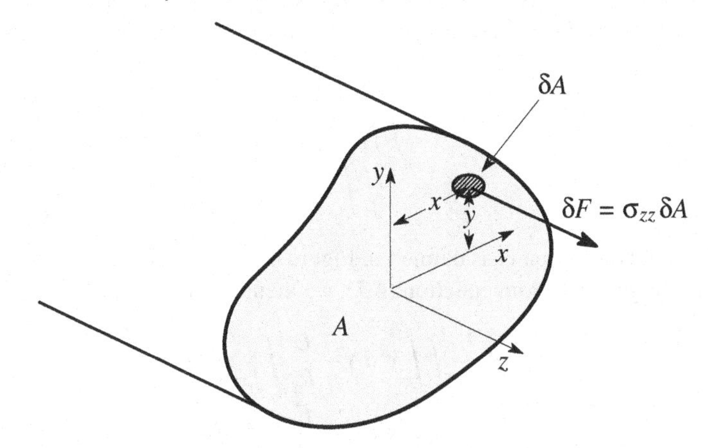


*Figure 4.5: Elemental force due to $\sigma_{zz}$ acting over $\delta A$*

The normal stress $\sigma_{zz}$ acting over the elemental area $\delta A$ causes an elemental force $\delta F = \sigma_{zz}\delta A$ in the $z$-direction, as shown in Figure 4.5. The total axial force can be found by summing these elemental forces over the cross section, with the result

$$F = \int\int_A \sigma_{zz} dA \tag{4.9}$$

$$= \frac{E}{R_x}\int\int_A ydA - \frac{E}{R_y}\int\int_A xdA \tag{4.10}$$

from (4.7). Notice that the summation leads to a double integral, since it is performed over an area. For example, the element of area $dA$ can be expressed in Cartesian coordinates as the elemental rectangle $dxdy$ and implies integration with respect to both $x$ and $y$ over the area defined by the cross section.

It is convenient to decompose the general bending problem into (i) a problem of pure bending ($F = 0$) and (ii) a problem with purely axial loading ($M_x = M_y = 0$) and then superpose the results. For the axial loading problem (ii), we assume that initially plane sections remain parallel after deformation, in which case the stress is uniform and given by

$$\sigma_{zz} = \frac{F}{A} . \tag{4.11}$$

Notice that the resultant of this uniform distribution must act through the centroid of the area $A$.

For the pure bending problem (i), the condition $F = 0$ can conveniently be satisfied by choosing the coordinate system such that

$$\iint_A x dA = 0 \;;\; \iint_A y dA = 0 \tag{4.12}$$

[see equation (4.10)], which is equivalent to choosing the origin of coordinates to coincide with the centroid of the section $A$.

The bending moments $M_x, M_y$ are determined by summing the moments of the elemental forces $\sigma_{zz}dA$ about the $x$ and $y$-axes respectively in Figure 4.5, with the result

$$M_x = \iint_A \sigma_{zz} y dA \tag{4.13}$$

$$M_y = -\iint_A \sigma_{zz} x dA \,, \tag{4.14}$$

where the sign convention is as defined in Figure 4.1.

Substituting for $\sigma_{zz}$ from equation (4.7), we then obtain

$$M_x = \frac{E}{R_x}\iint_A y^2 dA - \frac{E}{R_y}\iint_A xy dA$$

$$M_y = \frac{E}{R_y}\iint_A x^2 dA - \frac{E}{R_x}\iint_A yx dA \,.$$

Introducing the notation

$$I_x = \iint_A y^2 dA \;;\; I_y = \iint_A x^2 dA \;;\; I_{xy} = \iint_A xy dA \,, \tag{4.15}$$

these equations can be written

$$\frac{M_x}{E} = \frac{I_x}{R_x} - \frac{I_{xy}}{R_y} \tag{4.16}$$

$$\frac{M_y}{E} = \frac{I_y}{R_y} - \frac{I_{xy}}{R_x} \,. \tag{4.17}$$

### 4.1.5 Uncoupled problems

Notice that the first term in each of equations (4.16, 4.17) is the same as that obtained in the elementary theory of bending for symmetric beams. The quantity $I_{xy}$ (known as the *product second moment of area* or *product inertia*) introduces coupling between bending about the two axes. For example, a bending moment $M_x$ will generally produce some bending about the $y$ axis as well as the $x$-axis. However, in the special case where $I_{xy} = 0$, there is no coupling and the general bending problem reduces to the superposition of two applications of the elementary theory — one for each moment component $M_x, M_y$.

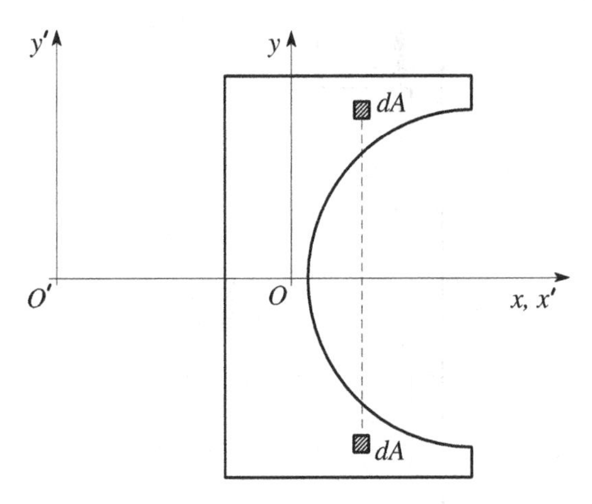


*Figure 4.6: Section for which $I_{xy} = 0$ due to symmetry*

We shall show in §4.4 below that all beam sections posess a pair of orthogonal axes about which the product inertia is zero, but a commonly occurring case is that in which either $Ox$ or $Oy$ is an axis of symmetry for the cross section. If this condition is satisfied, for every element of area $dA$ in the integral definition (4.15) there is an equal element $dA$ at the symmetric point for which one coordinate is the same and the other is equal and opposite. Thus, for example, the two elements $dA$ identified in the section of Figure 4.6 make equal and opposite contributions to $I_{xy}$ about any axis system that includes the axis of symmetry $Ox$. It follows that the product inertia is zero for this section about both of the systems $Oxy$, $O'x'y'$.

**Example 4.1**

*The channel section of Figure 4.7 is subjected to a bending moment with components $M_x = 8$ kNm, $M_y = 4$ kNm. Find the maximum tensile stress due to bending, if the appropriate second moments of area are $I_x = 8.87 \times 10^6$ mm$^4$, $I_y = 0.403 \times 10^6$ mm$^4$.*

Since $I_{xy} = 0$, there is no coupling and equations (4.16, 4.17) give[3]

$$\frac{E}{R_x} = \frac{M_x}{I_x} = \frac{8 \times 10^6}{8.87 \times 10^6} = 0.902 \text{ N/mm}^3$$

$$\frac{E}{R_y} = \frac{M_y}{I_y} = \frac{4 \times 10^6}{0.403 \times 10^6} = 9.926 \text{ N/mm}^3 \ .$$

Substitution in equation (4.7) then gives

$$\sigma_{zz} = 0.902y - 9.926x$$

in MPa, if $x, y$ are in mm.

[3] Notice that 1 MPa=1N/mm$^2$. Thus, if all quantities are expressed in the units of N and mm, the final expression for stress will be in MPa.

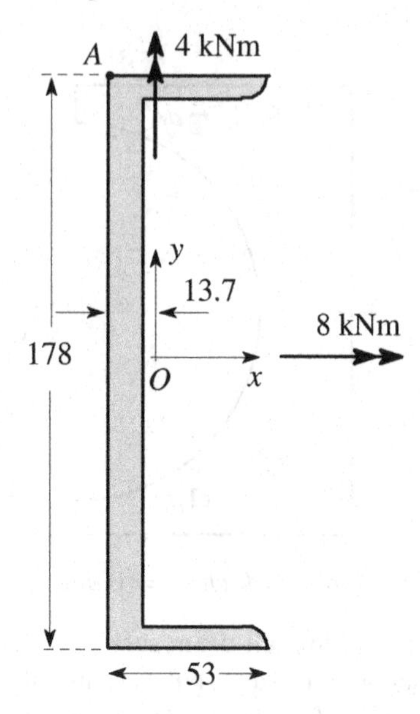


*Figure 4.7*

The maximum tensile stress will occur when $y$ is at its maximum and $x$ at its greatest negative value — i.e. at the point $A$ in Figure 4.7 defined by

$$x_A = -13.7 \text{ mm} \quad ; \quad y_A = 89 \text{ mm} .$$

(Remember that $x, y$ must be measured from the centroid.)

It follows that the maximum tensile stress is

$$\sigma_{\text{max}} = 0.902 \times 89 - 9.926 \times (-13.7) = 216.3 \text{ MPa} .$$

### 4.1.6 Coupled problems

If $I_{xy} \neq 0$, equations (4.16, 4.17) must be solved simultaneously for $R_x, R_y$ before substitution in equation (4.7). This can be done in symbolic terms, with the result

$$\frac{E}{R_x} = \frac{M_x I_y + M_y I_{xy}}{(I_x I_y - I_{xy}^2)} \tag{4.18}$$

$$\frac{E}{R_y} = \frac{M_y I_x + M_x I_{xy}}{(I_x I_y - I_{xy}^2)} , \tag{4.19}$$

after which equation (4.7) gives

$$\sigma_{zz} = \frac{(M_x I_y + M_y I_{xy})y - (M_y I_x + M_x I_{xy})x}{(I_x I_y - I_{xy}^2)} . \tag{4.20}$$

In particular, the general equation for the neutral axis ($\sigma_{zz}=0$) is

$$(M_x I_y + M_y I_{xy})y - (M_y I_x + M_x I_{xy})x = 0 . \tag{4.21}$$

We shall show in §4.4.2 below that the expression $(I_x I_y - I_{xy}^2)$, which appears in equations (4.18–4.20) can never be zero — in fact it is always positive — so neither the curvature nor the bending stress can be unbounded.

**Example 4.2**

*Figure 4.8 shows the cross section of a beam of length 2 m which is built in at one end, the other end being subjected to a transverse load of 3 kN in the direction shown. The centroid is located at the point O as shown and the appropriate second moments of area are $I_x = 1.33\times10^6$ mm$^4$, $I_y = 0.917\times10^6$ mm$^4$, $I_{xy} = 0.030\times10^6$ mm$^4$. Find the inclination of the neutral axis and hence the location and magnitude of the maximum tensile and compressive stresses.*

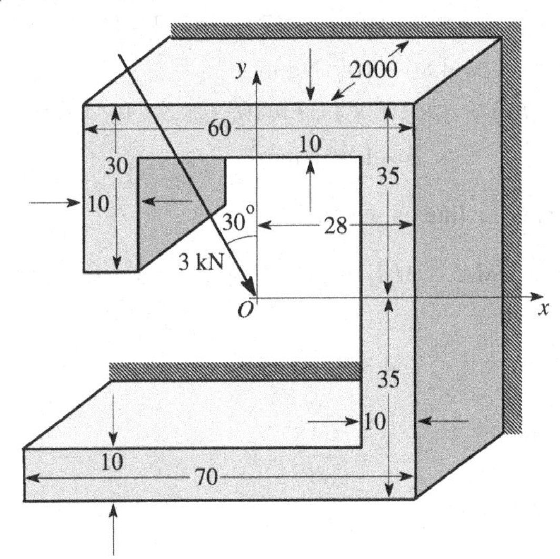


all dimensions in mm.

*Figure 4.8*

The maximum bending moment will occur at the built-in end and is

$$M = 3\text{ kN} \times 2\text{ m} = 6\text{ kNm}$$

about an axis inclined at 30° to the $x$-axis, as shown in Figure 4.9. This in turn can be resolved into components

$$M_x = M\cos 30^\circ = 5.2\text{ kNm} = 5.2\times10^6\text{ Nmm}$$

$$M_y = M\sin 30^\circ = 3\text{ kNm} = 3\times10^6\text{ Nmm} .$$

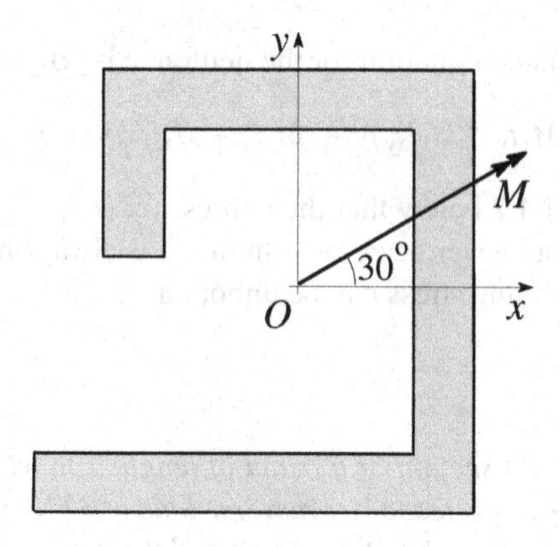


*Figure 4.9*

Evaluating the separate terms in (4.21), we have

$$\begin{aligned}(M_xI_y + M_yI_{xy}) &= 5.2 \times 10^6 \times 0.917 \times 10^6 + 3 \times 10^6 \times 0.03 \times 10^6 \\ &= 4.86 \times 10^{12}\ \text{Nmm}^5 \\ (M_yI_x + M_xI_{xy}) &= 3 \times 10^6 \times 1.33 \times 10^6 + 5.2 \times 10^6 \times 0.03 \times 10^6 \\ &= 4.15 \times 10^{12}\ \text{Nmm}^5\ .\end{aligned}$$

The neutral axis is the line on which

$$(M_xI_y + M_yI_{xy})y - (M_yI_x + M_xI_{xy})x = 0$$

— i.e.

$$4.86y - 4.15x = 0$$

and hence

$$\frac{y}{x} = \frac{4.15}{4.86} = 0.85\ .$$

This defines a line inclined at

$$\psi = \tan^{-1} 0.85 = 40.5^\circ$$

to the $x$-axis, as shown in Figure 4.10.

We also have

$$I_xI_y - I_{xy}^2 = (1.33 \times 0.917 - 0.03^2) \times 10^{12} = 1.219 \times 10^{12}\ \text{mm}^4\ .$$

The maximum tensile and compressive stresses occur at the points furthest from the neutral axis — i.e. at $A$ and $B$ in Figure 4.10. The coordinates of $A$ are

$$x_A = 28 - 60 = -32\ \text{mm}\ \ ;\ \ y_A = 70 - 35 = 35\ \text{mm}\ ,$$

whilst for $B$

$$x_B = 28\ \text{mm}\ \ ;\ \ y_B = -35\ \text{mm}\ .$$

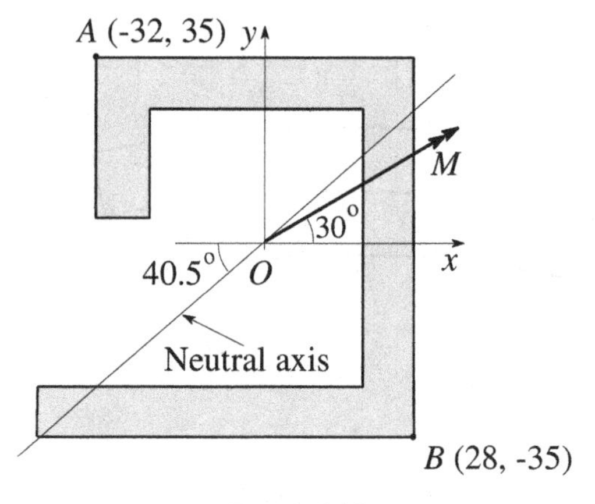


*Figure 4.10*

Substituting into equation (4.20), we therefore have

$$\sigma_A = \frac{4.86\times 10^{12}\times 35 - 4.15\times 10^{12}\times(-32)}{1.219\times 10^{12}}$$
$$= 248 \text{ MPa (maximum tensile stress)}$$

$$\sigma_B = \frac{4.86\times 10^{12}\times(-35) - 4.15\times 10^{12}\times 28}{1.219\times 10^{12}}$$
$$= -286 \text{ MPa (maximum compressive stress)}\,.$$

## 4.2 Displacements of the beam

In general, we expect the beam to have displacement components $u_x, u_y$ in both transverse directions and, as in the elementary theory, these are related to the corresponding radii of curvature through the equations

$$\frac{d^2u_y}{dz^2} = -\frac{1}{R_x} \quad ; \quad \frac{d^2u_x}{dz^2} = \frac{1}{R_y}\,. \tag{4.22}$$

The sign convention here is chosen to conform with that introduced in Figures 4.2, 4.3 for $R_x, R_y$, which is repeated here for convenience in the orthogonal views of Figure 4.11 *(a,b)*. Notice that $u_x, u_y$ are positive in the direction of the coordinate axes $x, y$ respectively and a positive value of the second derivative $d^2u/dz^2$ implies an increase of slope $du/dz$ as we move in the $z$-direction.

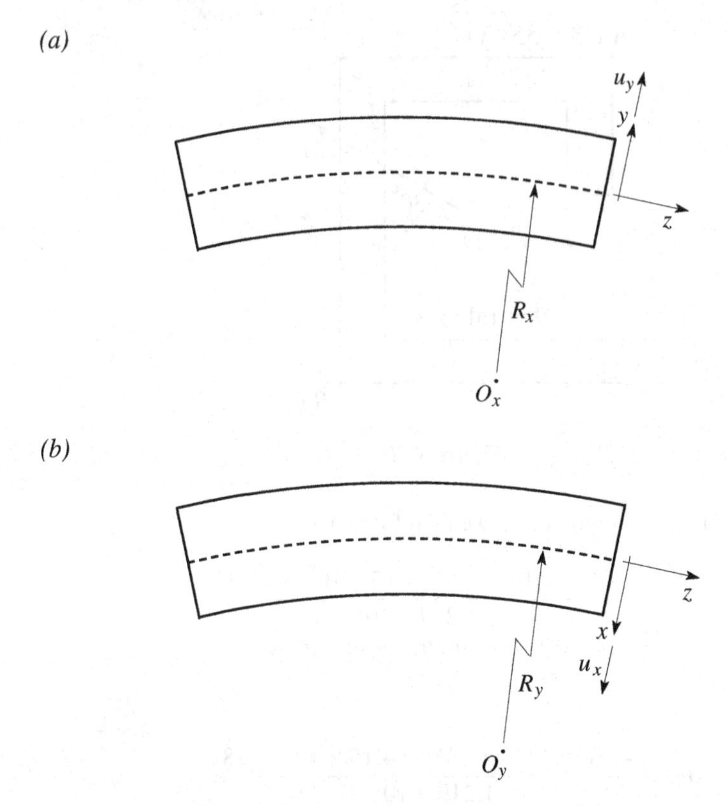


*Figure 4.11: Sign convention for radii of curvature and displacement components*

Substituting for $R_x, R_y$ from equations (4.18, 4.19), we obtain

$$\frac{d^2u_y}{dz^2} = -\frac{M_xI_y + M_yI_{xy}}{E(I_xI_y - I_{xy}^2)} \tag{4.23}$$

$$\frac{d^2u_x}{dz^2} = \frac{M_yI_x + M_xI_{xy}}{E(I_xI_y - I_{xy}^2)} \ . \tag{4.24}$$

If the problem is determinate, $M_x, M_y$ can be found from equilibrium considerations and equations (4.23, 4.24) can be integrated directly, as in the elementary theory. The displacement components $u_x, u_y$ can therefore be found by independent calculations. Notice however that if $I_{xy} \neq 0$, forces in the $x$-direction will tend to produce displacements in the $y$-direction and *vice versa*.

**Example 4.3**

*An L-section beam with second moments* $I_x = 1,512,500$ *mm*$^4$, $I_y = 412,500$ *mm*$^4$, $I_{xy} = -450,000$ *mm*$^4$ *and length* 2 *m is built in at one end and loaded by a vertical distributed load of* 2 *kN/m as shown in Figure 4.12. If the beam is made of steel with Young's modulus* 210 *GPa, find the displacement components at the free end B.*

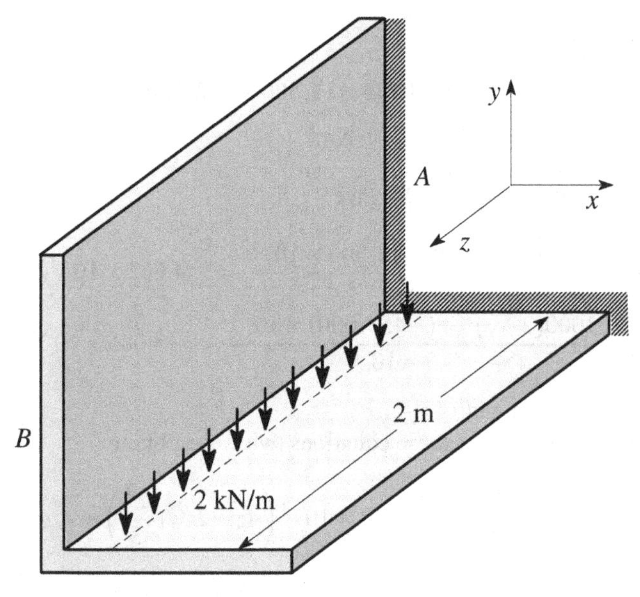


*Figure 4.12*

We cut the beam at a distance $z$ from $A$ and draw a free body diagram, as shown in Figure 4.13. Notice that $M_x$ acts in the negative $x$-direction on this cut, since the cut surface is a negative $z$-plane — i.e. the $z$-direction is the *inward* normal.

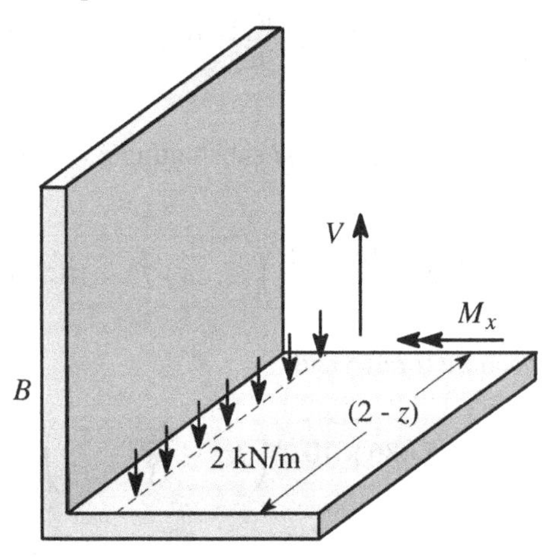


*Figure 4.13*

Equilibrium arguments then yield the results

$$M_x = 2000 \times (2-z) \times \frac{(2-z)}{2} = 1000(2-z)^2 \text{ Nm}$$

$$M_y = 0\,.$$

We also have

$$E(I_x I_y - I_{xy}^2) = 210 \times 10^9 (1{,}512{,}500 \times 412{,}500 - 450{,}000^2) \times 10^{-24}$$
$$= 88.5 \times 10^{-3}\ \mathrm{Nm}^6$$

and hence equations (4.23, 4.24) give

$$\frac{d^2 u_y}{dz^2} = -\frac{1000(2-z)^2 \times 412{,}500 \times 10^{-12}}{88.5 \times 10^{-3}} = -4.662 \times 10^{-3}(2-z)^2\ \mathrm{m}^{-1}$$
$$\frac{d^2 u_x}{dz^2} = \frac{1000(2-z)^2 \times (-450{,}000) \times 10^{-12}}{88.5 \times 10^{-3}} = -5.086 \times 10^{-3}(2-z)^2\ \mathrm{m}^{-1}\ ,$$

where $u_x, u_y, z$ are in metres.

Integrating the first of these equations twice, we obtain

$$\frac{du_y}{dz} = -4.662 \times 10^{-3}\left(4z - 2z^2 + \frac{z^3}{3}\right) + A$$

$$u_y = -4.662 \times 10^{-3}\left(2z^2 - \frac{2z^3}{3} + \frac{z^4}{12}\right) + Az + B\ ,$$

where $A, B$ are two arbitrary constants.

The beam is built in at $z=0$ and hence

$$u_y(0) = 0\ \ ;\ \ \frac{du_y}{dz}(0) = 0\ ,$$

from which $A=B=0$.

The end deflection is then obtained by substituting $z=2$ into the above expression, with the result

$$u_y(2) = -4.662 \times 10^{-3}\left(28 - \frac{16}{3} + \frac{16}{12}\right) = -49.73 \times 10^{-3}\ \mathrm{m} = -49.73\ \mathrm{mm}\ .$$

An exactly similar procedure for $u_x$ yields

$$u_x = -5.086 \times 10^{-3}\left(2z^2 - \frac{2z^3}{3} + \frac{z^4}{12}\right)$$

and hence

$$u_x(2) = -54.25\ \mathrm{mm}\ .$$

Notice that the beam experiences a horizontal displacement, even though the loading is purely vertical and in fact the horizontal displacement exceeds the vertical displacement. Further insight into this behaviour is provided by the arguments of §4.4.4 below.

## 4.3 Second moments of area

The procedure developed so far requires that the second moments of area be given. The properties of common structural sections such as I-beams, channels and angles are standardized and tabulated in reference works, but many engineering applications involve non-standard beam sections for which the second moments must be calculated.

The reader should be familiar with methods for evaluating $I_x, I_y$, from the elementary theory of bending. However, since we need to extend these methods to include the product inertia $I_{xy}$, we shall take the opportunity to give a brief review of the complete procedure.

The direct method is to evaluate the double integrals in equation (4.15). This process is demonstrated in Appendix B, but we seldom need to use it, since most engineering beam sections can be decomposed into rectangular and circular sub-areas, for which the appropriate properties are tabulated. However, these quantities need to be referred to a set of axes through the centroid of the whole section, which necessitates first finding the location of the centroid and then transferring the second moments to centroidal axes using the parallel axis theorem.

### 4.3.1 Finding the centroid

Consider the area $A$ of Figure 4.14, defined in the coordinate system $O'x'y'$. We construct a parallel set of axes $Oxy$ through the as yet unknown centroid $O$, defined such that

$$\iint_A x dA = 0 \quad ; \quad \iint_A y dA = 0 \,. \tag{4.25}$$

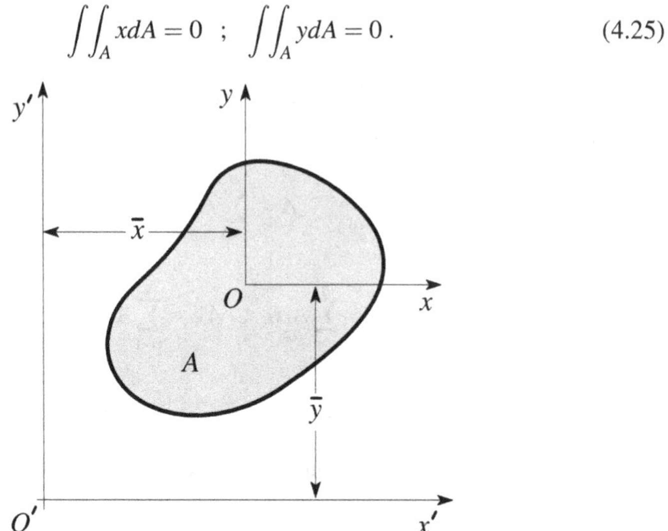


*Figure 4.14: Centroidal and non-centroidal axes*

From the geometry of Figure 4.14, the coordinates of a general point $x', y'$ can be written

$$x' = \bar{x} + x \ ; \ y' = \bar{y} + y , \tag{4.26}$$

where $\bar{x}, \bar{y}$ are the $x', y'$-coordinates of the centroid $O$. It follows that

$$\iint_A x' dA = \iint_A (\bar{x} + x) dA = \bar{x} \iint_A dA + \iint_A x dA , \tag{4.27}$$

since $\bar{x}$ does not depend on the position of the elemental area $dA$.

The second integral in equation (4.27) is zero from (4.25) and

$$\iint_A dA = A ,$$

so

$$\iint_A x' dA = A\bar{x} \quad \text{or} \quad \bar{x} = \frac{1}{A} \iint_A x' dA . \tag{4.28}$$

By a similar argument

$$\iint_A y' dA = A\bar{y} \quad \text{or} \quad \bar{y} = \frac{1}{A} \iint_A y' dA . \tag{4.29}$$

If the area $A$ consists of a number of sub-areas $A_1, A_2, A_3$, etc., whose areas and centroids are already known, we can write

$$\iint_A x' dA = \iint_{A_1} x' dA + \iint_{A_2} x' dA + \iint_{A_3} x' dA + \ldots$$

and hence

$$A\bar{x} = A_1\bar{x}_1 + A_2\bar{x}_2 + A_3\bar{x}_3 + \ldots ,$$

where $\bar{x}_1, \bar{x}_2, \bar{x}_3$ are the $x'$ coordinates of the centroids of the separate areas $A_1, A_2, A_3$ respectively. More generally, if

$$A = \sum_{i=1}^{n} A_i , \tag{4.30}$$

then

$$A\bar{x} = \sum_{i=1}^{n} A_i\bar{x}_i \ ; \ A\bar{y} = \sum_{i=1}^{n} A_i\bar{y}_i . \tag{4.31}$$

### 4.3.2 The parallel axis theorem

The second moment of area $I'_x$ about the *non-centroidal* axis $O'x'$ is defined as

$$I'_x = \iint_A y'^2 dA = \iint_A (y + \bar{y})^2 dA ,$$

from equation (4.26). Expanding the integrand and taking the constant $\bar{y}$ outside the integral sign, we obtain

$$I_x' = \int\int_A y^2 dA + 2\bar{y}\int\int_A y dA + \bar{y}^2 \int\int_A dA\,. \tag{4.32}$$

Now the second integral in (4.32) is zero from (4.25) and hence

$$I_x' = \int\int_A y^2 dA + \bar{y}^2 \int\int_A dA = I_x + A\bar{y}^2\,. \tag{4.33}$$

By a similar argument we also have

$$I_y' = I_y + A\bar{x}^2\,. \tag{4.34}$$

These results define the *parallel axis theorem* for second moments of area. They state that the second moment of area about any axis is equal to the corresponding second moment about a parallel axis through the centroid of the area plus the product of the area and the square of the distance between the two axes.

Since the terms $A\bar{x}^2, A\bar{y}^2$ are always positive, it follows that $I_y'$ is a minimum when $\bar{x}=0$ — i.e. when $O'y'$ coincides with $Oy$. In other words, *the centroidal axes give the minimum values of* $I_x', I_y'$. Notice that it is essential that the second moments $I_x, I_y$ on the right hand sides of equations (4.33, 4.34) be centroidal.

A similar theorem can be proved for the product inertia $I_{xy}$. Defining

$$I_{xy}' = \int\int_A x'y' dA = \int\int_A (x+\bar{x})(y+\bar{y}) dA\,,$$

we have on expanding the product

$$I_{xy}' = \int\int_A xy dA + \bar{x}\int\int_A y dA + \bar{y}\int\int_A x dA + \bar{x}\bar{y}\int\int_A dA\,.$$

The second and third integral are zero because of (4.25) and hence

$$I_{xy}' = I_{xy} + A\bar{x}\bar{y}\,. \tag{4.35}$$

It is *not* the case that $I_{xy}'$ is a maximum at $\bar{x}=\bar{y}=0$, since $\bar{x}\bar{y}$ can be positive or negative.

The parallel axis theorem can be used to determine the centroidal second moment of area for a composite area $A = \sum_{i=1}^{n} A_i$. Suppose the centroid of the sub-area $A_i$ is located at $\bar{x}_i, \bar{y}_i$ and that it has second moments of area $I_x^i, I_y^i, I_{xy}^i$ *about it's own centroidal axes*. It follows from equations (4.33–4.35) that the contribution of $A_i$ to the second moments about the communal centroidal axes through $O(\bar{x}, \bar{y})$ will be

$$I_x^i + A_i(\bar{y}_i - \bar{y})^2,\quad I_y^i + A_i(\bar{x}_i - \bar{x})^2,\quad I_{xy}^i + A_i(\bar{x}_i - \bar{x})(\bar{y}_i - \bar{y})\,,$$

since the distance between the communal centroidal axes and the individual centroidal axes are $(\bar{x}_i - \bar{x})$ and $(\bar{y}_i - \bar{y})$.

The second moments for the complete area will therefore be

$$I_x = \sum_{i=1}^{n} I_x^i + \sum_{i=1}^{n} A_i(\bar{y}_i - \bar{y})^2 \tag{4.36}$$

$$I_y = \sum_{i=1}^{n} I_y^i + \sum_{i=1}^{n} A_i(\bar{x}_i - \bar{x})^2 \tag{4.37}$$

$$I_{xy} = \sum_{i=1}^{n} I_{xy}^i + \sum_{i=1}^{n} A_i(\bar{x}_i - \bar{x})(\bar{y}_i - \bar{y}) \, . \tag{4.38}$$

These expressions are conveniently evaluated by constructing a table, as in the following example.

**Example 4.4**

*Determine the location of the centroid and the centroidal second moments of area for the beam section shown in Figure 4.15.*

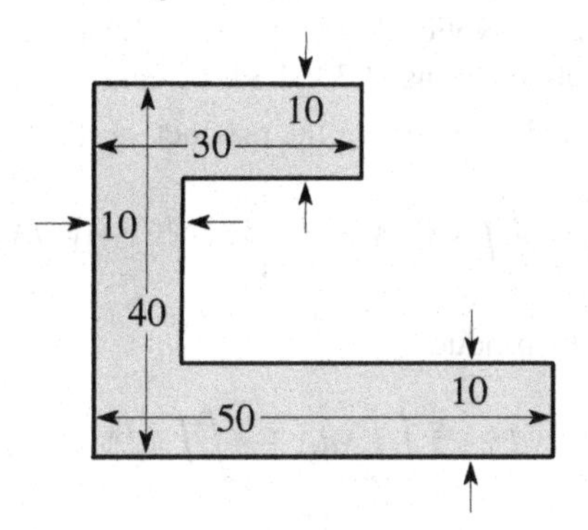


*all dimensions in mm.*

*Figure 4.15*

We first divide the area into sub-areas $A_1, A_2, A_3$, as in Figure 4.16.

For the rectangle $A_1$, we have

$$A_1 = 10 \times 50 = 500 \text{ mm}^2$$

and the centroid is located at the point $O_1$ with coordinates

$$\bar{x}_1 = 25 \;\; ; \;\; \bar{y}_1 = 5 \, ,$$

relative to the point $O'$.

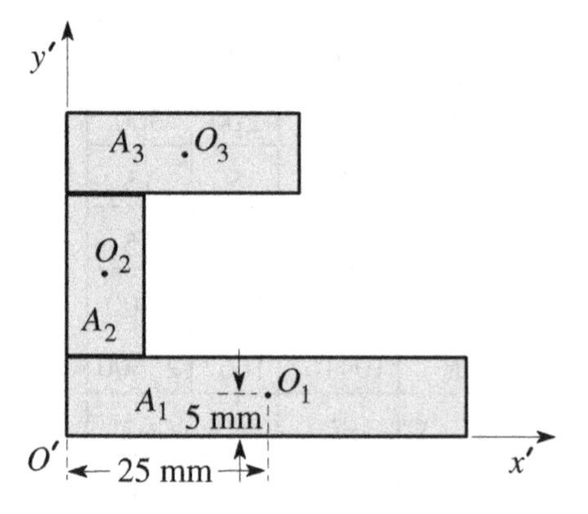


*Figure 4.16*

Similar arguments for the areas $A_2, A_3$ enable us to complete the following table:-

| $i$ | 1 | 2 | 3 |
|---|---|---|---|
| $A_i$ | 500 | 200 | 300 |
| $\bar{x}_i$ | 25 | 5 | 15 |
| $\bar{y}_i$ | 5 | 20 | 35 |

*Table 4.1*

It follows from equations (4.30, 4.31) that

$$A = 500 + 200 + 300 = 1000 \text{ mm}^2$$

$$A\bar{x} = 500 \times 25 + 200 \times 5 + 300 \times 15 = 18{,}000 \text{ mm}^3$$

$$A\bar{y} = 500 \times 5 + 200 \times 20 + 300 \times 35 = 17{,}000 \text{ mm}^3$$

and hence

$$\bar{x} = 18 \text{ mm} \quad ; \quad \bar{y} = 17 \text{ mm} .$$

The centroidal second moment of area for a rectangle $b \times h$ is $bh^3/12$ (see Appendix B), so we have

$$I_x^1 = \frac{50 \times 10^3}{12} = 4167 \text{ mm}^4 \quad ; \quad I_y^1 = \frac{10 \times 50^3}{12} = 104167 \text{ mm}^4 ,$$

where we note that the dimension cubed is that perpendicular to the required axis.

Using similar arguments for $A_2, A_3$, we can therefore extend Table 4.1 to read

| $i$ | 1 | 2 | 3 |
|---|---|---|---|
| $A_i$ | 500 | 200 | 300 |
| $\bar{x}_i$ | 25 | 5 | 15 |
| $\bar{y}_i$ | 5 | 20 | 35 |
| $I_x^i$ | 4167 | 6667 | 2500 |
| $I_y^i$ | 104167 | 1667 | 22500 |
| $(\bar{x}_i - \bar{x})$ | 7 | $-13$ | $-3$ |
| $(\bar{y}_i - \bar{y})$ | $-12$ | 3 | 18 |

*Table 4.2*

and hence, substituting the tabulated values into (4.36, 4.37), we have

$$\begin{aligned} I_x &= 4167 + 6667 + 2500 + 500 \times (-12)^2 + 200 \times 3^2 + 300 \times 18^2 \\ &= 184{,}333 \text{ mm}^4 \\ I_y &= 104167 + 1667 + 22500 + 500 \times 7^2 + 200 \times (-13)^2 + 300 \times (-3)^2 \\ &= 189{,}333 \text{ mm}^4 \;. \end{aligned}$$

For the product inertia, we note that the individual rectangular areas have zero product inertia about their own centroids by symmetry, so

$$\begin{aligned} I_{xy} &= 500 \times 7 \times (-12) + 200 \times (-13) \times 3 + 300 \times (-3) \times 18 \\ &= -66{,}000 \text{ mm}^4 \;, \end{aligned}$$

from (4.38).

### 4.3.3 Thin-walled sections

Many structural applications involve beams in which the thickness of the material is small in comparison with the linear dimensions of the cross section. Examples include structural steel sections, such as the channel of Figure 4.7 above, and also beams formed by bending metal plate such as the angle irons available in hardware stores and those used for supporting roadside signs.

Calculations in such cases can be simplified by describing the section in terms of the mean line and the wall thickness, and by neglecting the ratio of thickness to linear dimensions in comparison with unity. This is particularly advantageous in situations where the general shape is known, but the dimensions have yet to be determined from design considerations, since the approximation permits managable general expressions to be obtained in symbolic form.

**Example 4.5**

*Determine the second moments of area $I_x, I_y, I_{xy}$ for the unequal Z-section of Figure 4.17(a), for which $t \ll a$. The section is defined by the dimensions of the mean line ABCD which is equidistant between the two sides of the wall. The corresponding mean line is shown for clarity in Figure 4.17(b).*

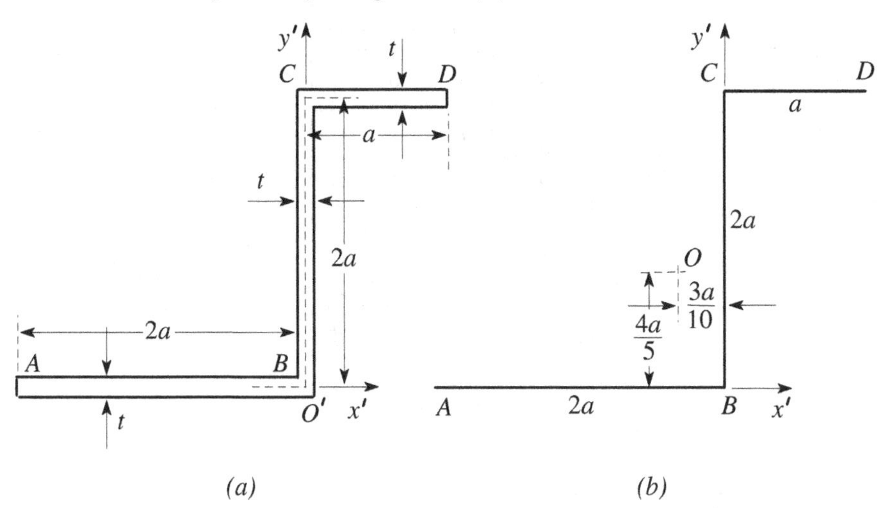


*Figure 4.17: (a) The unequal Z-section, (b) the equivalent mean line*

We choose to centre the coordinate system $O'x'y'$ on the point $B$, as shown in Figure 4.17 *(a)*. The area of each section is taken to be the product of the length of the mean line and the section thickness, which has the effect of replacing the actual corner section $CD$ of Figure 4.18 *(a)* by the overlapping rectangles of Figure 4.18 *(b)*. The error involved in this process is clearly of the order of $t/a$ and hence negligible.

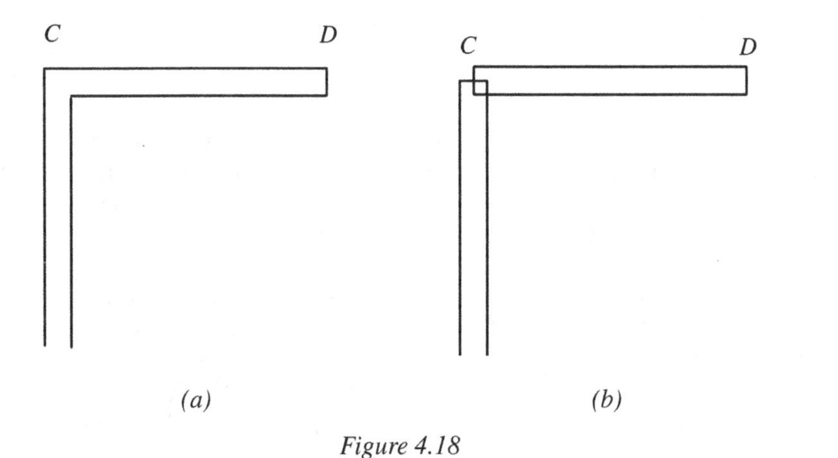


*Figure 4.18*

We identify the segments $AB$, $BC$, $CD$ as $A_1$,$A_2$,$A_3$ respectively, after which we can construct the first five rows of the table:-

| $i$ | 1 | 2 | 3 |
|---|---|---|---|
| segment | $AB$ | $BC$ | $CD$ |
| $A_i$ | $2at$ | $2at$ | $at$ |
| $\bar{x}_i$ | $-a$ | 0 | $\frac{a}{2}$ |
| $\bar{y}_i$ | 0 | $a$ | $2a$ |
| $I_x^i$ | 0 | $\frac{2a^3t}{3}$ | 0 |
| $I_y^i$ | $\frac{2a^3t}{3}$ | 0 | $\frac{a^3t}{12}$ |
| $(\bar{x}_i - \bar{x})$ | $-\frac{7a}{10}$ | $\frac{3a}{10}$ | $\frac{4a}{5}$ |
| $(\bar{y}_i - \bar{y})$ | $-\frac{4a}{5}$ | $\frac{a}{5}$ | $\frac{6a}{5}$ |

*Table 4.3*

Equations (4.30, 4.31) then give

$$A = 2at + 2at + at = 5at$$

$$A\bar{x} = -2a^2t + 0 + \frac{a^2t}{2} = -\frac{3a^2t}{2}$$

$$A\bar{y} = 0 + 2a^2t + 2a^2t = 4a^2t$$

and hence the coordinates of the centroid are

$$\bar{x} = -\frac{3a}{10} \quad ; \quad \bar{y} = \frac{4a}{5} ,$$

as shown in Figure 4.17(b).

We can now complete the remaining rows of the table, noting that the centroidal second moment of a thin rectangle about its mean line is negligible, since it is of the order $at^3 \ll a^3t$.

Using equations (4.36–4.38), we therefore obtain

$$I_x = \frac{2a^3t}{3} + 2at\left(-\frac{4a}{5}\right)^2 + 2at\left(\frac{a}{5}\right)^2 + at\left(\frac{6a}{5}\right)^2 = \frac{52a^3t}{15}$$

$$I_y = \frac{2a^3t}{3} + \frac{a^3t}{12} + 2at\left(-\frac{7a}{10}\right)^2 + 2at\left(\frac{3a}{10}\right)^2 + at\left(\frac{4a}{5}\right)^2 = \frac{51a^3t}{20}$$

and the product inertia is

$$I_{xy} = 2at\left(-\frac{7a}{10}\right)\left(-\frac{4a}{5}\right) + 2at\left(\frac{3a}{10}\right)\left(\frac{a}{5}\right) + at\left(\frac{4a}{5}\right)\left(\frac{6a}{5}\right) = \frac{11a^3t}{5} .$$

Notice how all these expressions reduce to a numerical multiplier on $a^3t$, since any other combinations of $a,t$ are small in the sense of $t \ll a$.

## 4.4 Further properties of second moments

In addition to the parallel axis theorem, there are relationships between the second moments of area about axes that are not parallel. These relations give further insight into the section properties and often enable us to predict how a beam will behave under given loading without performing more than the simplest calculations.

### 4.4.1 Coordinate transformation

Suppose we are given $I_x, I_y, I_{xy}$ in some coordinate system $Oxy$ and we wish to find the corresponding quantities $I'_x, I'_y, I'_{xy}$ in some other system centred on the same point, but rotated with respect to the first set through an angle $\theta$ as shown in Figure 4.19.

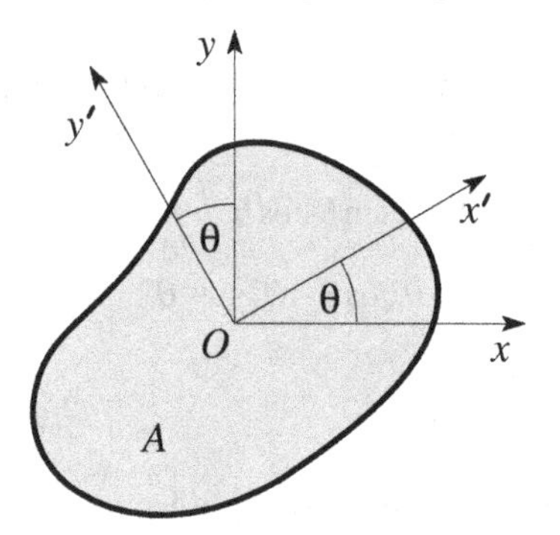


*Figure 4.19: Rotation of coordinate axes*

We first note that the vector $\boldsymbol{r}$ defining the position of a given element of area $dA$ can be written

$$\boldsymbol{r} = \boldsymbol{i}x + \boldsymbol{j}y = \boldsymbol{i}'x' + \boldsymbol{j}'y' , \tag{4.39}$$

where $\boldsymbol{i}, \boldsymbol{j}, \boldsymbol{i}', \boldsymbol{j}'$ are unit vectors in the $x, y, x', y'$-directions respectively.

Using this notation, we can write

$$\begin{aligned} x' &= \boldsymbol{r} \cdot \boldsymbol{i}' = (\boldsymbol{i}x + \boldsymbol{j}y) \cdot \boldsymbol{i}' = x(\boldsymbol{i} \cdot \boldsymbol{i}') + y(\boldsymbol{j} \cdot \boldsymbol{i}') \\ &= x\cos\theta + y\sin\theta . \end{aligned} \tag{4.40}$$

and by a similar argument,

$$y' = y\cos\theta - x\sin\theta . \tag{4.41}$$

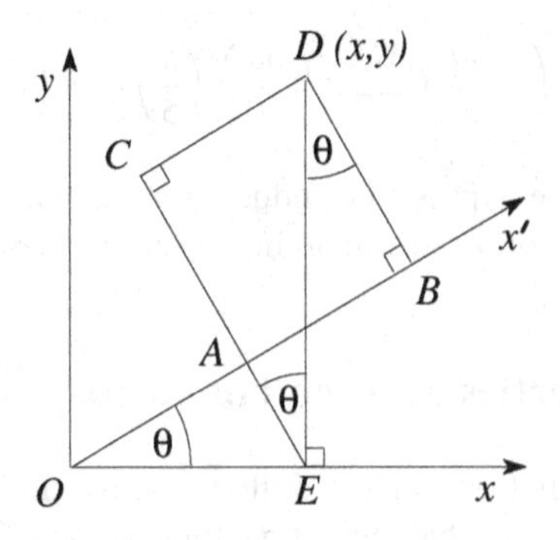


*Figure 4.20: Trigonometric construction for establishing equation (4.42)*

These results could also be proved using trigonometry as shown in Figure 4.20, since, for example, $x' = OB = OA + AB$; $OA = OE\cos\theta = x\cos\theta$; $AB = CD = DE\sin\theta = y\sin\theta$, leading to equation (4.40) as before.

Now, by definition

$$I'_x = \iint_A y'^2 dA = \iint_A (y\cos\theta - x\sin\theta)^2 dA \,, \tag{4.42}$$

using (4.41). Expanding the square in the integrand and separating terms, we then have

$$\begin{aligned} I'_x &= \cos^2\theta \iint_A y^2 dA - 2\sin\theta\cos\theta \iint_A xy dA + \sin^2\theta \iint_A x^2 dA \\ &= I_x\cos^2\theta - 2I_{xy}\sin\theta\cos\theta + I_y\sin^2\theta \,, \end{aligned} \tag{4.43}$$

using the definitions (4.15).

By a similar procedure, we find

$$\begin{aligned} I'_y &= \iint_A (x\cos\theta + y\sin\theta)^2 dA \\ &= I_y\cos^2\theta + 2I_{xy}\sin\theta\cos\theta + I_x\sin^2\theta \end{aligned} \tag{4.44}$$

$$\begin{aligned} I'_{xy} &= \iint_A (x\cos\theta + y\sin\theta)(y\cos\theta - x\sin\theta) dA \\ &= (I_x - I_y)\sin\theta\cos\theta + I_{xy}(\cos^2\theta - \sin^2\theta) \,. \end{aligned} \tag{4.45}$$

### 4.4.2 Mohr's circle of second moments

The perceptive reader will notice a similarity between equations (4.44, 4.45) and the stress transformation equations (2.1, 2.2), except for some sign differences. In fact the signs can be made the same by establishing the equivalences

$$\sigma_{xx} \to I_x \quad ; \quad \sigma'_{xx} \to I'_x \tag{4.46}$$

$$\sigma_{yy} \to I_y \quad ; \quad \sigma'_{yy} \to I'_y \tag{4.47}$$

$$\sigma_{xy} \to -I_{xy} \quad ; \quad \sigma'_{xy} \to -I'_{xy} \,. \tag{4.48}$$

It follows that we can draw a Mohr's circle for $I'_x, I'_{xy}$ as shown in Figure 4.21.

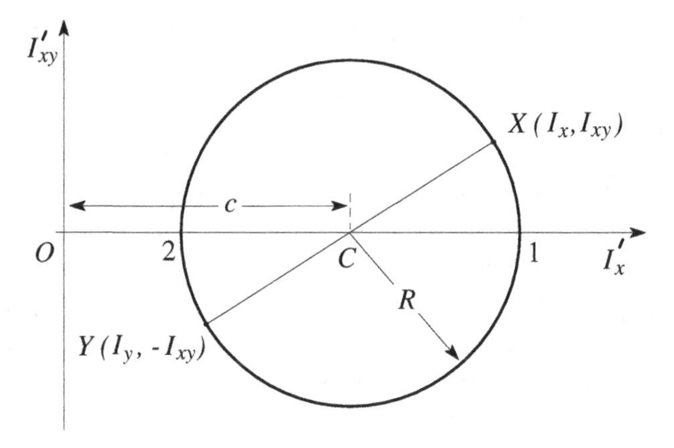


*Figure 4.21: Mohr's circle for second moments*

With the convention used here, clockwise rotation around the circle will correspond to clockwise rotation of the axes about which the second moments are calculated.

### Some important consequences

It follows immediately that:-

(i) There are two principal axes, labelled 1,2 in Figure 4.21.
(ii) About these axes, the product inertia $I_{xy} = 0$.
(iii) About these axes, $I_x$ is a maximum (point 1) or a minimum (point 2). We shall refer to axis 1 as the *stiff axis* and axis 2 as the *flexible axis* for the beam. They are respectively the axes about which it is hardest and easiest to bend the beam.
(iv) Since $I'_x > 0$ for all $\theta$ — it is the result of integrating a squared quantity [see equation (4.44)] — the circle must lie totally in the right half-plane.

As in §2.1.1, the distance to the centre of the circle and its radius are easily shown to be

$$c = \frac{I_x + I_y}{2} \quad ; \quad R = \sqrt{\left(\frac{I_x - I_y}{2}\right)^2 + I_{xy}^2} \tag{4.49}$$

[compare with equations (2.5, 2.6)]. Since the circle must lie completely in the right half-plane (see (iv) above), we must have $c > R$ and hence $c^2 > R^2$. Substituting from equations (4.49) for $c, R$, we therefore have

$$\left(\frac{I_x + I_y}{2}\right)^2 > \left(\frac{I_x - I_y}{2}\right)^2 + I_{xy}^2$$

and hence

$$I_x I_y - I_{xy}^2 > 0 \,, \tag{4.50}$$

which confirms that the denominator in equations (4.18–4.20) can never be zero or negative.

### Example 4.6

*Sketch the Mohr's circle of second moments for the unequal Z-section of Example 4.5 (Figure 4.17) and hence determine the principal second moments $I_1, I_2$ and the inclination of the principal axes.*

From Example 4.5, we have $I_x = 3.467a^3t$, $I_y = 2.55a^3t$, $I_{xy} = 2.2a^3t$. The Mohr's circle is found by plotting the points $X(I_x, I_{xy})$, $Y(I_y, -I_{xy})$ and then sketching the circle through these points as diameter, as shown in Figure 4.22.

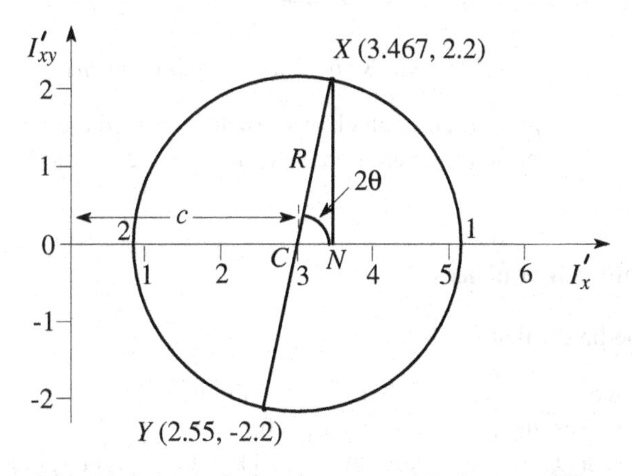


*Figure 4.22: Mohr's circle for the unequal Z-section of Figure 4.17*

The centre of the circle is defined by the value

$$c = \frac{(3.467 + 2.55)a^3t}{2} = 3.008a^3t$$

and the distance $CN$ is

$$CN = 3.467a^3t - 3.008a^3t = 0.459a^3t \,.$$

The radius $R$ is therefore

$$R = \left(\sqrt{0.459^2 + 2.2^2}\right) a^3t = 2.247a^3t \,.$$

It follows that the principal second moments are

$$I_1 = 3.008a^3t + 2.247a^3t = 5.255a^3t$$

$$I_2 = 3.008a^3t - 2.247a^3t = 0.761a^3t \,.$$

The inclination of the principal axes is defined by the angle $\theta$ where

$$\tan(2\theta) = \frac{NX}{CN} = \frac{2.2}{0.459} = 4.793\,,$$

giving

$$2\theta = 78.2^{\circ} \quad ; \quad \theta = 39.1^{\circ}\,.$$

Rotation in the Mohr's circle is in the same direction as that in the diagram of the beam section.[4] To get from the point $X$ to the point 1, we need to rotate around the circle clockwise by $78.2^{\circ}$ and hence the stiff axis 1 can be found by rotating the $x$-axis clockwise thorugh $39.1^{\circ}$, as shown in Figure 4.23. The flexible axis 2 is at right angles to the stiff axis and hence is inclined at $39.1^{\circ}$ clockwise from the $y$-axis or $50.9^{\circ}$ anticlockwise from the $x$-axis as shown.

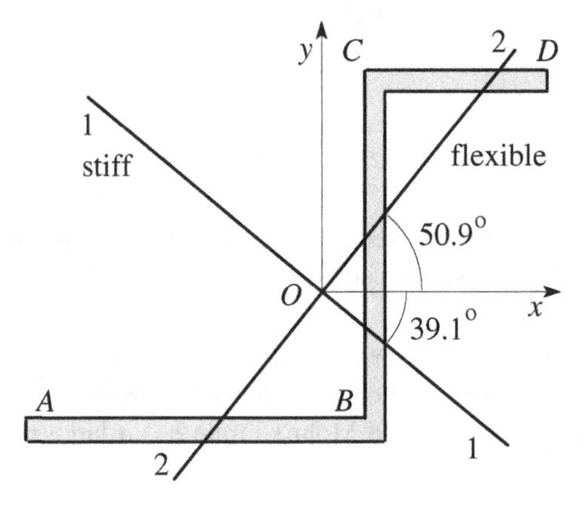


*Figure 4.23: Principal axes for the unequal Z-section of Figure 4.17*

### Mohr's circle of zero radius

An important class of beams includes all those whose cross sections define a Mohr's circle of radius $R=0$. In this case, the circle reduces to a point on the $I_x$ axis and it follows that

(i) the product inertia $I'_{xy} = 0$ for all $\theta$,
(ii) $I'_x$ is independent of $\theta$.

A beam of this class will therefore always bend only about the axis of the resultant bending moment. The neutral axis will be parallel to the axis of this moment and the flexural rigidity of the beam $EI$ will be the same for all axes.

Clearly an axisymmetric beam such as a solid or hollow circular cylinder will behave in this way, but there are other examples as well. For example, if any three or

[4] This is in contrast to the case of stress tranformation and results from the sign differences noted in connection with the relations (4.46–4.48) above.

more different axes can be found about which $I_x$ is the same, the Mohr's circle must reduce to a point, since a vertical line can only cut the circle in two points. This is the case if the section repeats itself every $120^o$ (or any other submultiple of $360^o$ greater than 2) as shown in Figure 4.24.

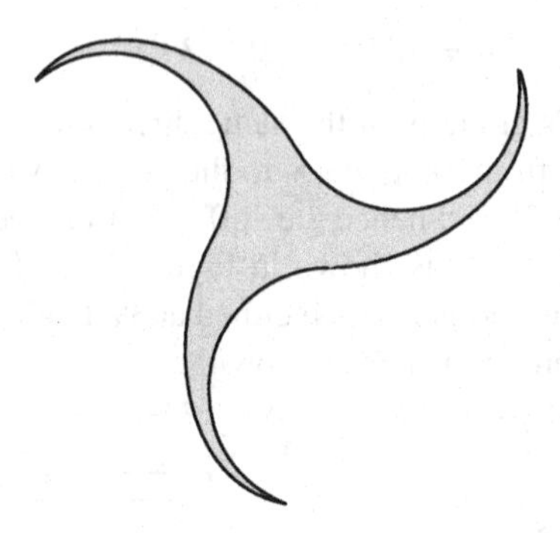

*Figure 4.24: Section with three-way symmetry*

It also applies to the case of a section with two planes of symmetry if the second moments about these two axes are equal — as in the case of the square section of Figure 4.25.

More generally, the coupling terms which cause the neutral axis to deviate from the moment axis in unsymmetrical bending depend on the product inertia $I_{xy}$ and are therefore most significant when the Mohr's circle has a large radius (see Problem 4.29).

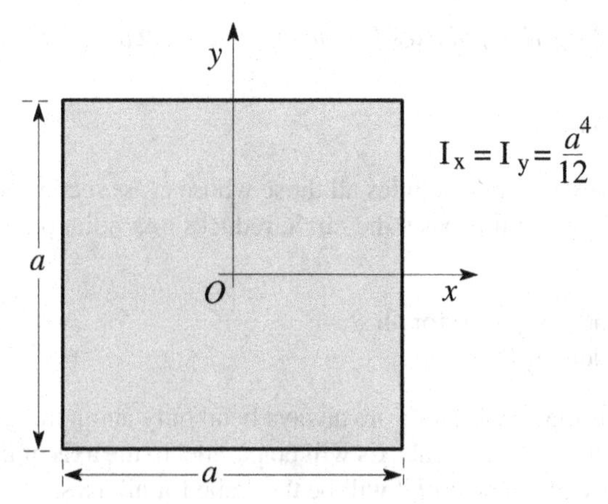


*Figure 4.25: The square cross section has the same second moment of area about all centroidal axes*

### 4.4.3 Solution of unsymmetrical bending problems in principal coordinates

If the principal second moments $I_1, I_2$ and the inclination of the principal axes are known, the bending problem is simplified by using the principal directions as coordinate directions, since then there will be no coupling between the two moment components and the problem can be solved by superposition as in §4.1.4. It is not generally cost effective to use this method if the principal second moments are *not* already known, since the extra effort involved in determining them, added to the greater complexity of the coordinates of the maximum stress points with inclined Cartesian axes makes for a longer solution than that using the simultaneous equations of §4.1.5. However, the principal second moments have been tabulated[5] for commonly occurring unsymmetrical structural steel sections, so civil engineers generally prefer to solve such problems using principal coordinates.

Denoting the moments about the principal axes $M_1, M_2$ and the corresponding radii of curvature $R_1, R_2$ respectively, equations (4.18, 4.19) reduce to

$$\frac{E}{R_1} = \frac{M_1}{I_1} \quad ; \quad \frac{E}{R_2} = \frac{M_2}{I_2} \tag{4.51}$$

and (4.7) then gives

$$\sigma_{zz} = \frac{M_1 y'}{I_1} - \frac{M_2 x'}{I_2} , \tag{4.52}$$

where $x'y'$ are aligned with the stiff and flexible axes 1,2, respectively.

#### Example 4.7

*Figure 4.26 (a) shows the location of the centroid and the inclination of the principal axes for an L5×3×1/4 unequal angle. The principal second moments are $I_1 = 5.70$ in$^4$, $I_2 = 0.85$ in$^4$ and the beam is loaded by a bending moment $M = 2000$ lb.in about the x-axis as shown. Find the maximum tensile and compressive stresses in the section.*

We first resolve the moment into components about the principal axes, obtaining

$$M_1 = M\cos(37.2^{\circ}) = 1593 \text{ lb.in} \quad ; \quad M_2 = -M\sin(37.2^{\circ}) = -1209 \text{ lb.in} ,$$

with the directions as defined in Figure 4.26 *(b)*.

---

[5] These tables generally include the values of $I_x, I_y, I_1, I_2$ and the inclination of the principal axes, but do not include the product inertia $I_{xy}$. In this case the only practical option is to solve the problem in principal coordinates. However, if $I_{xy}$ is also known, the author's personal opinion is that the method of §4.1.5 remains more efficient because the coordinates of the maximum stress points are then more easily determined.

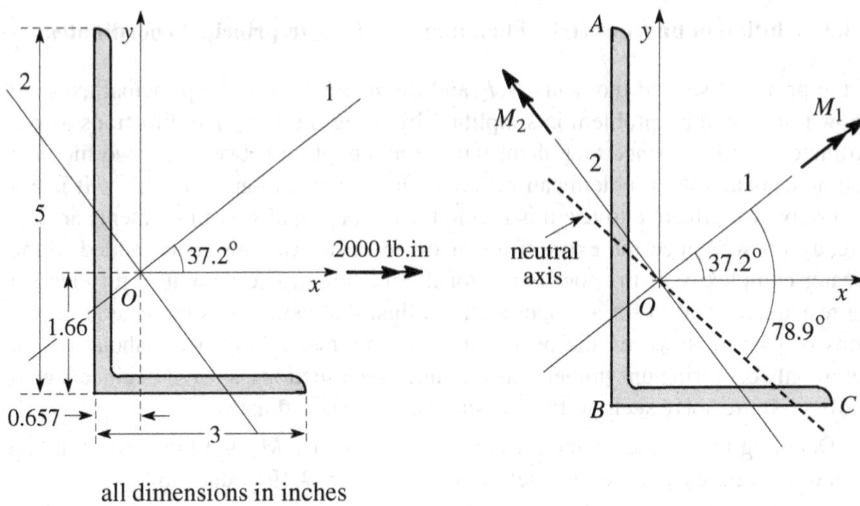


*Figure 4.26*

The stress distribution is then given by equation (4.52) as

$$\sigma_{zz} = \frac{1593}{5.70}y' - \frac{(-1209)}{0.85}x' = 279.5y' + 1422x'$$

(in psi if $x', y'$ are in inches) and hence the neutral axis is defined by the equation

$$\frac{y'}{x'} = -\frac{1422}{279.5} = -5.089\,.$$

This defines a line inclined at an angle

$$\psi = \tan^{-1} 5.089 = 78.9^{\circ}$$

clockwise from the $x'$-axis, as shown in Figure 4.26*(b)*.

The maximum tensile and compressive stresses will occur at the points $A, B$, which are the furthest from the neutral axis. In Cartesian coordinates $Oxy$, the points $A, B$ are defined by

$$x_A = -0.657 \text{ in} \quad ; \quad y_A = 5 - 1.66 = 3.34 \text{ in}$$

$$x_B = -0.657 \text{ in} \quad ; \quad y_B = -1.66 \text{ in.}$$

Hence, using (4.40, 4.41), the coordinates in terms of principal axes are

$$x'_A = -0.657\cos(37.2^{\circ}) + 3.34\sin(37.2^{\circ}) = 1.50 \text{ in}$$

$$y'_A = 3.34\cos(37.2^{\circ}) - (-0.657)\sin(37.2^{\circ}) = 3.06 \text{ in}$$

$$x'_B = -0.657\cos(37.2^{\circ}) - 1.66\sin(37.2^{\circ}) = -1.53 \text{ in}$$

$$y'_B = -1.66\cos(37.2^\circ) - (-0.657)\sin(37.2^\circ) = -0.93 \text{ in.}$$

Substituting these values in the equation for $\sigma_{zz}$, we then have

$$\sigma_A = 279.5(3.06) + 1422(1.50) = 2983 \text{ psi (maximum tensile stress)}$$

and

$$\sigma_B = 279.5(-0.93) + 1422(-1.53) = -2436 \text{ psi (maximum compressive stress)}\,.$$

### 4.4.4 Design estimates for the behaviour of unsymmetrical sections

Arguably, the greatest practical use of the coordinate transformation relations and Mohr's circle of second moments of area is to enable us to draw general conclusions that provide insight into the behaviour of unsymmetrical beam sections. We have already remarked on some of these properties in §4.4.2 and in this section we shall show how related arguments can be developed to permit quite accurate estimates to be made for the location of the principal axes and the neutral axis of bending, without doing any calculations.

We showed in §4.3.2 that out of the class of all axes parallel to a given direction, that through the centroid has the lowest value of

$$I = \iint_A n^2 dA\,, \tag{4.53}$$

where $n$ is the perpendicular distance from a given elemental area $dA$ to the axis. The Mohr's circle results show that if we now change the *direction* of the axis, the lowest value of $I$ corresponds to the flexible principal axis 2 in Figure 4.21. Thus, the flexible centroidal principal axis corresponds to the absolute minimum value of $I$ for axes of all inclinations and positions and we could reformulate the question of determining it as a minimization problem — i.e. *determine the axis for which $I$ as defined by equation (4.53) is a minimum.*

Now recall that the usual procedure for finding the 'best' straight line approximation to a set of experimental points is to choose that line which minimizes the sum of the squares of the distances $n_i$ of the individual points from the line — i.e. minimize

$$|E|^2 \equiv \sum_{i=1}^{N} n_i^2\,. \tag{4.54}$$

This is known as the 'least squares fit' and it is exactly similar to the minimization of (4.53) except that in the latter case we have elemental areas $dA$ instead of points. Thus, if we imagine dividing up the section into equal small areas and place one imaginary point at the centre of each, the flexible axis will be the best straight line fit to the resulting set of points. We can find this line quite well by eye, as shown in

Figure 4.27, provided the section is reasonably 'long and thin' — which is equivalent to the statement that the maximum and minimum second moments should be significantly different.[6]

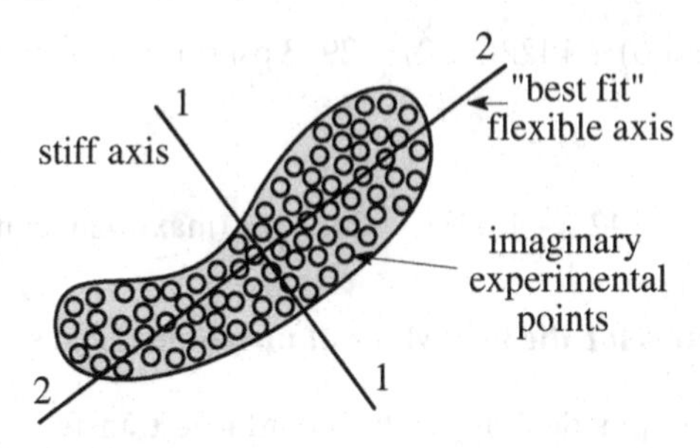


*Figure 4.27: Estimating the principal axes using the 'least squares fit' method*

Once we have estimated our best straight line — corresponding to the flexible axis — the stiff axis can be found by estimating the location of the centroid (which must lie on the flexible axis) and then drawing a line through it at right angles to the flexible axis as shown.

### Location of the neutral axis

We shall now show that the neutral axis will always lie between (i) the flexible principal axis and (ii) the axis about which the bending moment is applied (see Figure 4.28). The bigger the ratio between $I_1$ and $I_2$, the closer the neutral axis will get to the flexible axis.

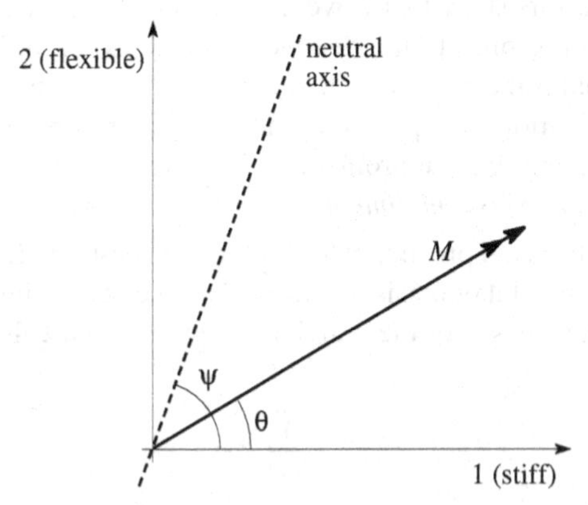


*Figure 4.28: Location of the neutral axis*

[6] If they are not, it doesn't matter too much from an engineering perspective where the principal axes are anyway, since the maximum product inertia is $(I_1 - I_2)/2$, from Figure 4.21 and hence the coupling effect will be weak — i.e. the elementary theory will give quite good answers.

To demonstrate this, we first resolve the applied moment $M$ into components about the principal axes — i.e.

$$M_1 = M\cos\theta \quad ; \quad M_2 = M\sin\theta \; . \tag{4.55}$$

Substitution into (4.52) then gives

$$\sigma_{zz} = \frac{My\cos\theta}{I_1} - \frac{Mx\sin\theta}{I_2} \tag{4.56}$$

and the neutral axis ($\sigma_{zz}=0$) is defined by the equation

$$\frac{y}{x} = \tan\psi = \frac{M\sin\theta}{I_2} . \frac{I_1}{M\cos\theta} = \frac{I_1}{I_2}\tan\theta \; . \tag{4.57}$$

Since $I_1 > I_2$ by definition, it follows that $|\tan\psi| > |\tan\theta|$ and hence that $|\psi| > |\theta|$. Furthermore, $|\tan\psi|$ increases with the ratio $I_1/I_2$ (i.e. the neutral axis moves closer to the flexible axis) for all $\theta$ except $\theta = 0$. In particular, a section with a very large ratio $I_1/I_2$ will tend to bend about the flexible axis, regardless of the direction of the applied moment. You can test this experimentally with a thin plastic ruler.

These results provide a convenient method of estimating the location of the neutral axis and hence the maximum stress points without doing any algebraic calculations, provided we can estimate the value of the ratio $I_1/I_2$. For a rectangular section $a \times b$, with $a > b$, this ratio is

$$\frac{I_1}{I_2} = \left(\frac{a}{b}\right)^2 , \tag{4.58}$$

so a plausible first estimate is to use the square of the ratio of maximum to minimum dimensions.

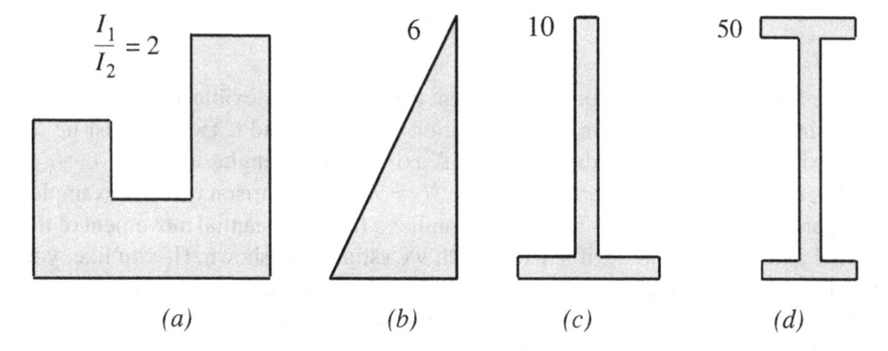


*Figure 4.29: Values of the ratio $I_1/I_2$ for some representative beam sections*

Figure 4.29 shows a variety of beam sections and the corresponding values of $I_1/I_2$. Most people are initially surprised that these values are so high, particularly for sections like Figure 4.29*(a)* that appear not too far from equi-axed. However, it is fairly easy to develop the ability to get a rough estimate at sight and this coupled with a visual 'least squares' estimate of the direction of the flexible axis permits us to place the neutral axis with some confidence. It is a good idea to do this as an initial step even when a full algebraic solution is intended, since the visual estimate provides a check against algebraic errors. You will be surprised to see how close your estimate is to the calculated value.

**Example 4.8**

*The beam section in Figure 4.30 is loaded by a bending moment M about the horizontal axis, as shown. Without doing any calculations, estimate the location of the principal axes of the section and the resulting neutral axis. Hence identify the points of maximum tensile and compressive stress in the section.*

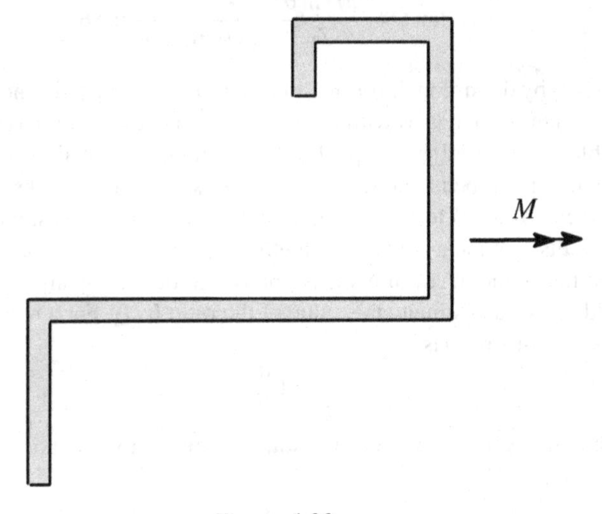


*Figure 4.30*

We first use the 'least squares' argument to estimate the flexible axis 2, as shown in Figure 4.31. We then estimate the location of the centroid $C$ (which must lie on the flexible axis) and hence draw in the stiff axis 1, at right angles to 2.

For this section, we estimate the ratio $I_1/I_2 \approx 7$, by comparison with the examples in Figure 4.29. Equation (4.57) therefore implies a fairly substantial movement of the neutral axis towards the flexible axis which we estimate as shown. (If you like, you can do a one line calculation using equation (4.57) to estimate $\psi$, but you still have to estimate where this is on the diagram, which we are not drawing strictly to scale, so probably a guess is good enough here.)

Finally, the maximum tensile and compressive stresses will occur at the points furthest from the neutral axis — i.e. at $A, B$ in Figure 4.31. To determine which is which, imagine that the moment arrow is rotated to line up with the neutral axis. A clockwise moment about this arrow will involve motion out of the paper) at $A$ and into the paper at $B$, so the maximum tensile stress will be at $A$ and the maximum compressive stress at $B$.

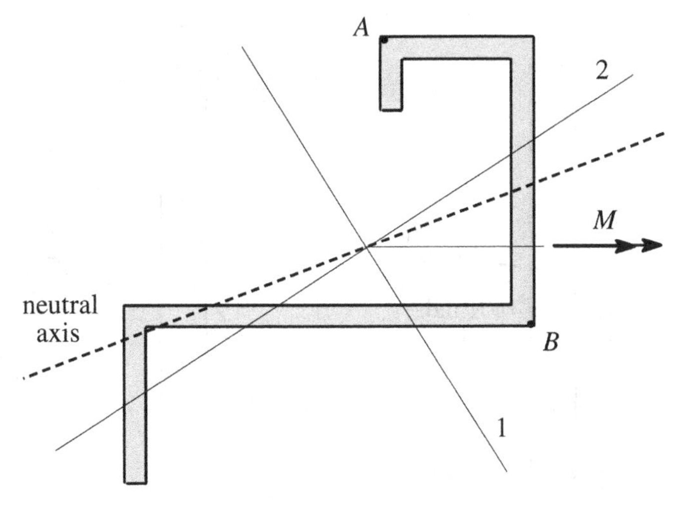


*Figure 4.31*

### 4.4.5 Errors due to misalignment

We have seen that that differences between the two principal second moments cause the neutral axis to deviate considerably from the moment axis towards the direction of the flexible principal axis. This effect can be particularly important if the section is designed to have a large value of $I_1$ about the axis where bending is anticipated, since small variations in the moment axis, due perhaps to misaligment of supports or loading, can then cause significant increase in stresses and deflections.

As an example, we consider a rectangular section of aspect ratio 10, subjected to a moment $M_0$, whose axis deviates from the stiff axis by the small angle 0.01 radians ($=0.6^{\circ}$), as shown in Figure 4.32*(a)*. Notice that the angle actually drawn in this figure is exaggerated by a factor of 10 for clarity.

The two principal second moments are

$$I_1 = \frac{10^3}{12} \;\; ; \;\; I_2 = \frac{10}{12} \; , \tag{4.59}$$

so that the ratio $I_1/I_2 = 100$.

The neutral axis is therefore defined by

$$\tan\psi = 100\tan(0.01) \approx 1 \; ,$$

from equation (4.57). Thus, the beam bends about an axis inclined at $45^{\circ}$ to the stiff axis, despite the fact that the moment is only very slightly misaligned (see Figure 4.32*(b)*).

We also calculate the maximum tensile stress, which will occur at $A$. We have

$$\sigma_A = \frac{M_0 \times 5 \times 12}{1000} - \frac{M_0}{100}\frac{(-0.5) \times 12}{10} = \frac{6.6M_0}{100} \; , \tag{4.60}$$

from (4.52).

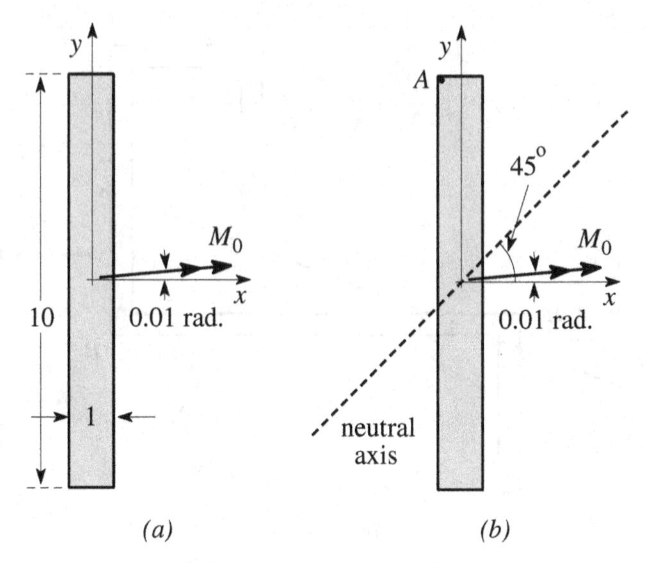


*Figure 4.32: Effect of moment misalignment*

Without the misalignment, the result would be $6M_0/100$, so the error is only 10% which is not too serious. However, the unwanted out of plane flexure of the beam could be a problem. For beams with a larger ratio between $I_1$ and $I_2$, misalignment could have more serious effects. This effectively places a limit on the extent to which beams can be tailored to give high strength and stiffness about a specified moment axis.

## 4.5 Summary

In this chapter, we have extended the elementary theory of elastic bending to beams of arbitrary cross section. All sections posess an orthogonal pair of principal axes which are respectively the stiffest and the most flexible axes for bending. The elementary theory of bending is exact when the bending moment is applied about one of the principal axes, but in all other cases it is non-conservative — the actual stresses will be larger than those predicted by the elementary theory and the errors can be quite large. These errors stem from the fact that the axis of bending (the neutral axis) generally deviates from the moment axis in unsymmetrical beams.

The engineering designer should pay particular attention to §4.4.4, in which methods are discussed for estimating the location of the principal axes of a section and the neutral axis of bending by eye. These methods enable us to judge when the deviation of the neutral axis (and hence the errors in the elementary bending prediction) will be large enough to warrant the extra complexity of a full unsymmetrical bending calculation. Situations where this happens are generally undesirable,[7] since

[7] Ask yourself why you are proposing to use an unsymmetrical section in bending. Legitimate reasons include the convenience of attachment of devices to the beam or geometrical constraints on the volume which it occupies.

they imply that significant bending is occurring about the flexible (and hence weaker) axis.

The stiffness of a beam in bending is governed by the second moments of area of the section. Methods have been presented for determining the second moments in a specific coordinate system. Mohr's circle can be used to transform these results to other axis systems and, in particular, to determine the principal values and the inclination of the principal axes.

Unsymmetrical bending effects are most pronounced when the ratio of principal second moments $(I_1/I_2)$ is large. Such beams may appear to be optimal when bending moments are expected only about one axis, but they are very sensitive to manufacturing or assembly imperfections or load misalignment.

## Further reading

W.B. Bickford (1998), *Advanced Mechanics of Materials,* Addison Wesley, Menlo Park, CA, pp. 182–204.
A.P. Boresi, R.J. Schmidt, and O.M. Sidebottom (1993), *Advanced Mechanics of Materials,* John Wiley, New York, 5th edn., Chapter 7.

## Problems

### Section 4.1.4

**4.1.** Figure P4.1 shows the cross section of a rectangular beam which is subjected to a bending moment of 500 Nm inclined at 25° to the $x$-axis. Find the location and magnitude of the maximum tensile stress due to bending.

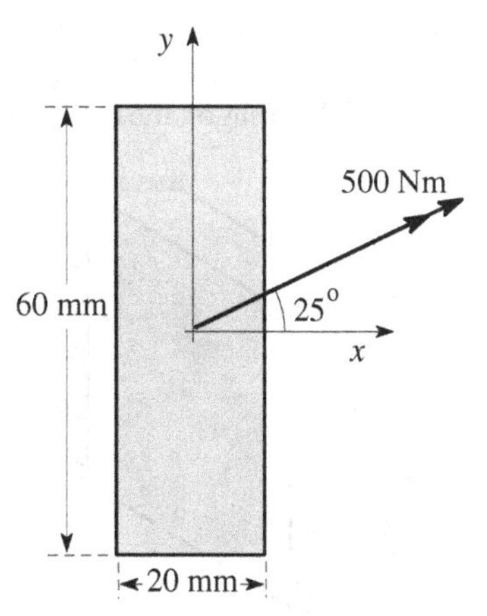


*Figure P4.1*

**4.2.** Figure P4.2 shows the dimensions of a W200×22 I-beam, for which $I_x = 20\times10^6$ mm$^4$, $I_y = 1.42\times10^6$ mm$^4$. The length of the beam is 10 m and it is simply supported at its ends. The loading consists of a downward vertical distributed load of 1000 N/m throughout the length of the beam (i.e. in the negative $y$-direction) and a horizontal concentrated force of 1000 N in the positive $x$-direction located at the mid-point of the beam. Find the magnitude and location of the maximum tensile stress in the beam.

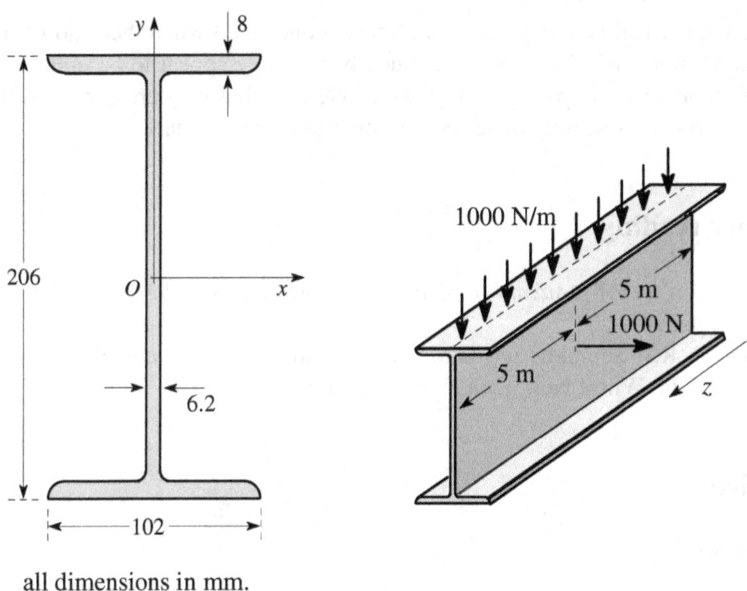


*Figure P4.2*

**4.3.** A cantilever beam of rectangular cross section is loaded by a force $F$ directed along the diagonal $AC$ of the section as shown in Figure P4.3. Show that the neutral axis in this case coincides with the other diagonal $BD$ for all values of the dimensions $a, b$ of the rectangle.

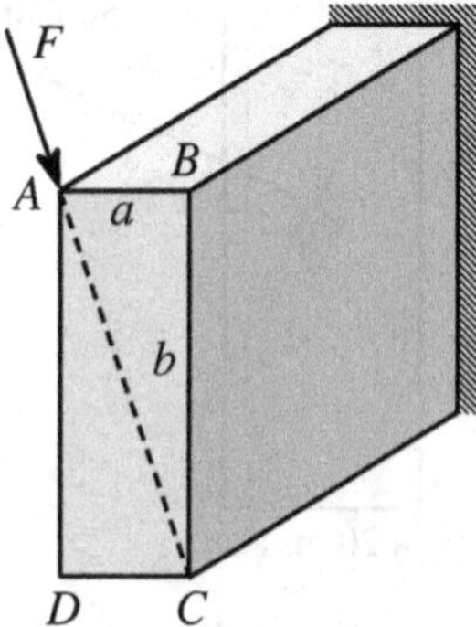


*Figure P4.3*

**4.4.** A beam of rectangular cross section $(a \times b)$ is loaded by a compressive axial force $F$ whose line of action passes through the point $(c, d)$ in centroidal coordinates $Oxy$. Find the range of values of $c, d$ for which the stress at all points in the section remains compressive.

**4.5.** The C200×20 channel section of Figure P4.5 has second moments of area $I_x = 15 \times 10^6$ mm$^4$, $I_y = 0.637 \times 10^6$ mm$^4$. It is loaded by the bending moments $M_x = 2200$ Nm and $M_y = -350$ Nm. Find the magnitude and location of the maximum tensile stress in the beam.

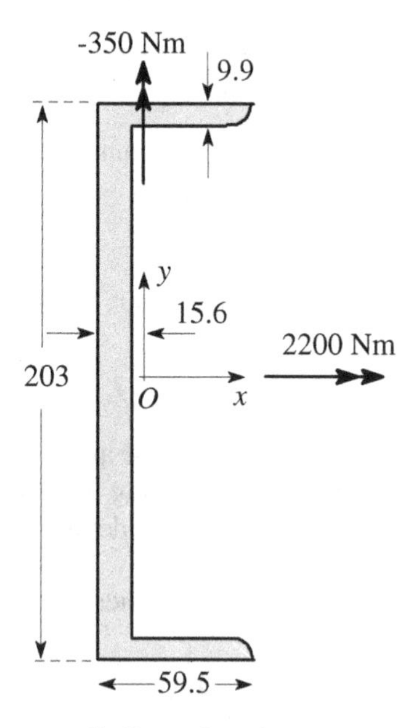


*Figure P4.5*

### Section 4.1.5

**4.6.** The L-section shown in Figure P4.6 has second moments $I_x = 1,512,500$ mm$^4$, $I_y = 412,500$ mm$^4$, $I_{xy} = -450,000$ mm$^4$ about centroidal axes $Oxy$. The centroid is located at the point (15,35) relative to the corner $C$ as shown.

The beam is subjected to a bending moment $M = 1000$ Nm about the horizontal axis $Ox$. Find the inclination of the neutral axis to $Ox$ and the location and magnitude of the maximum tensile stress in the section.

Compare your results with the value $My_{max}/I_x$ from the elementary bending theory.

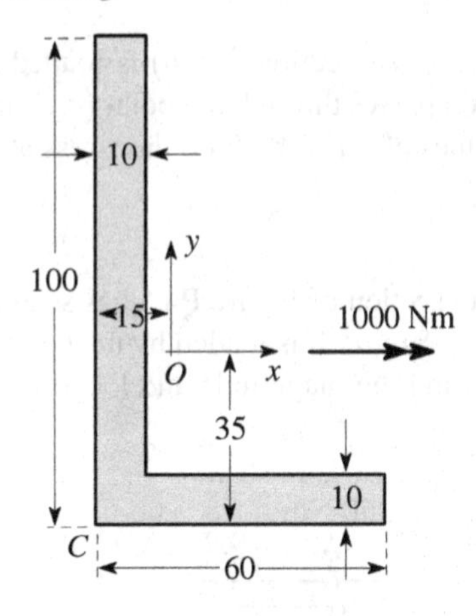


*Figure P4.6*

**4.7.** The second moments of area for the right-angle triangular section of Figure P4.7 are

$$I_x = \frac{a^3 b}{36} \quad ; \quad I_y = \frac{ab^3}{36} \quad ; \quad I_{xy} = \frac{a^2 b^2}{72}$$

and the centroid $O$ is located a distance $a/3$ above and $b/3$ to the left of $B$.

A beam of this section is loaded by a bending moment $M_0$ about the $x$-axis — i.e. $M_x = M_0$, $M_y = 0$. Show that the neutral axis is defined by the line joining $C$ to the mid-point of $AB$.

Comment on the implications of this result when $b \ll a$.

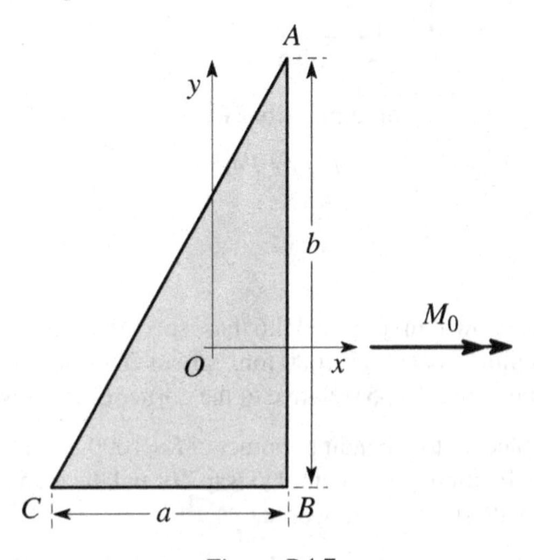


*Figure P4.7*

**4.8.** The Z-section of Figure P4.8 has second moments of area $I_x = 560,000\ \text{mm}^4$, $I_y = 290,000\ \text{mm}^4$, $I_{xy} = 300,000\ \text{mm}^4$. It is used for a beam of length 2m which is simply supported at its ends and loaded by a uniformly distributed vertical load of 1000 N/m. Find the location and magnitude of the maximum tensile stress.

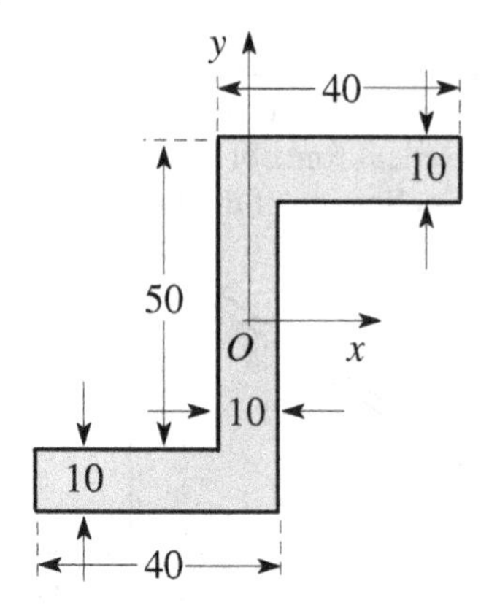


*Figure P4.8*

**4.9.** The beam section shown in Figure P4.9 has second moments of area $I_x = 317,600\ \text{mm}^4$, $I_y = 791,700\ \text{mm}^4$, $I_{xy} = -298,500\ \text{mm}^4$. The beam is subjected to a bending moment $M_0 = 1000$ Nm about the $x$-axis as shown. Find the equation defining the neutral axis and hence the location and magnitude of the maximum tensile stress.

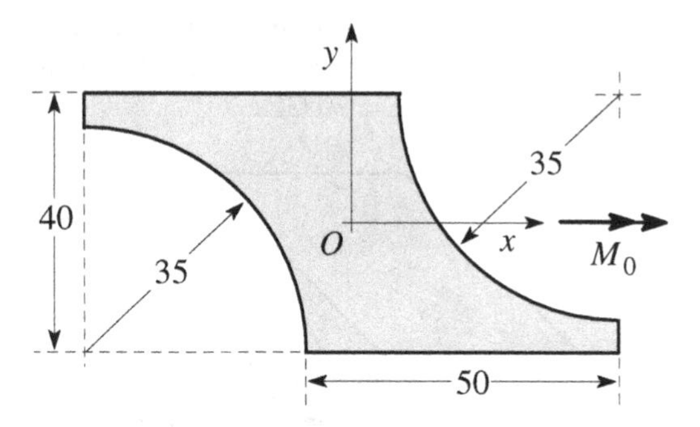


*Figure P4.9*

**Section 4.2**

**4.10.** Find the maximum horizontal and vertical components of the displacement for the beam of Problem 4.8, if the beam is made of steel for which $E = 210$ GPa.

**4.11.** A beam with the triangular cross section of Figure P4.7 is of length $L$ and made of material with Young's modulus $E$. It is built in at one end and loaded by a force $F$ in the $x$-direction at the other end. Find the two components of displacement at the loaded end.

**4.12.** Figure P4.12 shows an L-section steel beam ($E = 30 \times 10^6$ psi) with $I_x = 69.6$ in$^4$, $I_y = 11.6$ in$^4$, $I_{xy} = -16.2$ in$^4$. The beam is 12 feet long, is simply supported at its ends and is loaded by a vertical force of 1000 lbs as shown. Find the *horizontal* displacement at the point of application of the load.

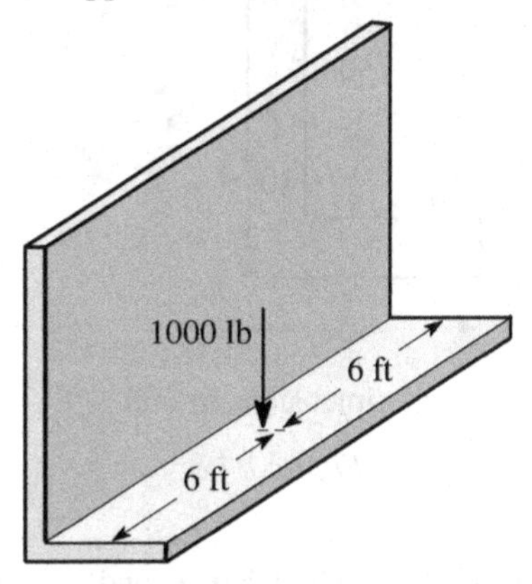


*Figure P4.12*

**4.13*.** Figure P4.13 shows a beam with the cross section of Figure P4.9 loaded by a uniformly distributed horizontal load of 1 kN/m. It is built in at $A$ and the end $B$ is free to move horizontally, but is restrained from moving vertically by a frictionless support. Find the vertical reaction and the horizontal displacement at $B$ if the beam is made of aluminium alloy for which $E = 70$ GPa.

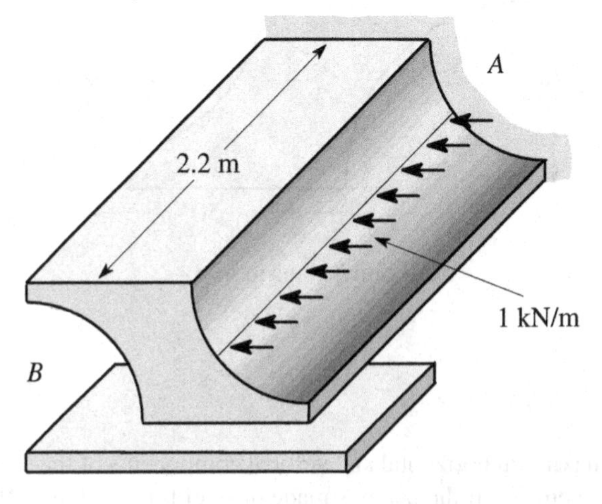


*Figure P4.13*

## Section 4.3

**4.14.** Calculate $I_x, I_y, I_{xy}$ for the symmetric[8] Z-section of Figure P4.8 above.

**4.15.** Calculate $I_x, I_y, I_{xy}$ for the section of Figure P4.9 above. You can use the results from Example B.1.

**4.16.** Calculate $I_x, I_y, I_{xy}$ for the L-section shown in Figure P4.16.

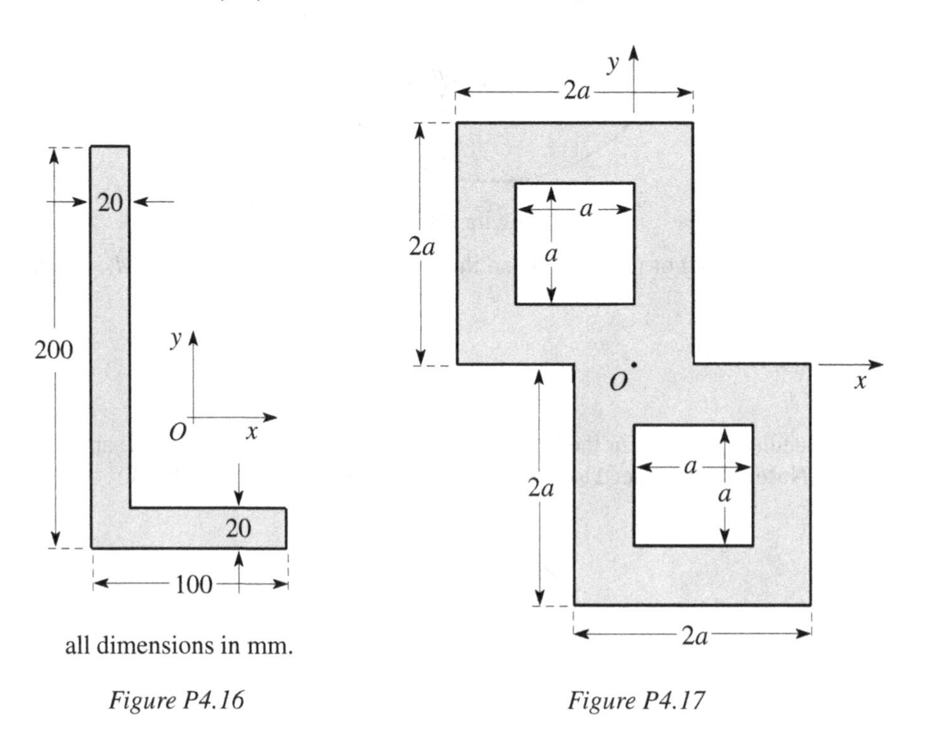


*Figure P4.16*

*Figure P4.17*

**4.17.** Find the product inertia $I_{xy}$ for the beam section shown in Figure P4.17. Notice that the centroid must be at $O$ by symmetry.

[8] Notice that this category of symmetry (reflection through the origin) does *not* guarantee $I_{xy} = 0$, by the argument of Figure 4.6, since symmetric points have equal and opposite values of both $x$ and $y$ coordinates, resulting in contributions to $I_{xy}$ of the same sign.

**4.18.** A beam of circular cross section, radius $3a$ has two eccentric circular holes of radius $a$ as shown in Figure P4.18.

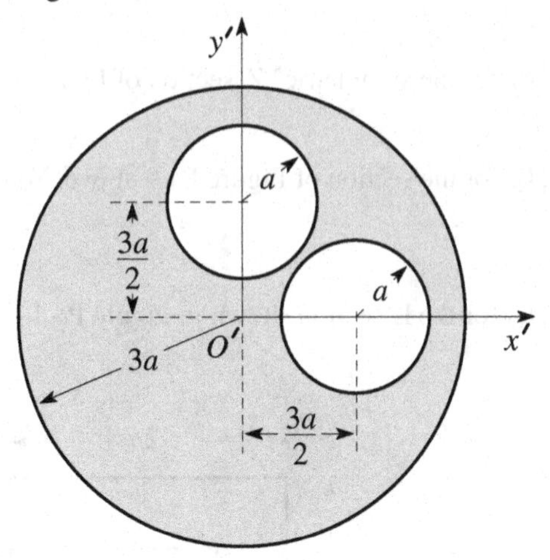


*Figure P4.18*

Find the centroid of the section and the second moments of area $I_x, I_y, I_{xy}$ about centroidal axes.

**Section 4.3.3**

**4.19.*** Calculate $I_x, I_y, I_{xy}$ for the S-section of Figure P4.19, using the '$t \ll a$' approximation. **Note:** You will need to use the integration method of Appendix B.

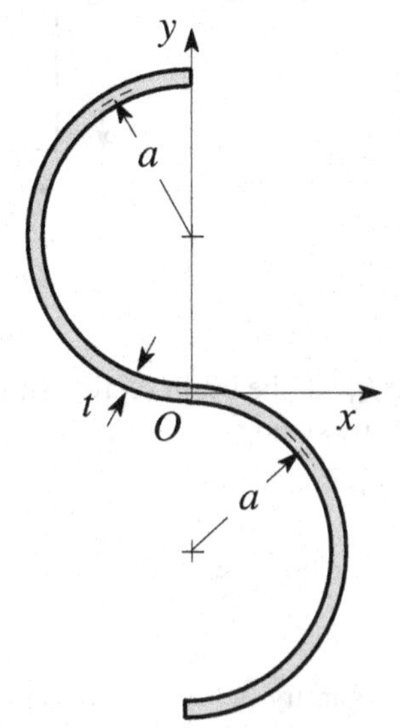


*Figure P4.19*

**4.20.** Calculate $I_x, I_y, I_{xy}$ for the section of Figure 4.15, using the '$t \ll a$' approximation and describing the section by the mean line. Which of the three second moments is the most in error by this method and by what percentage?

**4.21.** A beam of the section shown in Figure P4.21 is loaded by a bending moment $M_0$ about the $y$-axis. Calculate $I_x, I_y, I_{xy}$ using the '$t \ll a$' approximation and hence determine the inclination of the neutral axis.

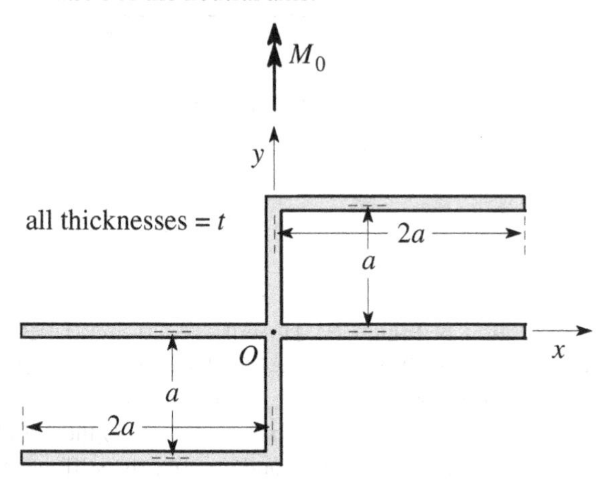


*Figure P4.21*

**4.22.** Find the location of the centroid and the second moments of area $I_x, I_y, I_{xy}$ for the thin-walled section of Figure P4.22.

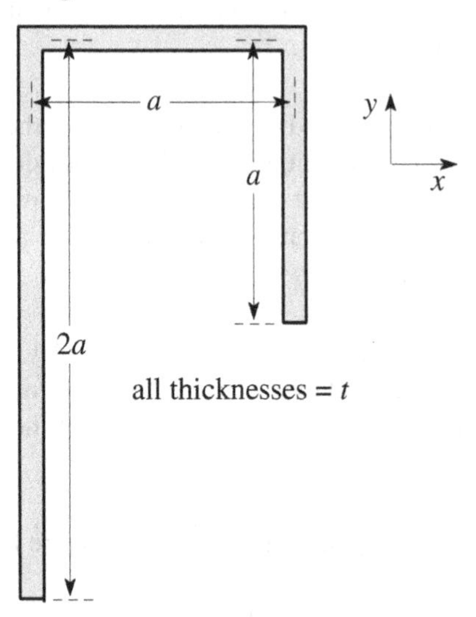


*Figure P4.22*

**4.23.** Find the location of the centroid and the second moments of area $I_x, I_y, I_{xy}$ for the section of Figure P4.23.

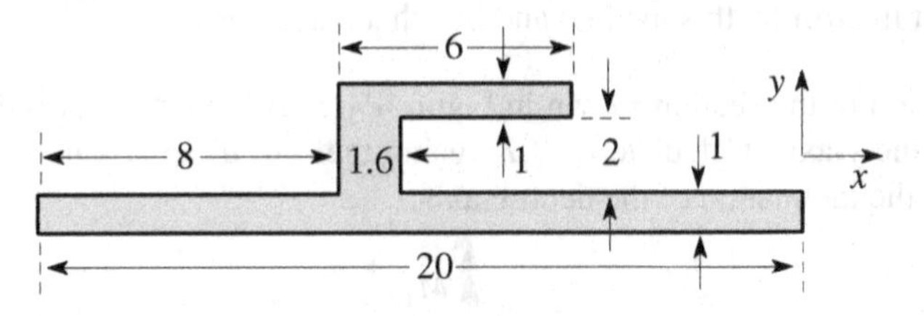


all dimensions in mm.

*Figure P4.23*

**Section 4.4**

**4.24.** For the cross section of Figure P4.21, with $a = 20$ mm, $t = 3$ mm, we obtain $I_x = 160 \times 10^3$ mm$^4$, $I_y = 256 \times 10^3$ mm$^4$, $I_{xy} = 96 \times 10^3$ mm$^4$. Sketch the Mohr's circle of second moments of area and hence determine the magnitudes of the principal second moments and the inclination of the principal axes to the $x$ and $y$ axes. Draw in the principal axes on a copy of the figure and identify the stiff and flexible axes as '1', '2', respectively.

**4.25.** Sketch Mohr's circle of second moments of area for the section of Figure P4.6, using the results for $I_x, I_y, I_{xy}$ given in problem 4.6. Hence determine the magnitudes of the principal second moments and the inclination of the principal axes to the $x$ and $y$ axes. Draw in the principal axes on a copy of the figure and identify the stiff and flexible axes as '1', '2', respectively.

**4.26.** Sketch Mohr's circle for second moments of area for the section of Figure P4.8, using the results for $I_x, I_y, I_{xy}$ given in problem 4.8. Hence determine the magnitudes of the principal second moments and the inclination of the principal axes to the $x$ and $y$ axes. Draw in the principal axes on a copy of the figure and identify the stiff and flexible axes as '1', '2', respectively.

**4.27.** Using the results for $I_x, I_y, I_{xy}$ given in problem 4.9, determine the magnitudes of the principal second moments and the inclination of the principal axes for the section of Figure P4.9. Draw in the principal axes on a copy of the figure and identify the stiff and flexible axes as '1', '2', respectively.

**4.28.** The second moments of area for the section of Figure P4.28 are

$$I_x = 26a^3t \;\; ; \;\; I_y = \frac{40a^3t}{3} \;\; ; \;\; I_{xy} = -14a^3t \, .$$

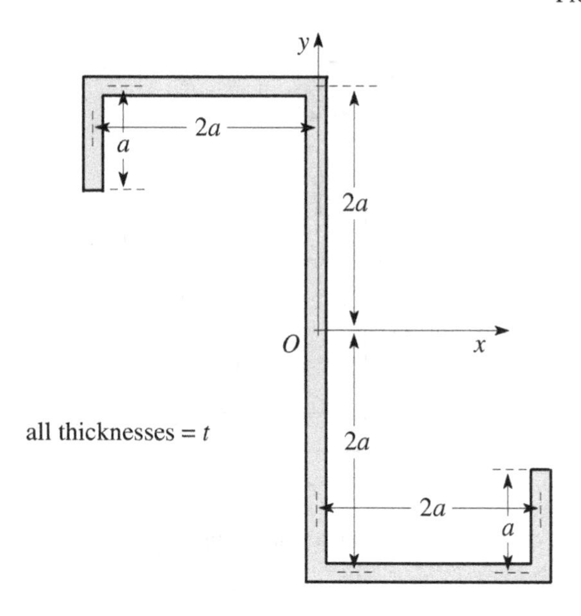


*Figure P4.28*

Find the principal second moments of area and the inclination of the principal axes. Draw in the principal axes on a copy of the figure and identify the stiff and flexible axes as '1', '2', respectively.

**4.29*.** A beam with second moments of area $I_x, I_y, I_{xy}$ is loaded by a bending moment, $M_0$ about the $x$-axis. Find an expression for the angle $\psi$ between the resulting neutral axis and the $x$-axis.

Devise a geometrical construction for the angle $\psi$ on the Mohr's circle of Figure 4.21 and hence show that for a given section, the maximum deviation between the moment axis and the neutral axis is

$$\psi_{max} = \sin^{-1}\left(\frac{I_1 - I_2}{I_1 + I_2}\right) .$$

Find the angle between the moment axis and the stiff axis (1) for this maximum deviation to occur.

## Section 4.4.3

**4.30.** Figure P4.30 shows the cross section of an unequal angle beam for which the principal second moments of area are $I_1 = 23.95 \times 10^6$ mm$^4$, $I_2 = 2.53 \times 10^6$ mm$^4$. The location of the centroid and the inclination of the principal axes are as shown. The beam is subjected to a bending moment of 10 kNm about the $x$-axis. Find the location and magnitude of the maximum tensile and compressive stress.

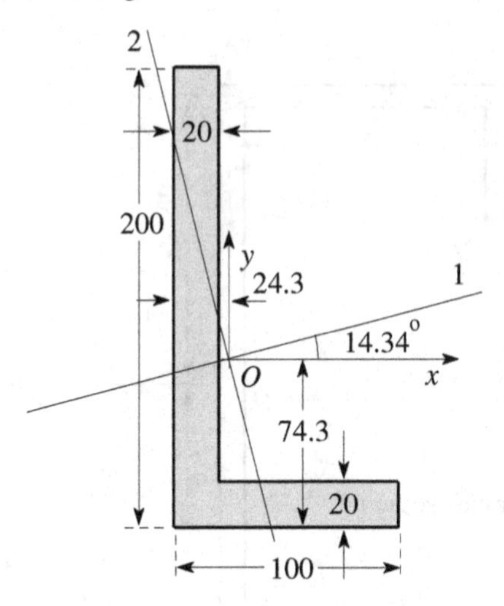


all dimensions in mm.

*Figure P4.30*

**4.31.** The beam of Problem 4.30 is subjected to a bending moment of 6 kNm about the $y$-axis. Find the location and magnitude of the maximum tensile and compressive stress.

**4.32.** The unequal Z-section of Figure 4.17 *(a)* has principal second moments $I_1 = 5.255a^3t, I_2 = 0.761a^3t$ and the inclination of the principal axes are as shown in Figure 4.23. The beam is loaded by a bending moment $M_0$ about the $x$-axis. Find the location and magnitude of the maximum tensile stress.

**Section 4.4.4**

**4.33.** Make a copy of the three beam sections in Figure P4.33 and use the 'least squares fit' method to estimate the location of the principal axes on each section. Label the stiff and flexible axes '1' and '2' respectively.

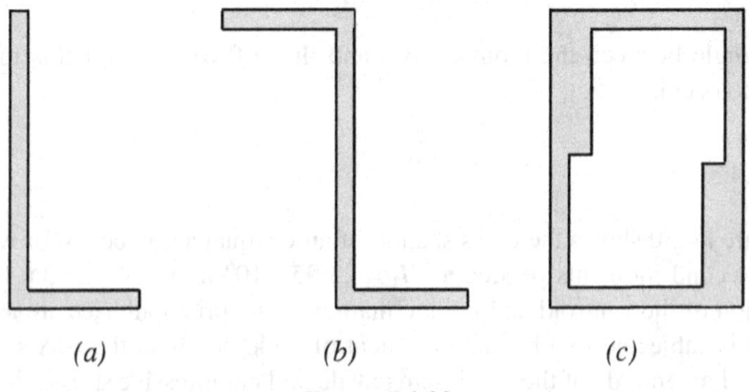


*Figure P4.33*

**4.34.** Make a copy of the beam section of Figure P4.17 and use the 'least squares fit' method to estimate the location of the principal axes on each section. Label the stiff and flexible axes '1' and '2' respectively.

**4.35.** Make a copy of the beam section of Figure 4.15 and use the 'least squares fit' method to estimate the location of the principal axes on each section. Label the stiff and flexible axes '1' and '2' respectively.

**4.36.** Make a copy of the beam section of Figure P4.22 and use the 'least squares fit' method to estimate the location of the principal axes on each section. Label the stiff and flexible axes '1' and '2' respectively.

**4.37.** A beam with the cross section of Figure P4.37 is loaded by a bending moment about the vertical axis as shown. Show on the figure your best estimate of (i) the location of the principal axes, (ii) the location of the neutral axis and (iii) the point $P$ where the maximum tensile stress will occur.

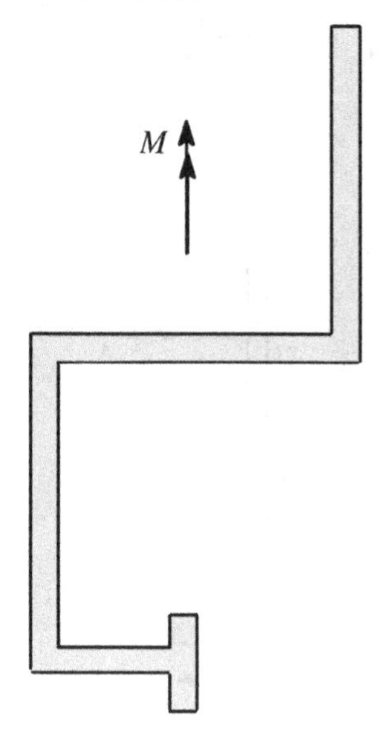


*Figure P4.37*

**4.38.** A beam with the S-section of Figure P4.19 is loaded by a bending moment about the $x$-axis. Show on the figure your best estimate of (i) the location of the principal axes, (ii) the location of the neutral axis and (iii) the point $P$ where the maximum tensile stress will occur.

**4.39.** Estimate the location of the principal axes for the beam section of Figure P4.6 using the 'least squares fit' method and hence find the approximate location of the neutral axis when the beam is loaded by a bending moment about the $x$-axis. If you previously solved Problem 4.6, compare your estimate for the inclination of the neutral axis with the exact calculation.

**4.40.** Estimate the location of the principal axes for the beam section of Figure P4.9 using the 'least squares fit' method and hence find the approximate location of the neutral axis when the beam is loaded by a bending moment about the $x$-axis. If you previously solved Problem 4.9, compare your estimate for the inclination of the neutral axis with the exact calculation.

## Section 4.4.5

**4.41.** The I-beam shown in Figure P4.41 is subjected to a bending moment which is intended to be aligned with the axis $Ox$, but which is actually inclined at $1^\circ$ towards $Oy$ because of a manufacturing error.

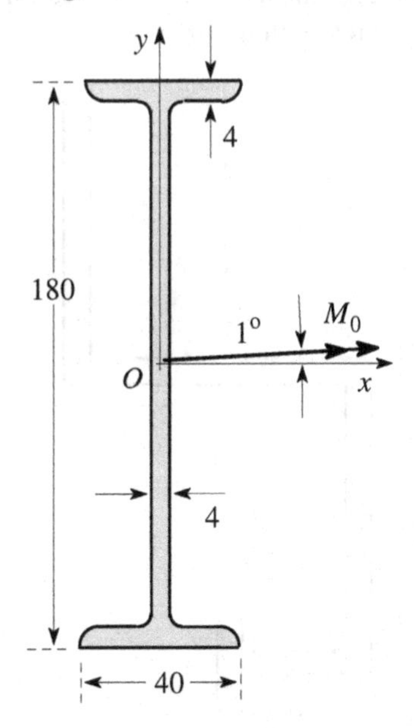


*all dimensions in mm.*

*Figure P4.41*

Find $I_x, I_y$ and hence the direction of the neutral axis, (which is also the axis about which bending occurs).

**4.42.** An I-beam section is to be used for a cantilever beam to support an end load whose direction may deviate from the vertical by $\pm 0.5^\circ$ due to manufacturing and assembly errors. Determine the maximum permissible ratio $I_1/I_2$ if the horizontal displacement is not to exceed 10% of the vertical displacement.

# 5

# Non-linear and Elastic-Plastic Bending

In this chapter, we shall consider the question of determining the stress distribution and curvature when a beam is subjected to bending moments sufficient to cause plastic deformation. The same analytical procedure can also be used for the bending of beams made of materials such as rubber, which have a non-linear elastic constitutive behaviour.

Engineering components are generally not expected to experience plastic deformation in normal service, but a calculation of the maximum load that can be carried in a single exceptional loading experience is often of interest from a safety viewpoint. A dramatic example is provided by studies of the crash behaviour of automobiles, which are based on elastic-plastic analysis of the structural elements of the vehicle.

We shall also find that when a beam is loaded into the plastic range and then released, it does not return to its original configuration and there are generally residual stresses remaining. This information forms the basis of analyses of simple metal forming processes, such as the forming of a curved bar by plastic bending of a bar that is initially straight.

## 5.1 Kinematics of bending

In the elementary theory of bending, it is customary to assume that initially plane sections remain plane. This assumption seems plausible, but the background to it is seldom discussed in any detail.

To fix ideas, suppose we have a very long beam, subjected to pure bending by the application of equal and opposite moments at the end. We imagine the beam as made up of a large number of identical slices, with initially parallel plane faces, as shown in Figure 5.1 *(a)*. The stress distribution near the ends of the beam will be affected by the precise way in which the moment is applied — i.e. by the traction distribution on the two end faces, but it is reasonable to assume that this end effect will decay as we get further from the ends and that there is some 'preferred' way in which the moment will be transmitted along the beam. This idea is known as *Saint-Venant's*

J.R. Barber, *Intermediate Mechanics of Materials*, Solid Mechanics and Its Applications 175, 2nd ed., DOI 10.1007/978-94-007-0295-0_5, © Springer Science+Business Media B.V. 2011

*principle*. Thus, sufficiently far from the ends of the beam, the stress distribution on all cross-sectional planes will be the same.

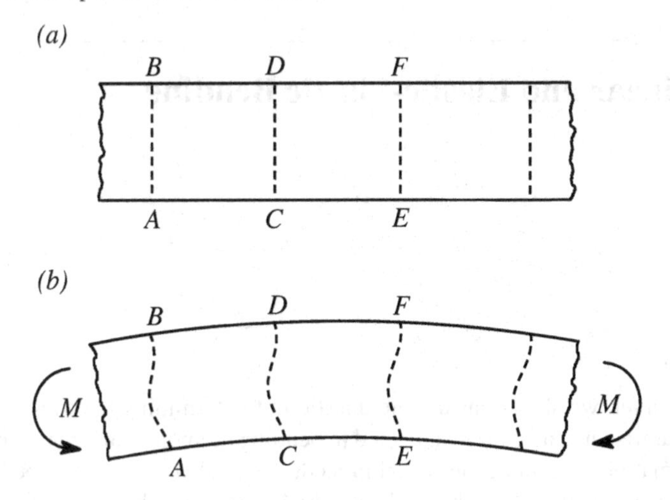


*Figure 5.1: Deformation of the beam: (a) the undeformed configuration; (b) after deformation*

In particular, the stress distribution across *AB* in Figure 5.1 *(a)* will be the same as that across *CD* and *EF*. It follows that the element *ABCD* is loaded in exactly the same way as the element *CDEF* and hence the deformed shapes of the two elements must be the same. This implies that the deformed shape of the initially straight line *AB* must be the same as that of *CD*.

Now the deformation of the elements must be one which preserves compatibility — in other words, adjacent slices must still be capable of stacking against each other without generating any gaps. We illustrate this in Figure 5.1 *(b)*. Each of the initially straight vertical lines must deform to the same shape and the most general deformation permitted in the beam is one in which (for example) *AB* deforms to an arbitrary new shape and *CD* deforms to the same shape, but with an arbitrary (small) rigid-body translation and rotation. The translation of *CD* to the right relative to *AB* will lead to a uniform tensile strain across the element *ABCD* and the rotation will lead to a tensile strain that varies linearly with $y$, regardless of the common deformed shape of *AB* and *CD*. To convince yourself of this, draw an arbitrary curve on a sheet of paper, copy the same curve onto a transparent sheet and experiment with various *small* rigid body displacements of one curve with respect to the other. You will find that the most general change in the distances between corresponding points on the curves due to the relative motion is a general linear function of $y$ and furthermore, that the multiplier on $y$ is proportional to the rigid-body relative rotation of the two curves in radians.

Similar considerations apply to the more general three-dimensional case and lead to the conclusion that regardless of the common deformed shape adopted by the various initially plane sections, the axial strain $e_{zz}$ must be a linear function of position in the section — i.e.

$$e_{zz} = C_0 + C_1 x + C_2 y \,, \tag{5.1}$$

where $C_0, C_1, C_2$ are unknown constants. This is all that is necessary to establish the bending theory. It is not necessary to make the more restrictive assumption that plane sections remain plane. To verify this, go back and review §4.1. You will find that we only used the assumption to arrive at equation (4.1), which is a reduced form of (5.1).

Notice that this kinematic argument makes no appeal to linearity in the stress-strain law, since it makes no reference to stress at all. Thus, it applies equally to any kind of material behaviour, including an elastic-plastic material loaded beyond the yield stress, a non-linear elastic material, or a generally anisotropic material.

As in §4.1, we can establish a relation between the constants $C_1, C_2$ and the radii of curvature of the beam, leading to the alternative expression

$$e_{zz} = C_0 - \frac{x}{R_y} + \frac{y}{R_x} \,. \tag{5.2}$$

It is sometimes convenient to define a Cartesian coordinate system aligned with the resultant bending axis. If $x', y'$ is rotated through $\theta$ anticlockwise from $x, y$, where

$$\tan\theta = -\frac{C_1}{C_2} \;\; ; \;\; \sin\theta = -\frac{C_1}{\sqrt{C_1^2+C_2^2}} \;\; ; \;\; \cos\theta = \frac{C_2}{\sqrt{C_1^2+C_2^2}} \,, \tag{5.3}$$

we have

$$x' = \frac{C_2 x}{\sqrt{C_1^2+C_2^2}} - \frac{C_1 y}{\sqrt{C_1^2+C_2^2}} \;\; ; \;\; y' = \frac{C_1 x}{\sqrt{C_1^2+C_2^2}} + \frac{C_2 y}{\sqrt{C_1^2+C_2^2}} \tag{5.4}$$

from (4.40, 4.41). It then follows from (5.1) and the second of (5.4) that

$$e_{zz} = C_0 + y'\sqrt{C_1^2+C_2^2} = C_0 + \frac{y'}{R} \,, \tag{5.5}$$

where the resultant curvature

$$\frac{1}{R} = \sqrt{\frac{1}{R_x^2} + \frac{1}{R_y^2}} \,. \tag{5.6}$$

## 5.2 Elastic-plastic constitutive behaviour

The uniaxial stress strain curve for a typical ductile material is shown in Figure 5.2. It exhibits a linear elastic region $OA$, beyond which the curve may be of quite complex shape. The yield stress $S_Y$ is defined as that stress beyond which some permanent deformation is produced — i.e. the stress-strain curve does not return to the point $O$ on unloading. Experimentally, we need to see a measurable level of permanent strain to be sure some yield has occurred, so the yield point $B$ is conventionally identified as the stress sufficient to leave a permanent tensile strain of 0.2% on unloading. The yield point $B$ is not necessarily identified with the limit of proportionality $A$, though for metals they are generally very close.

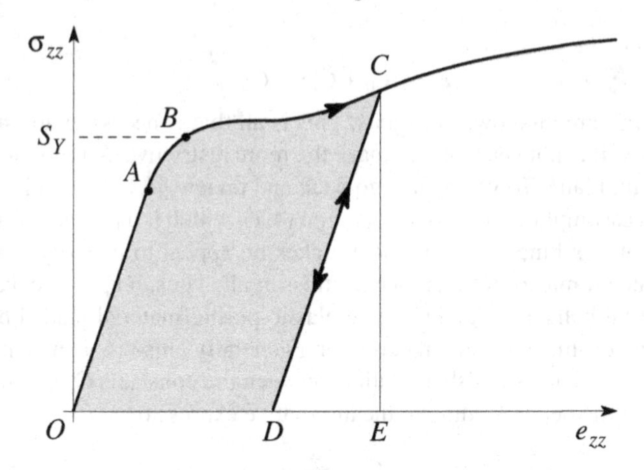


*Figure 5.2: Uniaxial stress strain curve for a typical ductile material*

Beyond the yield point, the stress continues to increase with strain, but the rate of increase (the slope of the curve) is orders of magnitude lower than that in the linear elastic range *OA*.

### 5.2.1 Unloading and reloading

If the stress is increased into the plastic range and then reduced, the material will generally unload along a straight line (*CD* in Figure 5.2) with the same slope as *OA*. Reloading from *D* will follow the same line *DC*, showing that the unloading/reloading process is elastic until the original maximum stress $\sigma(C)$ is exceeded. Thus a specimen which has been loaded and unloaded along the path *OABCD* behaves as a material with an increased yield stress $S'_Y = \sigma(C)$. In effect, it has become a new 'work-hardened' material. Remember that when we machine a specimen of a ductile material to perform a tensile test, we have no knowledge of the previous plastic deformation of the material, for example during the rolling process. We essentially measure just the current properties of the material.

During the loading process, work is done by the applied loads and the *specific work* (i.e. the work done per unit volume of material) is equal to the area under the stress-strain curve. In the elastic range, this work is stored in the material as strain energy (see §2.2.3) and is recovered on unloading. Beyond the yield point *B*, some of the work done is dissipated as heat and is not recoverable. For the loading path *OABCD*, the specific work done during loading is defined by the area *OABCE* and the work recovered during unloading is equal to the triangular area *CDE*. Thus, the shaded area *OABCD* in Figure 5.2 represents the energy dissipated as heat and *CDE* represents the elastic strain energy at point *C*. Notice that the elastic strain energy increases as a result of work hardening, so not all of the work done between *B* and *C* is dissipated as heat, but for most materials the rate of work hardening is sufficiently slow for this difference to be negligible.

### 5.2.2 Yield during reversed loading

So far we have discussed only those loading scenarios leading to positive (tensile) stresses, but many engineering applications involve alternating tensile and compressive stresses (see for example §2.3). For isotropic ductile materials, yielding in compression occurs at $\sigma_{zz} = -S_Y$ — i.e. at the same stress magnitude as in tensile loading. This result is implicit in both the Tresca and von Mises theories of ductile failure (§2.2.3).

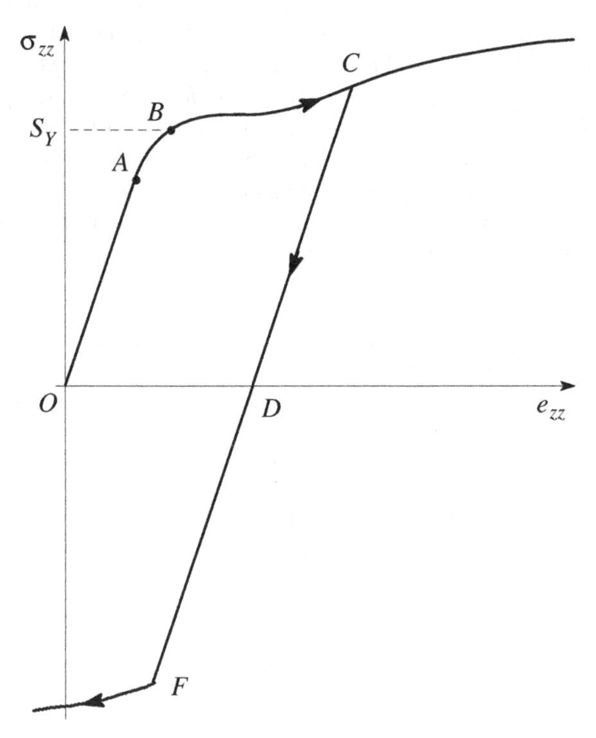


*Figure 5.3: Yield during reversed loading*

If the material has work hardened, this will generally imply a corresponding increase in compressive yield stress, so that negative loading following the scenario *OABCDF* in Figure 5.3 would be expected to involve compressive yield at $F$, with $CD = DF$. This is known as 'isotropic hardening', since it assumes that the material remains isotropic even after plastic deformation. However, significant anisotropic effects are often observed during plastic deformation, since the shape of the grains and the distribution of dislocations is influenced by the particular loading scenario. Also, a drop in the stress for reversed yield is often observed under uniaxial conditions. This is known as the Bauschinger effect.[1] The conditions at yield under complex

[1] See for example, F.A. McClintock and A.S. Argon (1966), *Mechanical Behaviour of Materials*, Addison-Wesley, Reading MA, §5.8

loading scenarios remain a subject of active research, since finite element methods now permit complex forming operations to be analysed and the constitutive law for the material in the plastic régime can place limits on the accuracy achievable.

### 5.2.3 Elastic-perfectly plastic material

In many applications, it is sufficient to use a simplified constitutive law with no work hardening — i.e. to assume that the stress remains at a constant yield stress $S_Y$ during plastic deformation, as shown in Figure 5.4, which also shows the assumed behaviour for unloading ($CD$) and for yield under reversed loading ($CF$). A material which behaves in this way is described as *elastic-perfectly plastic*. It represents quite a good approximation to Figures 5.2, 5.3 as long as the plastic strains are not too large and this in turn is likely to be true when the plastic zone is adjacent to material that is still elastic. For example, in the bending of beams, plastic strains sufficient to cause significant work hardening typically require the beam to be bent to a radius less than about five times the beam thickness.

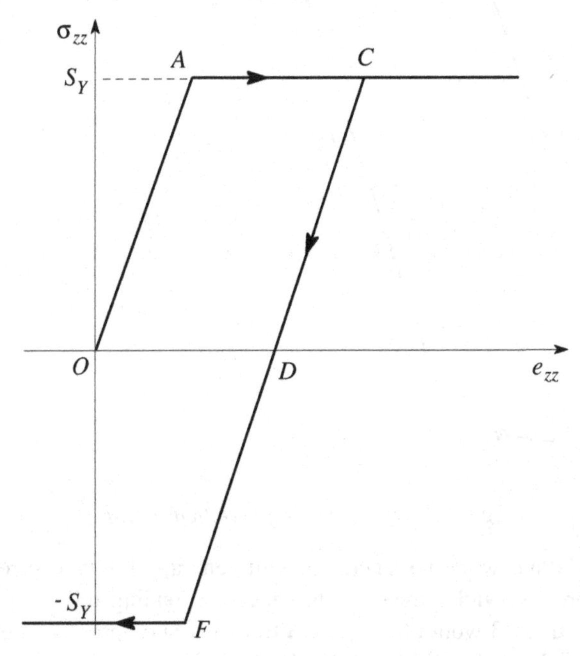


*Figure 5.4: Constitutive behaviour for an elastic-perfectly plastic material*

The loading portion $OAC$ of Figure 5.4 can be described by the equations

$$\sigma_{zz} = Ee_{zz} \quad ; \; |e_{zz}| < \frac{S_Y}{E} \tag{5.7}$$

$$= S_Y \,\text{sgn}(e_{zz}) \quad ; \; |e_{zz}| > \frac{S_Y}{E} \,, \tag{5.8}$$

where the expression $\mathrm{sgn}(e_{zz})$ takes the value $+1$ if $e_{zz} > 0$ and $-1$ if $e_{zz} < 0$, and is included to allow for the possibility of compressive loading $\sigma_{zz} < 0$, for which the corresponding strain will be negative.

## 5.3 Stress fields in non-linear and inelastic bending

We saw in §5.1 that the tensile strain $e_{zz}$ due to bending must be a linear function of $x, y$, as defined by equations (5.1, 5.2, 5.5). If the monotonic stress-strain relation is

$$\sigma_{zz} = f(e_{zz}) \; , \tag{5.9}$$

the stress will be given by

$$\sigma_{zz} = f(C_0 + C_1 x + C_2 y) = f\left(C_0 + \frac{y'}{R}\right) . \tag{5.10}$$

It follows that, as in elastic bending, there will be a neutral axis defined by the straight line

$$C_0 + C_1 x + C_2 y = C_0 + \frac{y'}{R} = 0 \tag{5.11}$$

on which the stress $\sigma_{zz} = 0$. Also, the stress is constant along any line parallel to the neutral axis. Indeed, a plot of the stress distribution along $y'$, perpendicular to the neutral axis, has the same form as the constitutive law except for a linear scaling of the axis.

These results apply for any function $f(e_{zz})$ in equation (5.9) and can therefore be used for beams made of non-linear elastic materials as well as for elastic-plastic bending.

For the special case of an elastic-perfectly plastic material, substitution of (5.5) into (5.7, 5.8) gives

$$\sigma_{zz} = EC_0 + \frac{Ey'}{R} \quad ; \quad \left| EC_0 + \frac{Ey'}{R} \right| < S_Y \tag{5.12}$$

$$= S_Y \,\mathrm{sgn}\left(C_0 + \frac{Ey'}{R}\right) \quad ; \quad \left| EC_0 + \frac{Ey'}{R} \right| > S_Y \, . \tag{5.13}$$

Thus, there will be an elastic region bounded by the two parallel lines

$$EC_0 + \frac{Ey'}{R} = \pm S_Y \; , \tag{5.14}$$

which are also parallel to and equidistant from the neutral axis, as shown in Figure 5.5.

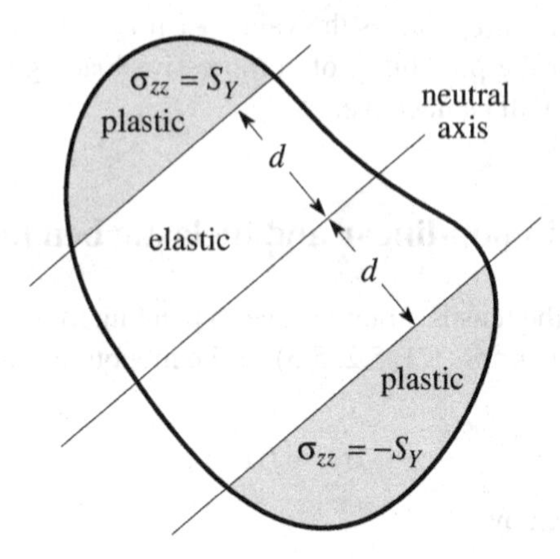


*Figure 5.5: Elastic and plastic zones in a beam cross section*

The perpendicular distance between each line and the neutral axis is

$$d = \frac{S_Y R}{E} . \tag{5.15}$$

Outside these lines, the material will have yielded and the stress will be equal to $\pm S_Y$, the positive sign being taken on one side and the negative on the other. At sufficiently small applied moments, $R$ will be large and the two bounding lines will lie outside the beam cross section, all of which will therefore lie in the linear elastic range. However, at some critical bending moment, denoted by $M_Y$, one or both of the bounding lines will touch the section and yielding will start. We refer to $M_Y$ as the *moment for first yield*. Further increase in moment will cause the bounding lines to move closer together, increasing the size of the plastic zone at the expense of the elastic zone.

The maximum moment that can be carried, $M_P$, occurs when the elastic zone has shrunk to zero. Equation (5.15) shows that this requires that the curvature be unbounded (i.e. that the radius of curvature $R$ be zero) so it represents local collapse of the beam. A point in a beam where this has happened is known as a *plastic hinge* (see §5.8 below). The moment $M_P$ is known as the *fully-plastic moment* for the beam section.

### 5.3.1 Force and moment resultants

If the constants $C_0, C_1, C_2$ defining the strain field are known, we can compute the force and moment resultants $F, M_x, M_y$ by substituting (5.10) into (4.9, 4.13, 4.14) — i.e.

$$F = \iint_A \sigma_{zz} dA \;\; ; \;\; M_x = \iint_A \sigma_{zz} y dA \;\; ; \;\; M_y = -\iint_A \sigma_{zz} x dA \tag{5.16}$$

and evaluating the integrals. In the more usual case where $F, M_x, M_y$ are specified, (5.16) provides three simultaneous equations which can be solved for the three unknowns $C_0, C_1, C_2$, after which substitution in (5.10) defines the complete stress distribution in the beam.

However, the problem is non-trivial because the non-linearity of the constitutive law (5.9) causes the resulting equations to be non-linear. In addition, we can no longer treat the three resultants separately and use superposition. For a general unsymmetrical section or a beam loaded by both an axial force and a bending moment, the neutral axis is likely to move and rotate as the loads increase and most cases are sufficiently complex to require a numerical solution.

## 5.4 Pure bending about an axis of symmetry

The solution is considerably simplified if:-

(i) there is no axial force,
(ii) the beam cross section is symmetric about the axis of the applied bending moment, and
(iii) the relation between stress and strain is the same in tension and compression,

since the neutral axis must then coincide with the axis of symmetry.

To prove this result, we first adopt it as a tentative hypothesis, in which case

$$e_{zz} = \frac{y}{R} . \tag{5.17}$$

Substituting for $\sigma_{zz}$ from (5.10) into (5.16), we then obtain

$$F = \iint_A f\left(\frac{y}{R}\right) dA \; ; \; M_x = \iint_A f\left(\frac{y}{R}\right) y dA \; ; \; M_y = -\iint_A f\left(\frac{y}{R}\right) x dA . \tag{5.18}$$

In view of the symmetry of the beam cross section, symmetrically disposed elements $dA$ with equal and opposite values of $y$ will make equal and opposite contributions to the first and third of these integrals as long as $f(y/R) = -f(-y/R)$ as required by condition (iii) above. Thus, the conditions $F = 0$, $M_y = 0$ are satisfied identically by the choice of the symmetry axis as neutral axis, justifying this initial assumption. The second of (5.18) then defines the relation between the bending moment $M_x$ and the radius of curvature of the beam $R$.

### Example 5.1

*A rectangular beam of (horizontal) width b and height h is made of a rubber for which the uniaxial constitutive law is approximated by the equation*

$$\sigma_{zz} = E\left(e_{zz} + \frac{e_{zz}^3}{e_0^2}\right) ,$$

*where $E, e_0$ are material constants. Find the relation between the bending moment $M$ and the radius of curvature $R$ for bending about a horizontal axis.*

*If the material fails at a tensile stress $\sigma = Ee_0$, find the largest moment that can be transmitted by the beam.*

Using equation (5.17) and the given constitutive law, we have

$$\sigma_{zz} = E\left(\frac{y}{R} + \frac{y^3}{R^3 e_0^2}\right) .$$

The second of (5.16) or (5.18) gives

$$M = E\int_{-h/2}^{h/2}\int_{-b/2}^{b/2}\left(\frac{y}{R} + \frac{y^3}{R^3 e_0^2}\right) ydxdy = Eb\int_{-h/2}^{h/2}\left(\frac{y}{R} + \frac{y^3}{R^3 e_0^2}\right) ydy$$
$$= Eb\left(\frac{h^3}{12R} + \frac{h^5}{160R^3 e_0^2}\right) ,$$

which defines the required relation between $M$ and $R$.

The maximum tensile stress occurs at $y = h/2$ and is

$$\sigma_{\max} = E\left(\frac{h}{2R} + \frac{h^3}{8R^3 e_0^2}\right) .$$

Thus, failure will occur when

$$\left(\frac{h}{2R} + \frac{h^3}{8R^3 e_0^2}\right) = e_0 \quad \text{or} \quad \left(\frac{h}{2Re_0}\right) + \left(\frac{h}{2Re_0}\right)^3 = 1 ,$$

which has the solution

$$\frac{h}{2Re_0} = 0.682 \quad \text{or} \quad R = 0.733\frac{h}{e_0} .$$

The bending moment at this limiting condition is

$$M_{\max} = Eb\left(\frac{h^2 e_0}{12 \times 0.733} + \frac{h^2 e_0}{160 \times (0.733)^3}\right) = 0.1296Ebh^3 e_0 .$$

### 5.4.1 Symmetric problems for elastic-perfectly plastic materials

The elastic-perfectly plastic constitutive law of equations (5.7, 5.8) has the same behaviour in tension and compression and hence satisfies condition (iii) above. As in Figure 5.5, there will be a central elastic region and symmetrically disposed plastic regions yielded in tension and compression respectively. The bending moment-curvature relation can be determined exactly as in the preceding example, but the discontinuity at $\sigma_{zz} = S_Y$ necessitates breaking the integral into three parts — one for the central elastic region and one for each of the plastic regions.

Some simplification follows from the fact that the stress is uniform in each plastic region, permitting the stress distribution there to be replaced by a force resultant through the local centroid. Figure 5.6 shows *(a)* a representative symmetric section and *(b)* the corresponding stress distribution when the beam is loaded into the plastic régime.

If the beam is bent to a radius $R$, the dimension $d$ is given by equation (5.15). The resultant bending moment can be written symbolically in the form

$$M = M_1 + M_2 + M_3 \equiv \iint_{A_1} \sigma_{zz} y dA + \iint_{A_2} \sigma_{zz} y dA + \iint_{A_3} \sigma_{zz} y dA \,, \tag{5.19}$$

where $A_1, A_2, A_3$ are the areas of the two plastic regions and the elastic region respectively, as shown in Figure 5.6*(a)*.

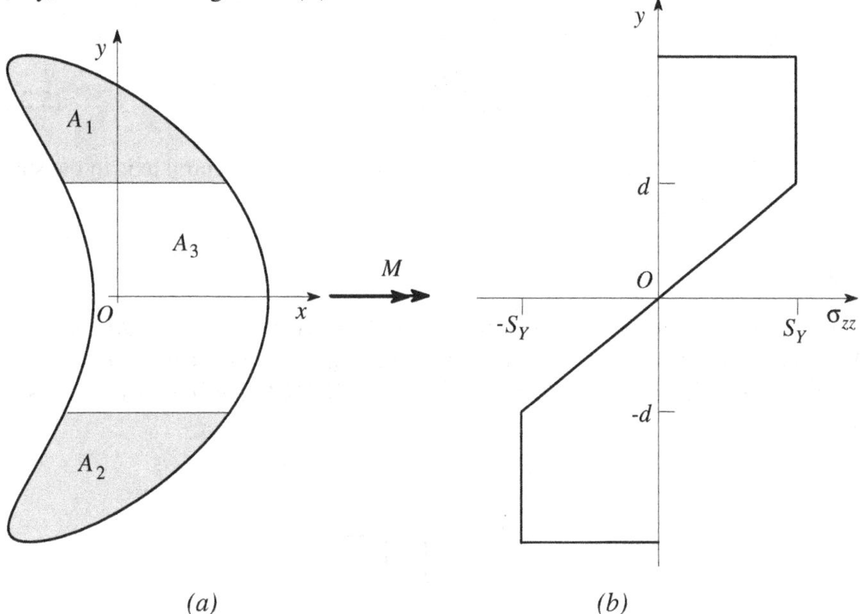


*Figure 5.6: Bending about an axis of symmetry: (a) elastic and plastic zones, (b) the corresponding stress distribution*

In $A_1$, the stress is uniform and equal to $S_Y$, so

$$M_1 = S_Y \iint_{A_1} y dA = S_Y A_1 \bar{y}_1 \,, \tag{5.20}$$

where $\bar{y}_1$ is the coordinate of the centroid of $A_1$. Similarly, in $A_2$, $\sigma_{zz} = -S_Y$, leading to

$$M_2 = -S_Y A_2 \bar{y}_2 \,, \tag{5.21}$$

but since the beam is symmetrical and loaded symmetrically, $\bar{y}_1 = -\bar{y}_2$, and hence $M_1 = M_2$ and $M_1 + M_2 = 2M_1$.

These results have a simple physical explanation. The stress is uniform over $A_1$ and hence the resultant is a force of magnitude $S_Y A_1$ acting through the centroid of $A_1$. As in §4.3.1, the value of $A_1\bar{y}_1$ for composite areas can conveniently be written as the sum of similar results for the component areas.

Turning now to the elastic region $A_3$, the expression for $M_3$ [the third integral in equation (5.19)] is exactly the same as the moment transmitted by an elastic beam of cross section $A_3$ bent to a radius $R$ — i.e.

$$M_3 = \frac{EI_3}{R} , \tag{5.22}$$

where $I_3$ is the second moment of area of the 'embedded elastic beam' of cross section $A_3$. Thus, the total moment can be written

$$M = \frac{EI_3}{R} + 2S_Y A_1 \bar{y}_1 \tag{5.23}$$

$$= S_Y \left( \frac{I_3}{d} + 2A_1\bar{y}_1 \right) , \tag{5.24}$$

using (5.15), where we recall that $d$ is the distance from the neutral axis to the start of the plastic region.

**Example 5.2**

*The I-beam of Figure 5.7 is loaded by a monotonically increasing bending moment $M$. Find the relation between $M$ and the radius of curvature $R$ if the material is elastic-perfectly plastic with Young's modulus $E=210$ GPa and yield stress $S_Y=400$ MPa.*

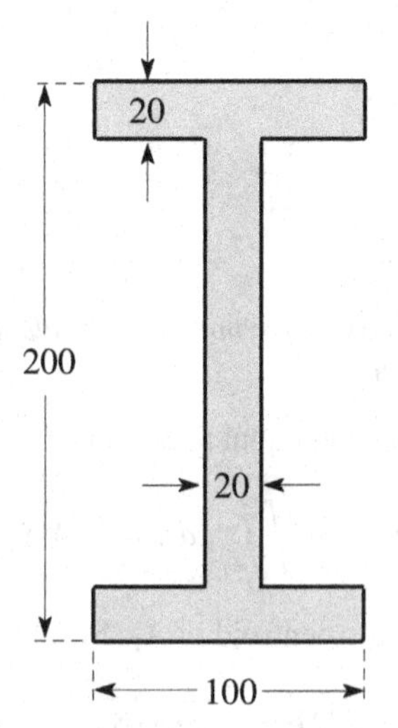


*Figure 5.7*

For low values of $M$, the beam will be entirely elastic and the relation between $M$ and $R$ is given by the elementary equation

$$M = \frac{EI}{R} ,$$

where $I$ is the second moment of area for the entire section, which we determine below.

After plastic deformation begins, we can distinguish the two qualitatively different states shown in Figures 5.8 *(a,b)*.

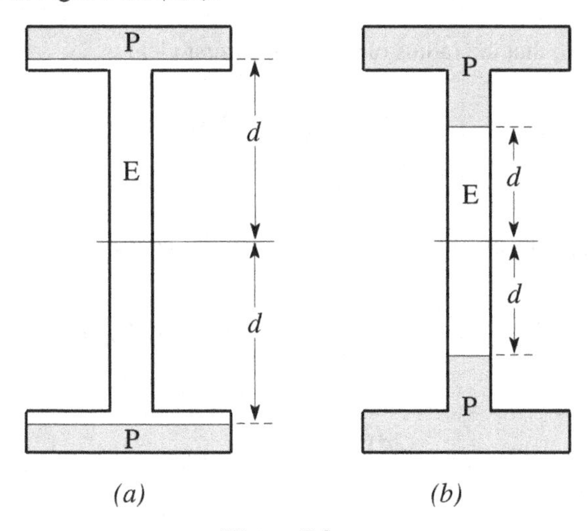


*Figure 5.8*

For case *(a)*, the second moment of area $I_3$ for the embedded elastic beam is

$$I_3 = \frac{(2d)^3 \times 0.1}{12} - \frac{0.16^3 \times 0.08}{12} = 0.0667d^3 - 27.3 \times 10^{-6} \text{ m}^4$$

and

$$A_1\bar{y}_1 = 0.1 \times (0.1 - d) \times \frac{(0.1 + d)}{2} = 0.5 \times 10^{-3} - 0.05d^2 \text{ m}^3 .$$

The value of $I$ for the full section, which we need to describe the solution in the elastic range, is obtained by setting $d = 0.1$ m in the expression for $I_3$ above — i.e.

$$I = 0.0667 \times (0.1)^3 - 27.3 \times 10^{-6} = 39.4 \times 10^{-6} \text{ m}^4 .$$

Substituting for $I_3$, $A_1\bar{y}_1$ in (5.24), we have

$$\begin{aligned} M &= S_Y(0.0667d^2 - 27.3 \times 10^{-6}d^{-1} + 10^{-3} - 0.1d^2) \\ &= 400 \times 10^6(10^{-3} - 0.0333d^2 - 27.3 \times 10^{-6}d^{-1}) \text{ Nm}. \end{aligned}$$

In particular, first yield corresponds to the value $d = 0.1$ and hence

$$M_Y = 157{,}600 \text{ Nm} = 157.6 \text{ kNm}.$$

The relation between $M$ and $R$ is obtained by substituting

$$d = \frac{S_Y R}{E} = 1.90 \times 10^{-3} R$$

and hence

$$M = 400 \times 10^6 (10^{-3} - 0.121 \times 10^{-6} R^2 - 14.33 \times 10^{-3} R^{-1}) \text{ Nm}.$$

We also note that the radius of curvature at first yield is

$$R_Y = \frac{0.1E}{S_Y} = 52.5 \text{ m},$$

from (5.15).

Case *(a)* applies as long as $0.08 < d < 0.1$ and hence the transition to case *(b)* occurs when

$$M = 400 \times 10^6 (10^{-3} - 0.0333 \times 0.08^2 - 27.3 \times 10^{-6} 0.08^{-1}) = 178100 \text{ Nm}.$$

For case *(b)*,

$$I_3 = \frac{(2d)^3 \times 0.02}{12} = 0.01333 d^3 \text{ m}^4$$

and

$$\begin{aligned} A_1 \bar{y}_1 &= 0.1 \times 0.02 \times 0.09 + (0.08 - d) \times 0.02 \times \frac{(0.08 + d)}{2} \\ &= 0.244 \times 10^{-3} - 0.01 d^2 \text{ m}^3 , \end{aligned}$$

where we have treated the flange and the section of web above $y = d$ as two separate areas for this computation.

Thus, using equation (5.24),

$$\begin{aligned} M &= S_Y (0.01333 d^2 + 0.488 \times 10^{-3} - 0.02 d^2 \\ &= 400 \times 10^6 (0.488 \times 10^{-3} - 6.67 \times 10^{-3} d^2) \text{ Nm} \\ &= 400 \times 10^6 (0.488 \times 10^{-3} - 24 \times 10^{-9} R^2) \text{ Nm}. \end{aligned}$$

The maximum bending moment $M_P$ occurs when $d = 0, R = 0$ and the section is fully plastic. We therefore have

$$M_P = 400 \times 10^6 \times 0.488 \times 10^{-3} = 195{,}200 \text{ Nm} = 195.2 \text{ kNm}.$$

The relation between $M$ and $R$ is illustrated in Figure 5.9.

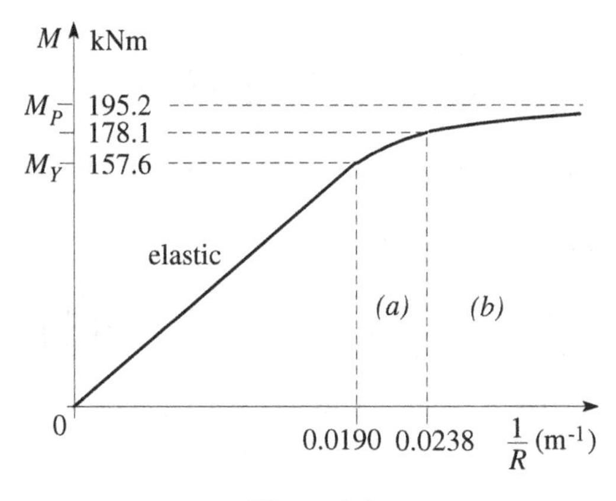


*Figure 5.9*

### 5.4.2 Fully plastic moment and shape factor

In the above example, Figure 5.9 shows that the curvature increases rapidly once we exceed the moment $M_Y$ for first yield and in fact the fully plastic moment $M_P$ is not very much larger than $M_Y$. We define the *shape factor*

$$f = \frac{M_P}{M_Y} \tag{5.25}$$

and for the I-beam of the example we have $f = 195.2/157.6 = 1.239$. In other words, a 24% increase in moment beyond first yield is sufficient to cause the beam to collapse completely. I-beams in general have relatively low values of shape factor because the flanges are responsible for the greater part of the moment transmitted in bending and the flange stresses are nowhere much below the yield stress when the outermost fibres yield.

From a design point of view, it is often sufficient to determine just $M_Y$ and $M_P$, since these define respectively the moment above which some permanent damage can be expected and the moment for total collapse of the beam. For the fully plastic case, equation (5.24) reduces to

$$M_P = 2S_Y A\bar{y}\,, \tag{5.26}$$

where $A$ is the area above the axis of symmetry and $\bar{y}$ is the coordinate of its centroid.

### Example 5.3

*Determine $M_Y, M_P$ and the shape factor $f$ for a solid circular section of diameter $d$, if the uniaxial yield stress for the material is $S_Y$.*

First yield will occur when the stress at $y = d/2$ is equal to $S_Y$, so

$$\frac{M_Y}{I} = \frac{2S_Y}{d}\,,$$

giving

$$M_Y = \frac{2S_Y}{d}\frac{\pi d^4}{64} = \frac{\pi S_Y d^3}{32} ,$$

where we have used the result that $I = \pi d^4/64$ for a circular cross section of diameter $d$.

For the fully plastic case, we have

$$A\bar{y} = \iint y dA = \int_0^{\pi}\int_0^{d/2} r\sin\theta r dr d\theta = \frac{2}{3}\left(\frac{d}{2}\right)^3 ,$$

for the area above the axis of symmetry, using the integration scheme of Figure 5.10.

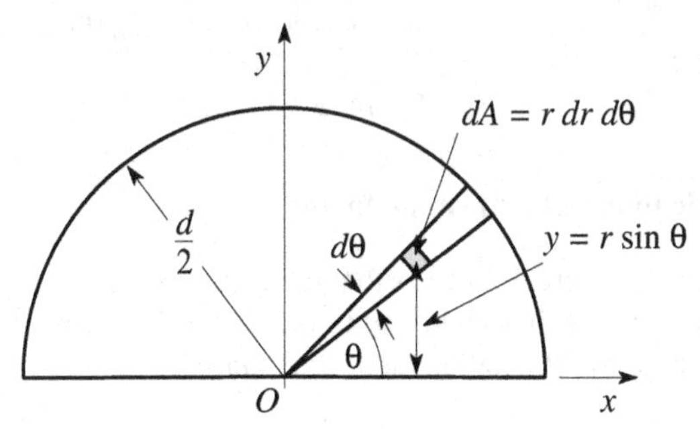


*Figure 5.10*

It then follows that

$$M_P = 2S_Y \times \frac{2}{3}\left(\frac{d}{2}\right)^3 = \frac{S_Y d^3}{6}$$

and the shape factor is

$$f = \frac{M_P}{M_Y} = \frac{S_Y d^3}{6}\frac{32}{\pi S_Y d^3} = 1.70 .$$

Notice that the shape factor is larger for the circle than for the I-beam, because the greater volume of material nearer to the axis of symmetry contributes significantly to the transmitted moment.

## 5.5 Bending of a symmetric section about an orthogonal axis

Figure 5.11 shows a beam section which is symmetric about the $y$-axis, but which is bent about the $x$-axis.

Suppose that the neutral axis is parallel to the $x$-axis, so that the strain and hence the stress is a function of $y$ only. It then follows from the third of equations (5.16) that $M_y = 0$, since symmetric elements with equal and opposite values of $x$ cancel each other in the integral. We therefore deduce that *the neutral axis will be parallel to the bending axis, when this is orthogonal to an axis of symmetry for the section.*

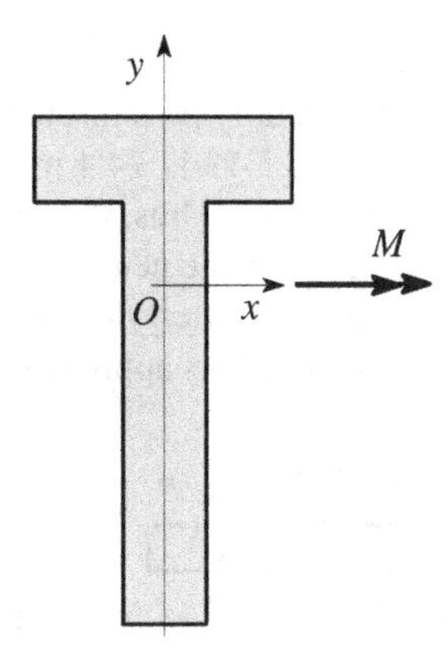


*Figure 5.11: Symmetric section bent about an orthogonal axis*

This argument is not subject to the other restrictions listed in §5.4 — in particular, it applies if there is an axial force and also if the monotonic stress-strain relation is different in tension and compression. However, the neutral axis will not generally pass through the centroid of the section and it will move (whilst remaining parallel to the bending axis) as the loading increases.

### 5.5.1 The fully plastic case

The simplest case is that of full plasticity, where every point in the section has yielded and is either at $S_Y$ or $-S_Y$ as shown in Figure 5.12.

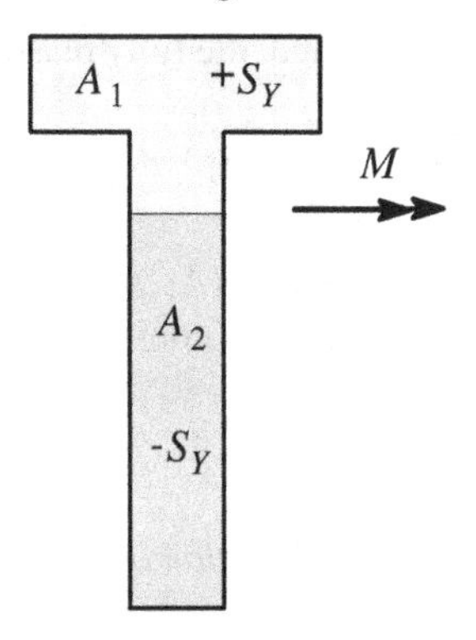


*Figure 5.12: Fully plastic state for bending of a symmetric section about an orthogonal axis*

The axial force is then

$$F = \iint_A \sigma_{zz} dA = \iint_{A_1} S_Y dA + \iint_{A_2} (-S_Y) dA = (A_1 - A_2) S_Y \ , \tag{5.27}$$

where $A_1, A_2$ are the areas above and below the neutral axis respectively. If there is no axial force, $F = 0$ and $A_1 = A_2$. In other words, the neutral axis for full plasticity divides the section into two equal areas. For sections that are not symmetric about the $x$-axis, this 'median line' does not generally pass through the centroid. For example, the centroid of the T-section of Figure 5.13 is 74.4 mm from the bottom, whereas the median line is 90 mm from the bottom. Thus, if we apply a gradually increasing bending moment $M$ to the T-section beam, the neutral axis will initially pass through the centroid, but it will migrate upwards after yielding starts, tending towards the median line as the fully plastic moment $M_P$ is approached.

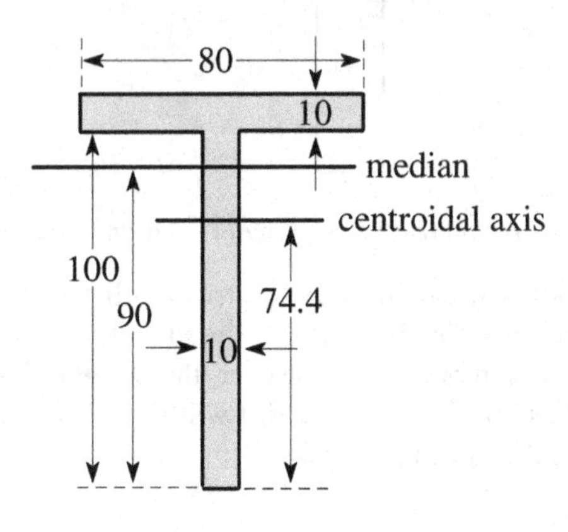


*Figure 5.13: Centroidal and median axes for a T-section*

Once the neutral axis has been found, the fully plastic moment is determined as

$$M_P = \iint_{A_1} S_Y y dA + \iint_{A_2} (-S_Y) y dA = S_Y (A_1 \bar{y}_1 - A_2 \bar{y}_2) , \qquad (5.28)$$

where $\bar{y}_1, \bar{y}_2$ define the centroids of the areas $A_1, A_2$. Since there is no axial force, any convenient origin may be used to define the coordinates $\bar{y}_1, \bar{y}_2$.

**Example 5.4**

*Find the fully plastic moment for the U-section of Figure 5.14, if the material yields at a stress $S_Y = 300$ MPa.*

The total area of the section is

$$A = 100 \times 100 - 80 \times 60 = 5200 \text{ mm}^2 ,$$

so the area above the median line must be $A/2 = 2600$ mm$^2$. The neutral axis is therefore 65 mm from the top of the section (since $65 \times 40 = 2600$) as shown in Figure 5.15.

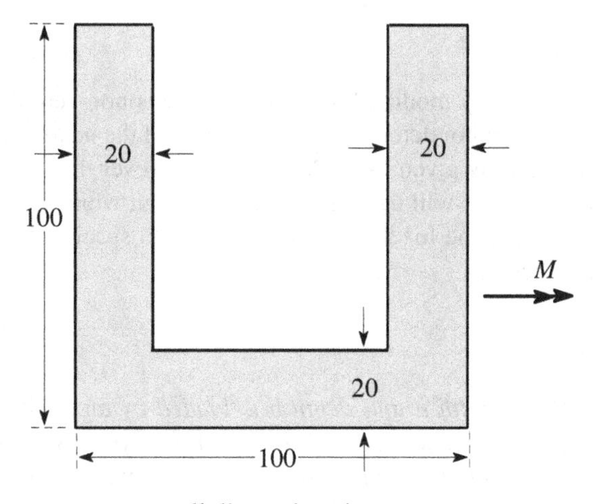


all dimensions in mm

*Figure 5.14*

We therefore have

$$A_1\bar{y}_1 = 65 \times 40 \times \frac{65}{2} = 84,500\ \text{mm}^3$$

$$A_2\bar{y}_2 = 20 \times 100 \times (-25) + 15 \times 40 \times \left(-\frac{15}{2}\right) = -54,500\ \text{mm}^3$$

and

$$M_P = 300 \times 10^6 (84500 + 54500) \times 10^{-9} = 41,700\ \text{Nm},$$

from equation (5.28).

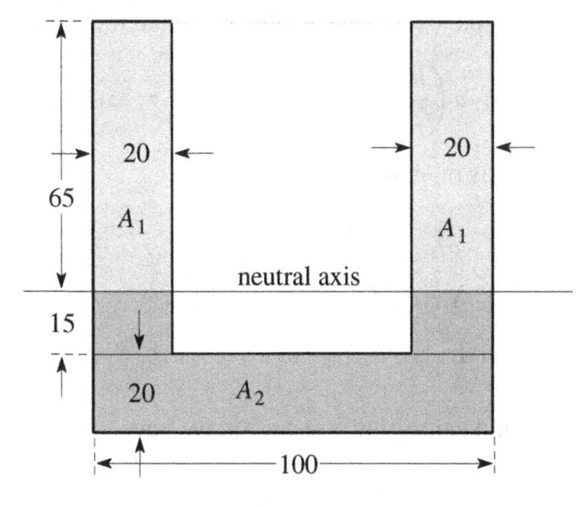


all dimensions in mm

*Figure 5.15*

### 5.5.2 Non-zero axial force

The method is only slightly modified if the axial force is non-zero. Equation (5.27) again provides a condition for determining the location of the neutral axis after which the fully plastic moment is given by (5.28). Notice however that if there is an axial force, the resultant moment will depend on the axis about which it is computed. An alternative description of the loading in such cases is to specify the resultant axial force and its line of action.

#### Example 5.5

*A rectangular beam of width $a$ and depth $b$ is loaded by an axial tensile force $F$ whose line of action is displaced a distance $d$ from the axis of symmetry, as shown in Figue 5.16. Find the collapse load $F_P$ and the dimension $c$ defining the location of the neutral axis and sketch their variation with $d$.*

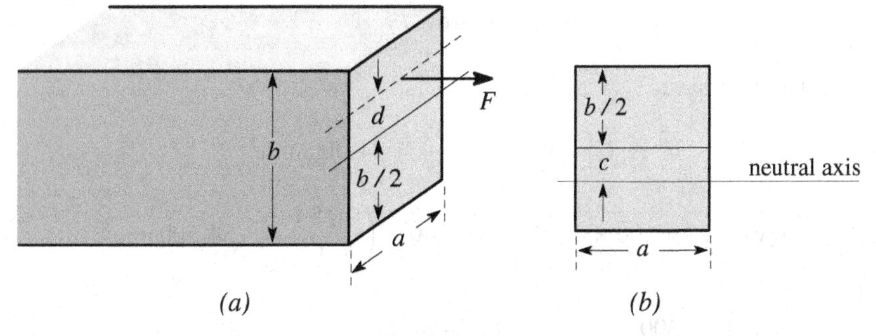


*Figure 5.16*

At collapse, the section above the neutral axis has yielded in tension and that below in compression, so the axial force is obtained from (5.27) as

$$F_P = S_Y a\left(\frac{b}{2}+c\right) - S_Y a\left(\frac{b}{2}-c\right) = 2S_Y ac\,.$$

The moment about the axis of symmetry is

$$\begin{aligned} M = F_P d &= S_Y a\left(\frac{b}{2}+c\right)\frac{\left(\frac{b}{2}-c\right)}{2} - S_Y a\left(\frac{b}{2}-c\right)\frac{\left(-\frac{b}{2}-c\right)}{2} \\ &= S_Y a\left(\frac{b^2}{4}-c^2\right)\,. \end{aligned}$$

Substituting for $F_P$ and cancelling a factor $S_Y$, we have

$$\frac{b^2}{4} - c^2 = 2cd\,,$$

with solution

$$c = \pm\sqrt{d^2 + \frac{b^2}{4}} - d\,.$$

If $d > 0$, only the positive square root[2] gives a physically reasonable result. For this case,

$$F_P = 2S_Y a\left(\sqrt{d^2 + \frac{b^2}{4}} - d\right)\,.$$

The dimensionless quantities

$$\frac{F_P}{S_Y ab}\ ;\ \frac{2c}{b}$$

both vary with the load offset $d$ as shown in Figure 5.17.

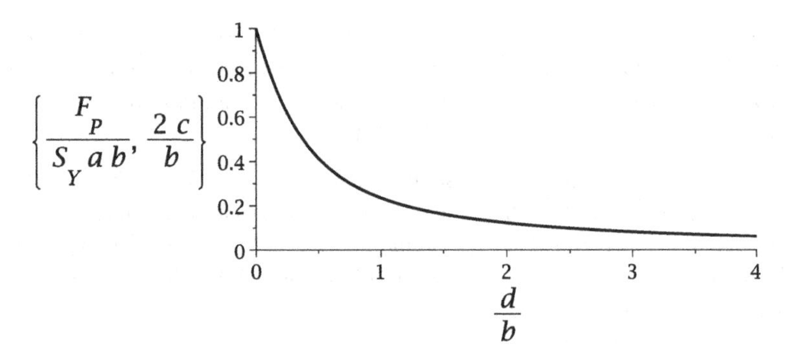


*Figure 5.17*

Notice that when $d = 0$, there is no offset of the load and we obtain $c = b/2$, which identifies the neutral axis with the lower edge of the cross section. This simply means that the whole section is above the neutral axis and hence is yielding in tension at $\sigma_{zz} = S_Y$, leading to a collapse load, $F_P = S_Y ab$.

For any non-zero value of $d$, however small, $c$ will be less than $b/2$ and there will be a region of compressive stress below the neutral axis.

### 5.5.3 The partially plastic solution

The problem of determining the stress field for applied moments between $M_Y$ and $M_P$ is rather more complicated. We first note that the beam will remain elastic as long as $\sigma_{zz} = My/I < S_Y$ for all values of $y$, measured from the centroidal axis. When this condition is violated, yield will start at the point in the section where $|y|$ is a maximum and this will generally be on one side of the axis only, since the section is unsymmetrical. This plastic zone will grow with $M$ until at some intermediate value a second plastic zone initiates on the other side of the axis. Both zones then grow at

[2] For $d < 0$, we would need to take the negative square root, but the solution would merely be a mirror image of that given here.

the expense of the elastic zone until eventually the section becomes fully plastic at $M_P$.

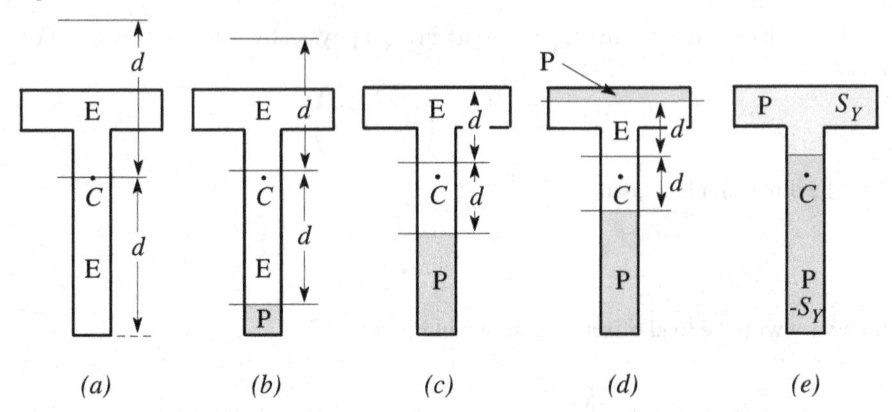


*Figure 5.18: Development of plastic zones for a symmetric section bent about an orthogonal axis: (a) first yield, (b) one plastic zone, (c) initiation of yield at top of section, (d) two plastic zones, (e) full plasticity*

This process is illustrated by the sequence of Figures 5.18 *(a–e)*. Notice how the plastic zones develop and also how the neutral axis moves away from the centroid $C$ as the moment is increased. Throughout the process, the position of the neutral axis and the dimension $d$ are determined in terms of the force $F$ and the moment $M_x$ using the first two of equations (5.16).

**Example 5.6**

*The T-beam of Figure 5.19 is loaded by a bending moment about the x-axis. Find an equation relating the bending moment M and the dimension c defining the extent of the plastic zone for the case illustrated where there is only one plastic zone. Also, find the value of c at which a second plastic zone will start to develop at the top of the section. The material yields at constant yield stress* $\pm S_Y$.

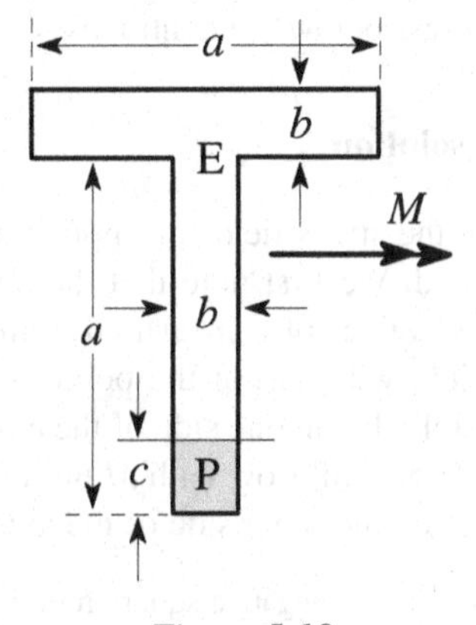


*Figure 5.19*

Since we don't yet know the location of the neutral axis, it is convenient to take the origin at the bottom of the section as shown in Figure 5.20 *(a)*. The material has yielded in compression ($\sigma_{zz} = -S_Y$) in $0 < y < c$ and the stress must vary linearly in the elastic range as shown in Figure 5.20 *(b)*. Thus

$$\begin{aligned}\sigma_{zz} &= -S_Y \quad &; \quad 0 < y < c \\ &= -S_Y + C(y-c) \quad &; \quad c < y < a+b\,,\end{aligned}$$

where $C$ is a constant to be determined. Notice how the second expression satisfies the condition that the stress in the elastic region just above $y = c$ should be on the point of yielding — in other words, the stress is a continuous function of $y$.

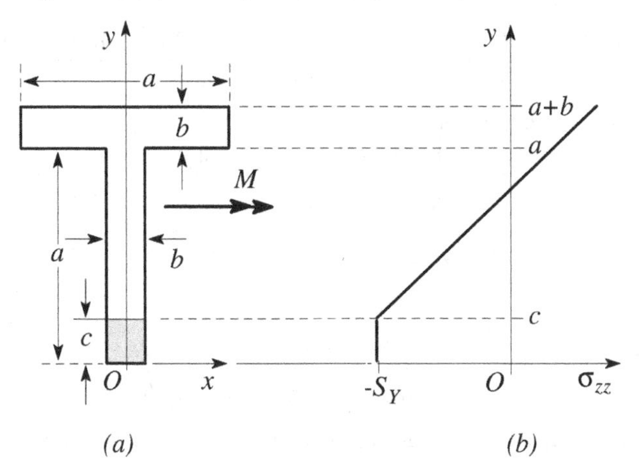


*Figure 5.20*

We can determine the value of $C$ from the condition that the total axial force be zero. From the first of (5.16), we have

$$F = \int\int_A \sigma_{zz} dA = -S_Y bc + \int_c^a [-S_Y + C(y-c)] b dy + \int_a^{a+b} [-S_Y + C(y-c)] a dy\,.$$

Performing the integrals and simplifying, we find

$$F = -2abS_Y + \frac{Cb}{2}(3a^2 + ab + c^2 - 4ac) = 0$$

and hence

$$C = \frac{4aS_Y}{3a^2 + ab + c^2 - 4ac}\,.$$

The neutral axis corresponds to the point $y = y_0$, where $\sigma_{zz} = 0$ and hence

$$-S_Y + C(y_0 - c) = 0\,.$$

Solving for $y_0$ and using the above value of $C$, we obtain

$$y_0 = c + \frac{S_Y}{C} = c + \frac{3a^2+ab+c^2-4ac}{4a} = \frac{3a^2+ab+c^2}{4a}\,.$$

The moment corresponding to this stress distribution can be found by taking moments about any convenient horizontal axis, since there is no axial force $F$. Thus, using the axis $y=0$ of Figure 5.20*(a)*, we have

$$M = \iint_A \sigma_{zz} y dA = \int_0^c (-S_Y) ybdy + \int_c^a [-S_Y + C(y-c)]bydy$$
$$+ \int_a^{a+b} [-S_Y + C(y-c)]aydy\,.$$

After performing the integrals and substituting for $C$, we find

$$M = \frac{S_Y ab(5a^3+6a^2b+12a^2c+5ab^2-9ac^2-3bc^2+4c^3-12abc)}{6(3a^2+ab+c^2-4ac)}\,.$$

This is the required relation between $M$ and $c$.

The second plastic zone will start when $\sigma_{zz}=S_Y$ at $y=a+b$ — i.e.

$$-S_Y + C(a+b-c) = S_Y\,.$$

Substituting for $C$ and simplifying, we obtain

$$a^2 - ab + c^2 - 2ac = 0\,,$$

with solution

$$c = a \pm \sqrt{a^2-(a^2-ab)} = a \pm \sqrt{ab}\,.$$

Only the negative square root makes sense here, so yield at the top of the section starts when

$$c = a - \sqrt{ab}\,.$$

It is clear from this example that the process is algebraically very complicated, even for a relatively simple beam section. The second stage of the process, where there are two plastic zones is equally complicated. The chances of making an algebraic error in such calculations is so large that they are only really practicable if performed symbolically using a language such as Mathematica or Maple or implemented numerically. Fortunately most of the important information for design against failure can be extracted from the limiting cases of first yield and full plasticity, which we have seen are easier to compute. The exception is the problem of residual stress and springback, which we discuss in §5.7 below and which unfortunately requires a solution for the stress field at the maximum applied load.

## 5.6 Unsymmetrical plastic bending

If the beam section has no axis of symmetry or if the bending moment is applied about an inclined axis, the problem is even more complicated and generally necessitates a numerical solution. However, the limiting solutions for first yield and full plasticity are still approachable algebraically.

For first yield, we perform an elastic analysis, using the techniques of Chapter 4. After finding the location and magnitude of the maximum stress $\sigma_{max}$, the yield moment $M_Y$ can then be determined by equating $\sigma_{max}$ to $\pm S_Y$ as appropriate.

For full plasticity, we know that the neutral axis will divide the section into two equal areas of tensile and compressive yield respectively, if there is no axial force. Because of the lack of symmetry, we do not know the inclination of the neutral axis, but we can obtain an equation to determine it by requiring that the resultant plastic moment be about the specified bending axis. One way to implement this condition is to take moments about an axis perpendicular to the applied moment axis and set the resultant to zero. This procedure leads to manageable algebraic calculations for thin-walled sections and for some other simple geometries.

### Example 5.7

*A beam of square cross section $a \times a$ is subjected to a bending moment about an axis inclined at an angle $\theta$ to the horizontal, as shown in Figure 5.21. The moment is sufficient to cause plastic collapse of the section. Assuming that the neutral axis is inclined at an angle $\alpha$ to the horizontal, find the relation between $\alpha$ and $\theta$ in the range $-\pi/4 < \theta < \pi/4$ and the corresponding value of the fully plastic moment $M_P$.*

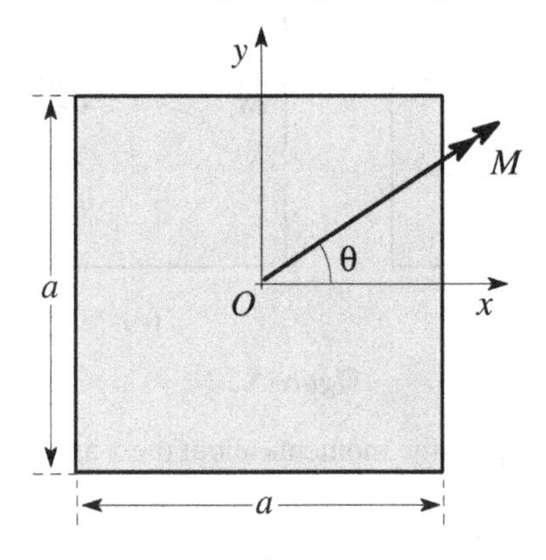


*Figure 5.21*

The inclined neutral axis must pass through the centre of the square, since any such line divides the section into two equal areas. If the axis is inclined at an angle $\alpha$, the stress at plastic collapse will be as shown in Figure 5.22 *(a)*.

For the purposes of calculation, it is convenient to consider this as the sum[3] of the distributions of Figure 5.22 *(b,c)*. The area of each of the triangles in Figure 5.22 *(c)* is

$$A_t = \frac{1}{2}\left(\frac{a}{2}\right)\left(\frac{a}{2}\tan\alpha\right) = \frac{a^2\tan\alpha}{8}$$

[3] Notice that this superposition is valid as long as we only use it to calculate force and moment resultants, but the deformations due to the stress fields of Figure 5.22 *(b,c) do not* generally sum to those due to Figure 5.22 *(a)*, since the constitutive law is non-linear.

and their centroids are at the points

$$\left(\frac{a}{3}, \frac{a\tan\alpha}{6}\right) \; ; \; \left(-\frac{a}{3}, -\frac{a\tan\alpha}{6}\right) .$$

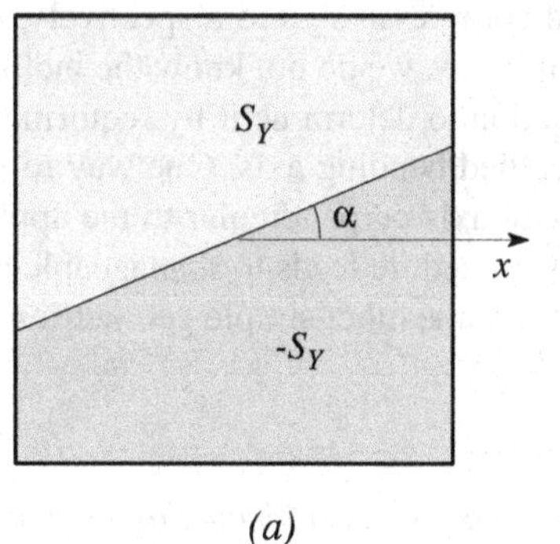


(a)

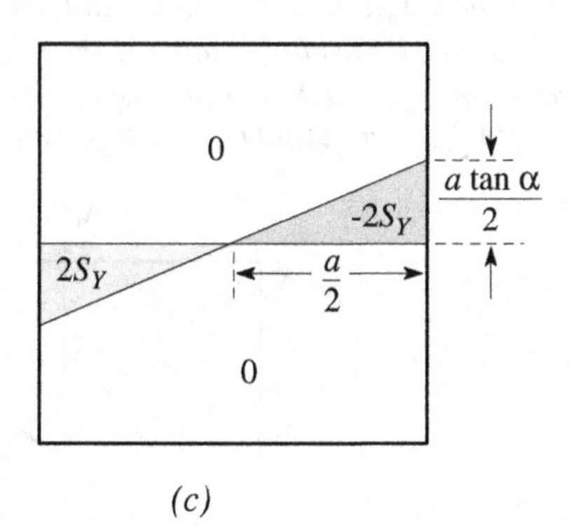


(b) (c)

*Figure 5.22*

We can therefore compute the moments about the $x$ and $y$ axes as

$$\begin{aligned} M_x &= 2S_Y(a)\left(\frac{a}{2}\right)\left(\frac{a}{4}\right) - 2(2S_Y)\left(\frac{a^2\tan\alpha}{8}\right)\left(\frac{a\tan\alpha}{6}\right) \\ &= a^3 S_Y\left(\frac{1}{4} - \frac{1}{12}\tan^2\alpha\right) \\ M_y &= 2(2S_Y)\left(\frac{a^2\tan\alpha}{8}\right)\left(\frac{a}{3}\right) \\ &= \frac{a^3 S_Y \tan\alpha}{6} . \end{aligned}$$

The angle of inclination of the applied moment axis can then be written

$$\tan\theta = \frac{M_y}{M_x} = \frac{2\tan\alpha}{(3 - \tan^2\alpha)} ,$$

with solution

$$\tan\alpha = \frac{\sqrt{1+3\tan^2\theta}-1}{\tan\theta} \; .$$

This is the required relation between $\alpha$ and $\theta$. It is valid only in the range $-\pi/4 < \alpha < -\pi/4$, since beyond this range the neutral axis will intersect the top and bottom of the section rather than the sides. However, the same result can be extended into other ranges by rotating the figure through 90°.

The variation of $\alpha$ with $\theta$ (in degrees) is illustrated in Figure 5.23 *(a)*. Notice that $\alpha = \theta$ at $\theta = -\pi/4$, 0, $\pi/4$ etc. due to symmetry.

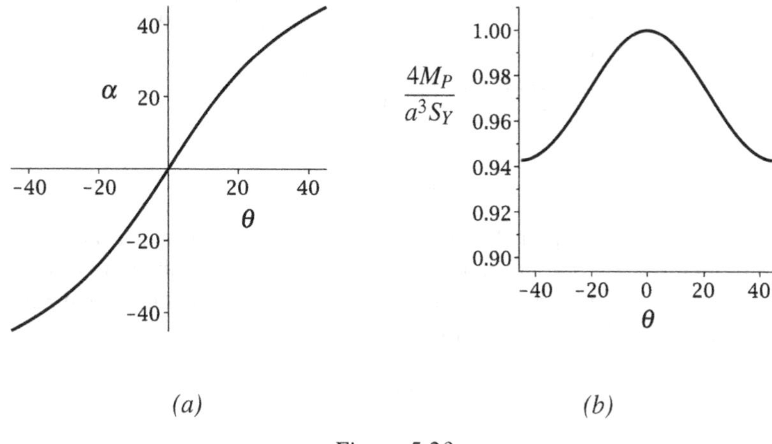


*(a)* *(b)*

*Figure 5.23*

The resultant fully plastic moment is

$$M_P = \sqrt{M_x^2 + M_y^2} = \frac{a^3 S_Y}{12}\sqrt{9 - 2\tan^2\alpha + \tan^4\alpha} \; .$$

This expression is shown in Figure 5.23 *(b)*. The moment $M_P$ varies between a maximum of $0.25a^3 S_Y$ at $\theta = 0$ and a minimum of $0.236a^3 S_Y$ at $\theta = \pm\pi/4$.

**Example 5.8**

*The thin-walled unsymmetrical U-section of Figure 5.24 is loaded by a moment about the horizontal axis. Find the fully plastic moment $M_P$ and the inclination of the corresponding neutral axis.*

There is no axial force and hence the areas above and below the neutral axis must be equal, giving

$$(a - d_1) + (2a - d_2) = d_1 + a + d_2$$

— i.e.

$$2(d_1 + d_2) = 2a \; .$$

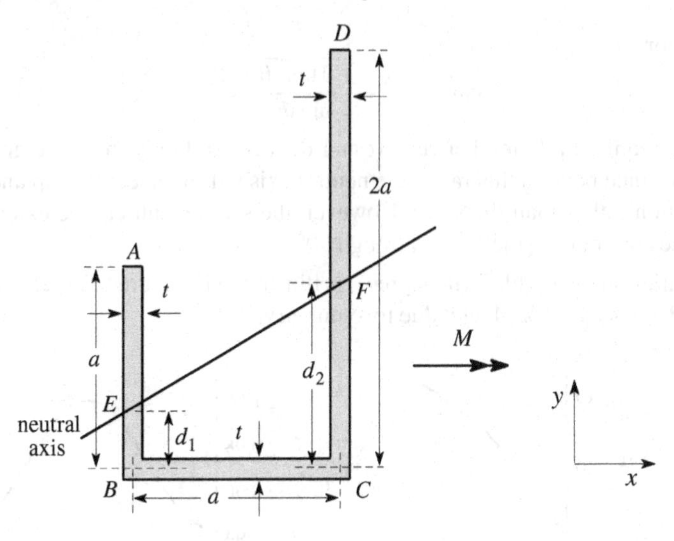


*Figure 5.24*

Since the moment is applied about the $x$-axis, the component $M_y$ must be zero. It is convenient to enforce this condition by taking moments about a vertical axis through the point $B$ in Figure 5.24, since then the forces acting on $AE$ and $EB$ have zero moment arm and make no contribution. We obtain

$$-S_Y(at)\frac{a}{2} - S_Y(d_2t)a + S_Y(2a-d_2)ta = 0$$

— i.e.

$$d_2 = \frac{3a}{4} .$$

We can then recover $d_1$ as

$$d_1 = a - d_2 = \frac{a}{4} .$$

This completes the solution for the neutral axis and shows that the horizontal bending moment causes the beam to bend about an axis that is inclined at an angle of $\tan^{-1}(0.5)$ to the horizontal. The neutral axis also completely defines the stress field at full yield and the plastic moment can therefore be obtained by taking moments about the horizontal axis passing through $BC$, giving

$$\begin{aligned} M_P &= S_Y(a-d_1)t\left(\frac{a+d_1}{2}\right) + S_Y(2a-d_2)t\left(\frac{2a+d_2}{2}\right) - S_Y d_1 t\left(\frac{d_1}{2}\right) \\ &\quad - S_Y d_2 t\left(\frac{d_2}{2}\right) \\ &= \frac{15 S_Y a^2 t}{8} , \end{aligned}$$

after substituting for $d_1, d_2$.

## 5.7 Unloading, springback and residual stress

If a beam is loaded into the plastic range and then unloaded, there will be some permanent residual curvature and the material will be left in a state of residual stress. We first note that the argument of §5.1 remains valid during the unloading process, so the strain is at all times a linear function of position in the section. As we start to reduce the bending moment, the magnitude of the strain will decrease and those parts of the cross section that have yielded will unload along a parallel elastic line such as $CD$ in Figures 5.2, 5.3. In most cases, the unloading process is completely elastic (i.e. no further yield occurs as the bending moment is reduced to zero) and hence the *change* in the stress field during unloading is identical to that produced in a corresponding elastic beam subjected to the same *change* of bending moment.

Figure 5.25 shows the evolution of the stress distribution for a symmetric beam bent about an axis of symmetry. We suppose the maximum bending moment $M_{max}$ lies in the range $M_Y < M_{max} < M_P$, so that the corresponding stress distribution has the form of Figure 5.25 *(a)*, with plastic zones at the top and bottom of the beam, separated by a central elastic region of depth $2d$. As the moment is reduced, the strain and hence the stress reduces by an amount proportional to the distance from the neutral axis, leading to a stress distribution of the form of Figure 5.25 *(b)*. If the moment is completely removed, the stress distribution must have zero resultant moment and hence must have regions of tensile and compressive stress on each side of the neutral axis, as shown in Figure 5.25 *(c)*.

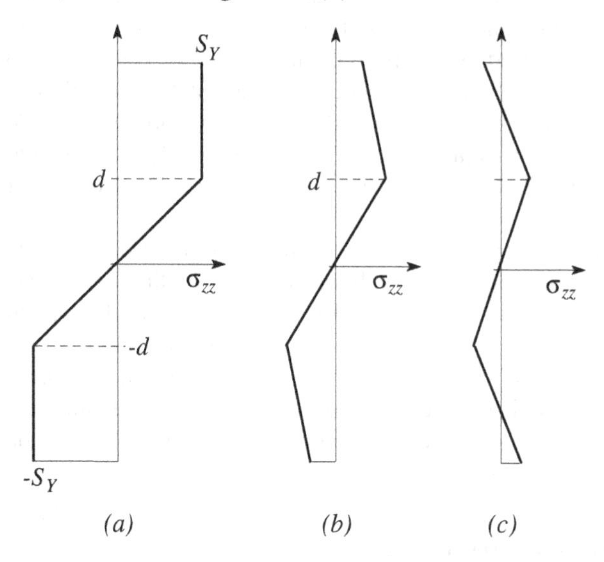


*Figure 5.25: Evolution of the stress distribution during loading and unloading: (a) maximum load, (b) partial unloading, (c) complete unloading*

In algebraic terms, the stress field at $M_{max}$ (Figure 5.25 *(a)*) can be written

$$\sigma_{zz} = S_Y \quad ; \quad y > d \tag{5.29}$$

$$= \frac{S_Y y}{d} \;\; ; \;\; -d < y < d \tag{5.30}$$

$$= -S_Y \;\; ; \;\; y < -d \tag{5.31}$$

and that for any stage of elastic unloading is

$$\sigma_{zz} = S_Y - \frac{M^* y}{I} \quad ; \; y > d \tag{5.32}$$

$$= \left(\frac{S_Y}{d} - \frac{M^*}{I}\right) y \; ; \; -d < y < d \tag{5.33}$$

$$= -S_Y - \frac{M^* y}{I} \quad ; \; y < -d \, , \tag{5.34}$$

where $M^* = M_{max} - M$ is the *reduction* in bending moment from the elastic-plastic state described by equations (5.29–5.31). In the special case of complete unloading [Figure 5.25 *(c)*], where we reduce the applied moment to zero, $M^* = M_{max}$ and hence the residual stress field is the difference between that which is obtained by elastic-plastic analysis at $M_{max}$ and that which would have been obtained at the same moment ($M_{max}$), *if the material had had a sufficiently high yield stress to remain elastic.* Students sometimes have difficulty with this concept, arguing that if the superposed stress field due to $M^*$ is one that could not have been applied to the original beam without causing yield, then some yielding should occur on unloading. However, yielding is governed by the instantaneous value of stress — not by its change from some previous value. Indeed, in the uniaxial stress-strain curve of Figure 5.3, the stress can be reduced by $2S_Y$ from point $C$ before yielding occurs at $F$, whereas an unstressed bar could only experience a change of $S_Y$ without yielding. For the bending problem, as long as the stress defined by equations (5.32–5.34) remains in the range $-S_Y < \sigma_{zz} < S_Y$ at all points for $0 < M^* < M_{max}$, the unloading process will remain elastic. This is always the case when bending occurs about an axis of symmetry.

An exactly similar process of superposition can be used for beams which are not bent about an axis of symmetry. In other words, we first solve the elastic-plastic problem as in §§5.5, 5.6 to establish the stress field under the maximum bending moment and then subtract the solution of an equivalent elastic problem to obtain the residual stress. Notice however that in this case, the neutral axis for the two superposed stress fields will be different. If the maximum bending moment is at or near to $M_P$, this can cause a small amount of additional yielding to occur near the neutral axis during unloading.

### 5.7.1 Springback and residual curvature

During unloading, the curvature is reduced and the magnitude of this change is given by the beam bending equation as

$$\frac{1}{R^*} = \frac{M^*}{EI} \, . \tag{5.35}$$

This reduction in curvature is known as 'springback'. The final radius of curvature $R_u$ on complete unloading can be determined as

$$\frac{1}{R_u} = \left(\frac{1}{R}\right)_{\max} - \frac{1}{R^*} = \frac{S_Y}{Ed} - \frac{M_{\max}}{EI} , \tag{5.36}$$

where $(1/R)_{\max}$ is the curvature at the maximum bending moment $M_{\max}$.

**Example 5.9**

*The T-beam of Figure 5.26 is made of an aluminium alloy with* $E = 80$ *GPa and* $S_Y = 200$ *MPa. It is bent to a radius of* 10 *m and then released. Find the final radius of curvature and the residual stress distribution.*

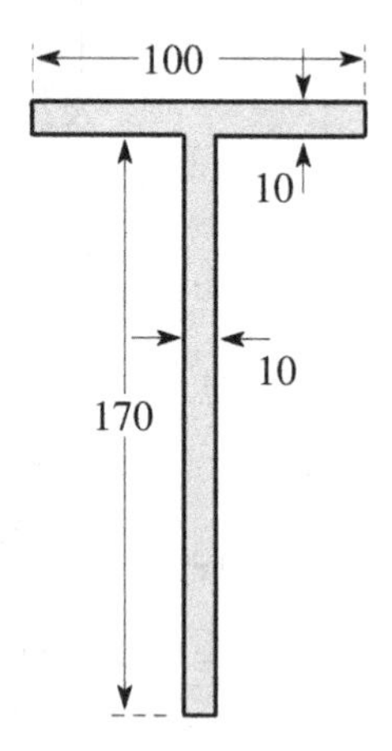


all dimensions in mm

*Figure 5.26*

We must first determine the applied moment $M_{\max}$ and the corresponding stress distribution. From equation (5.15), we have

$$d = \frac{S_Y R}{E} = \frac{200 \times 10^6 \times 10}{80 \times 10^9} = 0.025 \text{ m } = 25 \text{ mm.}$$

The location of the neutral axis is determined by the condition that the axial force $F = 0$. However, since $d$ is considerably less than the total depth of the section, it is likely that there is an upper plastic zone and that it extends down into the web as shown in Figure 5.27 *(a)*. In this case, the stress distribution will be as shown in Figure 5.27 *(b)*, where $c$ is an unknown dimension to be determined. The condition $F = 0$ then requires that

$$(0.1 \times 0.01 + c \times 0.01)S_Y - ([0.17 - 0.05 - c] \times 0.01)S_Y = 0 ,$$

since the elastic region makes self-cancelling contributions to the axial force. Solving for $c$, we have

$$c = 0.01 \text{ m.}$$

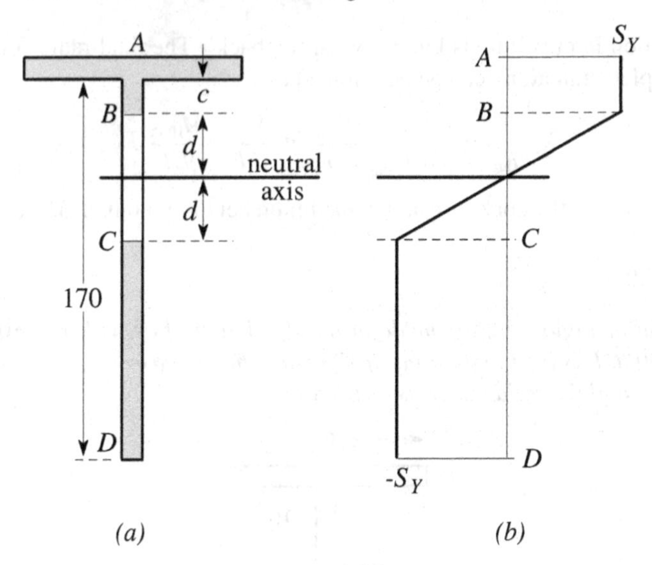


*Figure 5.27*

The corresponding bending moment is

$$\begin{aligned} M_{\max} &= S_Y(0.1\times 0.01\times 0.04 + 0.01\times 0.01\times 0.03) \\ &\quad - S_Y(0.11\times 0.01\times(-0.08)) + 2S_Y\left(\frac{0.01\times 0.025}{2}\right)\left(\frac{2}{3}\times 0.025\right) \\ &= 200\times 10^6\times 13.517\times 10^{-3} = 27{,}033 \text{ Nm}. \end{aligned}$$

For the unloading problem, we require the location of the centroid and the second moment of area $I_x$. Using the method of §4.3 and treating the web and the flange as areas $A_1, A_2$ respectively, we can construct the first three rows of Table 5.1 (in mm). Distances are measured from the bottom of the section.

| $i$ | 1 | 2 |
|---|---|---|
| $A_i$ | 1700 | 1000 |
| $\bar{y}_i$ | 85 | 175 |
| $I_x^i$ | $4.09\times 10^6$ | $8.3\times 10^3$ |
| $(\bar{y}_i - \bar{y})$ | $-33.3$ | 56.7 |

*Table 5.1*

Thus, the total area is

$$A = 1700 + 1000 = 2700 \text{ mm}^2$$

and the location of the centroid is defined by

$$A\bar{y} = 1700 \times 85 + 1000 \times 175 ,$$

giving

$$\bar{y} = 118.3 \text{ mm}.$$

This permits us to complete the last row of the table, after which we obtain

$$\begin{aligned} I_x &= 4.09 \times 10^6 + 8.3 \times 10^3 + 1700 \times 33.3^2 + 1000 \times 56.7^2 \\ &= 9.202 \times 10^6 \text{ mm}^4 = 9.202 \times 10^{-6} \text{ m}^4 . \end{aligned}$$

The residual curvature is therefore given by (5.36) as

$$\frac{1}{R_u} = \frac{1}{10} - \frac{27033}{80 \times 10^9 \times 9.202 \times 10^{-6}} \text{ m}^{-1}$$

from which

$$R_u = 15.8 \text{ m}.$$

Notice that this represents a substantial springback from the radius of 10 m at $M_{\max}$.

The final stress distribution is obtained by subtracting the distribution

$$\sigma_{zz} = \frac{M_{\max} y}{I} = \frac{27033 y}{9.202 \times 10^{-6}} = 2938 y \text{ MPa}$$

($y$ in mm) from the distribution of Figure 5.27 *(b)*, noting however that $y$ is measured from the elastic neutral axis. This gives the piecewise linear stress distribution sketched in Figure 5.28, which is completely determined by the stresses at the points $A, B, C, D$, in Figure 5.27 *(a)*, which are

$$\begin{aligned} \sigma_A &= \quad 200 - 2938(0.18 - 0.1183) \;= 18.7 \text{ MPa} \\ \sigma_B &= \quad 200 - 2938(0.16 - 0.1183) \;= 77.5 \text{ MPa} \\ \sigma_C &= -200 - 2938(0.11 - 0.1183) = -175.6 \text{ MPa} \\ \sigma_D &= \quad -200 - 2938(-0.1183) \quad = 148 \text{ MPa}. \end{aligned}$$

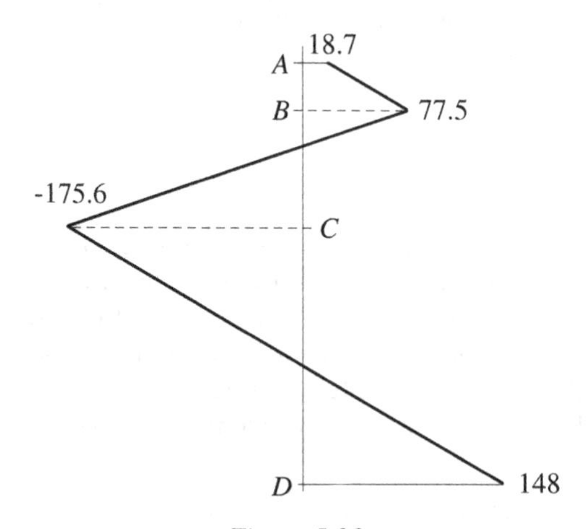


*Figure 5.28*

### 5.7.2 Reloading and shakedown

The results of the last section show that a beam that is plastically deformed and then unloaded will generally exhibit a state of residual stress opposite to that experienced at the most extreme loading condition. This residual stress is beneficial if the beam is subsequently loaded again in the same direction. Indeed, since the unloading process is completely elastic, it is clear that we could reload the beam to within an arbitrarily small value below the initial maximum moment $M_{max}$ without producing any additional plastic deformation. Perhaps more important from an engineering perspective is the fact that loading the deformed beam between zero and a moment *lower* than $M_{max}$ will result in maximum stresses lower than those that would be experienced by a beam without residual stresses. This means that the fatigue resistance of a beam subjected to zero to maximum loading can be enhanced by applying an initial overload in the direction of the maximum loading, sufficient to produce some plastic deformation. In effect, the residual stress induced shifts the mean value of the cyclic stresses closer to zero, without affecting the alternating component.

If the bending moment in a beam oscillates from zero to a maximum value in the range $M_Y < M < M_P$, plastic deformation will occur during the first loading cycle, but in subsequent cycles it will just reach the yield limit but not surpass it and no subsequent plastic deformation will occur. This is a common phenomenon in elastic-plastic structures subjected to cyclic loading and is known as *shakedown*. In effect, the structure tends to deform plastically early in the process in such a way as to develop a state of residual stress that discourages further plastic deformation. Indeed, there is a theorem, known as *Melan's theorem*[4], which proves that if any state of residual stress can be found that would be sufficient to inhibit plastic deformation during cyclic loading, the structure will in fact shake down until the behaviour is purely elastic. It is almost as though the structure would do its utmost to avoid its being subjected to repeated plastic deformation!

It is clear, however, that if the beam is plastically deformed under a bending moment $M$ and is then subjected to a bending moment of opposite sign, the residual stress from the first loading will *increase* the tendency for plastic deformation, so that a smaller moment is needed to cause first yield in the reverse direction. In most cases, if a beam is plastically deformed by a bending moment $M_1$, ($M_Y < M_1 < M_P$), elastic deformation will persist on reversed loading as long as

$$M_1 - 2M_Y < M < M_1 \; . \tag{5.37}$$

In other words, the *range* of bending moments under which the defomation is purely elastic is equal to $2M_Y$ as it would be for a beam without residual stress ($-M_Y < M < M_Y$), but the mean value is shifted in the direction of the original moment $M_1$. Equation (5.37) may overstate the elastic range under initial unloading if the beam is unsymmetric and $M_1$ is close to the fully plastic moment, but under cyclic loading the system will still shake down eventually as long as (5.37) is satisfied.

---

[4] P.S.Symonds (1951), Shakedown in continuous media, *ASME Journal of Applied Mechanics*, Vol.18, pp.85–89.

## 5.8 Limit analysis in the design of beams

We have seen in §§5.4–5.6 that the fully plastic moment $M_P$ is generally easier to calculate than $M_Y$ and exceeds $M_Y$ by a fairly modest percentage. Since all design calculations are subject to safety factors to allow for unforeseen overloads and other uncertainties, an alternative procedure is therefore to perform a fully plastic 'collapse' or 'limit' analysis of the problem and apply a somewhat larger safety factor.

### 5.8.1 Plastic hinges

Figure 5.29 shows a simply supported beam with a central load and the corresponding bending moment diagram.

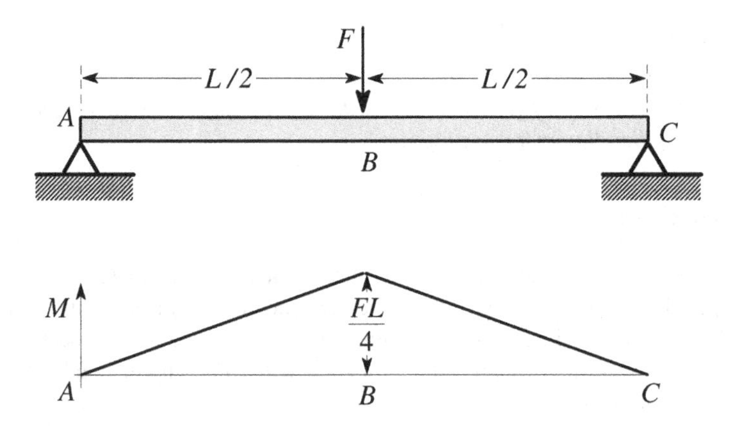


*Figure 5.29: A simply supported beam with a central load*

Failure will occur when the maximum bending moment $FL/4$ reaches $M_P$ and hence

$$F = \frac{4M_P}{L} .$$

Once this happens, unlimited bending of the beam will occur at $B$ without further increase in $F$ and the beam will collapse as shown in Figure 5.30*(a)*.

Some plastic deformation will occur at all points where $M > M_Y$ and this defines a range of points around $B$ as shown in Figure 5.30*(b)*. However, for most beam sections, the greater part of the deformation is concentrated near $B$ as in Figure 5.30*(a)* and the beam behaves as though it had a sticky hinge at $B$ which requires a moment $M_P$ to move it. We describe this process by saying that the beam develops a plastic hinge at $B$. Once the hinge is developed, the beam becomes a mechanism in the sense that unlimited further deformation can occur at the hinge without further increase in force or additional deformation elsewhere in the beam.

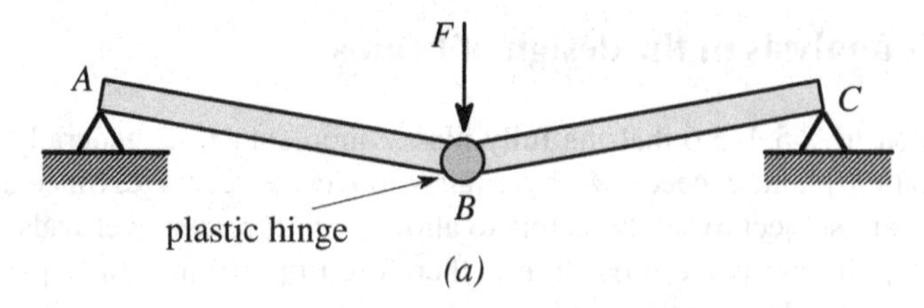


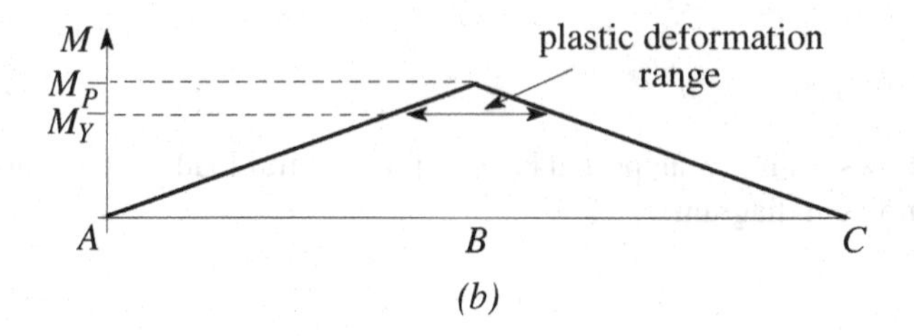


*Figure 5.30: (a) Mode of plastic collapse for the simply supported beam, (b) the corresponding bending moment diagram*

### 5.8.2 Indeterminate problems

For indeterminate problems, collapse requires that the beam or structure develop enough plastic hinges to remove the redundancy and convert the structure to a mechanism. It is usually fairly easy to determine where these hinges must occur. A good method is to sketch the probable shape of the shear force and/or bending moment diagrams. Remember that

$$\frac{dM}{dz} = V , \tag{5.38}$$

where $V$ is the shear force, so the maximum and minimum values of bending moment correspond to points where the shear force passes through zero. In particular, no plastic hinge can develop in the middle of a beam segment with no lateral load, since the shear force there must be constant.

Once a probable arrangement of plastic hinges is identified, the problem is reduced to one which can be solved using equilibrium considerations alone. This is almost always a great deal easier than the solution of the corresponding elastic indeterminate problem — for example using Castigliano's Second Theorem (§3.10).

**Example 5.10**

*The built-in beam of Figure 5.31 (a) is loaded by a uniformly distributed load $w$ per metre. The beam is made of steel with $S_Y = 300$ MPa and its cross section is as shown in Figure 5.31 (b). Find the value of $w$ at plastic collapse.*

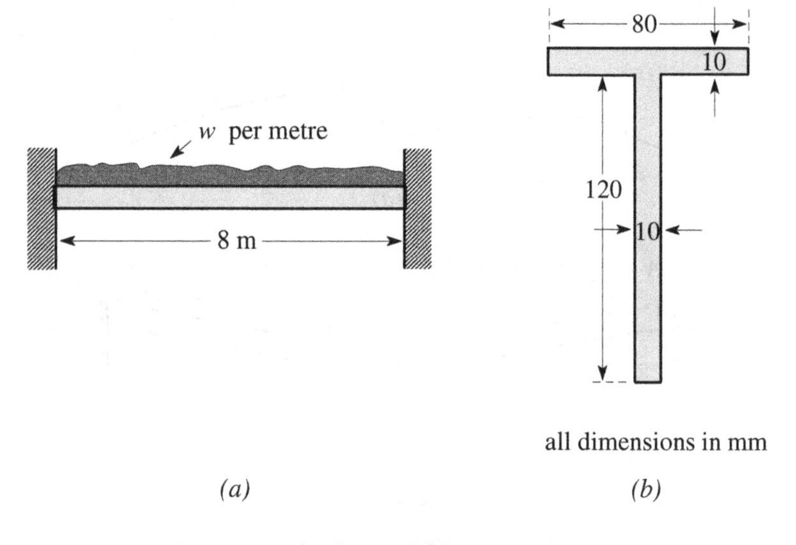


*Figure 5.31*

We first need to determine the fully plastic moment $M_P$. The total area of the section is

$$A = 80 \times 10 + 120 \times 10 = 2000\ \text{mm}^2\ ,$$

so the median line cuts the section at a distance 100 mm (1000 mm$^2$/10 mm) from the bottom of the section as shown in Figure 5.32.

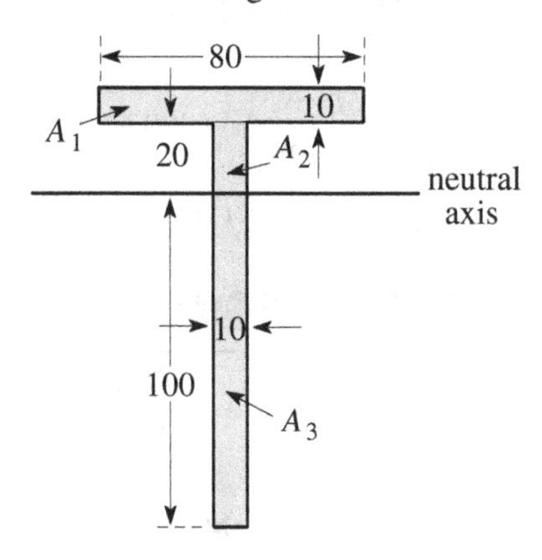


*Figure 5.32*

Taking moments about the bottom of the section, the fully plastic moment is then determined as

$$\begin{aligned} M_P &= 300 \times 10^6 (0.08 \times 0.01 \times 0.125 + 0.02 \times 0.01 \times 0.11 - 0.1 \times 0.01 \times 0.05) \\ &= 21600\ \text{Nm}. \end{aligned}$$

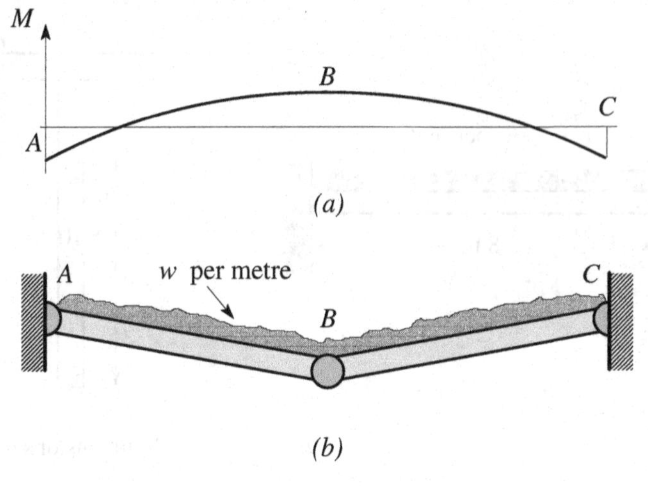


*Figure 5.33*

The bending moment diagram will be as shown in Figure 5.33*(a)*, so first yield will occur either at the supports or at the mid-point. However, the development of a single plastic hinge is not sufficient to cause the beam to collapse. The load can be increased further until the beam is converted to a mechanism by the establishment of plastic hinges at all three points *ABC* as shown in Figure 5.33*(b)*. Taking moments about *C* in the free-body diagram[5] for the segment *BC* (Figure 5.34) then yields the equilibrium equation

$$2M_P - 4w \times 2 = 0$$

and hence

$$w = \frac{2M_P}{(4\text{ m } \times 2\text{ m})} = 5400\text{ N/m.}$$

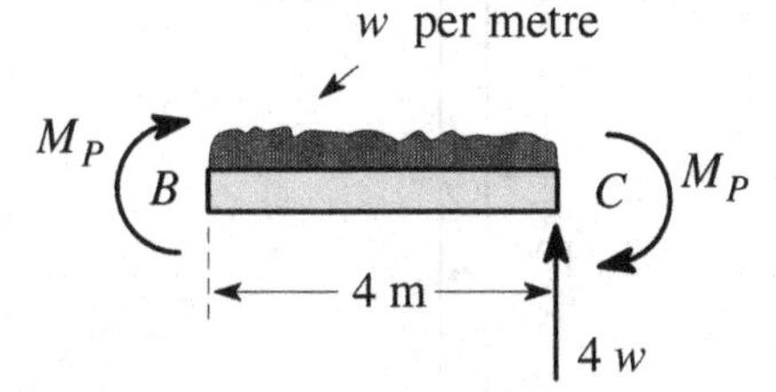


*Figure 5.34*

## 5.9 Summary

In this chapter, we have developed methods for determining the stress distribution and the curvature for beams that are loaded into the plastic range. The same methods can also be used for non-linear elastic materials. The strain is always a linear function

[5] Notice that the moments $M_P$ must oppose the relative rotations in Figure 5.33*(b)*.

of position in the section and the three constants describing this deformation can in principle always be found from three equilibrium equations defining the transmitted force and moments.

If a beam is bent about an axis of symmetry, this will also be the neutral axis if there is no axial force. Symmetric regions yielded in tension and compression are separated by an elastic region. If the moment axis is not an axis of symmetry, the neutral axis generally moves as plastic deformation progresses.

The maximum bending moment (the fully plastic moment) corresponds to the situation where the elastic zone has shrunk to zero and the beam then develops a plastic hinge. For a rectangular beam section, the shape factor $f = 1.5$, meaning that the load can be increased by 50% beyond first yield before collapse occurs. This 'residual strength' for an elastic-beam is a useful reserve in design to accommodate unlikely extreme loading events. However, more 'efficient' sections such as I-beams have values of $f$ much closer to unity, since most of the section area is concentrated at the same distance from the neutral axis. In such cases, a good design strategy is to calculate the fully plastic moment and use an appropriate safety factor. This method is particularly useful in designing complex indeterminate structures for strength.

If a beam is loaded into the plastic range and then unloaded, it will be left in a self-equilibrated state of residual stress and will have some residual curvature. This is important in forming operations involving bending, since the elastic recovery or 'springback' must be allowed for in designing appropriate forms and tools. Also, the tensile residual stresses may influence the fatigue life of formed components.

## Further reading

### Elastic-perfectly plastic stress distributions

W.F. Riley and W. Zachary (1989), *Introduction to Mechanics of Materials*, Wiley, New York, NY, Chapter 10.

### Nonlinear constitutive laws

J.M. Gere and S.P. Timoshenko (1997), *Mechanics of Materials,* PWS Publishing Company, Boston, MA, 4th edn., §6.10.

### Limit analysis

E.P. Popov (1998), *Engineering Mechanics of Solids* Prentice Hall, Upper Saddle River, NJ, 2nd edn., Chapter 20.
I.H. Shames and F.A. Cozzarelli (1992), *Elastic and Inelastic Stress Analysis*, Prentice Hall, Engelwood Cliffs, NJ, Chapter 10.
W.F. Riley and W. Zachary (1989), *loc. cit.*

## Problems

### Section 5.4

**5.1.** A beam of square cross section $a \times a$ is bent about a diagonal axis by a moment $M$, as shown in Figure P5.1. Find the relation between $M$ and the radius of curvature $R$, if the material obeys the constitutive law

$$\sigma_{zz} = E\left(e_{zz} + \frac{e_{zz}^3}{e_0^2}\right) .$$

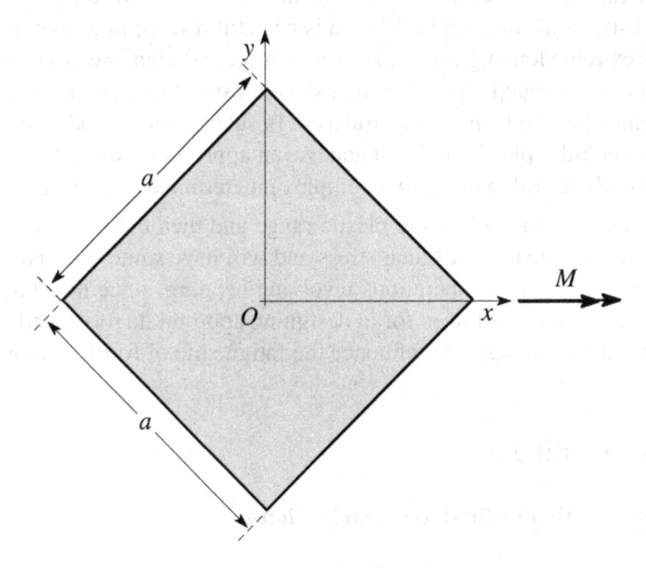


*Figure P5.1*

**5.2.** A certain polymer fails in tension at a stress $\sigma_0$ and its uniaxial constitutive law is approximated by the equation

$$\begin{aligned} \sigma_{zz} &= 0.6\sigma_0 e_{zz}^{0.2} \quad ; \; e_{zz} > 0 \\ &= -0.6\sigma_0 |e_{zz}|^{0.2} ; \; e_{zz} < 0 . \end{aligned}$$

It is used for a rectangular beam of width $b$ and height $h$. Find the maximum bending moment that can be transmitted by the beam without failure.

**5.3.** Find the fully plastic moment $M_P$ for the I-beam shown in Figure P5.3, if the uniaxial yield stress of the material is $S_Y = 100$ MPa.

At what percentage of the fully plastic moment will the flanges of the beam be fully plastic, whilst the web remains elastic?

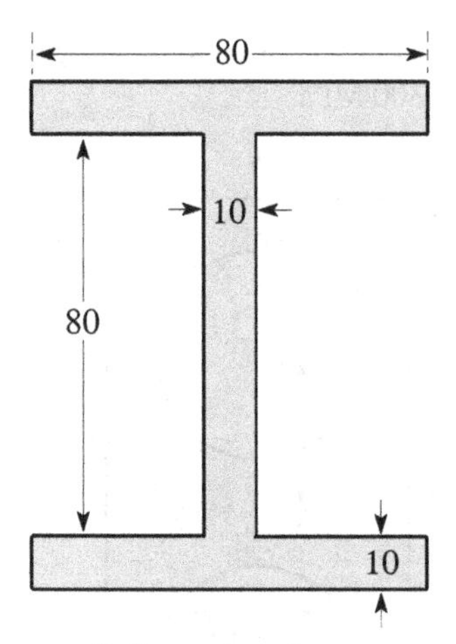


*Figure P5.3*

**5.4.** Find the applied moment for first yield $M_Y$, the fully plastic moment $M_P$, and the shape factor for a hollow circular cylindrical beam of outer radius $2a$ and inner radius $a$.

**5.5.** A rectangular beam of width $b$ and height $h$ is loaded by a bending moment $M$ about a horizontal axis. Find the relation between bending moment and radius of curvature $R$, if the material is elastic-perfectly plastic with Young's modulus $E$ and uniaxial yield stress $S_Y$.

**5.6.** The channel section of Figure P5.6 is bent by a moment $M$ which is just sufficient to cause the flanges to yield. Find the value of $M$ and the corresponding radius of curvature $R$, if $E = 210$ GPa and $S_Y = 350$ MPa.

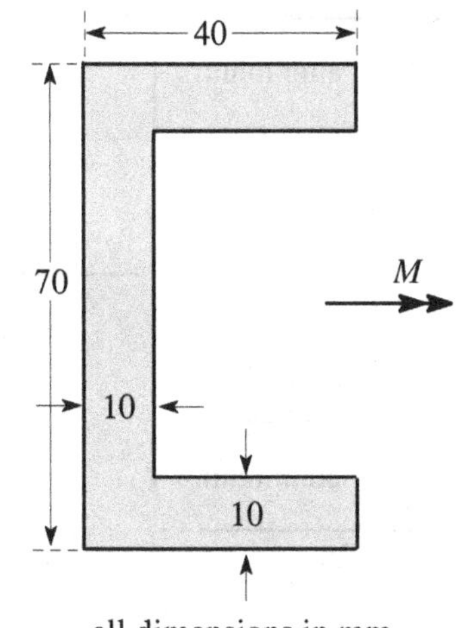


*Figure P5.6*

**5.7.** Find the fully plastic moment $M_P$ for the symmetric beam section of Figure P5.7, if the uniaxial yield stress is 300 MPa.

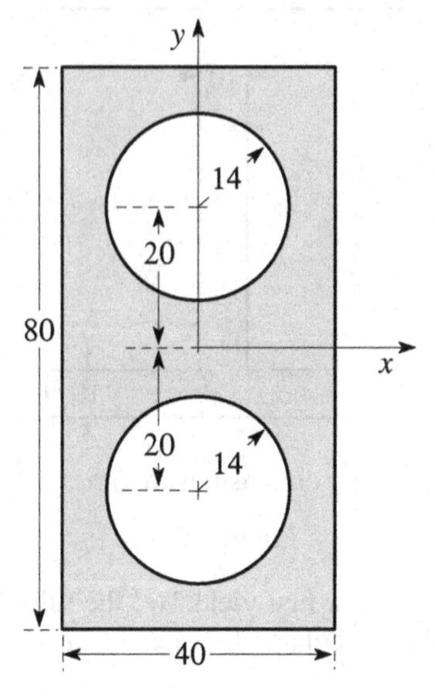


*Figure P5.7*

**5.8*.** Figure P5.8 shows a composite rectangular beam, manufactured from aluminium alloy ($E = 80$ GPa, $S_Y = 60$ MPa) and steel ($E = 210$ GPa, $S_Y = 300$ MPa). The beam is bent about the horizontal axis to a radius $R = 7$ m. Find the required bending moment $M$ and sketch the resulting stress distribution.

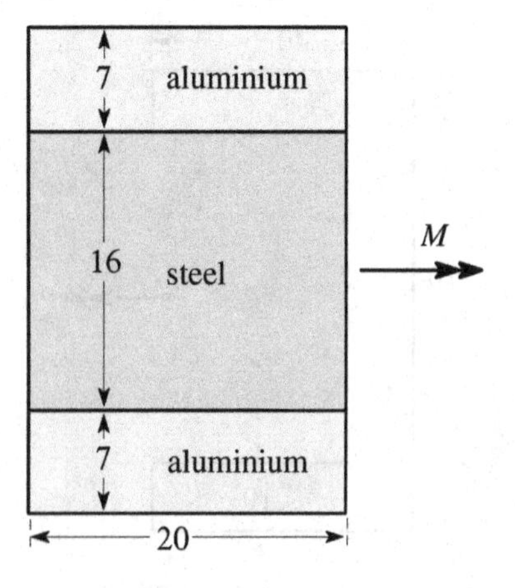


*Figure P5.8*

**5.9**.** It can be shown that the curvature of an elastic-plastic beam of square cross section $a\times a$ is

$$\frac{1}{R} = \frac{S_Y}{Ea}\left(\frac{3M}{M_P}\right) \quad ; \quad M < \frac{2M_P}{3}$$

$$= \frac{2S_Y}{Ea}\sqrt{\frac{1}{3(1-M/M_P)}} \quad ; \quad \frac{2M_P}{3} < M < M_P \,,$$

where $M_P = S_Y a^3/4$ is the fully plastic moment[6].

The beam is built in at $z=0$ and loaded by a force $F$ at $z=L$ as shown in Figure P5.9. If the force is just sufficient to cause the beam to become fully plastic at $z=0$, show that the downward end deflection is then

$$u(L) = \frac{40S_Y L^2}{27Ea} \,.$$

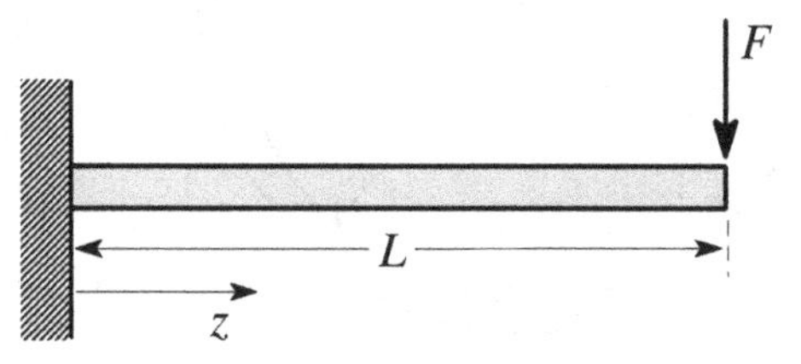


*Figure P5.9*

## Section 5.5

**5.10.** Find the fully plastic moment $M_P$ and the location of the corresponding neutral axis for the beam section of Figure P5.10, if the uniaxial yield stress is 200 MPa.

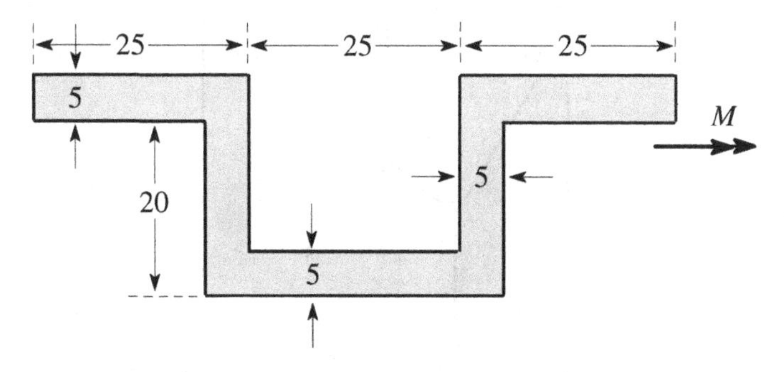


all dimensions in mm

*Figure P5.10*

[6] The proof of these results is a special case ($b=h=a$) of Problem 5.5.

**5,11.** Find the fully plastic moment and the shape factor for the beam section of Figure P5.11, if the uniaxial yield stress is 30 ksi.

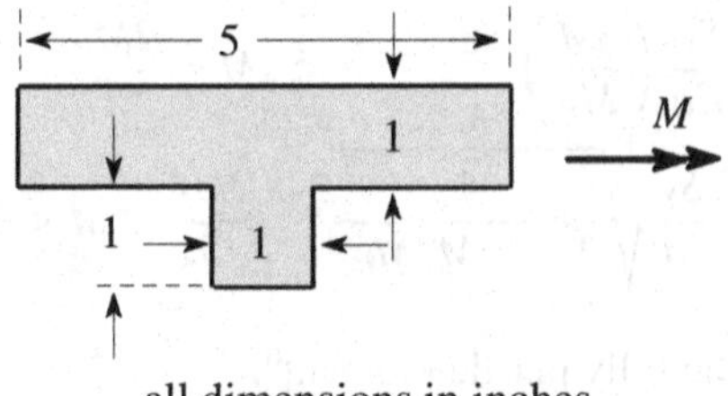


*Figure P5.11*

**5.12.** Find the fully plastic moment $M_P$ for the thin-walled semi-circular section of Figure P5.12, if the yield stress of the material is $S_Y$. Assume $t \ll a$.

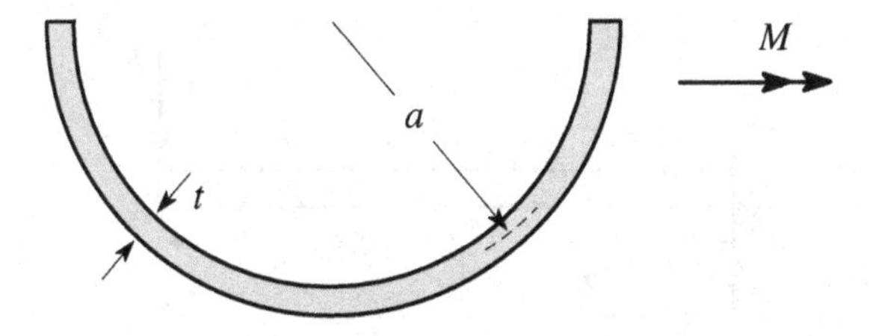


*Figure P5.12*

**5.13.** The equilateral triangular column of Figure P5.13 is loaded by an axial force $F$, whose line of action is a distance $h/3$ from one vertex. If the force is large enough to cause plastic collapse, find the location of the neutral axis.

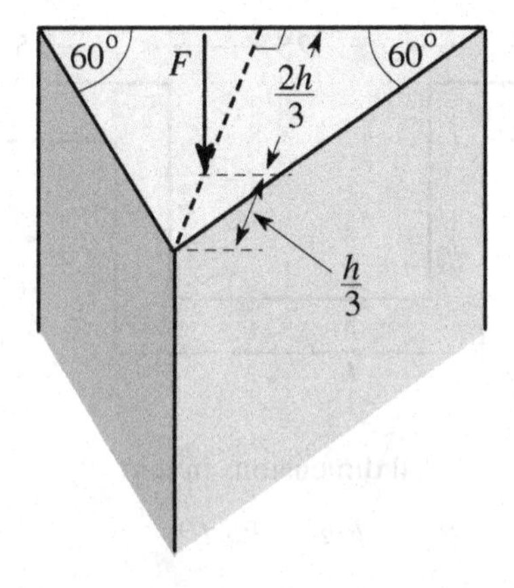


*Figure P5.13*

**5.14*.** Find the fully plastic moment $M_P$ for the T-section shown in Figure 5.13, if the uniaxial yield stress is $S_Y = 200$ MPa. Find also the moment at which a second plastic zone starts to be developed at the upper edge.

**5.15*.** The channel section of Figure P5.6 is bent to a radius of 16 m *about the vertical axis*. Find the location of the neutral axis, the extent of the plastic zone and the bending moment, if the relevant material properties are $E = 80$ GPa, $S_Y = 100$ MPa.

**5.16.** The beam of Problem 5.10 is bent to a radius of 1 m. Find the bending moment, the extent of the plastic zones and the location of the neutral axis ($E = 210$ GPa, $S_Y = 200$ MPa).

**5.17*.** A rectangular beam of width $b$ and height $h$ is made from a material with different elastic moduli in tension and compression — i.e.

$$\sigma_{zz} = E_1 e_{zz} \ ; \quad e_{zz} > 0$$
$$= E_2 e_{zz} \ ; \quad e_{zz} < 0 \, .$$

Find the location of the neutral axis and the relation between the applied moment $M$ and the radius of curvature $R$, if $E_1 = E$ and $E_2 = 2E$.

### Section 5.6

**5.18.** The I-beam of Figure P5.18 is to be bent to plastic collapse about an axis inclined at 45° as shown. Find the magnitude $M_P$ of the required resultant moment and its inclination $\theta$ to the horizontal axis, if the material has a uniaxial yield stress of 300 MPa. Comment on your results.

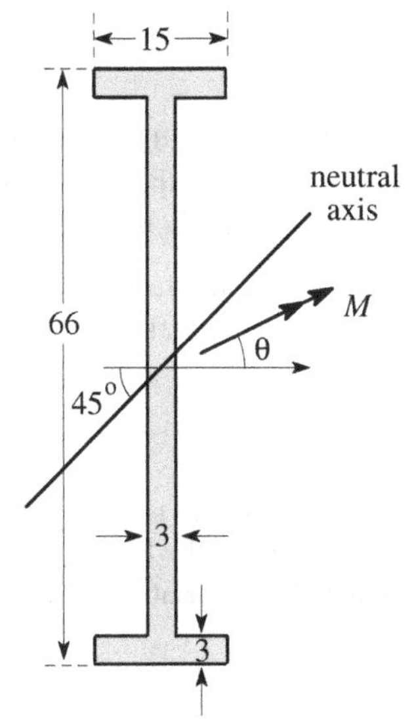


all dimensions in mm

*Figure P5.18*

**5.19.** The thin-walled unsymmetrical angle section shown in Figure P5.19 is bent in such a way that the neutral axis in the fully plastic state cuts the lower leg at a distance $d$ from the corner, where $d < a$. Show that the condition of zero axial force $F = 0$ then demands that the corresponding intercept on the vertical leg is at a distance $(1.5a - d)$ from the corner.

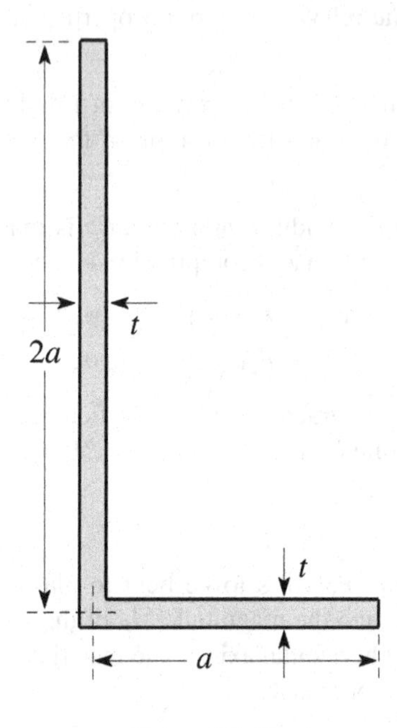


*Figure P5.19*

Determine $d$ for the case where the applied moment $M$ acts about the horizontal axis and hence determine the inclination of the neutral axis for this case. Assume that the thickness of the section $t \ll a$ throughout the calculation.

What interpretation do you place on the fact that all potential neutral axes that cut the section in only one place (chosen to make $F = 0$) will correspond to the same stress distribution?

**5.20.** The thin-walled I-beam of Figure P5.20 is bent by a fully plastic moment such that the neutral axis at angle $\alpha$ cuts through the flanges as shown. Find the inclination $\theta$ of the resultant moment and sketch the variation of $\alpha$ with $\theta$.

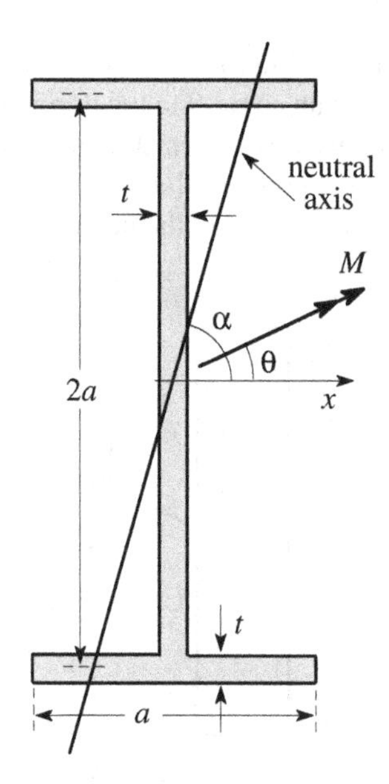


*Figure P5.20*

**5.21*.** The thin-walled section of Figure P5.21 is subjected to a bending moment about the horizontal axis. Find the fully plastic moment and the inclination of the corresponding neutral axis, if the yield stress of the material is $S_Y$.

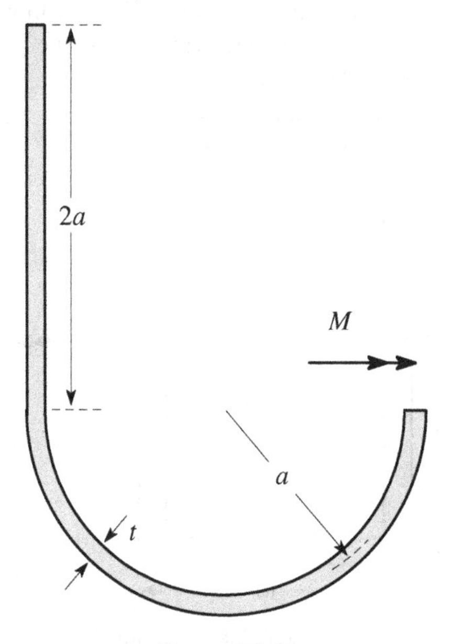


*Figure P5.21*

**5.22*.** The rectangular section of Figure P5.22 is loaded to plastic collapse by a moment about an axis inclined at an angle $\theta$ to the horizontal. Find the inclination of the neutral axis $\alpha$ and the fully plastic moment $M_P$ as functions of $\theta$. The yield stress for the material is 20 ksi. Sketch a graph of the resulting expressions. **Hint:** You will need to consider separately the cases where the neutral axis cuts (i) the sides and (ii) the top and bottom of the section.

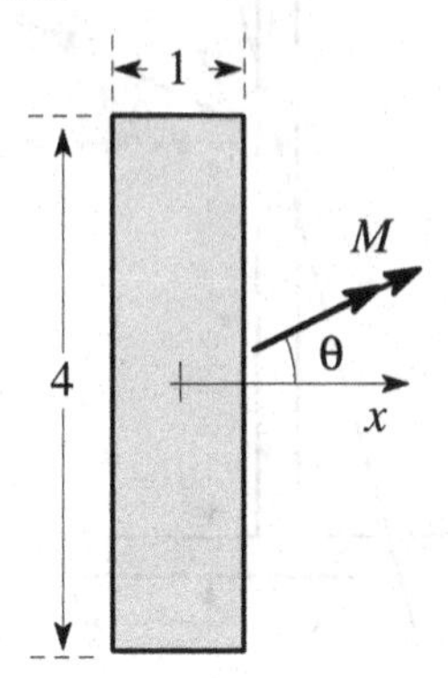


all dimensions in inches

*Figure P5.22*

## Section 5.7

**5.23.** The hollow rectangular section of Figure P5.23 is subjected to a bending moment just sufficient to yield the upper and lower walls. The moment is then released. Find the residual radius of curvature of the beam $R_u$ and the residual stress distribution, if $E = 210$ GPa and $S_Y = 350$ MPA.

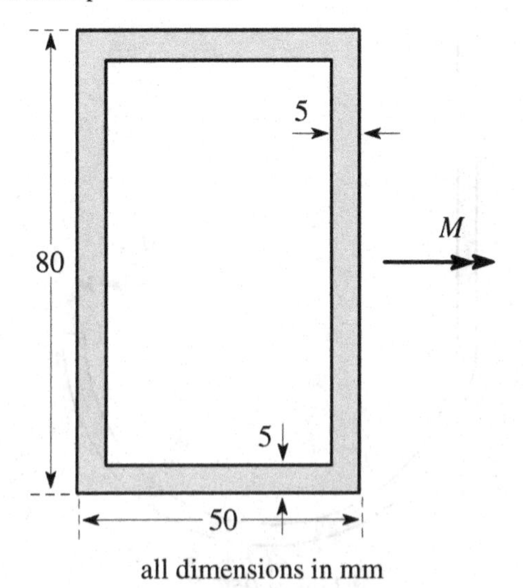


all dimensions in mm

*Figure P5.23*

**5.24.** A bar of solid circular cross section of radius $a$ is loaded by the fully plastic moment and then unloaded. Find the residual stress distribution in the bar, if the yield stress of the material is $S_Y$.

**5.25*.** In a manufacturing process, it is desired to form an initially straight bar of square cross section 25 mm × 25 mm to a curved bar of mean radius 1m. This is to be done by passing the bar between rollers, in the direction of the velocity $V$ in Figure P5.25, such that the central region is in a state of pure bending at a radius of curvature $R$. What must be the radius $R$, if the beam is to spring back to the required radius of 1m when the load is removed? Assume that the bar is of steel with $E = 210$ GPa and $S_Y = 210$ MPa.

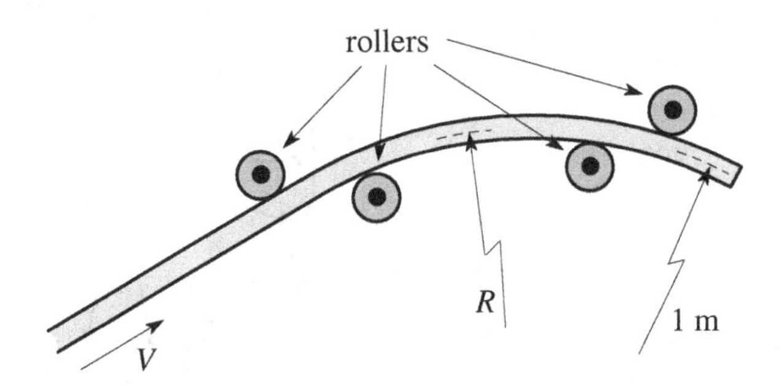


*Figure P5.25*

**5.26.** The section of Figure P5.11 is loaded until a second plastic zone is just about to develop at the upper edge. Find the radius of curvature at this state if $E = 30 \times 10^3$ ksi and $S_Y = 30$ ksi.

The beam is now unloaded. What will be the residual curvature?

**5.27.** A 1 mm diameter steel wire ($E = 210$ GPa, $S_Y = 210$ MPa) is close coiled into a helix around a 25 mm diameter rigid cylindrical drum. What will be the inside diameter of the coil when it is released from the drum.

**5.28*.** The T-section of Figure 5.13 is loaded by the fully plastic moment and then unloaded. Show that the assumption of elastic unloading leads to stresses exceeding the yield stress between the elastic and fully plastic neutral axes.

The actual unloading process will involve additional yield in a small central region, the rest of the section remaining elastic. Find the extent of this yield zone and the final residual stress distribution. Take care that your solution satisfies the condition of zero axial force ($F = 0$). The yield stress is $S_Y = 200$ MPa.

**5.29*.** A circular ring of radius 200 mm and square cross section 4 mm×4 mm is to be manufactured from steel of Young's modulus $E$ =200 GPa and tensile yield stress $S_Y$ =200 MPa.

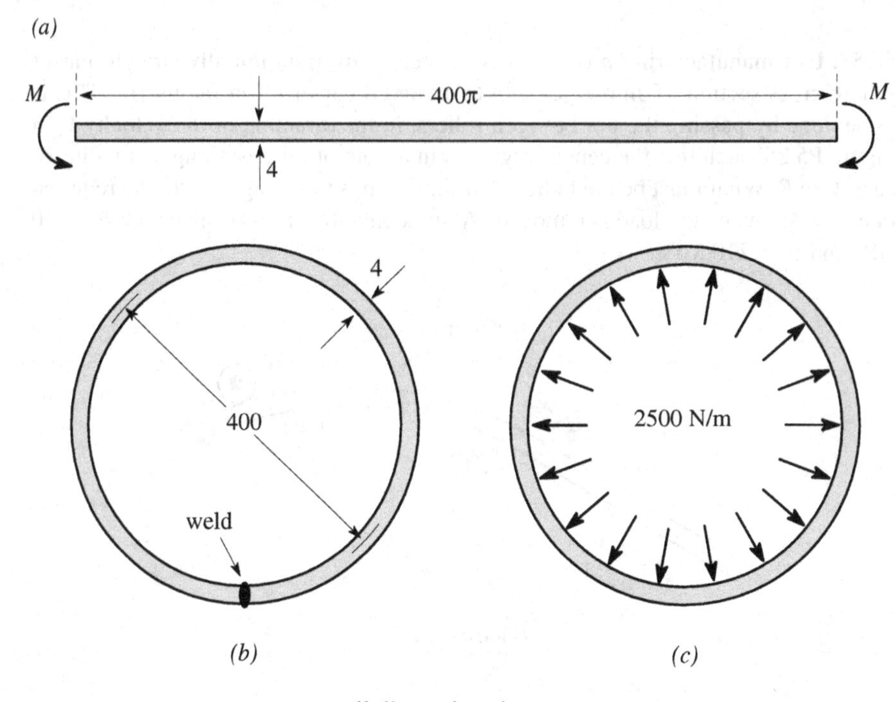


all dimensions in mm

*Figure P5.29*

(i) The first stage in the process is to take the initially straight and stress-free bar of cross section 4 mm×4 mm and length $400\pi$ mm of Figure P5.29 *(a)*, apply end moments $M$ sufficient to bend it into a circle and then weld the ends together, as shown in Figure P5.29 *(b)*.

Find the moment $M$ required and the corresponding bending stress distribution.

(ii) In an attempt to reduce the residual stress in the ring, a uniform radial force 2500 N/m per unit circumference is now applied, as shown in Figure P5.29 *(c)*, and then released.

Find the final stress distribution in the ring.

## Section 5.8

**5.30.** Figure P5.30 shows a beam on simple supports, loaded by a force $F$. Find the value of $F$ at plastic collapse if the fully plastic moment is $M_P$.

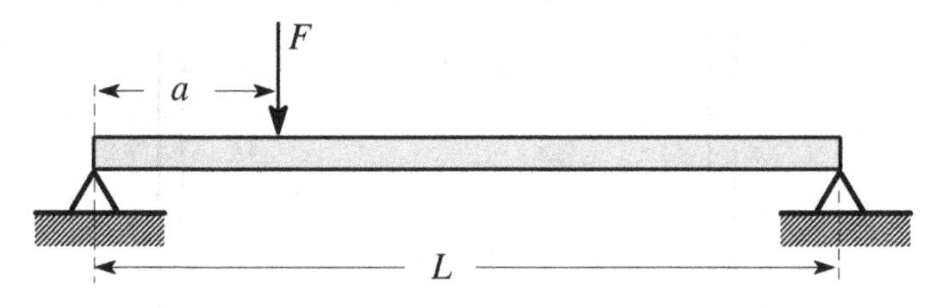


*Figure P5.30*

**5.31.** Figure P5.31 shows a beam on three simple supports subjected to a force $F$. If the fully plastic moment for the beam is $M_P$, find the value of $F$ that will cause plastic collapse.

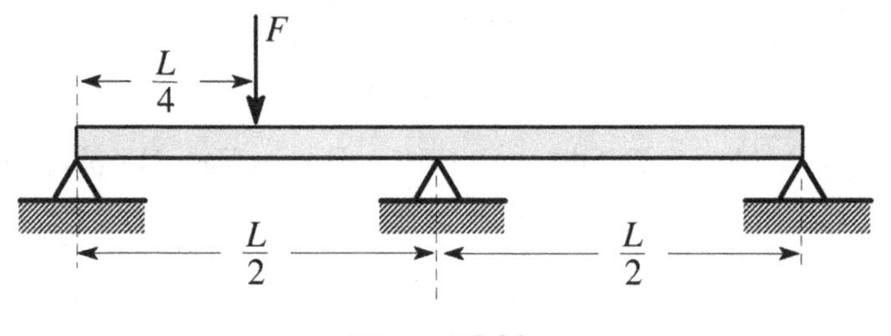


*Figure P5.31*

**5.32*.** Figure P5.32 shows a beam built in at one end, simply supported at the other and loaded by a distributed load $w_0$ per unit length over one half of its length $L$. Sketch the shape of the shear force and bending moment diagrams.

Show that a plastic hinge will form at some point $z = d$ in the range $0 < d < L/2$. Determine $d$ and the value of $w_0$ at collapse.

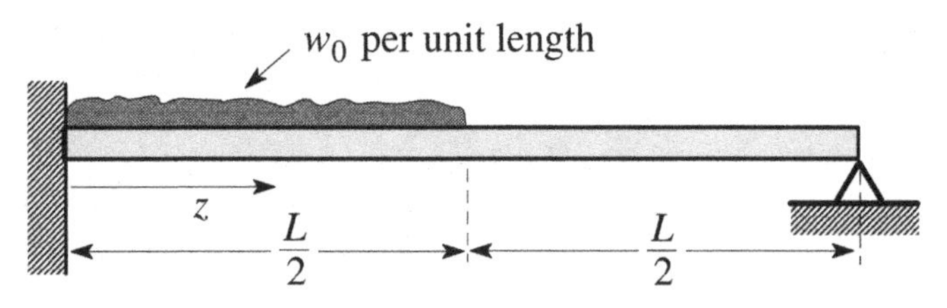


*Figure P5.32*

**5.33.** The U-frame structure of Figure P5.33 is fabricated from three identical beams of length $L$. It is built in at $A, D$ and loaded by a horizontal force $F$ at $B$. Find the collapse load if the fully plastic moment for all segments of the beam is $M_P$.

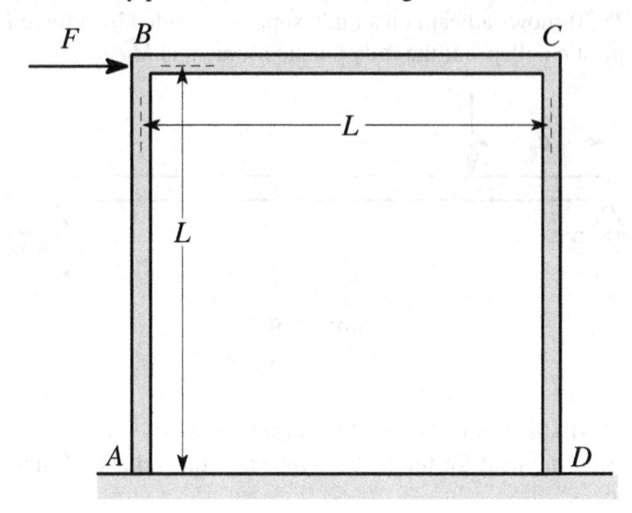


*Figure P5.33*

**5.34.** Re-solve Problem 5.33 if the supports at $A, D$ are replaced by pin joints.

# 6

# Shear and Torsion of Thin-walled Beams

In Chapters 4 and 5, we investigated the stresses developed in beams subjected to bending moments and, in some cases, an axial force — i.e. those forms of loading that give rise to normal stresses $\sigma_{zz}$ on transverse planes $z=c$, where $c$ is a constant. It is also worth noting that the combination of an axial force and a bending moment is equivalent to an equal axial force with a different line of action (see for example §5.5.2).

A similar relation exists between the two shear forces $V_x, V_y$ and the torque $T$, shown in Figure 6.1. All of these loads give rise to shear stresses on the transverse plane $A$ and in general the combination is statically equivalent to a resultant shear force whose magnitude and direction is determined by $V_x, V_y$, but whose line of action is modified by the torque $T$.

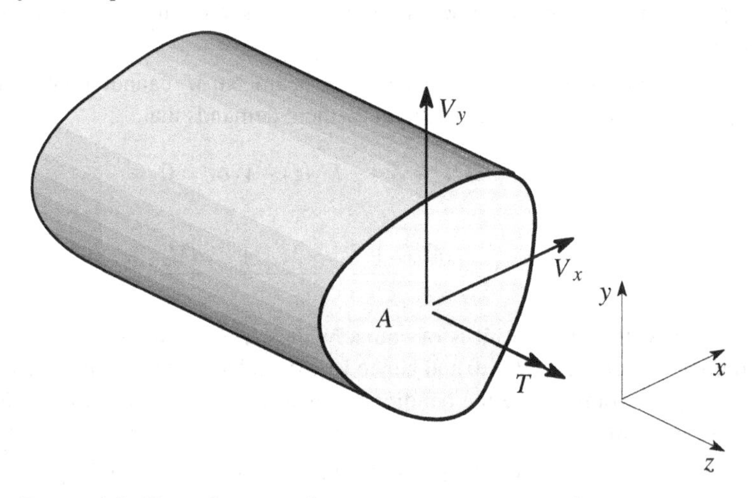


*Figure 6.1: Shear forces and torque on a transverse plane*

The shear stress distribution due to the loading of Figure 6.1 can be obtained for the elastic case, but the solution for general geometries is considerably more difficult than the bending theory and requires the solution of a partial differential equation

J.R. Barber, *Intermediate Mechanics of Materials*, Solid Mechanics and Its Applications 175, 2nd ed., DOI 10.1007/978-94-007-0295-0_6, 

with appropriate conditions at the boundaries of the section.[1] Fortunately, a more straightforward approximate theory can be developed, which approaches the exact theory in the important limiting case of thin-walled sections.

To introduce the subject, we shall first revisit the derivation of the elementary formula for the shear stress in a beam transmitting a shear force $V$. Later (in §6.3) we shall draw on the ideas of §4.4.3 to extend the argument to a general unsymmetrical section by choosing the coordinate system $Oxy$ to coincide with the principal axes of the section and using superposition.

## 6.1 Derivation of the shear stress formula

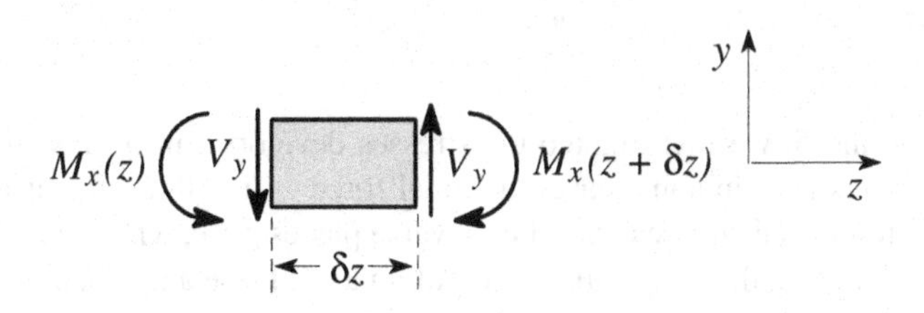


*Figure 6.2: Equilibrium of a beam segment*

Consider a small length $\delta z$ of a beam, transmitting a shear force $V_y$ and a bending moment $M_x$ as shown in Figure 6.2. As in Chapter 4, we define a coordinate system such that the $z$-axis points along the beam and $x, y$ are in the plane of the cross section. The $y$-axis points upwards in Figure 6.2 and hence the $x$-axis must point into the paper.[2] A positive moment $M_x$ is therefore clockwise on the positive $z$-surface, as shown.

We assume that $V_y$ is constant along the beam, but $M_x$ cannot be constant, because equilibrium of moments on the small element demands that

$$M_x(z+\delta z) - M_x(z) - V_y \delta z = 0$$

and hence

$$V_y = \frac{M_x(z+\delta z) - M_x(z)}{\delta z} = \frac{dM_x}{dz} . \tag{6.1}$$

This well known result is of course widely used to find the maximum bending moment (where $dM_x/dz = 0$ and hence $V_y = 0$) and to sketch shear force and bending moment diagrams, since the bending moment is the integral of the shear force and hence is the area under the shear force diagram up to the point under consideration.

We next consider the bending stresses on the element of length $\delta z$ shown in Figure 6.3.

---

[1] J.R.Barber (2010), *Elasticity*, Springer, Dordrecht, 3rd edn., Chapters 16,17.
[2] Compare with Figure 6.1.

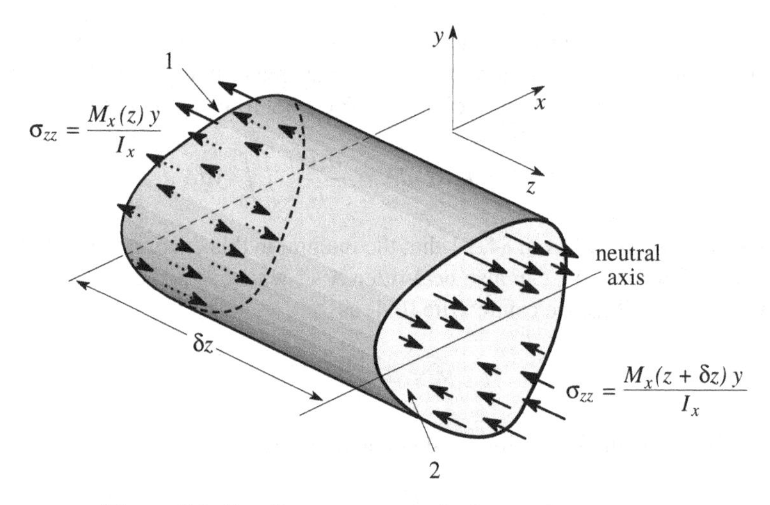


*Figure 6.3: Bending stresses on the beam element*

On the left end 1, we have

$$\sigma_{zz} = \frac{M_x(z)y}{I_x} , \tag{6.2}$$

but on the right end 2,

$$\sigma_{zz} = \frac{M_x(z+\delta z)y}{I_x} . \tag{6.3}$$

These stresses are statically equivalent to the moments in Figure 6.2 and hence the element is in equilibrium as long as equation (6.1) is satisfied. However, if we make a further cut longitudinally along the beam axis as shown in Figure 6.4*(a)*, the resulting element [Figure 6.4*(b)*] generally has different axial forces on its two ends because of the differing values of $\sigma_{zz}$.

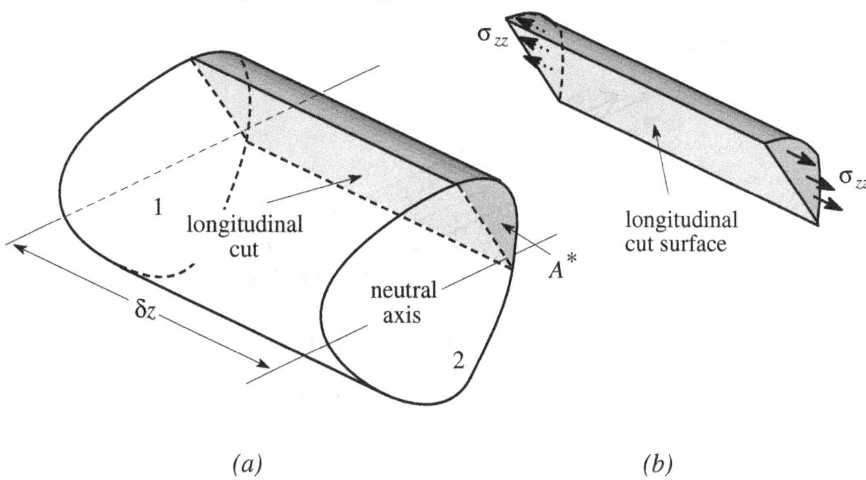


*Figure 6.4: (a) The longitudinal cut, (b) bending stresses on the section of the beam element removed*

We shall refer to the area shaded in Figure 6.4*(a)* as $A^*$ to distinguish it from the total area of the cross section $A$. Thus, $A^*$ is the area of that part of the cross section that is on one side of the longitudinal cut. The axial force on $A^*$ at section 1 is

$$F_1 = \iint_{A^*} \sigma_{zz} dA = \frac{M_x(z)}{I_x} \iint_{A^*} y dA \,, \tag{6.4}$$

using (6.2). We recall from §4.3.1 that the integral in this equation, which is the first moment of the area $A^*$, can also be written $A^*\bar{y}^*$ where $\bar{y}^*$ is the $y$-coordinate of the centroid of $A^*$. Thus, we can rewrite (6.4) as

$$F_1 = \frac{M_x(z)A^*\bar{y}^*}{I} \,. \tag{6.5}$$

A similar argument for section 2 using equation (6.3) gives

$$F_2 = \frac{M_x(z+\delta z)A^*\bar{y}^*}{I_x} \,. \tag{6.6}$$

The forces $F_1$ and $F_2$ will not be equal unless $A^*\bar{y}^* = 0$. If $A^*$ were the whole beam cross section, this condition would be satisfied by virtue of the choice of the centroid of the section as coordinate centre, but it will not generally be satisfied for that portion of the cross section removed by the longitudinal cut.[3] To keep the element in equilibrium, there must therefore be an extra axial force and the only place for this force to act is at the cut surface, since the axial forces on the transverse sections have been taken into account already. This means there must be a horizontal shear force $Q = F_2 - F_1$ on the cut, as shown in Figure 6.5. Thus

$$Q = F_2 - F_1 = \frac{[M_x(z+\delta z) - M_x(z)]A^*\bar{y}^*}{I_x} = \frac{V_y \delta z A^*\bar{y}^*}{I_x} \,, \tag{6.7}$$

using equation (6.1).

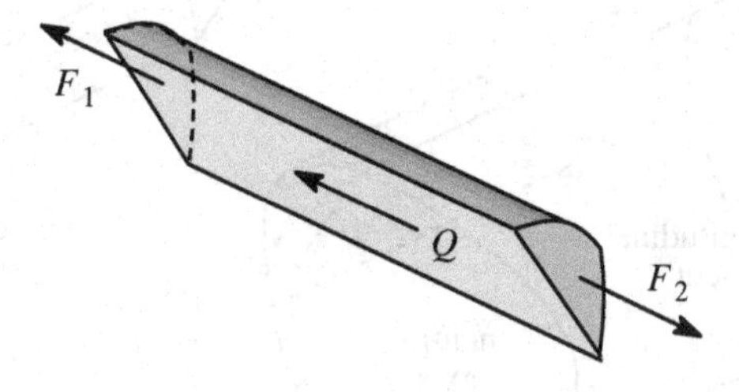


*Figure 6.5: Equilibrium of the element of Figure 6.4(b)*

The shear force *per unit length* of cut surface $q$, is obtained by dividing equation (6.7) by $\delta z$, i.e.

$$q = \frac{Q}{\delta z} = \frac{V_y A^*\bar{y}^*}{I_x} \,. \tag{6.8}$$

[3] In fact, it will be zero if and only if the centroid of $A^*$ lies on the neutral axis.

The quantity $q$ is sometimes referred to as the *shear flow*.

Up to this point, the above analysis is exact, within the limitations of the elementary beam theory, which itself can be shown to be exact as long as the shear force $V_y$ does not vary along the beam.[4] We can take a further *approximate* step if we assume that the shear stress is uniform across the width $t$ of the cut, in which case $q = \tau t$ and

$$\tau = \frac{q}{t} = \frac{V_y A^* \bar{y}^*}{I_x t} \, . \tag{6.9}$$

The stress calculated is a 'longitudinal' shear stress. It acts on the longitudinal cut surface as shown in Figure 6.6*(a)*. Also, $\tau$ is parallel with the beam axis, so it points away from the cut line $PQ$ at right angles. As discussed in §1.5.1, the stress matrix (1.2) is symmetric and there must therefore be an equal complementary shear stress pointing away from $PQ$ on the end face, as shown in Figure 6.6*(b)*.

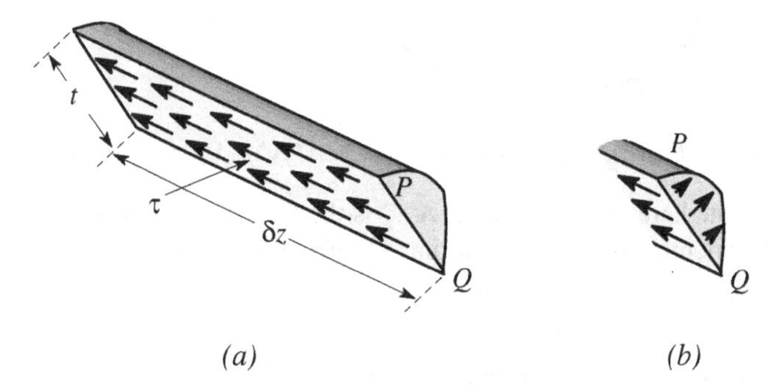


*Figure 6.6: (a) Longitudinal shear stress on the cut surface, (b) complementary shear stress on the transverse section*

Thus, $\tau$ is also the component of transverse shear stress at the cut pointing perpendicularly into the area $A^*$.

## Summary

A common error is to misinterpret the quantities $A^*, \bar{y}^*, t$ etc. appearing in the shear stress equations (6.8, 6.9), so we here list their definitions for reference. These definitions are illustrated in Figure 6.7.

$A^*$ is the area of the transverse section removed by the longitudinal cut (shown shaded in Figure 6.7).

$\bar{y}^*$ is the $y$-coordinate of the centroid of $A^*$ measured from the neutral axis of bending for the whole section.

$t$ is the total length of the cut $PQ$. Notice that $t$ appears in the formula because it defines the width of the area available to transmit the longitudinal shear force $Q$. Thus, if the cut passes through a hole in the section, only the segments through the solid walls will contribute to $t$.

[4] See, for example, J.R.Barber (2010), *Elasticity*, Springer, Dordrecht, 3rd edn., Chapter 17.

$V_y$ is the shear force in the $y$-direction.
$I_x$ is the centroidal second moment of area of the whole section about the $x$-axis.
$\tau$ is the component of transverse shear stress at the line of the cut in the direction pointing perpendicularly into the area $A^*$.

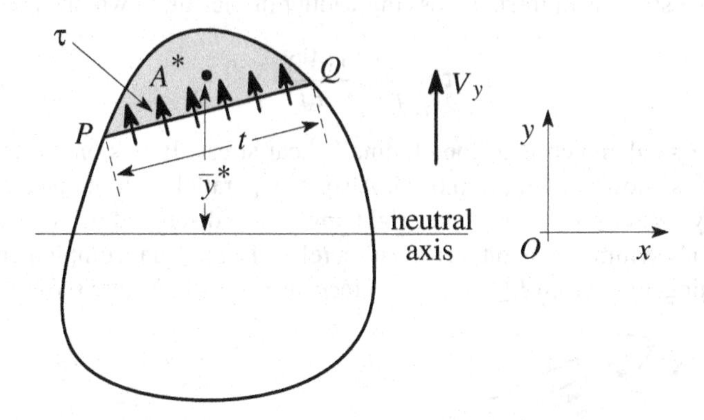


*Figure 6.7: Illustration of the definitions of $A^*, \bar{y}^*, t, V_y, \tau$ in equation (6.9).*

We also note here that the product $A^*\bar{y}^*$ is the first moment of the area $A^*$ and it is often conveniently obtained by splitting $A^*$ into several rectangular areas $A_1, A_2, ...$ and summing $A_1\bar{y}_1 + A_2\bar{y}_2 + ...$ as in §4.3.1.

### 6.1.1 Choice of cut and direction of the shear stress

In elementary treatments, the longitudinal cut is often taken to be parallel to the neutral axis of bending, but no part of the preceding argument depends upon this assumption. The results therefore apply for an arbitrary cut, which need not even be a plane. The only requirement is that it should separate the original beam cross section into two parts, one of which is identified as $A^*$.

It may require some explanation that the transverse shear stress on the end face has components, i.e. it has magnitude and direction. A shear stress by definition has to be parallel to the plane on which it acts, but in three dimensions, this still leaves it two orthogonal directions in which to have components. By taking different cuts through a given point, we can *estimate* both components of transverse shear stress, but remember that we have made the assumption that $\tau$ is uniform across $t$, so the results may not be very accurate, particularly if we have reason to believe that the distribution will be very non-uniform.

The most striking application of these arguments is in thin-walled open sections such as the I-beam. With the cut $BC$ in Figure 6.8 *(a)*, $A^*\bar{y}^*$ is large, $t$ is small and we get a substantial vertical shear stress, as we should expect. However, if we make a parallel cut $DE$ in the flange, as in Figure 6.8 *(b)*, $A^*\bar{y}^*$ may still be fairly large, but $t$ is now also large, so

$$\tau = \frac{V_y A^* \bar{y}^*}{I_x t}$$

will be small.

By contrast, if we make the cut $FG$ through the flange as in Figure 6.8 *(c)*, $A^*\bar{y}^*$ will still be fairly large and $t$ is now small, so that a significant shear stress is obtained. This stress has to point at the cut $FG$ from the rest of the section and is therefore a *horizontal* shear stress as shown in Figure 6.8 *(c)*. Thus we conclude that, in the flange, the horizontal component of shear stress will be substantial, but the vertical component will be small, giving a resultant that is predominantly horizontal.

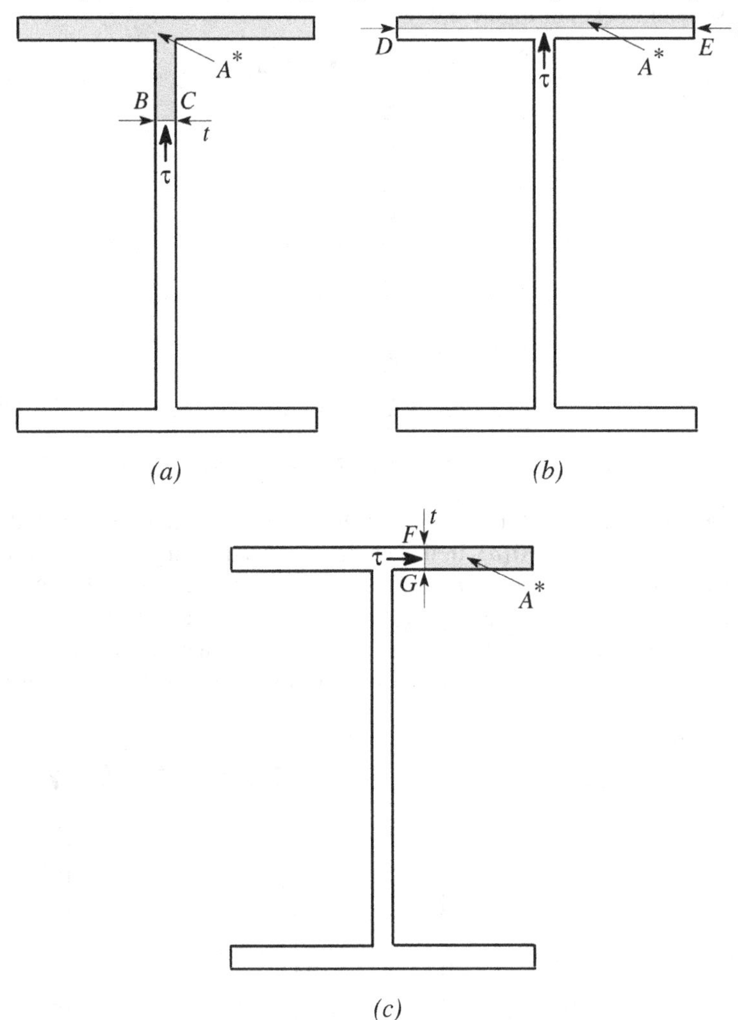


*Figure 6.8: Various cuts for the I-beam: (a) through the web, (b) horizontally through the flange, (c) vertically through the flange*

In general, when we have a thin-walled section, the shear stress component parallel to the wall will be significant because the corresponding value of $t$ is small, whereas that perpendicular to the wall will be negligible, because the corresponding $t$ is relatively large. In other words, the resultant shear stress tends to run parallel to the thin walls of the section. In the I-beam, this results in the distribution shown in Figure 6.9. This picture gives some idea of why $q$ is referred to as the shear flow, since the shear stresses appear to flow around the section like a fluid. However, notice that $q$ is not generally constant — it is a maximum at the neutral axis, where $A^*\bar{y}^*$ is maximum, and tends to zero at the ends of the section — i.e. at the ends of the flanges in the I-beam of Figure 6.9.

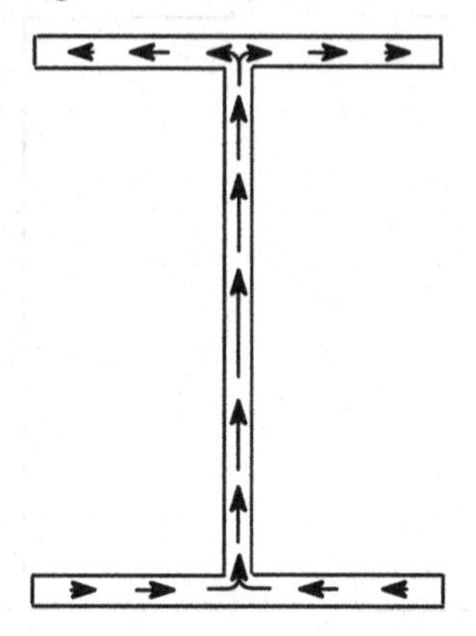

*Figure 6.9: Shear stress distribution in the I-beam*

Another unexpected result is obtained when the section bends back towards the neutral axis, as in Figure 6.10*(a)*. In this case, the preceding arguments lead to the stress distribution of Figure 6.10*(b)* in which the shear stress near the ends acts downwards on the section, despite the fact that the resultant is an upward shear force. Notice however that the regions with downward shear stress have lower stress values, being near to the free end (where the stress is always zero), so the resultant *is* an upward force.

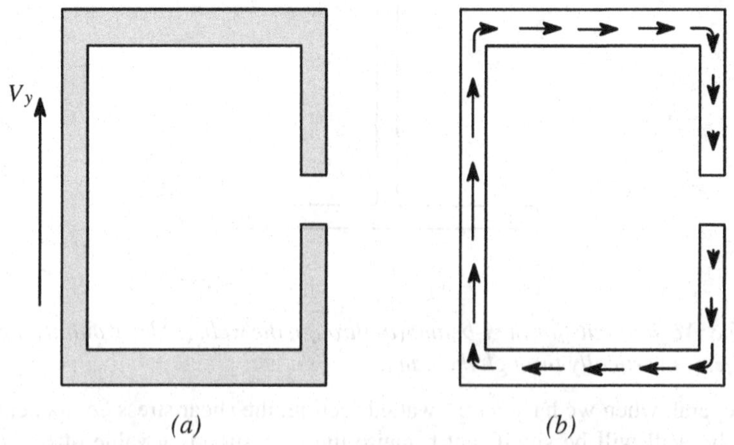


*Figure 6.10: Shear stress distribution for a slit rectangular tube*

In general, if the section crosses the neutral axis only once, the stress there will be in the same direction as $V_y$ and the shear flow will continue along the section, diminishing in magnitude towards the extremities of the section. At all points it remains locally parallel to the wall of the section.

## Example 6.1

*The I-beam of Figure 6.11 is loaded by a vertical shear force of 50 kN. The appropriate centroidal second moment of area for the section is $I_x = 39.4 \times 10^6$ mm$^4$. Find*

*(i) the complete distribution of shear stress in the section,*
*(ii) the maximum shear stress,*
*(iii) the average vertical shear stress,*
*(iv) the proportion of the shear force that is carried by the web,*
*(v) the average shear stress in the web.*

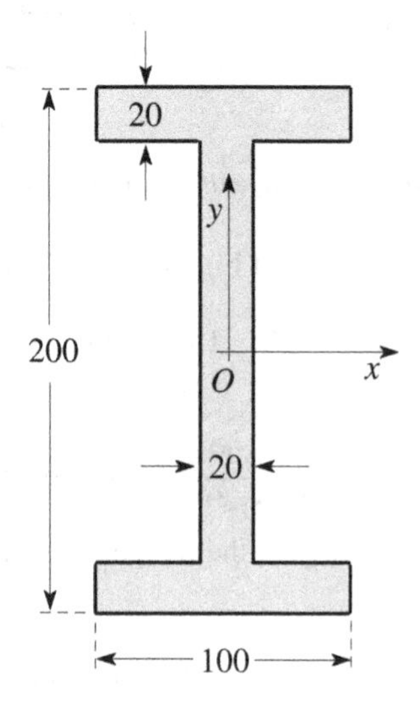


*Figure 6.11*

**(i)** To determine the complete shear stress distribution, we need to make two cuts at arbitrary points — one in the web and one in the flange as defined in Figure 6.12.

For the web, we use the cut *BC* of Figure 6.12 *(a)*, obtaining

$$\begin{aligned} A^*\bar{y}^* &= A_1\bar{y}_1 + A_2\bar{y}_2 = (80-y) \times 20 \times \frac{(80+y)}{2} + 100 \times 20 \times 90 \\ &= 244{,}000 - 20y^2 \text{ mm}^3 \ (y \text{ in mm}) \end{aligned}$$

and hence

$$\tau = \frac{50 \times 10^3 \times (244{,}000 - 20y^2)}{39.4 \times 10^6 \times 20} = 15.5 - 1.27 \times 10^{-3} y^2 \text{ MPa}\,.$$

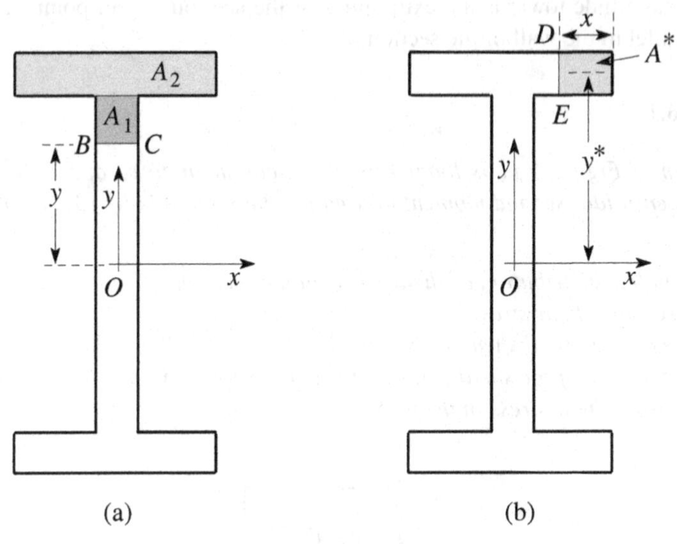


*Figure 6.12: (a) cut BC, (b) cut DE.*

For the flange, we use the cut $DE$ of Figure 6.12 *(b)*, obtaining

$$A^*\bar{y}^* = x \times 20 \times 90 = 1800x \text{ mm}^3 \ (x \text{ in mm})$$

and hence

$$\tau = \frac{50 \times 10^3 \times 1800x}{39.4 \times 10^6 \times 20} = 0.1142x \text{ MPa}\,.$$

Thus, the shear stress varies linearly in the flanges and quadratically in the web, as shown in Figure 6.13.

**(ii)** Since $t$ is the same at all sections, the maximum shear stress occurs in the web at the neutral axis ($y = 0$), where $A^*\bar{y}^*$ is maximum. We obtain

$$\tau_{\max} = 15.5 \text{ MPa}\,.$$

**(iii)** The average vertical shear stress is simply the shear force divided by the total area of the section which is

$$A = 2 \times 100 \times 20 + 160 \times 20 = 7200 \text{ mm}^2\,.$$

Hence

$$\tau_{\text{ave}} = \frac{50 \times 10^3}{7200} = 6.9 \text{ MPa}\,.$$

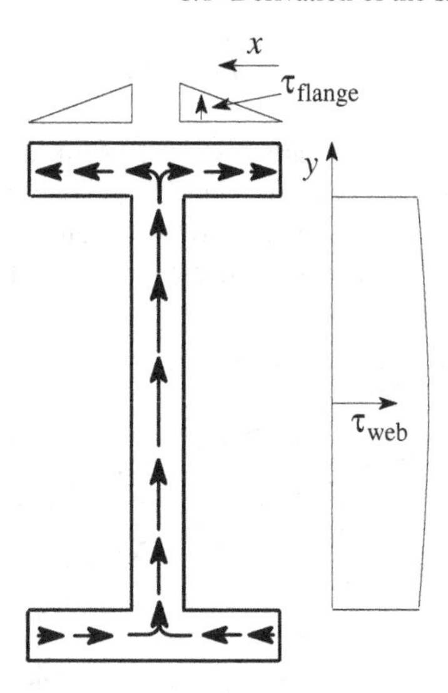


*Figure 6.13: Shear stress distribution in the I-beam*

**(iv)** The total shear force carried in the web can be obtained by summing (integrating) the shear stress over the web. We obtain

$$F = \int_{-80}^{80} (15.5 - 1.27 \times 10^{-3} y^2) 20 dy = 49,167 \text{ N (49 kN)} .$$

Thus the proportion of the shear force carried by the web is

$$\frac{49,167}{50,000} = 0.983 \ \ (98.3\%) .$$

**(v)** The average shear stress in the web is

$$\tau_{\text{ave}}^{\text{web}} = \frac{F}{A_{\text{web}}} = \frac{49167}{160 \times 20} = 15.4 \text{ MPa} .$$

Notice, from (iv) above, that almost all the vertical shear force is carried by the web. Also, there is only a modest variation in shear stress over the web. Even at the top of the web, where $y = 80$ mm, the shear stress is 11.4 MPa, from (i) above with $y = 80$ mm. For this reason, a reasonable estimate of the maximum shear stress in the I-beam can be obtained very quickly by just dividing the shear force by the web area — i.e.

$$\tau_{\text{max}} \approx \frac{50,000}{160 \times 20} = 15.6 \text{ MPa} ,$$

which is a very good quick approximation to the correct value obtained in (ii) above.

### 6.1.2 Location and magnitude of the maximum shear stress

In equation (6.9), only $A^*, \bar{y}^*$ and $t$ vary with position in the section and hence the maximum transverse shear stress occurs at the point where the ratio $A^*\bar{y}^*/t$ is a maximum. Elemental areas $dA$ make a positive contribution to

$$A^*\bar{y}^* = \iint_{A^*} y dA$$

if and only if $y > 0$ — i.e. if they are above the neutral axis. It follows that the maximum value of $A^*\bar{y}^*$ occurs when the cut is made at the neutral axis, in which case $A^*$ comprises that part of the section above the neutral axis. If the thickness $t$ is constant, as in Example 6.1, the maximum shear stress will therefore always occur at the neutral axis.

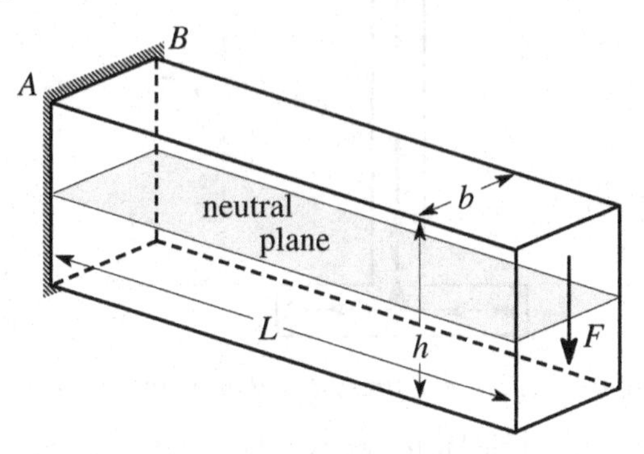


*Figure 6.14: Cantilever beam of rectangular cross section*

For a beam of rectangular cross section, the transverse shear stresses are always substantially smaller than the normal stresses due to bending. Consider, for example, the rectangular cantilever beam of Figure 6.14, loaded by an end force $F$. Elementary calculations show that the maximum bending stress occurs along the line $AB$ and is

$$\sigma_{max} = \frac{6FL}{bh^3} \tag{6.10}$$

and the maximum shear stress, calculated from equation (6.9) occurs at the neutral plane and is

$$\tau_{max} = \frac{3F}{2bh} \, . \tag{6.11}$$

Thus, $\tau_{max}$ will exceed $\sigma_{max}$ if and only if $L < h/4$ — i.e. if the beam is only a quarter as long as its height, which takes the problem well outside the range of application of beam theory, for which we require $L \gg h, b$.

Larger transverse shear stresses will be obtained if the section has a locally reduced thickness $t$, particularly if this occurs at or near to the neutral axis, as shown in Figure 6.15. However, it is still comparatively rare to encounter an application where the shear stresses due to shear loading are comparable with those due to bending. For this reason, it is frequently appropriate in design to neglect shear stresses altogether, or else to use a crude approximation for them such as $\tau \approx V_y/A$ or $\tau \approx V_y/A_{web}$. In Example 6.1, these approximations gave at least an order of magnitude estimate of the maximum shear stress and could therefore have been used to determine whether this magnitude was sufficiently high to justify the effort involved in a more exact calculation.

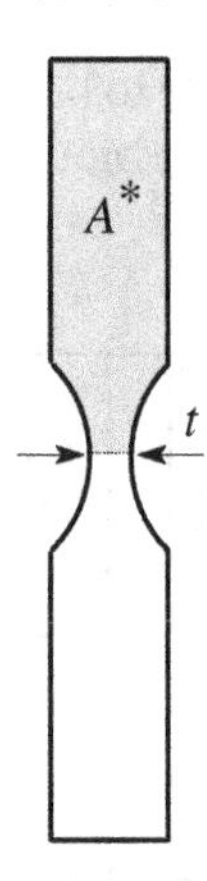


*Figure 6.15: Beam with a reduced thickness at the neutral axis*

### 6.1.3 Welds, rivets and bolts

There are not many applications where we would need to use a beam with a locally reduced thickness $t$, but a similar effect is produced when a beam is fabricated by welding three plates together as shown in Figure 6.16*(a)*. In this case, the thickness at the weld may be smaller than that of the rest of the section. Also, the stresses in the weld are of particular interest, since the weld material may contain defects and hence be weaker than the rest of the material.

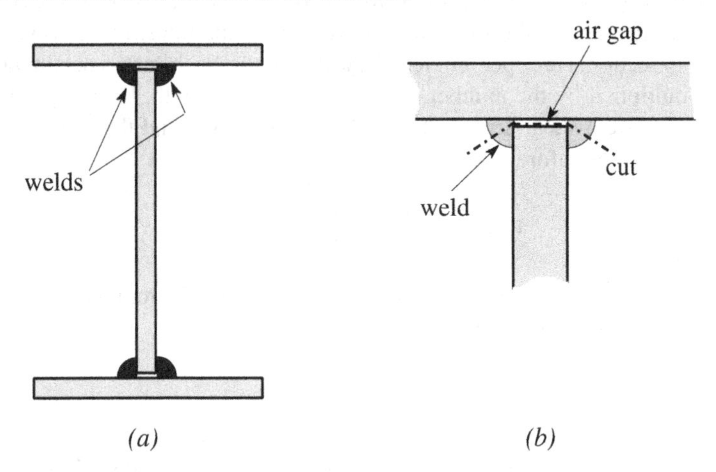


*Figure 6.16: A beam fabricated from three rectangular plates*

To determine the maximum shear stress in the weld, we make the *shortest* cut that would be sufficient to cause the flange to drop off, since the cut must separate the section into two parts in view of the argument of §6.1. This can be achieved by cutting through both welds as shown in Figure 6.16*(b)*.

Notice that the cut doesn't even have to be a plane or to be connected. The thickness $t$ is the total length along the cut line, not including air. Here we would get twice the smallest thickness through the weld.[5]

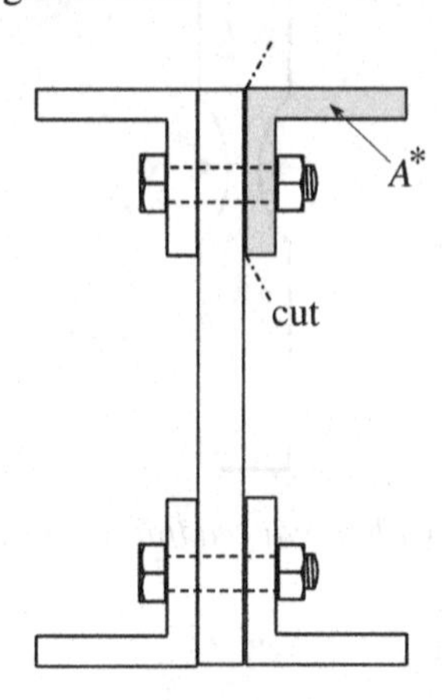


*Figure 6.17: A beam assembled with bolts*

A similar technique can be used if rivets or bolts are used to attach the flange instead of welds, as shown in Figure 6.17. This time, we imagine cutting through the bolts to make one angle drop off. However, we cannot define a thickness $t$ here, since the angle is joined to the rest of the section by a set of discrete areas. Instead, we must go back to the concept of shear flow $q$, which is the horizontal shear force per unit length transmitted through the bolts. This must be transmitted through the total bolt cross-sectional area per unit length, which in turn is the cross-sectional area of one bolt multiplied by the number of bolts $N$ per unit length. If the bolts are of diameter $d$, we have a shear area of $\pi N d^2/4$ per unit length and the average shear stress in the bolts is therefore

$$\tau_{\text{bolts}} = \frac{4q}{\pi d^2 N} = \frac{4V_y A^* \bar{y}^*}{\pi d^2 N I_x} , \tag{6.12}$$

where we note that $A^*$ and $\bar{y}^*$ relate to the section that would be removed by the cut, as in more conventional problems.

**Example 6.2**

*Figure 6.18 shows a W610×101 I-beam that has been stiffened against bending by the attachment of two 230×25 mm rectangular plates. The plates are secured to the I-beam by 10 mm diameter bolts with an axial spacing of 100 mm. The second moment of area of the I-beam without the stiffeners is $764\times10^6$ mm$^4$. Find the average shear stress in the bolts if the beam transmits a shear force of 50 kN.*

[5] In calculating the shear stress for this cut using equation (6.9), we can neglect the small contribution to $A^*\bar{y}^*$ from the fragments of weld adhering to the section removed.

We first calculate the second moment of area $I_x$ for the stiffened section, using the parallel axis theorem, with the result

$$I_x = 764 \times 10^6 + 2\left(\frac{230 \times 25^3}{12} + 230 \times 25 \times 314^2\right) = 1898 \times 10^6 \text{ mm}^4 .$$

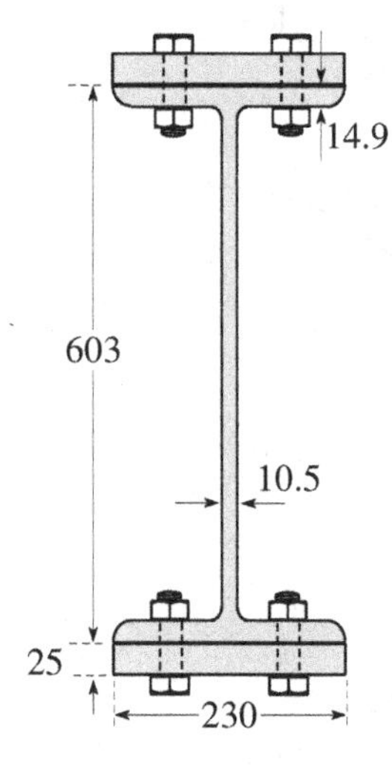


*Figure 6.18*

If we cut through the top flange bolts, the area cut off ($A^*$) is $230 \times 25$ and

$$A^* \bar{y}^* = 230 \times 25 \times 314 = 1.806 \times 10^6 \text{ mm}^3 .$$

The axial bolt spacing is 100 mm and there are two bolts at each axial location, so the number of bolts per unit length is

$$N = \frac{2}{100} = 0.02 \text{ bolts per mm} .$$

Substituting these results into equation (6.12), we obtain

$$\tau_{\text{bolts}} = \frac{4 \times 50,000 \times 1.806 \times 10^6}{\pi \times 10^2 \times 0.02 \times 1898 \times 10^6} = 30 \text{ MPa} .$$

### 6.1.4 Curved sections

If a thin-walled beam contains curved sections, the procedure for determining the transverse shear stress is the same, but integration will generally be needed in determining $I_x$ and $A^* \bar{y}^*$. We illustrate this with the example of the semicircular section of Figure 6.19, with radius $a$ and thickness $t$, where $t \ll a$.

To determine $I_x$, we consider a small element $dA = atd\theta$ shown shaded in Figure 6.19. Then

$$I_x = \iint_A y^2 dA = \int_{-\pi/2}^{\pi/2} (a\sin\theta)^2 at\,d\theta$$

$$= a^3 t \int_{-\pi/2}^{\pi/2} \sin^2\theta\,d\theta = \frac{\pi a^3 t}{2}\,. \tag{6.13}$$

Notice that $I_x$ is a property of the whole section, so the integral is performed over the entire semi-circle from end to end.

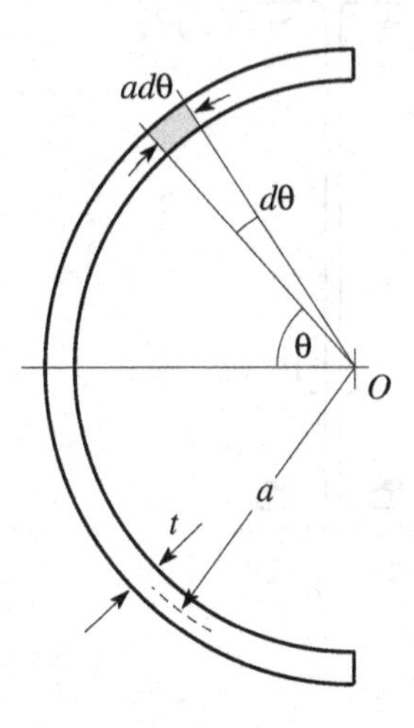


*Figure 6.19: Thin-walled semi-circular cross section*

We now make a cut at a general point $\theta = \phi$ to determine the shear stress, as shown in Figure 6.20.

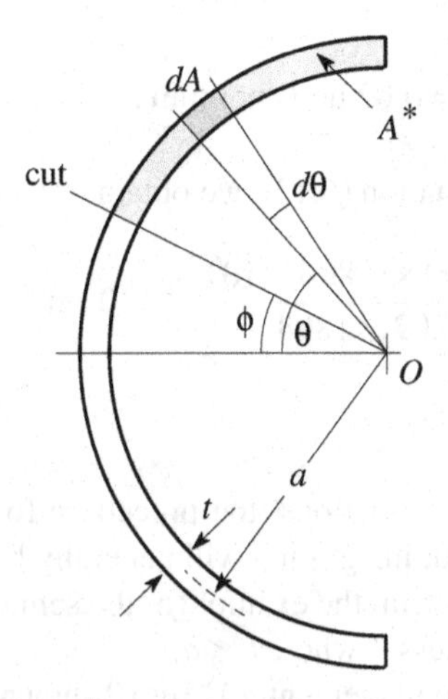


*Figure 6.20: Cut for determining the shear stress in the semi-circular section*

The quantity $A^*\bar{y}^*$ relates only to the section cut off, which is defined by $\phi < \theta < \pi/2$. These are therefore the limits of integration.[6] We obtain

$$A^*\bar{y}^* = \int_{\phi}^{\pi/2} ydA = \int_{\phi}^{\pi/2} (a\sin\theta)atd\theta$$
$$= a^2 t\cos\phi \; . \tag{6.14}$$

It is important to ensure that the limits of integration correspond to the edges of the area $A^*$ and that the lower limit is smaller than the upper limit.

The shear stress at the point $\theta = \phi$ is

$$\tau = \frac{V_y A^*\bar{y}^*}{I_x t} = \frac{V_y \times (a^2 t\cos\phi)}{(\pi a^3 t/2)\times t} = \frac{2V_y\cos\phi}{\pi a t} \; . \tag{6.15}$$

We see from equation (6.15) that the shear stress varies sinusoidally around the section from zero at the ends ($\phi = \pm\pi/2$) to a maximum at the neutral axis ($\phi = 0$), given by

$$\tau_{\max} = \frac{2V_y}{\pi a t} \; . \tag{6.16}$$

It is parallel to the section wall at all points, as shown in Figure 6.21.

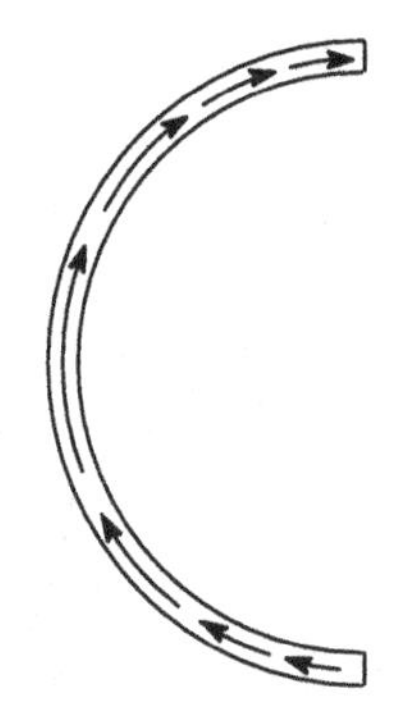

*Figure 6.21: Shear stress distribution for the semi-circular section*

## 6.2 Shear centre

The theory presented above gives a unique solution for the shear stress throughout the section for a given value of $V_y$. This implies that the line of action of $V_y$ is prescribed, since the distribution of $\tau$ over the section has a resultant in magnitude, direction *and* line of action.

[6] Notice that it is necessary to define two indefinite angles $\theta, \phi$ in order to calculate the stress at a general point. One of these angles ($\phi$) defines the point under consideration and the other ($\theta$) serves as a variable of integration in the summation for $A^*\bar{y}^*$. A common error is to use only one such variable and then use definite limits in the integration. At best, this will only yield the stress at a particular point.

If the actual shear force applied has a different line of action from that implied by the theoretical distribution, there will be some twisting of the section and a more complex shear stress distribution will be developed.

There is a unique point $C$ for any given beam section, such that if the line of action of $\boldsymbol{V}$ passes through $C$, the section does not twist. This point is called the *shear centre*. For a doubly-symmetric section, like an I-beam, the shear centre will coincide with the intersection between the two axes of symmetry. For sections with one axis of symmetry, the shear centre lies on that axis, but it's location must be found by determining the line of action of the shear force implied by the distribution of shear stress from the preceding analysis.

### 6.2.1 Finding the shear centre

To find the shear centre of a thin-walled section, we must determine the line of action of the resultant force due to the shear stress distribution $\tau$ of equation (6.9). The shear stress is everywhere locally parallel to the thin wall and it generates an elemental force $dF = \tau dA$ on the elemental area $dA$ shown in Figure 6.22*(a)*. We can write $dA = tdS$, where $dS$ is an element of the perimeter and hence

$$dF = \tau t dS = q dS \, . \tag{6.17}$$

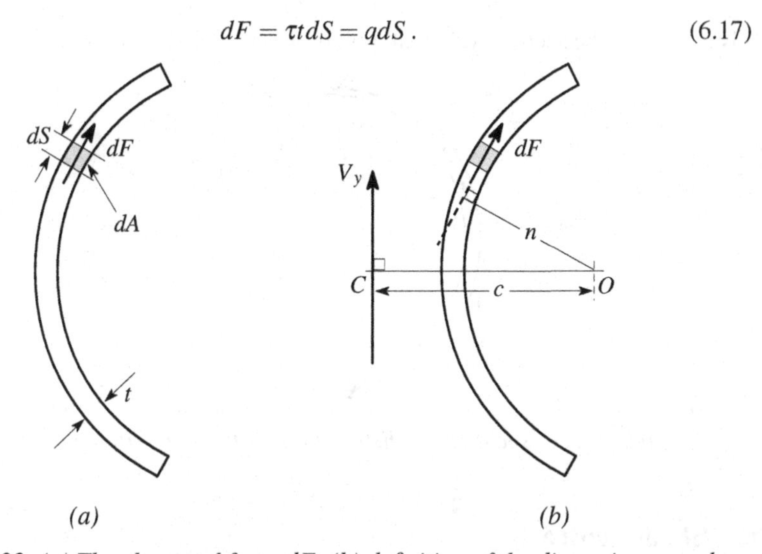


*Figure 6.22: (a) The elemental force dF, (b) definition of the dimensions c and n*

The shear force $V_y$ is the resultant of the elemental forces $dF$ and it must therefore have the same moment about any point $O$. We conclude that

$$V_y c = \int_S n dF = \int_S n \tau t dS = \int_S n q dS \, , \tag{6.18}$$

where the integration is performed over the entire perimeter of the section, $c$ defines the location of the shear centre and $n$ is the moment arm for the elemental force $dF$, i.e. the perpendicular distance from $O$ to the line of action of $dF$ (the local tangent to the mean line of the section), as shown in Figure 6.22*(b)*.

**Example 6.3**

*Find the distance c defining the shear centre for the semi-circular section of Figure 6.23.*

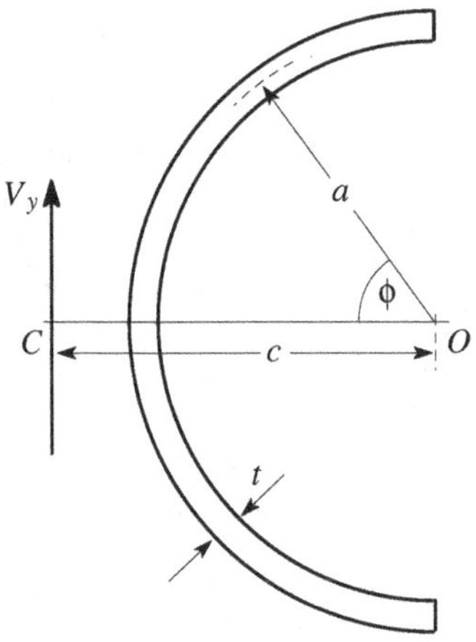


*Figure 6.23: Location of the shear centre for the semi-circular cross section*

The shear stress distribution for this section has already been determined in §6.1.4 above and is given by equation (6.15). In applying equation (6.18) to find $c$, it is convenient to take moments about the centre of the semi-circle, since the moment arm will then be constant and equal to the radius $a$. We therefore have

$$V_y c = \int\int_A (a)\tau dA = \int_{-\pi/2}^{\pi/2} (a) \left( \frac{2V_y \cos\phi}{\pi a t} \right) a t d\phi = \frac{2V_y a}{\pi} \int_{-\pi/2}^{\pi/2} \cos\phi d\phi = \frac{4V_y a}{\pi} ,$$

using (6.15) and hence

$$c = \frac{4a}{\pi} \approx 1.27a .$$

It follows that the shear centre $C$ is actually to the left of the section, as shown in Figure 6.23 — i.e. it is outside the semi-circle. This is surprising at first sight, but it is reasonable when we realize that the moment of the distribution about $C$ has to be zero, whereas for any point on the concave side of the section the integrand would be everywhere positive (positive $\tau$ and positive moment arm). In general, for open sections such as semi-circles, channels, open box sections etc., the shear centre is outside the section, but usually not very far outside. Some typical examples are shown in Figure 6.24.

Notice that the sections have to be *open* for this to be true. A different procedure is needed if they are closed[7] and generally the shear centre is then *inside* the section, as shown for example in Figure 6.25. This will be discussed in §6.6 below.

[7] A closed thin-walled section is one that completely encloses a volume. A closed section is rendered open by a cut at any point, even if the width of the cut is infinitesimal. Another way of defining this distinction is to note that a closed section is a multiply-connected shape, whereas an open section is simply-connected.

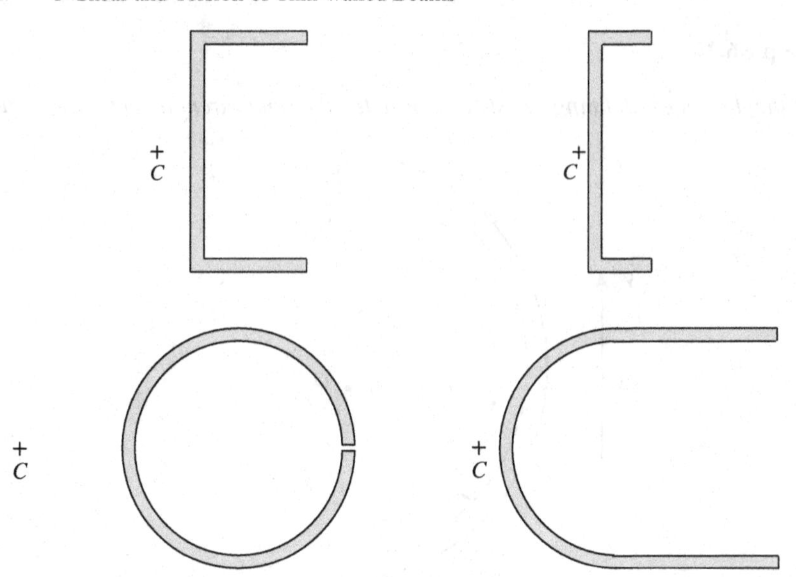


*Figure 6.24: Typical shear centre locations for open sections*

Thus, a major difference in shear behaviour is produced by slitting open a closed thin-walled section. You can examine this effect by comparing a loosely rolled cylindrical tube of paper with the same tube after it is taped shut along the edge.[8] The distinction arises because the open tube permits relative motion along the axis at the edge, whereas the closed tube does not.

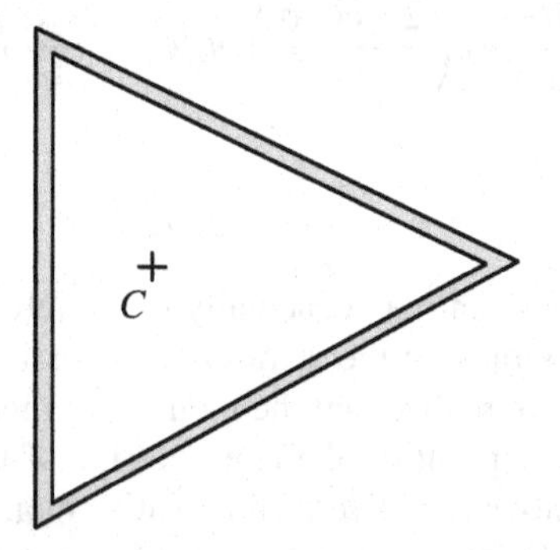


*Figure 6.25: Shear centre for a closed section*

**Example 6.4**

*Find the location of the shear centre for the thin-walled channel section of Figure 6.26.*

[8] For a more robust experiment, compare the cardboard tube from the centre of a roll of kitchen paper with a similar tube that has been slit along the axis, as in the third cross section of Figure 6.24.

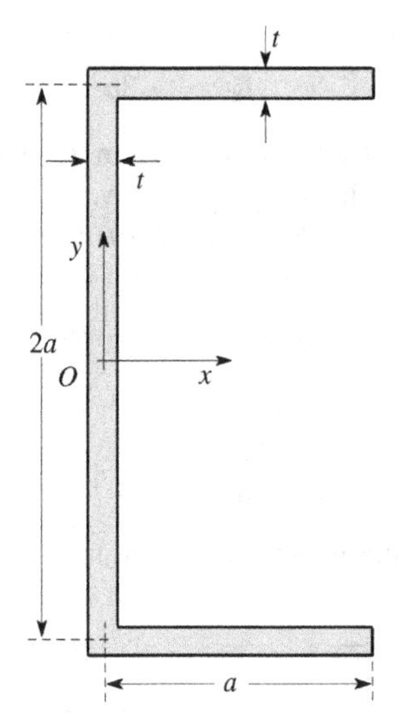


*Figure 6.26*

It is first necessary to determine the second moment of area which is

$$I_x = \frac{(2a)^3 t}{12} + 2(at)a^2 = \frac{8a^3 t}{3} \; .$$

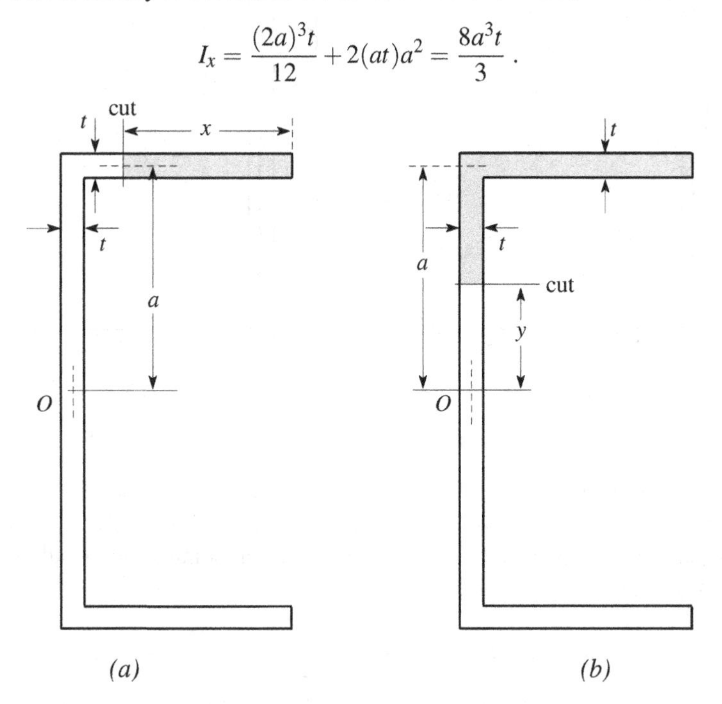


*Figure 6.27: Cuts for determining the shear stress (a) in the flange and (b) in the web*

If the beam is loaded by a vertical shear force $V_y$, the horizontal shear flow in the upper flange can be found using the cut of Figure 6.27 *(a)* and is

$$q = \frac{V_y(tx)a}{I_x} = \frac{3V_y x}{8a^2} ,$$

where $x$ is measured from the free end as shown in the figure. An equal and opposite stress is obtained in the lower flange by symmetry.[9]

For the web, we use the cut of Figure 6.27 *(b)*, obtaining

$$A^*\bar{y}^* = (at)a + \frac{(a-y)(a+y)t}{2} = \frac{3a^2t}{2} - \frac{y^2t}{2}$$

and hence

$$q = \frac{3V_y}{16a}\left(3 - \frac{y^2}{a^2}\right) ,$$

where $y$ is measured from the neutral axis. The stress distribution is illustrated in Figure 6.28 *(a)*.

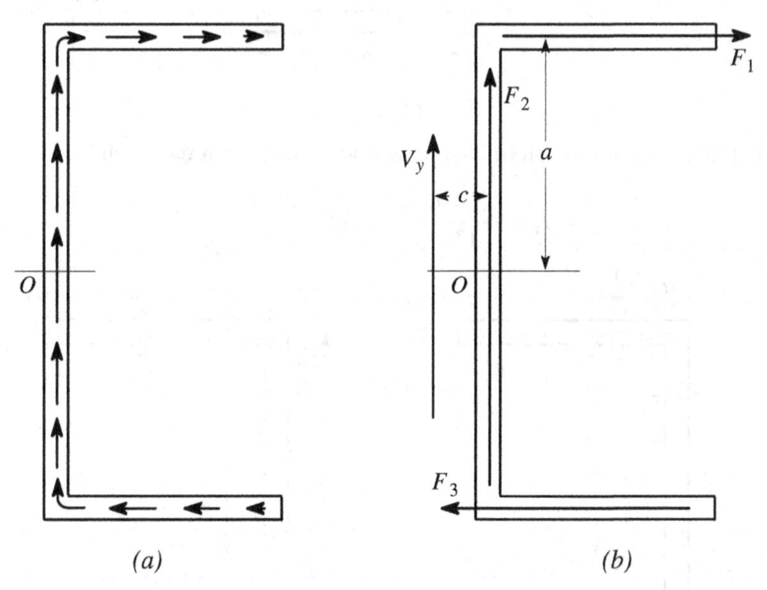


*Figure 6.28: (a) shear stress distribution in the channel section, (b) force resultants*

The channel section is made up of straight segments and all the elemental forces in any given segment will have the same line of action. We can therefore sum the forces separately for each segment by integration. For example, the resultant of the shear flow $q$ in the top flange is

$$F_1 = \int_0^a q dx = \int_0^a \left(\frac{3V_y x}{8a^2}\right) dx = \frac{3V_y}{16}$$

and it has the line of action shown in Figure 6.28 *(b)*.

[9] Alternatively, we could make a similar cut in the lower flange obtaining exactly the same expression, except that $\bar{y}^*$ would be $-a$, leading to a change of sign.

The resultant of the shear stress in the web, $F_2$ in Figure 6.28 *(b)*, must be equal to $V_y$, since this is the only place where the shear stress has a vertical component. We can confirm this by evaluating $F_2$ as

$$F_2 = \int_{-a}^{a} \frac{3V_y}{16a}\left(3-\frac{y^2}{a^2}\right)dy = \frac{3V_y}{16a}\left[3y-\frac{y^3}{3a^2}\right]_{-a}^{a} = \frac{3V_y}{16a}\left(6a-\frac{2a}{3}\right) = V_y\,.$$

The force $F_3$ in the lower flange is equal and opposite to $F_1$ by symmetry, so horizontal equilibrium is satisfied.

We can now determine the location of the shear centre by taking moments about the point $O$ in Figure 6.28 *(b)*. The force $V_y$ is statically equivalent to the system of forces $F_1, F_2, F_3$ and hence must have the same moment as this system about any point. It follows that

$$V_y \times c = F_1 \times a + F_2 \times 0 + F_3 \times a = \frac{3V_y a}{16} + \frac{3V_y a}{16}$$

and hence

$$c = \frac{3a}{8}\,.$$

Notice that by taking moments about $O$, we were able to make the moment arm for $F_2$ equal to zero. If we had recognized this from the beginning, we could have avoided calculating the shear stress in the web, thereby shortening the calculation. In general, when the problem asks only for the location of the shear centre, a careful choice of the point about which to take moments can often result in a saving of algebraic work.

## Experiment

It is difficult to load a channel section at the shear centre, since the latter occurs outside the section. It is easier to load it on the flange. This is equivalent to superposing a moment in the clockwise sense, as shown in Figure 6.29, and it will twist the section. You can test this experimentally by making a channel section beam from a piece of thin metal plate, fixing it at one end and loading the other end by a transverse force, as in Figure 6.29 *(a)*. You will find that the free end rotates clockwise under the action of the equivalent moment of Figure 6.29 *(b)*.

As a variant on this experiment, cut out a rectangular plate as in Figure 6.30 *(a)* and fold it along the dotted lines to form the beam of Figure 6.30 *(b)*. You will then be able to investigate the effect of loading outside the envelope of the section and in particular to locate the shear centre by finding the line of action of the force $F$ for which there is no twist.

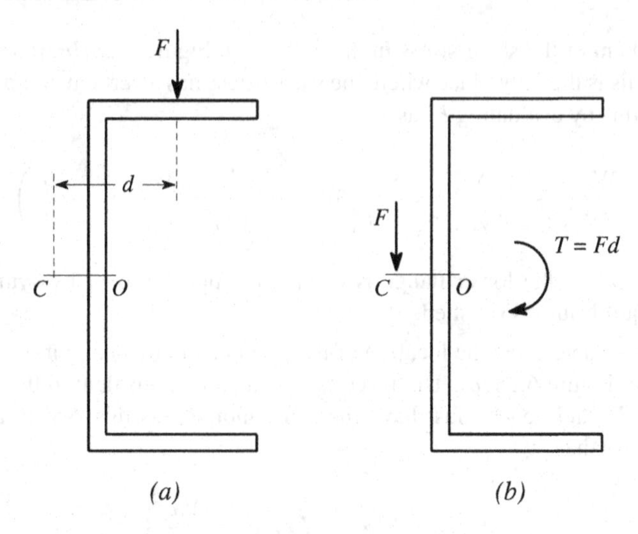


*Figure 6.29: (a) shear loading of the channel section, (b) equivalent loading system with the force acting through the shear centre and a torque*

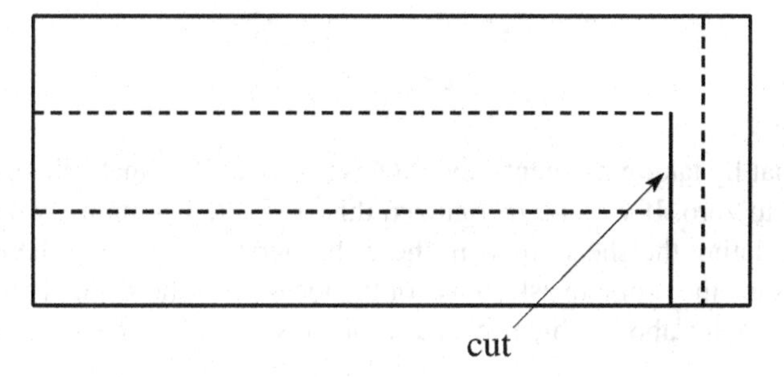


*Figure 6.30(a): An experimental model for finding the shear centre of a channel section*

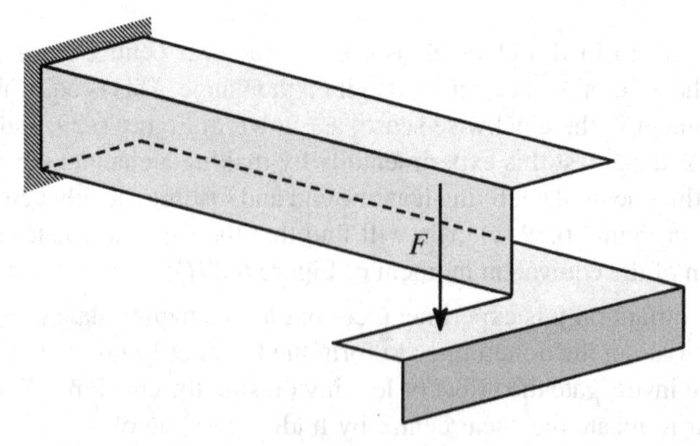


*Figure 6.30(b): Fold the plate to form the channel-shaped cantilever shown and explore the effect of the line of action of F on the twist of the section.*

## 6.3 Unsymmetrical sections

So far, we have considered only sections with at least one axis of symmetry. The generalization to unsymmetrical sections is routine, but leads to significantly more complicated algebra. The principal reason for performing these calculations is to determine the location of the shear centre, since the moment resulting from shear loading of an open section away from its shear centre [see e.g. Figure 6.29 *(b)*] can cause significant torsional stresses and twist, as will be discussed in §6.7 below. By contrast, the shear stresses associated with shear loading acting through the shear centre are often sufficiently small to be neglected, or estimated by simple approximations as discussed in §6.1.2.

### 6.3.1 Shear stress for an unsymmetrical section

The derivation of §6.1 can be generalized to unsymmetrical sections, using equation (4.20) in place of (6.2) to define the bending stress. We obtain

$$q = \tau t = \frac{(V_x I_x - V_y I_{xy})A^*\bar{x}^* - (V_x I_{xy} - V_y I_y)A^*\bar{y}^*}{(I_x I_y - I_{xy}^2)} , \tag{6.19}$$

where $(\bar{x}^*, \bar{y}^*)$ are the coordinates of the centroid of $A^*$ relative to the centroid of the whole section and we have used the relations

$$V_x = -\frac{dM_y}{dz} \;\; ; \;\; V_y = \frac{dM_x}{dz} , \tag{6.20}$$

obtained by similar arguments to equation (6.1), using the sign convention of Figures 4.1, 6.1.

If the coordinate system $Oxy$ is chosen to coincide with the principal axis system as in §4.4.3, the product inertia $I_{xy}$ will be zero and equation (6.19) reduces to

$$q = \tau t = \frac{V_x A^*\bar{x}^*}{I_y} + \frac{V_y A^*\bar{y}^*}{I_x} , \tag{6.21}$$

which is of course equivalent to the superposition of two applications of equation (6.9) about orthogonal axes.

### 6.3.2 Determining the shear centre

Suppose that the shear centre of the unsymmetrical thin-walled section of Figure 6.31 is at the point $C(a,b)$. As before, the shear flow $q$ defined by equation (6.19) corresponds to an elemental force $dF = \tau t dS = q dS$. The moment of the complete stress distribution about $O$ can then be equated to the moment of the two shear force components $V_x, V_y$, giving

$$aV_y - bV_x = \int_S nq dS . \tag{6.22}$$

For a particular problem, $q$ will be known from (6.19) and can be substituted into (6.22). When the integral is evaluated, the coordinates $a, b$ of the shear centre can be determined by equating coefficients of $V_x, V_y$. Alternatively, we can solve two separate problems — one with $V_x = 0$ to determine $a$ and another with $V_y = 0$ to determine $b$.

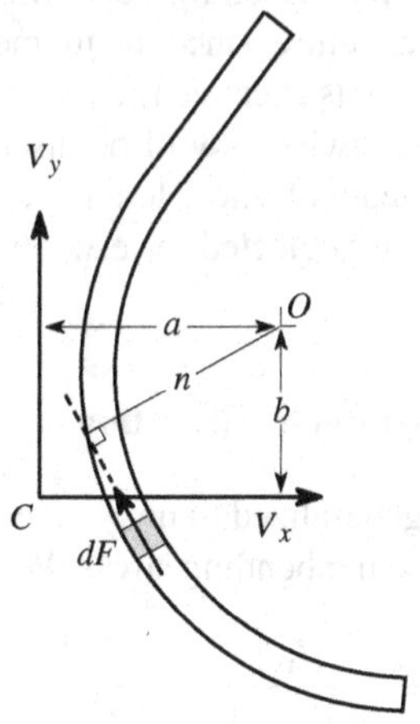


*Figure 6.31: Determining the shear centre for an unsymmetric section*

The procedure is algebraically tedious and will not be pursued here. More details and worked examples can be found in Cook and Young §9.8 or Bickford §4.4.

**Shear centre for angle sections**

One example for which the determination of the shear centre is very straightforward is the thin-walled angle section of Figure 6.32.

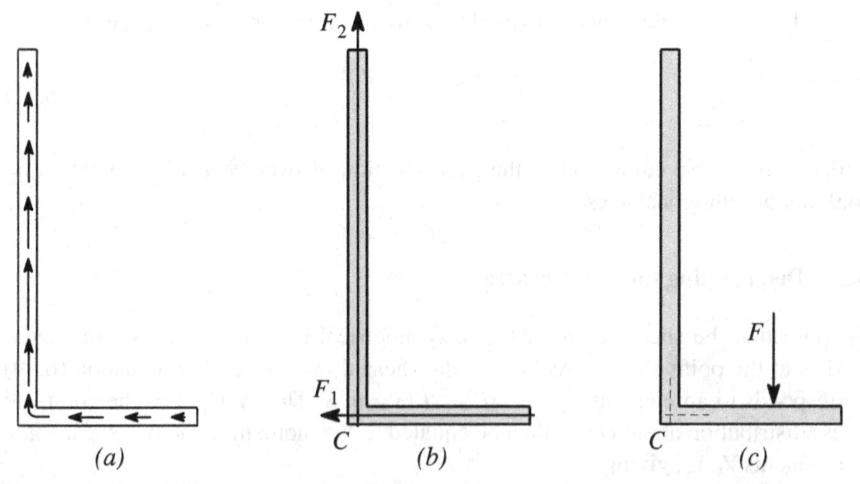


*Figure 6.32: (a) Shear stress distribution, (b) equivalent forces and shear centre location for an angle section, (c) loading by a force on the flange*

The shear stress distribution has to follow the walls of the section as in Figure 6.32 *(a)* and hence is equivalent to the two forces $F_1, F_2$ shown in Figure 6.32 *(b)*. The resultant of these two forces must act through the corner of the section $C$ which is therefore the shear centre.

Loading of an angle by a shear force on the lower flange, as shown in Figure 6.32 *(c)*, will therefore lead to clockwise twist of the section. Notice incidentally that vertical loading as in Figure 6.32 *(c)* implies that $F_1 = 0$ and hence that the shear stress in Figure 6.32 *(a)* changes direction at some point in the horizontal leg.

## 6.4 Closed sections

A closed section is one with one or more *enclosed* cavities. Another way of describing it is to say that a single longitudinal cut through the wall will not generally separate the section into two parts, but will simply render the section open, as in the case of the slit tube of Figure 6.33 *(b)*.

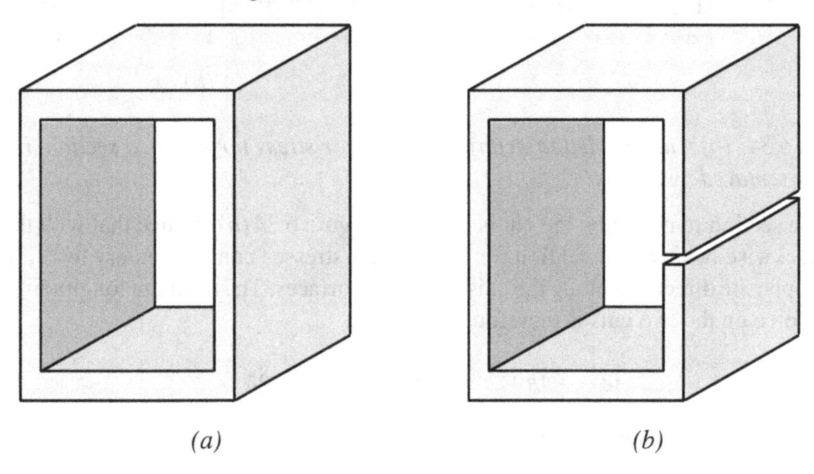


*Figure 6.33: (a) Closed section and (b) equivalent open section obtained by slitting the closed section along the axis*

In this section, we shall restrict attention to the case of a single enclosed cavity — e.g. a cylindrical tube or a box section. Other sections with several cavities (i.e. a closed section with continuous internal stiffeners) can be treated by similar methods.[10]

### 6.4.1 Determination of the shear stress distribution

Suppose we have an arbitrary closed thin-walled section transmitting a shear force $V_y$. We follow the same procedure as in §6.1, but in order to remove a piece of the

[10] See §6.6.2 below.

section to write an equilibrium equation, we now have to make *two* cuts — e.g. one at $A$ and one at $B$ in Figure 6.34 *(a)*. We shall adopt the convention that $\tau$ is positive in the anticlockwise sense, as shown, remembering that because of the thinness of the wall, the stress is constrained to be parallel to the wall at every point.

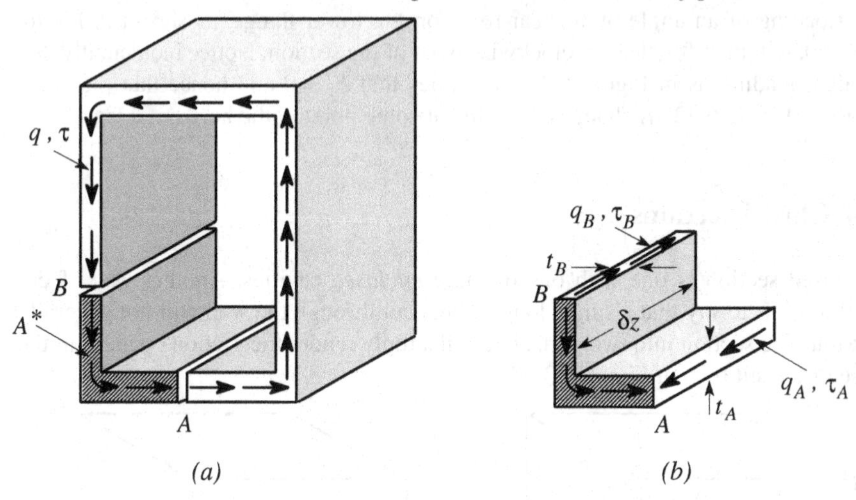


*Figure 6.34: (a) Cuts needed to determine the shear stress for a closed section, (b) section removed by the cut*

The section removed by the cut is shown in Figure 6.34 *(b)*. Notice that with the anticlockwise positive convention for $\tau$, the shear stresses complementary to $\tau_A, \tau_B$ act in opposite directions along the axis on the cut surfaces. The resultant longitudinal shear force on the two cuts is therefore

$$Q = \tau_B t_B \delta z - \tau_A t_A \delta z = (q_B - q_A)\delta z$$

and substitution into equation (6.7) yields

$$q_B = \frac{V_y A^* \bar{y}^*}{I_x} + q_A \,. \tag{6.23}$$

In using this result, it is important to adhere to the convention of Figure 6.34 *(a)* that $A^*$ is that part of the section traversed in passing clockwise from $A$ to $B$. However, $A^*\bar{y}^*$ for the complete section (shaded + unshaded) is zero and hence

$$(A^*\bar{y}^*)_{\text{shaded}} = -(A^*\bar{y}^*)_{\text{unshaded}} \,. \tag{6.24}$$

It follows that we can pass in the *anticlockwise* direction from $A$ to $B$, provided the area traversed $A^*$ is treated as negative in equation (6.23).

Equation (6.23) defines the shear flow $q$ throughout the section except for an arbitrary additive constant. Suppose we choose an arbitrary fixed point $A$ and write $q_A = C$, where $C$ is unknown. Equation (6.23) then takes the form

$$q_B = \frac{V_y A^* \bar{y}^*}{I_x} + C \tag{6.25}$$

and, since $B$ can be chosen to be any point, it completely determines the shear flow distribution except for the unknown constant $C$. Different values of $C$ will give different but parallel lines of action for the resultant shear force $V_y$, and these can be determined using the method of §6.2. Conversely, if the line of action of $V_y$ is prescribed, this can be used to determine $C$.

Thus, we see that there is a qualitative difference between the shear behaviour of open and closed thin-walled sections. For an open section, equation (6.9) corresponds to a unique line of action for the shear force, whereas for a closed section, equation (6.25) with an appropriate value of the constant $C$ describes the distribution for any line of action of $V_y$.

However, closed sections *do* have shear centres, in the sense of a point where $V$ must act in order for there to be no twist of the section. We shall return to this question in §6.6 below.

### Summary

The procedure for finding the shear flow and hence the shear stress in a thin-walled closed section loaded by a shear force with a prescribed line of action can be summarized as follows:-

(i) Choose any convenient point $A$ as a reference point and denote the shear flow there as $q_A = C$.
(ii) Use equation (6.25) to determine general expressions for $q$ in each segment of the section. Remember that the area $A^*$ is treated as positive if it is traversed in passing clockwise from $A$ to $B$ and negative if it is traversed anticlockwise.
(iii) Equate the moment of the distribution about any convenient point to the corresponding moment of the prescribed shear force and use the resulting equation to determine the constant $C$.

### Example 6.5

*The closed semi-circular section of Figure 6.35 is loaded by a vertical shear force $V_y$, whose line of action passes through the point $O$ as shown. Determine the complete distribution of shear stress in the section.*

We first determine the second moment of area $I_x$ for the section. We can use equation (6.13) for the semi-circular part of the section and add the contribution for the straight part, to obtain

$$I_x = \frac{\pi a^3 t}{2} + \frac{(2a)^3 t}{12} = 2.237 a^3 t \, .$$

It is convenient to choose the top corner of the section to be the point $A$, since segments starting at $A$ have a simple geometric shape. The shear flow in the straight portion is found from equation (6.25) by considering the shaded segment $AB_1$ of Figure 6.36*(a)*, giving

$$q_{B_1} = \frac{V_y(a-y)t(a+y)}{2(2.237a^3t)} + C = \frac{V_y}{4.474a}\left(1 - \frac{y^2}{a^2}\right) + C\,.$$

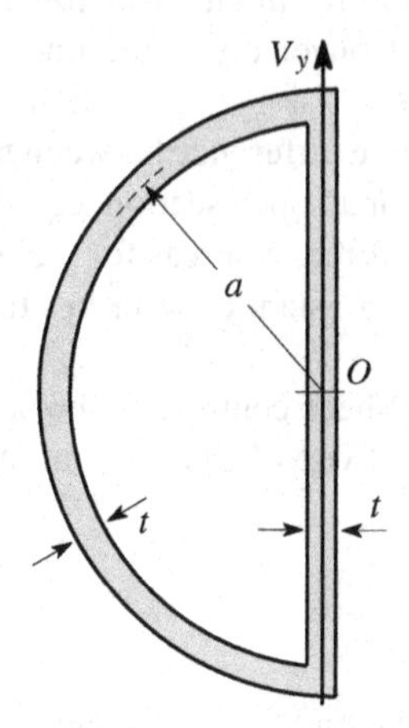


*Figure 6.35: Closed semi-circular section*

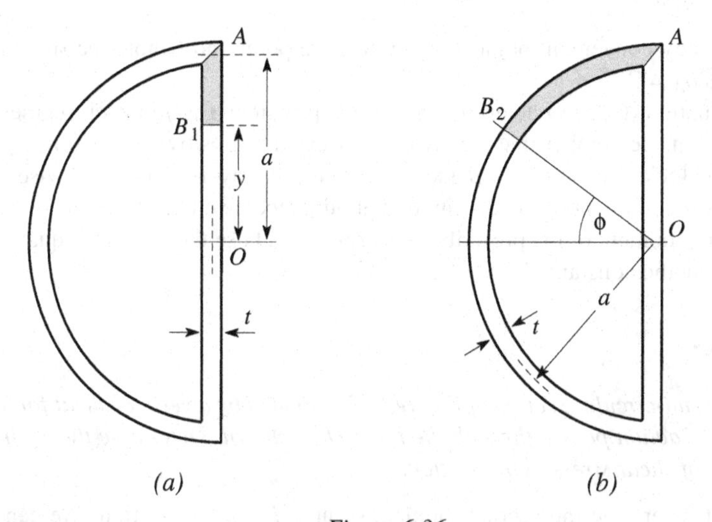


*Figure 6.36*

To find $q$ in the curved portion, we use the segment shaded in Figure 6.36*(b)*, but since it is traversed in passing anticlockwise from $A$ to $B$, the corresponding area $A^*$ must be treated as negative. We obtain

$$q_{B_2} = -\int_{\phi}^{\pi/2} \frac{V_y a \sin(\theta) a t d\theta}{2.237a^3t} + C = -\frac{V_y \cos\phi}{2.237a} + C\,.$$

The above equations completely define the shear flow $q$ and hence the shear stress, $\tau = q/t$, in the section, except for the unknown constant $C$. To determine $C$, we take moments about any convenient point, as in §6.2, to enforce the condition that the distribution is statically equivalent to the force $V$. The result is most easily obtained by taking moments about $O$, since the moment arm is then zero for both $q_{B_1}$ and $V_y$. The moment arm for $q_{B_2}$ is the radius $a$ and we obtain

$$V_y \times 0 = \int_{-\pi/2}^{\pi/2} a q_{B_2}(\phi) a d\phi = -\frac{V_y a}{2.237} \int_{-\pi/2}^{\pi/2} \cos\phi d\phi + \pi a^2 C$$

— i.e.

$$0 = \pi a^2 C - \frac{2V_y a}{2.237} ,$$

from which

$$C = \frac{2V_y}{2.237\pi a} = 0.285 \frac{V_y}{a} .$$

Substituting this value back into the expressions for shear flow, we obtain

$$q_{B_1} = \frac{V_y}{a} \left( 0.508 - 0.223 \frac{y^2}{a^2} \right) \quad ; \quad q_{B_2} = \frac{V_y}{a} (0.285 - 0.447 \cos\phi) ,$$

which completes the solution.

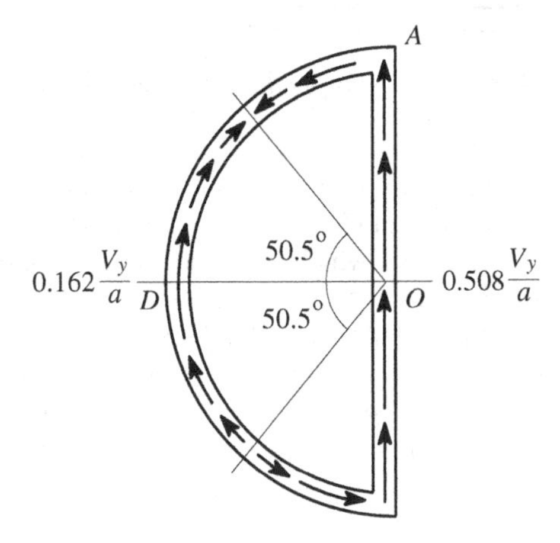


*Figure 6.37: Shear stress field due to the shear force $V_y$ passing through O*

The distribution is illustrated in Figure 6.37. Notice that the shear flow is negative and hence clockwise in the segment $-\phi_0 < \phi < \phi_0$, where

$$\phi_0 = \cos^{-1}\left(\frac{0.285}{0.447}\right) = 50.5^\circ .$$

The magnitude of $q$ reaches a maximum at the two points where the section crosses the neutral axis, being $0.508V/a$ at $O$ and $-0.162V/a$ at $D$.

## 6.5 Pure torsion of closed thin-walled sections

For a closed section, the shear flow can be non-zero even if the shear force $V$ is zero, in which case equation (6.25) reduces to

$$q_B = C\,. \tag{6.26}$$

In other words the shear flow is constant around the section. This corresponds to the case of a thin-walled closed section subjected to pure torsion.

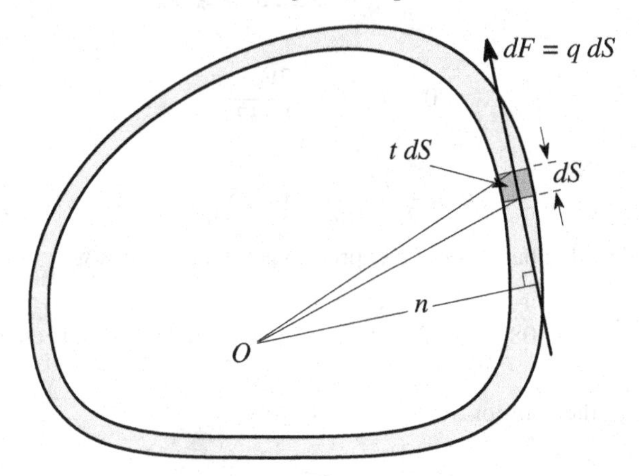


*Figure 6.38: Calculating the torque due to a given shear flow*

The torque $T$ is the same about any point if $V = 0$ and we can find it by taking moments exactly as in §§6.2, 6.4. Consider the contribution of the small element of area $tdS$ in Figure 6.38, where $dS$ is an element of length around the mean line of the section and $t$ is the *local* thickness, which is not necessarily constant. The elemental force associated with $tdS$ is $dF = \tau t dS = q dS$. We assume that $\tau$ is uniformly distributed over $tdS$ and hence the line of action of $dF$ is locally tangent to the mean line of the section — i.e. the line which is equidistant between the inner and outer edges of the wall. The moment of $dF$ about $O$ is therefore $ndF$ where $n$ is the normal distance from $O$ to the local tangent to the mean line. Summing the contribution for all such elements by integration, we have

$$T = \oint_S ndF = \oint_S qndS = C\oint_S ndS \tag{6.27}$$

from equation (6.26). Notice that $C$ can be taken out of the integral because it is constant.

The final integral in (6.27) can be given a physical interpretation. From Figure 6.39 *(a)*, we see that $ndS$ is the product of the base and the height of the triangle $OPQ$ and hence is twice the area of that triangle. Summing $ndS$ over all elements $dS$, we therefore obtain twice the area $A$ enclosed by the mean line of the section. In other words

$$\oint_S n dS = 2A$$

and hence

$$T = C \oint_S n dS = 2AC \,. \tag{6.28}$$

The mean line and the corresponding enclosed area $A$ are illustrated in Figure 6.39*(b)*. Notice however that since $t \ll a$, only a small error would be involved in using either the inner or the outer wall of the section to define $A$.

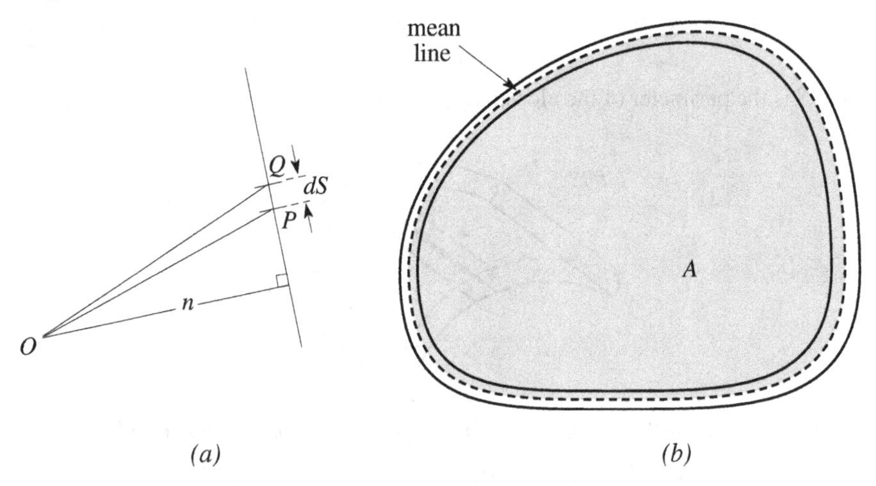


*Figure 6.39: (a) Contribution to the integral in equation (6.27) from the element dS, (b) Interpretation of the enclosed area A*

The shear flow and shear stress can be determined from (6.26, 6.28) as

$$q = C = \frac{T}{2A} \quad ; \quad \tau = \frac{q}{t} = \frac{T}{2At} \,. \tag{6.29}$$

Notice that the shear flow is constant around the section, but the shear stress will vary inversely with thickness $t$ if the latter is not constant. In particular, the maximum shear stress will occur where the thickness $t$ is a minimum.

### 6.5.1 Torsional stiffness

It is also useful to know the angle of twist $\theta$, and hence the torsional stiffness of the section. This is most conveniently found using an energy method. As in §3.3, we can equate the work done by the torque $T$ during loading to the stored strain energy. The work done is

$$W = \frac{1}{2} T\theta \,, \tag{6.30}$$

since the system is elastic and the torque will be everywhere proportional to the instantaneous angle of twist [see equation (3.30)].

The stored strain energy can be obtained from equation (3.33) as

$$U = \frac{1}{2G}\iiint_V \tau^2 dV \,, \tag{6.31}$$

where $V$ is the volume of material in the beam and we have used the relation $G = E/2(1+\nu)$. The shear stress is assumed to be uniform across the thickness $t$ and hence is constant in the volume element $dV = LtdS$ shown in Figure 6.40. The integral for the strain energy can therefore be written

$$U = \frac{L}{2G}\oint_S \tau^2 t dS \,, \tag{6.32}$$

where $S$ is the perimeter of the closed section.

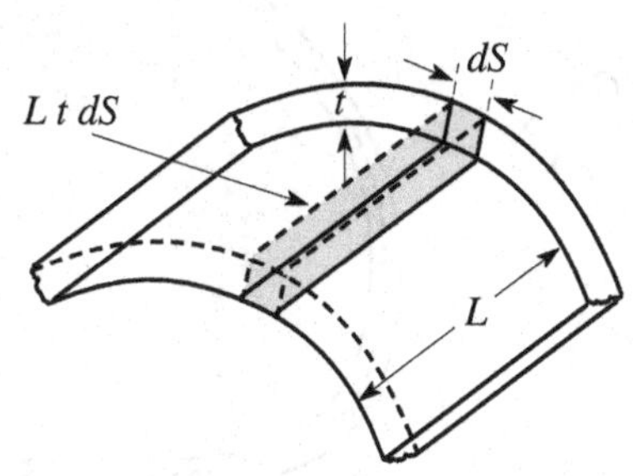


*Figure 6.40: Volume element for the integration of equation (6.31)*

Equating the work done and the strain energy $U$ and using (6.29) for the shear stress $\tau$, we then obtain

$$\frac{1}{2}T\theta = \frac{L}{2G}\oint_S \left(\frac{T}{2At}\right)^2 t dS = \frac{T^2 L}{8A^2 G}\oint_S \frac{dS}{t} \,, \tag{6.33}$$

from which

$$\theta = \frac{TL}{4A^2 G}\oint_S \frac{dS}{t} = \frac{TL}{GK} \,, \tag{6.34}$$

where

$$K = 4A^2 \Big/ \oint_S \frac{dS}{t} \tag{6.35}$$

is a measure of the torsional stiffness of the cross section. Notice that $K$ has the dimensions of length to power 4. It reduces to the polar moment of area $J$ in the special case where the section is axisymmetric.

In many problems, the thickness $t$ is constant and the integral in equations (6.34, 6.35) reduces to

$$\oint_S \frac{dS}{t} = \frac{S}{t} \,. \tag{6.36}$$

Alternatively, if the perimeter can be divided into segments $S_1, S_2, \ldots$ with thickness $t_1, t_2, \ldots$ respectively, we have

$$\oint_S \frac{dS}{t} = \frac{S_1}{t_1} + \frac{S_2}{t_2} + \ldots \,. \tag{6.37}$$

**Example 6.6**

*The symmetrical closed thin-walled section of Figure 6.41(a) is twisted by a torque of 20 Nm. Find the maximum shear stress and the twist of the beam per unit length if the material is steel (G = 80 GPa).*

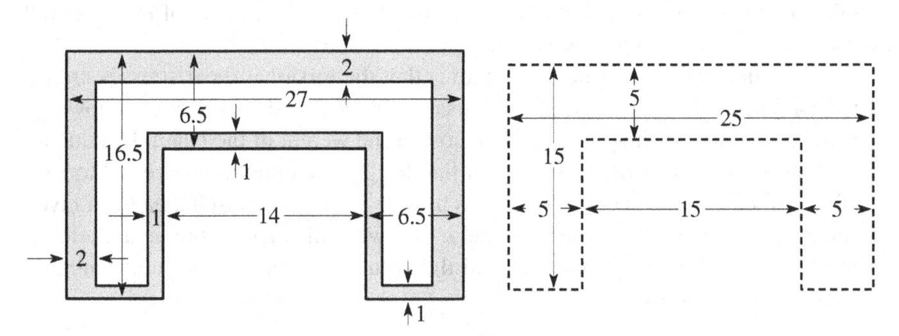


*Figure 6.41: (a) A closed thin-walled section, (b) the mean line*

We first determine the area enclosed by the mean line, which is

$$A = 5 \times 15 + 5 \times 15 + 5 \times 15 = 225 \text{ mm}^2 ,$$

from Figure 6.41 *(b)*. The maximum shear stress will occur where the thickness $t$ is a minimum ($t_{\min} = 1$ mm) and hence is

$$\tau = \frac{20{,}000}{2 \times 225 \times 1} = 44.5 \text{ MPa.}$$

The perimeter $S$ of the section comprises a segment of length $S_1 = 5 + 10 + 15 + 10 + 5 = 45$ mm of thickness 1 mm and a segment $S_2 = 15 + 25 + 15 = 55$ mm of thickness 2 mm. We therefore have

$$\oint_S \frac{dS}{t} = \frac{45}{1} + \frac{55}{2} = 72.5$$

and

$$K = \frac{4 \times 225^2}{72.5} = 2973 \text{ mm}^4 ,$$

from equation (6.35). The twist per unit length is

$$\frac{\theta}{L} = \frac{T}{GK} = \frac{20}{80 \times 10^9 \times 2973 \times 10^{-12}} = 0.084 \text{ rad/m} .$$

### 6.5.2 Design considerations in torsion

Equations (6.29, 6.34) are extremely simple and efficient to use for determining the shear stress and twist in a thin-walled closed section loaded in torsion. In the sprit of §1.2.2 they are therefore very useful as a design tool and there is also considerable justification for using them 'out of context' to get a rough estimate of the torsional behaviour of closed sections whose walls are not strictly thin.

Another interesting consideration here is that the torsional shear stresses are reduced and the stiffness increased by increasing the enclosed area $A$, which however does not contribute to either the material cost or the weight of the beam. If reducing weight and cost are primary concerns in the design, the optimal choice for torsion is a thin-walled circular section, since this has the greatest enclosed area for a given perimeter. For a cylinder of mean radius $a$ and wall thickness $t$, the area enclosed is $A=\pi a^2$, the perimeter is $S=2\pi a$ and the volume of material per unit length is $V=St=2\pi at$. It follows that

$$t=\frac{V}{2\pi a} \tag{6.38}$$

and the shear stress and stiffness are respectively

$$\tau=\frac{T}{2\pi a^2 t}=\frac{T}{Va} \tag{6.39}$$

$$K=\frac{4(\pi a^2)^2 t}{2\pi a}=2a^2 V\ . \tag{6.40}$$

Thus, for a given material volume $V$, the torsional shear stress decreases inversely with the radius $a$ of the tube and the stiffness increases with $a^2$.

Of course, there are practical reasons limiting the values of $a$ that can be used in particular applications. The beam has to fit into a given space and connect to other components at the ends. Also, the thickness for given $V$ is inverse with $a$ from (6.38) and for very thin sections, local buckling of the wall is a limiting failure mode. The reader can verify this by rolling up a piece of paper, taping it to make a closed section and then loading it in torsion. We shall discuss failure modes involving buckling in Chapter 12.

The contribution of the enclosed area $A$ to torsional stiffness can have some unexpected consequences when design changes are made. In my undergraduate days, a popular student pastime was to buy an old family saloon car and make something more sporty by removing the roof and parts of the door surrounds. However, many of these cars were built using 'monocoque' construction, in which there is no chassis. Instead, the car body is an integral part of the structure and front and rear axle assemblies are attached only through local sub-frames. In this case, the torsional stiffness between the front and rear of the car relies on the roof (and to some extent the doors) which render the car body a closed section. Removing the roof therefore reduced this torsional stiffness dramatically, resulting in a vehicle with quite bizarre handling characteristics, particularly on rough roads.

It is in fact quite a challenge to obtain sufficient torsional stiffness in an open car, even with a chassis. Closed sections can be used for the chassis members, but this is somewhat limited by space considerations. The Lotus Seven — an inexpensive kit car manufactured by Lotus in the 60s — solved this problem by boxing in the transmission tunnel.

## 6.6 Finding the shear centre for a closed section

The energy argument of §6.5.1 can be generalized to determine the location of the shear centre for a closed section. Remember that the shear centre is defined as the point through which $V_y$ must act if there is to be no twist.

We first apply the reciprocal theorem of §3.8 to the bar subjected to the two separate loads comprising (i) a torque $T$ and (ii) a force $V_y$ acting through the shear centre $C$. By the definition of $C$, $V_y$ produces no twist, so $T$ does no work against the deflections due to $V_y$. It follows by the reciprocal theorem that $V_y$ does no work against the deflections due to $T$ and hence that during the application of $T$, the shear centre does not move. Thus, *the shear centre is also the centre of rotation of the section when the beam is loaded in pure torsion.*[11]

Now a shear force with a more general line of action can be decomposed into an equal force $V_y$ through the shear centre and a torque $T = V_y d$, where $d$ is the perpendicular distance from $C$ to the line of action of $V_y$. With this decomposition, the shear force acting through $C$ will produce a displacement $u_C$ of $C$ but no twist, whereas the torque $T$ will produce a twist $\theta$ of the section, but no displacement of $C$. Thus, the work done when the two components are applied simultaneously is

$$W = \frac{1}{2}V_y u + \frac{1}{2}T\theta\,. \tag{6.41}$$

In other words, there will be no coupling terms between $V_y, u$ and $T, \theta$.

The twist $\theta$ due to $T$ must be in the same direction as $T$ and hence the second term in equation (6.41) must be positive for all $T$ and hence for all non-zero values of $d$. But the first term is independent of $d$ and hence $W$ *will be a minimum for a given magnitude and direction of* $V_y$*, if we choose the line of action of* $V_y$ *such that* $d=0$. In other words, the work done by $V_y$ is a minimum when its line of action passes through the shear centre. Thus, we can find the shear centre by writing the strain energy for a shear force of general line of action, using the expression for shear stress from equation (6.25), and requiring it to be a minimum.

From equation (6.32), we have

$$U = \frac{1}{2G}\oint_S \tau^2 t dS = \frac{1}{2G}\oint_S \frac{q^2 dS}{t} \tag{6.42}$$

and hence

$$\frac{\partial U}{\partial d} = \frac{1}{G}\oint_S q\frac{\partial q}{\partial d}\frac{dS}{t}\,. \tag{6.43}$$

This must be zero if $V_y$ acts through the shear centre $C$.

[11] Notice how these energy theorems are really quite useful!

Now changing the line of action of $V_y$ (without changing its direction) is equivalent to superposing a torque and we know from §6.5 that this merely changes $q$ everywhere by a constant. It follows that

$$\frac{\partial q}{\partial d} = \text{constant} \tag{6.44}$$

and this constant is non-zero (as is $G$), so the condition $\partial U/\partial d = 0$ simplifies to

$$\oint_S \frac{q dS}{t} = 0 , \tag{6.45}$$

which must be satisfied if $V_y$ is to act through the shear centre.

**Summary**

The procedure for finding the shear centre of a closed thin-walled section can therefore be summarized as follows:-

(i) Choose any convenient point $A$ as a reference point and denote the shear flow there as $q_A = C$.
(ii) Use equation (6.25) to determine general expressions for $q$ in each segment of the section. Remember that the area $A^*$ is treated as positive if it is traversed in passing clockwise from $A$ to $B$ and negative if it is traversed anticlockwise.
(iii) Substitute the result into condition (6.45), which then serves as an equation for the unknown constant $C$.
(iv) Substitute for $C$ into the original expressions for $q$ and take moments about any point as in §6.2 for open sections, to find the line of action of $V_y$, using equation (6.18).

### 6.6.1 Twist due to a shear force

If the shear force $V_y$ does not pass act through the shear center, it will cause the section to twist. If the location of the shear centre is known, this twist is easily found by decomposing the shear force into an equal force through the shear centre and an equivalent torque $T = V_y d$, where $d$ is the perpendicular distance from the shear centre to the line of action of $V_y$. In this representation, only the torque produces twist, which is therefore given by

$$\theta = \frac{TL}{4A^2 G} \oint_S \frac{dS}{t} , \tag{6.46}$$

from equation (6.34).

If the location of the shear centre is not known, a similar decomposition together with the relation (6.45) yields a general expression for the twist of the section. We first write the shear flow

$$q = q_1 + q_2 , \tag{6.47}$$

where $q_1$ is associated with the force $V_y$ through the shear centre and $q_2$ with the equivalent torque. It then follows that

$$\oint_S \frac{q_1 dS}{t} = 0 , \tag{6.48}$$

from (6.45) and

$$q_2 = \frac{T}{2A} , \tag{6.49}$$

from (6.29). We can therefore write

$$\oint_S \frac{q dS}{t} = \oint_S \frac{q_1 dS}{t} + \oint_S \frac{q_2 dS}{t} = \frac{T}{2A} \oint_S \frac{dS}{t} \tag{6.50}$$

and substitution in (6.46) yields

$$\theta = \frac{L}{2AG} \oint_S \frac{q dS}{t} . \tag{6.51}$$

Thus, if the line of action of $V_y$ is prescribed, we first use the procedure of §6.4.1 to determine $q$ throughout the section and then substitute into (6.51) to determine the twist $\theta$.

### Example 6.7

*The closed semi-circular section of Figure 6.35 is made of aluminium ($G$=30 GPa) and has dimensions $a$=30 mm, $t$=2 mm. Find the location of the shear centre. Hence or otherwise, find the twist per unit length when the beam is loaded by a shear force $V_y$=2 kN passing through the point O as shown in Figure 6.35*

In determining the location of the shear centre, the first two steps are identical with those of Example 6.5 above and lead to the results

$$q_{B_1} = \frac{V_y}{4.474a}\left(1 - \frac{y^2}{a^2}\right) + C$$

$$q_{B_2} = -\frac{V_y \cos\phi}{2.237a} + C ,$$

where the coordinates $y, \phi$ are defined in Figures 6.35*(a,b)* and $C$ is an unknown constant.

It follows that

$$\oint_S \frac{q dS}{t} = \int_{-a}^{a} \left[\frac{V_y}{4.474a}\left(1 - \frac{y^2}{a^2}\right) + C\right]\frac{dy}{t} + \int_{-\pi/2}^{\pi/2}\left(-\frac{V_y \cos\phi}{2.237a} + C\right)\frac{a d\phi}{t}$$

$$= \frac{2V_y}{4.474t} - \frac{2V_y}{3 \times 4.474t} + \frac{2aC}{t} - \frac{2V_y}{2.237t} + \frac{\pi a C}{t} = \frac{(\pi + 2)aC}{t} - \frac{4V_y}{6.711t} .$$

Notice that the integrals must be written so that the lower limit is smaller than the upper limit. It is *not* necessarily correct to write them such that they pass around the section always in the same sense (clockwise or anticlockwise), since all we are doing here is performing a summation.

To satisfy equation (6.45), we require

$$C = \frac{4V}{(\pi+2)(6.711a)} = 0.116\frac{V_y}{a} .$$

To find the shear centre, we equate the moment of the distribution about $O$ to that of $V_y$, as in §6.4.1. Only the stresses in the curved segment have a non-zero moment arm $(=a)$. The distance $c$ is defined to place the shear centre on the left of $O$ as in Figure 6.42, so the moment $V_y \times c$ is defined as clockwise positive. This convention has to be observed also in taking moments for the distribution and we obtain

$$V_y \times c = -\int_{-\pi/2}^{\pi/2} q_{B_1} a^2 d\phi = -\pi a^2 C + \frac{2V_y a}{2.237} .$$

Substituting for $C$ and dividing through by $V_y$, we find

$$c = -\pi a(0.116) + \frac{2a}{2.237} = 0.530a ,$$

which determines the location of the shear centre, shown in Figure 6.42. Notice that for a convex closed section, the shear centre is always inside the section. Also, since the torsional stiffness of closed sections is generally large (see §6.5.1), the exact location of the shear centre is of comparatively little importance, since not much twist will be produced by applying a shear force through a different point.

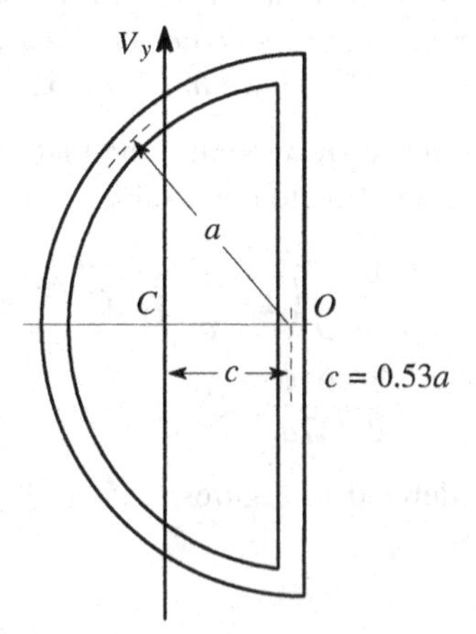


*Figure 6.42: Location of the shear centre for the closed semi-circular section*

We can use this result to determine the twist by noting that the shear force acting through $O$ is equivalent to an equal force through the shear centre (which produces no twist) and an equivalent anticlockwise torque

$$T = V_y c = 0.53 V_y a .$$

The area enclosed by the section is

$$A = \frac{\pi a^2}{2}$$

and

$$\oint_S \frac{dS}{t} = \frac{S}{t} = \frac{(\pi+2)a}{t} ,$$

so

$$\frac{\theta}{L} = \frac{T}{4A^2G}\oint_S \frac{dS}{t} = \frac{0.53V_y a(\pi+2)a}{\pi^2 a^4 Gt} = \frac{0.53(\pi+2)V_y}{\pi^2 a^2 Gt} .$$

Substituting the given values, we obtain

$$\frac{\theta}{L} = \frac{0.53(\pi+2)\times 2000}{\pi^2\times 30^2\times 10^{-6}\times 30\times 10^9\times 2\times 10^{-3}} = 0.010 \text{ rad/m} .$$

An alternative method for determining the twist per unit length is to substitute the shear flow corresponding to the line of action through $O$ directly into equation (6.51). We showed in Example 6.5 that $q$ is then given by the above expressions with $C = 0.285a$, so

$$\oint_S \frac{qdS}{t} = \frac{(\pi+2)aC}{t} - \frac{4V_y}{6.711t} = \frac{0.285(\pi+2)V_y}{t} - \frac{4V_y}{6.711t} = 0.869\frac{V_y}{t} .$$

We then have

$$\frac{\theta}{L} = \frac{0.869V_y}{2AGt} = \frac{0.869\times 2000}{\pi\times 30^2\times 10^{-6}\times 30\times 10^9\times 2\times 10^{-3}} = 0.010 \text{ rad/m}$$

as before.

### 6.6.2 Multicell sections

Figure 6.43 shows a closed thin-walled section with an internal partition which divides the enclosed area into two cells.

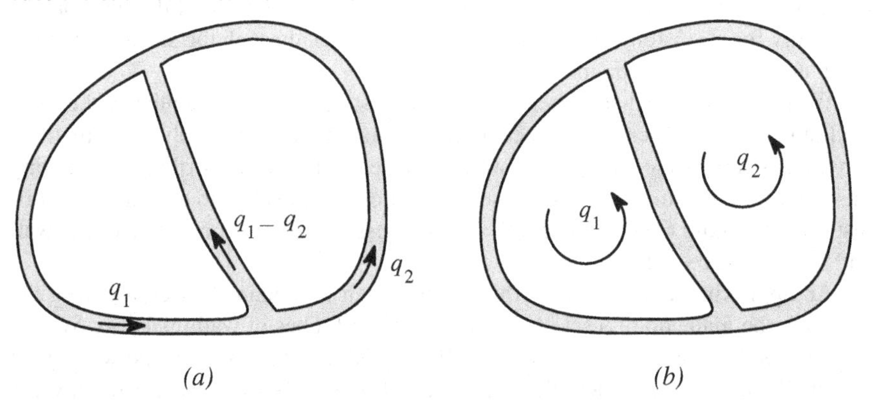


*Figure 6.43: (a) Multicell section, (b) equivalent shear flow*

If this section is loaded in torsion, the arguments of §6.5 show that the shear flow must be constant in each segment of the wall, but there is now the possibility that part of the shear flow passes down the partition as shown. Another way to think of this is to define a flow $q_i$ anticlockwise around each cell as shown in Figure 6.43 *(b)*, in which case superposition shows that the flow in the partition will be $q_1 - q_2$ upwards, as in Figure 6.43 *(a)*.

The equilibrium argument of §6.5 leads to the result

$$T = \sum_{i=1}^{N} 2A_i q_i \,, \tag{6.52}$$

where $A_i$ is the area enclosed by the $i$th cell and $N$ is the total number of cells, but for $N > 1$ this is insufficient to determine the shear flows $q_i$ in the individual cells for a given torque $T$. To complete the solution, we need to impose the compatibility condition that the individual cells twist through the same angle. Using equation (6.51), this yields

$$\frac{1}{A_1}\oint_{S_1}\frac{qdS}{t} = \frac{1}{A_2}\oint_{S_1}\frac{qdS}{t} = \ldots. \,, \tag{6.53}$$

which provides just sufficient additional equations to determine the $q_i$.

An essentially similar procedure can be used for a multicell section loaded by a shear force. The arguments of §6.4 enable us to determine how $q$ varies in any wall segment, except that an additional arbitrary constant is introduced by each partition. The compatibility conditions (6.53) provide the additional equations needed to determine these constants. For more details of this procedure, see Cook and Young, §§8.8, 9.9.

## 6.7 Torsion of thin-walled open sections

We have seen in §§6.4–6.6 that closed sections transmit shear forces and torques by essentially the same mechanism, generating a shear stress that is approximately uniform through the wall thickness. For open sections, this form of distribution can arise only for shear forces whose line of action passes through the shear centre and it can never arise from torsion alone.

We recall from §6.1 that the shear stress must always be tangential to the local wall of the section, because any local normal component would imply a complementary shear stress on the traction-free wall of the beam. The only distribution satisfying this condition in a thin-walled open section which is equivalent to a torque is that illustrated in Figure 6.44 *(a)*, where the shear stress passes along one side of the wall and back along the other.

An approximate treatment of this problem can be obtained by considering the section to be made up of a set of nested thin-walled tubes of wall thickness $\delta y$, one of which is illustrated in Figure 6.44 *(b)*. If the thickness $t$ is constant and we neglect the small curved regions near the ends, each of these tubes will have parallel walls and the area enclosed and the perimeter will be approximately

$$A = 2yb \ ; \ S = 2b \,, \tag{6.54}$$

respectively, where $y$ is measured normal to the mean line, as shown in Figure 6.44 *(b)*.

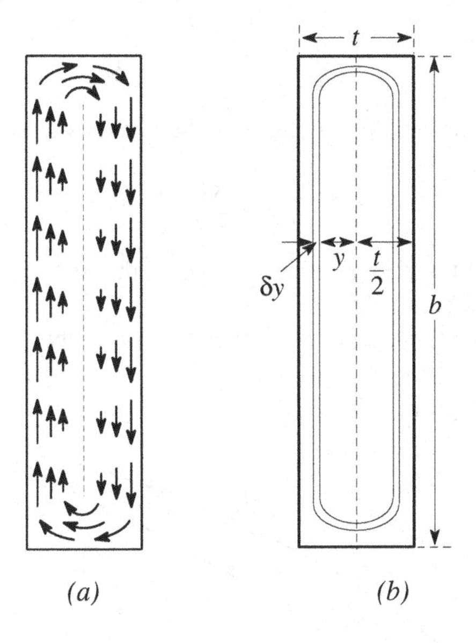


*Figure 6.44: (a) Shear stress distribution due to torsion in an open section, (b) elemental thin-walled tube contained within the section*

If a given tube carries an elemental torque $\delta T$, the twist per unit length will be

$$\frac{\theta}{L} = \frac{\delta T S}{4A^2 G \delta y} = \frac{\delta T}{8by^2 G \delta y} \,, \tag{6.55}$$

from (6.34) and hence

$$\delta T = \frac{8by^2 G \delta y \theta}{L} \,. \tag{6.56}$$

All the tubes must have the same twist and hence the total torque transmitted is

$$T = \int \delta T = \int_0^{t/2} \frac{8by^2 G \theta dy}{L} = \frac{bt^3 G \theta}{3L} \,. \tag{6.57}$$

Notice that this relation can be written

$$\theta = \frac{TL}{GK} \,, \tag{6.58}$$

where

$$K = \frac{bt^3}{3} \,. \tag{6.59}$$

Eliminating $\theta/L$ between equations (6.56, 6.57), we obtain

$$\delta T = \frac{24Ty^2\delta y}{t^3} \tag{6.60}$$

and hence the shear stress is given by (6.29) as

$$\tau = \frac{\delta T}{2A\delta y} = \frac{24Ty^2\delta y}{4yb\delta yt^3} = \frac{6Ty}{bt^3} \; . \tag{6.61}$$

Thus, the shear stress varies linearly through the thickness, reaching a maximum at $y = t/2$ of

$$\tau_{\max} = \frac{3T}{bt^2} \; . \tag{6.62}$$

This solution is inexact in regions near the ends of the section, but as long as $b \gg t$, the error involved will be small.

We have developed the argument for the case where the thickness $t$ is constant, but it is easily extended to sections where $t$ is a function of distance $\xi$ along the section, as shown in Figure 6.45.

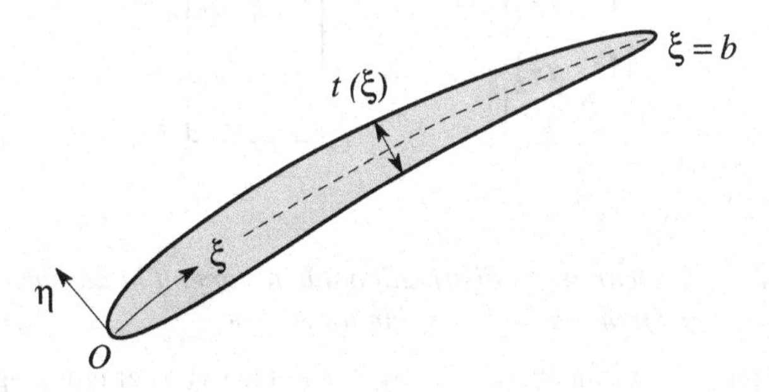


*Figure 6.45: Thin-walled open section with variable thickness*

We find that

$$K = \frac{1}{3}\int_0^b t^3 d\xi \tag{6.63}$$

and the maximum shear stress is

$$\tau_{\max}(\xi) = \frac{Tt(\xi)}{K} \; . \tag{6.64}$$

Thus, the absolute maximum shear stress occurs at the thickest part of the section.

**Example 6.8**

*The steel angle beam of Figure 6.46 is loaded by a torque of 10 Nm. Find the maximum shear stress and the twist per unit length ($G_{\text{steel}} = 80$ GPa).*

The perimeter of the section is 120 mm, so

$$K = \frac{120 \times 5^3}{3} = 5000 \text{ mm}^4 \; ,$$

from equation (6.59). The twist per unit length is therefore

$$\frac{\theta}{L} = \frac{T}{GK} = \frac{10}{80 \times 10^9 \times 5000 \times 10^{-12}} = 0.025 \text{ rad/m}$$

and the maximum shear stress is

$$\tau_{\max} = \frac{3T}{bt^2} = \frac{3 \times 10{,}000}{120 \times 5^2} = 10 \text{ MPa}.$$

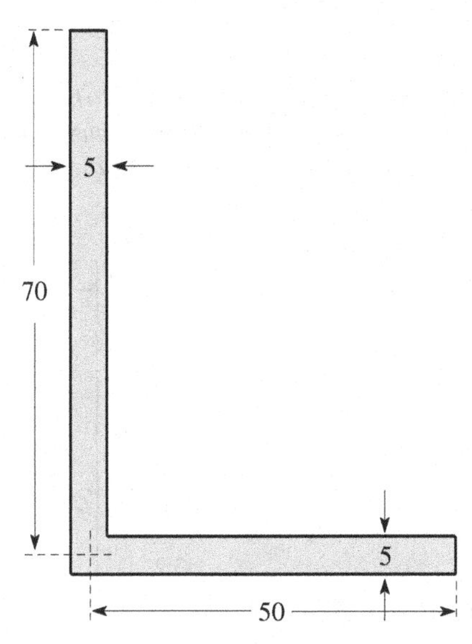


all dimensions in mm

*Figure 6.46*

Alternatively, we could find the maximum shear stress from equation (6.64) which gives

$$\tau_{\max} = \frac{Tt}{K} = \frac{10{,}000 \times 5}{5000} = 10 \text{ MPa}$$

as before.

Notice that this torque gives a modest shear stress (typical steels have yield stresses over 100 MPa), but produces quite a substantial twist ($\approx 14^{\circ}$ per metre). This illustrates the fact that open sections are surprisingly flexible in torsion. Notice also how easy these equations are to use.

### 6.7.1 Loading of an open section away from its shear centre

These results can be combined with those of §§6.1–6.3 to determine the shear stresses and the twist for an open section loaded by a shear force whose line of action does not pass through the shear centre. The procedure is:-

(i) Find the shear stress distribution due to an equal shear force acting through the shear centre (§6.1).
(ii) Find the location of the shear centre (§6.2).
(iii) Find the equivalent torque due to the offset of the prescribed line of action of the shear force from the shear centre.
(iv) Find the stresses and the twist due to this torque (§6.7).
(v) Superpose the results from (i) and (iv).

**Example 6.9**

*A steel beam with the channel section of Figure 6.47(a) is loaded by a shear force of 10 kN with the line of action shown. Find the maximum shear stress and the twist per unit length. The section has a second moment of area $I_x = 0.81 \times 10^6$ mm$^4$ and $G = 80$ GPa for steel.*

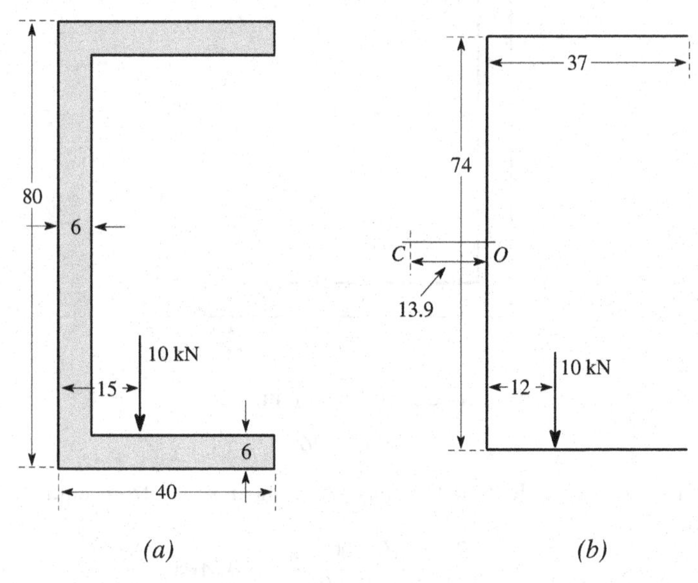


*Figure 6.47: (a) The channel section, (b) the mean line and location of the shear centre*

The mean line of the thin-walled section has the dimensions of Figure 6.47*(b)* and the section is similar to Figure 6.26 with $a = 37$ mm and $t = 6$ mm. The first two steps in the above procedure can therefore be taken from Example 6.4 (§6.2). Referring to Figure 6.28 and the associated calculations, we note that the maximum shear stress due to an equal shear force through the shear centre occurs in the web at the neutral axis ($y = 0$) and is

$$\tau_{max} = \frac{9V_y}{16at} = \frac{9 \times 10{,}000}{16 \times 37 \times 6} = 25.3 \text{ MPa}.$$

Also, from Example 6.4, the shear centre is located at a distance

$$c = \frac{3a}{8} = 13.9 \text{ mm}$$

to the left of $O$, as shown in Figure 6.47*(b)*.

The offset of the shear force is therefore equivalent to a clockwise torque

$$T = 10{,}000 \times (13.9 + 12) \times 10^{-3} = 259 \text{ Nm}.$$

The channel section has a constant thickness $t = 6$ mm and perimeter

$$b = 37 + 74 + 37 = 148 \text{ mm},$$

so

$$K = \frac{bt^3}{3} = \frac{148 \times 6^3}{3} = 10.7 \times 10^3 \text{ mm}^4$$

and the twist per unit length is

$$\frac{\theta}{L} = \frac{T}{GK} = \frac{259}{80 \times 10^9 \times 10.7 \times 10^{-9}} = 0.30 \text{ rad/m } (17.3^\circ\text{/m}).$$

The torque generates a maximum shear stress

$$\tau = \frac{Tt}{K} = \frac{259 \times 10^3 \times 6}{10.7 \times 10^3} = 145 \text{ MPa},$$

which occurs at the edges of the wall all around the section. The maximum shear stresses due to the shear force and the equivalent torque are shown in Figures 6.48 *(a,b)* respectively.

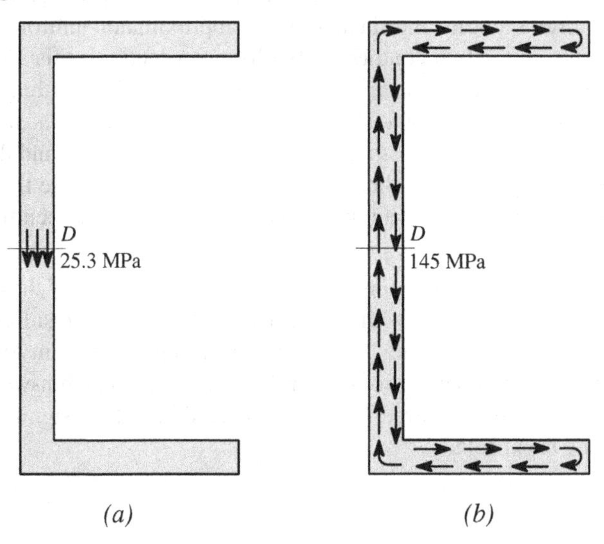


*Figure 6.48: Magnitude and location of the maximum shear stress due to (a) an equal shear force through the shear centre and (b) the torque due to the offset of the shear force*

Superposing these stress fields, it is clear that the maximum shear stress will occur at $D$, where they act in the same direction, giving

$$\tau_{max} = 25 + 145 = 170 \text{ MPa}.$$

This example shows that the offset of the line of action of the shear force has a dramatic effect on the behaviour of the beam in shear. It increases a modest maximum shear stress of 25 MPa by a factor of 7 and produces an unacceptable amount of twist per unit length for most practical applications. Thus, with open sections, it is important to ensure that shear forces are applied as near to the shear centre as possible. This is generally easier to achieve with symmetric sections such as I-beams, rather than with channels, for which the shear centre lies outside the section envelope.

## 6.8 Summary

In this chapter, we have developed methods of estimating the shear stress distribution in beams due to transmitted shear forces and torques. For any beam cross section, there exists a unique point known as the shear centre, through which the shear force must act if there is to be no twist of the beam.

When the shear force acts through the shear centre, the shear stresses are usually small compared with the bending stresses and they can often be neglected or approximated as $\tau = V_y/A_{web}$, except for built-up or welded sections or in situations where high accuracy is required.

For thin-walled sections, the shear stress must always be parallel to the edges of the wall. For closed sections, the shear stress is approximately uniform through the wall thickness for all loading conditions, but for open sections, this situation arises only when the shear force acts through the shear centre. When thin-walled open sections are loaded away from the shear centre or in torsion, the shear stress varies linearly through the wall thickness and the maximum shear stress and the twist may be quite large. Furthermore, the shear centre can be located outside the envelope of the section, making it difficult to apply the load through the shear centre.

By contrast, the shear centre for closed thin-walled sections always lies within the envelope of the section. The torsional stiffness and strength of these sections is very easily determined with a few lines of calculation, using the results of §6.5. You will generally find that the section is sufficiently strong and stiff in torsion for the exact location of the shear centre to be of little concern. Closed thin-walled sections are therefore much to be preferred in design applications where significant shear or torsional loading is expected.

## Further reading

### Sections 6.1–6.3

W.B. Bickford (1998), *Advanced Mechanics of Materials,* Addison Wesley, Menlo Park, CA, §§4.1–4.4.
A.P. Boresi, R.J. Schmidt, and O.M. Sidebottom (1993), *Advanced Mechanics of Materials,* John Wiley, New York, 5th edn., §§8.1–8.3.
R.D. Cook and W.C. Young (1985), *Advanced Mechanics of Materials*, Macmillan, New York, §§9.5–9.8.

### Section 6.4, 6.6

W.B. Bickford (1998), *loc. cit.*, §§4.5–4.7.
R.D. Cook and W.C. Young (1985), *loc. cit.*, §9.9.

### Section 6.5

W.B. Bickford (1998), *loc. cit.*, §§3.4, 3.5.
R.D. Cook and W.C. Young (1985), *loc. cit.*, §8.7.

### Section 6.7

J.R. Barber (2010), *Elasticity*, Springer, Dordrecht, 3rd edn., Chapter 16.
S.P. Timoshenko and J.N. Goodier (1970), *Theory of Elasticity*, McGraw-Hill, New York, 3rd edn., Chapter 10.

## Problems

### Section 6.1

*In Problems 6.1–6.10, assume that the shear force $V_y$ acts through the shear centre of the section, as defined in §6.2.*

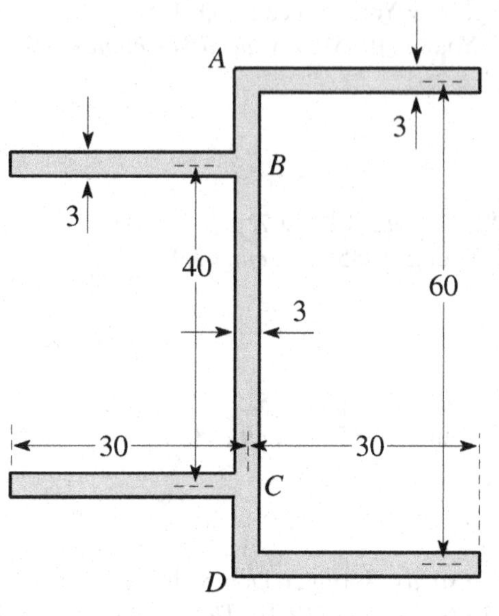


*Figure P6.1*

**6.1.** The second moment of area for the symmetric beam section of Figure P6.1 is $I_x = 0.288 \times 10^6$ mm$^4$. The beam is loaded by a vertical shear force of 4kN. Find the maximum shear stress and compare it with the average value $V_y/A_{\rm web}$, where $A_{\rm web}$ is the area of the vertical part $AD$ of the section only.

**6.2.** Figure P6.2 shows the dimensions of a W36×230 I-beam ($I_x = 15 \times 10^3$ in$^4$) that is loaded by a vertical shear force of 75,000 lbs. Find the maximum shear stress.

**6.3.** The wide flange I-beam of Figure P6.3 has second moment of area $I_x = 7a^3t/12$ and is loaded by a vertical shear force $V_y$. Find (i) the shear stress at $A$, (ii) the shear stress at $O$ and (iii) the average shear stress in the web.

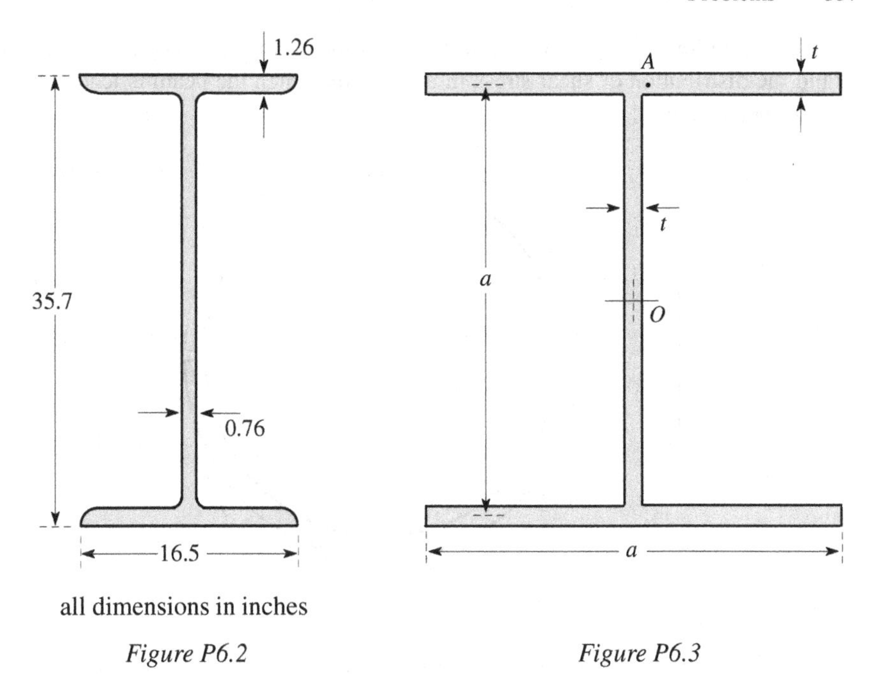


*Figure P6.2*

*Figure P6.3*

**6.4.** Figure P6.4 shows the cross section of a thin-walled beam subjected to a vertical shear force $V_y$. Find the distribution of shear stress in the section.

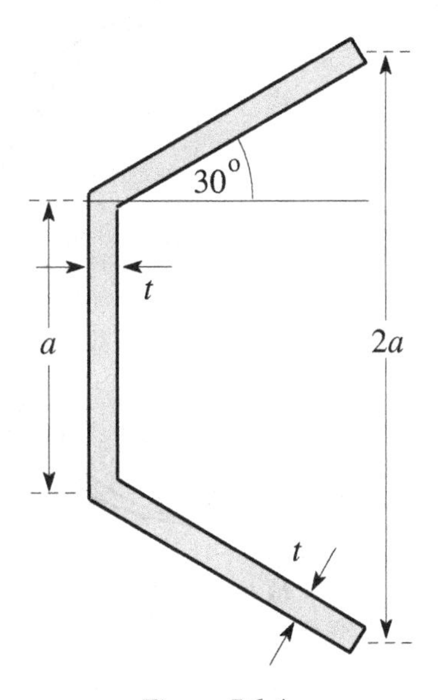


*Figure P6.4*

**6.5.** The slit cylindrical tube of Figure P6.5 has mean radius $a$ and wall thickness $t$. Find the distribution of shear stress in the section, when the beam is loaded by a vertical shear force $V_y$.

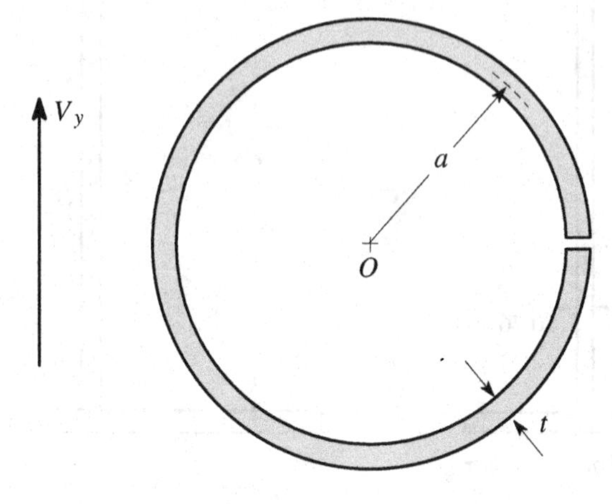


*Figure P6.5*

**6.6.** The crescent beam section of Figure P6.6 has mean radius $a$ and thickness $t = t_0 \cos\theta$. Find the distribution of shear stress in the section, when the beam is loaded by a vertical shear force $V_y$.

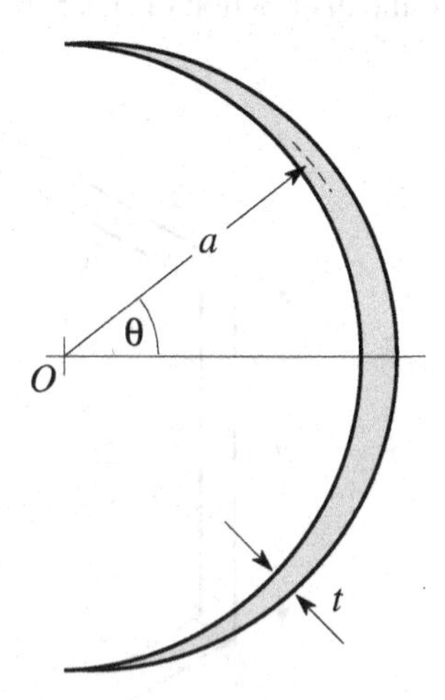


*Figure P6.6*

**6.7.** Figure P6.7 shows an I-beam fabricated by welding together three rectangular plates. Find the shear stress in the welds when the vertical shear force is 800 kN.

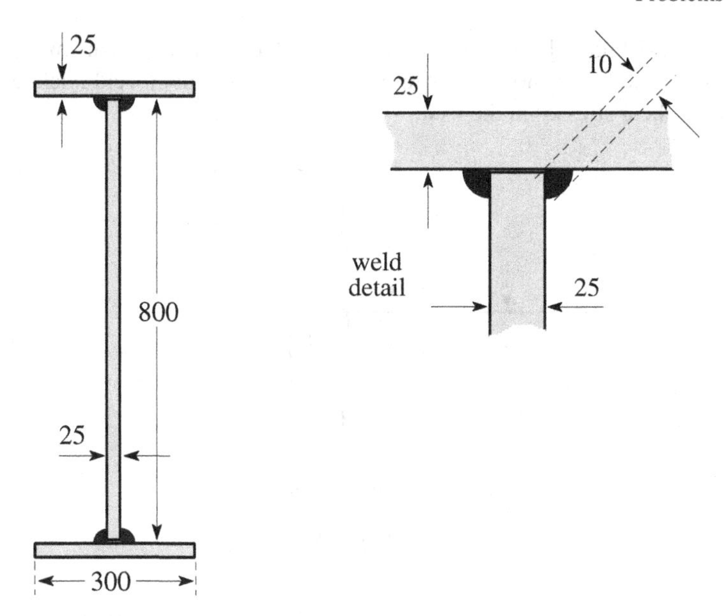


all dimensions in mm

*Figure P6.7*

**6.8.** An I-beam is fabricated by bolting four angle irons to a rectangular plate, using 1 inch diameter bolts, as shown in Figure P6.8. Determine the number of bolts needed per unit length if the average shear stress in the bolts is to be equal to the maximum shear stress at the neutral plane when the beam transmits a vertical shear force.

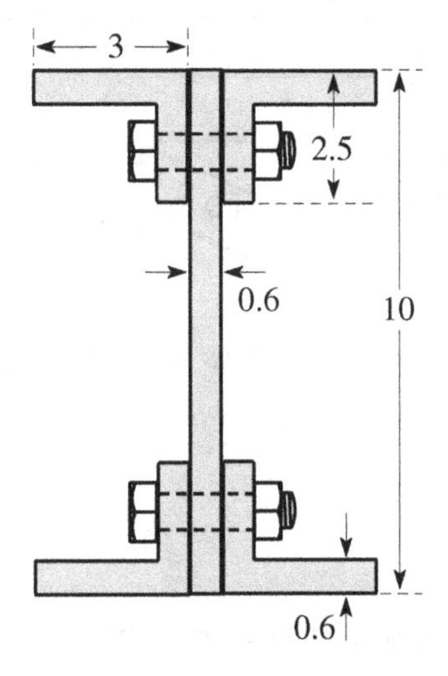


all dimensions in inches

*Figure P6.8*

**6.9.** The beam of Figure P6.1 is fabricated from a channel and two bars by welding, details of the weld being shown in Figure P6.9. If the minimum weld thickness is 1 mm, find the shear stresses in the weld.

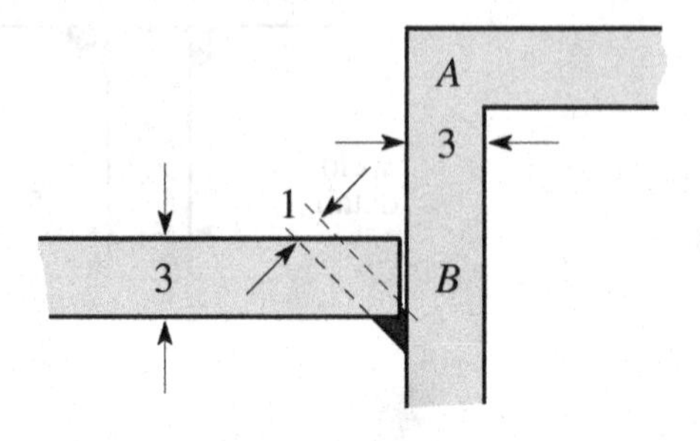


all dimensions in mm

*Figure P6.9*

**6.10.** A team of industrial engineers wishes to attach a small bracket to the I-beam of Figure P6.2 in order to mount a video camera to record factory operations. Details of the bracket and its attachment are shown in Figure P6.10 It is proposed to use 1/8 inch diameter screws with an axial spacing of 6 inches. Estimate the average shear stress in the screws. You may neglect the effect of the bracket on the second moment of area of the beam.

Would you advise the engineers to relocate the bracket and if so, to where?

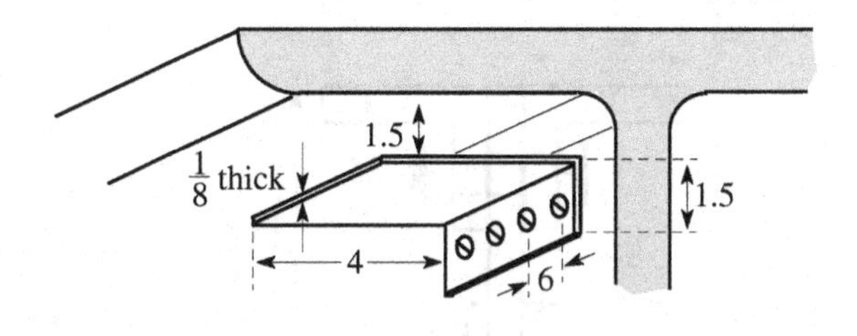


all dimensions in inches

*Figure P6.10*

## Section 6.2

**6.11.** Find the location of the shear centre for the beam of Figure P6.1.

**6.12.** Find the location of the shear centre for the slit box section shown in Figure P6.12. All segments have thickness $t$.

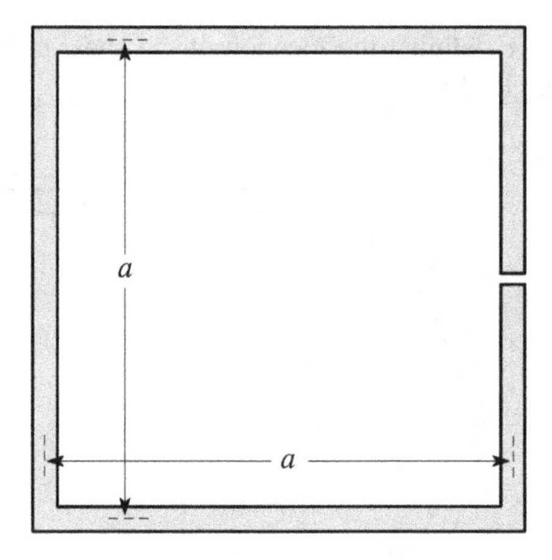


*Figure P6.12*

**6.13.** Find the location of the shear centre for the slit cylindrical tube of Figure P6.5.

**6.14.** Find the location of the shear centre for the crescent section of Figure P6.6.

**6.15.** Find the location of the shear centre for the beam section of Figure P6.4.

**6.16.** Find the location of the shear centre for the beam section shown in Figure P6.16.

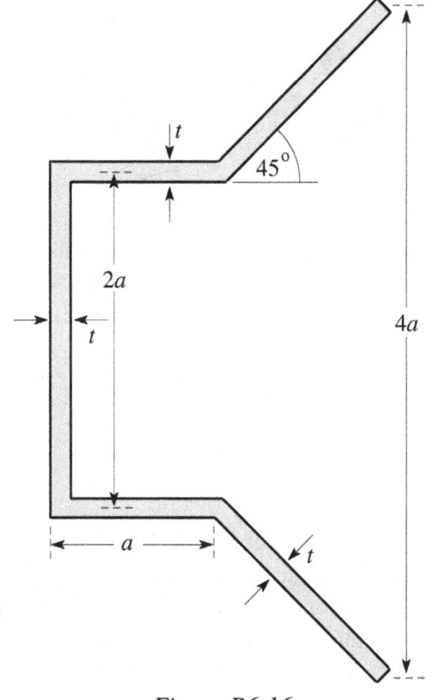


*Figure P6.16*

**6.17.** Find the distance $c$ defining the location of the shear centre $C$ for the beam section shown in Figure P6.17.

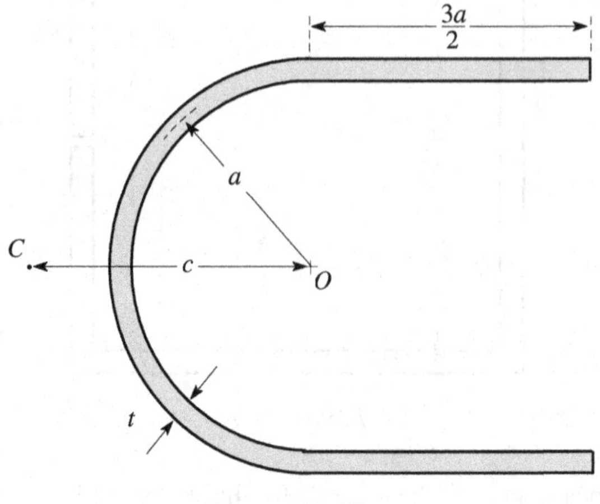


*Figure P6.17*

**6.18.** Find the distance $c$ defining the location of the shear centre $C$ for the beam section shown in Figure P6.18.

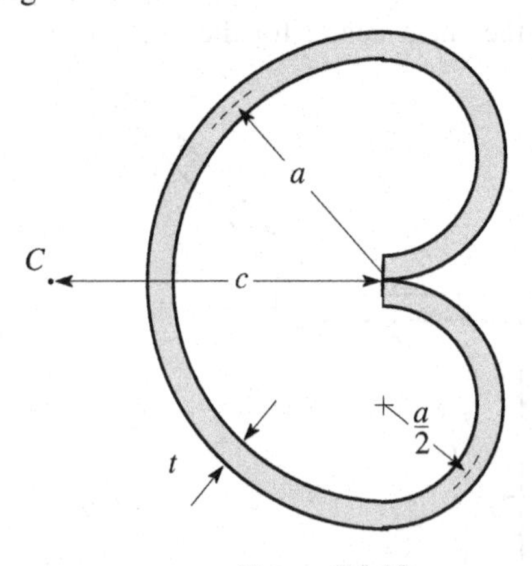


*Figure P6.18*

## Section 6.4

**6.19.** The closed thin-walled section of Figure P6.19 has an equilateral triangular cross section of side $a$, but the vertical side is of thickness $2t$, compared with $t$ for the two inclined walls. The section transmits a vertical shear force $V_y$, whose line of action passes through the point $O$. Find the shear stress distribution in the section. In particular, find the maximum shear stress and identify the points where the shear stress is zero.

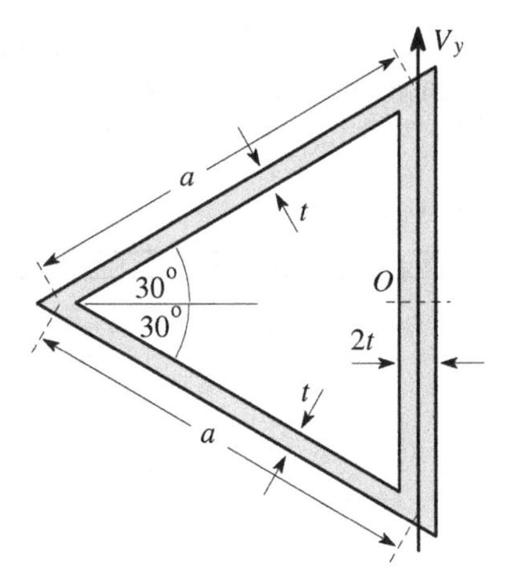


*Figure P6.19*

**6.20.** The outer and inner circular walls of a closed thin-walled circular tube have slightly different centres, as shown in Figure P6.20, so that the thickness of the section varies according to

$$t = t_0(1 - \varepsilon \cos\theta) ,$$

where $\varepsilon < 1$. The tube is loaded by a vertical shear force $V_y$ acting through the point $O$. Find the maximum shear stress in the section.

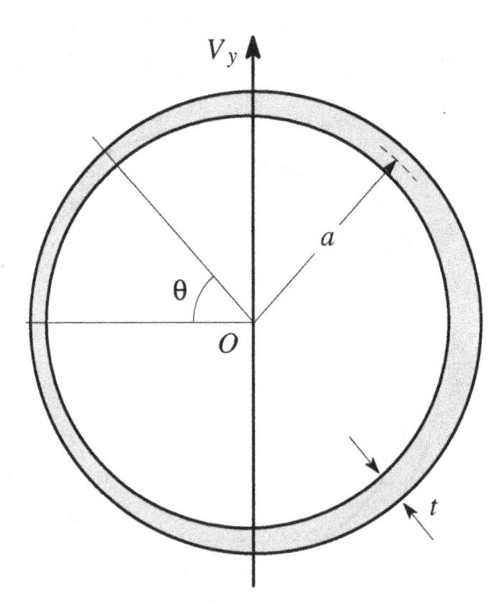


*Figure P6.20*

**6.21.** The closed thin-walled section of Figure P6.21 transmits a vertical shear force $V_y$, whose line of action passes through the point $O_1$. Find the shear stresses at the points $A$ and $B$. **Note:** By symmetry, the shear centre for this section must be at $C$.

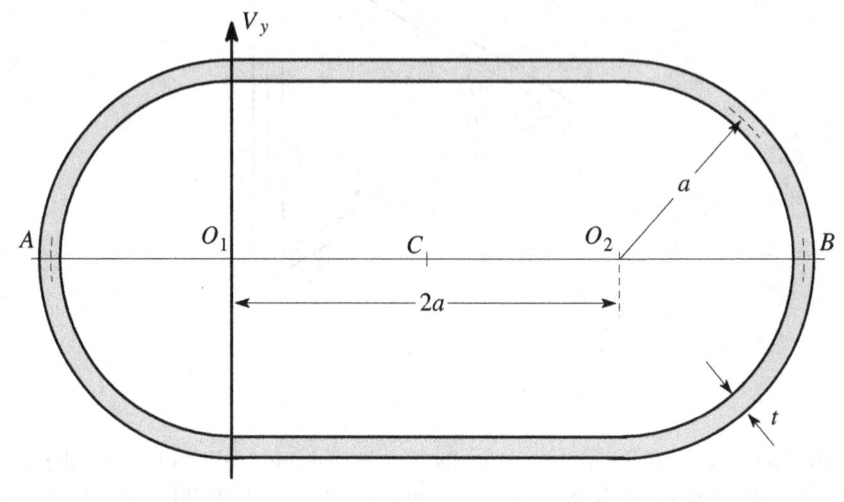


*Figure P6.21*

**6.22.** The closed circular section of Figure P6.22 is loaded by a vertical shear force $V_y$ whose line of action is such that the shear stress at $B$ is zero. Find the shear stress at $A$ and the distance $d$ defining the line of action. Does this mean that the same distance $d$ defines the shear centre for the open section of Figure P6.5?

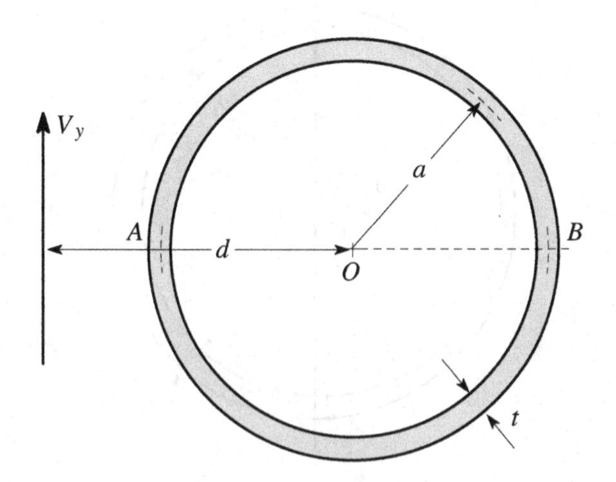


*Figure P6.22*

**6.23.** The unsymmetrical square box section of Figure P6.23 transmits a vertical shear force $V_y$ whose line of action passes through the middle of the thicker left-hand section as shown. Find the maximum shear stress in the section.

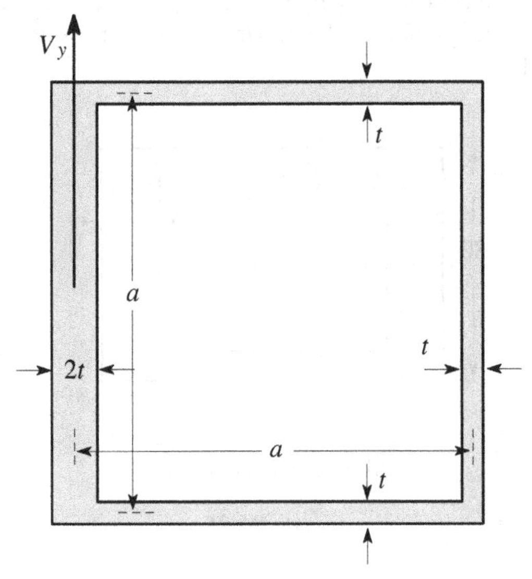


*Figure P6.23*

**6.24.** The closed thin-walled section of Figure P6.24 transmits a vertical shear force $V_y$, whose line of action passes through the point $O$. Find the distribution of shear stress in the section.

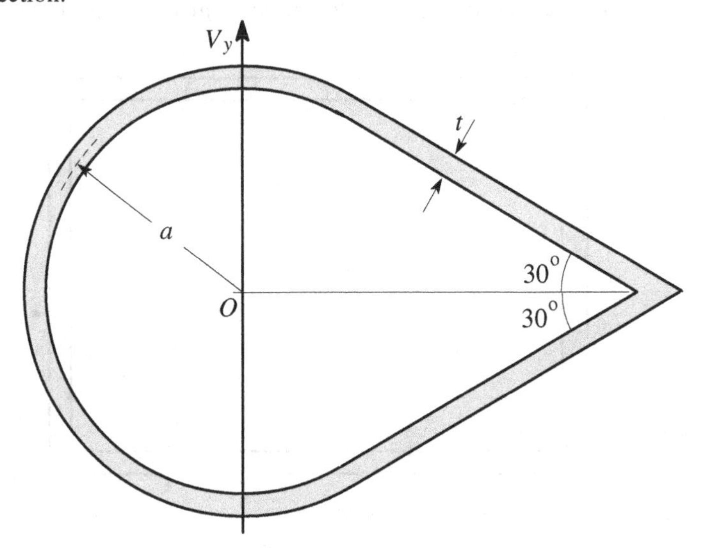


*Figure P6.24*

## Section 6.5

**6.25.** The symmetric box section of Figure P6.25 is loaded by a torque of 4 kNm. Find the location and magnitude of the maximum shear stress and the twist per unit length if $G=80$ GPa.

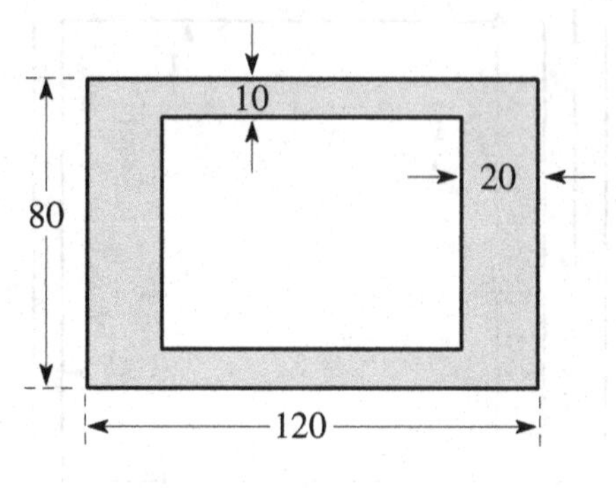


*Figure P6.25*

**6.26.** The hollow thin-walled box section of Figure P6.26 has sides $a,b$ and wall thickness $t$ and is subjected to a torque $T$. Find the shear stress distribution and the torsional stiffness $K$. Express the results in terms of the perimeter of the section $p=2(a+b)$ and the ratio $r=a/b$ between the sides. Hence show that a square section $(r=1)$ has the greatest stiffness and strength for a given volume of material.

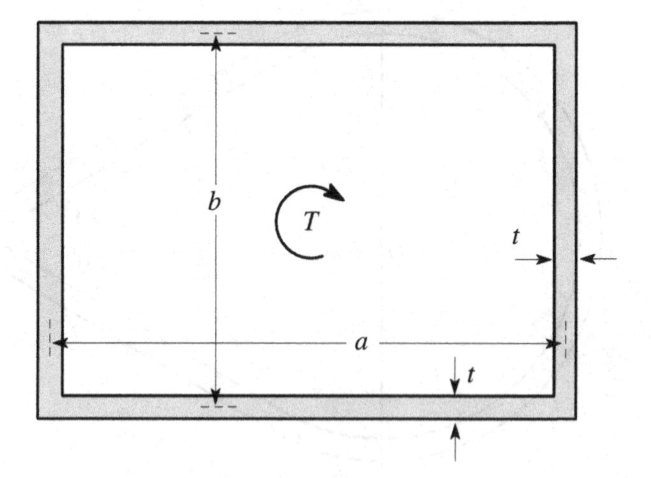


*Figure P6.26*

**6.27.** A beam with the section of Figure P6.24 is loaded by a torque $T$. Find the maximum shear stress and the twist per unit length, if the material has shear modulus $G$.

**6.28.** Find the torsional stiffness $K$ for the closed thin-walled beam section of Figure P6.28. The wall thickness is everywhere equal to $t$.

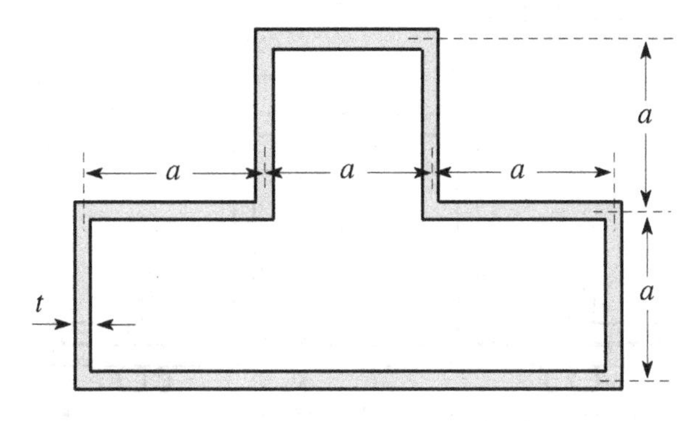


*Figure P6.28*

**6.29.** Find the torsional stiffness $K$ for the closed thin-walled beam section of Figure P6.29. The wall thickness is everywhere equal to $t$.

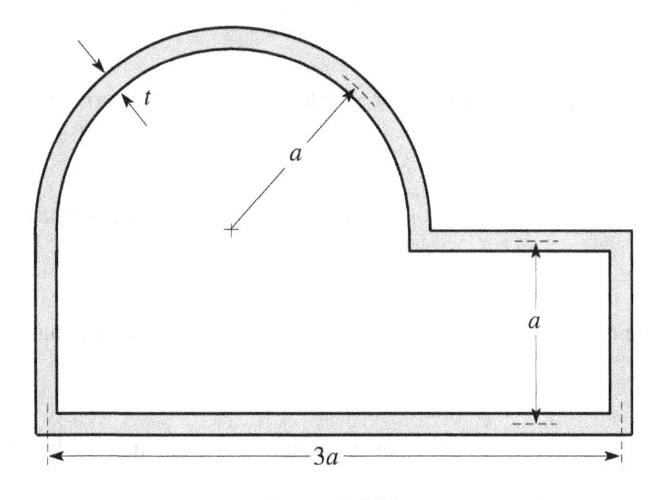


*Figure P6.29*

**6.30.** A beam with the section of Figure P6.20 is loaded by a torque $T$. Find the maximum shear stress and the twist per unit length, if the material has shear modulus $G$.

**6.31.** A beam with the section of Figure P6.21 is loaded by a torque $T$. Find the maximum shear stress and the twist per unit length, if the material has shear modulus $G$.

**6.32.** The torsional stiffness of a hat section is to be increased by bolting a flat plate to it, creating the closed section of Figure P6.32. The bolts are of 3 mm diameter and have an axial spacing of 50 mm. Estimate the shear stress in the bolts when the beam is loaded by a torque of 10 Nm.

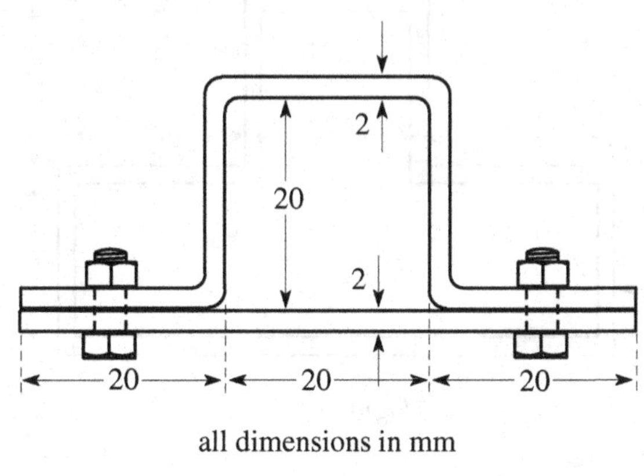


*Figure P6.32*

## Section 6.6

**6.33.** Find the twist per unit length for Problem 6.19, if the material has shear modulus $G$.

**6.34.** Find the twist per unit length for Problem 6.20, if the material has shear modulus $G$.

**6.35.** Find the twist per unit length for Problem 6.21, if the material has shear modulus $G$.

**6.36.** Find the twist per unit length for Problem 6.24, if the material has shear modulus $G$.

**6.37.** Find the location of the shear centre for the triangular section of Figure P6.19.

**6.38.** Find the location of the shear centre for the section of Figure P6.20.

**6.39.** Find the location of the shear centre for the unsymmetrical box section of Figure P6.23.

**6.40.** Find the location of the shear centre for the section of Figure P6.24.

**6.41.** Find the location of the shear centre for the closed thin-walled section of Figure P6.41.

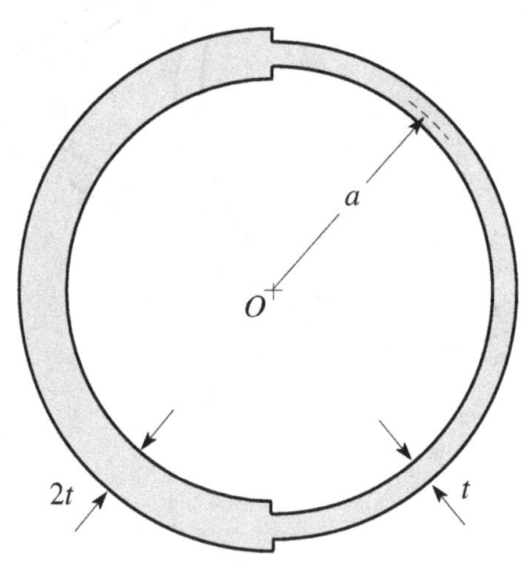


*Figure P6.41*

**6.42.** The two-cell, thin-walled beam section of Figure P6.42 is loaded by a torque $T$. Find the location and magnitude of the maximum shear stress and the torsional stiffness $K$ for the section. The wall thickness is everywhere equal to $t$. Compare your results with those obtained when there is no internal partition, as in Figure P6.28.

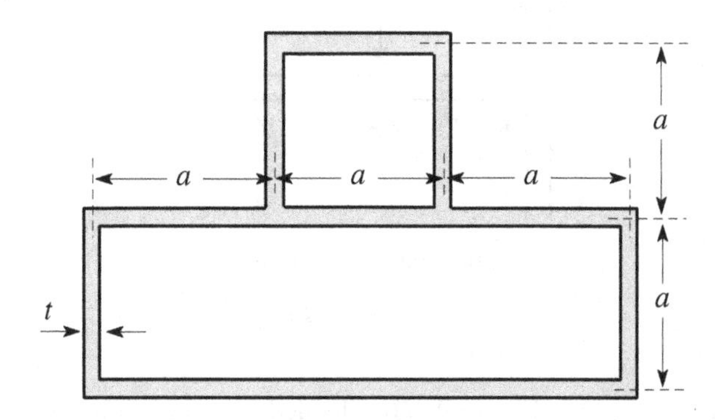


*Figure P6.42*

**6.43.** The two-cell, thin-walled beam section of Figure P6.43 is loaded by a torque $T$. Find the location and magnitude of the maximum shear stress and the torsional stiffness $K$ for the section. The wall thickness is everywhere equal to $t$.

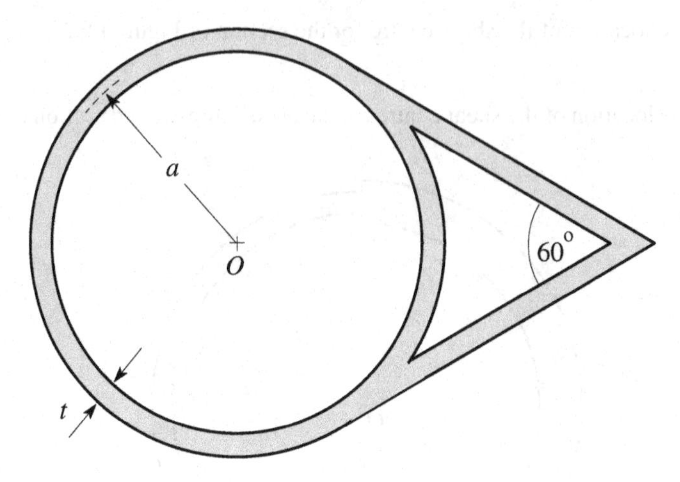


*Figure P6.43*

**Section 6.7**

**6.44.** Find the torsional stiffness $K$ for the cresent section of Figure P6.6.

**6.45.** The T-beam of Figure P6.45 transmits a shear force of 20 kN with the line of action shown. Find the maximum shear stress and the twist per unit length, if the material is steel with $G = 80$ GPa.

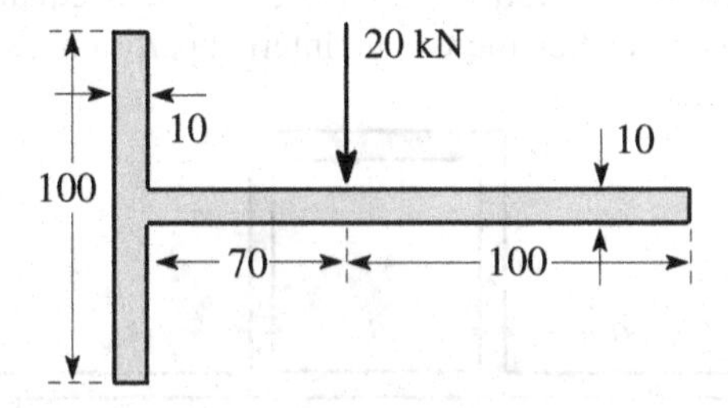


all dimensions in mm

*Figure P6.45*

**6.46.** The unequal angle of Figure P6.46 is loaded by a shear force of 6 kN with the line of action shown. Find the twist per unit length, if the material of the beam is steel with $G = 80$ GPa.

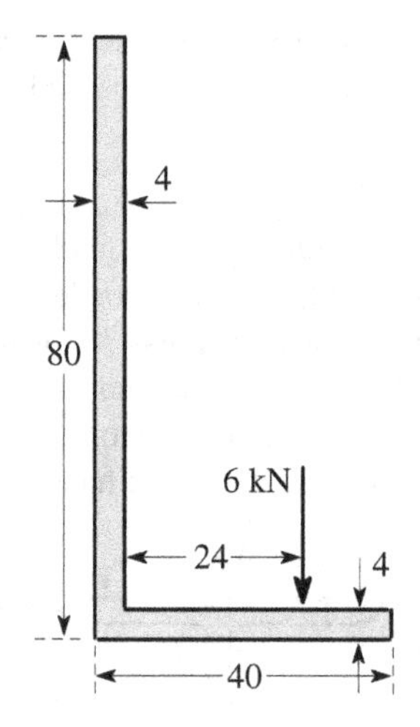


all dimensions in mm

*Figure P6.46*

**6.47*.** Use the results of §6.6 to find the twist per unit length for Problem 6.22. Use your result to determine the line of action of $V_y$ for which the *open* section of Figure P6.5 has no twist. Comment on the implication of your result for the statement that the shear centre is that point through which $V_y$ must act in order that there be no twist.

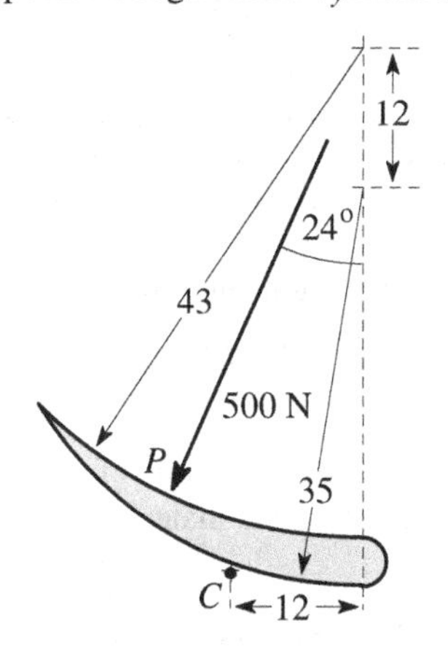


all dimensions in mm

*Figure P6.48*

**6.48*.** A titanium alloy turbine blade is idealized by the section of Figure P6.48. The leading edge is a semicircle of radius 2 mm and the inner and outer edges of the trailing section are cylindrical surfaces of radius 43 mm and 35 mm respectively, with centres as shown. The shear centre is located at $C$ and the blade is loaded by a 500 N force normal to the inner surface at $P$ as shown. Find the maximum shear stress due to torsion and the twist of the blade per unit length. The shear modulus for the alloy is 43 GPa.

**6.49.** The W200×22 I-beam of Figure P6.49 is made of steel ($G=80$ GPa) and is 15 m long. One end is built in and the other is loaded by two forces $F$ constituting a torque, as shown. Find the torsional stiffness $K$ for the beam section and hence determine the value of $F$ needed to cause the end to twist through an angle of 20°. Does the result surprise you? How large an angle of twist[12] could a person of average strength produce in such a beam using his or her hands only?

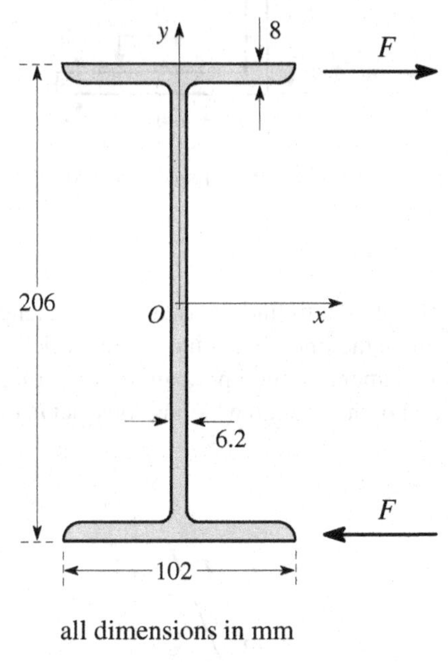


*Figure P6.49*

**6.50.** The beam of Figure P6.1 is loaded by a torque of 40 Nm. Find the maximum shear stress and the twist per unit length, if the material is steel with $G=80$ GPa.

**6.51.** The shear force in Problem 6.2 is displaced by 1 inch from the plane of symmetry. Find the percentage increase in the maximum shear stress caused by this offset.

**6.52.** Use the results of §6.7 to answer Problem 1.19 for the angle iron and compare the results with experiment.

[12] If you want to impress your unsuspecting colleagues with your engineering expertise (or your immense strength), ask them to guess the answer to this question, along the lines of the questions in §1.2, and then perform the experiment.

# 7

# Beams on Elastic Foundations

Many engineering applications involve a relatively rigid structure supported by a more flexible distributed 'foundation'. An important class of examples arises in civil engineering, where buildings or other structures are supported on a soil base. Other less obvious examples include steel components supported on extended rubber bushings, floating structures (where the support is provided by the buoyancy force) and the transmission of load between bones through an intervening cartilidge or other softer tissue layer. The roughness and/or lack of conformity between contacting bodies can also simulate an intervening softer foundation. In all these cases, the foundation has the effect of distributing the load over a more extended area. For example, the weight of a locomotive transmitted to the track through the relatively concentrated wheel/rail contact will be distributed over quite an extended region as a result of the compliance of the ballast and soil foundation. An important objective in analyzing such problems is to determine the extent of this distribution in order to choose appropriate properties for the foundation materials.

A linear elastic foundation is any form of distributed support in which the displacement is a linear function of the applied load. In general, if we apply a localized load to a foundation, the maximum displacement will occur under the load, but adjacent regions of the surface will also be displaced, as shown in Figure 7.1. You can demonstrate this effect by pressing the sharp edge of a ruler into a rubber eraser resting on a plane surface. The extent of this *non-local* displacement depends on the geometry of the foundation and also its material properties. Displacements tend to be more localized if the elastic modulus of the foundation material increases with depth[1] as in the case of many soils, or if the foundation comprises a thin layer of flexible material resting on or bonded to a rigid base.[2]

[1] C.R. Calladine and J.A. Greenwood (1978), Line and point loads on a non-homogeneous incompressible elastic half-space, *Quarterly Journal of Mechanics and Applied Mathematics,* Vol.31, pp.507–529.

[2] K.L. Johnson (1985), *Contact Mechanics*, Cambridge University Press, Cambridge, §5.8.

J.R. Barber, *Intermediate Mechanics of Materials*, Solid Mechanics and Its Applications 175, 2nd ed., DOI 10.1007/978-94-007-0295-0_7, © Springer Science+Business Media B.V. 2011

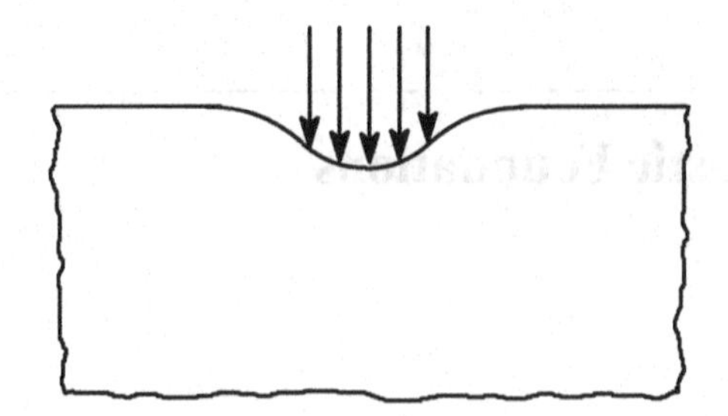

*Figure 7.1: Surface displacements due to a localized load*

A limiting case is the *Winkler foundation*, in which the displacement $u(z)$ at a point defined by the coordinate $z$ depends only upon the local force per unit length $p(z)$ — i.e.

$$u(z) = \frac{p(z)}{k} , \tag{7.1}$$

where $k$ is known as the *modulus* or the *stiffness* of the foundation. The modulus is the support per unit length of beam, per unit displacement, and it therefore has the same dimensions as a stress (force/length$^2$). The Winkler foundation acts essentially like a bed of unconnected springs, as shown in Figure 7.2. Equation (7.1) leads to significant simplification in the analysis of problems and hence is often used as an approximation in situations where the foundation displacement is not strictly local. We shall discuss ways of determining whether this is an appropriate simplification in §7.3 below.

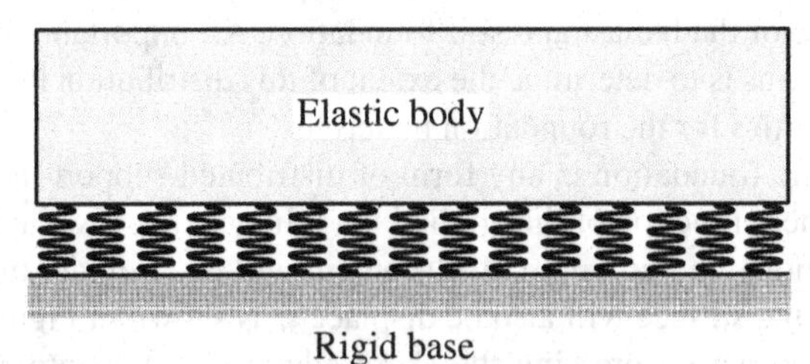


*Figure 7.2: Schematic representation of the Winkler foundation*

## 7.1 The governing equation

Figure 7.3 *(a)* shows a uniform beam of flexural rigidity $EI$, supported by a Winkler foundation and subjected to a distributed load $w(z)$ per unit length. The forces acting on a small element of beam of length $\delta z$ are shown in Figure 7.3 *(b)*. The force corresponding to the distributed load is $w(z)\delta z$ and the support provides a downward force $p(z)\delta z$ opposing the upward displacement $u(z)$.

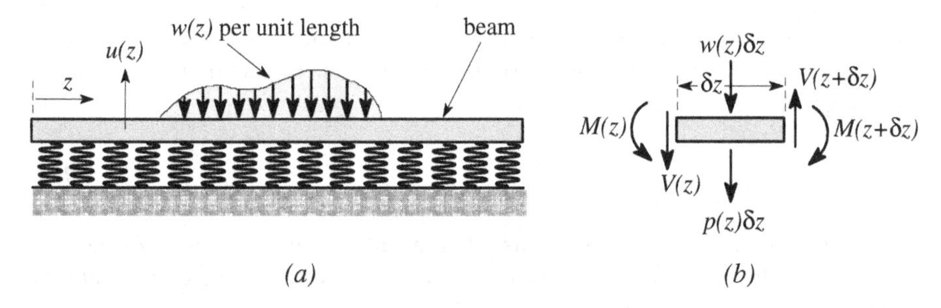


*Figure 7.3: (a) A beam supported on a Winkler foundation, (b) equilibrium of a beam element*

Equilibrium of vertical forces therefore requires that the shear force $V$ satsify the equation

$$V(z+\delta z)-V(z)-w(z)\delta z-p(z)\delta z=0$$

and hence

$$\frac{V(z+\delta z)-V(z)}{\delta z}=w(z)+p(z)\ .$$

Allowing $\delta z\to 0$ and using (7.1), we then obtain

$$\frac{dV}{dz}=w(z)+p(z)=w(z)+ku(z)\ . \tag{7.2}$$

As in §6.1, moment equilibrium requires

$$\frac{dM}{dz}=V \tag{7.3}$$

and the bending moment $M$ and the displacement $u$ are related by the bending equation (1.17)

$$M=-EI\frac{d^2u}{dz^2}\ . \tag{7.4}$$

Eliminating $M,V$ between equations (7.2–7.4), we obtain

$$EI\frac{d^4u}{dz^4}+ku=-w\ , \tag{7.5}$$

which is the governing equation for an elastic beam supported by a Winkler foundation.

### 7.1.1 Solution of the governing equation

The applied load $w$ will generally be a known function of $z$ and hence (7.5) constitutes an inhomogeneous ordinary differential equation for the unknown displacement $u(z)$. Once this equation is solved, the bending moment can be recovered by substitution in (7.4).

The general solution of (7.5) can be written as the sum of a *particular solution* and the general *homogeneous solution.* The particular solution is *any* function that satisfies (7.5) and the homogeneous solution is the general solution of the corresponding homogeneous equation

$$EI\frac{d^4u}{dz^4} + ku = 0 \, . \tag{7.6}$$

Since (7.6) is a fourth order equation, its solution will contain four arbitrary constants which permit the satisfaction of two boundary conditions at each end of the beam, as in elementary beam problems.

The particular solution and the homogeneous solution can be given a physical interpretation. The particular solution corresponds to a state in which the correct load $w(z)$ is applied, without regard to the end conditions, which will therefore generally not be those required in the problem. Hence, we must superpose forces and/or moments at the ends so as to restore the correct end conditions. This is the function of the general homogeneous solution, which describes all possible states of the same beam loaded *at the ends only* — after all, (7.6) is simply (7.5) with $w(z)$ equal to zero. We shall first examine the homogeneous solution in some detail, since it affords considerable insight into the general behaviour of beams on elastic foundations. In particular, we shall find that the deformation tends to be quite localized near the loads, in contrast to the behaviour of beams on discrete supports.

## 7.2 The homogeneous solution

If the beam has no distributed load $w(z)$, the displacement is given by equation (7.6), whose form suggests the existence of solutions of the form

$$u(z) = Ae^{bz} \, , \tag{7.7}$$

where $A, b$ are constants. Substitution into (7.6) yields

$$EIb^4Ae^{bz} + kAe^{bz} = 0 \, ,$$

showing that (7.7) will be a solution of (7.6) if and only if

$$b^4 = -\frac{k}{EI} \, . \tag{7.8}$$

This equation has no real roots, but it has four complex roots which can be written

$$b = (\pm 1 \pm i)\beta \, ,$$

where

$$\beta = \sqrt[4]{\frac{k}{4EI}} \, . \tag{7.9}$$

This solution is easily verified by substitution back into (7.8).

It follows that the general solution of the homogeneous equation (7.6) can be written

$$u(z) = A_1 e^{(1+i)\beta z} + A_2 e^{(1-i)\beta z} + A_3 e^{(-1+i)\beta z} + A_4 e^{(-1-i)\beta z} , \tag{7.10}$$

where $A_1, A_2, A_3, A_4$ are four independent complex constants. However, the displacement must be a real function, so we must have $A_2 = \bar{A}_1, A_4 = \bar{A}_3$, where the overbar denotes the complex conjugate. After some algebraic manipulations, the most general real function of the form (7.10) can be written

$$u(z) = B_1 e^{\beta z} \cos(\beta z) + B_2 e^{\beta z} \sin(\beta z) + B_3 e^{-\beta z} \cos(\beta z) + B_4 e^{-\beta z} \sin(\beta z) . \tag{7.11}$$

An alternative form which is sometimes more convenient for beams of finite length is

$$\begin{aligned} u(z) = {} & C_1 \cosh(\beta z) \cos(\beta z) + C_2 \sinh(\beta z) \sin(\beta z) + C_3 \cosh(\beta z) \sin(\beta z) \\ & + C_4 \sinh(\beta z) \cos(\beta z) . \end{aligned} \tag{7.12}$$

Notice that (7.11) and (7.12) are equivalent because of the identities

$$\cosh(\beta z) = \frac{e^{\beta z} + e^{-\beta z}}{2} \quad ; \quad \sinh(\beta z) = \frac{e^{\beta z} - e^{-\beta z}}{2} .$$

In fact, (7.12) can be obtained from (7.11) by writing

$$B_1 = \frac{C_1 - C_4}{2} \; ; \; B_2 = \frac{C_2 + C_3}{2} \; ; \; B_3 = \frac{C_1 + C_4}{2} \; ; \; B_4 = \frac{C_3 - C_2}{2} .$$

The reader uncomfortable with the complex algebra leading to equations (7.11, 7.12) is encouraged to verify them by direct substitution into (7.6).

### 7.2.1 The semi-infinite beam

Figure 7.4 shows a semi-infinite beam, $z > 0$, loaded by a force $F_0$ and a moment $M_0$ at the end $z = 0$. The displacement must be bounded as $z \to \infty$ and hence only the exponentially decaying terms should be retained from (7.11), giving

$$u(z) = B_3 e^{-\beta z} \cos(\beta z) + B_4 e^{-\beta z} \sin(\beta z) . \tag{7.13}$$

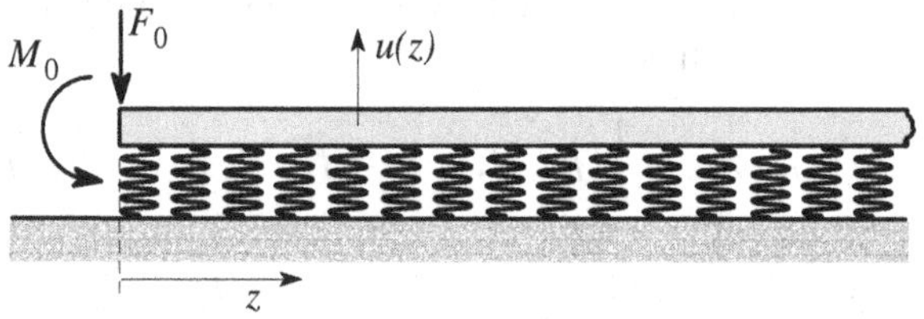

*Figure 7.4: The semi-infinite beam loaded only at the end*

A more rigorous argument showing that the exponentially growing terms should be dropped can be obtained by considering a beam of finite length $L$, where $\beta L \gg 1$. In this case, all four terms in (7.11) should be retained in order to satisfy the two boundary conditions at each end, but it will be found that the coefficients $B_1, B_2 \to 0$ as $\beta L \to \infty$, for any boundary conditions on the end $z=L$. We shall consider finite beam problems further in §7.6 below.

Progressive derivatives of equation (7.13) with respect to $z$ provide corresponding expressions for the slope $\theta$, bending moment $M$, and shear force $V$ as functions of $z$. We have

$$\theta \equiv \frac{du}{dz} = -B_3\beta\, e^{-\beta z}[\cos(\beta z) + \sin(\beta z)] + B_4\beta e^{-\beta z}[\cos(\beta z) - \sin(\beta z)] \tag{7.14}$$

$$M = -EI\frac{d^2u}{dz^2} = -2B_3EI\beta^2 e^{-\beta z}\sin(\beta z) + 2B_4EI\beta^2 e^{-\beta z}\cos(\beta z) \tag{7.15}$$

$$V = \frac{dM}{dz} = -2B_3EI\beta^3 e^{-\beta z}[\cos(\beta z) - \sin(\beta z)] - 2B_4EI\beta^3 e^{-\beta z}[\cos(\beta z) + \sin(\beta z)]\,. \tag{7.16}$$

It is convenient to define the four functions

$$f_1(x) = e^{-x}\cos x\ ;\ \ f_2(x) = e^{-x}\sin x\ ;\ \ f_3(x) = e^{-x}(\cos x + \sin x)\ ;$$
$$f_4(x) = e^{-x}(\cos x - \sin x)\,, \tag{7.17}$$

which permit equations (7.13–7.16) to be written in the compact form

$$u(z) = B_3 f_1(\beta z) + B_4 f_2(\beta z) \tag{7.18}$$
$$\theta(z) = -B_3\beta f_3(\beta z) + B_4\beta f_4(\beta z) \tag{7.19}$$
$$M(z) = -\frac{B_3 k}{2\beta^2} f_2(\beta z) + \frac{B_4 k}{2\beta^2} f_1(\beta z) \tag{7.20}$$
$$V(z) = -\frac{B_3 k}{2\beta} f_4(\beta z) - \frac{B_4 k}{2\beta} f_3(\beta z)\,, \tag{7.21}$$

where we have used (7.9) to eliminate the flexural rigidity $EI$ in favour of the modulus $k$ in equations (7.20, 7.21). It can be verified by substitution that the functions $f_1, f_2, f_3, f_4$ satisfy the relations

$$f_1 = \frac{1}{2}(f_3 + f_4)\ ;\ f_2 = \frac{1}{2}(f_3 - f_4)\ ;\ f_3 = f_1 + f_2\ ;\ f_4 = f_1 - f_2 \tag{7.22}$$

and

$$\frac{df_1}{dx} = -f_3\ ;\ \frac{df_2}{dx} = f_4\ ;\ \frac{df_3}{dx} = -2f_2\ ;\ \frac{df_4}{dx} = -2f_1\,. \tag{7.23}$$

Equations (7.18–7.21) can be used to determine the constants $B_3, B_4$ for any two boundary conditions at the end $z=0$. For the case illustrated in Figure 7.4, we have

$$F_0 = V(0) = -\frac{B_3 k}{2\beta} - \frac{B_4 k}{2\beta} \quad ; \quad M_0 = M(0) = \frac{B_4 k}{2\beta^2}$$

and hence

$$B_3 = -\frac{2\beta F_0}{k} - \frac{2\beta^2 M_0}{k} \quad ; \quad B_4 = \frac{2\beta^2 M_0}{k} \, .$$

Substituting into (7.18–7.21) and using (7.22) to simplify the resulting expressions, we obtain

$$u(z) = -\frac{2\beta F_0}{k} f_1(\beta z) - \frac{2\beta^2 M_0}{k} f_4(\beta z) \qquad (7.24)$$

$$\theta(z) = \frac{2\beta^2 F_0}{k} f_3(\beta z) + \frac{4\beta^3 M_0}{k} f_1(\beta z) \qquad (7.25)$$

$$M(z) = \frac{F_0}{\beta} f_2(\beta z) + M_0 f_3(\beta z) \qquad (7.26)$$

$$V(z) = F_0 f_4(\beta z) - 2M_0 \beta f_2(\beta z) \, . \qquad (7.27)$$

**Example 7.1**

*A semi-infinite steel beam of second moment of area $I_x = 0.5 \times 10^6$ mm$^4$ is supported on an elastic foundation of modulus $k = 10$ MPa and loaded by a downward force $F_0 = 10$ kN at the end, as shown in Figure 7.5. Find the slope and deflection at the free end and the location and magnitude of the maximum bending moment ($E_{\text{steel}} = 210$ GPa).*

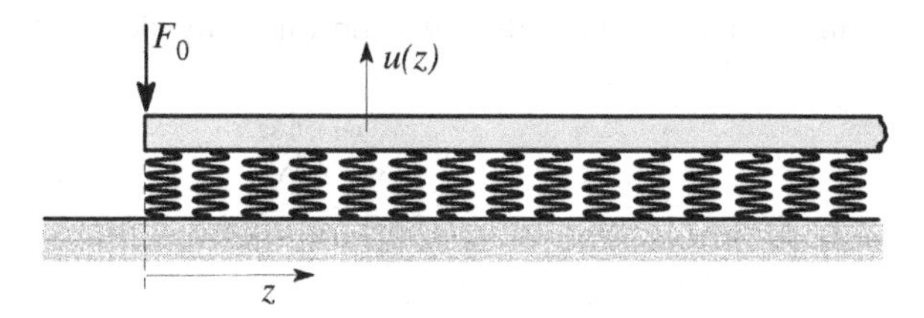


*Figure 7.5: The semi-infinite beam loaded by an end force*

We first use equation (7.9) to calculate

$$\beta = \sqrt[4]{\frac{k}{4EI}} = \sqrt[4]{\frac{10 \times 10^6}{4 \times 210 \times 10^9 \times 0.5 \times 10^{-6}}} = 2.209 \text{ m}^{-1} \, .$$

The complete solution can be obtained by substituting $F_0 = 10$ kN, $M_0 = 0$ into equations (7.24–7.27). In particular, the end displacement and slope are

$$u(0) = -\frac{2\beta F_0}{k} = -\left(\frac{2 \times 2.209 \times 10 \times 10^3}{10 \times 10^6}\right) = -4.4 \times 10^{-3} \text{ m}$$

— i.e. 4.4 mm downwards, and

$$\theta(0) = \frac{2\beta^2 F_0}{k} = \frac{2\times 2.209^2\times 10\times 10^3}{10\times 10^6} = 9.8\times 10^{-3}\text{ radians } (0.56^{\circ})\,.$$

To find the maximum bending moment, we first note that

$$\frac{dM}{dz} = V = -\left(-\frac{2\beta F_0}{k}\right)\frac{k}{2\beta}f_4(\beta z) = F_0 e^{-\beta z}[\cos(\beta z) - \sin(\beta z)]\,,$$

using (7.24). The maximum bending moment occurs when $dM/dz=0$ and hence

$$\cos(\beta z) - \sin(\beta z) = 0\,.$$

This equation has roots at

$$\beta z = \frac{\pi}{4}, \frac{5\pi}{4}, \frac{9\pi}{4}, \ldots$$

In view of the exponential decay, the magnitude of $M$ will be greatest at the first root, where

$$z = \frac{\pi}{4\beta} = \frac{\pi}{4\times 2.209} = 0.356\text{ m}\,.$$

At this point, we have

$$M_{\max} = \frac{F_0 e^{-\pi/4}}{\sqrt{2}\beta} = 0.322\frac{F_0}{\beta} = \frac{0.322\times 10\times 10^3}{2.209} = 1457\text{ Nm}\,.$$

Figure 7.6 shows the deformed shape of the beam and the bending moment as a function of $z$. Notice that the disturbance decays rapidly with $z$, but the trigonometric functions in equations (7.13–7.16) cause the decay to be oscillatory, so there are actually regions in which the beam is displaced upwards[3] by the downward force $F_0$.

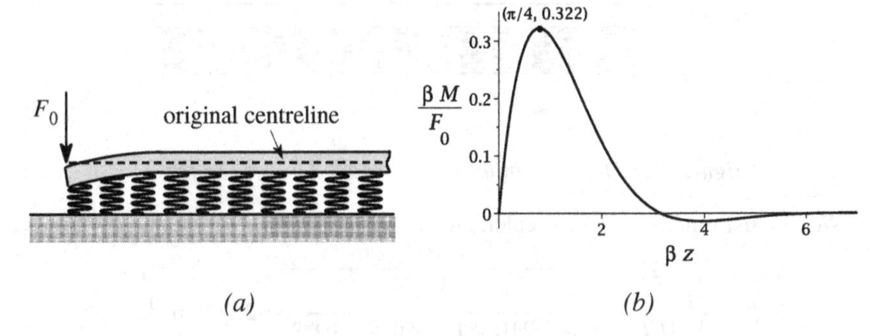


*Figure 7.6: (a) Deformed shape of the beam and (b) the bending moment distribution*

[3] If the beam is simply resting on the support, rather than being attached to it, an upward displacement will cause the beam to separate from the foundation and the restoring force $p(z)$ will then be locally zero, rather than being given by equation (7.1). A foundation of this kind that exerts a restoring force only for displacements in one direction is known as a *unilateral support*. However, in problems with downward loading, the weight of the beam will probably be sufficient to maintain contact in the upward displacement regions.

## 7.3 Localized nature of the solution

The rapid decay of the disturbance observed in the above example is typical of end loading problems for beams on elastic foundations. Figure 7.7 shows the four functions $f_1, f_2, f_3, f_4$, of equations (7.17) and all have esentially decayed to negligible values at $x=4$. In view of equations (7.18–7.21), we conclude that for any end loading of the beam, the deformation will be restricted to a region at the end defined approximately by

$$0 < z < 4l_0 \, , \tag{7.28}$$

where

$$l_0 \equiv \frac{1}{\beta} = \sqrt[4]{\frac{4EI}{k}} \tag{7.29}$$

has the dimensions of length and functions as a characteristic decay length for the beam/support system. For Example 7.1, the decay length was $l_0 = 1/2.209 = 0.453$ m.

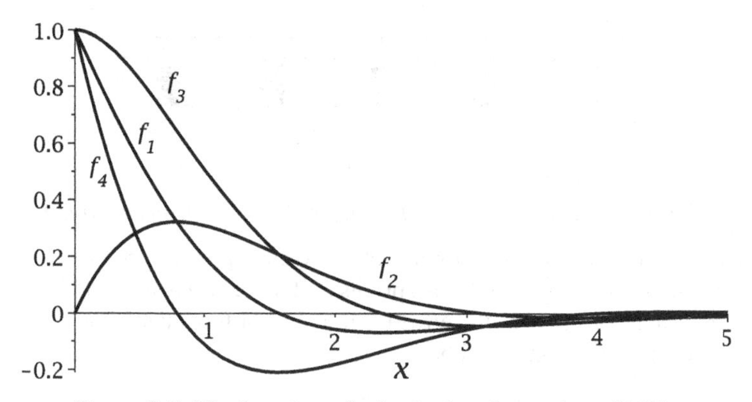


*Figure 7.7: The functions $f_1, f_2, f_3, f_4$, of equations (7.17)*

In any practical problem involving a beam on an elastic foundation, the very first step should always be to calculate $l_0$ from the properties of the beam and the foundation. Comparison of this length with other length scales in the problem then provides important information about the behaviour of the system and can often be used to justify a simpler solution procedure. If the length of the beam $L \gg l_0$ (i.e. if $\beta L \gg 1$), it follows that end loads on the beam will have only local effects and there will be a central region, distant from the ends, where the beam will be unaffected by the end conditions. This condition is often satisfied in practice and it simplifies the resulting problem, since effects at the two ends can then be analyzed separately, using equations (7.18–7.21).

Similar localization occurs when loads are applied over a restricted central region of the beam. If the loads are at least $4l_0$ from either end of the beam, their effects will not be influenced by conditions at the ends and it is sufficient to solve the simpler problem in which the same loads are applied to an infinite beam (see §7.4 below).

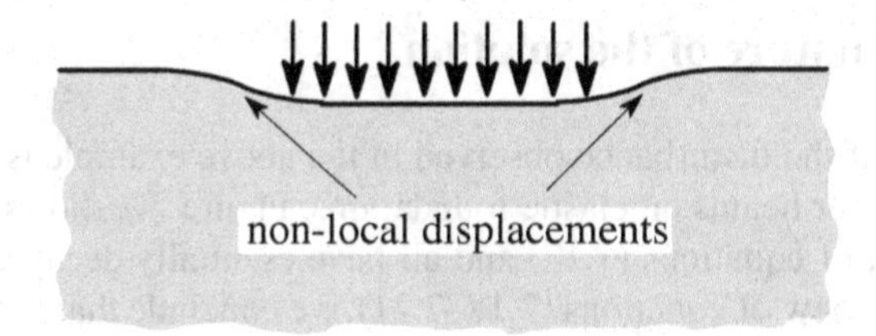


*Figure 7.8: Estimation of the significance of non-local effects*

We can also use an estimate of $l_0$ to determine whether a more realistic 'non-local' foundation can be approximated by an equivalent Winkler foundation in a particular application (see Figure 7.1 and the associated discussion). Suppose a uniform load is applied to the non-local foundation over an extended region, as shown in Figure 7.8. The Winkler approximation is reasonable if the regions of significant non-local displacements adjacent to the load extend a distance significantly smaller than $l_0$.

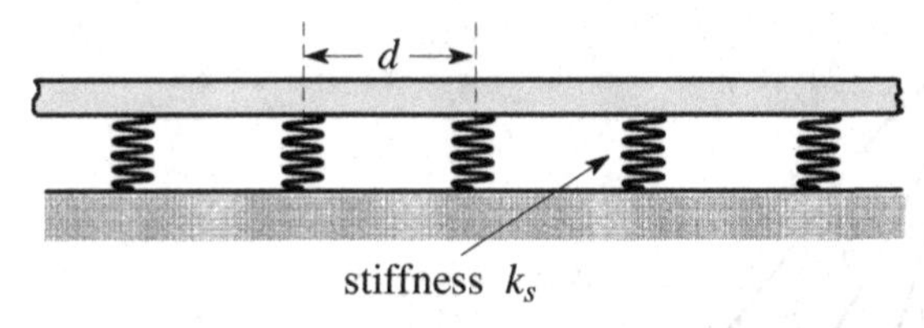


*Figure 7.9: A beam supported on discrete springs with spacing d*

Another case of some interest concerns a beam supported on a set of discrete springs, as shown in Figure 7.9. If the springs each have stiffness $k_s$ and the distance between adjacent springs is $d$, the effective modulus of the foundation (the stiffness associated with a unit length) will be

$$k = \frac{k_s}{d} , \tag{7.30}$$

giving

$$l_0 = \sqrt[4]{\frac{4EId}{k_s}} . \tag{7.31}$$

We can then argue that representation of the discrete springs as a continuous foundation will be reasonable if $d \ll l_0$ and hence

$$\sqrt[4]{\frac{d^3 k_s}{4EI}} \ll 1. \tag{7.32}$$

## 7.4 Concentrated force on an infinite beam

In view of the preceding discussion, it is of interest to consider the problem of Figure 7.10*(a)*, in which an infinite beam on an elastic foundation is loaded by a concentrated force $F_0$. We can locate the origin at the point of application of the force without loss of generality.

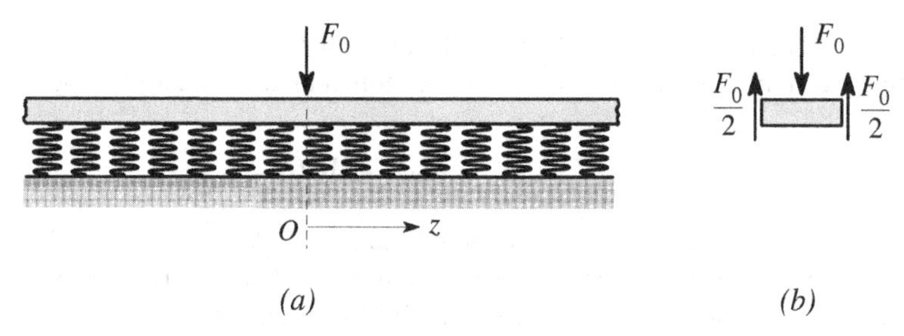


*Figure 7.10: (a) Infinite beam loaded by a concentrated force, (b) equilibrium of an infinitesimal element at O*

Strictly speaking, this is an inhomogeneous problem in which the load $w(z) = F_0\delta(z)$, where $\delta(z)$ is the Dirac delta function. However, we can reduce the problem to a homogeneous problem for the semi-infinite beam $z>0$ using a symmetry argument.

The problem is symmetrical about $O$ and hence the slope $\theta$ must be zero at $O$. Furthermore, equilibrium of an infinitesimal beam element immediately under the load [Figure 7.10 *(b)*] shows that the shear force immediately to the right of the force must be $F_0/2$. Thus the region $z>0$ has end conditions

$$\theta(0) = 0 \quad ; \quad V(0) = \frac{F_0}{2} .$$

Substituting into equations (7.19, 7.21), we obtain the two simultaneous equations

$$-B_3 + B_4 = 0 \quad ; \quad -\frac{B_3 k}{2\beta} - \frac{B_4 k}{2\beta} = \frac{F_0}{2}$$

for $B_3, B_4$, with solution

$$B_3 = B_4 = -\frac{F_0\beta}{2k} .$$

It follows from (7.18, 7.20) that

$$u(z) = -\frac{F_0\beta}{2k} e^{-\beta z}[\cos(\beta z) + \sin(\beta z)] = -\frac{F_0\beta}{2k} f_3(\beta z) \tag{7.33}$$

$$M(z) = -\frac{F_0}{4\beta} e^{-\beta z}[\cos(\beta z) - \sin(\beta z)] = -\frac{F_0}{4\beta} f_4(\beta z) \tag{7.34}$$

in $z>0$. The symmetry of the problem requires that the same solution can be used in $z<0$ with $z$ replaced by $-z$. Thus, expressions that are correct for all values of $z$ can be written

$$u(z) = -\frac{F_0\beta}{2k} f_3(\beta|z|) \quad ; \quad M(z) = -\frac{F_0}{4\beta} f_4(\beta|z|) . \tag{7.35}$$

The slope $\theta$ and shear force $V$ are odd functions of $z$ (e.g. $V(z) = -V(-z)$) and can be written

$$\theta(z) = \frac{F_0\beta^2}{k} \mathrm{sgn}(z) f_2(\beta|z|) \quad ; \quad V(z) = \frac{F_0}{2} \mathrm{sgn}(z) f_1(\beta|z|) , \tag{7.36}$$

using (7.3, 7.21), where $\mathrm{sgn}(z)=1$ for $z>0$ and $-1$ for $z<0$.

The maximum displacement and bending moment occur under the load at $z=0$ and are

$$u_{\max}=u(0)=-\frac{F_0\beta}{2k} \quad ; \quad M_{\max}=M(0)=-\frac{F_0}{4\beta} \,. \tag{7.37}$$

The deformed shape of the beam is shown in Figure 7.11. As in Example 7.1 [Figure 7.6*(a)*], the decay of the disturbance is oscillatory, so there are regions where the downward force causes an upward (albeit small) displacement.

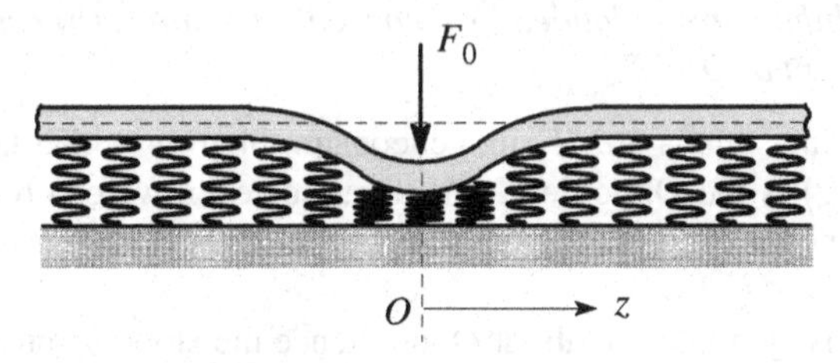


*Figure 7.11: Deformation of the beam under a concentrated force*

### 7.4.1 More general loading of the infinite beam

The preceding solution can be extended to problems involving several concentrated forces, using superposition. For example, if forces $F_1, F_2, F_3$ act at the points $z=a_1, a_2, a_3$ respectively, as shown in Figure 7.12, the resulting displacement will be

$$u(z)=-\frac{F_1\beta}{2k}f_3(\beta|z-a_1|)-\frac{F_2\beta}{2k}f_3(\beta|z-a_2|)-\frac{F_1\beta}{2k}f_3(\beta|z-a_3|)\,,$$

from (7.35). A similar expression can be written for the moment $M(z)$.

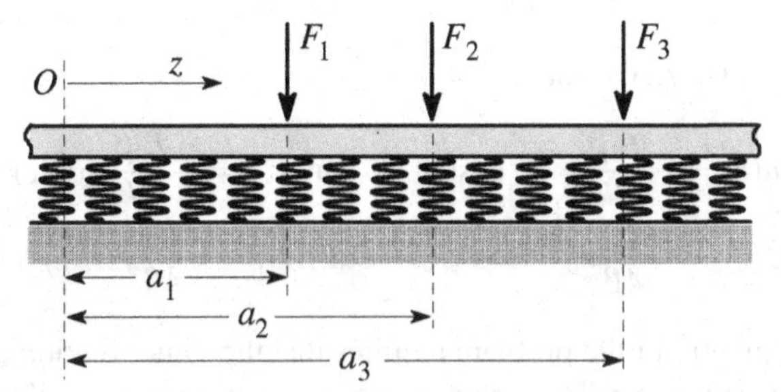


*Figure 7.12: Infinite beam with several concentrated forces*

The same method can be extended to the case of a distributed load $w(z)$ in $a<z<b$ by considering it to consist of a set of concentrated forces[4] $w(z')\delta z'$, as shown in Figure 7.13.

[4] Notice that this is another case where we need to introduce a second variable $z'$ for the purposes of integration (*cf* §6.1.4).

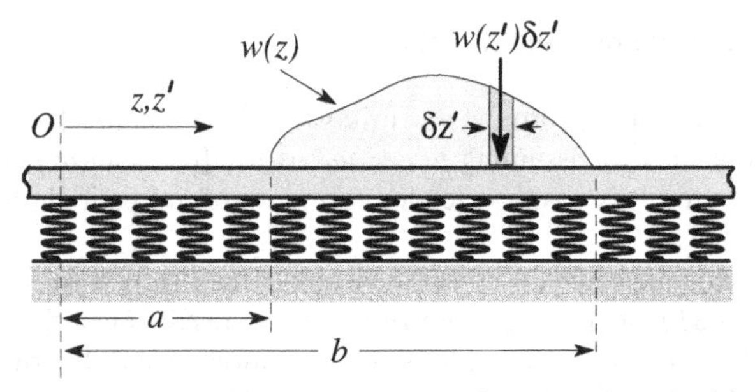


*Figure 7.13: Infinite beam with a distributed load $w(z)$*

We obtain

$$u(z) = -\frac{\beta}{2k}\int_a^b w(z')f_3(\beta|z-z'|)dz' . \tag{7.38}$$

Be careful with the absolute value expression $|z-z'|$, particularly when evaluating the displacement at a point in the range $a<z<b$. In fact, in this case it is safer to split the range of integration to make the sign of $(z-z')$ explicit. We then obtain

$$u(z) = -\frac{\beta}{2k}\int_a^z w(z')f_3[\beta(z-z')]dz' - \frac{\beta}{2k}\int_z^b w(z')f_3[\beta(z'-z)]dz' \; ; \; a<z<b ,$$

since$|z-z'|=(z'-z)$ when $z'>z$.

Equation (7.38) defines a general solution for the displacement of an infinite beam subjected to a distributed load $w(z)$, but this is not generally the most efficient way of solving such problems, since the integrals tend to be algebraically complicated. Instead, we usually account for a distributed load by seeking a particular solution of equation (7.5), as will be discussed in the next section.

## 7.5 The particular solution

If the applied load $w(z)$ is an elementary function, it is usually possible to determine a particular solution to equation (7.5) by assuming a displacement function $u(z)$ of similar form. Special cases that can always be solved in this way include polynomial, trigonometric and exponential functions.

### Polynomials

If $w(z)$ is a polynomial of degree $n$ in $z$, a particular solution can always be obtained by assuming $u(z)$ to be another polynomial of the same degree, substituting into equation (7.5) and equating coefficients. For polynomials of degree 3 and below, the first (derivative) term in (7.5) degenerates to zero and we obtain the simple result

$$u(z) = -\frac{w(z)}{k} . \tag{7.39}$$

**Trigonometric and exponential functions**

If the applied load is of the form $w(z) = w_0 \cos(az)$ [or $w_0 \sin(az)$], a particular solution can be obtained by assuming $u(z) = u_0 \cos(az)$ [or $u_0 \sin(az)$] and equating coefficients. This result is easily extended to more general periodic loading using Fourier series.

A similar technique can be used for the exponential function $w(z) = w_0 \exp(az)$. The more advanced reader will recognize that these results open the way to more general procedures for determining particular solutions, using Fourier or Laplace transforms.

### 7.5.1 Uniform loading

If a beam is subjected to a uniform load $w(z) = w_0$ throughout its length, the particular solution

$$u(z) = -\frac{w_0}{k} \tag{7.40}$$

satisfies equation (7.5) and corresponds to a state in which the beam simply moves downwards as a rigid body to compress the support until an equal and opposite distributed support force is generated. In this case there will be no shear force or bending moment in the beam and the particular solution will be the complete solution of the problem if the ends of the beam are free and unloaded.

Equation (7.40) will not be the complete solution if the ends of the beam are supported or if the load does not extend over the whole length of the beam. In such cases, the solution is obtained by superposing the homogeneous solution of §7.2 to correct the boundary conditions at the ends or at the ends of the loaded region.

**Example 7.2**

*Figure 7.14 shows a semi-infinite beam of flexural rigidity $EI$ supported on an elastic foundation of modulus $k$ and simply supported at the end $z=0$. The beam is subjected to a uniformly distributed load $w_0$ per unit length. Find the maximum bending moment in the beam.*

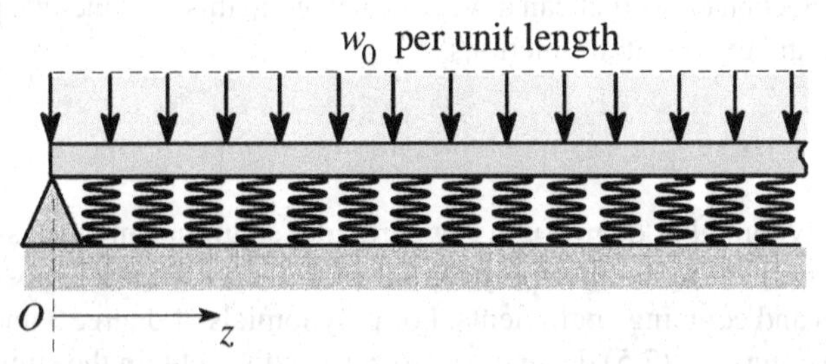


*Figure 7.14: Semi-infinite beam with a uniform load $w_0$ per unit length, supported at the end*

If there were no force at the support, the beam would fall until $u(z) = -w_0/k$, as in Figure 7.15*(a)*. To correct the end condition, we must superpose an upward force $F_0$ at $z=0$, as in Figure 7.15*(b)*. The force must be just sufficient to produce an upward displacement $u(0) = w_0/k$, so as to re-establish contact with the support. Notice that no moment is applied at the end, since this is a simple support.

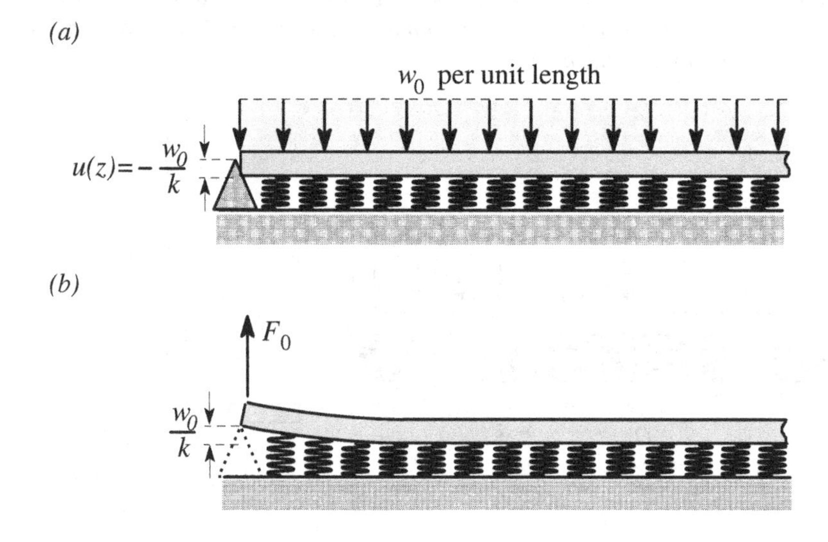


*Figure 7.15: (a) The particular solution, (b) force at the support*

In Example 7.1, we found that a downward force $F_0$ at the end of a similar semi-infinite beam produced a displacement $u(0) = -2\beta F_0/k$, so having regard to the sign difference between Figures 7.5 and 7.15*(b)*, we conclude that

$$F_0 = \frac{w_0}{k}\frac{k}{2\beta} = \frac{w_0}{2\beta} .$$

The particular solution produces no bending moment, so the maximum bending moment is that due to the end force $F_0$ which from Example 7.1 is of magnitude

$$|M_{\max}| = 0.322\frac{F_0}{\beta} = 0.161\frac{w_0}{\beta^2} .$$

As before, it occurs at the point $z = \pi/4\beta = \pi l_0/4$.

### 7.5.2 Discontinuous loads

If only part of the beam is loaded, we can make a cut in the beam at the discontinuity, develop a particular solution for each segment and then apply loads to the ends of each segment sufficient to restore continuity of displacement and slope.

**Example 7.3**

*A long T-section beam is supported on a Winkler foundation of modulus $k = 10$ MPa and subjected to a uniformly distributed load $w(z) = w_0 = 50$ kN/m in $z > 0$ only, as shown in Figure 7.16(a). The cross section of the beam and the location of the centroid C are shown in Figure 7.16(b). The appropriate second moment of area is $I_x = 1.17 \times 10^{-6}$ m$^4$ and $E = 210$ GPa. Find the location and magnitude of the maximum tensile stress in the beam.*

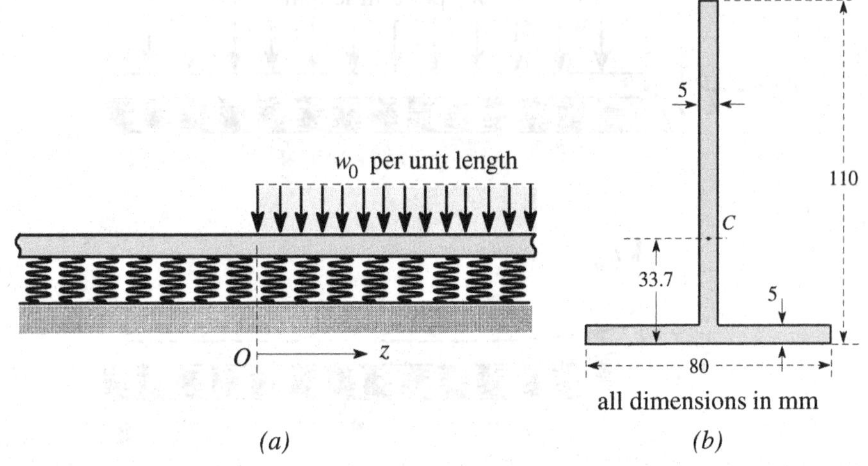


*Figure 7.16: (a) Infinite beam with a discontinuous load, (b) cross section of the beam*

We first calculate

$$\beta = \sqrt[4]{\frac{k}{4EI}} = \sqrt[4]{\frac{10 \times 10^6}{4 \times 210 \times 10^9 \times 1.17 \times 10^{-6}}} = 1.786 \text{ m}^{-1} .$$

Suppose we now make a cut in the beam at $z = 0$, permitting the segment $z > 0$ to deflect to

$$u(z) = -\frac{w_0}{k} = -\frac{50 \times 10^3}{10 \times 10^6} = -5 \times 10^{-3} \text{ m (5 mm downwards)}.$$

The segment $z < 0$, being unloaded, will have no displacement and both beams will move as rigid bodies, involving no bending moment and no loads on the cut ends. The displacement produced by this process has the discontinuous form shown in Figure 7.17.

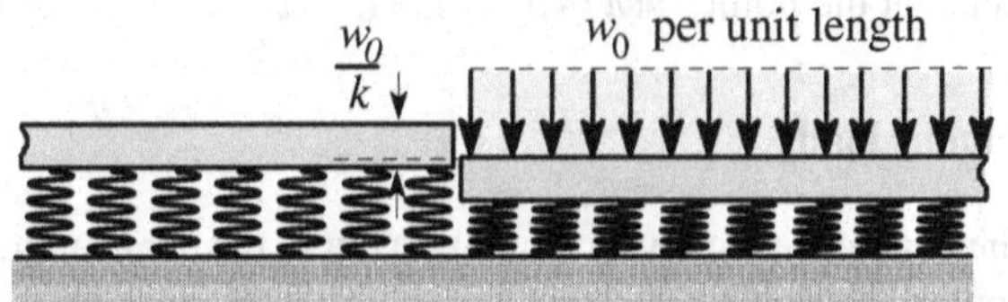


*Figure 7.17: Displacement of the cut beam*

We now apply equal and opposite forces and moments to the cut ends of the two segments, chosen so as to close the gap whilst preserving continuity of slope at $z=0$. Since the two segments have similar properties, it is clear that equal and opposite forces $F_0$ will produce equal slopes at the ends, as shown in Figure 7.18.

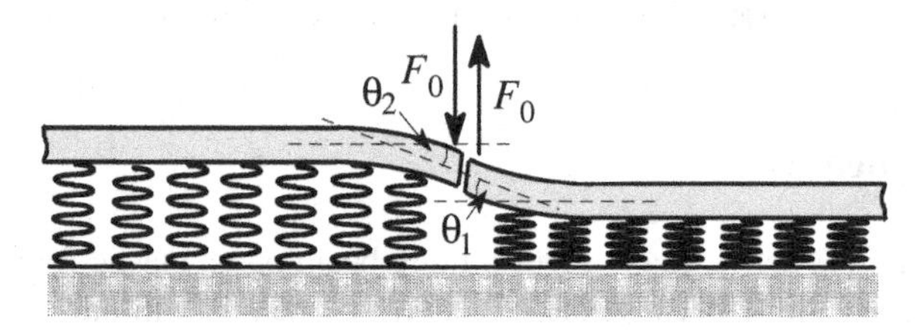


*Figure 7.18: Slopes $\theta_1 = \theta_2$ for all values of $F_0$*

Thus, no moments are required at the cut to preserve continuity of slope. Furthermore, the forces produce equal and opposite displacements on the two ends, each of which must therefore have magnitude $w_0/2k$ to restore continuity of displacement. This is half of the end displacement in the problem of Figure 7.15*(b)* and hence the required force is

$$F_0 = -V(0) = \frac{w_0}{4\beta} = \frac{50 \times 10^3}{4 \times 1.786}\ \text{N} = 7.0\ \text{kN}\,.$$

No bending moment is generated at $z=0$. It is a point of inflexion in the deformed beam at which $d^2u/dz^2=0$. The maximum bending moment therefore occurs either in $z>0$ or $z<0$. Both beam segments are deformed in a similar manner — the load $w_0$ in $z>0$ merely causes a superposed rigid body displacement as in Figure 7.17. The maximum bending moment in each segment is therefore the same and, as in Example 7.1, we have

$$|M_{\max}| = 0.322\frac{F_0}{\beta} = \frac{0.322 \times 7}{1.786} = 1.25\ \text{kNm}\,.$$

The maximum positive moment occurs at

$$z = -\frac{\pi l_0}{4} = -\frac{\pi}{4\beta} = -0.44\ \text{m}$$

(in the unloaded segment) and the equal maximum negative moment occurs in the loaded segment at $z=0.44$ m.

The bending stresses at the top and bottom of the section at $z=-0.44$ m in the unloaded segment are given by

$$\sigma_{\text{top}} = \frac{My_{\max}}{I} = \frac{1.25 \times 10^6 \times (110 - 33.7)}{1.17 \times 10^6} = 81.5\ \text{MPa}$$

$$\sigma_{\text{bottom}} = -\frac{1.25 \times 10^6 \times 33.7}{1.17 \times 10^6} = -36.0\ \text{MPa}\,.$$

These values are reversed in sign at $z=0.44$ m in the loaded segment. We conclude that the maximum tensile stress occurs in the unloaded segment at the top of the section and is 81.5 MPa. As a check on the sign convention, we note from Figure 7.18 that the curvature of the beam will place the upper part of the section in tension on the left and in compression on the right of the discontinuity.

## 7.6 Finite beams

We argued in §7.3 that in many cases a finite beam on an elastic foundation can be treated as semi-infinite because the effects from one end will have decayed before the other end is reached, leading to essentially uncoupled homogeneous problems for the conditions at the respective ends. However, this decoupling requires that $\beta L \gg 1$, where $L$ is the length of the beam. If this condition is not satisfied, all four constants in the homogeneous solution (7.11) must be retained and determined from four simultaneous equations, representing the two conditions at each end of the beam.

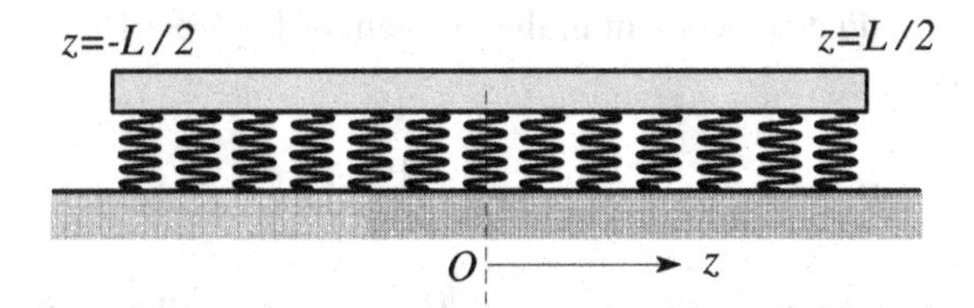


*Figure 7.19: Coordinate system for the finite beam*

In this case, it is generally more efficient to take advantage of the symmetry of the finite beam by moving the origin to the mid-point, as shown in Figure 7.19, and using the hyperbolic form (7.12) of the homogeneous solution, which we here restate as

$$u(z) = C_1 g_1(\beta z) + C_2 g_2(\beta z) + C_3 g_3(\beta z) + C_4 g_4(\beta z) , \tag{7.41}$$

where

$$\begin{aligned} g_1(x) &= \cosh(x)\cos(x) \; ; \quad g_2(x) = \sinh(x)\sin(x) \; ; \\ g_3(x) &= \cosh(x)\sin(x) \; ; \quad g_4(x) = \sinh(x)\cos(x) \, . \end{aligned} \tag{7.42}$$

Notice that $g_1, g_2$ are even functions of $x$ and $g_3, g_4$ are odd functions, so in a symmetric or antisymmetric problem, only two of the four arbitrary constants need to be included. The derivatives of these functions satisfy the relations

$$\frac{dg_1}{dx} = g_4 - g_3 \; ; \quad \frac{dg_2}{dx} = g_3 + g_4 \; ; \quad \frac{dg_3}{dx} = g_1 + g_2 \; ; \quad \frac{dg_4}{dx} = g_1 - g_2 \tag{7.43}$$

and

$$\frac{d^2 g_1}{dx^2} = -2g_2 \; ; \quad \frac{d^2 g_2}{dx^2} = 2g_1 \; ; \quad \frac{d^2 g_3}{dx^2} = 2g_4 \; ; \quad \frac{d^2 g_4}{dx^2} = -2g_3 \, . \tag{7.44}$$

We can therefore write general expressions for the slope, bending moment and shear force in the form

$$\begin{aligned}\theta(z) &= C_1\beta\,[g_4(\beta z) - g_3(\beta z)] + C_2\beta\,[g_3(\beta z) + g_4(\beta z)] \\ &\quad + C_3\beta\,[g_1(\beta z) + g_2(\beta z)] + C_4\beta\,[g_1(\beta z) - g_2(\beta z)] \end{aligned} \tag{7.45}$$

$$M(z) = \frac{C_1 k}{2\beta^2} g_2(\beta z) - \frac{C_2 k}{2\beta^2} g_1(\beta z) - \frac{C_3 k}{2\beta^2} g_4(\beta z) + \frac{C_4 k}{2\beta^2} g_5(\beta z) \tag{7.46}$$

$$\begin{aligned}V(z) &= \frac{C_1 k}{2\beta}[g_3(\beta z) + g_4(\beta z)] - \frac{C_2 k}{2\beta}[g_4(\beta z) - g_3(\beta z)] \\ &\quad - \frac{C_3 k}{2\beta}[g_1(\beta z) - g_2(\beta z)] + \frac{C_4 k}{2\beta}[g_1(\beta z) + g_2(\beta z)]\,. \end{aligned} \tag{7.47}$$

The results are still considerably more algebraically complex than those for long beams and there is justification in practical problems for using the semi-infinite beam results as an order of magnitude approximation, even for values of $\beta L$ as low as 3.

**Example 7.4**

*A beam of length $L$ and flexural rigidity $EI$ is supported on a foundation of modulus $k$ and also by two simple supports at the ends. It is loaded by a uniformly distributed load $w_0$ per unit length as shown in Figure 7.20. Find the reactions at the supports and the bending moment at the mid-point. Show that these results reduce to those for a simply-supported beam with no elastic foundation when the beam is sufficiently short.*

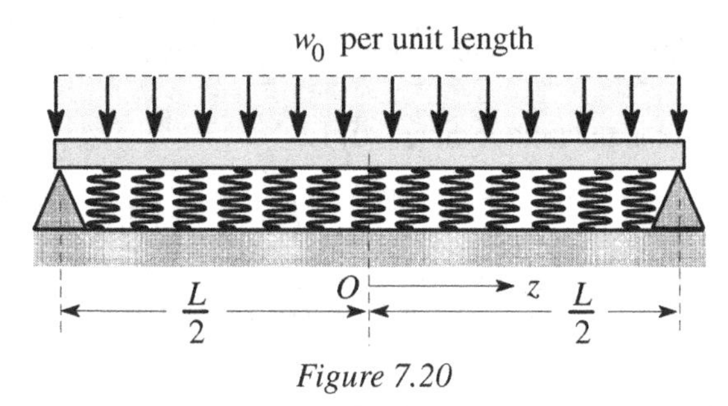


*Figure 7.20*

We construct the solution as the sum of the particular solution (7.40) and the homogeneous solution (7.41), i.e.

$$u(z) = -\frac{w_0}{k} + C_1 g_1(\beta z) + C_2 g_2(\beta z) + C_3 g_3(\beta z) + C_4 g_4(\beta z)\,.$$

However, the problem is symmetric about the origin and hence the solution must satisfy the condition $u(z) = u(-z)$, which requires that the constants $C_3 = C_4 = 0$.

We therefore have

$$u(z) = -\frac{w_0}{k} + C_1 g_1(\beta z) + C_2 g_2(\beta z)$$

and

$$M(z) = \frac{C_1 k}{2\beta^2} g_2(\beta z) - \frac{C_2 k}{2\beta^2} g_1(\beta z) ,$$

from (7.47).

The end conditions require

$$u\left(\frac{L}{2}\right) = u\left(-\frac{L}{2}\right) = 0 \ ; \ M\left(\frac{L}{2}\right) = M\left(-\frac{L}{2}\right) = 0$$

and hence

$$C_1 \cosh\left(\frac{\beta L}{2}\right) \cos\left(\frac{\beta L}{2}\right) + C_2 \sinh\left(\frac{\beta L}{2}\right) \sin\left(\frac{\beta L}{2}\right) = \frac{w_0}{k}$$

$$C_1 \sinh\left(\frac{\beta L}{2}\right) \sin\left(\frac{\beta L}{2}\right) - C_2 \cosh\left(\frac{\beta L}{2}\right) \cos\left(\frac{\beta L}{2}\right) = 0 ,$$

using the definitions (7.42).

The solution of these simultaneous equations is

$$C_1 = \frac{2w_0 \cosh\left(\frac{\beta L}{2}\right) \cos\left(\frac{\beta L}{2}\right)}{k\left[\cosh(\beta L) + \cos(\beta L)\right]} \ ; \ C_2 = \frac{2w_0 \sinh\left(\frac{\beta L}{2}\right) \sin\left(\frac{\beta L}{2}\right)}{k\left[\cosh(\beta L) + \cos(\beta L)\right]} ,$$

where we have used the identity

$$\cosh^2\frac{x}{2}\cos^2\frac{x}{2} + \sinh^2\frac{x}{2}\sin^2\frac{x}{2} = \frac{1}{2}(\cosh x + \cos x) .$$

The bending moment at the mid-point ($z=0$) is

$$M(0) = -\frac{C_2 k}{2\beta^2} = -\frac{w_0 \sinh\left(\frac{\beta L}{2}\right) \sin\left(\frac{\beta L}{2}\right)}{\beta^2\left[\cosh(\beta L) + \cos(\beta L)\right]} .$$

For the support reactions, we first note that

$$V(z) = \frac{C_1 k}{2\beta}\left[g_3(\beta z) + g_4(\beta z)\right] - \frac{C_2 k}{2\beta}\left[g_4(\beta z) - g_3(\beta z)\right] ,$$

from (7.47). Substituting for $C_1, C_2$, using the definitions (7.42) and setting $z=L/2$ we have, after some simplification,

$$V\left(\frac{L}{2}\right) = \frac{w_0}{2\beta}\left(\frac{\sinh(\beta L) + \sin(\beta L)}{\cosh(\beta L) + \cos(\beta L)}\right) ,$$

which is also the upward force at the support.

In the limit $\beta L \ll 1$,

$$\sinh(\beta L), \sin(\beta L) \to \beta L \;\; ; \;\; \cosh(\beta L), \cos(\beta L) \to 1 \;\; ;$$

$$\sinh\left(\frac{\beta L}{2}\right), \sin\left(\frac{\beta L}{2}\right) \to \frac{\beta L}{2}$$

and we obtain

$$M(0) \to -\frac{w_0\left(\frac{\beta L}{2}\right)^2}{2\beta^2} = -\frac{w_0 L^2}{8}$$

$$V\left(\frac{L}{2}\right) \to \frac{w_0(2\beta L)}{2\beta(2)} = \frac{w_0 L}{2} \; .$$

Both these results agree with the elementary bending theory for a beam on simple supports with a uniformly distributed load. The elementary theory provides quite a good approximation up to values of $\beta L$ that are not strictly small. For example, with $\beta L = 1$, the exact solution gives $M(0) = -0.120 w_0 L^2$ which differs from the elementary prediction by only 5%.

## 7.7 Short beams

We saw in the preceding example that when $\beta L \ll 1$, the solution reduced to that predicted by the elementary bending theory. In effect, the elastic deformations associated with bending were very small and the corresponding support forces generated by the foundation were negligible.

This will always be the case for a sufficiently short beam on an elastic foundation if it is also kinematically supported, i.e. if there are rigid discrete supports at the ends or elsewhere, sufficient to prevent rigid body motion if the elastic foundation were removed. The solution is then very simple. Once we have verified that the beam parameters satisfy[5] the condition $\beta L \ll 1$, we just ignore the elastic foundation and solve the remaining conventional beam problem.

If the beam is not kinematically supported, but still satisfies the condition $\beta L \ll 1$, the foundation forces can be determined by assuming that the beam translates vertically and rotates as a rigid body. The unknown parameters defining this rigid body motion are then found by requiring that the complete beam be in equilibrium.

An important special case concerns beam-like structures floating on the surface of a liquid.

[5] For practical purposes, $\beta L < 1$ is generally sufficient to ensure that the approximation will be sufficiently accurate.

**Example 7.5**

*Figure 7.21(a) shows a uniform rectangular beam of length L and weight W floating on the surface of a liquid and loaded by a concentrated force F. Assuming that the beam is short ($\beta L \ll 1$), find the maximum bending moment in the beam. How large a force F is required to cause one end of the beam to lift out of the liquid.*

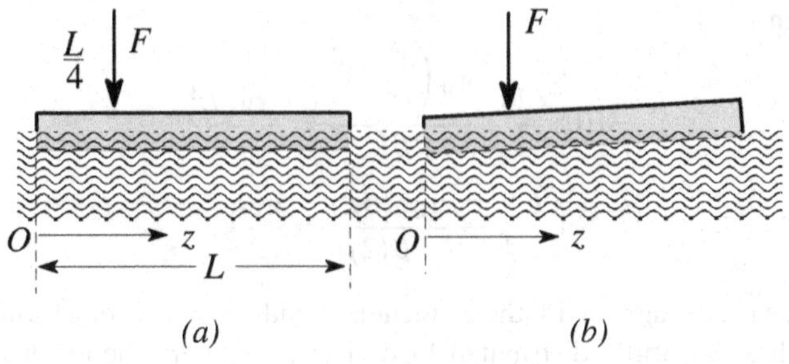


*Figure 7.21*

We first treat the beam as a rigid body, in which case its displacement must be a linear function of $z$ — i.e.

$$u(z) = A + Bz\,,$$

where $A, B$ are unknown constants. Figure 7.21 *(b)* shows the displaced configuration of the beam under load.

It follows that the buoyancy force must be

$$p(z) = k(A + Bz)\,,$$

where $k$ is a constant. If the beam is to be in equilibrium, we must have

$$F + W + \int_0^L p(z)dz = 0 \;\;;\;\; \frac{FL}{4} + \frac{WL}{2} + \int_0^L zp(z)dz = 0\,,$$

where we follow the convention of defining $u(z)$ positive upwards and hence $p(z)$ positive downwards [see Figure 7.3 *(a)*]. Substituting for $p(z)$ and evaluating the integrals, we obtain

$$kAL + \frac{kBL^2}{2} = -F - W \;\;;\;\; \frac{kAL^2}{2} + \frac{kBL^3}{3} = -\frac{FL}{4} - \frac{WL}{2}\,,$$

with solution

$$A = -\frac{5F}{2kL} - \frac{W}{k} \;\;;\;\; B = \frac{3F}{kL^2}$$

and

$$p(z) = -\frac{F}{L}\left(\frac{5}{2} - \frac{3z}{L}\right) - \frac{W}{L}\,.$$

The beam will lift from the liquid if the predicted value of $p > 0$ and this will occur first at $z = L$ if $F > 2W$.

The last term $(-W/L)$ in the expression for $p(z)$ simply cancels the self-weight $w(z)=W/L$ and hence makes no contribution to the bending moment in the beam. The remaining loads are shown in Figure 7.22*(a)* and we can sketch the shear force and bending moment diagrams as in Figure 7.22*(b)*.

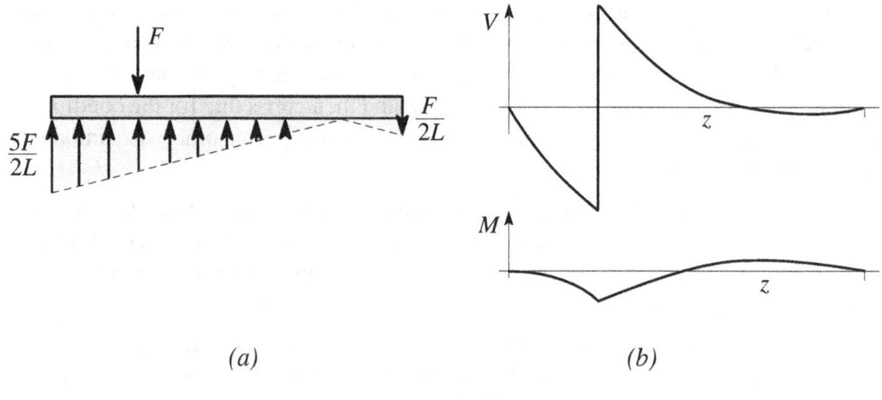


*Figure 7.22*

It is clear from the bending moment diagram that the maximum bending moment will occur under the load at $z=L/4$ and is

$$|M_{\max}| = \int_0^{L/4} zp(z)dz = \frac{F}{2}\int_0^{L/4}\left(\frac{5z}{2} - \frac{3z^2}{L}\right) = \frac{FL^2}{32} \;.$$

## 7.8 Summary

In this chapter, we have considered problems in which a loaded beam is supported by an elastic foundation that provides a restoring force proportional to the local displacement. The theory of beams on elastic foundations has traditionally been used in civil engineering to model structures and railway tracks on soil foundations. However, this formalism is also a useful way of describing systems such as thin metal components bonded to rubber, stiff fibres in a relatively flexible matrix or cartiledge layers (the foundation) attached to bones (the beam).

The resulting deformation is localized, the displacement decaying exponentially with distance from the loaded region at a rate which depends upon the foundation modulus and the flexural rigidity of the beam. The first stage in any engineering design problem of this kind should always be to calculate the decay rate $\beta$, since the nature of the solution depends qualitatively on its relation to the length of the beam or to the length of the loaded region. The length scale associated with this decay also provides criteria to assess whether a non-local support or a system of periodic discrete supports can reasonably be approximated by a uniform local (Winkler) foundation.

For continuously loaded beams, the solution can be constructed as the sum of a particular solution, and a homogeneous solution which describes the localized deformation associated with the end conditions. A general particular solution can be written as an integral [equation (7.38)], but for many practical loading scenarios, the result is more easily obtained directly from the governing differential equation (7.5).

For long beams, where $\beta L \gg 1$, the load is predominantly carried by the foundation and bending effects are important only in a region near the supports or a discontinuity in loading. The problem is then conveniently treated by finding a particular solution for similar loading of an infinite beam and then correcting for the conditions at each end separately, using the results for the end loading of an otherwise unloaded semi-infinite beam (7.11).

For short beams ($\beta L \ll 1$) the foundation has little influence on the deformation if the beam is kinematically supported. Otherwise, the beam compresses the foundation essentially as a rigid body and the arbitrary constant(s) defining this rigid-body motion can be determined from kinematic and equilibrium arguments.

For design purposes, reasonable accuracy can be expected using the long beam approximation for $\beta L > 3$ and the short beam approximation for $\beta L < 1$. For beams of intermediate length, there is interaction between the conditions at the two ends of the beam, requiring the solution of four simultaneous equations, though some simplification can often be achieved by locating the origin of coordinates at the midpoint of the beam and using symmetry.

## Further reading

A.P. Boresi, R.J. Schmidt, and O.M. Sidebottom (1993), *Advanced Mechanics of Materials,* John Wiley, New York, 5th edn., §§10.1–10.6.
R.D. Cook and W.C. Young (1985), *Advanced Mechanics of Materials*, Macmillan, New York, §§5.1–5.6.
A.C. Ugural and S.K. Fenster (1995), *Advanced Strength and Applied Elasticity*, Prentice-Hall, Eaglewood Cliffs, NJ, 3rd edn., Chapter 9.

## Problems

### Sections 7.1–7.4

**7.1.** A semi-infinite beam of flexural rigidity $EI$ is supported on an elastic foundation of modulus $k$ and simply supported at the end. A moment $M_0$ is applied to the beam at the support. Find the reaction induced at the support and the slope at the end.

**7.2.** A semi-infinite steel beam of second moment of area $I = 5 \times 10^6$ mm$^4$ is supported on an elastic foundation of modulus $k = 10$ MPa and loaded by a moment of 20 kNm at the end. Find the slope and deflection at the end ($E_{steel} = 210$ GPa).

**7.3.** A long steel I-beam of second moment of area $I_x = 1.2\times 10^6\ \text{mm}^4$ is supported on an elastic foundation of modulus 1.2 MPa. Find the force required to deflect the end of the beam by 5 mm ($E_{\text{steel}} = 210$ GPa).

**7.4.** A long steel I-beam of second moment of area $I_x = 3.0\times 10^6\ \text{mm}^4$ is supported on an elastic foundation of modulus 4.0 MPa. An end force is applied so as to cause an end deflection of 2 mm. Find the force required and the resulting maximum bending moment in the beam ($E_{\text{steel}} = 210$ GPa).

**7.5.** Figure P7.5 shows the cross section of an oil tanker. Obtain an order of magnitude estimate of the flexural rigidity of the section by approximating it by a 70 ft diameter hollow cylinder with a 2 inch thick steel wall ($E = 30\times 10^6$ psi). Hence estimate the decay length $l_0$ for the tanker, considered as a beam on an elastic foundation. Would a typical tanker be 'long'or 'short' according to the criteria of §7.3? The density of sea water is 62 lbs/ft$^3$.

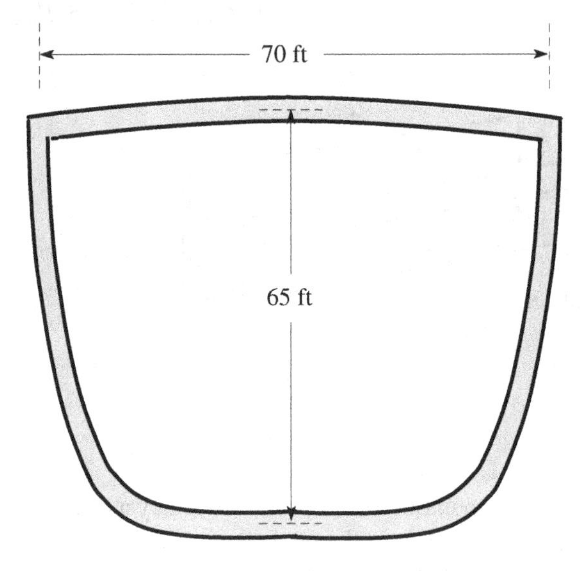


*Figure P7.5*

**7.6.** Figure P7.6 shows the cross section of a semi-infinite $C254\times 45$ steel channel which rests on an elastic foundation of modulus $k = 10$ MPa and is loaded by a downward force of 20 kN at the end. The second moment of area of the channel for bending about the horizontal axis is $I_x = 1.64\times 10^6\ \text{mm}^4$ and the centroid $C$ is located at a distance 18 mm from the bottom of the section. Find the deflection at the free end and the location and magnitude of the maximum tensile stress in the beam. ($E_{\text{steel}} = 210$ GPa)

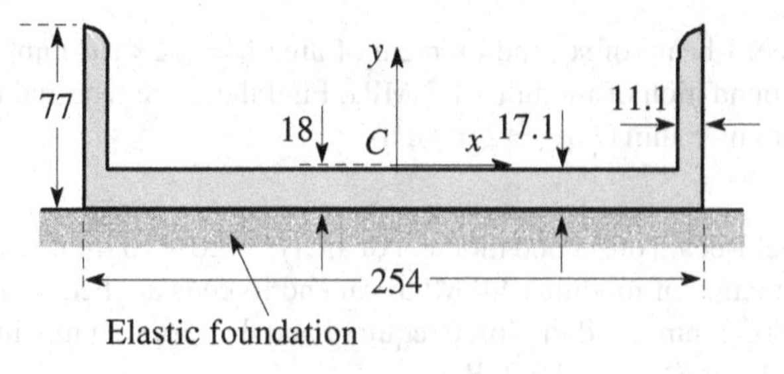


all dimensions in mm

*Figure P7.6*

**7.7.** A mounting assembly involves a 2 mm steel strip bonded to a 4 mm rubber layer, which in turn is supported by a rigid base, as shown in Figure P7.7. Estimate the modulus of the rubber foundation, assuming that the out-of-plane stress $\sigma_{xx}$ is always zero. Hence, estimate the decay length $l_0$ for the assembly, considered as a steel beam on a rubber foundation ($E_{\mathrm{rubber}} = 1.4$ MPa, $\nu_{\mathrm{rubber}} = 0.5$, $E_{\mathrm{steel}} = 210$ GPa, $\nu_{\mathrm{steel}} = 0.3$).

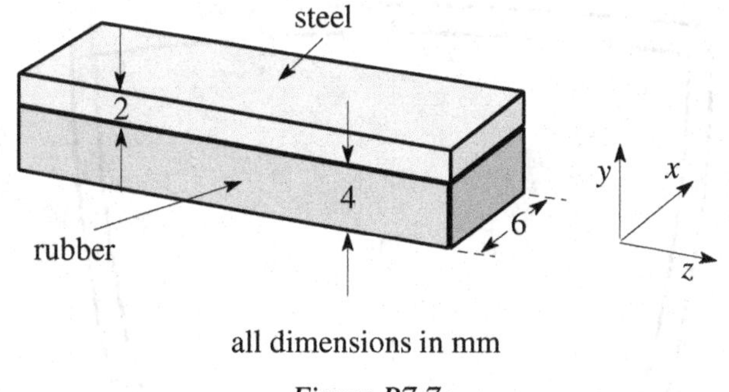


all dimensions in mm

*Figure P7.7*

**7.8.** An infinite beam on an elastic foundation is loaded by a concentrated force $F_0$. What will be the percentage error in the predictions of (i) the maximum deflection and (ii) the maximum bending moment, if the modulus $k$ of the foundation is over-estimated by 50%?

**7.9.** An air duct consists of an aluminium thin-walled tube of diameter 1 ft and wall thickness 1/8 in. It is supported by a set of steel cables of diameter 1/8 in and length 4 feet, with an axial spacing of 3 feet, as shown in Figure P7.9. Find the characteristic length $l_0$ for this system and comment on whether it can reasonably be analyzed as a beam on an elastic foundation ($E_{\mathrm{steel}} = 30 \times 10^6$ psi, $E_{\mathrm{aluminium}} = 11 \times 10^6$ psi).

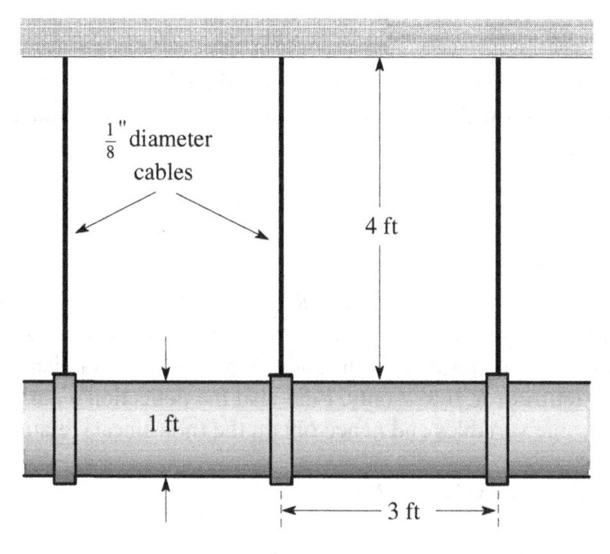


*Figure P7.9*

**7.10.** Figure P7.10 shows a long steel beam of $80 \times 80$ mm square cross section supported on discrete springs of stiffness 1 kN/mm, spaced 100 mm apart. Can this support reasonably be considered as an elastic foundation?

The beam is loaded by a force of 20 kN as shown. Find the maximum tensile stress in the beam and the maximum force carried by any individual spring. ($E_{steel} =$ 210 GPa).

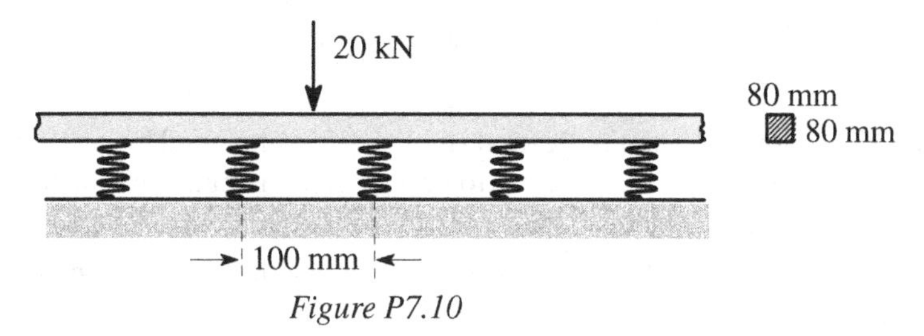


*Figure P7.10*

**7.11.** In Problem 7.10 additional springs are provided to reduce the axial spacing to 50 mm, other parameters remaining the same. Find the maximum tensile stress in the beam and the maximum force carried by any individual spring.

**Section 7.5**

**7.12.** An infinite beam of flexural rigidity $EI$ is supported by an elastic foundation of modulus $k$ and is subject to a sinusoidal load

$$w(z) = w_0 \sin(az) ,$$

where $a$ is a constant. Find the displacement $u$ and the bending moment $M$ as functions of $z$ and hence show that for a given value of the load amplitude $w_0$, the maximum bending moment occurs when

$$a = \sqrt[4]{\frac{k}{EI}} .$$

**7.13**.** A rigid cylinder of large radius $R$ is pressed against an infinite beam of flexural rigidity $EI$ supported by an elastic foundation of modulus $k$, as shown in Figure P7.13. Find an expression for the force $F$ required to establish a contact area of length $2a$ in terms of $R, \beta, k, a$ only. Find also the deflection $u_0$ at the mid-point in terms of the same variables and hence obtain the (non-linear) relation between $F$ and $u_0$.

What happens at low values of $F$?

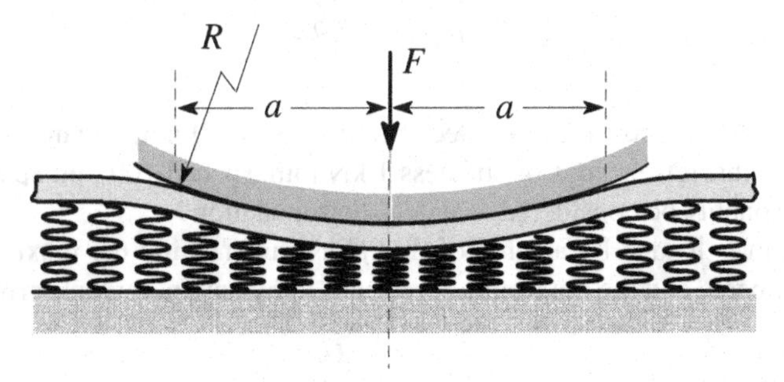


*Figure P7.13*

**7.14*.** An infinite beam of flexural rigidity $EI$ is supported by an elastic foundation of modulus $k$ and is loaded by a force $F_0$, which is uniformly distributed over the region $-L<z<L$ — i.e. $w(z)=F_0/2L$ in this range.

Find expressions for the displacement and the bending moment as functions of $z$ (i) in $-L<z<L$ and (ii) in $z>L$.

**Hint:** This problem can be solved by integration, using equation (7.38), but an easier method is to represent the loading as the superposition of (i) a load $w(z) = F_0/2L$ in $-L<z<\infty$ and (ii) a negative load $w(z) = -F_0/2L$ in $L<z<\infty$. These separate problems are both similar to that treated in Example 7.3, except for a change of origin.

**7.15.** Write a computer program to locate the maximum bending moment for Problem 7.14 (by finding the points where $V=0$).

Run the program for various values of $\beta L$ and plot a graph of $M_{max}/M_0$ as a function of $\beta L$, where $M_0$ is the value of $M_{max}$ for $L=0$ — i.e. $M_0$ is the moment under a concentrated force $F_0$ and is given by equation (7.37).

Note that there are many places where $V = 0$, both under the load and in the unloaded region. Your program must determine at which one the magnitude of the bending moment is greatest.

**7.16.** Figure P7.16 shows an infinite beam of flexural rigidity $EI$, supported by an elastic foundation of modulus $k$ and subjected to a linearly varying load

$$w(z) = w_0 z$$

in $z > 0$. Find the bending moment in the beam as a function of $z$. Hence show that the maximum bending moment occurs at $O$ and find its magnitude.

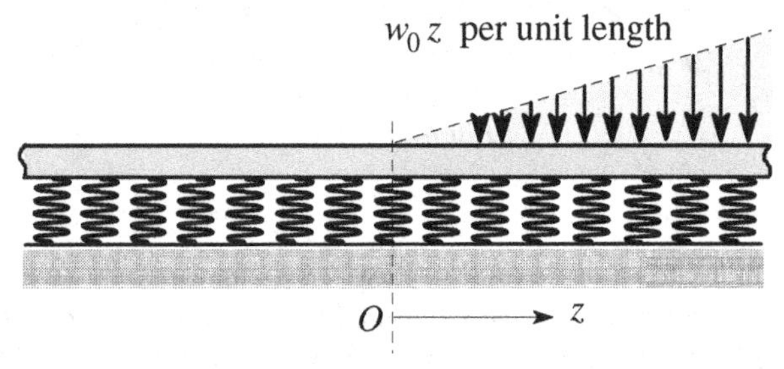


*Figure P7.16*

**7.17.** A semi-infinite beam of flexural rigidity $EI$ is supported by an elastic foundation of modulus $k$ and built in at the end, as shown in Figure P7.17. The beam is subjected to a distributed load

$$w(z) = w_0 e^{-az} .$$

Find the bending moment at the support in terms of $w_0, a, \beta, k$ only.

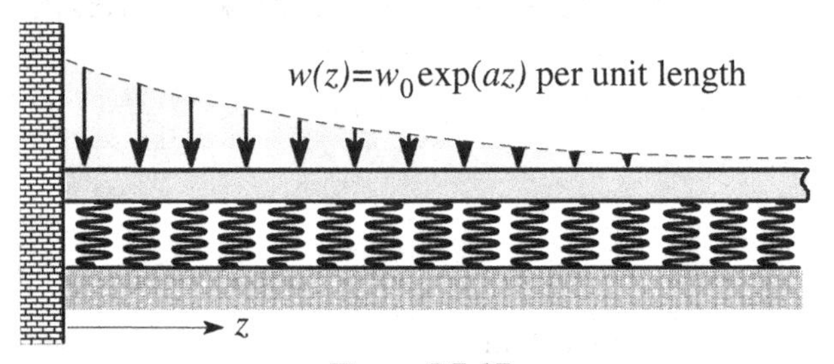


*Figure P7.17*

**7.18**.** An infinite beam of flexural rigidity $EI$ is supported on an elastic foundation whose modulus is $k_0$ in $z > 0$ and $2k_0$ in $z < 0$. The beam is subjected to a uniformly distributed load $w_0$ per unit length, as shown in Figure P7.18. Find the location and magnitude of the maximum bending moment.

**Hint:** Split the beam at $z=0$ and then determine the moment and shear force needed there to re-establish continuity of slope and deflection. Notice however that $\beta$ takes different values on the two sides, so the problem is not symmetrical, in contrast to Example 7.3.

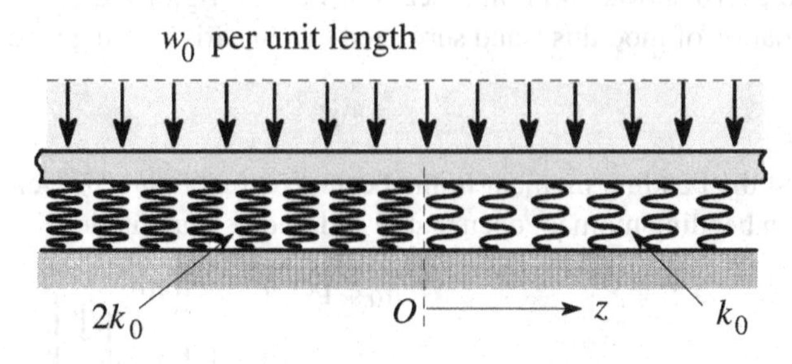


*Figure P7.18*

**7.19.** A semi-infinite beam of flexural rigidity $EI$ is supported by an elastic foundation of modulus $k$ and loaded by a concentrated force $F$ a distance $a$ from the end, as shown in Figure P7.19. Find the deflection under the load, the deflection at the end and the bending moment under the load.

**Hint:** Use the solution for a force on an infinite beam, make a cut at the point $z=-a$ and correct for the end conditions.

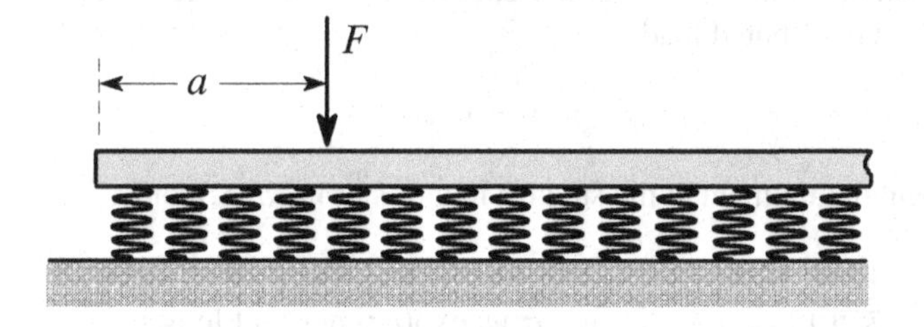


*Figure P7.19*

**Sections 7.6, 7.7**

**7.20.** A rectangular wooden plank of width 12 in, thickness 2 in and length 30 ft floats on the surface of a lake. The wood has specific gravity 0.8 and Youngs modulus $E=1.6\times10^6$ psi and the density of water is 60 lb/ft$^3$. A man weighing 150 lbs steps carefully onto the plank at one end. Will he get his feet wet?

**7.21.** Figure P7.21 shows a finite beam of flexural rigidity $EI$ loaded by a concentrated force $F_0$ at the mid-point. The beam is supported on an elastic foundation of modulus $k$, but is otherwise unsupported. Find the displacement and the bending moment at the mid-point.

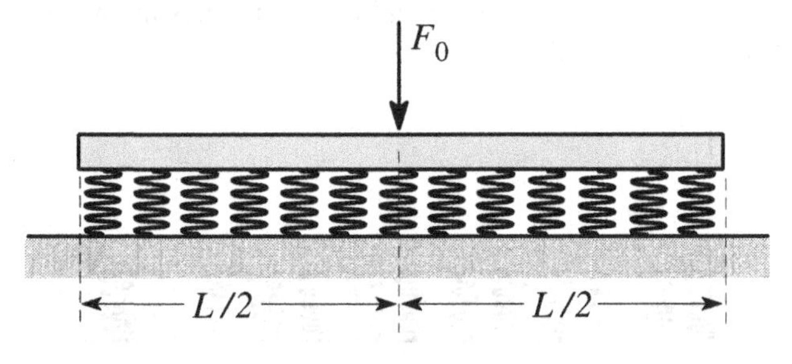


*Figure P7.21*

**7.22.** Figure P7.22 shows a beam of length 2 m and flexural rigidity $EI = 2 \times 10^6$ Nm$^2$, subjected to a uniform load $w_0 = 4$ kN/m. The beam is built in at both ends and supported on an elastic foundation of modulus $k = 0.8$ MPa. Find the bending moments at the support and at the mid-point and the maximum deflection.

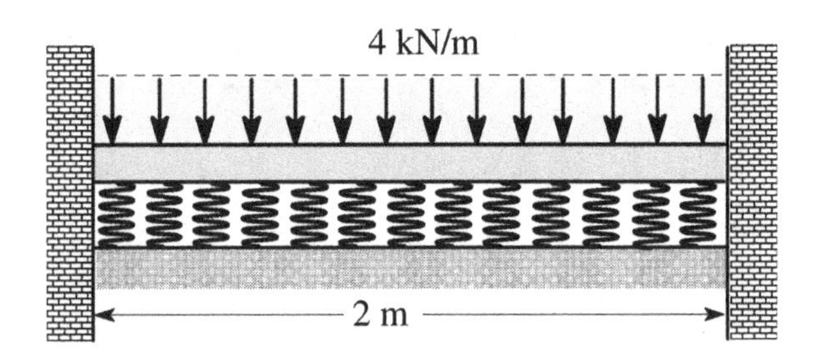


*Figure P7.22*

**7.23.** The oil tanker of Problem 7.5 is 500 feet long and when fully laden it rides in still water so that the lowest point of the hull is 50 feet below the water line. In a gale, the tanker has to traverse waves of peak-to-trough height of 18 feet and wavelength 50 feet. Estimate the maximum tensile stress generated in the hull. Will the stresses be worse when cutting through the waves at right angles, or at an angle?

**7.24.** A finite beam of length $L$ and flexural rigidity $EI$ is built in at its ends and supported on a foundation of modulus $k$. It is subjected to a uniform load $w(z) = w_0$. Find the deflection at the centre and the restraining moments at the supports for the case where $\beta L = 1$.

**7.25.** A beam of length $L$ is pinned at one end and supported on an elastic foundation of modulus $k$, such that $\beta L \ll 1$. The beam is loaded by a force $F$ at the end, as shown in Figure P7.25. Find the location and magnitude of the maximum bending moment and the deflection at the loaded end.

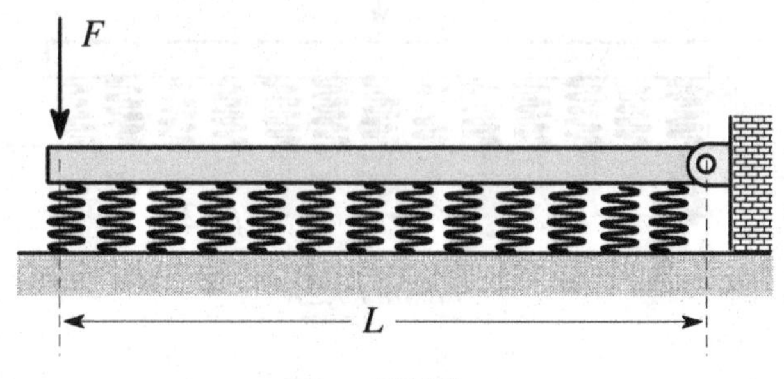


*Figure P7.25*

# 8

# Membrane Stresses in Axisymmetric Shells

A *shell* is any structure or part of a structure in which one local dimension is much smaller than the other two. Shells abound in all branches of engineering and in the natural world. Examples include gas tanks, thin-walled pipes and air ducts, grain hoppers, car bodies, aircraft wings and fuselages, window panes, eggshells, wine-glasses, concrete roofs and the human skull.

A shell which is plane in the undeformed state is known as a *plate*. Plates are generally much less efficient than curved shells as structural elements, being more flexible and less capable of transmitting transverse loads (i.e. loads normal to the plane) without failure. You can verify this by pushing against a plate glass window or the plane side of a metal filing cabinet. You will find that it is quite easy to produce an observable deflection. The same material formed into a curved shell (e.g. a car body panel or a drinking glass) will sustain an equal transverse load without observable deformation.

The basic idea behind the analysis of shells is that the stresses can't vary much across the thin dimension and hence we can approximate the stress distribution *across the thickness* by a Taylor series. When the shell is very thin, only the first term of this series will be important — this is the 'membrane theory', in which the stresses are assumed to be uniform through the thickness. In many problems — particularly those involving shells with discontinuities or built-in boundary conditions — it is necessary to include both the first and second terms of the series. This constitutes the 'bending theory of shells'.

It might be helpful to draw an analogy with beams, which of course are structures in which *two* dimensions are much smaller than the third. The 'membrane theory of beams' assumes that the axial stress $\sigma_{zz}$ is uniform across the section and hence corresponds only to the transmission of an axial force passing through the centroid. The addition of one extra term in the Taylor series permits the stress to vary linearly across the section and defines the bending theory of beams.

In this chapter, we shall restrict attention to the membrane theory of shells, which therefore implies that bending moments are everwhere negligible. It follows that *derivative* of the bending moment must also be everywhere negligible and hence

J.R. Barber, *Intermediate Mechanics of Materials*, Solid Mechanics and Its Applications 175, 2nd ed., DOI 10.1007/978-94-007-0295-0_8, © Springer Science+Business Media B.V. 2011

there can be no shear force or transverse shear stress, by extension of equation (6.1). In other words, the only non-zero stresses are in the local (tangential) plane of the shell and are uniform through the thickness.

Despite these restrictions, we shall find that the curvature of a membrane shell permits it to transmit a wider range of loads than either a straight beam or a plate under similar restrictions, making the shell a useful and versatile lightweight structural member. We shall illustrate this here for the particular case of axisymmetric shells under axisymmetric loading.

## 8.1 The meridional stress

Figure 8.1 shows the cross section of an axisymmetric shell subjected to various kinds of axisymmetric loads. In particular, the shell may contain a liquid of density $\rho_l$ and a gas at pressure $p$, there may be resultant axial forces $F$ from the attached ductwork and reactions $R$ at the support, and the density $\rho_s$ of the shell material itself may cause significant loading.

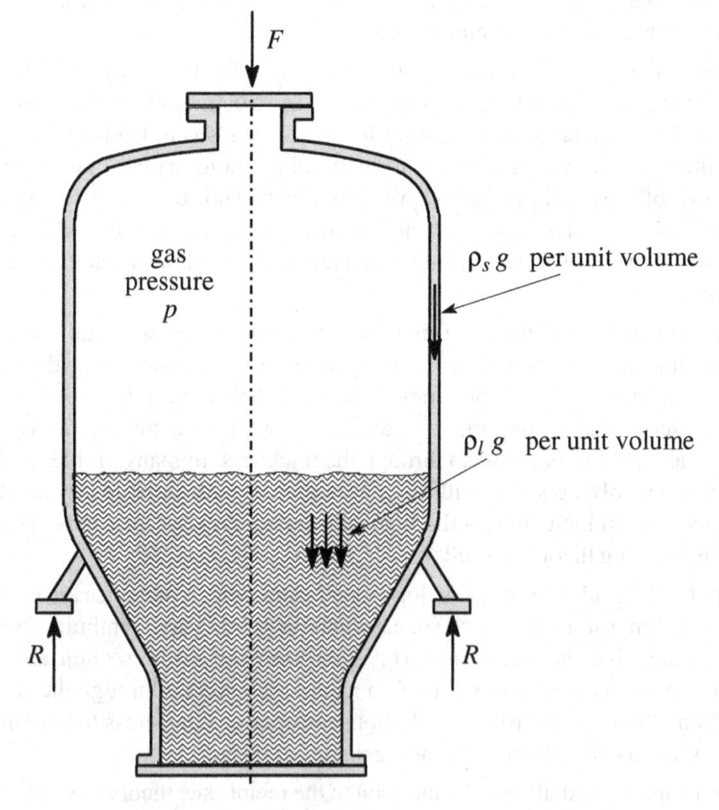


*Figure 8.1: Loading of an axisymmetric shell*

Suppose we now separate the shell into two parts by making a cut perpendicular to the axis of symmetry. The free body diagram of one such part is shown in Figure 8.2. We anticipate a membrane stress $\sigma_l$ at the cut as shown. Notice that the cut is made perpendicular to the shell wall,[1] so as to expose the membrane stress which is tangential to the wall.

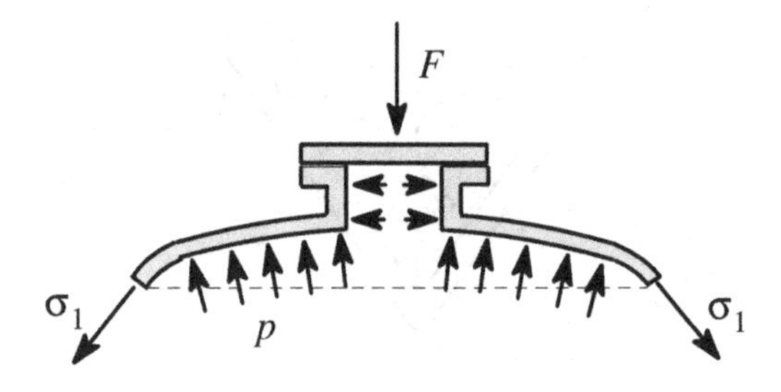


*Figure 8.2: Free body diagram exposing the meridional stress $\sigma_l$*

The segment of shell in Figure 8.2 must be in equilibrium under the action of the forces shown and we can obtain an equilibrium equation by summing the forces in the axial direction. If the shell and the applied loads are axisymmetric, the membrane stress $\sigma_l$ will be the same all around the cut, so this equation will be sufficient to determine $\sigma_l$, which is known as the *meridional stress* or the *longitudinal stress*.[2]

**Example 8.1**

*Figure 8.3 shows a conical tank, supported at the top and filled with water of density $\rho$ to a height $h$ above the vertex. The cone has semi-angle $\alpha$ and the wall thickness is $t$. Find the meridional stress $\sigma_l$ as a function of $z$ and identify the location and magnitude of the maximum value. The weight of the shell may be neglected.*

To find $\sigma_l$ at height $z$, we make the cut $A-A$ shown in Figure 8.3, taking care to go through the shell at right angles. The simplest procedure is to 'cut' through the water as well, leaving the free-body diagram of Figure 8.4. Three forces act on the system shown in this figure, resulting from:-

(i) the meridional stress $\sigma_l$ acting on the cut shell surface;
(ii) the hydrostatic pressure $\rho g(h-z)$ due to the head of water above $A-A$ acting on the water surface $A-A$; and
(iii) the weight $W$ of the cone of water left below $A-A$ and shown in Figure 8.4.

[1] One way to remember this is to imagine actually making the cut with a hacksaw. The cut to make is the one that requires the least effort, because it goes through the minimum wall thickness.

[2] Because on a globe regarded as a shell, $\sigma_l$ would act in the direction of a line of longitude, also known as a meridian.

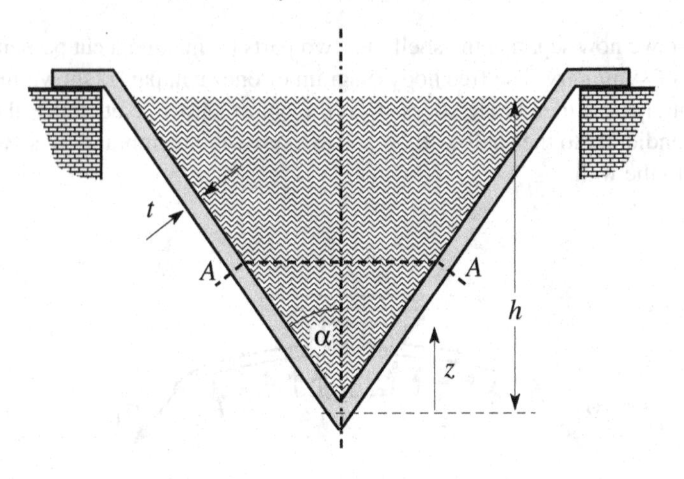


*Figure 8.3: The conical tank*

To determine $W$, we note that the volume of a cone of height $b$ and base radius $a$ is $V = \pi a^2 b/3$. Here, $b = z$ and $a = z\tan\alpha$, giving

$$V = \frac{1}{3}\pi(z\tan\alpha)^2 z = \frac{1}{3}\pi z^3 \tan^2\alpha$$

and

$$W = \rho g V = \frac{1}{3}\pi\rho g z^3 \tan^2\alpha \; .$$

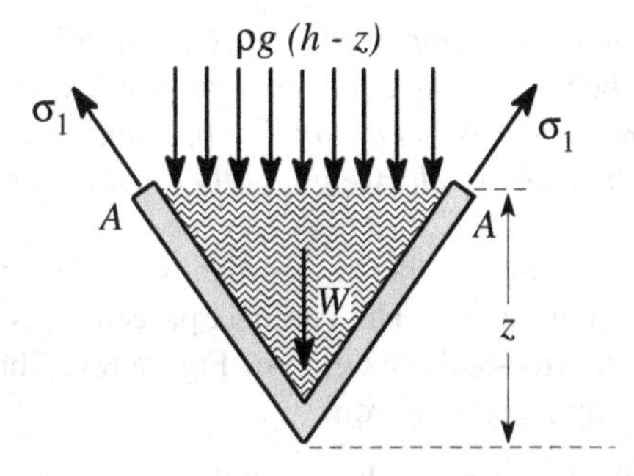


*Figure 8.4: Free-body diagram of the system below the cut A–A*

The stress $\sigma_1$ acts on the area of the cut through the shell wall, which is of thickness $t$ and circumferential length $2\pi a = 2\pi z\tan\alpha$. Finally, the water pressure $\rho g(h-z)$ acts over a circular area at $A-A$ of radius $a = z\tan\alpha$ whose area is therefore $\pi(z\tan\alpha)^2$.

The three forces must sum to zero and hence

$$\sigma_1(2\pi z t\tan\alpha)\cos\alpha - \rho g(h-z)\pi(z\tan\alpha)^2 - \frac{1}{3}\pi\rho g z^3\tan^2\alpha = 0 \, , \qquad (8.1)$$

where the factor $\cos\alpha$ in the first term results from resolution of the force due to $\sigma_1$ into the vertical (axial) direction.

Solving this equation for the meridional stress $\sigma_1$, we obtain

$$\sigma_1 = \frac{\rho g z \tan\alpha}{2t\cos\alpha}\left(h - \frac{2}{3}z\right) .$$

We note that $\sigma_1$ is zero at the vertex $z=0$ and elsewhere is always positive. The maximum value occurs when

$$\frac{d\sigma_1}{dz} = 0$$

which requires

$$h - \frac{4}{3}z = 0 \quad \text{or} \quad z = \frac{3h}{4} .$$

At this point, we have

$$\sigma_1^{\max} = \frac{3\rho g h^2 \tan\alpha}{16t\cos\alpha} .$$

### 8.1.1 Choice of cut

In Example 8.1, we made the cut in such a way as to leave a conical volume of water in the free-body diagram of Figure 8.4, but any other axisymmetric cut could have been used, two alternatives being illustrated in Figure 8.5 *(a,b)*.

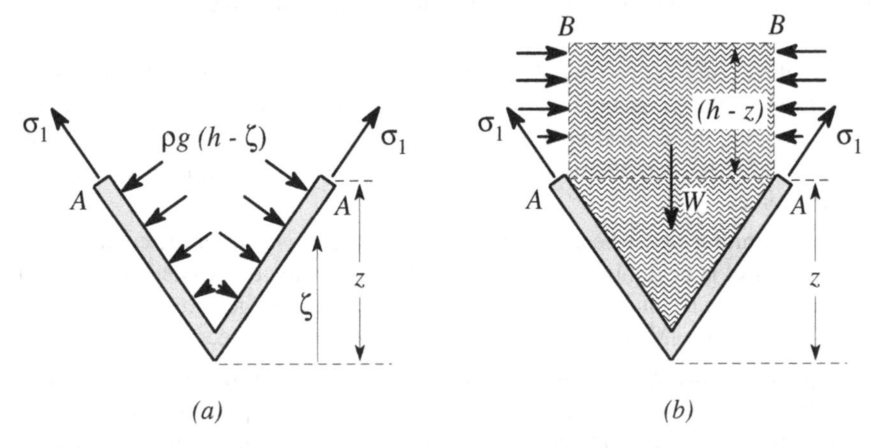


*Figure 8.5: Alternative cuts for determining* $\sigma_1$

In Figure 8.5 *(a)*, we show the shell segment only, implying a cut which passes between the water and the shell surface. In this case, the pressure exerted by the water on the shell must be shown, since it results in an external force on the shell segment. The resultant axial force due to this pressure can be found by considering the element of shell surface between height $\zeta$ and $\zeta+\delta\zeta$ above the vertex, multiplying by the local pressure $\rho g(h-\zeta)$, resolving the resulting elemental force into the axial direction and summing by integration. The final result will be found to be equal to the sum of terms (ii) and (iii) in the original derivation, but the calculation is significantly more tedious, which is why we used the cut of Figure 8.4.

Another alternative cut, shown in Figure 8.5 *(b)*, passes through the shell and then goes vertically upwards to the water surface, so as to include an additional cylindrical volume of water above the plane $AA$ in the free-body diagram. This cylinder is subjected to pressure from the rest of the water acting on the cylindrical surface $AB$, but since these forces are horizontal, they do not feature in the vertical equilibrium equation. Thus, the only two terms in the equilibrum equation are those due to $\sigma_1$ and to the weight of water in the free-body diagram [(i) and (iii) of the preceding solution]. However, the weight of water $W$ is now increased by the weight of the cylinder $AB$ and it is easily verified that this is equal to the missing term (ii). The volume of the cylinder is

$$V_{\rm cyl} = \pi(z\tan\alpha)^2(h-z) ,$$

so its weight is

$$W_{\rm cyl} = \pi\rho g(z\tan\alpha)^2(h-z) ,$$

which is exactly equal to the second term in the equilibrium equation (8.1).

In general, any axisymmetric cut can be used to determine the meridional stress, provided it cuts through the shell at only one location. All cuts will yield the same equilibrium equation, but a careful choice will sometimes simplify the calculations. A good 'default option' is to make a plane cut perpendicular to the axis,[3] as in Figure 8.4.

The three cuts defined by the free-body diagrams of Figure 8.4 and Figure 8.5 *(a,b)* yield the same final equilibrium equation because the volumes of water involved — i.e. the conical volume in Figure 8.4 and in the cone and the cylinder in Figure 8.5 *(b)* — are themselves in equilibrium under the action of gravity and pressure forces.

In the limiting case where the density and hence the weight of the contents of a pressure vessel is negligible, the pressure will be the same everywhere and we can conclude that the resultant of a uniform pressure acting on a general curved surface is the same for all surfaces terminating on the same closed curve.

In particular, the uniform pressure $p$ acting on the shell surface of Figure 8.6 *(a)* is statically equivalent to a force $pA_c$, where $A_c$ is the area of the plane surface $A-A$ and the line of action of this force passes through the centroid of $A_c$ and is perpendicular to the plane $A-A$. This result can be useful outside the context of the membrane theory of shells. For example, we can use it to deduce that the distributed load on the curved beam of Figure 8.6 *(b)* is statically equivalent to a vertical force $2w_0R$ acting through the centre $O$.

[3] Except in passing through the shell itself, where the cut must be normal to the local shell surface.

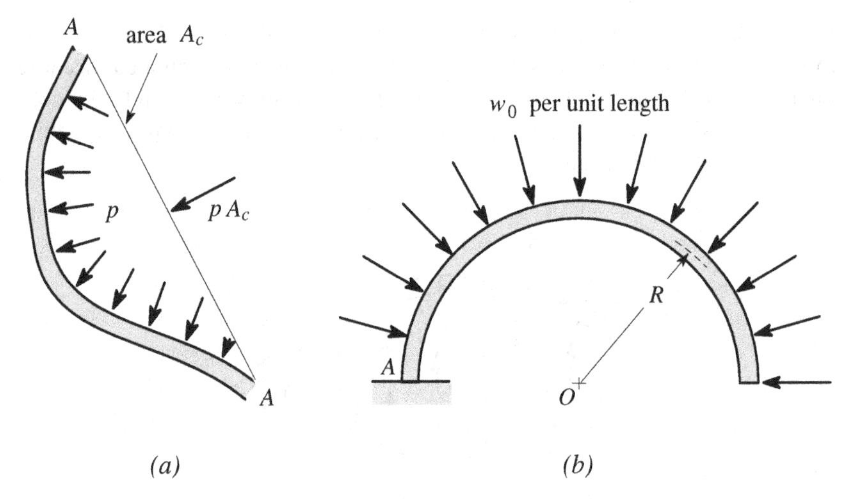


*Figure 8.6: (a) uniform pressure on a general shell segment, (b) uniform normal loading on a curved beam*

## 8.2 The circumferential stress

Since the system is axisymmetric, there can be no shear stresses on the meridional surface and hence $\sigma_1$ is a principal stress. The other principal stress $\sigma_2$ in the local tangent plane to the shell is known as the *circumferential stress* or the *hoop stress*. On a globe, it would act in the direction of a line of lattitude — i.e. in the East-West direction.

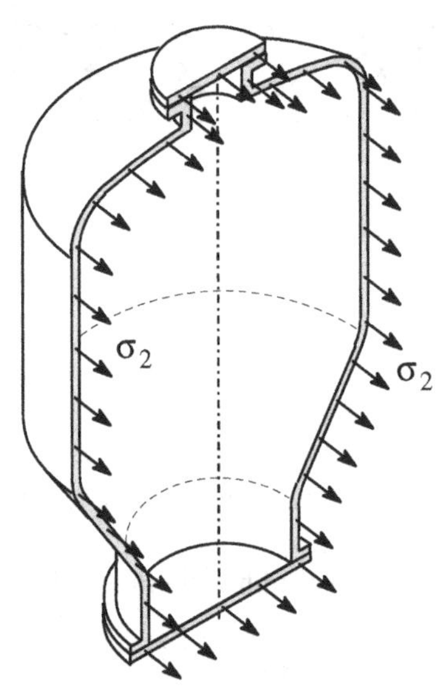


*Figure 8.7: The circumferential stress $\sigma_2$*

We can expose $\sigma_2$ by making any plane cut that includes the axis of symmetry, as shown in Figure 8.7, but we cannot generally determine it from such a cut because $\sigma_2$ is not necessarily uniform over the cut surface. However, we can find it by considering the equilibrium in the through-thickness direction of the small rectangular element of shell shown in Figure 8.8.

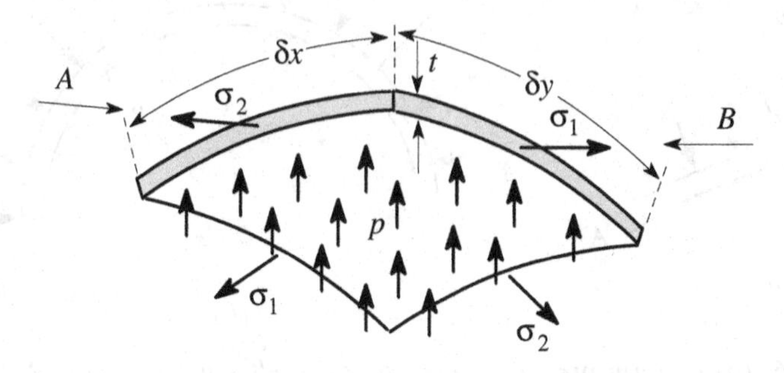


*Figure 8.8: Equilibrium of a small rectangular element of shell*

We suppose that the shell element is loaded only by an internal pressure $p$, acting normal to the shell, and the membrane stresses $\sigma_1, \sigma_2$. The pressure exerts a force $p\delta x\delta y$ in the direction normal to the shell. To determine the forces due to $\sigma_1, \sigma_2$, we first draw edge views of the element looking in directions $A, B$, as shown in Figure 8.9 *(a,b)*, respectively. In general, the curvature of the shell will be different in the two views, so we denote the corresponding radii by $R_1, R_2$, respectively.

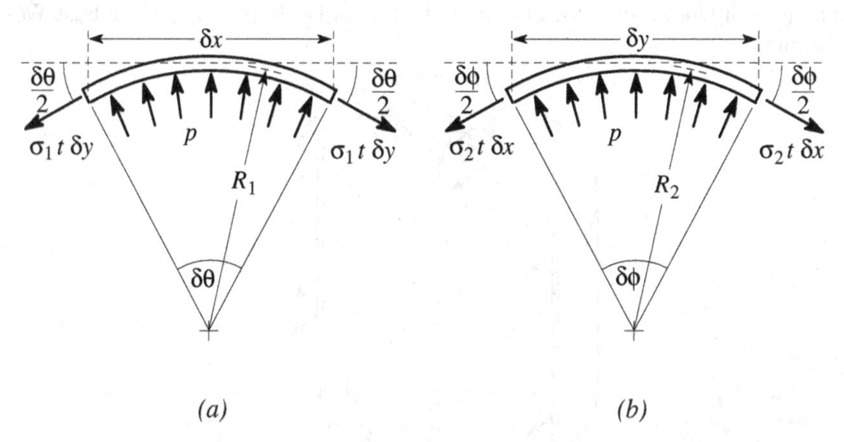


*Figure 8.9: View of the shell element in (a) direction A and (b) direction B*

As a result of the curvature, the length $\delta x$ in Figure 8.9 *(a)* subtends an angle

$$\delta\theta = \frac{\delta x}{R_1}$$

at the centre of curvature. The stress $\sigma_1$ acting on the edges of the element corresponds to forces $\sigma_1 t\delta y$ on each edge in Figure 8.9*(a)* and these forces are not horizontal in the figure because of the curvature of the shell. They therefore have a resultant in the direction normal to the shell (vertically downwards in the figure) equal to

$$2\sigma_1 t\delta y \sin\left(\frac{\delta\theta}{2}\right) \approx \sigma_1 t\delta y\delta\theta = \frac{\sigma_1 t\delta x\delta y}{R_1} .$$

The approximation involved in replacing $\sin(\delta\theta/2)$ by $\delta\theta/2$ will be exact in the limit $\delta x \to 0$. A similar argument applied to Figure 8.9*(b)* shows that the forces $\sigma_2 t\delta x$ due to stress $\sigma_2$ have a resultant in the same direction (normal to the shell) of

$$\frac{\sigma_2 t\delta x\delta y}{R_2} .$$

Summing the forces along the common normal for the element $\delta x\delta y$, we therefore obtain

$$\frac{\sigma_1 t\delta x\delta y}{R_1} + \frac{\sigma_2 t\delta x\delta y}{R_2} - p\delta x\delta y = 0 ,$$

and hence

$$\frac{\sigma_1}{R_1} + \frac{\sigma_2}{R_2} = \frac{p}{t} . \tag{8.2}$$

The meridional stress $\sigma_1$ can always be calculated by using axial equilibrium arguments, as in §8.1, so equation (8.2) can be used to determine the circumferential stress $\sigma_2$.

### 8.2.1 The radii of curvature

Notice that the shell element has two radii of curvature[4] $R_1, R_2$, which appear in the two orthogonal views on $A$ and $B$ respectively (see Figure 8.9). These radii are generally different. For example, for the cylindrical shell of Figure 8.10*(a)*, the corresponding element [Figure 8.10*(b)*] will appear curved when looking in direction $B$ along the axis, giving $R_2 = a$. However, it will look straight when looking in direction $A$, so $R_1 = \infty$.

[4] Even a general (non-axisymmetric) element of shell has two orthogonal principal curvatures and corresponding principal radii $R_1, R_2$. This can be established by a transformation of coordinates mathematically analogous to those used in §2.1.1 and §4.4.1. For the axisymmetric shell, the principal directions must always align with the circumferential and axial directions.

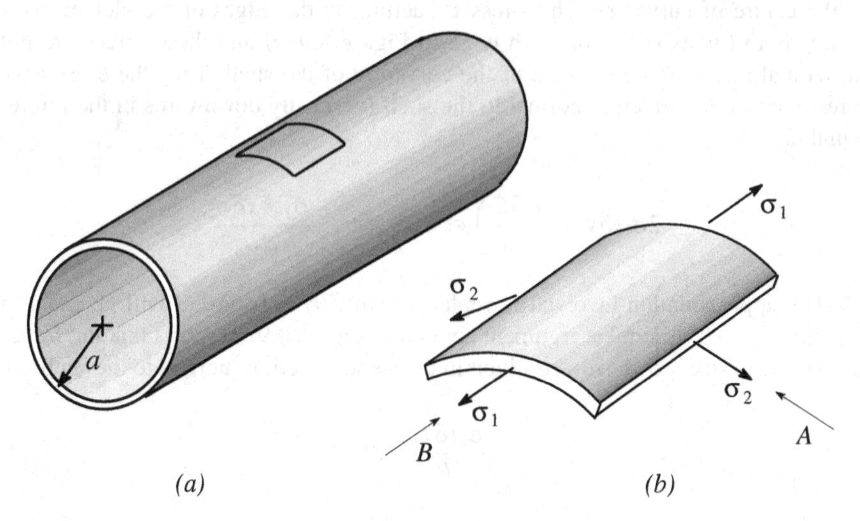


*Figure 8.10: (a) A cylindrical shell and (b) the corresponding rectangular shell element*

More generally, the radius $R_1$ is that which is seen in a cross-sectional view of an axisymmetric shell — i.e. when the shell is sectioned on a plane including the axis of symmetry. To determine the other radius $R_2$, we first note that both centres of curvature must lie on the common normal to the local piece of shell. Also, the axisymmetry demands that the centre $O_2$ must lie on the axis of symmetry. To find $O_2$ and hence $R_2$, we therefore extend the normal until it cuts the axis, as shown for various cases in Figure 8.11.

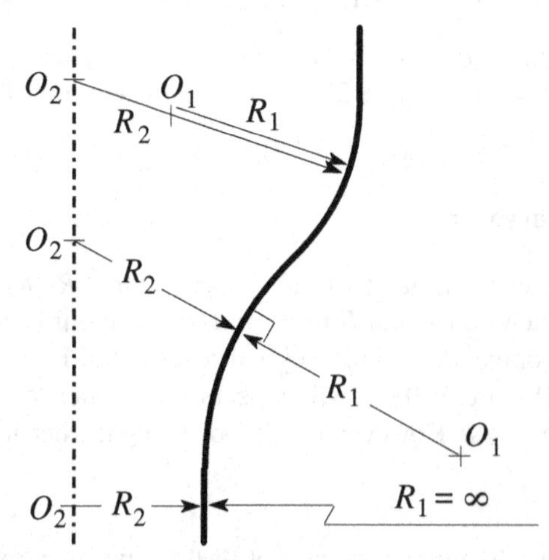


*Figure 8.11: Radii of curvature for the shell*

Take care not to confuse $R_2$ with the local cross-sectional radius $r$, shown in Figure 8.12*(a)*. In determining the meridional stress, $r$ will be the radius of the cross-sectional cut (e.g. $A-A$ in Figure 8.4) and it corresponds to a line perpendicular to the axis of symmetry. By contrast, the radius $R_2$ is perpendicular to the local tangent to the shell.

A particular case that demonstrates this distinction is the sphere. Everyone would agree from symmetry that a sphere of radius $a$ has its principal radii equal at all points — i.e. that $R_1 = R_2 = a$. However, if we think of the sphere as a special case of an axisymmetric shell, we would immediately determine $R_1 = a$, but the radius $r$ would then vary depending on whether we were looking near the poles or near the equator [see Figure 8.12 *(b)*].

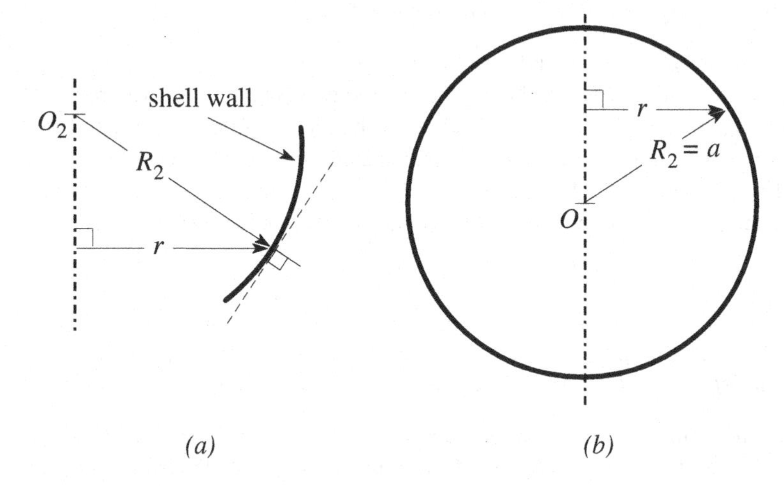


*Figure 8.12: (a) Distinction between the cross-sectional radius r and the circumferential radius of curvature of the shell $R_2$, (b) Radii r and $R_2$ for the sphere*

By contrast, if we construct the normal to the shell at any point and define $R_2$ as the distance along the normal to the axis of symmetry, we see that this leads to the correct conclusion that $R_2 = a$ and that the centres $O_1, O_2$ for both radii lie at the centre of the sphere for all points on the surface.

### 8.2.2 Sign conventions

The formula

$$\frac{\sigma_1}{R_1} + \frac{\sigma_2}{R_2} = \frac{p}{t} \tag{8.2}$$

was obtained under the assumptions that $\sigma_1, \sigma_2$ are tensile positive and that the centres $O_1, O_2$ corresponding to the radii $R_1, R_2$ are both on the same side of the shell as the pressure $p$.

Sometimes the two centres are on opposite sides of the shell, in which case the surface is locally saddle-shaped, as shown in Figure 8.13. Consider, for example the inside surface of a doughnut. This kind of curvature is sometimes referred to as *anticlastic*.

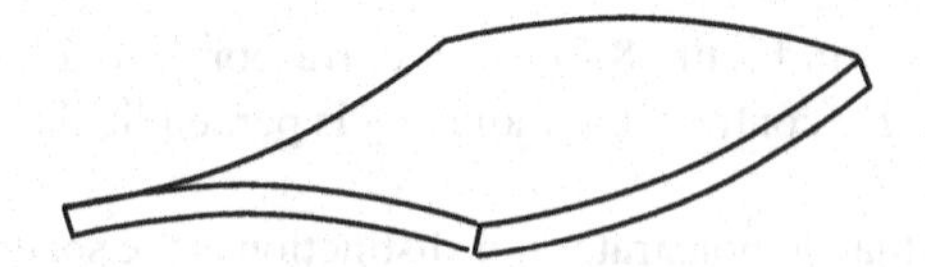

*Figure 8.13: Shell element with anticlastic curvature*

In addition, the pressure — if there is one – may be on the outside of the shell, as in the case of the shell of a submarine subjected to water pressure.

By referring to the derivation of equation (8.2), it is easily verified that it applies to such cases provided we follow the conventions:-

(i) The radius $R_2$ is always taken to be positive.
(ii) The radius $R_1$ is taken to be negative if the two centers of curvature $O_1, O_2$ are on the opposite sides of the shell (i.e. if the local region of shell has anticlastic curvature) and positive if they are on the same side of the shell.
(iii) The pressure $p$ is taken as positive if it acts on the inside surface of the shell (nearer to the axis) and negative if it acts on the outside surface.

**Example 8.2**

*Find the circumferential membrane stress $\sigma_2$ for the conical tank of Figure 8.3 as a function of height $z$. Find also the maximum shear stress in the tank.*

For the conical tank, we have $R_1 = \infty$, since a side view of the cone exhibits straight sides. To determine $R_2$, we draw a normal to the shell surface at height $z$ as shown in Figure 8.14, and extend it until it cuts the axis of symmetry.

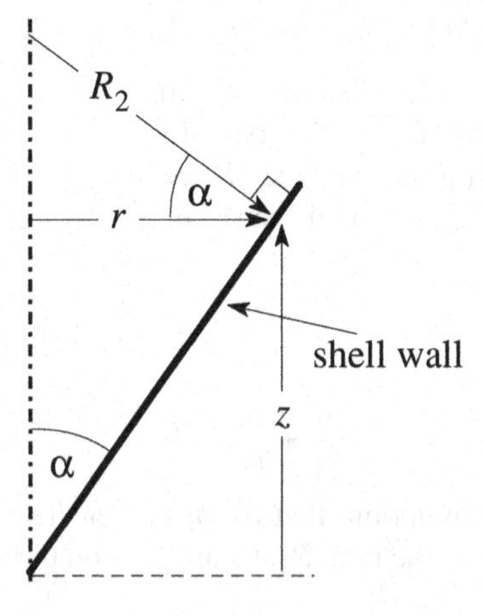


*Figure 8.14*

From the geometry of this figure, we note that the cross-sectional radius

$$r = z \tan \alpha$$

and hence

$$R_2 = \frac{r}{\cos\alpha} = \frac{z\tan\alpha}{\cos\alpha} .$$

We also note that the internal pressure at height $z$ is

$$p = \rho g(h-z) .$$

Using these values in equation (8.2), we have

$$\frac{\sigma_2\cos\alpha}{z\tan\alpha} + \frac{\sigma_1}{\infty} = \frac{\rho g(h-z)}{t}$$

and hence

$$\sigma_2 = \frac{\rho g(h-z)z\tan\alpha}{t\cos\alpha} .$$

It follows that the maximum value of $\sigma_2$ occurs at $z=h/2$ and is

$$\sigma_2^{\max} = \frac{\rho g h^2\tan\alpha}{4t\cos\alpha} ,$$

which is greater than the maximum meridional stress

$$\sigma_1^{\max} = \frac{3\rho g h^2\tan\alpha}{16t\cos\alpha}$$

found in §8.1 above.

In this example, $R_1=\infty$ and hence $\sigma_1$ does not enter into the equation. However, in general, we have to find $\sigma_1$ first by the method of §8.1, so that it can be eliminated in equation (8.2) to determine $\sigma_2$.

The maximum shear stress at any given point in the shell is the largest of

$$\left|\frac{\sigma_1-\sigma_2}{2}\right| , \left|\frac{\sigma_2-\sigma_3}{2}\right| , \left|\frac{\sigma_3-\sigma_1}{2}\right| ,$$

where $\sigma_1,\sigma_2,\sigma_3$ are the three principal stresses (see §2.1.2). In membrane stress problems, $\sigma_1,\sigma_2$ are the two membrane stresses and $\sigma_3$ is the stress on the local tangent plane to the shell. This generally varies through the shell thickness, since one side is loaded by fluid pressure, whilst the other is traction-free. However, the fact that the shell is thin-walled guarantees that $|\sigma_3| \ll |\sigma_1|,|\sigma_2|$. For example, in computing $\sigma_1$, the pressure force acting over the area $\pi r^2$ is resisted by the force due to $\sigma_1$ acting over the area $2\pi rt$, showing that $\sigma_1$ is greater than $p$ in the ratio of at least $r/2t$. We can therefore approximate $\sigma_3$ as zero in computing the maximum shear stress.

In the present example, both membrane stresses are everywhere tensile, so the maximum shear stress is

$$\tau_{\max} = \frac{\max(\sigma_1,\sigma_2)-0}{2} .$$

The maximum membrane stress is $\sigma_2^{\max}$ (see above) and hence

$$\tau_{\max} = \frac{\rho g h^2\tan\alpha}{8t\cos\alpha} .$$

## 8.3 Self-weight

So far, we have neglected the weight of the shell itself, which is usually a reasonable approximation for thin-walled shells subjected to pressure loading. However, there are cases (notably in civil engineering applications) where a significant part of the loading is associated with the self-weight of the structure.

The meridional stress $\sigma_1$ is determined exactly as in §8.1, except that the weight of the shell is included as an additional term in the equilibrium equation. In this calculation, it is sometimes useful to note that the volume of material in a shell of uniform thickness is approximately equal to the product of the mean surface area of the shell and the thickness.

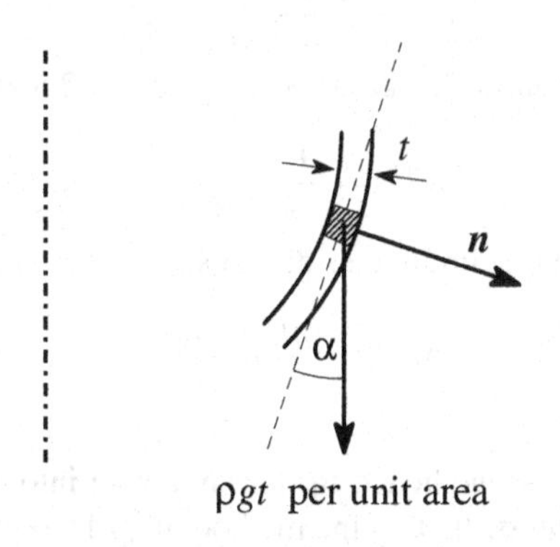


*Figure 8.15: Self-weight of the shell has a component equivalent to an additional pressure*

The self-weight of the shell also plays a rôle in the determination of the circumferential stress $\sigma_2$, since, in general, it has a component in the direction normal to the shell and hence modifies equation (8.2). For example, the small element of shell shown in Figure 8.15 is subject to a gravitational force of $\rho g t$ per unit area, where $t$ is the local thickness. This force has a component $\rho g t \sin\alpha$ in the direction of the outward normal $\boldsymbol{n}$, and hence has the same effect as an additional internal pressure of magnitude $\rho g t \sin\alpha$, where $\alpha$ is the angle between the local tangent to the shell and the vertical, as shown in Figure 8.15. It is important to check the direction of this component to make sure whether the resolved component is in the direction of the outward or the inward normal to the shell, since this will determine the sign of the additional term in equation (8.2).

**Example 8.3**

*Figure 8.16 shows a spherical roof shell of mean radius a and uniform thickness t, loaded only by its own weight. Find the membrane stresses as functions of the polar angle $\phi$ if the material has density $\rho_s$.*

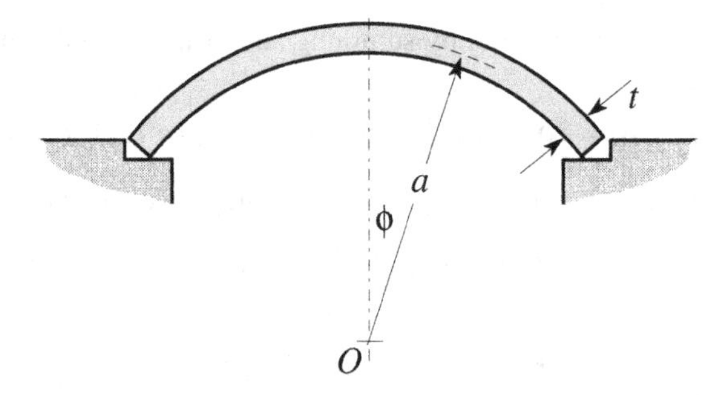


*Figure 8.16: Spherical roof shell*

To determine $\sigma_1$, we cut out the section of shell above the polar angle $\phi$, illustrated in Figure 8.17 *(a)*. The cross-sectional radius $r$ is then

$$r = a \sin\phi$$

and the meridional stress acts on the area of the cut through the shell thickness, which is $2\pi rt$ (the product of circumference and thickness). Axial equilibrium therefore requires that

$$\sigma_1(2\pi rt)\sin\phi + W = 2\pi\sigma_1 at\sin^2\phi + W = 0\,,$$

where $W$ is the weight of the shell segment and the factor $\sin\phi$ in the first term is required to resolve the force due to $\sigma_1$ into the vertical direction.

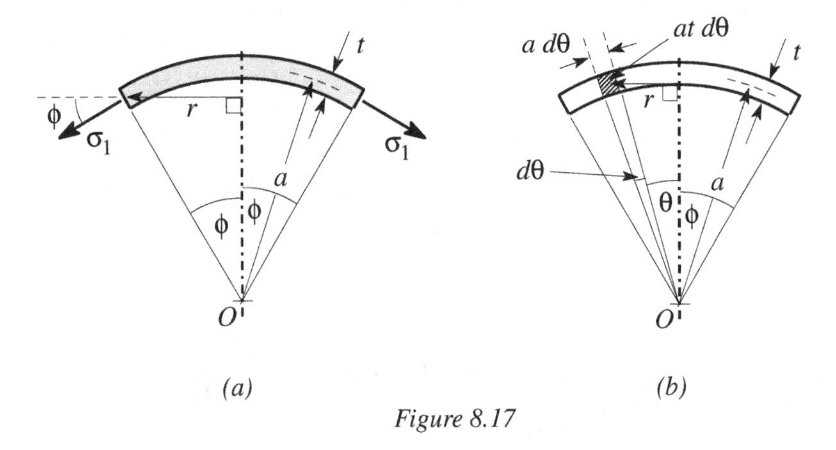


*Figure 8.17*

To find the weight $W$ we need to perform an integration. We first note that a small angle $d\theta$ at the centre of the sphere corresponds to a segment of shell of length $ad\theta$, as shown in Figure 8.17 *(b)*. This in turn defines an annular ring of material of radius $r=a\sin\theta$ and cross-sectional area $atd\theta$, shown shaded in Figure 8.17 *(b)*. The weight of this ring is

$$dW = 2\pi ratd\theta\rho_s g = 2\pi\rho_s ga^2 t\sin\theta d\theta$$

and the total weight $W$ is obtained by summing such rings by integration, with the result

$$W = \int_0^{\phi} dW = 2\rho_s g\pi a^2 t \int_0^{\phi} \sin\theta d\theta = 2\rho_s g\pi a^2 t(1-\cos\phi) .$$

Substituting this result into the equilibrium equation, we then have

$$2\pi\sigma_1 a t \sin^2\phi + 2\rho_s g\pi a^2 t(1-\cos\phi) = 0$$

and hence

$$\sigma_1 = -\frac{\rho_s g a(1-\cos\phi)}{\sin^2\phi} = -\frac{\rho_s g a}{(1+\cos\phi)} ,$$

since $\sin^2\phi = 1-\cos^2\phi$. Notice that $\sigma_1$ is compressive, as we should expect, and that it is bounded at the top, $\phi = 0$, where $\sigma_1 = -\rho_s g a/2$.

The circumferential stress $\sigma_2$ can be determined from equation (8.2). In the present case there is no internal pressure $p$, but as in Figure 8.15, the self-weight, $\rho_s g t$ per unit area has a component along the local normal to the shell. The direction of this force is shown in Figure 8.18 and it has a component along the *inward* normal of $\rho_s g t \cos\phi$. This simulates an *external* pressure, which is treated as negative, following the conventions of §8.2.2.

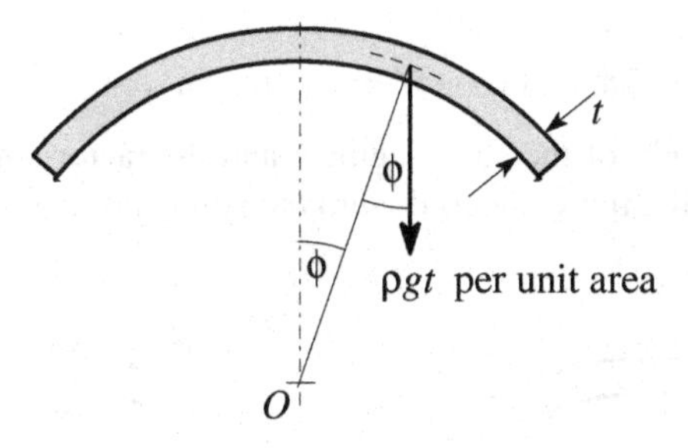


*Figure 8.18*

For the sphere, the two radii are $R_1 = R_2 = a$, so equation (8.2) takes the form

$$\frac{\sigma_1}{a} + \frac{\sigma_2}{a} = -\frac{\rho_s g t \cos\phi}{t} ,$$

from which

$$\sigma_2 = -\rho_s g a \cos\phi - \sigma_1 = \rho_s g a \left( -\cos\phi + \frac{1}{(1+\cos\phi)} \right)$$

$$= \rho_s g a \left( \frac{1-\cos\phi-\cos^2\phi}{1+\cos\phi} \right) .$$

It follows that $\sigma_2$ also tends to $-\rho_s g a/2$ as $\phi \to 0$. In fact, we shall always find that $\sigma_1 = \sigma_2$ at any point where a shell crosses the axis of symmetry without a discontinuity in slope[5] or a concentrated force.

[5] An axisymmetric shell with slope continuity must be locally spherical at a point where it crosses the axis, since all diameters at this point are equivalent.

The circumferential stress $\sigma_2$ is compressive near $\phi=0$, but becomes tensile if $\cos\phi+\cos^2\phi<1$, which corresponds to the range $\phi>51.8^{\mathrm{o}}$. This is a factor of importance if the shell is to be constructed from a material like concrete, which has a higher strength in compression than in tension.

## 8.4 Relative magnitudes of different loads

In the preceding sections, we have considered examples of shells loaded by internal pressure, the weight of the contents and the self-weight of the shell. In many practical applications, all these forms of loading will be present, but usually the membrane stresses will be dominated by one of them and the others can conveniently be neglected, as discussed in §1.3.

The weight of a contained fluid causes the pressure to increase with depth, so that at depth $h$, $p=\rho gh$. If the surface of the fluid is exposed to atmospheric pressure, the weight of the contents are therefore responsible for the principal pressure loading of the structure and must be included in the analysis. However, in some cases a vessel may contain a liquid above which there is a gas (usually vapour) at above atmospheric pressure. A typical example is a boiler delivering pressurized steam. The loading from the gas pressure will usually be significantly larger than that from the weight of the fluid, which can therefore be neglected. To test whether this approximation is legitimate, calculate the maximum pressure change $\rho gh$, between the top and bottom of the contained fluid and compare it with the gas pressure.

### Example 8.4

*A spherical boiler of diameter 6 ft is full to the 4 ft mark with water and delivers steam at a pressure of 200 psi. Is it necessary to include the weight of the water in a membrane stress calculation? The density of water is 60 lb/ft*$^3$.

The head of water between the bottom of the boiler and the water surface is

$$\rho gh=\frac{60\times 4\times 12}{12^3}=1.67\ \text{psi}\,.$$

This is less than 1% of the steam pressure (200 psi), so the weight of the water can reasonably be neglected.

A shell is by definition thin and hence the volume of shell material is orders of magnitude smaller than that of the contained volume. It follows that the self-weight of the shell will be significantly smaller than that of any liquid contents, even allowing for the fact that the shell material will typically have a density several times larger than that of the contents. After all, a full saucepan or water bottle is considerably heavier than an empty one. We have seen that the pressure due to fluid weight is $\rho gh$. In view of equation (8.2), this leads to membrane stresses of the order $\rho gha/t$, where $a$ is a representative shell radius.

By contrast, the self-weight of the shell will typically yield stresses of the order $\rho_s gl$, where $\rho_s$ is the density of the shell material and $l$ is a representative shell dimension. The results of the preceding examples confirm this general behaviour. For the conical shell with a contained fluid, we obtained

$$\sigma_1^{\max} = \frac{3\rho g h^2 \tan\alpha}{16t\cos\alpha} \;\; ; \;\; \sigma_2^{\max} = \frac{\rho g h^2 \tan\alpha}{4t\cos\alpha} \, ,$$

both of which contain the factor $\rho g h^2/t$ (since for the cone, radius $a$ is proportional to height $h$). The spherical roof shell of Example 8.3 yields the membrane stresses

$$\sigma_1 = -\frac{\rho_s g a}{(1+\cos\phi)} \;\; ; \;\; \sigma_2 = \rho_s g a \left(\frac{1-\cos\phi-\cos^2\phi}{1+\cos\phi}\right) ,$$

both of which contain the factor $\rho_s g a$, where $a$ is the radius of the sphere.

In view of these considerations, shell self-weight can almost always be neglected when there is a contained liquid, but this hypothesis can always be tested in particular cases by comparing representative estimates of $\rho_s g l$ and $\rho g h a/t$. A similar comparison can be made when a heavy shell is subjected to pressure loading $p$, using $pa/t$ as a representative membrane stress.

**Example 8.5**

*A concrete chimney of height 20 m, mean radius 0.8 m and wall thickness 100 mm is subject to a maximum pressure difference between inside and outside of 0.05 atm (5 KPa). The density of concrete is 2300 kg/m$^3$. Can either of the two forms of loading be neglected in a membrane stress calculation?*

We compute

$$\frac{pa}{t} = \frac{5\times 10^3 \times 0.8}{0.1} = 40\times 10^3 \text{ Pa}$$

$$\rho_s g h = 2300\times 9.81\times 20 = 450\times 10^3 \text{ Pa}$$

and conclude that in this case the loading from self-weight is dominant, at least near the base of the chimney, so the pressure loading can reasonably be neglected, as long as a 10% error is acceptable.

## 8.5 Strains and Displacements

Once the stresses are known, the strains and hence the deformation of the shell can be calculated. However, the predicted displacements may be constrained either by a support or by a point of connection between two segments of shell with different radii. In such cases, the shell experiences local bending.

Since the shell is axisymmetric, only radial and axial displacements are possible. Axial displacements $u_z$ are seldom of importance unless the change in contained volume in the shell due to internal pressure is required or the shell is prevented by the support system from axial expansion or contraction. The existence of a radial displacement $u_r$ implies that the local diameter and hence the local circumference of the shell has increased, as shown in Figure 8.19. It follows that there must be a circumferential (hoop) strain $e_{\theta\theta}$ or $e_2$. In the undeformed view, the radius in the cross section is $r$. After deformation, it is $r+u_r$, since all points move a distance $u_r$ outwards. Hence, the *new* circumference is $2\pi(u_r+r)$ and the circumferential strain is

$$e_2 = \frac{2\pi(u_r + r) - 2\pi r}{2\pi r} = \frac{u_r}{r}\,. \tag{8.3}$$

But, by Hooke's law,

$$e_2 = \frac{\sigma_2}{E} - \frac{\nu\sigma_1}{E} \tag{8.4}$$

and hence

$$u_r = \frac{(\sigma_2 - \nu\sigma_1)r}{E}\,. \tag{8.5}$$

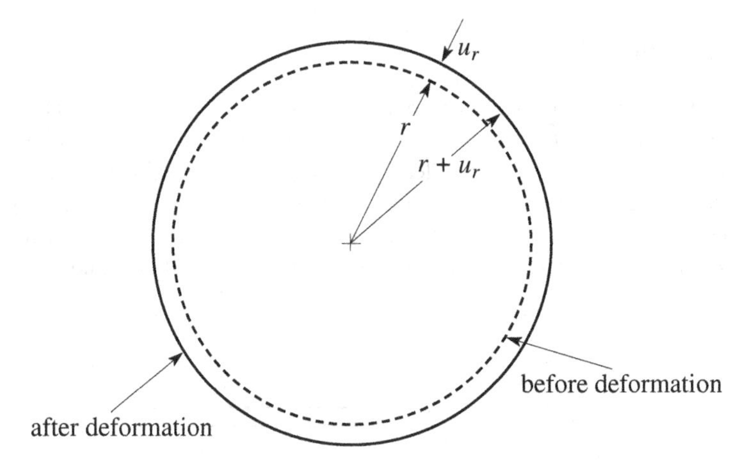


*Figure 8.19: Circumferential strain due to radial displacement $u_r$*

**Example 8.6**

*A spherical tank has mean radius 1 m, wall thickness 5 mm, and is made of steel for which $E=210$ GPa, $\nu=0.3$. It is completely full of water at atmospheric pressure. An additional 0.01 m$^3$ of water is then forced into the tank under pressure. Find the resulting internal pressure if the water can be treated as incompressible and the weight of the water and the tank can be neglected.*

Suppose the final internal pressure in the tank is $p$. Cutting the tank on a diameter, keeping half the water in each half, we can write the equilibrium equation

$$2\pi \times 1 \times 0.005 \times \sigma_1 - \pi \times 1^2 \times p = 0$$

and hence

$$\sigma_1 = 100p\,.$$

In view of the symmetry of the sphere, we also have $\sigma_2 = 100p$.

The radial displacement at all points is therefore

$$u_r = \frac{(\sigma_2 - \nu\sigma_1)r}{E} = \frac{100p(1-0.3)\times 1}{210\times 10^9} = \frac{p}{3\times 10^9}\ \text{m}\,,$$

where $p$ is in Pa. The increase in volume of the tank due to this displacement is

$$\Delta V = \frac{4\pi}{3}(1+u_r)^3 - \frac{4\pi}{3} \approx 4\pi u_r = \frac{4\pi p}{3\times 10^9}\ \text{m}^3\,,$$

but we require $\Delta V = 0.01\ \text{m}^3$, so the final pressure must be

$$p = \frac{3\times 10^9 \times 0.01}{4\pi} = 2.4\times 10^6\ \text{Pa}\quad (2.4\ \text{MPa}).$$

### 8.5.1 Discontinuities

Radial displacements are mainly of concern at the *ends* of a shell or at points of discontinuity. For example, suppose a cylindrical shell is built-in to a rigid wall and subjected to an internal pressure $p$. Clearly we get a non-zero value for $u_r$, since the stresses are non-zero, but at the end $u_r$ is constrained to be zero. In practice, the shell will deform as shown in Figure 8.20 and *bending stresses* will be developed at the end. We shall show how to determine the transitional shape and the bending stresses in Chapter 9.

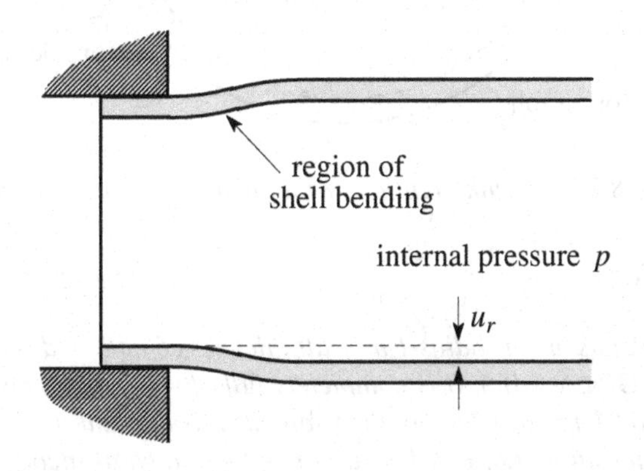


*Figure 8.20: Cylindrical pressure vessel built-in to a rigid wall at one end*

A less obvious example concerns the cylinder with hemispherical ends and internal pressure $p$, shown in Figure 8.21.

In the cylindrical section, the methods of §§8.1, 8.2 give

$$\sigma_1 = \frac{pa}{2t}\ ;\ \sigma_2 = \frac{pa}{t}$$

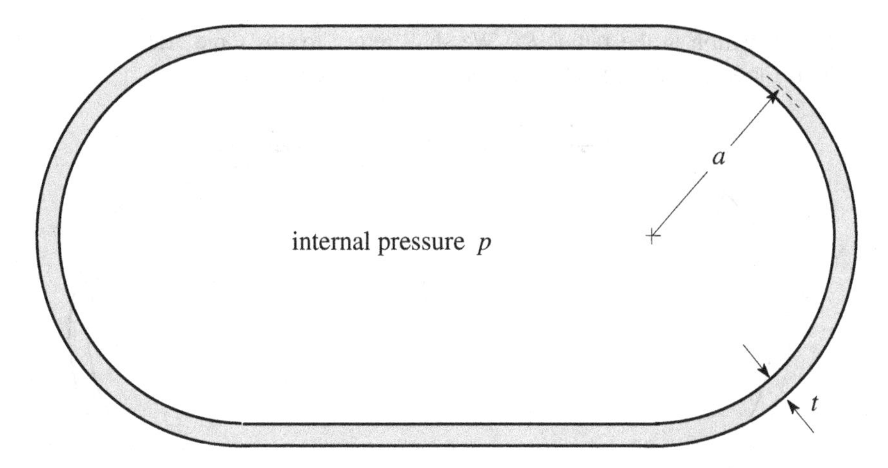


*Figure 8.21: Cylindrical pressure vessel with hemispherical ends*

and hence

$$u_r = \frac{a}{E}\left(\frac{pa}{t} - \frac{\nu pa}{2t}\right) = \frac{pa^2(2-\nu)}{2tE} \ .$$

For the spherical ends, we find

$$\sigma_1 = \frac{pa}{2t} \ ; \ \sigma_2 = \frac{pa}{2t}$$

and hence

$$u_r = \frac{a}{E}\left(\frac{pa}{2t} - \frac{\nu pa}{2t}\right) = \frac{pa^2(1-\nu)}{2tE} \ .$$

Thus, the spherical ends have less radial displacement than the cylinder. The deformed shape will be as shown in Figure 8.22. The predictions of the membrane theory are shown as a solid line, but clearly there must again be a transition region, shown dotted, in which bending stresses will be important.

More generally, the construction for $R_2$ in Figure 8.11 shows that this radius will be continuous as long as there are no sudden changes in the tangent plane of the shell (i.e. no kinks, as would occur for example at a cone/cylinder transition). Under these conditions, equilibrium considerations show that the meridional stress $\sigma_1$ will also be continuous as long as the thickness remains constant. However, equation (8.2) can be written

$$\sigma_2 = R_2\left(\frac{p}{t} - \frac{\sigma_1}{R_1}\right) , \tag{8.6}$$

showing that there will generally be a discontinuity in $\sigma_2$, and hence in the radial displacement $u_r$, if $R_1$ is discontinuous, as is clearly the case at the transition from a cylinder ($R_1 = \infty$) to a sphere. This discontinuity can be avoided by allowing a more gradual transition in the radius $R_1$. We shall return to this topic after introducing methods for determining bending stresses in the next chapter.

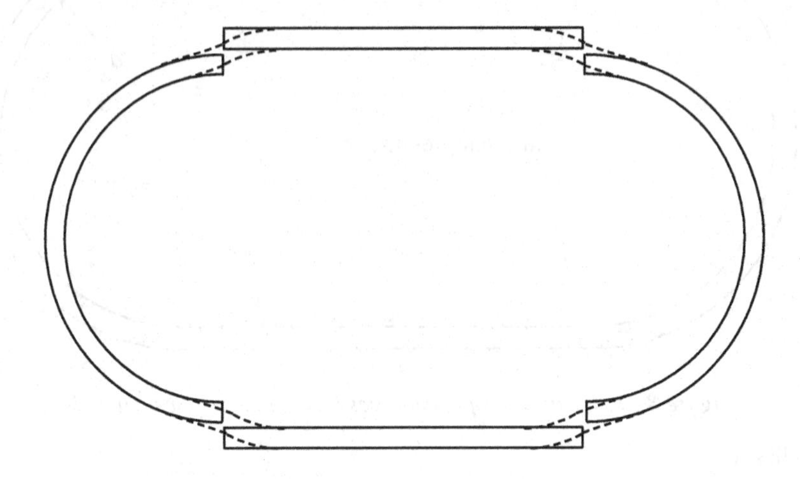

*Figure 8.22: Deformation under pressure of the vessel of Figure 8.21*

## 8.6 Summary

Shells are strong, stiff, lightweight structures that in most cases can transmit applied loads by in-plane membrane stresses (i.e. stresses which are uniform through the thickness). By contrast, flat plates always experience bending under lateral loading and are therefore relatively flexible and weak.

For axisymmetric shells, the meridional membrane stress can always be found by making an appropriate transverse cut, drawing the resulting free-body diagram and enforcing axial equilibrium. The circumferential stress can then be determined from the membrane equilibrium equation (8.2).

Once the membrane stresses are known, the strains and hence radial displacements can be calculated using Hooke's law. The membrane theory of shells predicts the occurrence of discontinuities in radial displacement wherever the loading or the shell radius is discontinuous. In these regions, localized shell bending will occur and higher stresses will be obtained. This topic is addressed in Chapter 9.

## Further reading

W.B. Bickford (1998), *Advanced Mechanics of Materials,* Addison Wesley, Menlo Park, CA, §§8.1–8.4.
A.H. Burr (1981), *Mechanical Analysis and Design*, Elsevier, New York, §§8.2–8.4.
R.D. Cook and W.C. Young (1985), *Advanced Mechanics of Materials,* Macmillan, New York, §§6.1–6.4.

W. Flügge (1973) *Stresses in Shells*, Springer-Verlag, New York, 2nd edn., §2.2.
S.P. Timoshenko and S. Woinowsky-Krieger (1959) *Theory of Plates and Shells*, McGraw-Hill, New York, 2nd edn., §§105–108.
A.C. Ugural and S.K. Fenster (1995), *Advanced Strength and Applied Elasticity*, Prentice-Hall, Eaglewood Cliffs, NJ, 3rd edn., §§13.9–13.11.

## Problems

### Sections 8.1, 8.2

**8.1.** A spherical fuel oil tank of radius $R$ sits on an annular ring support, as shown in Figure P8.1. The oil density is $\rho$ and the oil level is a distance $bR$ from the crown of the tank.

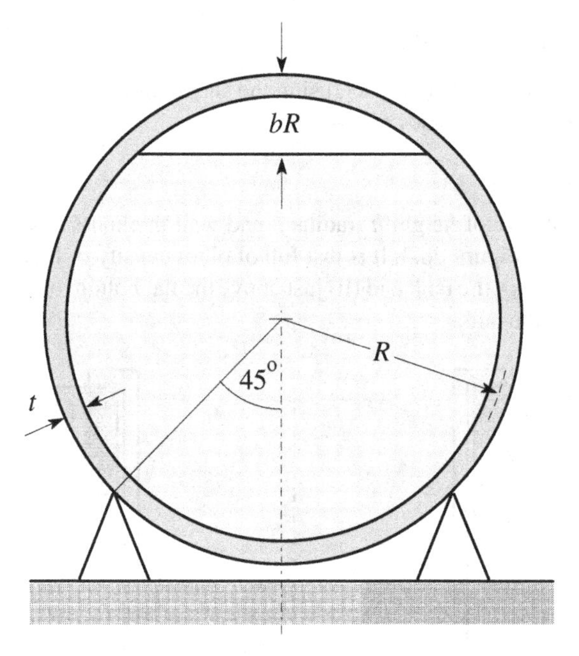


*Figure P8.1*

Find the meridional stress just above the support ring and hence find the value of $b$ for which this stress is numerically a maximum. The self-weight of the tank can be neglected.

**8.2.** Figure P8.2 shows the cross section of an axisymmetric chemical process vessel which is required to contain gas at a pressure of 1 MPa. The wall thickness is 10 mm everywhere. The self-weight of the shell can be neglected.

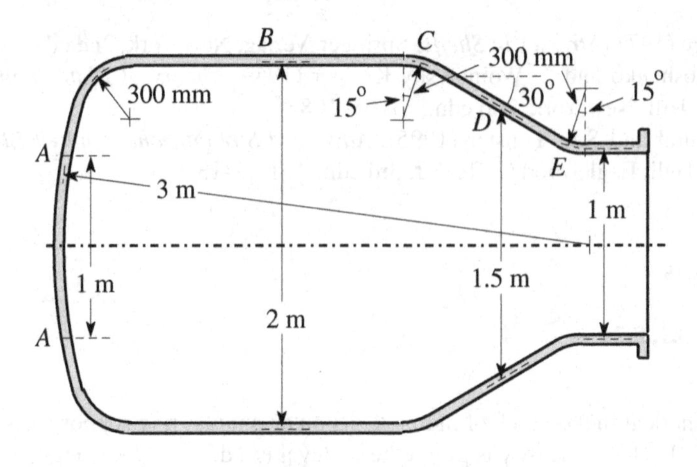


*Figure P8.2*

For each of the points *A, B, C, D, E*:-

(i) Find the radii of curvature $R_1, R_2$, using the sign convention of §8.2.2.
(ii) Find the membrane stresses $\sigma_1, \sigma_2$.

**8.3.** A cylindrical tank of height $h$, radius $a$ and wall thickness $t$ is supported from the top, as shown in Figure P8.3. It is just full of oil of density $\rho$. Find the membrane stresses (i) at the top of the tank and (ii) just above the flat bottom of the tank. Neglect the self-weight of the tank.

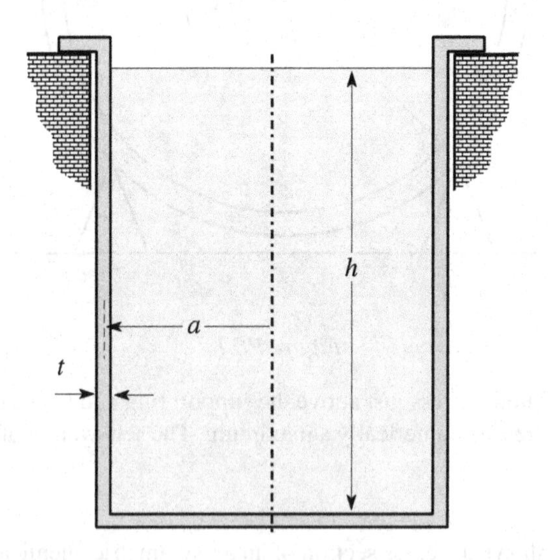


*Figure P8.3*

**8.4.** A lightweight crash helmet consists of a 6 in radius hemispherical aluminium shell of wall thickness $3/32$ in, padded on the inside with a 2 in layer of expanded polystyrene. In an impact, a maximum force of 2000 lbs is exerted on the helmet, uniformly distributed over a circular region of diameter 2 in as shown in Figure P8.4. Assuming the balancing force from the padding to consist of a uniform pressure over the entire hemispherical internal surface, find the maximum shear stress in the shell.

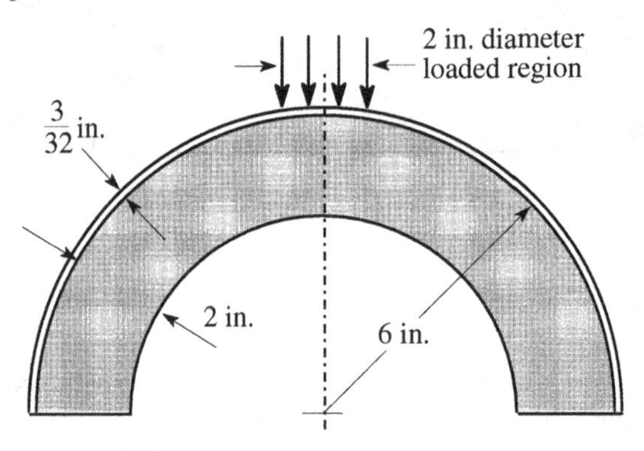


*Figure P8.4*

**8.5.** Figure P8.5 shows a design for the end of a cylindrical tank which is to contain air at 250 psi. Find the maximum tensile membrane stress in the tank.

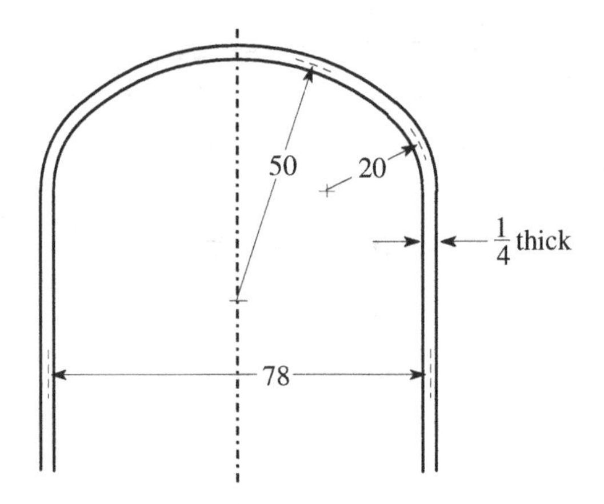


*Figure P8.5*

**8.6.** A hemispherical dome is partially submerged in water of density $\rho$, as shown in Figure P8.6. Find the membrane stresses at the point $A$, defined by the angle $\phi$. You can assume that the pressure inside the dome and above the water surface is atmospheric pressure $p_0$ and that the stresses due to self-weight of the shell are negligible.

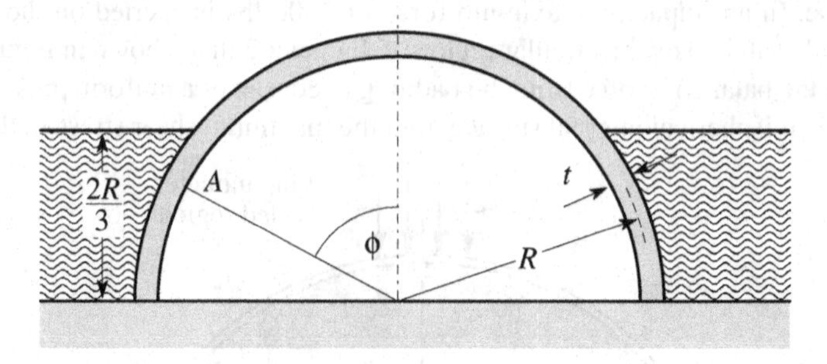


*Figure P8.6*

**8.7.** Figure P8.7 shows an aircraft fuselage, which can be assumed to be approximately axisymmetric. It comprises two conical segments of semi-angle 18° and 6° respectively, smoothly connected to a spherical nose and an intermediate segment of radius $R_1 = 15$ m and maximum diameter 6 m.

Estimate the maximum membrane stress and the maximum shear stress in the fuselage when the cabin pressure is 100 kPa and the outside air pressure is 30 kPa. The thickness of the shell is 4 mm everywhere.

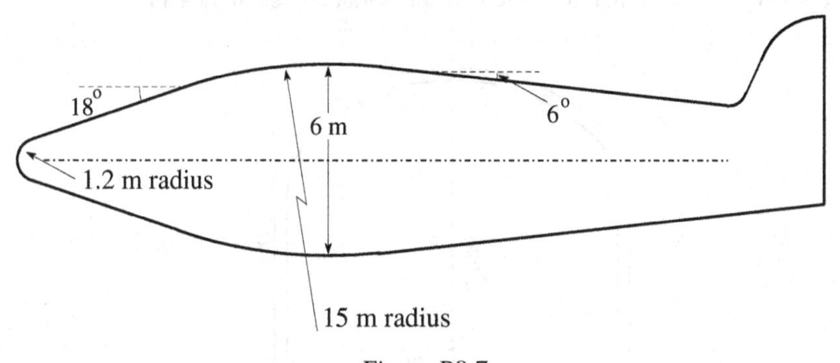


*Figure P8.7*

**8.8*.** Figure P8.8 shows the cross section of a toroidal pressure vessel, the toroid being essentially a tube of radius $a$ whose centreline is a circle of radius $b$. The wall thickness is $t(\ll a, b)$ and the vessel contains gas at pressure $p$.

Neglecting the weight of the gas and the shell, find the membrane stresses at a general point $Q$ defined by the coordinate $\phi$.

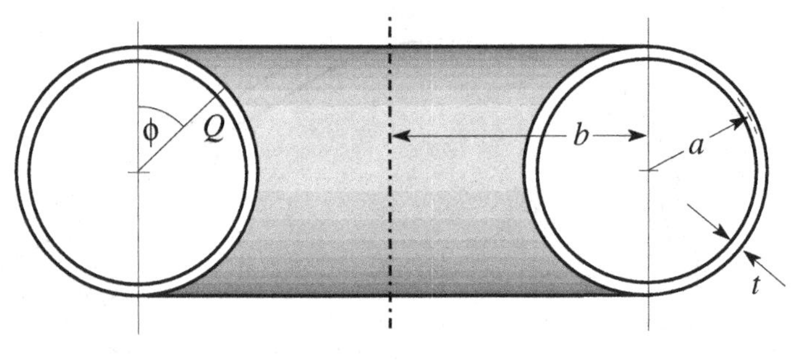


*Figure P8.8*

**8.9.** The thin-walled sphere in Figure P8.9 has a mean radius $a$ and wall thickness $t$. It is compressed by equal and opposite forces $F$ on a diameter. There is no internal pressure. Find the membrane stresses at $A$.

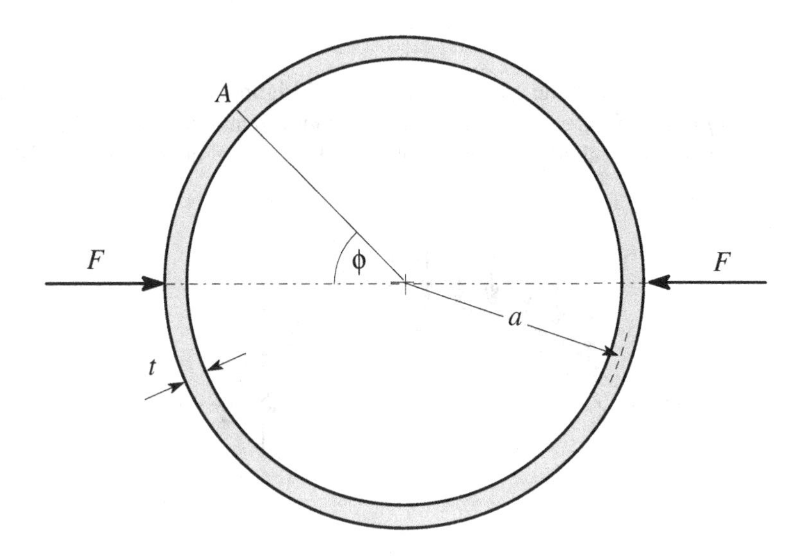


*Figure P8.9*

**8.10.** The thin-walled pressure vessel of Figure P8.10 is of uniform wall thickness $t$ and consists of a conical section of semi-angle $30^{\circ}$ and a spherical section of mean radius $a$ such that slope and radius are continuous at the transition. It is subjected to an *external* pressure $p_0$.

Find the magnitude and location of the maximum values of the two membrane stresses $\sigma_1, \sigma_2$ and of the stress difference $|\sigma_1 - \sigma_2|$.

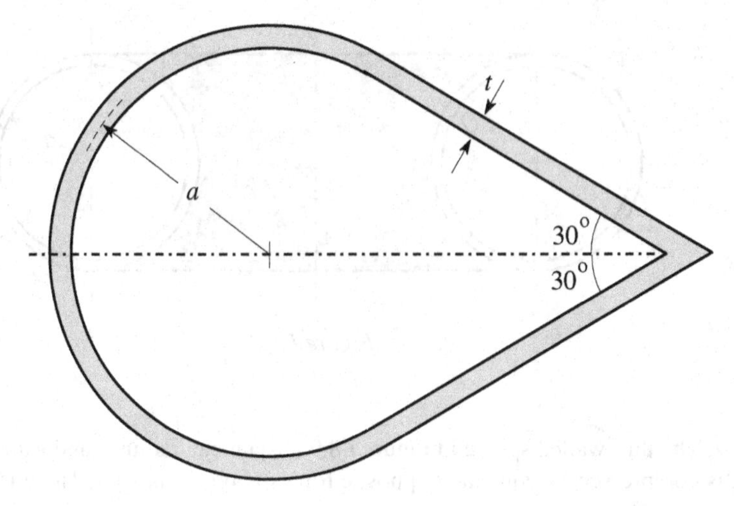


*Figure P8.10*

**8.11.** Snow accumulation on the spherical roof shell of Figure P8.11 causes a loading of $w_0$ per unit *projected*[6] area. The shell has mean radius $a$ and thickness $t$. Find the membrane stresses in the shell as functions of the polar angle $\phi$, neglecting the effect of shell self-weight.

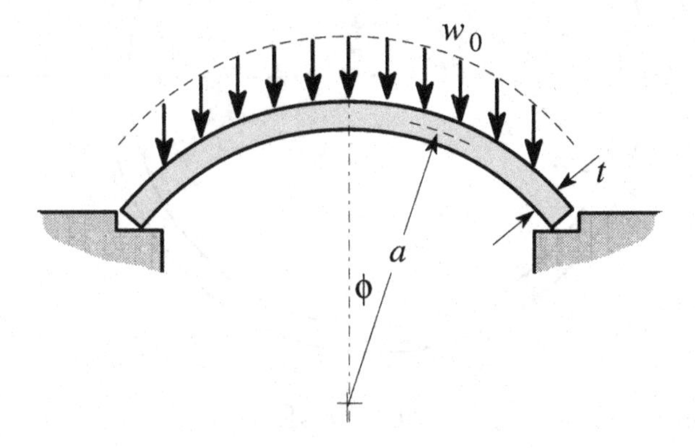


*Figure P8.11*

**Sections 8.3, 8.4**

**8.12.** Figure P8.12 shows a proposed design for the upper section of an axisymmetric cooling tower, which is to be made of concrete of density 2300 kg/m$^3$.

[6] In other words, the *vertical* thickness of the snow layer is uniform.

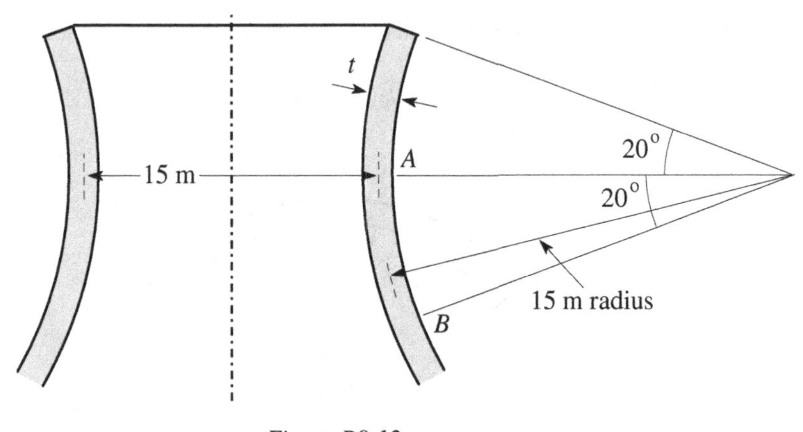


*Figure P8.12*

Find the minimum value of the shell thickness $t$ if the tower must be capable of sustaining an internal pressure 2kPa at $A$ and 5kPa at $B$ without inducing tensile membrane stresses at these points. With your choice of $t$, do you expect there to be tensile stresses elsewhere in the tower and if so where?

**8.13.** Figure P8.13 shows a closed spherical container of mean radius $a$ whose wall thickness $t$ ($t \ll a$) is chosen to cause it to float just half submerged in a fluid of density $\rho$. Develop expressions for the membrane stresses in the submerged portion of the sphere as a function of $\phi$. There is no internal pressure.

**Note:** The volume of a sphere of radius $a$ is $\frac{4}{3}\pi a^3$, whilst its surface area is $4\pi a^2$.

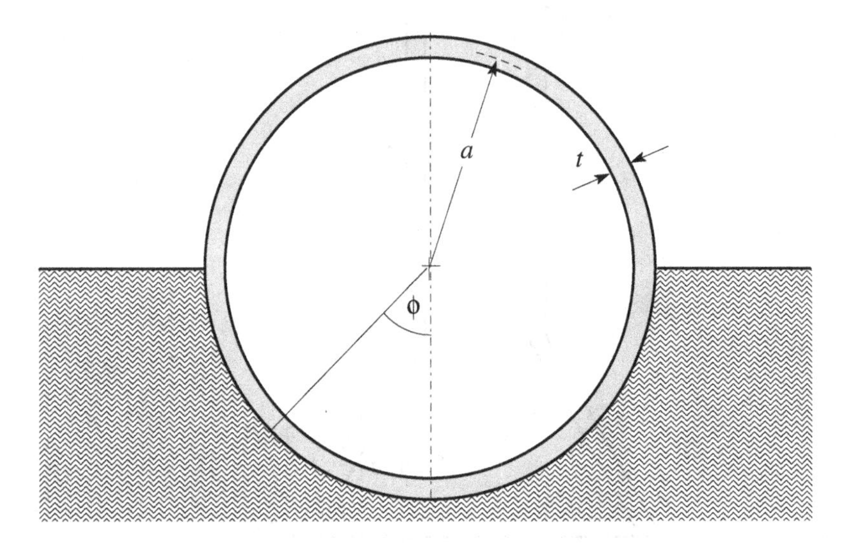


*Figure P8.13*

**8.14.** Figure P8.14 shows a spherical roof shell whose thickness $t$ varies with $\phi$, being given by

$$t = t_0(4 - 3\cos\phi)\ ,$$

where $t_0$ is a constant.

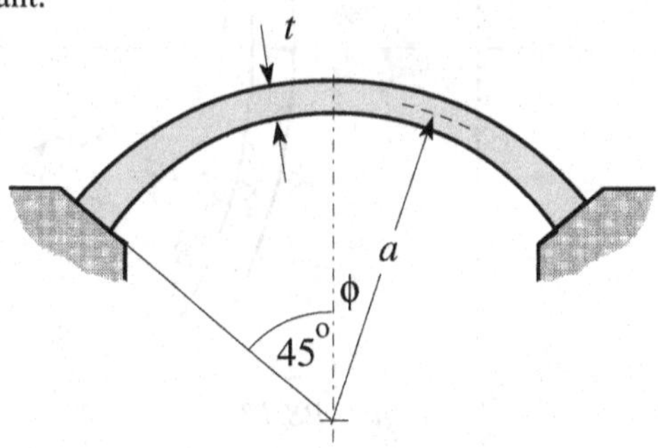


*Figure P8.14*

Calculate the membrane stresses in the roof due to self-weight alone, if it is made from a material of density $\rho$. Compare the values at $\phi = 0$ and $45^o$ and comment on your results.

**8.15.** Figure P8.15 shows the cross section of a concrete chimney, loaded only by its own weight. The density of concrete is 2300 kg/m$^3$. Find the membrane stresses at the base of the chimney.

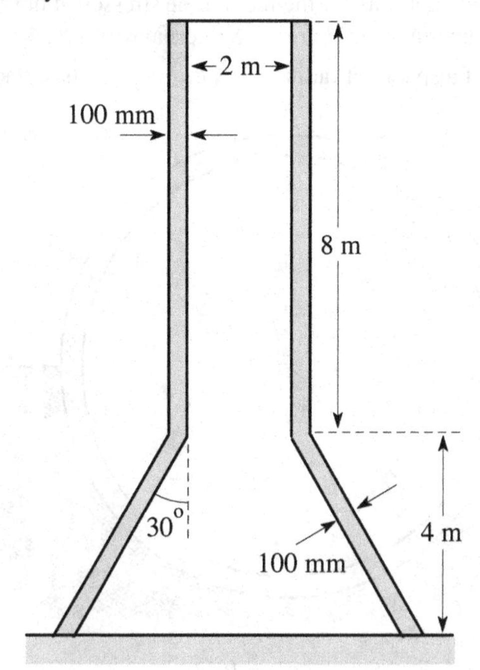


*Figure P8.15*

**8.16.** Figure P8.16 shows the approximate dimensions of a champagne bottle, manufactured from glass of density $\rho_g = 2400$ kg/m$^3$. The bottle is to contain liquid of density 1050 kg/m$^3$ at a pressure of 30 kPa above atmospheric.

Which of the following quantities are significant enough to require inclusion in a membrane stress calculation:-

(i) internal pressure,
(ii) weight of contents,
(iii) weight of the glass.

Estimate the maximum membrane stress in the bottle. Why is it necessary to use an increased wall thickness at the points $A-A$?

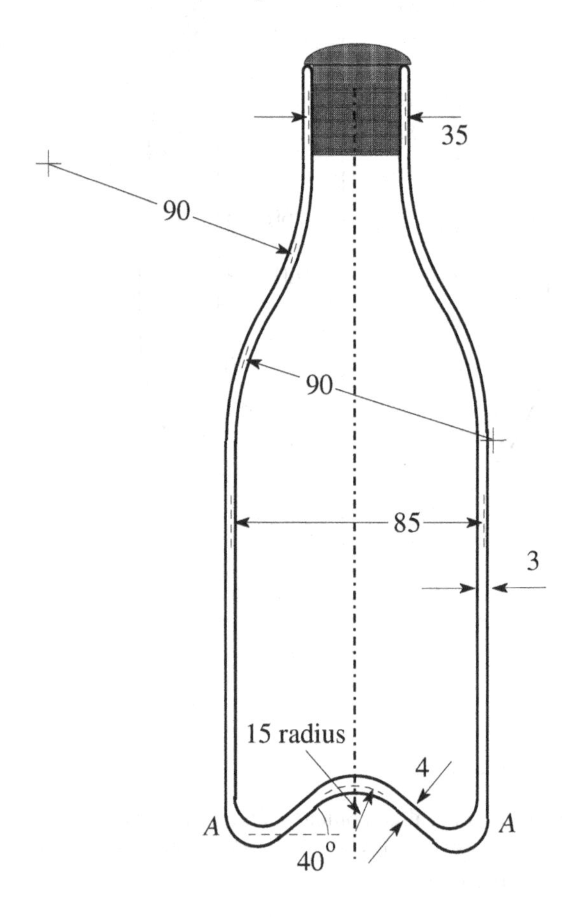


all dimensions in mm

*Figure P8.16*

**8.17.** Figure P8.17 shows the cross section of a vacuum flask for the temporary storage of liquified gasses. The space $A$ between the inner and outer walls is evacuated and the exposed surfaces are subject to atmospheric pressure of 100 kPa. The vessel is made of glass of density 2300 kg/m$^3$ and the wall thickness is everywhere 3 mm, except at $B$ where the vessel must be stiffened against bending.

Find the maximum tensile membrane stress in the wall, excluding the region $B$.

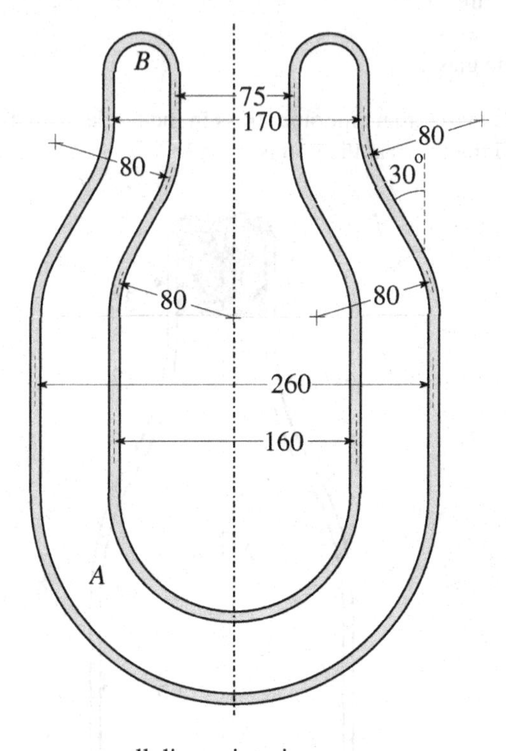


*Figure P8.17*

**Section 8.5**

**8.18.** Figure P8.18 shows a thin-walled conical shell of thickness $t$, semi-angle $\alpha$ and height $h$. It is loaded by self-weight only, the material of the shell having density $\rho_s$, Young's modulus $E$ and Poisson's ratio $\nu$.

Determine the membrane stresses $\sigma_1, \sigma_2$ at the base of the shell and hence find the radial displacement $u_r$ at the base of the shell.

**Note:** The surface area of a cone of semi-angle $\alpha$ and height $h$ is $\pi h^2 \tan\alpha / \cos\alpha$.

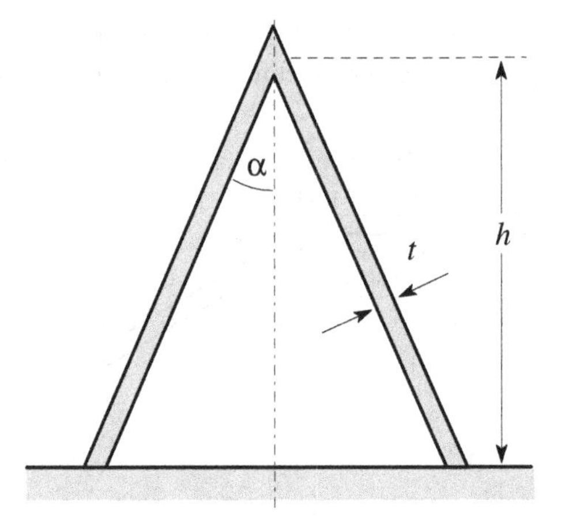


*Figure P8.18*

**8.19.** For the toroidal transition between the cylinder and the cone shown in Figure P8.19, find the membrane stresses $\sigma_1, \sigma_2$ at the points $A_1, A_2, B_1, B_2$, due to an internal pressure of 1 MPa, if the wall thickness is 10 mm.

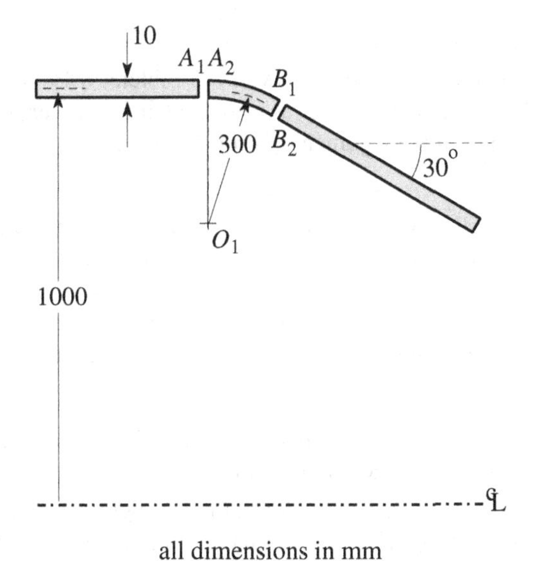


*Figure P8.19*

Hence find the circumferential strain and the radial displacement at these points, if the material is steel ($E = 210$ GPa, $\nu = 0.3$).

Sketch the distorted shapes of the shell sections at the transition and show how local bending will re-establish continuity of displacement.

**8.20.** Repeat the calculation of Problem 8.19 for the points $A_1$,$A_2$, assuming there is no toroidal transition — i.e. the cone is directly joined to the cylinder as shown in Figure P8.20.

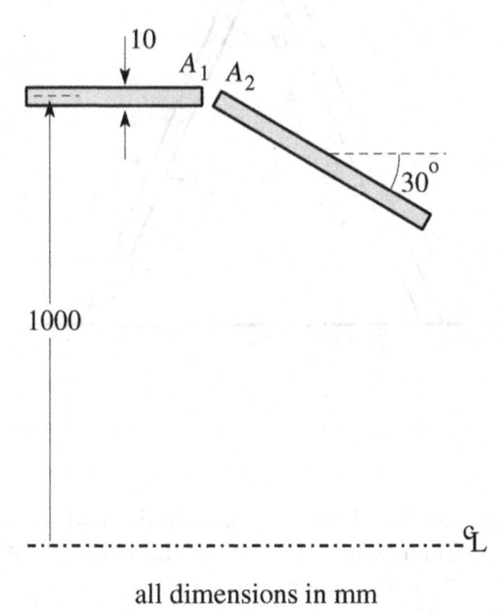


*Figure P8.20*

Comment on the relation between the answers to Problems 8.19, 8.20. Will the transition reduce the local bending stresses, as we would intuitively expect, or not? If so, explain what is missing in the argument of Problem 8.20.

**8.21.** A cylindrical shell of radius $a$, thickness $t$ is to be capped by a hemispherical shell of radius $a$ and thickness $t_1$. It is proposed to eliminate the bending effects at the discontinuity, discussed in §8.5.1, by choosing $t_1$ so as to cause the membrane radial displacements $u_r$ in the cylinder and the hemisphere to be equal. What is the required ratio $t_1/t$, if the material has Young's modulus $E$ and Poisson's ratio $\nu$?

**8.22.** A spherical shell of mean radius $a$ and thickness $t$ rotates at angular velocity $\Omega$ about a diameter. There is no internal pressure.

Find the membrane stresses as functions of polar angle and hence find the radial displacement at the equator. The material has density $\rho$, Young's modulus $E$ and Poisson's ratio $\nu$.

# 9

# Axisymmetric Bending of Cylindrical Shells

We saw in the last chapter that the equilibrium equations for an axisymmetric shell can be satisfied by membrane stresses $\sigma_1, \sigma_2$, which are uniform through the thickness of the shell. However, the radial displacements predicted by the membrane theory generally involve discontinuities at the supports or wherever there is a discontinuity in radius or thickness. In such cases, continuity is restored by localized bending of the shell. In the present chapter, we shall investigate the axisymmetric bending of cylindrical shells, but important features of the results carry over to other axisymmetric shell bending problems.

## 9.1 Bending stresses and moments

As in the bending of beams, shell bending involves stresses that vary linearly through the shell thickness. In combination with the membrane stresses, they constitute the first two terms in a Taylor series expansion of the variation of normal stress through the thickness.

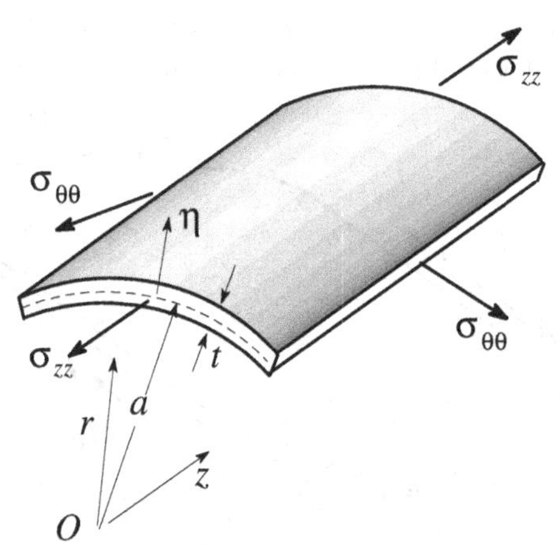


*Figure 9.1: Coordinate system and stresses for a shell element*

J.R. Barber, *Intermediate Mechanics of Materials*, Solid Mechanics and Its Applications 175, 2nd ed., DOI 10.1007/978-94-007-0295-0_9, 

Figure 9.1 shows the longitudinal stress $\sigma_{zz}$ and the circumferential stress $\sigma_{\theta\theta}$ acting on an element of cylindrical shell of mean radius $a$. To decouple membrane and bending stresses, it is convenient to define a coordinate $\eta$ in the direction of the outward normal and measured from the mid-plane of the shell. In other words, the distance from the cylinder axis is $r=a+\eta$. This is equivalent to the location of the origin at the centroid in beam bending problems. The bending theory is then based on the assumption that the local stress field can be written in the form

$$\sigma_{zz} = \sigma_1 + \frac{s_1\eta}{t} \tag{9.1}$$

$$\sigma_{\theta\theta} = \sigma_2 + \frac{s_2\eta}{t} , \tag{9.2}$$

where the constant terms $\sigma_1, \sigma_2$ are the membrane stresses already defined in Chapter 8, and the linearly varying terms represent the bending stresses.

The stress $\sigma_{\theta\theta}$ in equation (9.2) has a resultant bending moment $M_\theta$ per unit length of shell, given by

$$M_\theta = \int_{-t/2}^{t/2} \sigma_{\theta\theta}\eta d\eta = \int_{-t/2}^{t/2} \left(\sigma_2 + \frac{s_2\eta}{t}\right)\eta d\eta$$
$$= \frac{s_2 t^2}{12} . \tag{9.3}$$

We can use this result to eliminate $s_2$ in equation (9.2), obtaining

$$\sigma_{\theta\theta} = \sigma_2 + \frac{12M_\theta\eta}{t^3} . \tag{9.4}$$

A similar procedure with $\sigma_{zz}$ gives

$$\sigma_{zz} = \sigma_1 + \frac{12M_z\eta}{t^3} , \tag{9.5}$$

where $M_z$ is the bending moment per unit circumference. Notice that $M_\theta, M_z$ are bending moments *per unit length* and hence they have units of force (e.g. Nm per m $\equiv$ N). They act on the edges of the shell element in the sense defined in Figure 9.2.

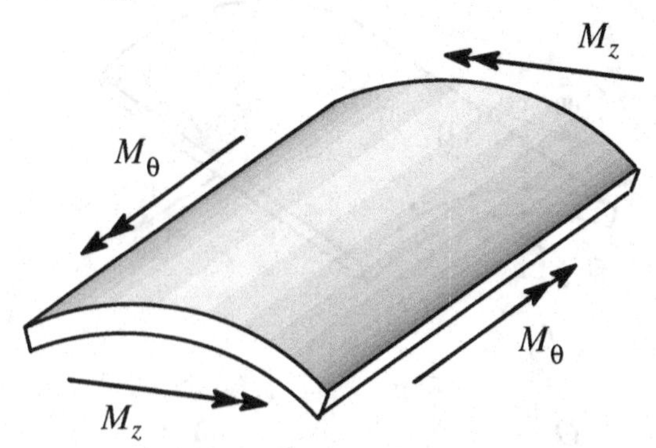


*Figure 9.2: Bending moments acting on the shell element*

The sign convention is chosen such that positive moments generate tensile stresses at the outside of the shell, where $\eta > 0$. The maximum bending stresses occur at $\eta = \pm t/2$ and are

$$\sigma_{\max} = \pm\frac{6M}{t^2} \; . \tag{9.6}$$

The bending stresses are additive to the membrane stresses [see equations (9.1, 9.2)], so generally we shall find that the maximum total stress in a shell will occur in a region of localized bending.

## 9.2 Deformation of the shell

A general axisymmetric deformation of a cylindrical shell can be characterized by the radial and axial displacements, $u_r$ and $u_z$ respectively, of the shell mean surface as functions of the axial co-ordinate $z$. These displacements are illustrated in Figure 9.3*(a)*, where the horizontal straight line represents the position of the shell mean surface before deformation. The point $A$ on this line moves to $A'$ as a result of combined membrane and bending deformations.

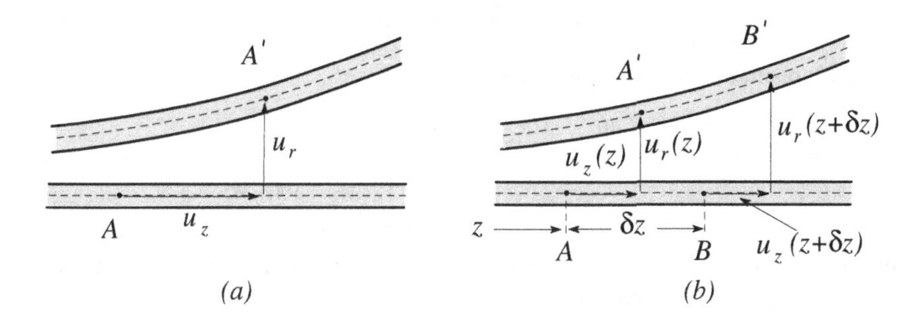


*Figure 9.3: (a) Definition of displacement components, (b) deformation of the shell element AB*

The cross section of the cylinder remains circular and, as in §8.5, we conclude that the circumferential strain $e_{\theta\theta} = u_r/a$. Hooke's law then gives

$$\begin{aligned}\frac{u_r}{a} = e_{\theta\theta} &= \frac{\sigma_{\theta\theta} - \nu\sigma_{zz}}{E} \\ &= \frac{\sigma_2 - \nu\sigma_1}{E} + \frac{12(M_\theta - \nu M_z)\eta}{Et^3} \; .\end{aligned} \tag{9.7}$$

We argued in Example 8.2 that the radial stress $\sigma_{rr}$ is negligible compared with $\sigma_{\theta\theta}, \sigma_{zz}$ and hence the corresponding radial strain,

$$e_{rr} = \frac{\partial u_r}{\partial r}$$

is also negligible.[1] It follows that $u_r$ cannot be a function of $\eta$, so the term varying with $\eta$ in equation (9.7) must be zero, leading to the results

$$M_\theta = \nu M_z \tag{9.8}$$

$$u_r = \frac{(\sigma_2 - \nu\sigma_1)a}{E} . \tag{9.9}$$

As in the elementary bending theory, we assume that plane transverse sections remain plane and perpendicular to the local shell wall. The deformation of a typical element $AB$ of length $\delta z$ is shown in Figure 9.3 *(b)*. The mean line increases in length by the amount

$$u_z(z+\delta z) - u_z(z)$$

and the section at $B$ rotates anticlockwise relative to that at $A$ through the angle

$$u_r'(z+\delta z) - u_r' ,$$

where $u_r' \equiv du_r/dz$. The longitudinal strain is therefore

$$e_{zz} = \frac{u_z(z+\delta z) - u_z(z)}{\delta z} - \frac{[u_r'(z+\delta z) - u_r']\eta}{\delta z}$$

and in the limit $\delta z \to 0$ we have

$$e_{zz} = \frac{du_z}{dz} - \eta \frac{d^2 u_r}{dz^2} .$$

Using this result and Hooke's law, we obtain

$$\begin{aligned} \frac{du_z}{dz} - \eta \frac{d^2 u_r}{dz^2} &= \frac{\sigma_{zz} - \nu\sigma_{\theta\theta}}{E} \\ &= \frac{(\sigma_1 - \nu\sigma_2)}{E} + \frac{12(M_z - \nu M_\theta)\eta}{Et^3} , \end{aligned} \tag{9.10}$$

from equations (9.4, 9.5). This equation must be satisfied for all $z, \eta$, so we can equate coefficients of $\eta$, obtaining

$$\frac{du_z}{dz} = \frac{(\sigma_1 - \nu\sigma_2)}{E} \tag{9.11}$$

$$\frac{d^2 u_r}{dz^2} = -\frac{12(M_z - \nu M_\theta)}{Et^3} . \tag{9.12}$$

Furthermore, we can eliminate $M_\theta$ from (9.12) using (9.8), with the result

$$M_z = -D\frac{d^2 u_r}{dz^2} , \tag{9.13}$$

where

$$D = \frac{Et^3}{12(1-\nu^2)} \tag{9.14}$$

is known as the *stiffness* of the shell.

[1] It might be argued that the membrane stresses $\sigma_1, \sigma_2$ will generate a non-negligible Poisson's ratio strain $e_{rr}$. However, a more rigorous treatment including this effect shows that the resulting modification of equation (9.8) is still small of order $t/a$ (see Problem 9.1).

## 9.3 Equilibrium of the shell element

As in §8.2, we can write an equilibrium equation for the forces acting on the shell element in the normal direction, but when bending is included this equation will also include transverse shear forces. Figure 9.4 *(a)* shows the appropriate forces acting on the element $a\delta\theta\delta z$. Notice that the lengths $\delta x, \delta y$ in Figure 8.8 are here replaced by $\delta z, a\delta\theta$ respectively. We also include transverse shear forces $V(z)$ *per unit length* on the curved edges of the element. There can be no shear forces on the straight edges because the deformation is axisymmetric.

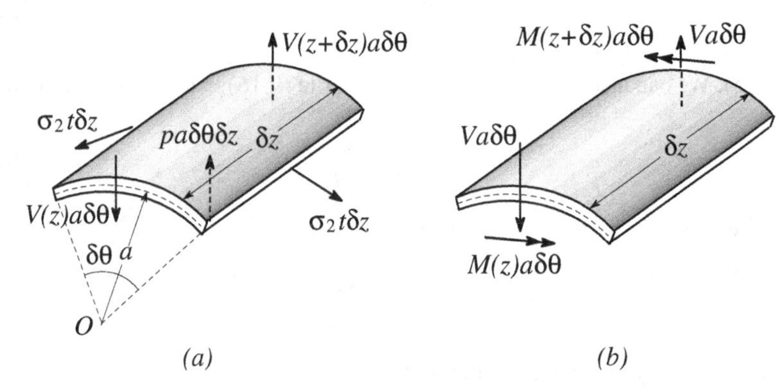


*Figure 9.4: Equilibrium of (a) forces and (b) moments on the shell element*

Summing the forces in the direction of the outward normal, we obtain

$$[V(z+\delta z) - V(z)]\, a\delta\theta + pa\delta\theta\delta z - \sigma_2 t\delta z\delta\theta = 0\,.$$

Here, the second term arises from the internal pressure $p$ acting on the area $a\delta\theta\delta z$ and the last term is the resultant of the two forces $\sigma_2 t\delta z$ and is obtained exactly as in the derivation of equation (8.2).

Dividing through by $a\delta\theta\delta z$ and taking the limit as $\delta z \to 0$, we obtain

$$\frac{dV}{dz} = \frac{\sigma_2 t}{a} - p\,. \tag{9.15}$$

A second equation is obtained by considering the moment equilibrium of the element under the action of the bending moments and shear forces shown in Figure 9.4 *(b)*. We obtain

$$\frac{dM_z}{dz} = V\,. \tag{9.16}$$

Alternatively, this result can be considered as another case of equation (6.1), regarding the shell element as a beam of unit width.

## 9.4 The governing equation

These equations can be expressed in terms of the radial displacement $u_r$, to obtain the governing equation for shell deformation. We first note that

$$\sigma_2 = \frac{Eu_r}{a} + \nu\sigma_1 , \tag{9.17}$$

from equation (9.9) and hence

$$\frac{dV}{dz} = \frac{Eu_r t}{a^2} + \frac{\nu\sigma_1 t}{a} - p , \tag{9.18}$$

from (9.15). We can also use equation (9.13) to write (9.16) in terms of $u_r$, obtaining

$$V = \frac{dM_z}{dz} = -D\frac{d^3u_r}{dz^3} . \tag{9.19}$$

Eliminating $V$ between (9.18, 9.19), we then obtain

$$-D\frac{d^4u_r}{dz^4} = \frac{Eu_r t}{a^2} + \frac{\nu\sigma_1 t}{a} - p .$$

This is conveniently rearranged in the form

$$\frac{d^4u_r}{dz^4} + 4\beta^4 u_r = 4\beta^4\left(\frac{pa^2}{Et} - \frac{\nu\sigma_1 a}{E}\right) , \tag{9.20}$$

where we define

$$\beta^4 = \frac{3(1-\nu^2)}{a^2t^2} \tag{9.21}$$

and we have used (9.14) to simplify the multiplying constants. Equation (9.20) is the governing equation for axisymmetric bending of a circular cylindrical shell of radius $a$ and thickness $t$.

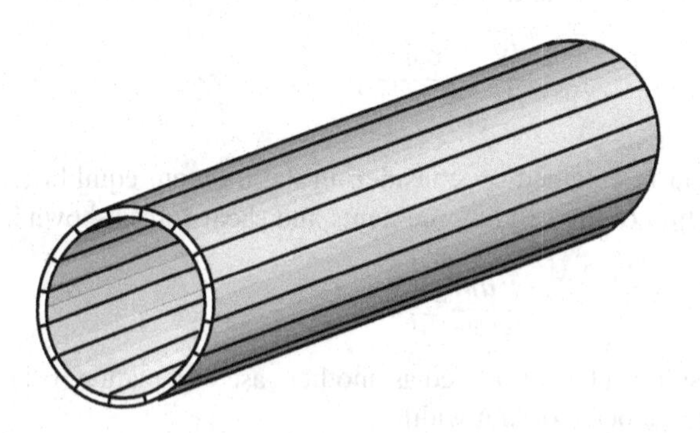

*Figure 9.5: The shell considered as a set of connected beams*

The attentive reader should immediately recognize this equation as having the same form as equation (7.5) for the bending of a beam on an elastic foundation. To understand the reason for this similarity, it might be helpful to regard the shell as a set of strips, as shown in Figure 9.5. Each strip can be considered as an elastic beam whose displacement in the radial direction is opposed by the elastic action of the circumferential stress $\sigma_2$.

All of the solution methods[2] developed in Chapter 7 can be applied to the solution of equation (9.20). However, the most important conclusion to be drawn from this equation, as in §7.3, is that the displacement due to radial loading of the shell will be localized, decaying exponentially with distance from the load. As before, we define a characteristic decay length

$$l_0 = \frac{1}{\beta} = C\sqrt{at} \,, \tag{9.22}$$

where the numerical factor

$$C = \sqrt[4]{\frac{1}{3(1-\nu^2)}} \tag{9.23}$$

lies in the very restricted range $0.76 < C < 0.82$ for $0 < \nu < 0.5$. Thus, the characteristic length is significantly smaller than the radius $a$, since $t \ll a$. As an example, a steel shell of radius 1 m and wall thickness 10 mm will have a characteristic length of 55 mm.

It follows that all practical shells are 'long' in the sense of Chapter 7 and can therefore be analyzed using the infinite or semi-infinite solutions of §§7.4, 7.2.1.

An alternative form for the right hand side of equation (9.20) can be obtained by noting that the membrane theory predicts that the circumferential stress $\sigma_2 = pa/t$ for the cylinder and hence the predicted membrane displacement is

$$u_r^m = \frac{(\sigma_2 - \nu\sigma_1)a}{E} = \frac{pa^2}{Et} - \frac{\nu\sigma_1 a}{E} \,,$$

using (8.5). We can substitute this result into (9.20), obtaining

$$\frac{d^4 u_r}{dz^4} + 4\beta^4 u_r = 4\beta^4 u_r^m \,. \tag{9.24}$$

This shows that the membrane solution (8.5) gives the exact value for the radial displacement[3] as long as the loading is a polynomial of third degree or below in $z$.

---

[2] Readers who have not studied Chapter 7 will find it helpful to review §§7.2–7.5 in parallel with the present derivations.

[3] A similar result holds for a beam on an elastic foundation subjected to polynomial loading and corresponds to the case where $u = w(z)/k$ and the loading is transmitted directly to the foundation.

### 9.4.1 Solution strategy

The first stage in the analysis of any axisymmetric shell is to find the 'membrane solution', using the methods of Chapter 8. We can then identify the regions of the shell (typically the ends of the shell or discontinuities of radius, thickness or loading) at which the membrane theory predicts discontinuities of radial displacement or slope. The bending theory is then used to construct a local solution, treating the shell as infinite or semi-infinite.

The equilibrium equation used to determine the meridional membrane stress $\sigma_1$ is unaffected by the presence of local bending, so this value carries over into the bending calculation. By contrast, equation (8.2) is modified by the presence of shear forces [see equation (9.15) above] and hence the circumferential stress $\sigma_2$ generally deviates from the value predicted by the membrane theory in a region of local bending. However, once $\sigma_1$ is determined, we can solve (9.20) or (9.24) for $u_r$ with appropriate end conditions. The local value of $\sigma_2$ can then be recovered from equation (9.17).

It is convenient to summarize these steps as follows:-

(i) Find the axial membrane stress $\sigma_1$ as in §8.1 by making a cut perpendicular to the axis, writing an equilibrium equation, and summing forces along the axis.
(ii) Use this result and the internal pressure $p$ to determine the right-hand side of equation (9.20), which may be a function of $z$.
(iii) Use the techniques of Chapter 7 to obtain the general solution of equation (9.20), noting that for uniform or low order polynomials we can find a simple particular solution and that the homogeneous solution is defined by equation (7.11) or (7.12). Usually the shell can be considered semi-infinite, so we only need the exponentially decaying terms and these can be written in the form (7.18).
(iv) Find the constants from the end conditions, using (9.13, 9.19) if necessary to convert force or moment end conditions to conditions on $u_r$.
(v) Once $u_r$ is known, $\sigma_2$ can be determined from (9.17).
(vi) The bending moment $M_z$ is given by (9.13) and the maximum stresses are then given by

$$\sigma_z^{\max} = \sigma_1 \pm \frac{6M_z^{\max}}{t^2} \; ; \quad \sigma_\theta^{\max} = \sigma_2 \pm \frac{6\nu M_z^{\max}}{t^2} \; . \tag{9.25}$$

**Example 9.1**

*A cylindrical steel tube of mean diameter 50 mm and wall thickness 4 mm is built in to a rigid wall at one end and loaded by an axial force of 60 kN, as shown in Figure 9.6. Find the maximum tensile stress in the tube. There is no internal pressure and the elastic properties for steel are $E=210$ GPa, $\nu=0.3$.*

The axial load of 60 kN will produce a meridional stress $\sigma_1$ given by the equilibrium equation

$$\pi \times 50 \times 4 \times \sigma_1 = 60 \times 10^3$$

and hence

$$\sigma_1 = \frac{60 \times 10^3}{\pi \times 50 \times 4} = 95.5 \text{ MPa}\,.$$

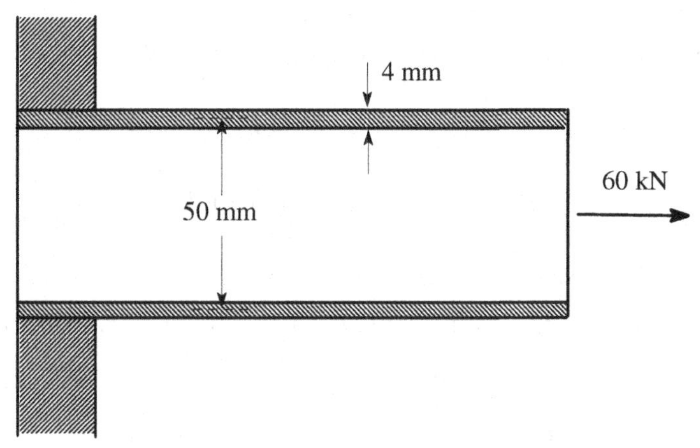


*Figure 9.6*

The membrane solution gives $\sigma_2 = 0$, from equation (8.2), since there is no internal pressure and $R_1 = \infty$ for the cylinder. Thus the membrane radial displacement will be

$$u_r^m = -\frac{\nu \sigma_1 a}{E} = -\frac{0.3 \times 95.5 \times 10^6 \times 0.025}{210 \times 10^9} = -3.41 \times 10^{-6} \text{ m}\,.$$

The negative sign indicates that the displacement is radially inwards — the diameter of the tube gets smaller as a result of Poisson's ratio strains.

This radial displacement is prevented at the built-in support, where $u_r = 0$ and $du_r/dz = 0$. There will therefore be a region of local bending.

We first calculate

$$\beta = \sqrt[4]{\frac{3(1-\nu^2)}{a^2 t^2}} = \sqrt[4]{\frac{3 \times 0.91}{25^2 \times 4^2}} = 0.129 \text{ mm}^{-1} \quad (129 \text{ m}^{-1})\,,$$

corresponding to a characteristic length, $l_0 = 7.78$ mm.

The membrane displacement $u_r^m$ is constant (independent of $z$) and hence a suitable particular integral of equation (9.24) is $u_r = u_r^m$. A sufficiently general solution of this equation can therefore be written

$$u_r = u_r^m + B_3 f_1(\beta z) + B_4 f_2(\beta z)\,,$$

using equation (7.18) from §7.2 for the homogeneous solution. The corresponding derivative is

$$\theta(z) \equiv \frac{du_r}{dz} = -B_3 \beta f_3(\beta z) + B_4 \beta f_4(\beta z)\,,$$

from (7.19) and the boundary conditions $u_r = du_r/dz = 0$ therefore require that

$$B_3 = -u_r^m \;\; ; \;\; -B_3 + B_4 = 0 \, ,$$

with solution

$$B_3 = B_4 = 3.41 \times 10^{-6}\,\text{m} \, .$$

The bending moment $M_z$ is given by equation (9.13, 7.23) as

$$M_z = -D\frac{d^2u_r}{dz^2} = -2D\beta^2[B_3 f_2(\beta z) - B_4 f_1(\beta z)] \, ,$$

where

$$D = \frac{Et^3}{12(1-\nu^2)} = \frac{210 \times 10^9 \times 4^3 \times 10^{-9}}{12 \times 0.91} = 1230\,\text{Nm} \, .$$

The maximum value of $M_z$ occurs at $z = 0$ and is

$$M_z^{\max} = 2D\beta^2 B_4 = 2 \times 1230 \times 129^2 \times 3.41 \times 10^{-6} = 138.6\,\text{Nm per m} \, .$$

The corresponding maximum bending stress is

$$\sigma_{\max} = \frac{6M_z^{\max}}{t^2} = \frac{6 \times 138.6}{4^2} = 52.0\,\text{MPa}.$$

This stress is additive to the membrane stress $\sigma_1 = 95.5$ MPa. The deformed shape of the tube is as shown in Figure 9.7, from which we conclude that the maximum tensile stress (the most extended fibres) occur at the outside of the shell near $z = 0$, since this is the outer radius of the curvature in the cross-sectional view. At this point, we have

$$\sigma_{zz} = 95.5 + 52.0 = 147.5\,\text{MPa} \, .$$

Alternatively, we can use equation (9.5) to obtain

$$\sigma_{zz} = \sigma_1 + \frac{M_z \eta}{t^3} \, ,$$

noting that $M_z$ is positive, to determine the maximum stress to be at $\eta = t/2$, where

$$\sigma_{zz} = 95.5 + \frac{12 \times 138.6 \times 2}{4^3} = 147.5\,\text{MPa} \, .$$

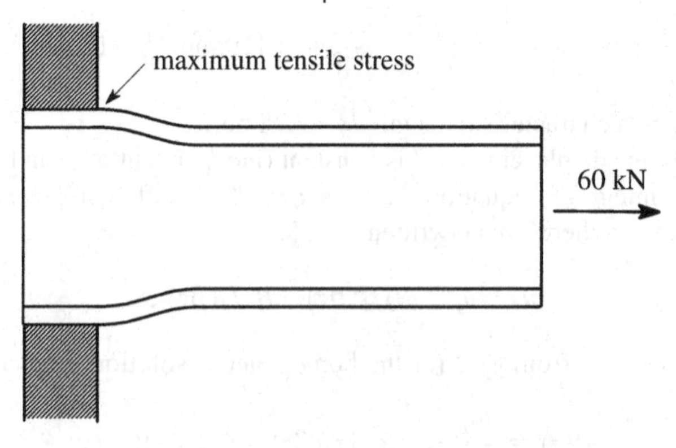


*Figure 9.7: Deformed shape of the tube*

A bending moment $M_\theta$ will also be induced in the circumferential direction and in some situations this can define the maximum tensile stress, when combined with a substantial circumferential membrane stress $\sigma_2$. We first calculate $\sigma_2$, using equation (9.17). At the support, $u_r = 0$ and we have

$$\sigma_2 = \nu\sigma_1 = 0.3 \times 95.5 = 28.7 \text{ MPa} .$$

The circumferential bending moment at the support is found from equation (9.8) as

$$M_\theta^{\max} = \nu M_z^{\max} = 0.3 \times 138.6 = 41.6 \text{ Nm per m}$$

and hence

$$\sigma_{\theta\theta}^{\max} = 28.7 + \frac{6 \times 41.6}{4^2} = 44.3 \text{ MPa} .$$

Once again, the maximum occurs at the outer surface of the shell, but it is smaller than $\sigma_{zz}^{\max}$.

## 9.5 Localized loading of the shell

Figure 9.8 *(a)* shows a cylindrical shell subjected to an axisymmetric radial load, $F_0$ per unit circumference. This problem is exactly analogous to the infinite beam problem of §7.4 and we conclude by analogy with equation (7.35) that the radial displacement will be given by

$$u_r = \delta f_3(\beta|z|) , \tag{9.26}$$

where $\delta$ is the displacement under the load.

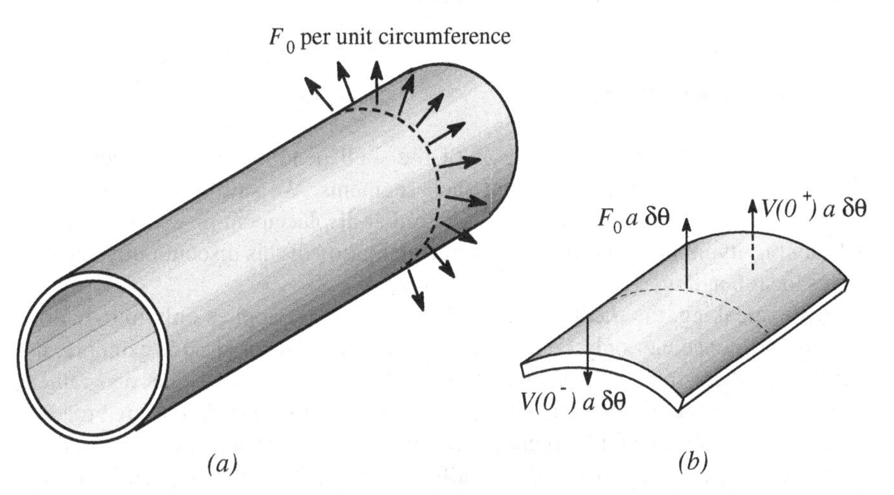


*Figure 9.8: (a) Cylindrical shell subjected to an axisymmetric radial load; (b) Equilibrium of a shell element*

To find the relation between $F_0$ and $\delta$, we note that

$$V = -D\frac{d^3u_r}{dz^3} = -4D\delta\beta^3 f_1(\beta|z|)\text{sgn}(z) ,$$

from equations (9.19, 9.26, 7.23), where $\text{sgn}(z)=1$ for $z>0$ and $-1$ for $z<0$. Figure 9.8 *(b)* shows a small element of shell including the load $F_0$ and we conclude that

$$F_0 = -V(0+) + V(0-) = 8\beta^3 D\delta ,$$

which permits us to rewrite (9.26) in the form

$$u_r = \frac{F_0}{8\beta^3 D} f_3(\beta|z|) . \tag{9.27}$$

The corresponding axial bending moment, $M_z$ is

$$M_z = -D\frac{d^2u_r}{dz^2} = \frac{F_0}{4\beta} f_4(\beta|z|) . \tag{9.28}$$

As in §7.4.1, we can generalize these results by superposition to obtain the solution for an arbitrary axisymmetric load acting over the segment $a<z<b$. We obtain

$$u_r = \frac{1}{8\beta^3 D}\int_a^b p(z')f_3(\beta|z-z'|)dz' \tag{9.29}$$

$$M_z = \frac{1}{4\beta}\int_a^b p(z')f_4(\beta|z-z'|)dz' , \tag{9.30}$$

where we note that a distributed axisymmetric load is equivalent to an internal pressure $p$ that is a function of the axial coordinate $z'$.

## 9.6 Shell transition regions

One of the most important applications of the shell bending theory concerns the design of transitions between different shell segments. We saw in §8.5.1 that the membrane theory predicts a discontinuity in radial displacement $u_r$ whenever there is a discontinuity in the radius $R_1$, and the reconciliation of this discontinuity always involves local bending.

Strictly speaking, the theory developed in this chapter applies only to cylindrical shells and hence can only be used for the cylindrical segment, for example, in the cylinder/sphere transition of Figures 8.21, 8.22. There is a distinct and rather more complicated theory for the bending of spherical shells.[4] The results are broadly similar to those for the cylinder (bending effects are localized and decay away from the loaded region), but the functions are affected by the fact that the shell becomes

[4] See for example, S.P. Timoshenko and S. Woinowsky-Krieger (1959), *Theory of Plates and Shells*, McGraw-Hill, New York, 2nd edn., §§129–132.

locally inclined to the axis and has a cross-sectional radius $r$, which varies with distance from the loaded region.

However, we have seen that shell bending effects are extremely localized near the discontinuity and in many cases these complicating factors do not reach significant proportions within the decay length, permitting the cylindrical shell bending theory to be used as an approximation. The appropriateness of this approximation can be assessed by calculating the characteristic length from equation (9.22) and determining the change in radius and shell inclination at the end of the bending zone. For example, Figure 9.9 shows a toroidal shell segment connecting a cylinder and a cone. The decay length in the cylinder is

$$l_0 = 0.78\sqrt{at} = 0.78\sqrt{500 \times 10} = 55 \text{ mm}\,.$$

The region affected by bending is therefore of the order of 55 mm to each side of the transition between the cylindrical and toroidal segments. On the toroidal side, the angle subtended at the centre $O_1$ is therefore

$$\alpha = \frac{55}{100} = 0.55 \text{ radians } = 31.6^{\text{o}}$$

and the *change* in cross-sectional radius, $r$ in this range is

$$100(1 - \cos 31.6^{\text{o}}) = 14.8 \text{ mm}\,.$$

Thus, in this region, the toroidal segment is sufficiently close to a cylinder to justify the use of the cylindrical bending theory, at least to obtain a first approximation to the magnitude of the bending stresses.

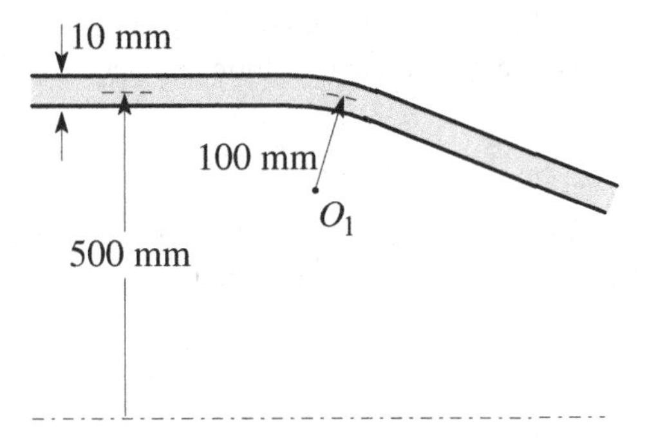


*Figure 9.9: Transition from a cylinder to a toroidal shell segment*

To determine the bending stresses associated with a transition between two shell segments, we first calculate the membrane stresses and displacements to determine the discontinuity in $u_r$ predicted at the transition. The reconciliation of this discontinuity is then effected by applying equal and opposite forces and moments to the two segments, as in Example 7.3. As in that case, the maximum bending moment occurs at the two points $\beta z = \pm\pi/4$.

**Example 9.2**

*Find the maximum tensile and compressive stresses for the toroidal to cylindrical transition of Figure 9.9, if the internal pressure is 2 MPa.*

We first calculate the membrane stresses and displacements. For both toroidal and cylindrical segments at the transition, we have

$$\sigma_1 = \frac{p \times \pi a^2}{2\pi a t} = \frac{2 \times 500}{2 \times 10} = 50 \text{ MPa} .$$

For the cylinder,

$$\sigma_2 = \frac{pa}{t} = \frac{2 \times 500}{10} = 100 \text{ MPa} ,$$

whereas for the toroidal segment ($R_1 = 100$ mm, $R_2 = 500$ mm),

$$\sigma_2 = R_2 \left( \frac{p}{t} - \frac{\sigma_1}{R_1} \right) = 500 \left( \frac{2}{10} - \frac{50}{100} \right) = -150 \text{ MPa} .$$

The membrane displacements are

$$\begin{aligned} u_r^m = \frac{(\sigma_2 - \nu\sigma_1)a}{E} &= \frac{(100 - 0.3 \times 50)500}{210 \times 10^3} = 0.202 \text{ mm (cylinder)} \\ &= \frac{(-150 - 0.3 \times 50)500}{210 \times 10^3} = -0.393 \text{ mm (toroidal segment)} . \end{aligned}$$

The bending solution must therefore reconcile a discontinuity $\delta = 0.202 + 0.393 = 0.595$ mm in $u_r$, half of which will be accommodated on each side of the transition. The bending moment will be zero at the transition and hence the bending displacement will be

$$u_r^b = \frac{\delta}{2} f_1(\beta z) ,$$

by analogy with equation (7.24), where $\beta = 1/l_0 = 18.2 \text{ m}^{-1}$ (the value of $l_0 = 55$ mm was calculated above). The stiffness of the shell is

$$D = \frac{Et^3}{12(1-\nu^2)} = \frac{210 \times 10^9 \times 10^3 \times 10^{-9}}{12 \times 0.91} = 19{,}230 \text{ Nm}$$

and the axial bending moment is therefore

$$M_z = -D \frac{d^2 u_r}{dz^2} = -D\delta\beta^2 f_2(\beta z) .$$

The maximum value of this expression occurs at $\beta z = \pi/4$ and is

$$M_z^{max} = 0.322 D\delta\beta^2 = 0.322 \times 19{,}230 \times 0.595 \times 10^{-3} \times 18.2^2 = 1258 \text{ Nm per m} .$$

The corresponding maximum bending stress is

$$\sigma = \frac{6M_z^{\max}}{t^2} = \frac{6 \times 1258}{10^2} = 75.5 \text{ MPa} ,$$

giving a maximum axial stress

$$\sigma_{zz}^{\max} = 50 + 75.5 = 125.5 \text{ MPa} .$$

We should also check to ensure that the circumferential stress $\sigma_{\theta\theta}$ nowhere exceeds this maximum value. We recall that the circumferential membrane stress

$$\sigma_2 = \frac{Eu_r}{a} + \nu\sigma_1 ,$$

from equation (9.17). For most of the shell, this will be *lower* than the limiting value of 100 MPa in the cylindrical segment, since the transition restrains the shell from expansion. However, the oscillatory nature of the decaying function $f_1(x)$ ensures that there is an axial location at which $u_r > u_r^m$ and hence $\sigma_2 > 100$ MPa. The maximum $u_r$ will occur where

$$\theta(z) = \frac{\beta\delta}{2} f_3(\beta z) = 0$$

and hence $\cos(\beta z) + \sin(\beta z) = 0$. The first solution of this equation is at $\beta z = 3\pi/4$, where

$$u_r^b = -\frac{\delta}{2} f_1\left(\frac{3\pi}{4}\right) = -\frac{0.595 \times (-0.067)}{2} = 0.020 \text{ mm} .$$

We therefore have,

$$\sigma_2^{\max} = \frac{210 \times 10^3(0.202 + 0.020)}{500} + 0.3 \times 50 = 108.2 \text{ MPa} .$$

In addition, the circumferential bending moment at this point is

$$M_\theta = \nu\beta^2 D f_2\left(\frac{3\pi}{4}\right) = 0.3 \times 19{,}230 \times 0.595 \times 10^{-3} \times 18.2^2 \times 0.067 = 76 \text{ Nm per m} ,$$

contributing an additional stress

$$\sigma_{\theta\theta}^b = \frac{6 \times 76}{10^2} = 4.6 \text{ MPa} .$$

However, it is clear that the maximum tensile stress in the shell is the axial stress $\sigma_{zz}^{\max} = 125$ MPa.

### 9.6.1 The cylinder/cone transition

The toroidal shell of Example 9.2 is often used to form a transition between a cylindrical and a conical shell segment, as shown in Figure 9.10. In this section, we shall discuss the criteria that might be used to determine an appropriate transition radius, $R$.

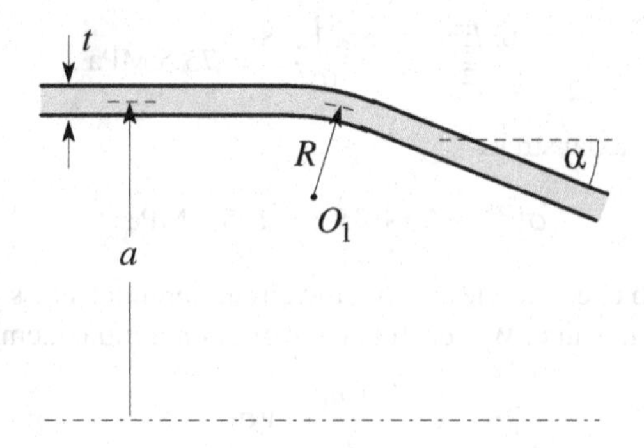


*Figure 9.10: Toroidal transition from a cylinder to a cone*

If $R$ is small, the circumferential membrane stress $\sigma_2$ will be negative (compressive) in the toroidal segment and the displacement discontinuity $\delta$ to be reconciled by bending will be relatively large. In this case, the largest tensile stress in the shell will be the axial stress $\sigma_{zz}$, as in Example 9.2, and it will be dominated by the bending term $6M_z/t^2$.

Increasing $R$ will reduce the bending stresses, but the total stress cannot be lower than the membrane stresses $\sigma_1, \sigma_2$ and these are dominated by the circumferential stress $\sigma_2$ in the cylindrical and conical segments. Thus, there is nothing to be gained by increasing $R$ beyond the point at which the maximum total axial stress is equal to the largest value of $\sigma_2$. Depending on the semi-angle of the cone, this break-even point occurs in the range $R/a = 0.5 \sim 0.6$ (see Problem 9.7).

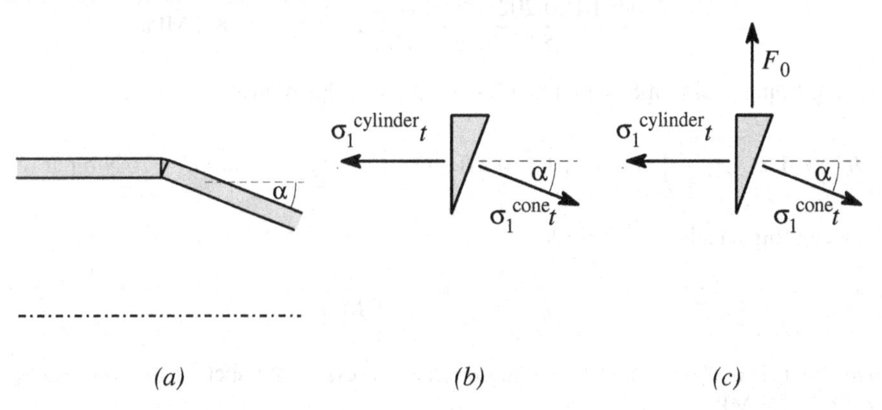


*Figure 9.11: (a) Abrupt cylinder/cone transition; (b) loading of the connecting element; (c) Equilibrium restored by a radial force*

At the other extreme, we could make an abrupt transition from the cylinder to the cone, as shown in Figure 9.11 *(a)*. In this case, the small triangular element connecting the segments will be subject to the loading shown in Figure 9.11 *(b)* and it cannot be in equilibrium. The meridional stresses are calculated in the usual way as

$$\sigma_1^{\text{cylinder}} = \frac{pa}{2t} \;\; ; \;\; \sigma_1^{\text{cone}} = \frac{pa}{2t\cos\alpha} \tag{9.31}$$

and axial equilibrium is satisfied.[5] To restore radial equilibrium, we would need to add a force $F_0$ per unit circumference as in Figure 9.11 *(c)*, where

$$F_0 = \sigma_1^{\text{cone}} t \sin\alpha = \frac{pa\tan\alpha}{2} \; . \tag{9.32}$$

In the actual shell, this force is not present and the transition zone therefore tends to pull radially inwards under the influence of the membrane stresses until additional circumferential stresses are developed sufficient to oppose the unbalanced radial force.

The bending stresses near the discontinuity can be estimated by first assuming that the force $F_0$ is provided. In this case, there will be some bending associated with the discontinuity in $u_r^m$ between the cylinder and the cone, but it will be small. We then remove the force $F_0$, which is equivalent to applying an *inward* concentrated radial load of the same magnitude. We have already solved this 'corrective' problem in §9.5 and the maximum bending moment produced is

$$M_z(0) = \frac{F_0}{4\beta} = \frac{pa\tan\alpha}{8\beta} \; ,$$

corresponding to a maximum bending stress of

$$\sigma_b^{\max} = \frac{6M}{t^2} = \frac{3pa\tan\alpha}{4\beta t^2} \; .$$

Unless $\alpha$ is very small, this will be substantially larger than the membrane stresses, which are of order $pa/t$.

One interpretation of the function of the toroidal transition is that is serves to distribute the force $F_0$ of equation (9.32) over the toroidal region, thereby reducing the maximum bending moment it produces. Thus, we might regard the toroidal segment as a sequence of infinitesimal changes of $\alpha$ at each of which an infinitesimal radial force is required. As long as $\alpha$ is not too large, the change in $\alpha$ per unit axial length is $d\alpha/dz = 1/R$ and the required inward radial force corresponds to pressure

$$p(z) = -\frac{pa}{2R} \; , \tag{9.33}$$

where the negative sign reflects the fact that the required distributed force is equivalent to an *external* pressure. The bending moment due to this load can be written down as

$$M_z(z) = -\frac{pa}{8\beta R}\int_0^{R\alpha} f_4(\beta|z-z'|)dz' \; , \tag{9.34}$$

using equation (9.30). This method can also be used for transitions that are not toroidal (i.e. that have variable radius $R$) always provided the maximum value of

[5] As indeed it must be, since this condition was used to determine equations (9.31).

$\alpha$ is not too large. From a design perspective, a good rule of thumb is that a transition will have a significant effect in moderating bending stresses if its length (e.g. the arc length $R\alpha$ for the toroidal transition of Figure 9.10) is about twice the decay length $l_0$.

### 9.6.2 Reinforcing rings

If a discontinuous change in slope $\alpha$ is unavoidable in a shell design, the stress concentration at the transition can be reduced by using a reinforcing ring, as shown in Figure 9.12 *(a)*. The ring resists the tendency for the shell to pull in locally at the discontinuity under the influence of the axial loading of the shell and thereby reduces both the circumferential membrane stress $\sigma_2$ and the local bending stresses.

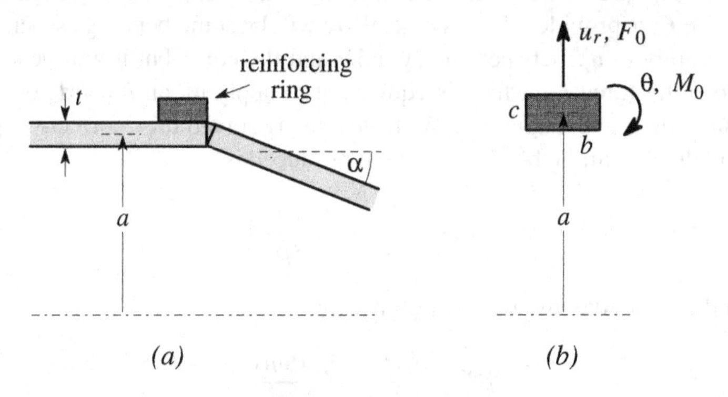


*Figure 9.12: (a) Cone/cylinder transition reinforced by a ring; (b) deformation of the ring due to a radial force and a moment*

A rigorous analysis of the geometry of Figure 9.12 *(a)* requires that the ring be treated as an elastic member whose rotation $\theta$ and radial expansion $u_r$ are functions of applied moment and radial load. For example, if the ring has a rectangular cross section $b\times c$ as shown in Figure 9.12 *(b)*, elementary calculations[6] show that

$$u_r = \frac{F_0 a^2}{Ebc} \quad ; \quad \theta = \frac{12M_0 a^2}{Eb^3c} \, , \tag{9.35}$$

where $F_0, M_0$ are respectively the radial force per unit circumference and the moment per unit circumference.

However, if the ring is fairly substantial compared with the shell ($b,c \gg t$), it can reasonably be approximated as rigid, in which case both shell segments have built in end conditions [$u_r(0) = \theta(0) = 0$].

Reinforcing rings can also be used to reduce the circumferential membrane stress in a cylindrical shell by restraining radial expansion. Figure 9.13 shows the deformed shape of a shell stiffened in this way.

[6] see Problems 9.12, 9.13 below.

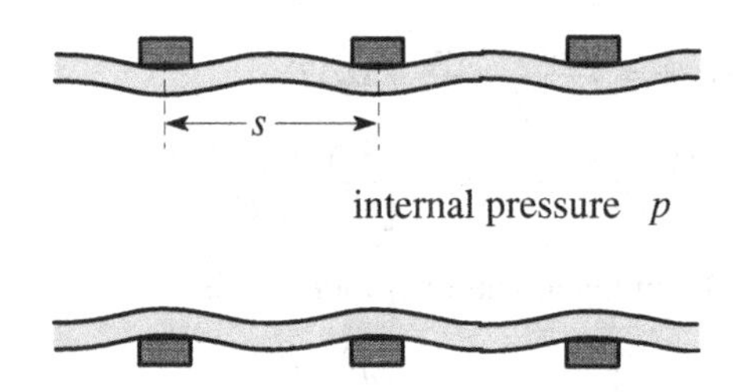


*Figure 9.13: Thin cylindrical shell stiffened by rings*

To be effective, the rings must be sufficiently closely spaced to ensure that the maximum radial displacement between rings is significantly lower than the unrestricted membrane displacement $u_r^m$. It can be shown[7] that this condition is satisfied if $\beta s < 1$ and hence the spacing $s < 0.78\sqrt{at}$.

## 9.7 Thermal stresses

Thin-walled vessels are often used for chemical processes and as components of thermal machines (e.g. boilers and heat exchangers). In these applications, additional loading will generally result from temperature gradients in the shell. Two distinct forms of thermal loading can be identified in an axisymmetric shell:-

(i) If there is heat flow through the wall thickness (e.g. as in a heat exchanger tube) there will be a radial temperature gradient which will induce additional shell bending stresses.
(ii) Even if the temperature is uniform through the thickness, axial temperature gradients will affect the membrane radial displacement $u_r^m$ in equation (9.24) and hence change the bending stress distribution.

These effects can be characterized by a temperature function of the form

$$\Delta T(\eta,z) = T_0(z) + \eta T_1(z) \ . \tag{9.36}$$

With this notation, the outward heat flux per unit area through the shell wall is

$$q(z) = -K\frac{\partial T}{\partial \eta} = -KT_1(z) \ , \tag{9.37}$$

where $K$ is the thermal conductivity of the shell material.

We recall from equations (1.14–1.16) that the effect of a local increase of temperature $\Delta T$ is to produce normal strains in all three directions of magnitude $\alpha\Delta T$. We can therefore generalize the preceding derivations by adding in these thermal strains at each point where Hooke's law was invoked — notably in equations (9.7, 9.10). The new version of (9.7) is

[7] The problem is analogous to Problem 7.23.

$$\frac{u_r}{a} = e_{\theta\theta} = \frac{\sigma_{\theta\theta} - \nu\sigma_{zz}}{E} + \alpha\Delta T$$
$$= \frac{\sigma_2 - \nu\sigma_1}{E} + \alpha T_0 + \left[\frac{12(M_\theta - \nu M_z)}{Et^3} + \alpha T_1\right]\eta\,.$$

The condition that $u_r$ be independent of $\eta$ then yields

$$M_\theta = \nu M_z - \frac{Et^3\alpha T_1}{12} \tag{9.38}$$
$$u_r = \left(\frac{\sigma_2 - \nu\sigma_1}{E} + \alpha T_0\right)a\,. \tag{9.39}$$

Equation (9.10) is generalized to

$$\frac{du_z}{dz} - \eta\frac{d^2u_r}{dz^2} = \frac{\sigma_{zz} - \nu\sigma_{\theta\theta}}{E} + \alpha T$$
$$= \frac{(\sigma_1 - \nu\sigma_2)}{E} + \alpha T_0 + \left[\frac{12(M_z - \nu M_\theta)}{Et^3} + \alpha T_1\right]\eta\,. \tag{9.40}$$

Equating coefficients of $\eta$ and using (9.38), we then have

$$\frac{du_z}{dz} = \frac{(\sigma_1 - \nu\sigma_2)}{E} + \alpha T_0 \tag{9.41}$$
$$M_z = -D\left[\frac{d^2u_r}{dz^2} + \alpha(1+\nu)T_1\right]. \tag{9.42}$$

A similar derivation to that in §9.4 then leads to the results

$$\sigma_2 = \frac{Eu_r}{a} + \nu\sigma_1 - E\alpha T_0 \tag{9.43}$$
$$V = -D\left[\frac{d^3u_r}{dz^3} + \alpha(1+\nu)\frac{dT_1}{dz}\right] \tag{9.44}$$

$$\frac{d^4u_r}{dz^4} + 4\beta^4 u_r = 4\beta^4 u_r^m - \alpha(1+\nu)\frac{d^2T_1}{dz^2}\,, \tag{9.45}$$

where the membrane displacement is

$$u_r^m = \frac{pa^2}{Et} - \frac{\nu\sigma_1 a}{E} + \alpha T_0 a\,. \tag{9.46}$$

We see from these results that the term $T_0(z)$ in equation (9.36) affects the bending solution only through the membrane displacement $u_r^m$. Once this has been calculated, the solution procedure is identical to that described in §9.4. In particular, bending stresses will be induced only at ends or discontinuities if $T_0$ is a constant or a linear function of $z$.

By contrast, the through-thickness temperature gradient $T_1(z)$ influences both the governing equation (9.45) and the moment-curvature relation (9.42) and it will always give rise to thermally induced bending stresses.

**Example 9.3**

*A long cylindrical heat exchanger tube of radius $a$ and thickness $t$ transmits a constant heat flux $q_0$ per unit area throughout its length. The temperature is independent of the axial coordinate $z$. Find the maximum bending stress in the tube if the material has Young's modulus $E$, Poisson's ratio $\nu$, thermal conductivity $K$, and coefficient of thermal expansion $\alpha$. There is no other loading of the shell.*

In this case, we have

$$T_1(z) = -\frac{q_0}{K} ,$$

from equation (9.37) and there is no axial temperature variation ($T_0=0$), so the membrane displacement $u_r^m=0$, from (9.46).

Substituting for $T_1$ into (9.45), we find that the right hand side of this equation is zero. It follows that $u_r=0$ is a particular integral of this equation and it is also the complete solution, since the tube is 'long', i.e the ends are far enough away not to influence the solution. We then have

$$M_z = \frac{\alpha(1+\nu)q_0 D}{K} ,$$

from (9.42) and

$$M_\theta = \frac{\alpha(1+\nu)\nu q_0 D}{K} + \frac{Et^3\alpha q_0}{12K} ,$$

from (9.38).

Substituting for $D$ and simplifying, we have

$$M_z = M_\theta = \frac{\alpha q_0 E t^3}{12K(1-\nu)}$$

and

$$\sigma_{\max} = \frac{6M}{t^2} = \frac{\alpha q_0 E t}{2K(1-\nu)} .$$

## 9.8 The ASME pressure vessel code

The consequences of pressure vessel failure can be extremely serious, particularly where toxic and/or high temperature fluids are involved. Designs are therefore usually based on standardized procedures described in the American Society of Mechanical Engineers' pressure vessel code.[8] Conformity to the code is also usually a requirement for obtaining liability insurance for the design. The recommendations of the code have evolved over many years, but in the area of stress analysis they are largely based on shell bending calculations such as those described in the present chapter.

[8] *ASME Boiler and Pressure Vessel Code — An American National Standard,* American Society of Mechanical Engineers, New York.

All common shell configurations are described in the code, but it is dangerous to use it as a substitute for thinking through the calculation process for a design. This is particularly true if the design is in any respect unconventional. It is a sobering thought that many code revisions have their origins in serious accidents involving vessels that were considered by their designers and insurers to be within the code at that time.

## 9.9 Summary

In this chapter, we have developed methods for determining the bending stresses in cylindrical shells subjected to axisymmetric radial, axial or thermal loading. The resulting governing equation has the same form as that for the bending of a beam on an elastic foundation and, as in that case, bending stresses are generally associated with discontinuities of geometry or loading.

Shell bending effects are important at the junction between two shells or at a point of localized loading, such as a support. Bending stresses are typically comparable with (and additive to) the membrane stresses and hence give stress concentration factors of the order of 2.

The effects are extremely localized, with an axial decay length significantly smaller than the shell radius. In design, the first step is to calculate the characteristic decay length given by equation (9.22). This enables us to determine whether a particular feature can be analyzed in isolation (as is usually the case).

Bending stresses at discontinuities can be reduced by using appropriate geometric transitions. The ASME Pressure Vessel Code provides detailed guidance for these cases, but a good rule of thumb is to use a transition whose axial length is twice the characteristic decay length.

## Further reading

A.P. Boresi, R.J. Schmidt, and O.M. Sidebottom (1993), *Advanced Mechanics of Materials,* John Wiley, New York, 5th edn., §10.7.

R.D. Cook and W.C. Young (1985), *Advanced Mechanics of Materials,* Macmillan, New York, §§6.1–6.4.

S.P. Timoshenko and S. Woinowsky-Krieger (1959), *Theory of Plates and Shells,* McGraw-Hill, New York, 2nd edn., §§114–119.

# Problems

## Sections 9.1–9.4

**9.1.** Develop a more rigorous version of the analysis of equations (9.7–9.8) noting that

(i) The circumferential strain

$$e_{\theta\theta} = \frac{u_r}{r} = \frac{u_r}{a+\eta}$$

and this expression should be used in place of $u_r/a$ in the left-hand side of equation (9.7).

(ii) The membrane stresses $\sigma_1, \sigma_2$ will generate Poisson's ratio strains

$$e_{rr} = \frac{\partial u_r}{\partial r} = \frac{\partial u_r}{\partial \eta} = -\frac{\nu(\sigma_1+\sigma_2)}{E} \; .$$

Show that this leads to the modified equation

$$(M_\theta - \nu M_z) = -\frac{t^3(1+\nu)\sigma_2}{12a}$$

in place of (9.8), and explain why this correction is negligible in view of the assumption $t \ll a$.

**9.2.** Figure P9.2 shows the design for the attachment of an 8 in diameter pipe to a cylindrical tank. The end of the pipe and the aperture in the tank are stiffened sufficiently for this region to afford a rigid (built-in) support to the end of the pipe.

Find the maximum tensile stress in the pipe when the internal pressure is 300 psi ($E = 30\times 10^6$ psi, $\nu = 0.3$).

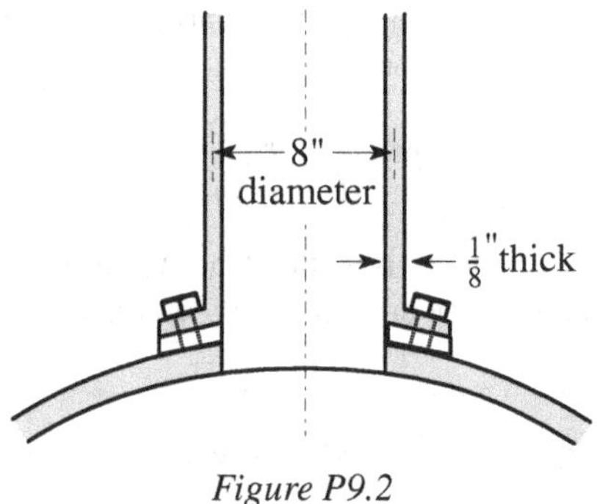


*Figure P9.2*

**9.3.** The vertical cylindrical water tank of Figure P9.3 is made from galvanized steel ($E = 30 \times 10^6$ psi, $\nu = 0.3$) of thickness $1/16$ inch. The bottom of the tank is a flat steel plate which is sufficiently rigid in extension to prevent radial displacement at the bottom, but which affords no restraint against rotation, $\theta$.

Find the location and magnitude of the maximum radial displacement of the tank when it is just full of water of density 62.5 lbs per $\text{ft}^3$. Find also the maximum tensile stress in the tank.

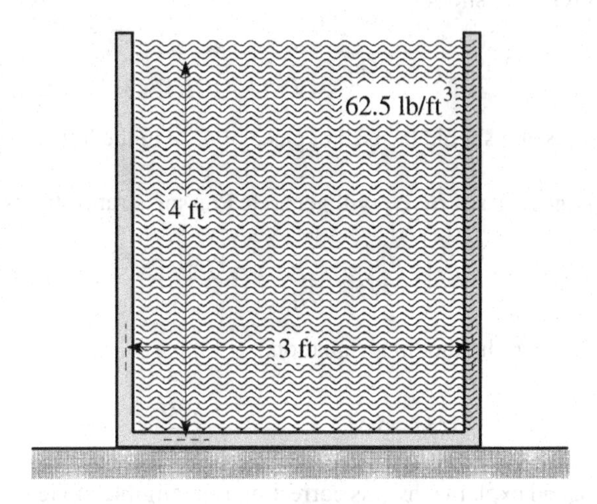


*Figure P9.3*

**9.4.** A cylindrical pressure vessel of radius 700 mm and wall thickness 8 mm has rigid end plates that can be considered to give built-in end conditions to the shell. Find the maximum tensile stress, if the internal pressure is 1.2 MPa and the material is steel ($E = 210$ GPa, $\nu = 0.3$).

**9.5*.** The analysis of the bending of circular plates shows that a moment $M_0$ per unit circumference applied to the edge of a plate of radius $a$ and thickness $t$, as shown in Figure P9.5, produces a local rotation[9]

$$\theta = \frac{M_0 a}{D(1+\nu)},$$

where the stiffness of the plate $D$ is defined by equation (9.14). A similar plate, simply supported around the edge and loaded by a uniform pressure $p_0$ experiences an edge rotation

$$\theta_1 = \frac{p_0 a^3}{8D(1+\nu)}.$$

[9] See S.P. Timoshenko and S. Woinowsky-Krieger (1959) *Theory of Plates and Shells*, McGraw-Hill, New York, 2nd edn., §15, §16.

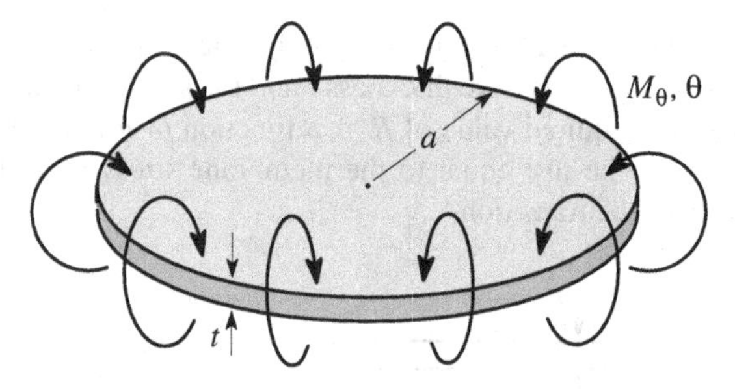


*Figure P9.5*

Two such plates are used to close the ends of a cylindrical shell of the same radius, thickness and material properties. Find an expression for the bending moment $M_0$ at the shell/plate junction when the vessel is subjected to an internal pressure $p_0$. Assume that the end plates completely resist radial expansion at the junction.

**9.6.** A rigid conical wedge is driven into a cylindrical tube by an axial force $F_0$, as shown in Figure P9.6. The wedge angle $\alpha \ll 1$ and the coefficient of friction at the contact point is $f$.

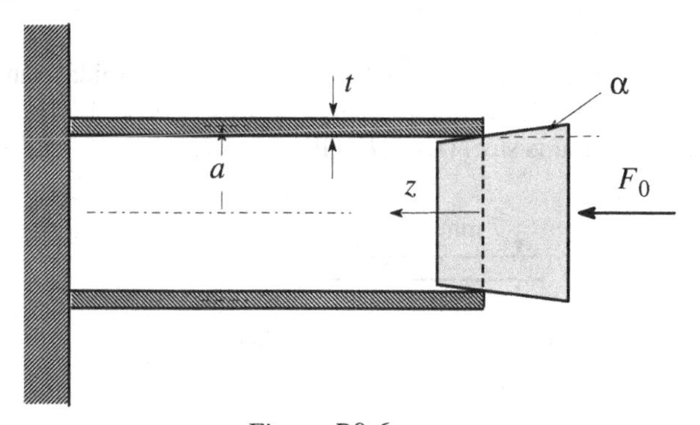


*Figure P9.6*

Assuming that contact occurs only at the end of the tube as shown, find an expression for the radial displacement of the tube $u_r$ as a function of $z$ and $F_0$. Hence show that there is a minimum angle $\alpha$ for which contact is restricted to the end and find it's value. The cylinder has a radius $a$ and thickness $t$ and the material has Young's modulus $E$ and Poisson's ratio $\nu$.

**Sections 9.5, 9.6**

**9.7.** A compressed air storage tank comprises a cylindrical segment and two hemispherical ends, each of radius 300 mm and wall thickness 4 mm. Find the maximum tensile stress in the tank when the internal pressure is 25 bar (1 bar = 100 kPa, $E=210$ GPa, $\nu=0.3$).

**9.8.** A toroidal shell of radius $R$ is joined to a cylindrical shell of radius $a$, as shown in Figure P9.8. Both segments have thickness $t$ and the shell is loaded by an internal pressure $p_0$. Find the required value of $R$ as a function of $a, t, p_0$, if the maximum axial stress $\sigma_{zz}^{\max}$ is to be just equal to the membrane stress $\sigma_2$ in the cylindrical segment, distant from the transition.

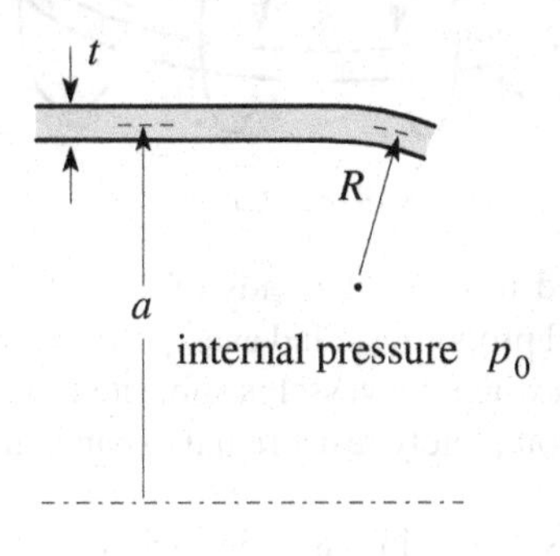


*Figure P9.8*

**9.9.** Find the maximum tensile stress for the cylindrical/toroidal transition of Figure P9.9 if the vessel is loaded by an internal pressure of 800 kPa. The shell thickness is 7 mm and the material is steel ($E = 210$ GPa, $\nu = 0.3$).

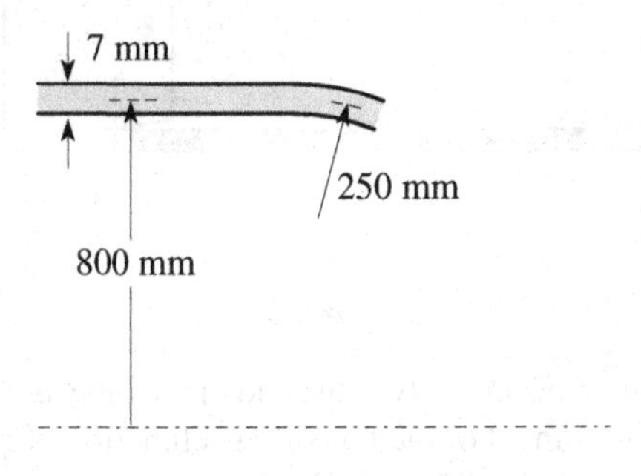


*Figure P9.9*

**9.10.** To reduce the bending stresses at a transition between a cylindrical shell of radius $a$ and a hemispherical shell, a transition region is proposed in which the radius

$$R_1 = \frac{ba}{z} ,$$

as shown in Figure P9.10.

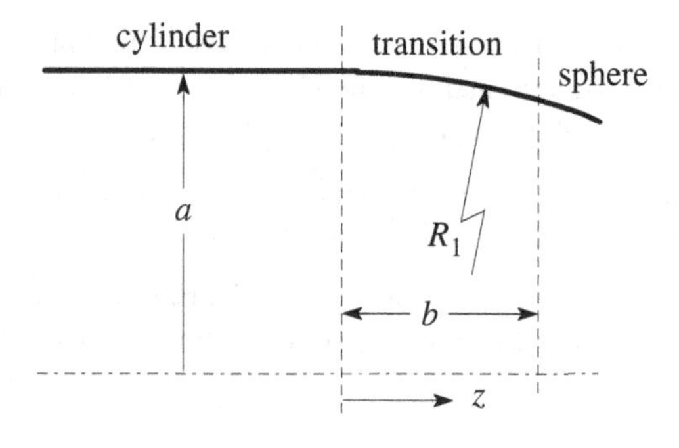


*Figure P9.10*

Find the bending moment at $z=0$, using the argument of §9.6.1 [equation (9.33)], assuming that $\beta b \gg 1$.

Discuss the relative benefits to be gained from such a transition.

**9.11.** A long shell of radius 500 mm and thickness 2 mm is locally stiffened by a reinforcing ring of 15 mm×10 mm rectangular cross section, as shown in Figure P9.11. The shell is subjected to an internal pressure $p_0=400$ kPa. Both the ring and the shell are made of steel for which $E=210$ GPa, $\nu=0.3$.

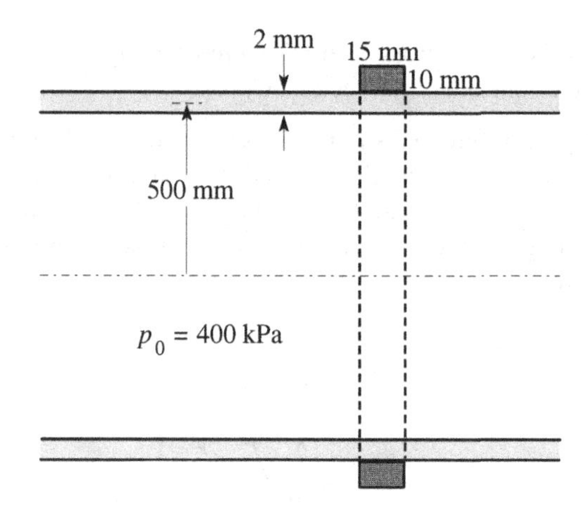


*Figure P9.11*

(i) Find the membrane displacement $u_r^m$ in the unstiffened shell;

(ii) Assuming the ring can be treated as rigid, find the radial force $F_0$ per unit circumference exerted on the ring by the shell;

(iii) Use equation (9.35) to determine the radial displacement of the ring due to this force. Does your result support the 'rigid' assumption used in (ii)?

**9.12.** Figure P9.12 shows a transition from a cylindrical shell of radius $a$ to a conical shell of semi-angle $\alpha$, stiffened by a reinforcing ring. Both shell segments have thickness $t$ and are made of the same material as the ring.

(i) Find the unbalanced radial force $F_0$ per unit circumference that must be resisted by the ring if the shell is loaded by an internal pressure $p_0$;

(ii) Estimate the required cross-sectional area of the ring if the radial displacement at the transition is to be limited to 10% of the membrane displacement $u_r^m$ in the cylindrical segment.

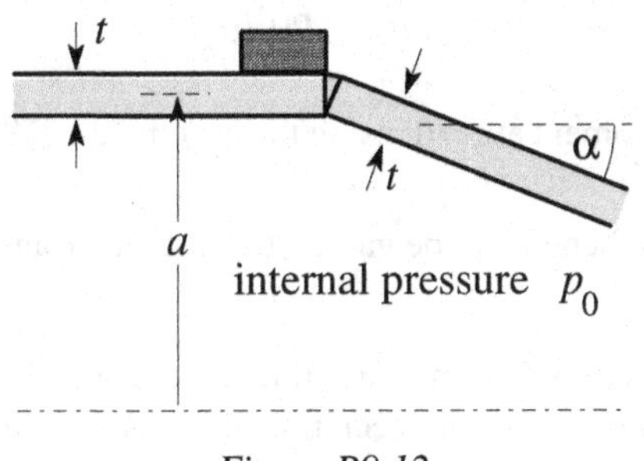


*Figure P9.12*

**9.13.** Figure P9.13 shows the cross section of a reinforcing ring whose mean radius (i.e. the radius to the centroid of the section, $C$) is $a$.

Determine the circumferential strain $e_{\theta\theta}$ as a function of the coordinates $\xi, \eta$ if the mean radius increases by $u_r$ whilst the section remains undistorted. Hence determine the radial force per unit circumference required to achieve this deformation, if the material has Young's modulus $E$. Assume that the dimensions of the cross section are small compared with $a$.

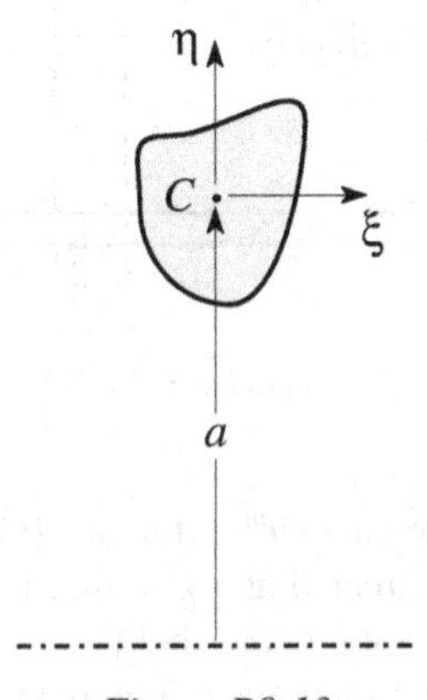


*Figure P9.13*

**9.14.** The ring cross section of Figure P9.13 is now rotated in the plane of the figure about the centroid $C$ through an angle $\theta$. As before, determine the circumferential strain $e_{\theta\theta}$ as a function of the coordinates $\xi, \eta$. Hence determine the moment $M_0$ per unit circumference that must be applied to the ring to achieve this rotation. Compare your result with equation (9.35) in the special case of a rectangular section $b \times c$.

**Section 9.7**

**9.15.** A heat exchanger tube of radius 60 mm and thickness 3 mm is made of copper ($E$= 121 GPa, $\nu$=0.33, $\alpha$=17×10$^{-6}$ per °C, $K$=381 W/m°C). The temperatures of the outside and inside surfaces of the tube $T_{out}, T_{in}$ vary with axial distance $z$ from the end according to the equations

$$T_{out} = 150\ ^{\circ}\mathrm{C} \quad ; \quad T_{in} = 30 + 120e^{-\lambda z}\ ^{\circ}\mathrm{C}\ ,$$

where $\lambda = 0.8\ \mathrm{m}^{-1}$ and $z$ is measured in metres.

The end of the tube is free of load and there is no internal or external pressure. Find the maximum tensile stress in the tube.

**9.16.** A long cylindrical shell of radius 200 mm and thickness 6 mm has a rigid (built-in) end at $z=0$. The temperature of the shell varies linearly with $z$ according to the equation

$$T(z) = 100z\ ^{\circ}\mathrm{C}\ ,$$

where $z$ is measured in metres. There is no internal pressure.

Find the maximum tensile stress in the shell if the relevant material properties are $E$=72 GPa, $\nu$=0.32, $\alpha$=22×10$^{-6}$ per °C.

**9.17.** The central section $-b<z<b$ of a long cylindrical shell of radius $a$ and thickness $t$ is heated to a mean temperature $T_0$, the rest of the shell being at zero temperature. Find the radial displacement of the shell and the axial bending moment $M_z$ at the point $z=0$.

# 10

# Thick-walled Cylinders and Disks

In this chapter, we shall consider a class of problems in which the stresses and displacements depend only on the radius $r$ in the cylindrical coordinate system $(r, \theta, z)$. The problems are therefore *axisymmetric* (i.e. there is no variation with the angle $\theta$) nor is there any variation with the distance $z$ along the axis.

Practical applications include a thick-walled cylinder loaded by internal or external pressure, a cylindrical grinding wheel loaded as a result of centrifugal acceleration during rotation, an automotive brake disk with an axisymmetric temperature distribution due to frictional heating, or the contact stresses developed when a cylindrical wheel is shrunk or pressed onto a cylindrical shaft.

## 10.1 Solution method

These applications are of considerable importance in their own right, but the study of axisymmetric problems has the additional advantage of introducing the reader to the methods used in the more advanced theories of elasticity, thermoelasticity and plasticity in a conveniently simple context.

The solution depends on the following three physical requirements:

(i) Each particle of the body must satisfy Newton's second law. This imposes a condition on the stress field in the body, since the tractions on the imaginary surface defining a particle are actually stress components. If there are no accelerations, this is an *equilibrium condition*.

(ii) The stresses and strains are related by a *constitutive law* for the material, which can be determined by performing appropriate experiments on idealized specimens. In the linear elastic régime, this will be Hooke's law [equations (1.14–1.16)], but we shall also consider problems in the plastic régime for elastic-perfectly plastic materials (§5.2.3), for which the constitutive law is the yield criterion for the material (§2.2.3).

(iii) The deformation must be kinematically possible, which places constraints on the admissible strain distributions. This is sometimes known as the *compatibility condition*.

J.R. Barber, *Intermediate Mechanics of Materials*, Solid Mechanics and Its Applications 175, 2nd ed., DOI 10.1007/978-94-007-0295-0_10, 

### 10.1.1 Stress components and the equilibrium condition

Figure 10.1 shows a small element of material in the cylindrical polar coordinate system $(r,\theta,z)$. For axisymmetric loading, there can be no shear stresses acting on the faces of this element and the only non-zero stress components are the normal stresses $\sigma_{rr}, \sigma_{\theta\theta}, \sigma_{zz}$, which are independent of $\theta$ and $z$, but which generally vary with radius $r$. This is recognized in the figure by the inclusion of the difference $\delta\sigma_{rr}$ between the normal stresses on the inner and outer curved surfaces of the element.

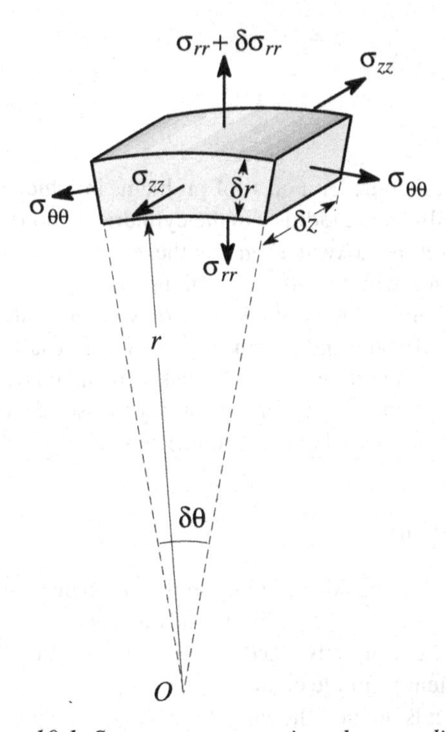


*Figure 10.1: Stress components in polar coordinates*

The stress component $\sigma_{rr} + \delta\sigma_{rr}$ acts over a 'rectangular' curved surface of area $(r+\delta r)\delta\theta\delta z$ and therefore exerts a radially-outwards force $(\sigma_{rr} + \delta\sigma_{rr})(r+\delta r)\delta\theta\delta z$, if $\delta\theta$ is sufficiently small. The forces acting on the other faces of the element can be found in the same way. Summing the forces in the radial direction, we can write Newton's law in the form

$$(\sigma_{rr} + \delta\sigma_{rr})(r+\delta r)\delta\theta\delta z - \sigma_{rr} r\delta\theta\delta z - 2\sigma_{\theta\theta}\delta r\delta z \sin\left(\frac{\delta\theta}{2}\right) = \rho a_r r\delta r\delta\theta\delta z\,, \tag{10.1}$$

where $\rho$ is the density of the material and $a_r$ is the outward radial acceleration of the element. In interpreting this equation, notice that

(i) The forces $\sigma_{\theta\theta}\delta r\delta z$ associated with the stress component $\sigma_{\theta\theta}$ are not co-linear because of the curvature of the element and they therefore have a resultant in the inward radial direction.[1]

(ii) The volume of the element is $r\delta r\delta\theta\delta z$ and hence its mass is $\rho r\delta r\delta\theta\delta z$.

The only state of motion leading to an axisymmetric stress field is that of uniform rotation about the axis at speed $\Omega$, in which case

$$a_r = -\Omega^2 r\,. \tag{10.2}$$

We now substitute (10.2) into (10.1), divide both sides of the equation by $r\delta r\delta\theta\delta z$, and take the limit as the small quantities $\delta r, \delta\theta \to 0$, obtaining the *differential equation of motion*

$$\frac{d\sigma_{rr}}{dr} + \frac{(\sigma_{rr} - \sigma_{\theta\theta})}{r} = -\rho\Omega^2 r\,. \tag{10.3}$$

In the special case where the body is not rotating ($\Omega = 0$), this reduces to the *equilibrium equation*

$$\frac{d\sigma_{rr}}{dr} + \frac{(\sigma_{rr} - \sigma_{\theta\theta})}{r} = 0\,. \tag{10.4}$$

### 10.1.2 Strain, displacement and compatibility

For axisymmetric loading, the most general deformation that can occur in the plane involves only a radial displacement $u_r$ that is a function of $r$ only. The normal strains $e_{rr}, e_{\theta\theta}$ can be written in terms of $u_r$ by considering the kinematics of the small element $ABCD$ in Figure 10.2, which deforms to the configuration $A'B'C'D'$ as a result of the radial displacement $u_r$. Notice that the arc $\delta\theta$ cannot change, since a sequence of elements around the circumference must add up to $2\pi$ and in axisymmetry the deformation of all elements is identical.

Writing the two strains as (extension)/(original length) for this element and then proceeding to the limit as $\delta r, \delta\theta \to 0$ we obtain

$$e_{rr} = \lim_{\delta r \to 0} \frac{u_r(r+\delta r) - u_r(r)}{\delta r} = \frac{du_r}{dr} \tag{10.5}$$

$$e_{\theta\theta} = \lim_{\delta\theta \to 0} \frac{(r+u_r)\delta\theta - r\delta\theta}{r\delta\theta} = \frac{u_r}{r}\,. \tag{10.6}$$

---

[1] A similar argument was used in the derivation of the equilibrium equation (8.2) for membrane stresses in §8.2.

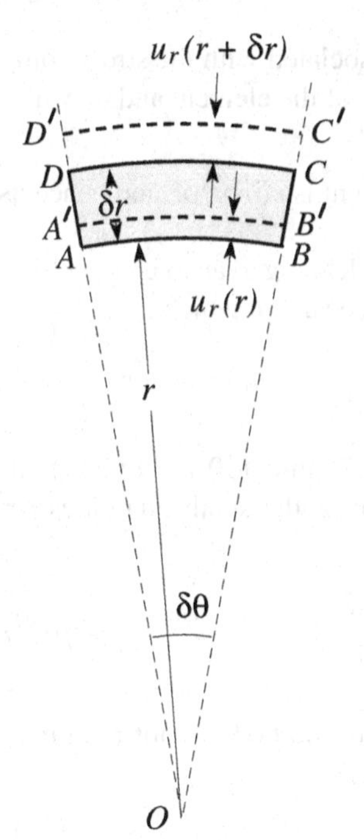


*Figure 10.2: Deformation associated with a radial displacement*

The *two* strains $e_{rr}, e_{\theta\theta}$ are defined in terms of *one* displacement component $u_r$ and we can therefore eliminate $u_r$ between equations (10.5, 10.6), obtaining

$$e_{rr} = \frac{d}{dr}(re_{\theta\theta}) = e_{\theta\theta} + r\frac{de_{\theta\theta}}{dr} ,$$

or

$$\frac{de_{\theta\theta}}{dr} + \frac{(e_{\theta\theta} - e_{rr})}{r} = 0 . \tag{10.7}$$

This is the *compatibility equation* for strains. It defines the condition that the strain components must satisfy if the particles of the body are to fit together in the deformed state.[2]

### 10.1.3 The elastic constitutive law

The equilibrium equation (10.4) [or the equation of motion (10.3)] relates the two stress components $\sigma_{rr}, \sigma_{\theta\theta}$, whereas the compatibility equation (10.7) relates the two strain components $e_{rr}, e_{\theta\theta}$. We need two equations to solve for two unknowns and we can therefore solve this problem in terms of either stress or strain if the relation between stress and strain (also known as the *constitutive law*) is known for

[2] For a discussion of compatibility equations in more general problems of elasticity, see J.R. Barber (2010), *Elasticity*, Springer, Dordrecht, Netherlands, 3rd edn., §2.2.

the material. Alternatively, we can use the strain-displacement relations (10.5, 10.6) and the constitutive law to express the stresses as functions of the single displacement component $u_r$, which is then determined from equation (10.3) or (10.4).

Notice that equations (10.3–10.7) are essentially mathematical relations, whereas the constitutive law is a property of the real material that must be found by an experiment such as a tensile test.

If the stresses are sufficiently small, most materials behave elastically and obey Hooke's law [equations (1.14–1.16)], which in polar coordinates can be written in the form

$$e_{rr} = \frac{\sigma_{rr}}{E} - \frac{\nu\sigma_{\theta\theta}}{E} - \frac{\nu\sigma_{zz}}{E} + \alpha T \tag{10.8}$$

$$e_{\theta\theta} = \frac{\sigma_{\theta\theta}}{E} - \frac{\nu\sigma_{zz}}{E} - \frac{\nu\sigma_{rr}}{E} + \alpha T \tag{10.9}$$

$$e_{zz} = \frac{\sigma_{zz}}{E} - \frac{\nu\sigma_{rr}}{E} - \frac{\nu\sigma_{\theta\theta}}{E} + \alpha T \,, \tag{10.10}$$

where $T$ should here be interpreted as the *change* in temperature from a datum at which the nominal geometry is defined.[3] Notice that each stress component produces a strain in its own direction, but also a 'Poisson's ratio' strain of opposite sign in the two orthogonal directions.

**Plane stress and plane strain**

The only state in which the stresses are truly independent of $z$ is that in which expansion in the $z$-direction is prevented, so that $e_{zz}=0$. This is known as a state of *plane strain*. Equation (10.10) can then be solved for $\sigma_{zz}$, with the result

$$\sigma_{zz} = \nu(\sigma_{rr} + \sigma_{\theta\theta}) - E\alpha T \,. \tag{10.11}$$

Substituting this result into (10.8, 10.9), we obtain

$$e_{rr} = \frac{(1-\nu^2)\sigma_{rr}}{E} - \frac{\nu(1+\nu)\sigma_{\theta\theta}}{E} + \alpha(1+\nu)T \tag{10.12}$$

$$e_{\theta\theta} = \frac{(1-\nu^2)\sigma_{\theta\theta}}{E} - \frac{\nu(1+\nu)\sigma_{rr}}{E} + \alpha(1+\nu)T \,. \tag{10.13}$$

These are the two-dimensional constitutive relations for axisymmetric plane strain.

Alternatively, if the cylinder is very short in the axial direction and the $z$-surfaces are free of traction, it can be argued that $\sigma_{zz}$ will always be small and hence can be neglected. This leads to an approximate solution appropriate for a thin disk and is known as a state of *plane stress*. Setting $\sigma_{zz}=0$, we obtain

$$e_{rr} = \frac{\sigma_{rr}}{E} - \frac{\nu\sigma_{\theta\theta}}{E} + \alpha T \tag{10.14}$$

$$e_{\theta\theta} = \frac{\sigma_{\theta\theta}}{E} - \frac{\nu\sigma_{rr}}{E} + \alpha T \,, \tag{10.15}$$

[3] In §1.5.4, we made this explicit by using the notation $\Delta T$.

which are the two-dimensional constitutive relations for axisymmetric plane stress.

The perceptive reader should immediately ask the question "How thin must the disk be for the plane stress approximation to be appropriate?" Clearly the answer cannot depend on the particular choice of units used to measure the thickness and hence must be stated in terms of an appropriate dimensionless ratio between the thickness and another significant dimension of the disk. In particular, the thickness needs to be small in comparison with the the inner and outer radii of the disk.

Comparing the relations (10.12, 10.13) and (10.14, 10.15), we notice that they are both linear equations connecting the same variables. They differ only in the multiplying constants. Thus, the mathematical solution procedure is the same for both plane stress and plane strain problems.

## 10.2 The thin circular disk

In this section, we shall develop elastic solutions for the thin circular disk using the plane stress approximation — i.e. using the constitutive law of equations (10.14, 10.15). The results will be applicable to thin rotating or stationary disks with arbitrary axisymmetric temperature distributions. Typical applications include automotive brake disks, grinding wheels and turbine blade disks.

We first solve equations (10.14, 10.15) for $\sigma_{rr}, \sigma_{\theta\theta}$ and use (10.5, 10.6) to express the results in terms of $u_r$, obtaining

$$\begin{aligned}\sigma_{rr} &= \frac{E}{(1-\nu^2)}(e_{rr}+\nu e_{\theta\theta}) - \frac{E\alpha T}{(1-\nu)} \\ &= \frac{E}{(1-\nu^2)}\left(\frac{du_r}{dr}+\frac{\nu u_r}{r}\right) - \frac{E\alpha T}{(1-\nu)} \end{aligned} \tag{10.16}$$

$$\begin{aligned}\sigma_{\theta\theta} &= \frac{E}{(1-\nu^2)}(e_{\theta\theta}+\nu e_{rr}) - \frac{E\alpha T}{(1-\nu)} \\ &= \frac{E}{(1-\nu^2)}\left(\frac{u_r}{r}+\nu\frac{du_r}{dr}\right) - \frac{E\alpha T}{(1-\nu)} \,. \end{aligned} \tag{10.17}$$

Substituting these expressions into the equation of motion (10.3), we obtain the ordinary differential equation

$$\frac{d^2u_r}{dr^2}+\frac{1}{r}\frac{du_r}{dr}-\frac{u_r}{r^2} = -\frac{\rho\Omega^2(1-\nu^2)r}{E}+\alpha(1+\nu)\frac{dT}{dr} \,, \tag{10.18}$$

for the radial displacement, $u_r$.

The general solution of this equation can be obtained using the identity

$$\frac{d^2u_r}{dr^2}+\frac{1}{r}\frac{du_r}{dr}-\frac{u_r}{r^2} = \frac{d}{dr}\left[\frac{1}{r}\frac{d}{dr}(ru_r)\right] \,,$$

which can be verified by expanding the right hand side, using differentiation by parts.

Rewriting (10.18) with this result, we have

$$\frac{d}{dr}\left[\frac{1}{r}\frac{d}{dr}(ru_r)\right] = -\frac{\rho\Omega^2(1-\nu^2)r}{E} + \alpha(1+\nu)\frac{dT}{dr} \tag{10.19}$$

and hence, after integration,

$$\frac{1}{r}\frac{d}{dr}(ru_r) = -\frac{\rho\Omega^2(1-\nu^2)r^2}{2E} + \alpha(1+\nu)T + C\,,$$

where $C$ is an arbitrary constant. Multiplying both sides of the equation by $r$ and integrating again yields

$$ru_r = -\frac{\rho\Omega^2(1-\nu^2)r^4}{8E} + \alpha(1+\nu)\int Trdr + \frac{Cr^2}{2} + D$$

and hence

$$u_r = -\frac{\rho\Omega^2(1-\nu^2)r^3}{8E} + \frac{\alpha(1+\nu)}{r}\int Trdr + \frac{Cr}{2} + \frac{D}{r}\,. \tag{10.20}$$

The stresses can then be recovered by substituting for $u_r$ in equations (10.16, 10.17). We obtain

$$\sigma_{rr} = -\frac{(3+\nu)\rho\Omega^2r^2}{8} - \frac{E\alpha}{r^2}\int rTdr + A + \frac{B}{r^2} \tag{10.21}$$

$$\sigma_{\theta\theta} = -\frac{(1+3\nu)\rho\Omega^2r^2}{8} + \frac{E\alpha}{r^2}\int rTdr - E\alpha T + A - \frac{B}{r^2}\,, \tag{10.22}$$

where $A, B$ are two new arbitrary constants related to $C, D$ through

$$A = \frac{EC}{2(1-\nu)} \quad ; \quad B = -\frac{ED}{(1+\nu)}\,,$$

which are introduced in order to avoid unnecessary algebraic complication in problems where only the stresses are required. The equivalent form of equation (10.20) using the constants $A, B$ is

$$u_r = -\frac{(1-\nu^2)\rho\Omega^2r^3}{8E} + \frac{\alpha(1+\nu)}{r}\int rTdr + \frac{A(1-\nu)r}{E} - \frac{(1+\nu)B}{Er}\,. \tag{10.23}$$

Equations (10.21–10.23) define the general solution for the problem of the axisymmetric circular disk. The temperature distribution $T$ and the rotational speed $\Omega$ will generally be known and the constants of integration $A, B$ are determined from one condition at each of the inner and outer radii of the disk. If the disk is solid, there is no inner radius, but a second equation for $A, B$ is obtained from the condition that the stresses must be bounded at the origin. We shall illustrate the solution procedure in two examples.

**Example 10.1 — Bursting speed of a grinding wheel**

*Figure 10.3 shows a grinding wheel of 150 mm inside diameter and 600 mm outside diameter. It is bonded to a cylindrical rigid shaft at the inside diameter and is unloaded at the outside diameter. The wheel has an effective modulus $E = 10$ GPa, $\nu = 0.3$ and the density is 2500 kg/m$^3$. Find the maximum safe rotational speed if the maximum permissible tensile stress is 10 MPa. (The maximum permissible rotational speed or the 'bursting speed' of a grinding wheel is specified by the manufacturer.)*

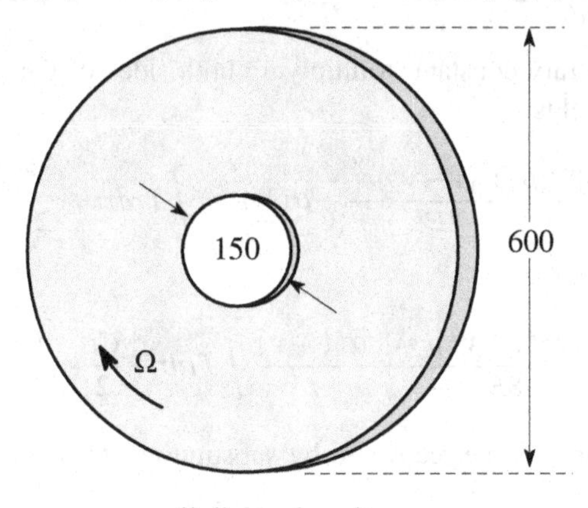


all dimensions in mm

*Figure 10.3: Grinding wheel*

In this problem, the principal loading is that due to rotation. Additional stresses may result from heating of the grinding wheel during operation, but we shall neglect this effect (dropping out the terms involving temperature) in the following calculation.

The curved outer edge of the wheel, corresponding to $r = 300$ mm, is unloaded. Referring back to Figure 10.1, we see that the stress component $\sigma_{rr}$ acts on the curved surface of the element[4] and hence we conclude that $\sigma_{rr} = 0$ at $r = 300$ mm. This condition and equation (10.21) gives the equation

$$0 = -\frac{3.3 \times 2500 \times \Omega^2 \times 0.3^2}{8} + A + \frac{B}{0.3^2} \; .$$

A second equation is obtained from the fact that the wheel is bonded to a rigid shaft at the inner edge $r = 75$ mm. This requires that the radial displacement of the wheel $u_r$ be zero at $r = 75$ mm, since otherwise a gap would open between the wheel and the shaft. Thus,

$$0 = -\frac{0.91 \times 2500 \times \Omega^2 \times 0.075^3}{8 \times 10 \times 10^9} + \frac{0.7 \times A \times 0.075}{10 \times 10^9} - \frac{1.3 \times B}{10 \times 10^9 \times 0.075} \; ,$$

[4] Notice that the *circumferential* stress $\sigma_{\theta\theta}$ does not act on the exposed curved surface of the disk and hence will not generally be zero at an unloaded edge.

from equation (10.23).

Solving these two simultaneous equations for $A, B$, we obtain

$$A = 89.87\Omega^2 \ ; \ B = 0.2653\Omega^2$$

and hence

$$\sigma_{rr} = \left(89.87 - 1031r^2 + \frac{0.2653}{r^2}\right)\Omega^2$$

$$\sigma_{\theta\theta} = \left(89.87 - 593.8r^2 - \frac{0.2653}{r^2}\right)\Omega^2 ,$$

from (10.21, 10.22), where the stresses will be in Pa if $r$ is in metres and $\Omega$ in radians per second.

The variation of these two stress components through the disk is illustrated in Figure 10.4, where we note that the maximum tensile stress is the value of $\sigma_{rr}$ at the inner radius $r = 0.075$ m, given by $131.2\Omega^2$. Setting this equal to the permissible stress of 10 MPa, we obtain

$$131.2\Omega_{max}^2 = 10 \times 10^6$$

and hence the maximum permissible rotational speed is

$$\Omega_{max} = \sqrt{\frac{10 \times 10^6}{131.2}} = 276 \text{ rad/s} = \frac{276 \times 60}{2\pi} = 2636 \text{ rpm} .$$

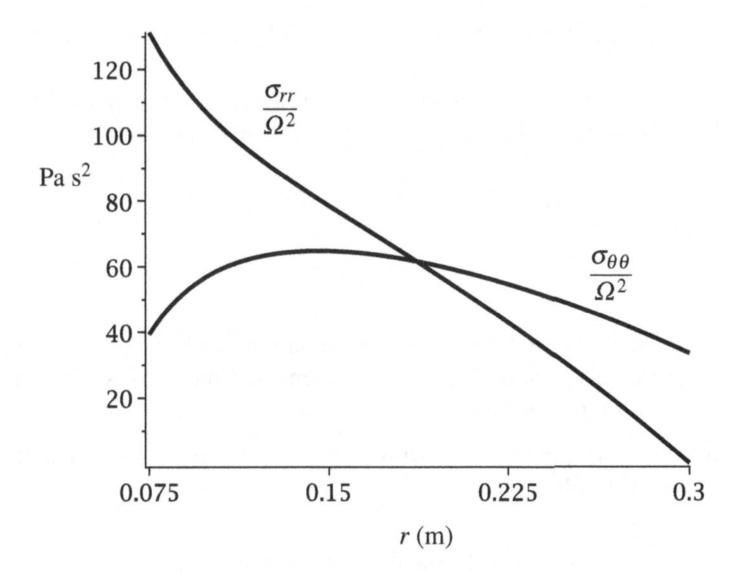


*Figure 10.4: Stress distribution in the grinding wheel*

**Example 10.2 — Automotive brake disk**

*An automotive brake disk is idealized as a uniform disk of diameter 10 in., as shown in Figure 10.5 (a). It is made of cast iron for which* $E=30\times10^6$ *psi and* $\alpha=7\times10^{-6}$ *per* $^{\circ}$*F. After a brake application, the temperature distribution, illustrated in Figure 10.5 (b), can be described by the equation* $T=(100+4r^2)$ $^{\circ}$*F, where r is the radial position in inches. Neglecting stresses due to rotation, find the stress distribution in the disk and determine the location and magnitude of the maximum shear stress.*

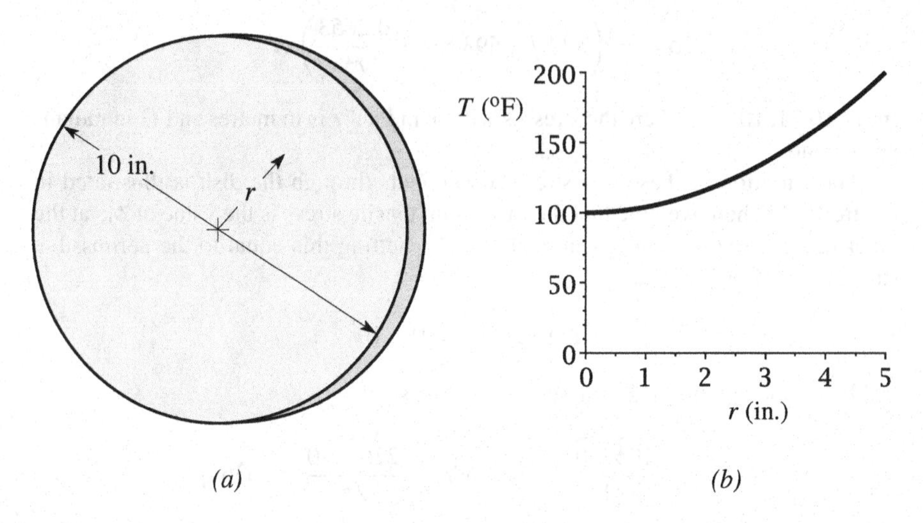


*Figure 10.5: The automotive brake disk*

General expressions for the stresses are given by equations (10.21, 10.22) as before. Substituting for $T$ and performing the integrations, we obtain

$$\sigma_{rr} = -210(50+r^2)+A+\frac{B}{r^2}$$
$$\sigma_{\theta\theta} = -210(50+3r^2)+A-\frac{B}{r^2}\ .$$

The disk has no central hole and hence one boundary condition is lost. However, an additional condition is obtained from the requirement that the stresses be bounded at $r=0$, which in turn implies that $B=0$.

At the outside radius, there is no traction and hence the stress component $\sigma_{rr}=0$ as in Example 10.1. It follows that

$$A = 210(50+5^2) = 15,750 \text{ psi}\ .$$

The complete stress field is therefore

$$\sigma_{rr} = 15750 - 210(50 + r^2) = 5250 - 210r^2$$
$$\sigma_{\theta\theta} = 15750 - 210(50 + 3r^2) = 5250 - 630r^2 ,$$

which is illustrated in Figure 10.6. Notice that the two stress components are equal at $r=0$. This is always the case in axisymmetric problems for the disk without a central hole, since all diameters are equivalent at this point.[5]

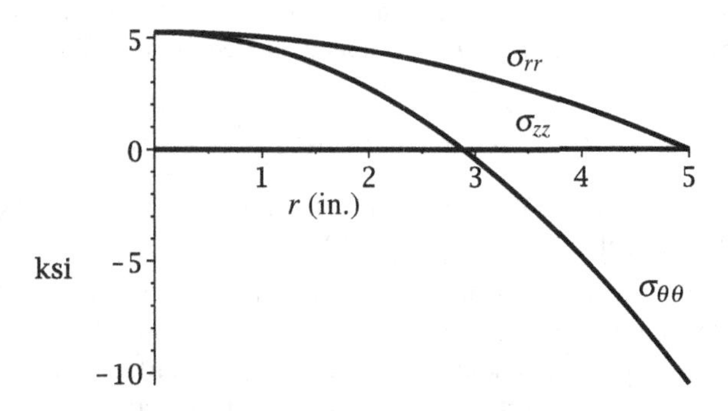


*Figure 10.6: Stress distribution in the brake disk*

We see from Figure 10.6 that the maximum normal stress occurs at $r=0$ (where both stress components are 5250 psi tensile) and at the outer radius, where $\sigma_{\theta\theta}$ is 10,500 psi and compressive (negative). The axisymmetry of the problem ensures that that there is no shear stress on any of the faces of the element in Figure 10.1 and hence that $\sigma_{rr}, \sigma_{\theta\theta}, \sigma_{zz}$ are principal stresses. Note that $\sigma_{zz}=0$ for the disk, because of the plane stress assumption. The maximum shear stress is half of the greatest difference between principal stresses (see §2.1). All three stresses are shown in Figure 10.6 and it is clear that the greatest difference between them occurs at the outside radius ($r=5$ inches) and is

$$\tau_{\max} = \frac{\sigma_{zz} - \sigma_{\theta\theta}}{2} = 5250 \text{ psi} .$$

## Discussion

These examples are illustrative of the kinds of design calculations that can be performed for circular disks. Notice that we considered only loading due to rotation in the grinding wheel and only thermoelastic loading in the brake disk, even though both components experience both rotation and heating during service. To explain why these choices were made, we first remark that

(i) the terms in the stress equations (10.21, 10.22) due to rotation are proportional to $\rho\Omega^2 r^2$;

[5] The same argument leads to the conclusion that the membrane stresses $\sigma_1, \sigma_2$ are equal on the axis for an axisymmetric shell which is continuous across the axis (see Example 8.3).

(ii) the thermal stress terms are proportional to $E\alpha T$;
(iii) a body that is subjected to a uniform increase in temperature increases in size, but is unstressed unless the boundaries are restrained in some way.

When a car is moving at 60 mph (88 ft/s), the wheels (and hence the brake disks) will be rotating at about 70 rad/s (670 rpm), where we have assumed a tyre outer radius of 1.25 ft. This is a low rotational speed and we therefore anticipate a small contribution to the stresses from the rotational terms. To confirm this, calculate $\rho\Omega^2r^2$ using the largest (outer) radius for $r$. The density of steel is about $7.4\times10^{-4}$ slug/in$^3$, so

$$\rho\Omega^2r^2 = 7.4\times10^{-4}\times70^2\times5^2 = 91 \text{ psi}\,.$$

This is small compared with the failure strength of typical steels ($\approx$ 30 ksi) and with the thermal stresses calculated in Example 10.2.

The grinding wheel is heated at the outer radius as a result of the grinding process. Grinding wheels are made of ceramic cutting materials bonded by a filler and will generally have low thermal conductivity. Thus, only the surface layers of the wheel can be expected to experience a significant increase in temperature. This might cause significant local circumferential stresses ($\sigma_{\theta\theta}$), but they will be compressive since the material wants to expand and is being prevented from doing so. Brittle materials are much stronger in compression than in tension. We also see from Example 10.1 above that the maximum stresses due to rotation occur at the centre of the wheel rather than the outside, so the questions of possible failure due to rotational stresses and due to local thermal stresses at the outer edge are independent of each other.

Preliminary thinking of this kind (perhaps supplemented by a few simple estimates) is essential when deciding which stress calculations to perform for design purposes.

## 10.3 Cylindrical pressure vessels

Figure 10.7 shows a thick-walled cylindrical vessel of inside radius $a$ and outside radius $b$, subjected to internal pressure $p_0$. Vessels of this kind are only needed when the pressure to be contained is very large. Boilers, gas tanks etc. generally operate at pressures small enough to permit the wall thickness to be small in comparison with the radius, in which case the simpler membrane theory of Chapter 8 can be used. Applications of thick-walled vessels include processes requiring extremely high pressure, such as the manufacture of synthetic diamonds, and testing machines to determine the properties of materials under high hydostatic pressure.[6]

[6] An important case is the determination of the viscosity of lubricants. In gearing and rolling contact bearings, the lubricant is squeezed to very high pressure and successful lubrication for such applications depends on the lubricant having a high viscosity at high pressures.

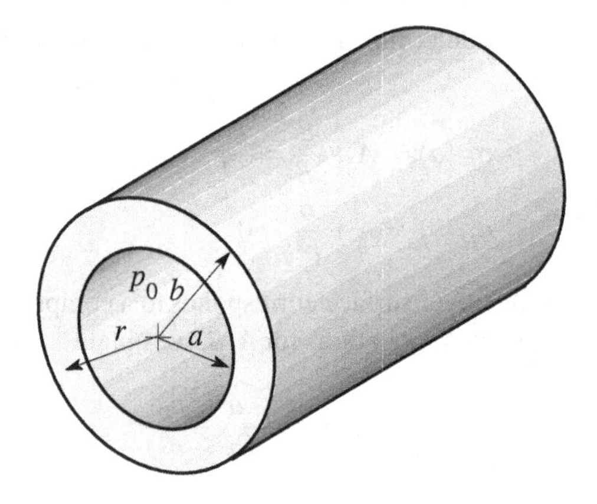


*Figure 10.7: Cylindrical vessel with internal pressure $p_0$*

If the ends of the vessel are prevented from moving, the plane strain constitutive relations (10.12, 10.13) are appropriate. These differ from the corresponding plane stress relations (10.14, 10.15) only in the multiplying constants and hence, mathematically, a problem in plane strain simply looks like a plane stress problem for a material with different material properties. In fact, it is easily verified that equations (10.12, 10.13) can be obtained from (10.14, 10.15) by making the substitutions

$$E = \frac{E'}{(1-\nu'^2)} \; ; \; \nu = \frac{\nu'}{(1-\nu')} \; ; \; \alpha = \alpha'(1+\nu') \tag{10.24}$$

and then dropping the primes.

It follows that the stresses and displacements in the cylinder can be obtained by making the same substitutions in equations (10.21–10.23) and are

$$\sigma_{rr} = -\frac{(3-2\nu)\rho\Omega^2 r^2}{8(1-\nu)} - \frac{E\alpha}{(1-\nu)r^2}\int rTdr + A + \frac{B}{r^2} \tag{10.25}$$

$$\sigma_{\theta\theta} = -\frac{(1+2\nu)\rho\Omega^2 r^2}{8(1-\nu)} + \frac{E\alpha}{(1-\nu)r^2}\int rTdr - \frac{E\alpha T}{(1-\nu)} + A - \frac{B}{r^2} \tag{10.26}$$

$$u_r = -\frac{(1-2\nu)(1+\nu)\rho\Omega^2 r^3}{8E(1-\nu)} + \frac{\alpha(1+\nu)}{(1-\nu)r}\int rTdr + \frac{A(1-2\nu)(1+\nu)r}{E} - \frac{(1+\nu)B}{Er} \, . \tag{10.27}$$

For the problem of the thick-walled cylinder, we might also be interested in the axial stress $\sigma_{zz}$, which can be obtained from (10.11) as

$$\begin{aligned} \sigma_{zz} &= \nu(\sigma_{rr} + \sigma_{\theta\theta}) - E\alpha T \\ &= -\frac{\nu\rho\Omega^2 r^2}{2(1-\nu)} - \frac{E\alpha T}{(1-\nu)} + 2\nu A \, . \end{aligned} \tag{10.28}$$

If there are no thermal or rotational effects, only the terms involving the constants $A, B$ will remain and these are determined by imposing boundary conditions at the inner and outer surfaces, i.e.

$$\sigma_{rr}(a) = A + \frac{B}{a^2} = -p_0$$
$$\sigma_{rr}(b) = A + \frac{B}{b^2} = 0\,.$$

Notice that the pressure on the inner surface corresponds to a compressive stress and is therefore negative. Solving these equations for $A, B$, we obtain

$$A = \frac{a^2 p_0}{(b^2 - a^2)} \;;\; B = -\frac{a^2 b^2 p_0}{(b^2 - a^2)} \tag{10.29}$$

and hence

$$\sigma_{rr} = \frac{a^2 p_0}{(b^2 - a^2)}\left(1 - \frac{b^2}{r^2}\right) \;;\; \sigma_{\theta\theta} = \frac{a^2 p_0}{(b^2 - a^2)}\left(1 + \frac{b^2}{r^2}\right)\,. \tag{10.30}$$

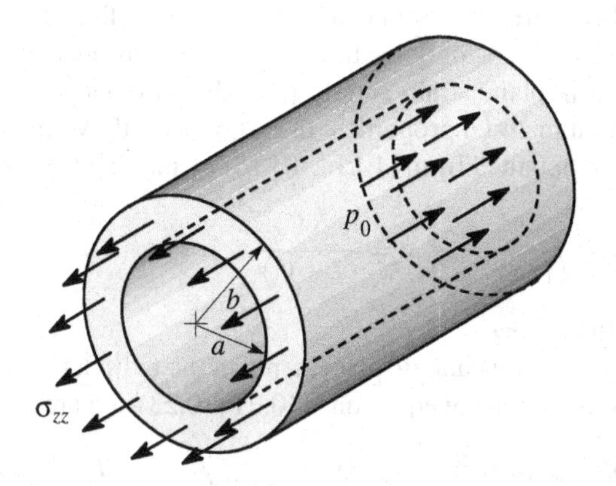


*Figure 10.8: Free-body diagram for determining $\sigma_{zz}$*

If the ends of the cylinder are prevented from moving, the axial stress is given by equation (10.28) as

$$\sigma_{zz} = 2\nu A = \frac{2\nu p_0 a^2}{(b^2 - a^2)}$$

and it is constant, i.e. it doesn't vary with radius $r$. However, we can generalize the solution to other end conditions by superposing a state of uniaxial tension, which will change $\sigma_{zz}$ by a constant without affecting the other two stress components (though it will change the radial displacement $u_r$, because of Poisson's ratio strains). The resulting axial stress will still be independent of $r$ and hence it can be found by considering the axial equilibrium of the cylinder. For example, if the vessel has closed ends and there are no external loads applied, the axial equilibrium of the free-body diagram of Figure 10.8 yields the equation

$$\pi(b^2 - a^2)\sigma_{zz} - \pi a^2 p_0 = 0 \,,$$

which states that the force exerted by the pressure $p_0$ on the end face (of area $\pi a^2$) is balanced by the force exerted by the axial stress over the annular cross section of the vessel [area $\pi(b^2-a^2)$]. It follows that

$$\sigma_{zz} = \frac{a^2 p_0}{(b^2 - a^2)} \,. \tag{10.31}$$

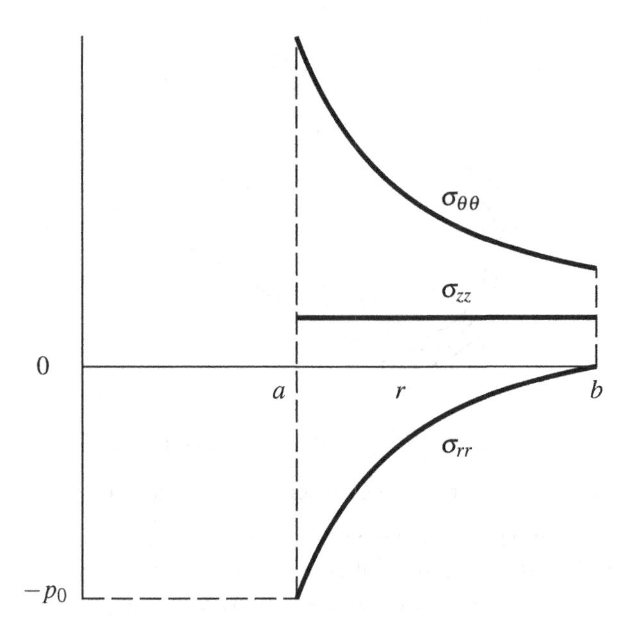


*Figure 10.9: Stress distribution in the cylindrical vessel*

The three stress components $\sigma_{rr}, \sigma_{\theta\theta}, \sigma_{zz}$ are presented graphically in Figure 10.9. Notice that the maximum tensile stress is $\sigma_{\theta\theta}$ at the inner radius, $r=a$, and is

$$\sigma_{\max} = \sigma_{\theta\theta}(a) = \frac{a^2 p_0}{(b^2 - a^2)} \left(1 + \frac{b^2}{a^2}\right) \,,$$

from equation (10.30).

Also, the maximum difference between principal stresses is $(\sigma_{\theta\theta} - \sigma_{rr})$ at the same location leading to

$$\tau_{\max} = \frac{\sigma_{\theta\theta}(a) - \sigma_{rr}(a)}{2} = \frac{2b^2 p_0}{(b^2 - a^2)} \,.$$

Thus, ductile failure is most likely to initiate at the inner radius $r=a$.

## 10.4 Composite cylinders, limits and fits

There are situations where it is advantageous to use an assembly of two concentric cylinders or disks, as shown in Figure 10.10. Usually, the central hole (inside diameter) of the outer cylinder is made slightly smaller than the outside diameter of the inner cylinder, so as to ensure that there is a firm fit between the components. Assembly can be achieved by forcing the cylinders together axially (usually using a lubricant) or by heating the outer cylinder[7] and/or cooling the inner one sufficiently for them to be assembled without force.

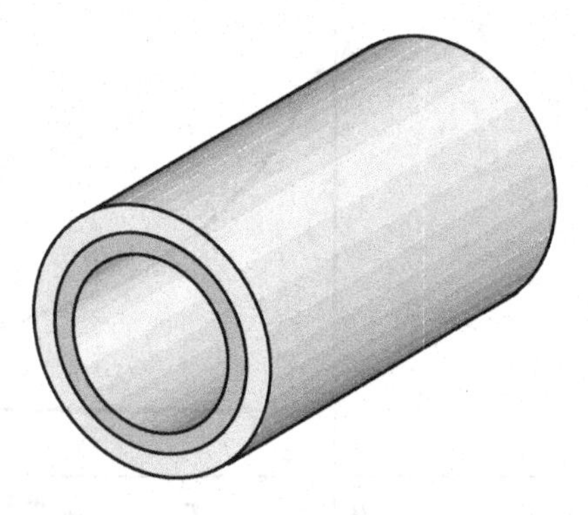

*Figure 10.10: A composite cylinder*

Advantages of using a composite cylinder assembly include

(i) It is a convenient way of mounting gears, pulleys and wheels on shafts. If the contact pressure at the interface after assembly is sufficiently large, it may not even be necessary to use a key to transmit torque, since high friction forces can be transmitted.
(ii) The two cylinders could be of different materials. For example, a corrosion-resistant (but possibly brittle) material could be used for the inner cylinder (to be in contact with a contained corrosive fluid) and a structurally strong material for the outer cylinder.
(iii) The pre-stress developed by the assembly process is beneficial and permits a composite cylinder with given inner and outer diameters to carry a larger internal pressure than a monolithic cylinder of the same material.

### 10.4.1 Solution procedure

Each of the two components of the cylinder or disk can be analyzed using the methods developed above. The stress and displacement fields will still be given by equations (10.21–10.23) or (10.25–10.27), but the arbitrary constants $A, B$ will generally

[7] Notice that when the outer cylinder is heated to a uniform temperature, all the dimensions, including the diameter of the hole, get bigger.

be different in the two components and will be denoted by $A_1, B_1$ for the inner and $A_2, B_2$ for the outer cylinder. Of course, the material constants may also differ and will be denoted similarly as $E_1, \nu_1, \alpha_1, E_2, \nu_2, \alpha_2$, respectively.

There are now four unknown constants and we require four equations to determine them. Two of these equations come from the conditions at the inner and outer radii, as before. Two additional conditions must be satisfied at the interface $r=c$.

(i) By Newton's third law, the contact pressure $p_c$ must act equally on both contacting surfaces, so that

$$p_c = -\sigma_{rr_1}(c) = -\sigma_{rr_2}(c) \tag{10.32}$$

(see Figure 10.11).

(ii) The fact that the surfaces are in contact imposes a kinematic constraint on the displacements at the interface. Suppose the inner radius of the outer cylinder is $c$ and the outer radius of the inner cylinder is $c+\delta$ where $\delta \ll c$. We describe $\delta$ as the *radial interference*. If the two cylinders now experience different radial displacements $u_{r_1}, u_{r_2}$, the final radii will be $c+u_{r_2}(c)$ and $c+\delta+u_{r_1}(c)$ and these must be equal (i.e. the two surfaces must be at the same place), if the cylinders are to be in contact. Thus

$$c+u_{r_2}(c) = c+\delta+u_{r_1}(c) \tag{10.33}$$

or

$$u_{r_2}(c) - u_{r_1}(c) = \delta \, . \tag{10.34}$$

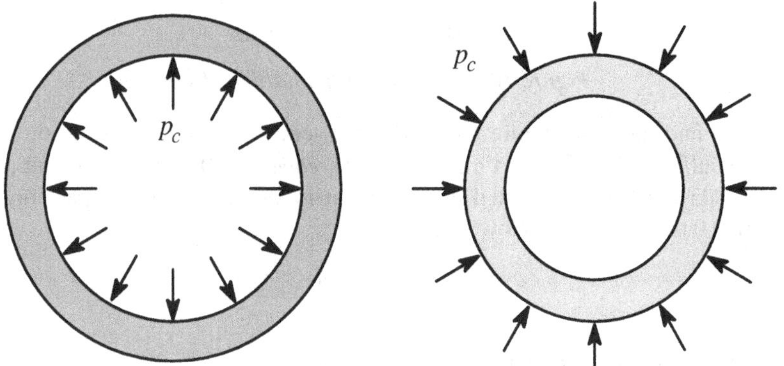


*Figure 10.11: Free-body diagram for the composite cylinder*

Usually the conditions at the inner and outer surfaces of the composite cylinder and the radial interference $\delta$ will be known and we shall wish to determine the interface pressure $p_c$ and the stress field in the cylinders. However, it is considerably easier to solve problems in which $p_c$ is prescribed instead of $\delta$, since then the problems for the two cylinders are uncoupled and can be solved separately. We can take advantage of this by pretending that $p_c$ is prescribed in all problems (writing it as a symbol in the resulting equations) and then eliminating or solving for it at the end. We shall illustrate this procedure with the following example.

**Example 10.3 — Wheel and tyre assembly**

*A wheel consisting of a solid steel disk of diameter 10 in. carries a hardened steel tyre of thickness 0.15 in., as shown in Figure 10.12. The radial interference at the interface before assembly is 0.012 in. and assembly is to be achieved by heating the tyre until it can be slipped over the wheel with a radial clearance of 0.005 in.*

*Find the temperature to which the tyre must be heated, and the maximum tensile stress in the tyre after the assembly has cooled down. Assume that* $E=30\times10^6$ *psi,* $\nu=0.3, \alpha=7\times10^{-6}$ *per °F for both components.*

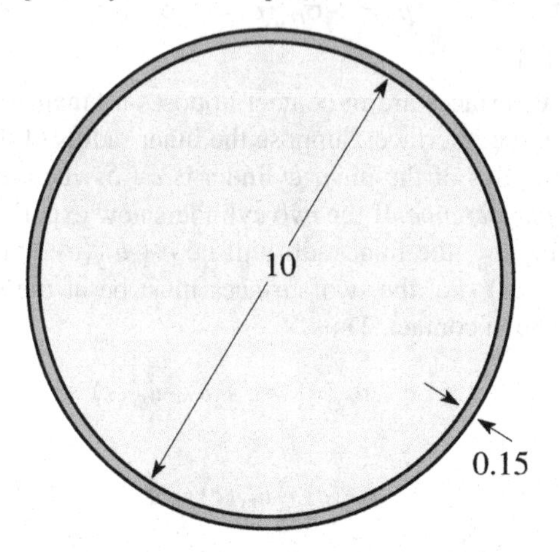


all dimensions in inches

*Figure 10.12: Wheel with a hardened steel tyre*

To find the temperature required for assembly, we note that the tire must expand sufficiently to convert a radial *interference* of 0.012 in. to a radial *clearance* of 0.005 in. Thus, the radial displacement $u_r$ due to thermal expansion must be 0.012+0.005 = 0.017 in. Now

$$e_{\theta\theta} = \alpha\Delta T = \frac{u_r}{r} = \frac{0.017}{5}$$

and hence the required temperature difference is

$$\Delta T = \frac{0.017}{7\times10^{-6}\times5} = 486\,^{\circ}\mathrm{F}\,.$$

Notice that the *radius* of the wheel is 5 in. (Take care not to confuse diameter with radius or radial interference with diametral interference.)

To find the stress state after assembly, we denote the final interface pressure by $p_c$. It then follows from equation (10.21) that $A_1=-p_c$, since there are no thermal or rotational effects, and $B_1$ must be zero to preserve continuity at $r=0$. Thus, the radial displacement at the surface of the disk is

$$u_{r_1} = \frac{A_1(1-\nu_1)r}{E_1} = -\frac{0.7 \times 5 \times p_c}{30 \times 10^6} = -0.1167 \times 10^{-6} p_c ,$$

from equation (10.23).

For the tyre, we have $\sigma_{rr} = -p_c$ at $r=5$ in and $\sigma_{rr}=0$ at $r=5.15$ in and hence

$$A_2 + \frac{B_2}{5^2} = -p_c \;\; ; \;\; A_2 + \frac{B_2}{5.15^2} = 0 ,$$

with solution

$$A_2 = 16.42 p_c \;\; ; \;\; B_2 = -435.5 p_c .$$

Substituting into (10.23), we find that the radial displacement at the inner radius of the tyre $(r=5)$ is

$$u_{r_2} = \frac{A_2(1-\nu_2)r}{E_2} - \frac{(1+\nu_2)B_2}{E_2 r} = \frac{A_2 \times 0.7 \times 5}{30 \times 10^6} - \frac{B_2 \times 1.3}{30 \times 10^6 \times 5} = 5.69 \times 10^{-6} p_c .$$

Finally, we substitute the above expressions for $u_{r1}, u_{r2}$ into (10.34), obtaining

$$5.69 \times 10^{-6} p_c + 0.1167 \times 10^{-6} p_c = 0.012$$

and hence $p_c = 2066$ psi.

The maximum tensile stress in the tyre occurs at the inner radius and is

$$\sigma_{\theta\theta} = A_2 - \frac{B_2}{5^2} = 33.84 p_c = 70 \text{ ksi} .$$

## Discussion

This example illustrates a calculation procedure which can be adapted to most problems involving composite cylinders. However, in the spirit of §1.2.2, we note that an acceptable result can often be obtained by a considerably shorter calculation. A clue to this is provided by the fact that the two radial displacements are very different in magnitude $(5.69 \gg 0.1167)$, so that the smaller of them can be neglected without affecting the result. In physical terms, this is equivalent to saying that the central disk is so much stiffer than the tyre that $u_{r_1}$ can be neglected and the problem treated as one of assembling a tyre over a rigid disk. Furthermore, the tyre is thin relative to its radius (i.e. $0.15 \ll 5$) and hence a thin-walled cylinder solution can be used for it, which amounts to assuming that $\sigma_{\theta\theta} \gg \sigma_{rr}$ and is uniform across the thickness (see Chapter 8).

With these idealizations, we obtain

$$u_{r_2} \approx \delta = 0.012 \text{ in.}$$

$$e_{\theta\theta} = \frac{u_{r_2}}{r} \approx \frac{0.012}{5} = 0.0024$$

$$\sigma_{\theta\theta} = E e_{\theta\theta} \approx 30 \times 10^6 \times 0.0024 = 72 \text{ ksi} ,$$

which is within 3% of the more exact result. An even better approximation can be obtained by using the average tyre radius, 5.075 inches, in calculating $e_{\theta\theta}$, giving a stress of 71 ksi.

Of course, this problem involves a fairly thin outer cylinder and the thin-walled approximation can be expected to be less accurate when the thickness is greater. This question is further explored in Problem 10.12.

### 10.4.2 Limits and fits

Example 10.3 shows that a fairly small difference between the inside diameter of the tyre and the outside diameter of the wheel leads to quite substantial stresses in the tyre. In practice, these dimensions can only be guaranteed in a manufacturing operation within some finite limits or *tolerances* and the components made to the design will show some statistical scatter. It is therefore important to specify the acceptable tolerances on the two mating dimensions and to choose these so that they are achievable by a practicable manufacturing operation.

Shrink or force fits are widely used to assemble gears and other components onto cylindrical shafts, so sets of standards have been established defining appropriate amounts of interference $\delta$ and tolerance limits on the dimensions for various categories of fits.[8]

If gears and shafts are randomly assembled, situations may arise where the largest permissible shaft is assembled to the smallest permissible hole and this might result in too large a tensile hoop stress at the inner radius of the gear. The opposite case of a small shaft assembled to a large hole will give a loose fit which might not develop sufficient contact pressure to maintain integrity under load. This problem can be alleviated to some extent, at the cost of additional organizational costs, by sorting the manufactured components into size ranges and assembling predominantly large shafts to large holes etc.

## 10.5 Plastic deformation of disks and cylinders

The general procedure outlined in §10.1 can be used for disks and cylinders of inelastic materials if the appropriate inelastic constitutive relations are used in place of equations (10.8–10.10). In this section, we shall consider the special case of a material that behaves elastically up to the yield stress and thereafter yields at constant stress (i.e. without work hardening). This idealization of ductile material behavior was used in Chapter 5 in the analysis of elastic-plastic bending and its justification is discussed in §5.2.3. The corresponding uniaxial tensile stress-strain relation for both loading and unloading is shown in Figure 5.4.

If a disk or cylinder of an elastic-plastic material is subjected to monotonically increasing load, the stresses will initially be given by the preceding elastic analysis

[8] See for example, American Standard Limits for Cylindrical Parts ANSI B4.2-1978, American Society of Mechanical Engineers, New York, or ISO 17.040.10 Limits and Fits at http://www.iso.org

until a load is reached at which yielding commences, usually at the inner or outer radius. Further increase of load will cause a plastic zone to grow, starting from the point of first yield. Eventually, the plastic zone will extend over the entire body and this fully-plastic state defines the maximum load the body can sustain without total collapse. As in the analysis of elastic-plastic bending, the limiting cases of first yield and full plasticity are considerably simpler than the intermediate condition that involves both elastic and plastic zones.

In cylinders and disks, at least two of the three principal stresses $\sigma_{rr}, \sigma_{\theta\theta}, \sigma_{zz}$ are non-zero and hence we need to supplement the uniaxial stress-strain relation with information about the influence of the other stress components on the conditions at yield. The general question of yield criteria is discussed in §2.2.3. For the present analysis, we shall adopt *Tresca's yield criterion*,[9] which states that the maximum shear stress $\tau_{max}$ during yielding is equal to a critical value $\tau_Y$. More specifically, we assume that the material behaves elastically as long as $\tau_{max} < \tau_Y$ and thereafter yields with $\tau_{max} = \tau_Y$.

In Chapter 2, we showed that the maximum shear stress is

$$\tau_{max} = \max\left(\frac{|\sigma_1 - \sigma_2|}{2}, \frac{|\sigma_2 - \sigma_3|}{2}, \frac{|\sigma_3 - \sigma_1|}{2}\right),$$

where $\sigma_1, \sigma_2, \sigma_3$ are the three principal stresses. If yielding occurs in uniaxial tension at a stress $S_Y$, we conclude that $\tau_Y = S_Y/2$. For the present axisymmetric problem, the principal stresses are $\sigma_{rr}, \sigma_{\theta\theta}, \sigma_{zz}$ and hence Tresca's yield criterion can be expressed in the form

$$\max\left(|\sigma_{rr} - \sigma_{\theta\theta}|, |\sigma_{\theta\theta} - \sigma_{zz}|, |\sigma_{zz} - \sigma_{rr}|\right) = S_Y . \tag{10.35}$$

If we knew which of the three stress differences was the greatest, this equation and the equation of motion (10.3) would constitute two equations for the two unknown stress components $\sigma_{rr}, \sigma_{\theta\theta}$ in any region that is yielding. Recall that the third stress component $\sigma_{zz}$ is zero under plane stress conditions and can be determined by axial equilibrium arguments for plane strain.

One approach is to use trial and error. We tentatively assume that a given stress difference is the greatest and solve the plasticity problem under this assumption. We can then check the final stress distribution to see whether our initial assumption is confirmed. In the worst case, we could try all possible combinations until a consistent solution is obtained, but a better approach is to start with the solution of the corresponding elastic solution and determine the stress difference governing first yield. In most cases, this will also govern the deformation in the plastic zone in the subsequent elastic-plastic problem. The elastic solution also tells us the location at which yield first occurs and hence shows where the plastic zone will develop at loads above that for first yield.

[9] The advantage of using Tresca's rather than von Mises' yield criterion is that the former leads to linear differential equations that can be solved analytically.

### 10.5.1 First yield

For first yield, the deformation is everywhere elastic and the solution proceeds exactly as in the preceding sections, except that the load is unknown and must be expressed symbolically. We then take the further step of identifying the maximum stress difference and equating it to $S_Y$, to determine the load and the location at which yield first occurs. This is best illustrated by example. Consider the thick cylinder loaded by internal pressure, for which the stress field is illustrated in Figure 10.9 above. We note that the greatest stress difference (i.e. the greatest distance between any two of the three lines) occurs at the inner radius $r=a$ and is the difference between $\sigma_{\theta\theta}$ and $\sigma_{rr}$. It follows that yield will start at the inner radius when

$$\sigma_{\theta\theta} - \sigma_{rr} = S_Y \; . \tag{10.36}$$

Substituting for the stresses from equations (10.30) and setting $r=a$ for the inner radius, we find that this occurs at a pressure $p_Y$ given by

$$\frac{a^2 p_Y}{(b^2-a^2)}\left(1+\frac{b^2}{a^2}\right) - \frac{a^2 p_Y}{(b^2-a^2)}\left(1-\frac{b^2}{a^2}\right) = S_Y \; .$$

The internal pressure for first yield is therefore

$$p_Y = \frac{S_Y}{2}\left(1-\frac{a^2}{b^2}\right) \; . \tag{10.37}$$

### 10.5.2 The fully-plastic solution

If the internal pressure is increased beyond $p_Y$, we expect yield to occur at the inner radius and further increase in pressure will cause the region of plastic deformation to extend outwards. Eventually, the whole cylinder will be plastically deformed and no further increase in pressure can occur without collapse. We shall denote the pressure at which this occurs (the fully plastic pressure) by $p_P$.

To solve the fully plastic problem, we tentatively assume that plasticity is governed by the same criterion [equation (10.36)] as at first yield. We can then eliminate $\sigma_{\theta\theta}$ between equations (10.4, 10.36) to obtain

$$\frac{d\sigma_{rr}}{dr} - \frac{S_Y}{r} = 0 \, ,$$

which has the general solution

$$\sigma_{rr} = S_Y \ln(r) + C \, , \tag{10.38}$$

where $C$ is an arbitrary constant of integration.

At the fully-plastic state, the internal pressure is $p_P$ and the external pressure is zero, giving the two boundary conditions

$$\sigma_{rr}(a) = S_Y \ln(a) + C = -p_P$$
$$\sigma_{rr}(b) = S_Y \ln(b) + C = 0 .$$

Solving these equations for $C, p_P$, we find

$$C = -S_Y \ln(b) \quad ; \quad p_P = S_Y \ln(b/a)$$

and substituting for $C$ into (10.38, 10.36), we obtain the complete stress field as

$$\sigma_{rr} = -S_Y \ln(b/r)$$
$$\sigma_{\theta\theta} = S_Y - S_Y \ln(b/r) .$$

These expressions as well as that for $\sigma_{zz}$ from (10.31) are shown in Figure 10.13, for the case where $b=2a$. Notice that the curves for $\sigma_{\theta\theta}$ and $\sigma_{rr}$ are parallel and separated by a distance $S_Y$, as demanded by the yield condition (10.36). We also notice that, for the case illustrated, $(\sigma_{\theta\theta} - \sigma_{rr})$ is indeed the greatest stress difference throughout the cylinder, as we assumed at the beginning. However, if $a/b<0.45$, the curves for $\sigma_{\theta\theta}$ and $\sigma_{zz}$ cross near $r=a$ and the solution is not valid in this range. We shall discuss the resolution of this difficulty in §10.5.4 below.

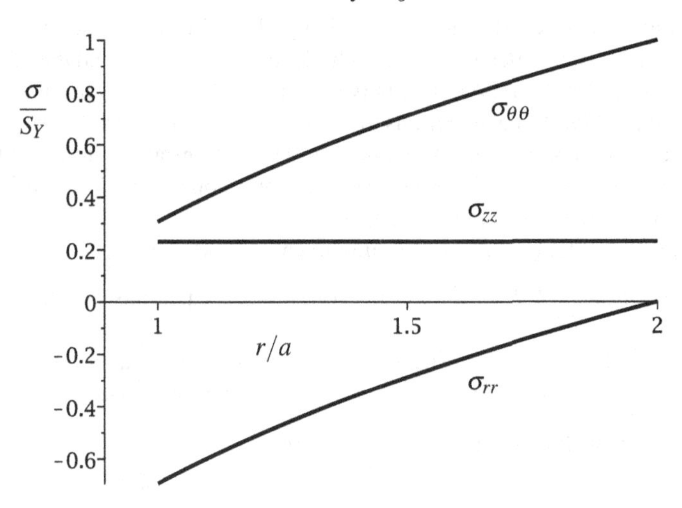


*Figure 10.13: Stress field in the fully-plastic state*

It is instructive to compare the internal pressure, $p_P$ for full plasticity with that for first yield, $p_Y$. We have

$$\frac{p_P}{p_Y} = \frac{2\ln(b/a)}{(1-a^2/b^2)} .$$

This ratio is plotted in Figure 10.14. For a thin-walled cylinder, $a/b$ is close to unity and a comparatively small additional pressure is sufficient to cause complete collapse once yield starts to occur. The curve is shown dashed for $a/b<0.45$, since the solution loses validity in this range.

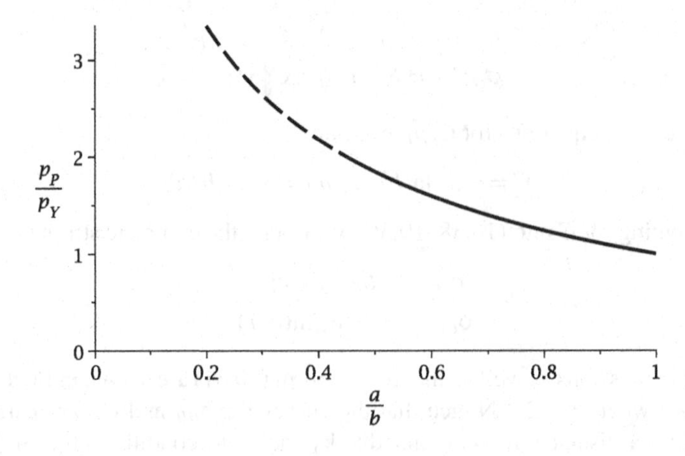


*Figure 10.14: The ratio $p_P/p_Y$ as a function of the radius ratio $a/b$*

### 10.5.3 Elastic-plastic problems

If the loading is intermediate between the first yield and fully-plastic limits, we can anticipate a solution in which part of the cylinder or disk is elastic and the rest plastic. The radius $c$ of the circular boundary between these zones is an additional unknown to be determined. Additional equations to determine this radius and the arbitrary constants in the solution will come from continuity conditions at the interface between the elastic and plastic zones. In fact, the procedure is somewhat similar to that used for composite elastic cylinders in §10.4.

In a typical problem, there are four unknowns comprising:-

- the two constants $A, B$ in the elastic zone — see equations (10.20–10.22) for plane stress or (10.25–10.27) for plane strain;
- a single constant $C$ in the plastic zone — see for example, equation (10.38); and
- the unknown radius $c$ defining the boundary of the plastic zone.

To determine them, we have the four conditions:-

- specified pressure or displacement at the inner and outer radii $r=a,b$;
- continuity of radial stress, $\sigma_{rr}$ at $r=c$ as in equation (10.32);
- continuity of *circumferential* stress $\sigma_{\theta\theta}$ at $r=c$.

This last condition requires some comment, since it does not apply in problems for composite elastic cylinders and disks. However, here we are dealing with a single cylinder or disk, part of which has yielded and part of which is still elastic. If the load is increased slightly, the plastic zone will extend, indicating that the material just inside the elastic zone must have been on the point of yielding. Since the yield criterion usually[10] involves the circumferential stress $\sigma_{\theta\theta}$, this implies that this stress component must be continuous across the elastic-plastic boundary.

[10] but see §10.5.4.

**Example 10.4**

*A thin solid disk of outer radius a rotates at speed $\Omega$. The outer edge of the disk is traction-free and thermal effects are negligible. The material of the disk is ductile with uniaxial yield stress $S_Y$, Young's modulus E, Poisson's ratio $\nu$ and density $\rho$.*

*Find the rotational speed for first yield $\Omega_Y$ and for complete plastic failure $\Omega_P$. Find also the radius c defining the boundary between the elastic and plastic zones for $\Omega_Y < \Omega < \Omega_P$.*

**First yield**

For the elastic solution we have

$$\sigma_{rr} = -\frac{(3+\nu)\rho\Omega^2 r^2}{8} + A + \frac{B}{r^2} ,$$

from equation (10.21). As long as the whole disk is elastic, we can argue that the constant $B$ must be zero to give bounded stresses at the centre, as in the automotive brake disk of Example 10.2. The remaining constant $A$ is determined from the condition that the outer edge of the disk be traction-free ($\sigma_{rr} = 0$ at $r = a$). We obtain

$$A = \frac{(3+\nu)\rho\Omega^2 a^2}{8}$$

and hence the elastic solution is defined by

$$\begin{aligned} \sigma_{rr} &= \frac{(3+\nu)\rho\Omega^2(a^2 - r^2)}{8} \\ \sigma_{\theta\theta} &= \frac{\rho\Omega^2}{8}\left[(3+\nu)a^2 - (1+3\nu)r^2\right] , \end{aligned}$$

from (10.21, 10.22).

This stress distribution is illustrated in Figure 10.15 for the case where $\nu = 0.3$. Both stresses are tensile throughout the disk and, since the third principal stress $\sigma_{zz} = 0$ for the disk, the greatest stress difference in the elastic solution is $(\sigma_{\theta\theta} - \sigma_{zz})$. First yield will therefore occur at the centre at a speed $\Omega_Y$ given by

$$\sigma_{\theta\theta} = \frac{\rho\Omega_Y^2(3+\nu)a^2}{8} = S_Y ,$$

i.e.

$$\Omega_Y = \sqrt{\frac{8S_Y}{(3+\nu)\rho a^2}} .$$

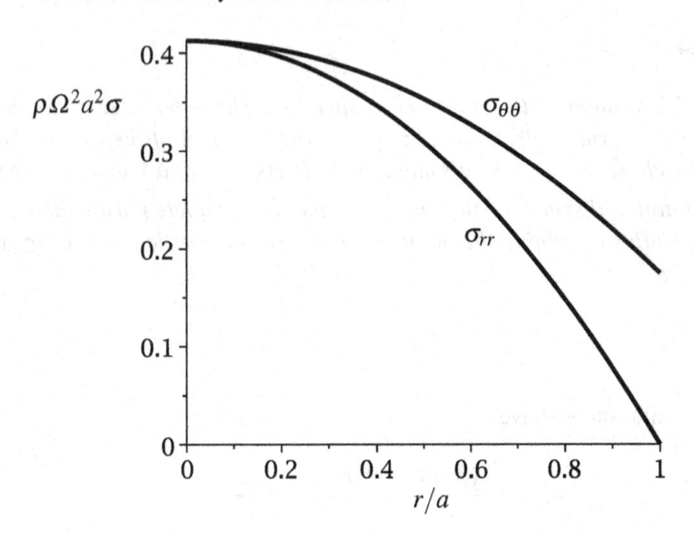


*Figure 10.15: Elastic stress distribution for the rotating solid disk ($\nu=0.3$)*

**The plastic zone**

If the speed is increased beyond $\Omega_Y$, a plastic zone will develop at the centre. In other words, the region $0 \le r < c$ will be plastic and the region $c < r < a$ will be elastic, where $c$ is an unknown radius to be determined.

In the plastic zone $0 \le r < c$, the elastic stress distribution of Figure 10.15 leads us to expect yielding to be governed by the condition $\sigma_{\theta\theta} - \sigma_{zz} = S_Y$ and hence

$$\sigma_{\theta\theta} = S_Y ,$$

since $\sigma_{zz} = 0$.

Substituting this result into the equation of motion (10.3), we obtain

$$\frac{d\sigma_{rr}}{dr} + \frac{\sigma_{rr}}{r} = \frac{S_Y}{r} - \rho\Omega^2 r ,$$

whose general solution is

$$\sigma_{rr} = S_Y - \frac{\rho\Omega^2 r^2}{3} + \frac{C}{r} ,$$

where $C$ is an arbitrary constant. The plastic zone includes the centre $r=0$ and hence we must have $C=0$ for the stresses to be bounded, leaving

$$\sigma_{rr} = S_Y - \frac{\rho\Omega^2 r^2}{3} .$$

**The elastic zone**

In the elastic zone $c<r<a$, the stresses are given by equations (10.21, 10.22) as

$$\sigma_{rr} = -\frac{(3+\nu)\rho\Omega^2 r^2}{8} + A + \frac{B}{r^2}$$
$$\sigma_{\theta\theta} = -\frac{(1+3\nu)\rho\Omega^2 r^2}{8} + A - \frac{B}{r^2} \ .$$

In contrast to the first yield solution, the constant $B$ is not here required to be zero, since the centre $r=0$ is not included in the elastic range $c<r<a$.

The two constants $A, B$ and the unknown radius $c$ are determined from the traction-free condition $\sigma_{rr}(a)=0$ at the outer radius and the two continuity conditions $\sigma_{rr}(c^+)=\sigma_{rr}(c^-)$, $\sigma_{\theta\theta}(c^+)=\sigma_{\theta\theta}(c^-)$ at $r=c$. Substituting for the appropriate stresses, we obtain

$$-\frac{(3+\nu)\rho\Omega^2 a^2}{8} + A + \frac{B}{a^2} = 0$$
$$-\frac{(3+\nu)\rho\Omega^2 c^2}{8} + A + \frac{B}{c^2} = S_Y - \frac{\rho\Omega^2 c^2}{3}$$
$$-\frac{(1+3\nu)\rho\Omega^2 c^2}{8} + A - \frac{B}{c^2} = S_Y \ .$$

Eliminating $A, B$ between these equations, we obtain

$$\left(\frac{c}{a}\right)^4 - 2\left(\frac{c}{a}\right)^2 + \frac{3}{(1+3\nu)}\left[(3+\nu) - \frac{8S_Y}{\rho\Omega^2 a^2}\right] = 0 \, ,$$

which has the solution

$$\left(\frac{c}{a}\right)^2 = 1 - \sqrt{\frac{8}{(1+3\nu)}\left(\frac{3S_Y}{\rho a^2\Omega^2} - 1\right)} \, , \qquad (10.39)$$

where we have taken the negative square root, since physically meaningful results require that $0<c<a$.

Complete plastic failure corresponds to the case $c=a$ (i.e. the plastic zone extends over the whole disk) and hence

$$\Omega_P = \sqrt{\frac{3S_Y}{\rho a^2}} \ .$$

The ratio

$$\frac{\Omega_P}{\Omega_Y} = \sqrt{\frac{3(3+\nu)}{8}}$$

increases with $\nu$, but is less than 1.15 even at the maximum possible value of $\nu=0.5$. Thus, the speed at complete plastic failure is never more than 15% higher than that at first yield. From a design perspective, most of the interesting results can be obtained from the elastic 'first yield' analysis and the fully plastic analysis. In the present case, the latter can of course be obtained directly by imposing the edge condition $\sigma_{rr}=0$ at $r=a$ in the expression for the stresses in the plastic zone.

For this reason, the complexities of an elastic-plastic solution are not justified in most practical applications. One exception is Problem 10.21, which describes a test method for determining the properties of soils.

### 10.5.4 Other failure modes

There is no guarantee that failure of the cylinder or disk will involve plastic deformation at all radii. For example, an annular disk with a large ratio between inner and outer radii (such as Problem 10.22) and loaded by internal pressure will fail as soon as the internal pressure reaches the yield stress in compression and before yielding has progressed to the outer radius. The innermost layer of material will then yield in compression, since there is no restraint to axial motion ($\sigma_{zz}=0$), and we shall see a thin layer of material extruded in the axial direction, much as a layer of viscous liquid could be squozen out from the interface between two disks during a shrink fit.

Problems exhibiting this kind of behavior can generally be identified by the fact that the attempt to solve the fully-plastic problem runs into difficulties. For example, the fully-plastic solution for the thick-walled cylinder in §10.5.2 above breaks down if it predicts $\sigma_{\theta\theta}<\sigma_{zz}$ at the inner radius. This is again an indication that we might expect axial extrusion of material at the inner radius. However, the cylinder differs from the disk in that there is nowhere for the material to go. In fact, the assumption that plane sections remain plane demands that stresses resisting this axial motion will be developed. As a result, the axial stress distribution $\sigma_{zz}$ will deviate from equation (10.31) (which was developed from elastic arguments) sufficiently to ensure that axial motion of material does not occur. In effect, $\sigma_{zz}$ will decrease at the inner radius, sufficiently to leave $(\sigma_{\theta\theta}-\sigma_{rr})$ as the greatest stress difference, and increase at the outer radius in order to preserve axial equilibrium. This process can continue until the whole cylinder is plastic.

### 10.5.5 Unloading and residual stresses

If a cylinder or disk is loaded into the plastic range and then unloaded, the unloading process will generally move the stress state inside the yield envelope and hence be elastic. In extreme cases, complete unloading might cause the state to reach the yield envelope at a different point and hence cause additional plastic deformation, but this possibility can be detected by performing an elastic analysis and checking the predicted final stress state.

If unloading is complete (i.e. if the final tractions, rotational speed and temperatures are returned to zero), the final residual stress field is obtained from that at the maximum load by subtracting the field that would have been obtained at the maximum load had the system remained elastic. The reader will recall that a similar procedure was used in §5.7 to determine the residual stress distribution in a beam loaded into the plastic range and then released.

The more general case where the loading is reduced, but not necessarily to zero, can be treated by finding the residual stress and then superposing the elastic stresses associated with the final loading. Alternatively, we can start from the stress field at the maximum load and superpose an elastic field corresponding to the *difference* between the maximum and the final load. Notice however that the elastic field varies with the square of the rotational speed, so if the speed changes from $\Omega_1$ to $\Omega_2$, the corresponding terms in this differential field will contain $\Omega_2^2-\Omega_1^2$.

### Example 10.5 — The rotating disk

*A solid disk of radius a starts from rest, accelerates up to the fully-plastic rotational speed $\Omega_P$ and then returns to rest. Determine the residual stress field in the disk, if the material has Young's modulus E, Poisson's ratio $\nu$ and tensile yield stress $S_Y$.*

The analysis of this problem during the loading phase has already been performed in Example 10.4 above. In particular, the elastic solution is

$$\sigma_{rr} = \frac{(3+\nu)\rho\Omega^2(a^2-r^2)}{8} \quad ; \quad \sigma_{\theta\theta} = \frac{\rho\Omega^2}{8}\left[(3+\nu)a^2-(1+3\nu)r^2\right]$$

and the fully plastic solution is

$$\sigma_{rr} = S_Y - \frac{\rho\Omega_P^2 r^2}{3} \quad ; \quad \sigma_{\theta\theta} = S_Y \; ,$$

with

$$\Omega_P = \sqrt{\frac{3S_Y}{\rho a^2}} \; .$$

To find the residual stresses we must subtract, from the fully-plastic solution, the stresses that would be produced at a rotational speed of $\Omega_P$ *if the material had remained elastic* — i.e. if there had been no plastic deformation. This in turn is obtained by substituting $\Omega=\Omega_P$ in the elastic equations above. Thus, the residual stresses are

$$\begin{aligned}
\sigma_{rr} &= S_Y - \frac{\rho\Omega_P^2 r^2}{3} - \frac{(3+\nu)\rho\Omega_P^2(a^2-r^2)}{8} \\
\sigma_{\theta\theta} &= S_Y - \frac{\rho\Omega_P^2}{8}\left[(3+\nu)a^2-(1+3\nu)r^2\right] \; .
\end{aligned}$$

After substituting for $\Omega_P$, these expressions simplify to

$$\sigma_{rr} = -\frac{S_Y(1+3\nu)}{8}\left(1-\frac{r^2}{a^2}\right) \quad ; \quad \sigma_{\theta\theta} = \frac{S_Y(1+3\nu)}{8}\left(\frac{3r^2}{a^2}-1\right) \; .$$

The stresses (including $\sigma_{zz}=0$) are illustrated in Figure 10.16. Notice that the maximum stress difference occurs at the outer edge $(r=a)$ and is

$$\sigma_{\theta\theta}(a) - \sigma_{rr}(a) = \frac{S_Y(1+3\nu)}{4} \; .$$

Since this value is less than $S_Y$, we conclude that there will be no further yielding during unloading.

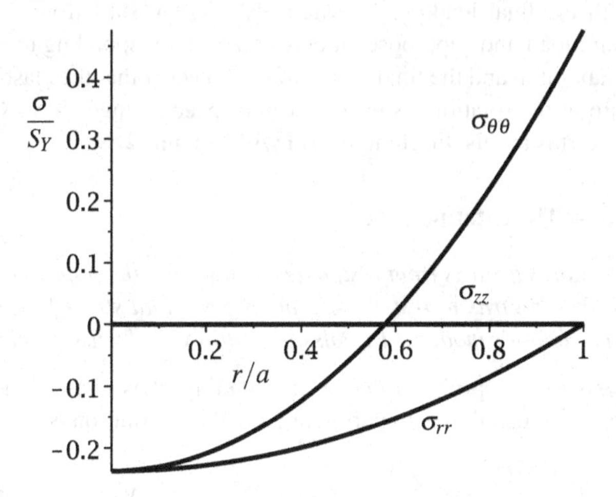


*Figure 10.16: Residual stresses in the disk after rotation to near $\Omega_P$ ($\nu = 0.3$)*

## 10.6 Summary

In this chapter, we have developed methods for determining the stresses in cylinders and disks under axisymmetric loading. Applications include thick-walled pressure vessels, rotating wheels and disks, and stresses associated with shrink and press fits.

The stresses must satisfy Newton's second law, the strains must satisfy a kinematic compatibility relation and the stresses and strains are related through the constitutive behavior of the material. These factors can be combined to define the governing ordinary differential equation for a material with any constitutive law. Radial traction or displacement can be prescribed at the inner and outer radii. The method was illustrated here for elastic and elastic-plastic materials, including the effects of rotation and thermal expansion.

The internal pressure required to cause yielding throughout a thick walled pressure vessel is significantly larger than that for first yield and a single overload condition leaves a state of residual stress that will prevent plastic deformation under more moderate loading. Similar favourable residual stress states can be developed by shrinking one thick cylinder onto another.

## Further reading

W.B. Bickford (1998), *Advanced Mechanics of Materials,* Addison Wesley, Menlo Park, CA, Chapter 6.

A.P. Boresi, R.J. Schmidt, and O.M. Sidebottom (1993), *Advanced Mechanics of Materials,* John Wiley, New York, 5th edn., Chapter 11.
A.H. Burr (1981), *Mechanical Analysis and Design*, Elsevier, New York, §§8.6–8.14.

## Problems

### Sections 10.1—10.3

**10.1.** A rotating disk of radius $b$ has a small central hole of radius $a$ ($\ll b$). If all the surfaces of the disk are traction-free, show that the hole causes a stress concentration of 2 — i.e. that the maximum tensile stress is twice as large as that in a similar disk without a hole, rotating at the same speed.

**10.2.** The turbine disk of Figure P10.2 comprises a solid central disk of diameter 800 mm and thickness 50 mm, to which are attached 50 blades, each of mass 0.4 kg. The centres of mass of the blades lie on a circle of radius 430 mm. Find the maximum tensile stress in the disk when the rotational speed is 10,000 rpm. The disk is made of steel for which $E = 210$ GPa, $\nu = 0.3$ and $\rho = 7700$ kg/m$^3$.

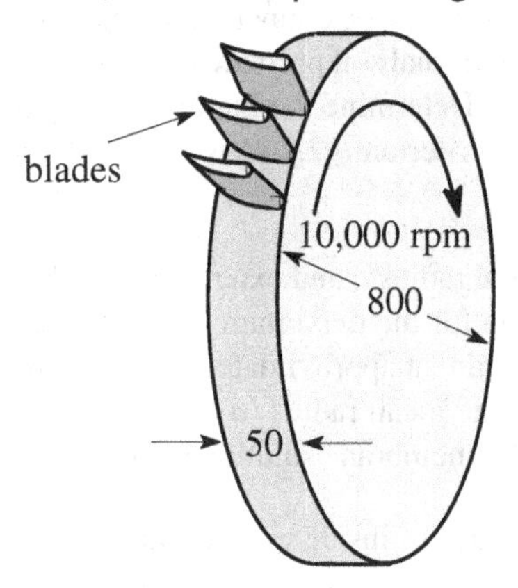


*Figure P10.2*

**10.3.** During a test on an automotive disk brake, it was found that the temperature $T$ of the disk could be approximated by the expression

$$T = \frac{5T_0}{4}(1-e^{-\tau}) - \frac{T_0 r^2}{2a^2}(1-e^{-\tau}) + \frac{T_0 r^4}{8a^4}\tau e^{-\tau} ,$$

where $\tau = t/t_0$ is a dimensionless time, $a$ is the radius of the disk, $r$ is the distance from the axis, $t$ is time and $T_0, t_0$ are constants. The disk is continuous at $r = 0$ and all its surfaces can be assumed to be traction-free.

Derive expressions for the radial and circumferential thermal stresses as functions of radius and time, and hence show that the maximum value of the stress difference $(\sigma_{\theta\theta}-\sigma_{rr})$ at any given time always occurs at the outer edge of the disk.

**10.4.** A long hollow cylinder has outside diameter 25 inches and inside diameter 15 inches. It is made of steel ($E=30\times 10^6$ psi, $\nu=0.3$) and is loaded by an *external* pressure of 600 psi. The ends of the vessel are prevented from moving axially. Find the maximum compressive stress in the cylinder and the change in diameter of the central hole.

**10.5.** The fuel rods of a nuclear reactor consist of solid uranium cylinders of diameter 70 mm. During operation, a typical rod experiences a temperature distribution approximated by the equation

$$T(r) = 600 - 0.1r^2 \ {}^{\circ}\mathrm{C} ,$$

where $r$ is in mm. Find the maximum stress in the fuel rod if the outer surface is traction-free and plane strain conditions can be assumed. The properties of uranium are $E=172$ GPa, $\nu=0.28$, $\alpha=11\times 10^{-6}$ per °C.

**10.6.** A concrete drainpipe has inside diameter 30 inches and wall thickness 4 inches. When dry, it is loaded by external soil pressure of 10 psi and when full, the maximum internal pressure is 30 psi. Determine the factor of safety against failure if concrete obeys the modified Mohr criterion (§2.2.4) with $S_t=0.6$ ksi, $S_c=5.0$ ksi. Assume that the axial stress $\sigma_{zz}=0$.

**10.7.** A cylinder of internal radius $a$ and external radius $b$ is loaded by internal pressure $p$. Find an expression for the maximum tensile stress. Then use the membrane theory of Chapter 8 to obtain an approximate solution of the same problem, treating the radius of the shell as the mean radius $(a+b)/2$. What is the maximum value of the ratio $b/a$ for which the membrane solution is in error by less than 10%?

**10.8.** A thick elastic cylinder of inside radius $a$ and outside radius $2a$ is loaded by internal pressure $p$. At the outer radius, radial displacement $u_r$ is prevented by a rigid reinforcing sleeve. Also, the cylinder is prevented from expanding in the axial direction ($e_{zz}=0$).

Find the three stress components $\sigma_{rr}, \sigma_{\theta\theta}, \sigma_{zz}$ and sketch a graph of the resulting functions of radius. Where, and at what internal pressure, will yield first occur if the material has a yield stress $S_Y$ in uniaxial tension and obeys Tresca's yield criterion? Comment on the results for the case where Poisson's ratio $\nu=0.5$.

**10.9*.** A cylinder is fabricated from a composite material with the fibres laid down in the circumferential direction, such that the constitutive law is

$$e_{rr} = \frac{\sigma_{rr}}{E_1} \ \ e_{\theta\theta} = \frac{\sigma_{\theta\theta}}{E_2} ,$$

where $E_2 = 4E_1$. (Note that Poisson's ratio is zero). Use these expressions and the equilibrium and compatibility conditions to find the differential equation satisfied by the radial stress $\sigma_{rr}$ for the case where there is no rotation and no temperature variation. Solve this equation using the identity

$$r^{-i-j}\frac{d}{dr}\left[r^j\frac{d}{dr}\left(r^i f\right)\right] \equiv \frac{d^2 f}{dr^2} + \frac{(2i+j)}{r}\frac{df}{dr} + \frac{i(i+j-1)f}{r^2} \; .$$

(You will need to find the appropriate values of $i, j$ to make the right hand side correspond to your differential equation and then use the equivalent left hand side form.)

Use your solution to find the stresses in a cylinder of inside radius $a$ and outside radius $2a$, subjected to an internal pressure $p$.

**10.10*.** Use the plane stress constitutive relations (10.14, 10.15) to eliminate the strains $e_{rr}, e_{\theta\theta}$ from the compatibility equation (10.7). This equation and the equation of motion (10.3) then constitute two equations for the two unknown stress components $\sigma_{rr}, \sigma_{\theta\theta}$. Eliminate $\sigma_{\theta\theta}$ between these two equations to obtain a single second order differential equation for $\sigma_{rr}$ and then use the identity in Problem 10.9 to obtain the general solution for $\sigma_{rr}$. Verify[11] that your final expression agrees with equation (10.21).

## Section 10.4

**10.11.** The steel turbine disk of Problem 10.2 is to be drilled with a central hole and shrunk on to a 100 mm diameter steel shaft with an interference such that the interface pressure after assembly is 100 MPa.

Find the required radial interference $\delta$ and the rotational speed in rpm at which the disk will come loose from the shaft.

**10.12.** A wheel of inner radius $a$ and outer radius $na$ is shrunk on to a disk of radius $a$ with radial interference $\delta$. The material has Young's modulus $E$ and Poisson's ratio $\nu$. Using the plane stress equations, find the interface pressure after assembly and the maximum tensile stress, which will be $\sigma_{\theta\theta}$ at the inner radius of the wheel.

Obtain an approximate solution to the same problem assuming the disk is rigid and using the thin-walled cylinder equations, as in the 'Discussion' following Example 10.3. Sketch graphs of the results by each method as functions of $n$ and identify

[11] This is an alternative derivation of the general solution. These two methods of solution are available for all elasticity problems. As in this problem, we can use the constitutive law to express the compatibility equation in terms of stresses and then solve the equilibrium and compatibility equations simultaneously. Alternatively, as in the text development, we can use the strain-displacement relations and the constitutive law to express the stresses in terms of the displacements and hence develop a statement of the equilibrium equation in terms of displacements (see J.R. Barber (2010), *Elasticity*, Springer, Dordrecht, Netherlands, 3rd edn., §2.3). When we work in terms of displacements, the compatibility equation is automatically satisfied.

the range of values of $n$ for which the approximate result is within $\pm30\%$ of the more exact solution.

**10.13.** A steel cylinder of outside diameter 180 mm is shrunk on to a tungsten carbide liner of outside diameter 60 mm and wall thickness 7.5 mm. If the stress in the liner is to remain compressive when an internal pressure of 110 MPa is applied to it, what must be the initial circumferential stress due to interference at the inside surface of the liner?

Assume that axial stresses $\sigma_{zz}$ are negligible. For tungsten carbide, $E=600$ GPa, $\nu=0.21$; for steel, $E=210$ GPa, $\nu=0.3$.

**10.14.** In the ANSI listings of limits and fits, the tightest fit suggested for mounting a high grade cast iron gear on a steel shaft is FN2. For a nominal shaft diameter of 4.5 in, the tolerances permit the hole diameter to be anywhere in the range 4.5000 to 4.5014 in and the shaft external diameter to be in the range 4.5030 to 4.5039 in.

Suppose the gear outside diameter is 8 in and we have the bad luck to combine the smallest permitted hole with the largest permitted shaft. What will be the maximum tensile stress in the cast iron?

For cast iron, $E=15\times10^6$ psi, $\nu=0.25$; for steel, $E=30\times10^6$ psi, $\nu=0.3$.

**10.15.** Each axle of an electric freight locomotive supports a load of 44,200 lb on two wheels with a 42 in effective diameter (see Figure P10.15). A gear of 63 teeth and 31.5 in pitch diameter on the axle is driven by a steel pinion of 16 teeth and 8 in pitch diameter on the steel shaft of an electric motor hung between the wheels. The teeth are 1.12 in high on a root (dedendum) diameter of 7 in.

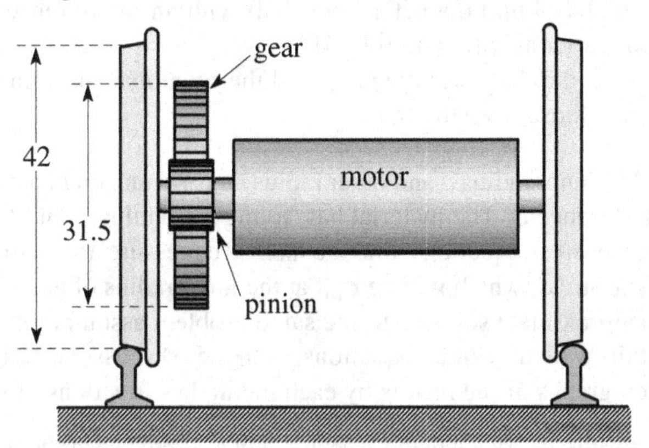


*Figure P10.15*

The pinion is bored with a hole of 5 in diameter and is 6 in long. It is proposed to press this onto the motor shaft without keying. The minimum coefficient of friction at the interference fit is expected to be 0.1, which is low because of the lubricant used in the pressing operation.

The maximum tractive force which can be developed at the wheel-rail contact corresponds to 35% adhesion — i.e. to a friction coefficient of 0.35 between the wheel and the rail.

(i) Determine the interference pressure, the diametral interference and the assembly force required if there is to be a safety factor of 3.33 against slipping in service.
(ii) What is the maximum tensile stress and the maximum shear stress in the pinion?
(iii) If a shrink fit is used instead of pressing, what change of temperature is required if the pinion is to be slipped over its shaft with a diametral clearance of 0.003 in. (For steel, $E=30\times10^6$ psi, $\nu=0.3$, $\alpha=6.3\times10^{-6}$ per $^\circ$ F.)

**10.16.** Figure P10.16 shows the cross section of a compound cylindrical pressure vessel, fabricated by shrinking the outer cylinder onto the inner cylinder, so as to leave a favourable state of prestress. Both cylinders are made of the same material. It is desired to optimize the interference fit such that when the vessel is loaded by a gradually increasing internal pressure, the yield condition is reached simultaneously at the inner surface of the inner cylinder and at the inner surface of the outer cylinder. What should be the value of the interface pressure after assembly if the uniaxial yield stress for the material is $S_Y$?

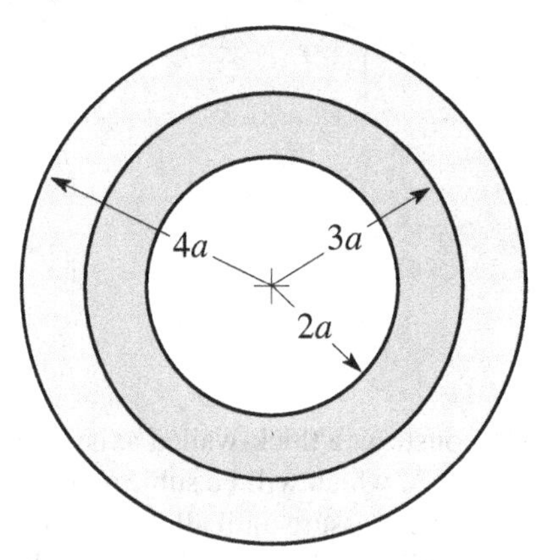


*Figure P10.16*

## Section 10.5

**10.17.** A thin disk of inner radius $a/3$ and outer radius $a$ rotates about its axis at speed $\Omega$. The inner radius is traction free and the outer radius carries turbine blades which exert an average tensile traction $\sigma_1$ due to centrifugal effects when the rotational speed is $\Omega_1$.

Show that the disk becomes fully plastic at a speed given by

$$\Omega^2 = \frac{2S_Y}{\rho a^2} \Big/ \left(\frac{26}{27} + \frac{3\sigma_1}{\rho a^2 \Omega_1^2}\right) ,$$

where $S_Y$ is the yield stress in uniaxial tension and the material obeys Tresca's yield criterion.

**10.18*.** Due to defects in the microstructure, a certain material has the biaxial yield surface shown in Figure P10.18, where $S_Y$ is the yield stress in uniaxial tension.

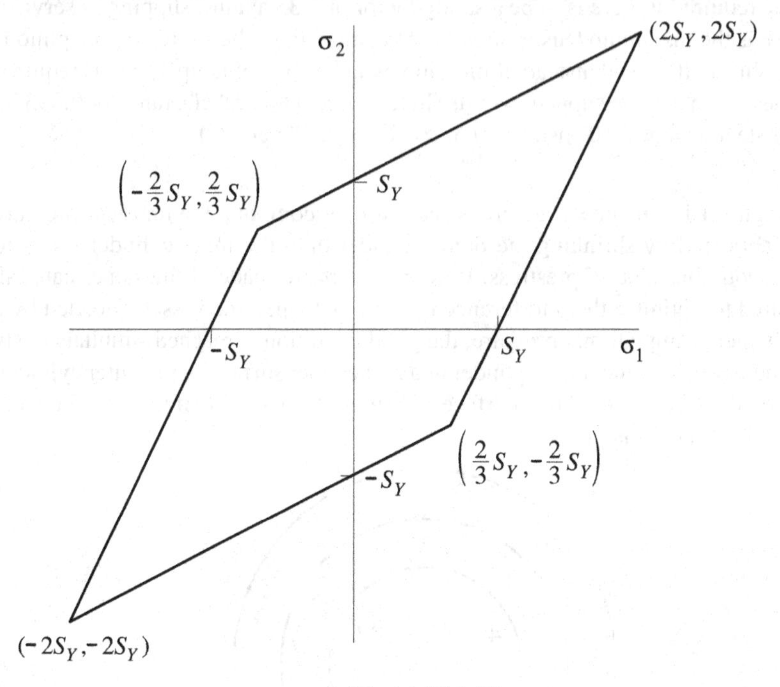


*Figure P10.18*

The material is used to construct a thick-walled axisymmetric pressure vessel of inner radius $a$ and outer radius $b$, which will be subjected to internal pressure $p$. The vessel is tested by increasing the pressure until all the material has yielded — i.e. the conditions are fully plastic. Find the value of $p$ needed to achieve complete plasticity and the corresponding distribution of radial and circumferential stress.

Assume that the axial stress $\sigma_{zz}$ (which is the third principal stress $\sigma_3$) has no effect on the yield surface in the range in question.

**10.19.** A thin circular disk of inner radius $a$ and outer radius $na$ carries blading at its circumference which imposes there a tensile traction of $ka^2\Omega^2$, where $\Omega$ is the rotational speed. Find expressions for (i) the speed at which first yield occurs and (ii) the speed at which the disk becomes fully plastic. Find the ratio of these values for the case where $n=2$ and $\nu=0.3$. Comment on your result.

**10.20.** Find the rotational speed at which the turbine disk of Problem 10.2 becomes fully plastic, if the steel has a uniaxial yield stress $S_Y = 700$ MPa.

**10.21***. An infinite solid consists of an incompressible elastic-plastic material which has modulus of rigidity $G$, and which yields at a shear stress $\tau_Y$. The solid contains a long cylindrical hole of radius $b$ which contains fluid at pressure $p$. Find the radial and circumferential stresses in the solid for the case where $p$ is sufficient for yielding to have progressed to a radius $c$.

Find the radial displacement at $r=c$ and hence show that

$$\frac{p}{\tau_Y} = 1 + \ln\left[\left(\frac{G}{\tau_Y}\right)\left(\frac{\Delta V}{V}\right)\right] ,$$

where $V$ is the original volume of the hole and $\Delta V$ is the change in its volume due to the pressure $p$.

**10.22.** A thin steel annular disk of internal radius 2 in and external radius 8 in is subjected to internal pressure at the inner radius, the outer edge being traction-free. Obtain expressions for the stresses in the disk when the pressure $p$ is sufficiently large for there to be a plastic zone of outer radius $c$. The appropriate material properties are $E=30\times 10^6$psi, $\nu=0.3, S_Y = 100\times 10^3$psi.

Show that the maximum value of $p$ that can be applied is $p=S_Y$ and that in this condition, part of the disk remains in the elastic state. Describe the resulting failure in physical terms.

**10.23.** A pressure vessel in the form of a long circular cylinder with closed ends is subjected to an external pressure $p$. The internal radius of the cylinder is $a$ and the external radius is $2a$. Show that if yielding is governed by the Tresca criterion and if the value of $p$ is large enough, yielding occurs up to a radius $c$ given by

$$S_Y \ln\left(\frac{c}{a}\right) - p + \frac{1}{2}S_Y\left(1 - \frac{c^2}{4a^2}\right) = 0 ,$$

where $S_Y$ is the tensile yield stress of the material.

Sketch the distribution of radial and circumferential stress through the cylinder wall for the case where $c = 1.2a$.

**10.24.** A rigid cylindrical container of radius $a$ is sealed when it is just full of water of density $\rho$ at pressure $p_0$. It is then rotated about its axis at speed $\Omega$. Treating water as an elastic-plastic material which yields at zero shear stress, but which has a bulk modulus $K$, find the pressure distribution in the water. Neglect gravitational effects. At what speed would you expect your solution to cease to be valid?

**Note**: The bulk modulus is defined such that

$$e \equiv e_{rr} + e_{\theta\theta} + e_{zz} = -\frac{p}{K}$$

where $p$ is the hydrostatic pressure.

**10.25.** A long solid cylinder of radius $a$ is rapidly heated around its curved surface producing an instantaneous temperature distribution

$$T = T_0 + T_1 \frac{r^2}{a^2}$$

If $T_1$ is large enough for some yielding to occur, find the radius of the elastic region and the instantaneous stress distribution in the cylinder. Assume that the material obeys Tresca's yield criterion and that the tensile yield stress $S_Y$ is independent of temperature.

**10.26.** A long circular cylinder of inside radius $a$ and outside radius $na$ has closed ends. The material may be assumed to be elastic-perfectly plastic, obeying Tresca's yield criterion, with a yield stress $S_Y$ in uniaxial tension. An external pressure just sufficient to cause yielding throughout the cylinder is applied and then removed. Find the distribution of residual stress and hence the maximum value of $n$ if there is to be no yielding on unloading.

**10.27.** A cylindrical pressure vessel of inside radius $a$, outside radius $2a$ and closed ends is subjected to an internal pressure $p$. Show that yield will occur throughout the vessel if $p = S_Y \ln 2$.

If this pressure is applied and then released, show that the residual circumferential stress at the inner radius is

$$-\left(\frac{8\ln 2}{3} - 1\right) S_Y \, .$$

# 11

# Curved Beams

The classical theory of the bending of beams is strictly exact if the axis of the beam is straight, the loads are applied only at the ends and the cross section is uniform along the length, which is much larger than any other linear dimension. However, the resulting equations are so simple that engineers routinely apply them outside this restrictive context and this usage is appropriate as long as we recognize that the resulting approximation may underestimate the bending stresses.

A particular case in which this occurs is that where the axis of the beam is curved. Important practical applications include chain links, crane hooks, pipe bends and curved segments of machine tool frames.

## 11.1 The governing equation

Figure 11.1 *(a)* shows a segment of a curved beam subtending a small angle $\delta\theta$ at its centre of curvature $O$ and transmitting a bending moment $M$. The resulting deformation is illustrated in Figure 11.1 *(b)*, where the reference frame has been chosen so as to leave the left end of the segment fixed.

As in the elementary analysis of straight beams, a convenient starting point is the assumption[1] that plane transverse sections remain plane during the deformation. The right end of the segment will rotate through some angle $\delta\phi$. This in turn causes the subtended angle in the deformed segment to increase to $\delta\theta+\delta\phi$ and the centre of curvature to move to $O'$ as shown.

[1] As in §5.1, we can establish a more rigorous starting point by remarking that transverse plane sections must all deform to the same shape, but will in general suffer different rigid body translation and rotation. However, if the material is isotropic, it follows from symmetry that plane sections remain plane.

J.R. Barber, *Intermediate Mechanics of Materials*, Solid Mechanics and Its Applications 175, 2nd ed., DOI 10.1007/978-94-007-0295-0_11, © Springer Science+Business Media B.V. 2011

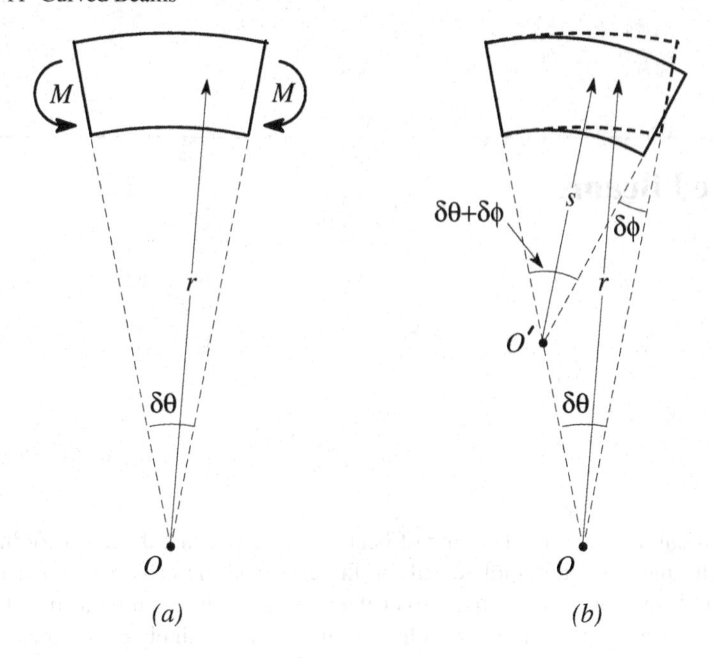


*Figure 11.1: (a) Curved beam loaded by a bending moment M; (b) the resulting deformation*

We denote the radius of a given point as $r$ in the undeformed state and $s$ in the deformed state. The circumferential strain can therefore be written

$$e_{\theta\theta} = \frac{s(\delta\theta + \delta\phi) - r\delta\theta}{r\delta\theta} \rightarrow \frac{s}{r}\left(1 + \frac{d\phi}{d\theta}\right) - 1 \tag{11.1}$$

in the limit where $\delta\theta \rightarrow 0$.

We anticipate the existence of a neutral surface at some radius $r = \hat{r}$ at which the circumferential strain is zero. The corresponding radius $\hat{s}$ after deformation is therefore (by definition) given by

$$\hat{s}\left(1 + \frac{d\phi}{d\theta}\right) = \hat{r} \,. \tag{11.2}$$

From the geometry of Figure 11.1 *(b)*, the distance $OO'$ is

$$d = r - s = \hat{r} - \hat{s}$$

and hence

$$s = r - d = r - \hat{r} + \hat{s} \,. \tag{11.3}$$

The strain (11.1) can therefore be written

$$e_{\theta\theta} = \left(1 - \frac{\hat{r}}{r} + \frac{\hat{s}}{r}\right)\left(1 + \frac{d\phi}{d\theta}\right) - 1 = \left(1 - \frac{\hat{r}}{r}\right)\frac{d\phi}{d\theta} \,, \tag{11.4}$$

using (11.2).

If the stress components $\sigma_{rr}, \sigma_{zz}$ can be assumed to be negligible, we then have

$$\sigma_{\theta\theta} = Ee_{\theta\theta} = E\left(1 - \frac{\hat{r}}{r}\right)\frac{d\phi}{d\theta} \ . \tag{11.5}$$

This stress acts on the cross section of the beam in the circumferential direction, as shown in Figure 11.2 and it follows that the total circumferential force is

$$F = \iint_A \sigma_{\theta\theta} dA = E\frac{d\phi}{d\theta}\iint_A \left(1 - \frac{\hat{r}}{r}\right) dA \ , \tag{11.6}$$

where the integral is performed over the cross-sectional area $A$ of the beam.

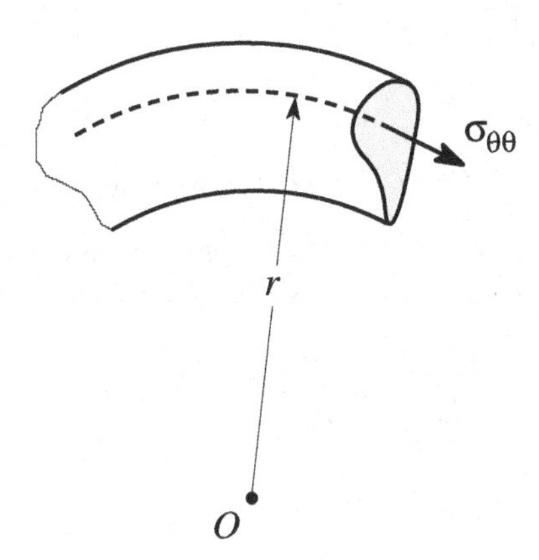


*Figure 11.2: The circumferential stress,* $\sigma_{\theta\theta}$

As with straight beams, it is convenient to treat circumferential (axial) loading and bending separately and then superpose the resulting stress distributions. We therefore focus initially on the case where the circumferential force is zero, in which case

$$\iint_A \left(1 - \frac{\hat{r}}{r}\right) dA = 0$$

and hence

$$A = \hat{r}\iint_A \frac{dA}{r} \ , \tag{11.7}$$

which serves to determine the radius $\hat{r}$ defining the location of the neutral surface.

### 11.1.1 Rectangular and circular cross sections

Figure 11.3 *(a)* shows a curved beam of rectangular cross section, with inner radius $a$, outer radius $b$ and thickness $t$.

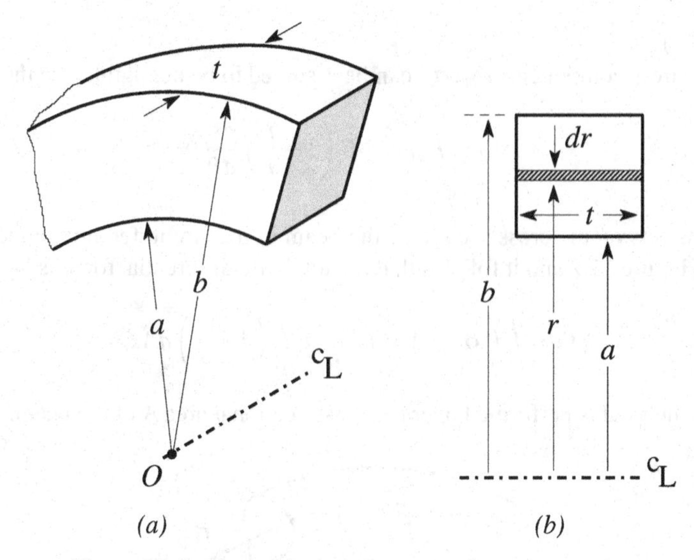


*Figure 11.3: Curved beam of rectangular cross section*

The integral in equation (11.7) for this cross section can be evaluated using the rectangular elemental area $dA = tdr$ shown shaded in Figure 11.3 *(b)*.

We then have

$$\iint_A \frac{dA}{r} = \int_a^b \frac{tdr}{r} = t\ln(b/a) . \tag{11.8}$$

The cross-sectional area is

$$A = (b-a)t$$

and hence the neutral radius $\hat{r}$ is given by

$$\hat{r} = \frac{(b-a)t}{t\ln(b/a)} = \frac{(b-a)}{\ln(b/a)} . \tag{11.9}$$

If the beam is only slightly curved, so that $a, b \gg (b-a)$ and $b/a$ is close to unity, it can be shown that $\hat{r} \to (a+b)/2$. In other words, the neutral surface approaches the mid-plane of the section as in the elementary theory.

Figure 11.4 *(a)* shows a curved beam of circular cross section of radius $a$, whose axis defines an arc of radius $R$. To evaluate the integral in equation (11.7), we use the shaded elemental area of Figure 11.4 *(b)*, obtaining[2]

$$\iint_A \frac{dA}{r} = \int_0^a \int_0^{2\pi} \frac{\rho d\theta d\rho}{(R+\rho\cos\theta)} = \int_0^a \frac{2\pi\rho d\rho}{\sqrt{R^2-\rho^2}}$$
$$= 2\pi(R-\sqrt{R^2-a^2}) . \tag{11.10}$$

[2] see for example I.S. Gradshteyn and I.M. Ryzhik (1980), *Tables of Integrals, Series and Products,* Academic Press, New York, 3.613.

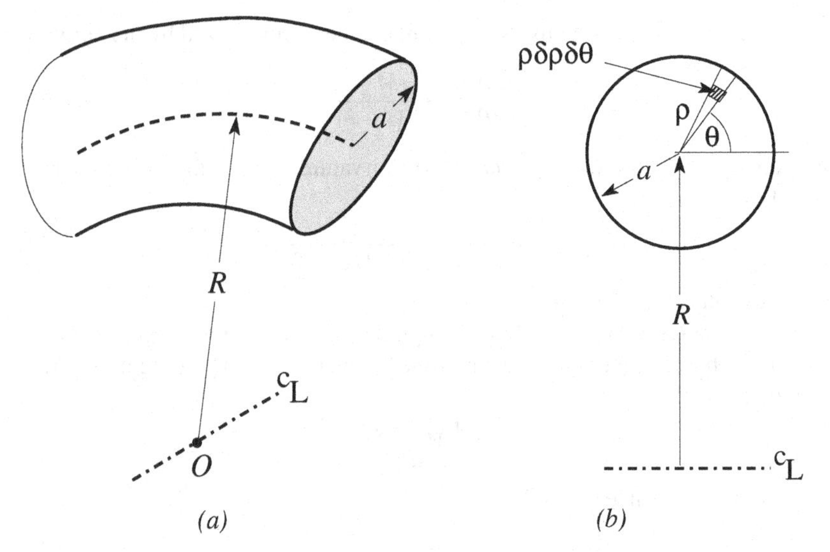


*Figure 11.4: Curved beam of circular cross section*

The cross-sectional area of the circle is $\pi a^2$ and the neutral radius is therefore

$$\hat{r} = \frac{\pi a^2}{2\pi(R - \sqrt{R^2 - a^2})} = \frac{a^2}{2(R - \sqrt{R^2 - a^2})} \; , \tag{11.11}$$

from (11.7).

### 11.1.2 The bending moment

Referring back to Figures 11.1, 11.2, we can develop an expression for the bending moment $M$ by taking moments about the curvature axis $O$. We obtain

$$\begin{aligned} M &= \iint_A \sigma_{\theta\theta} r dA = E\frac{d\phi}{d\theta}\iint_A \left(1 - \frac{\hat{r}}{r}\right) r dA \\ &= E\frac{d\phi}{d\theta}\left(\iint_A r dA - \iint_A \hat{r} dA\right) , \end{aligned}$$

using (11.5).

The first integral is equal to $A\bar{r}$, where $\bar{r}$ is the distance from $O$ to the centroid of $A$, so the moment expression can be written in the condensed form

$$M = EA(\bar{r} - \hat{r})\frac{d\phi}{d\theta} \; . \tag{11.12}$$

The stress due to the bending moment $M$ can then be obtained by eliminating $d\phi/d\theta$ between equations (11.5, 11.12), with the result

$$\sigma_{\theta\theta} = \frac{M}{A(\bar{r} - \hat{r})}\left(1 - \frac{\hat{r}}{r}\right) \; . \tag{11.13}$$

The deformation caused by the moment can be characterized by the derivative

$$\frac{d\phi}{d\theta} = \frac{M}{EA(\bar{r}-\hat{r})} . \tag{11.14}$$

Alternatively, we note that the *change* of curvature, $\Delta\kappa$ of the neutral surface is defined by

$$\Delta\kappa \equiv \frac{1}{\hat{s}} - \frac{1}{\hat{r}} = \frac{M}{EA\hat{r}(\bar{r}-\hat{r})} , \tag{11.15}$$

from equations (11.2, 11.14).

As in Chapter 3 (§3.2), we can compute the strain energy stored in the beam segment $\delta\theta$ by equating it to the work done by the moment $M$ during its application. We obtain

$$\delta U = \frac{1}{2}M\delta\phi = \frac{1}{2}M\frac{d\phi}{d\theta}\delta\theta = \frac{M^2\delta\theta}{2EA(\bar{r}-\hat{r})} ,$$

from (11.14) and hence

$$\frac{dU}{d\theta} = \frac{M^2}{2EA(\bar{r}-\hat{r})} . \tag{11.16}$$

This result permits us to use the energy methods of Chapter 3 in curved beam problems (see for example Problems 11.10, 11.11).

Equations (11.12–11.16) contain the factor $(\bar{r}-\hat{r})$, which is often small in comparison with $\bar{r}$. It is important to take a sufficient number of significant digits in the calculation of these quantities to ensure that accuracy is not lost. This will be clarified by the following example.

**Example 11.1**

*A curved beam of 1 in square cross section and inner radius 2 in subtends an angle of* $90^{\circ}$ *at the centre, as shown in Figure 11.5. Find the stresses at the inner and outer radii when the beam is subjected to a bending moment of 100 lb ft. Find also the relative rotation of the ends of the beam, if the material is steel with* $E = 30\times10^6$ *psi.*

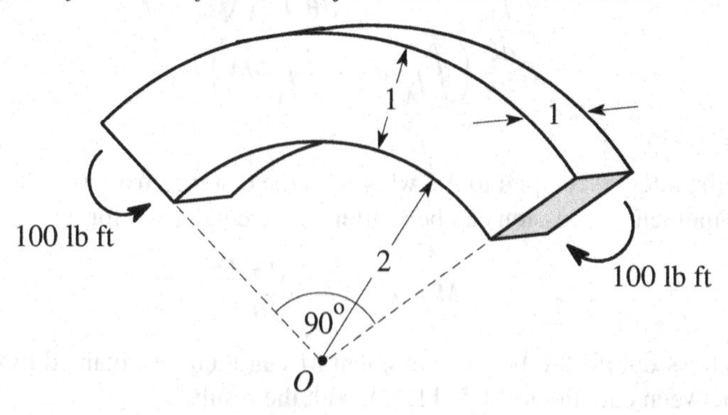


*Figure 11.5*

The inner and outer radii are $a=2$ in, $b=3$ in, respectively, so the neutral radius can be found from equation (11.9) as

$$\hat{r} = \frac{(b-a)}{\ln(b/a)} = \frac{1}{\ln(1.5)} = 2.46630 \text{ in.}$$

The centroidal radius is

$$\bar{r} = 2.5 \text{ in.}$$

Notice that we have to take a larger number of significant digits in $\hat{r}, \bar{r}$ than usual because the stress and deformation expressions involve the difference $(\bar{r}-\hat{r})$ between quantities of comparable magnitude.

The stress at the outer radius ($b=3$ in) can now be found from equation (11.13) as

$$\begin{aligned}\sigma_{\theta\theta}(b) &= \frac{M}{A(\bar{r}-\hat{r})}\left(1-\frac{\hat{r}}{b}\right) = \frac{100\times 12}{1(2.5-2.46630)}\left(1-\frac{2.46630}{3}\right)\\ &= 6335 \text{ psi (tensile).}\end{aligned}$$

A similar calculation for the inner radius, $a=2$ in, yields

$$\sigma_{\theta\theta}(a) = \frac{100\times 12}{1(2.5-2.46630)}\left(1-\frac{2.46630}{2}\right) = -8302 \text{ psi (compressive).}$$

Notice that the magnitude of the stress at the inner radius exceeds that at the outer radius. This is due to the stress concentrating effect of the curvature and is more pronounced when the ratio $a/b$ is smaller.

The deformation of the beam is defined by equation (11.14), from which we obtain

$$\frac{d\phi}{d\theta} = \frac{M}{EA(\bar{r}-\hat{r})} = \frac{100\times 12}{30\times 10^6\times 1(2.5-2.46630)} = 1.187\times 10^{-3}\,.$$

The relative rotation of the ends is therefore

$$\delta\phi = 1.187\times 10^{-3}\times 90^{\circ} = 0.107^{\circ}\,.$$

It is instructive to compare the maximum stresses with the predictions of the elementary bending theory for straight beams, which gives

$$\sigma_{\max} = \frac{Mc}{I} = \frac{100\times 12\times 0.5}{1^4/12} = 7200 \text{ psi.}$$

The elementary theory predicts equal and opposite stresses at the inner and outer radii, so it underestimates the stress at the inner radius and overestimates it at the outer radius, in each case by about 14%.

### 11.1.3 Composite cross sections

The preceding analysis applies to any cross section which is symmetric about a plane normal to the curvature axis of the beam. For a general cross section, we must first determine the neutral radius $\hat{r}$ from equation (11.7) and the centroidal radius $\bar{r}$ using the methods of §4.3. The circumferential stress distribution and the moment-curvature relation are then given by equations (11.13–11.15).

For composite areas

$$A = A_1 + A_2 + A_3 + \ldots ,$$

the integral in (11.7) can be written as the sum of a series of separate integrals, as in §4.3. We obtain

$$\frac{A}{\hat{r}} = \iint_A \frac{dA}{r} = \iint_{A_1} \frac{dA}{r} + \iint_{A_2} \frac{dA}{r} + \iint_{A_3} \frac{dA}{r} + \ldots \qquad (11.17)$$

The results of §11.1.1 therefore permit us to write down the integral for any area made up of rectangular and/or circular segments.

### 11.1.4 Axial loading

The analysis so far has been restricted to the case of pure bending, where the axial force $F$ defined by equation (11.6) is zero. In this section, we shall consider the opposite case, where there is an axial force, but no bending moment.

Figure 11.6*(a)* shows a segment of curved beam loaded by an axial force $F$ on its ends. The segment is clearly not in equilibrium, since the two forces are not co-linear, so we conclude that a curved beam cannot support a uniform axial load along its length without additional loading. The simplest loading scenario leading to a uniform axial force is that shown in Figure 11.6*(b)*, where the forces $F$ are balanced by a uniformly distributed radial load of $w$ per unit length. If the subtended angle $\delta\theta$ is sufficiently small, the equilibrium equation for this figure is

$$wR\delta\theta - 2F\frac{\delta\theta}{2} = 0$$

and hence

$$w = \frac{F}{R} .$$

This loading leads to an axisymmetric state of stress and deformation and is in fact analogous to that in the membrane theory of axisymmetric shells (Chapter 8). The subtended angle $\delta\theta$ must remain the same after deformation and hence the circumferential strain is

$$e_{\theta\theta} = \frac{u_r}{r} ,$$

where $u_r$ is the radial displacement.[3]

[3] *cf* equation (8.3).

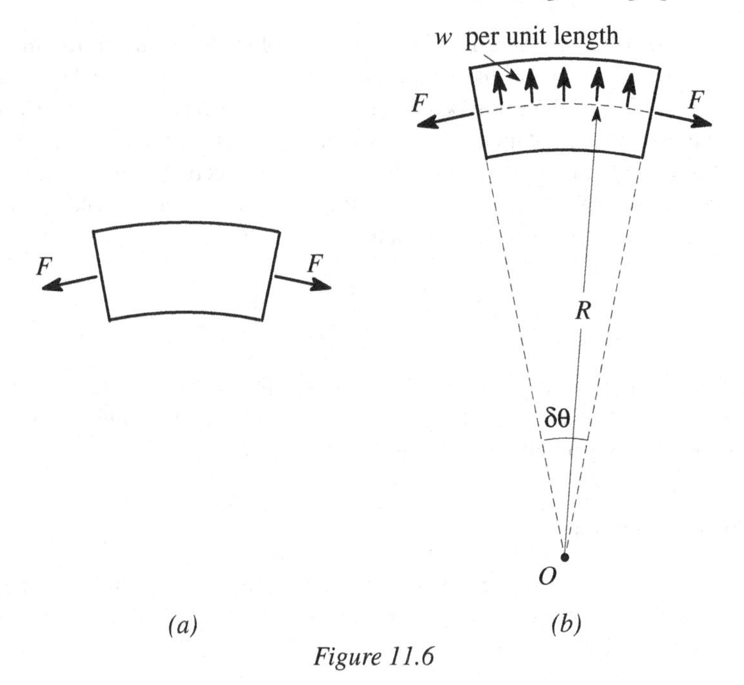


*Figure 11.6*

Neglecting Poisson's ratio effects from the stress components $\sigma_{rr}, \sigma_{zz}$, we then have

$$\sigma_{\theta\theta} = Ee_{\theta\theta} = \frac{Eu_r}{r} \tag{11.18}$$

corresponding to an axial force

$$F = \iint_A \sigma_{\theta\theta} dA = Eu_r \iint_A \frac{dA}{r} = \frac{EAu_r}{\hat{r}} , \tag{11.19}$$

using (11.7). Eliminating $u_r$ between equations (11.18, 11.19) gives

$$\sigma_{\theta\theta} = \frac{F\hat{r}}{Ar} . \tag{11.20}$$

We can find the line of action of the force $F$ by taking moments about the axis of curvature $O$. We find

$$M_O = \iint_A \sigma_{\theta\theta} r dA = Eu_r \iint_A dA = EAu_r .$$

It follows that the force passes through a point a distance

$$\frac{M_O}{F} = EAu_r \frac{\hat{r}}{EAu_r} = \hat{r}$$

from $O$. In other words, the force for this simple mode of deformation passes through the neutral surface of the bending analysis.

This represents the most natural decomposition of the load into a force and a moment, since the force then produces no rotation of transverse planes and the moment causes no extension of the fibre located at the line of action of the force. However, simpler expressions are obtained if we consider the stress field due to a force whose line of action passes through the centroid ($r=\bar{r}$) of the section. This loading can be regarded as the superposition of a force $F$ acting through $r=\hat{r}$ and a bending moment $M=F(\bar{r}-\hat{r})$. The resulting stress field is then

$$\sigma_{\theta\theta} = \frac{F\hat{r}}{Ar} + \frac{F(\bar{r}-\hat{r})}{A(\bar{r}-\hat{r})}\left(1-\frac{\hat{r}}{r}\right) = \frac{F}{A} . \tag{11.21}$$

Thus, if the line of action of the force passes through the centroid, the stress distribution is uniform and equal to $F/A$ as in the elementary bending theory, but the deformation will then involve a non-zero value of $d\phi/d\theta$.

### More general loading

If there is an axial force, but no radial load, the bending moment and the transverse shear force in the beam must vary along the beam. The corresponding beam segment is shown in Figure 11.7.

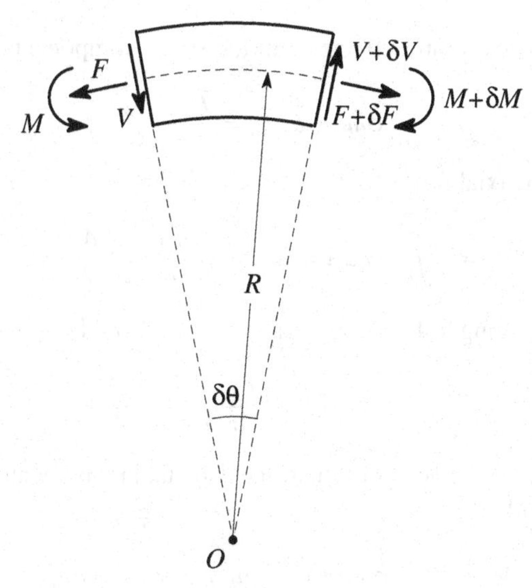


*Figure 11.7*

Resolving forces in horizontal and vertical directions and taking moments about the centre of curvature $O$, we obtain the three equilibrium equations

$$\delta F + V\delta\theta = 0 \;\; ; \;\; \delta V - F\delta\theta = 0 \;\; ; \;\; R\delta F + \delta M = 0$$

and hence, in the limit $\delta\theta \to 0$,

$$\frac{dF}{d\theta} = -V \;\; ; \;\; \frac{dV}{d\theta} = F \;\; ; \;\; \frac{dM}{d\theta} = R\frac{dF}{d\theta} = -RV \; . \tag{11.22}$$

The first two equations can be combined to yield

$$\frac{d^2F}{d\theta^2} + F = 0 \;\; ; \;\; \frac{d^2V}{d\theta^2} + V = 0 \tag{11.23}$$

and hence the most general solution for $F, V, M$ in the absence of distributed loads is

$$F = A\cos\theta + B\sin\theta \; ; \; V = A\sin\theta - B\cos\theta \; ; \; M = AR\cos\theta + BR\sin\theta + C, \tag{11.24}$$

where $A, B, C$ are three constants to be determined from the end loading conditions.

The shear force $V$ will generally cause initially plane transverse sections to deform out of plane and hence the preceding analysis is strictly not appropriate to this case. However, as with straight beams, the stresses associated with the shear force are generally fairly small and the simpler results can be used for general loading without much loss in accuracy.

### Example 11.2

*Figure 11.8 shows a U-shaped curved beam with a symmetric channel cross section, loaded by two opposed forces of magnitude 2kN. Find the neutral radius $\hat{r}$ associated with this cross section and the maximum tensile stress at the section $A-A$.*

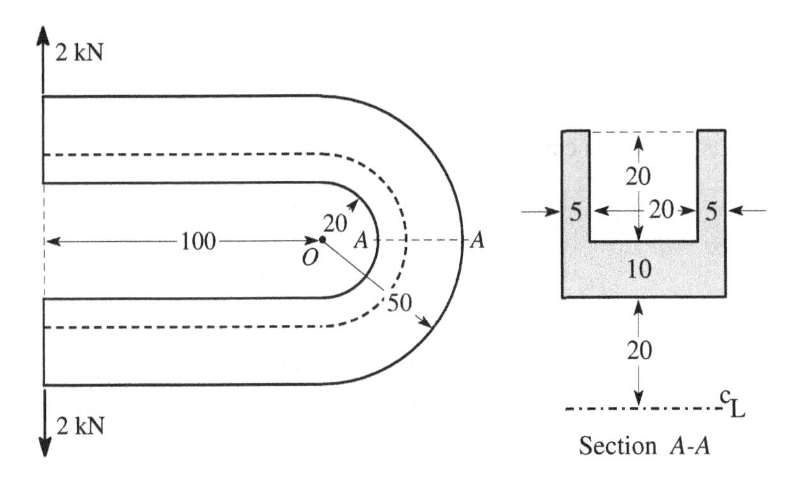


all dimensions in mm

*Figure 11.8*

The cross section is conveniently regarded as the difference between the solid (30×30) square and the dotted (20×20) square in Figure 11.9.

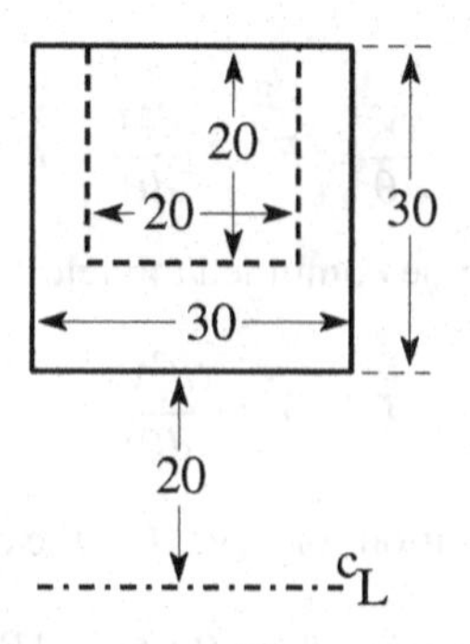


*Figure 11.9*

For the 30 × 30 square, ($A_1$), we have

$$A_1 = 900 \text{ mm}^2,\ a = 20 \text{ mm},\ b = 50 \text{ mm},\ \bar{r}_1 = \frac{a+b}{2} = 35 \text{ mm},\ t = 30 \text{ mm}$$

and

$$\iint_{A_1} \frac{dA}{r} = t\ln(b/a) = 27.4887 \text{ mm},$$

from (11.8).

For the 20 × 20 square ($A_2$),

$$A_2 = -400 \text{ mm}^2,\ a = 30 \text{ mm},\ b = 50 \text{ mm},\ \bar{r}_2 = 40 \text{ mm},\ t = 20 \text{ mm}$$

and

$$\iint_{A_2} \frac{dA}{r} = -10.2165 \text{ mm}.$$

These results are conveniently tabulated as in §4.3 as

| $i$ | 1 | 2 |
|---|---|---|
| $A_i$ | 900 | –400 |
| $\bar{r}_i$ | 35 | 40 |
| $\iint_{A_i} \frac{dA}{r}$ | 27.4887 | -10.2165 |

We then have

$$\begin{aligned} A &= A_1 + A_2 = 900 - 400 = 500 \text{ mm}^2 \\ A\bar{r} &= 900 \times 35 - 400 \times 30 = 15500 \text{ mm}^3 \\ \bar{r} &= \frac{15500}{500} = 31 \text{ mm} \\ \iint_A \frac{dA}{r} &= 27.4887 - 10.2165 = 17.2722 \text{ mm}. \end{aligned}$$

and the neutral radius, $\hat{r}$ is therefore

$$\hat{r} = A \Big/ \iint_A \frac{dA}{r} = \frac{500}{17.2722} = 28.9483 \text{ mm}.$$

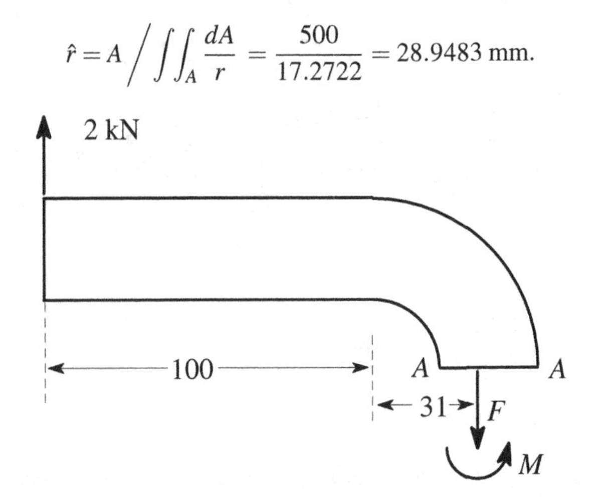


all dimensions in mm

*Figure 11.10*

Figure 11.10 shows a free-body diagram of the beam sectioned at $A$–$A$, at which there will be an axial force $F = 2$ kN and a bending moment, referred to the centroidal radius $\bar{r} = 31$ mm of

$$M = 2(100 + 31) = 262 \text{ Nm (kNmm)}.$$

The direction of the bending moment in Figure 11.10 is opposite to that in Figure 11.1 *(a)*, so we must take $M = -262$ Nm in equation (11.13). Also, it is clear from Figure 11.7 that the maximum tensile stress will occur at the inner radius, $r = 20$ mm and its magnitude is

$$\sigma_{\theta\theta}^{\max} = \sigma_{\theta\theta}(20) = \frac{F}{A} + \frac{M}{A(\bar{r} - \hat{r})}\left(1 - \frac{\hat{r}}{r}\right)$$
$$= \frac{2 \times 10^3}{500} + \frac{-262 \times 10^3}{500(31 - 28.9483)}\left(1 - \frac{28.9483}{20}\right) = 118.3 \text{ MPa}.$$

## 11.2 Radial stresses

In developing equation (11.5), we assumed that $\sigma_{\theta\theta}$ was the only stress and in particular that the radial stress $\sigma_{rr}$ was negligible. In practice, significant radial stresses can be developed in curved beams, but since their contribution to equation (11.5) is multiplied by Poisson's ratio which is fairly small, the above analysis generally gives a good approximation to the circumferential stresses in the beam.

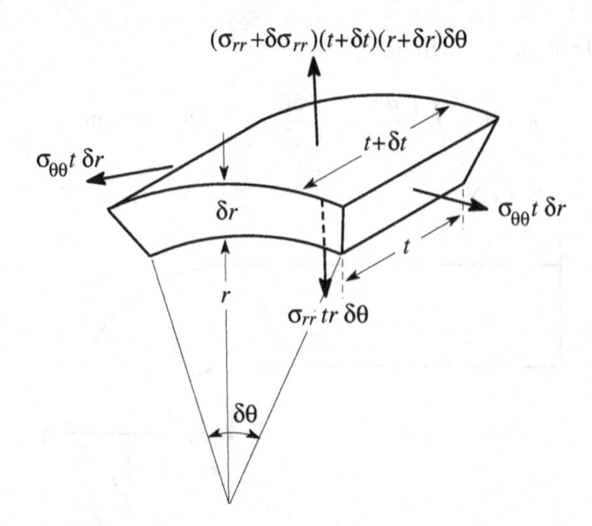


*Figure 11.11: Radial equilibrium of a small beam element*

We can estimate the distribution of radial stress by using an equilibrium argument. We first need to modify the equilibrium equation (10.4) to include the effect of a variable thickness in the axial direction. Figure 11.11 shows the small element of the beam contained between the radial surfaces $r$ and $r+\delta r$. The equilibrium equation for this element, analogous to equation (10.1) with no acceleration term, is

$$(\sigma_{rr}+\delta\sigma_{rr})(r+\delta r)(t+\delta t)\delta\theta-\sigma_{rr}rt\delta\theta-2\sigma_{\theta\theta}t\delta r\sin\left(\frac{\delta\theta}{2}\right)=0\,.$$

Dividing throughout by $r\delta r\delta\theta$ and proceeding to the limit as $\delta r$, $\delta\theta\to 0$, we obtain the differential equation of equilibrium

$$\frac{d}{dr}(t\sigma_{rr})+\frac{(\sigma_{rr}-\sigma_{\theta\theta})t}{r}=0\,, \tag{11.25}$$

which clearly reduces to (10.4) if $t$ is independent of $r$.

Equation (11.25) can be expressed in the form

$$\frac{d}{dr}(tr\sigma_{rr})=t\sigma_{\theta\theta}\,.$$

The radial stress must be zero at the inner radius $r=a$ and hence we can integrate this relation to obtain

$$\sigma_{rr}=\frac{1}{tr}\int_a^r t\sigma_{\theta\theta}dr\,, \tag{11.26}$$

which permits $\sigma_{rr}$ to be determined, once $\sigma_{\theta\theta}$ is known.

**Example 11.3**

*Find the maximum radial stress $\sigma_{rr}$ for the square section curved beam of Example 11.1.*

We already found the radii $\hat{r}, \bar{r}$ for this beam in Example 11.1 above as

$$\hat{r} = 2.46630 \text{ in} \quad ; \quad \bar{r} = 2.5 \text{ in}$$

and hence the general expression for $\sigma_{\theta\theta}$ is

$$\begin{aligned}\sigma_{\theta\theta}(r) &= \frac{M}{A(\bar{r}-\hat{r})}\left(1-\frac{\hat{r}}{r}\right) = \frac{100\times 12}{1(2.5-2.46630)}\left(1-\frac{2.46630}{r}\right)\\ &= 35608 - \frac{87821}{r} \text{ psi},\end{aligned}$$

where $r$ is in inches.

For this geometry, $t$ is constant and equal to 1 in, so substituting for $\sigma_{\theta\theta}$ into equation (11.26), we obtain

$$\sigma_{rr} = \frac{1}{r}\int_2^r\left(35608 - \frac{87821}{r}\right)dr = 35608\left(1-\frac{2}{r}\right) - \frac{87821}{r}\ln(r/2)\,.$$

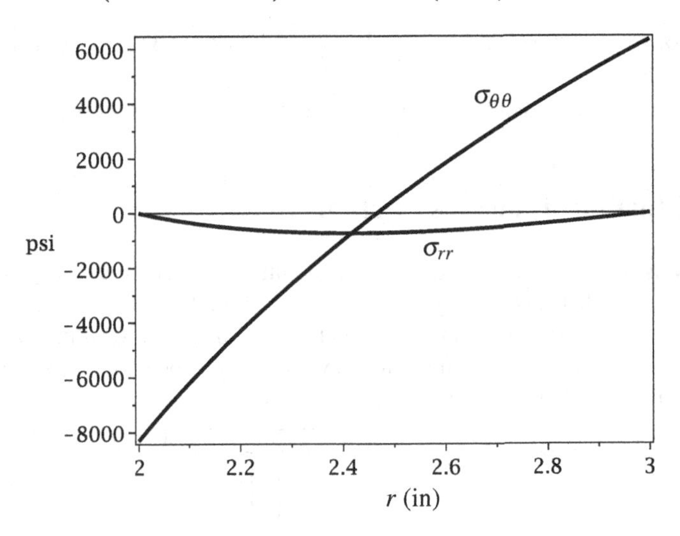


*Figure 11.12: Stress distribution in the square section beam*

The stresses $\sigma_{rr}, \sigma_{\theta\theta}$ are shown as functions of $r$ in Figure 11.12. The radial stress is zero at the inner and outer radii, as required by the traction-free boundary condition, and the maximum value occurs when

$$\frac{d\sigma_{rr}}{dr} = \frac{87821}{r^2}\ln(r/2) - \frac{16605}{r^2} = 0$$

and hence

$$r = 2\exp\left(\frac{16605}{87821}\right) = 2.416 \text{ in}.$$

The maximum radial stress is therefore

$$\sigma_{rr}^{\max} = 35608\left(1-\frac{2}{2.416}\right) - \frac{87821}{2.416}\ln(2.416/2) = -738 \text{ psi (compressive)}.$$

This is small compared with the maximum circumferential stress (6335 psi tensile at $r=3$ and 8302 psi compressive at $r=2$) and can generally be neglected. However, it is more significant for severely curved beams ($a/b \ll 1$) or for beams with a reduced thickness $t$ near the neutral surface.

The radial stress must necessarily be zero at the inner and outer radii and reach a maximum somewhere near the mean radius. The derivative $d\sigma_{rr}/dr$ in that region is generally fairly small, so for a quick estimate of the magnitude of the maximum radial stress, it is often sufficiently accurate to evaluate it at the neutral or centroidal radii instead of calculating the exact maximum location. In the present example, we have

$$\sigma_{rr}(\bar{r}) = 35608\left(1-\frac{2}{2.5}\right) - \frac{87821}{2.5}\ln(2.5/2) = 717 \text{ psi},$$

whilst

$$\sigma_{rr}(\hat{r}) = 35608\left(1-\frac{2}{2.4663}\right) - \frac{87821}{2.4663}\ln(2.4663/2) = 730 \text{ psi},$$

both of which are sufficiently close to the actual maximum of 738 psi.

## 11.3 Distortion of the cross section

The analysis in this chapter is based on the assumption that the cross section of the beam remains undistorted during bending deformation. Unfortunately, this is a non-conservative approximation and it can lead to serious errors for thin-walled cross sections, such as I-beams and thin-walled cylinders (pipe bends), where the stiffness is overestimated and the maximum stress underestimated.

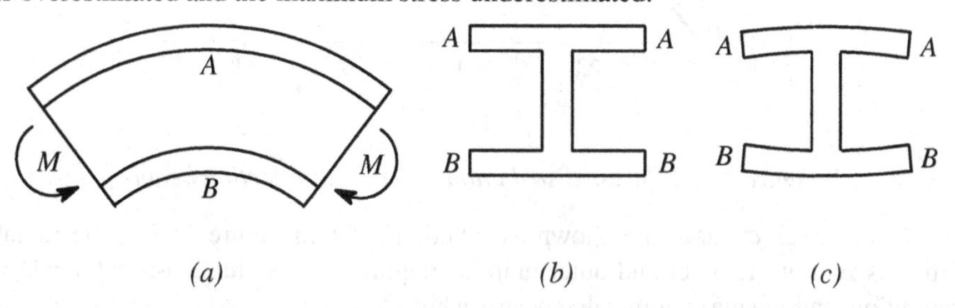


*Figure 11.13: Distortion of the cross section for a curved beam of I-section*

To understand the mechanism of section distortion during bending, it is helpful to consider the strain energy associated with the assumed deformation. Figure 11.13 *(a)* shows a curved beam whose thin-walled I-beam cross section is shown in Figure 11.13 *(b)*. If the beam is subjected to a bending moment $M$, the outer flange will be loaded in tension, whilst the inner flange is loaded in compression. Now suppose that the beam flanges were to distort, as shown in Figure 11.13 *(c)*, the tips of the outer flange $A$ bending inwards and those of the inner flange $B$ bending outwards. As a result of this deformation, the final radius of regions $A$ will be reduced, so that the tensile stress is reduced, resulting in a decrease in the strain energy $U$ stored in the beam. A similar effect is predicted for the outward motion of the inner flanges at $B$. Of course, there will be additional strain energy associated with the curvature of the flanges in Figure 11.13*(c)*, but for a sufficiently thin flange, this will be small and the net effect of the distortion will be a reduction of total strain energy. The stationary potential energy theorem (§3.5) therefore tells us to expect the beam section to deform during bending as shown in Figure 11.13*(c)*. If the bending moment is reversed in direction, the distortion is also reversed — i.e. the flanges $A$ will bend radially outwards and the flanges $B$ will bend inwards.

This effect is clearly non-conservative, since by reducing the magnitude of $\sigma_{\theta\theta}$ at $A$ and $B$, the distortion reduces the corresponding bending moment. If the bending moment is prescribed, additional curvature change ($\Delta\kappa$) will occur, leading to increased bending stresses at the middle of the flanges, where radial motion is restrained by the web.

For thin-walled circular tubes (pipe bends), the same effect causes the section to become oval during bending, as shown in Figure 11.14. The direction of ovalization depends on the direction of the applied moment.

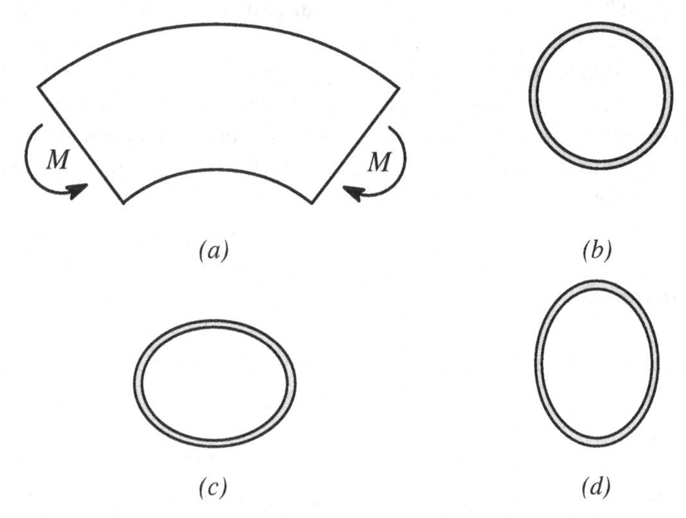


*Figure 11.14: (a) Curved thin-walled tube loaded in bending, (b) undeformed cross section, (c) deformed cross section for positive M, (d) deformed cross section for negative M*

The most straightforward way to estimate the distortion of a thin-walled cross section and its effect on the maximum stress is to approximate the expected deformed shape by an appropriate function and use the Rayleigh-Ritz method (§3.6 above). More details of this procedure can be found in Cook and Young (1985), §§10.5, 10.6.

## 11.4 Range of application of the theory

The analysis of curved beams is somewhat more complicated than that for straight beams and hence it makes sense to use the simpler theory if the radius of curvature is large compared with the radial thickness of the beam. A comparison of the two theories for a beam of rectangular cross section shows that the error involved in using the elementary theory is less than 10% as long as $b/a < 1.3$, where $a, b$ are the inner and outer radii respectively, as shown in Figure 11.3. Another way of expressing this condition is that the radial thickness of the beam $(b-a)$ should be less than 30% of the inner radius.

The elementary theory is much more successful in predicting the deflections in a curved beam, particularly if the neutral radius $\hat{r}$ is used to define the 'length' of the beam (see Problem 11.3). For example, the error in the predicted stiffness of a beam of rectangular cross section, compared with the curved beam result, is less than 10% for $b/a < 14$.

The curved beam theory is itself approximate, since equation (11.3) assumes that the radial displacement $u_r$ is independent of $r$ and we showed in §11.2 that the radial stress $\sigma_{rr}$ (and hence the strain $e_{rr}$) is non-zero. Fortunately, an exact solution to the bending problem can be obtained for the rectangular cross section[4] and a comparison with the curved beam theory[5] shows that the latter predicts a maximum stress within 10% of the exact value as long as $b/a < 11$.

We conclude that the curved beam theory is useful in the range $1.3 < b/a < 11$. However, in a design problem, it is usually sensible to use the elementary theory first to obtain an estimate of the maximum stress even in this range. If the result is very low relative to the allowable stress $\sigma_{all}$ (e.g. if $\sigma_{max}/\sigma_{all} < 0.2$), the extra effort involved in a curved beam calculation is not justified.

## 11.5 Summary

In this chapter, we have shown that the stress distribution in a curved beam due to a bending moment differs from that in a straight beam of similar cross section, and the neutral surface is displaced towards the axis of curvature. The inner radius of the beam acts as a stress concentration, causing an increase in the local bending stress.

Radial stresses as well as circumferential stresses are obtained and thin-walled cross sections may experience significant distortion.

The elementary bending theory gives a good estimate of the maximum stresses and the deformation except when the radius of curvature of the beam is comparable with the cross-sectional dimensions.

---

[4] J.R. Barber (2010), *Elasticity*, Springer, Dordrecht, Netherlands, 3rd edn., Chapter 10.

[5] R.D. Cook and W.C. Young (1985), *Advanced Mechanics of Materials,* Macmillan, New York, §10.3.

## Further reading

A.P. Boresi, R.J. Schmidt, and O.M. Sidebottom (1993), *Advanced Mechanics of Materials,* John Wiley, New York, 5th edn., Chapter 9.
R.D. Cook and W.C. Young (1985) *Advanced Mechanics of Materials,* Macmillan, New York, Chapter 10.

## Problems

**11.1.** Figure P11.1 shows a curved bar of solid circular cross section, with diameter 30 mm. The axis of the bar describes an arc with a radius 40 mm and the bar is loaded by a bending moment of 300 Nm. Find the maximum tensile stress. What would be the percentage error involved in using the elementary bending equation $\sigma = My/I$?

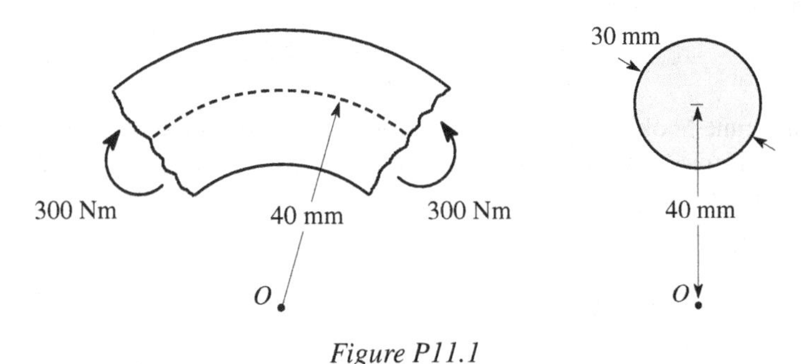


*Figure P11.1*

**11.2.** Find the maximum bending moment that can be transmitted by the curved beam of Figure P11.1 if the allowable tensile stress in the material is $\sigma_{\rm all} = 120$ MPa.

**11.3.** The elementary bending theory predicts that the ends of an initially straight beam of length $L$ will experience a relative rotation

$$\Delta\phi = \frac{ML}{EI} \ ,$$

when subjected to a bending moment $M$ [see equation (3.9)].

Use this result to estimate the relative rotation of the ends of the curved beam of Figure 11.5, assuming that the 'length' of the beam is the mean length, $L = \bar{r}\Delta\theta$, where $\Delta\theta$ is the subtended angle (in radians). Repeat the estimate using $L = \hat{r}\Delta\theta$. Comment on your results.

**11.4.** Figure P11.4 shows a U-shaped curved beam with a T-shaped cross section, loaded by two opposed forces of magnitude 500 lb. Find the maximum circumferential stress $\sigma_{\theta\theta}^{\max}$.

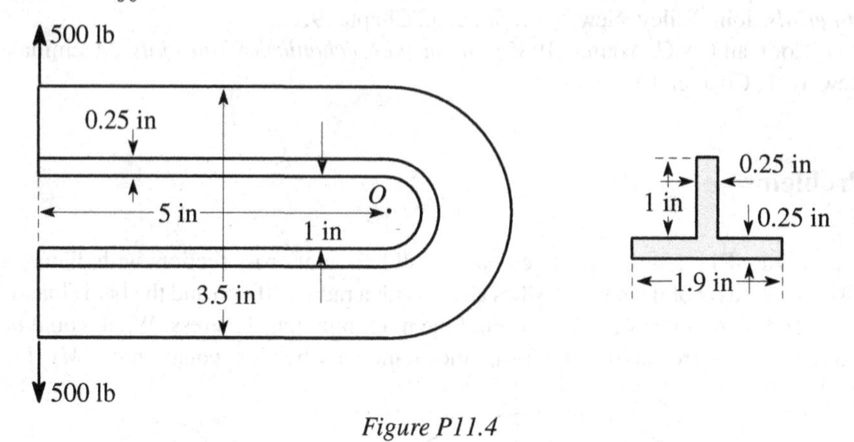


*Figure P11.4*

**11.5.** The crane hook in Figure P11.5 is loaded by a force of 8000 lb. Find the circumferential stress at the point $A$, where the hook has the trapezoidal section shown with $a=1$ in., $b=4$ in., $c=2$ in., $d=1$ in.

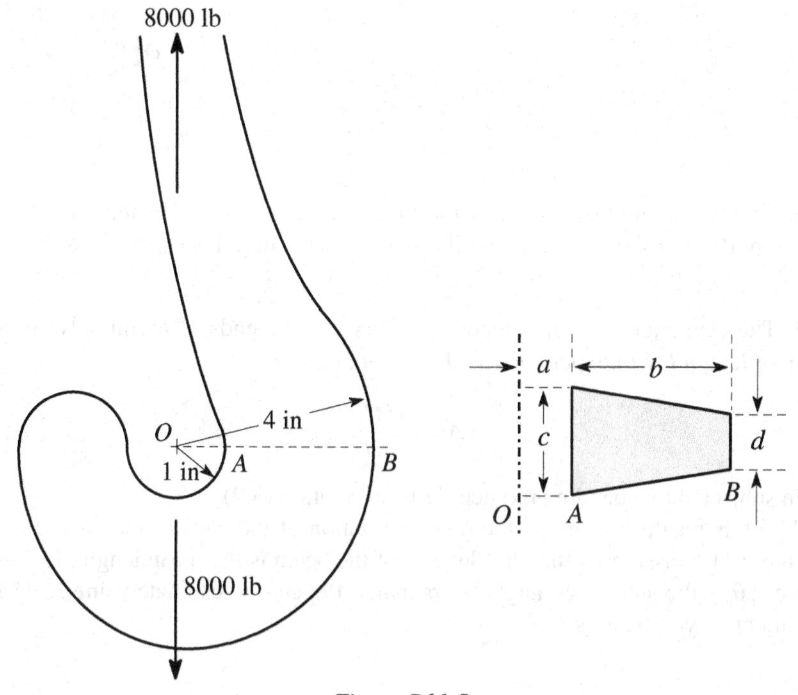


*Figure P11.5*

You may use without proof the following results for the trapezoidal section:-

$$A = \frac{(c+d)(b-a)}{2} \; ; \; \bar{r} = \frac{(2a+b)c+(a+2b)d}{3(c+d)} \; ;$$

$$\frac{A}{\hat{r}} = (d-c) + \frac{(bc-ad)}{(b-c)} \ln(b/a) \, .$$

**11.6.** Use integration to derive the expressions for $A, \bar{r}, A/\hat{r}$ for the trapezoidal section given in Problem 11.5.

**11.7.*** In a curved beam, the stress concentrating effect of the inner radius can be reduced by making the thickness there larger than that at the outer radius. This is the rationale for using the trapezoidal section in Figure P11.5. If the ratio between outer and inner radii is $b/a=2$, find the ratio $c/d$ for the trapezoidal section, if the circumferential stresses at the inner and outer radii are to be equal and opposite. The results given in Problem 11.5 may be used without proof.

**11.8.** Figure P11.8 shows part of the frame of a press. Find the maximum tensile stress at $A$, when a vertical load of 20 tons is applied as shown.

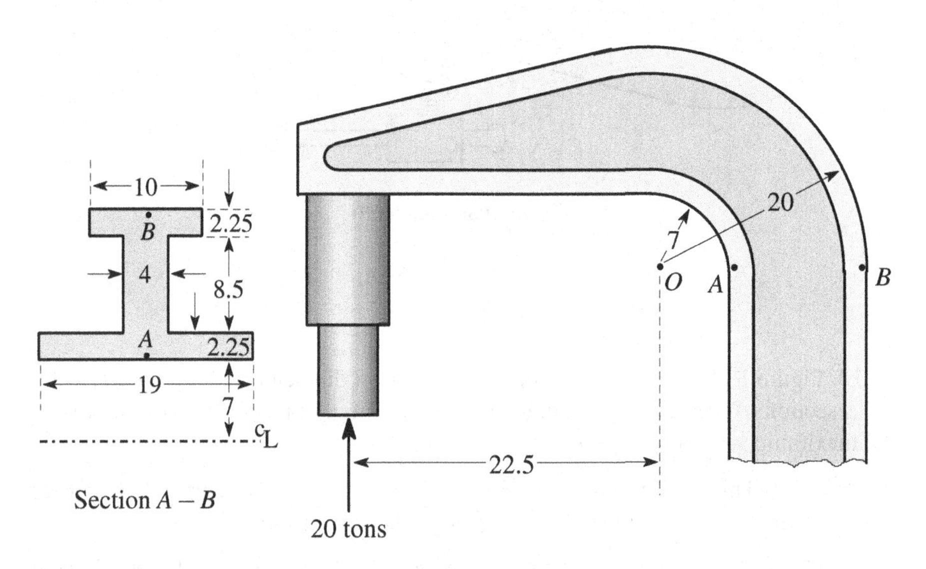


*Figure P11.8*

**11.9.** The C-clamp of Figure P11.9 is manufactured from a steel with a tensile yield stress $S_Y = 380$ MPa. The frame has the T-section shown. Find the maximum clamping force $P$, if there is to be a factor of safety of 2.5 against yielding at the point $A$.

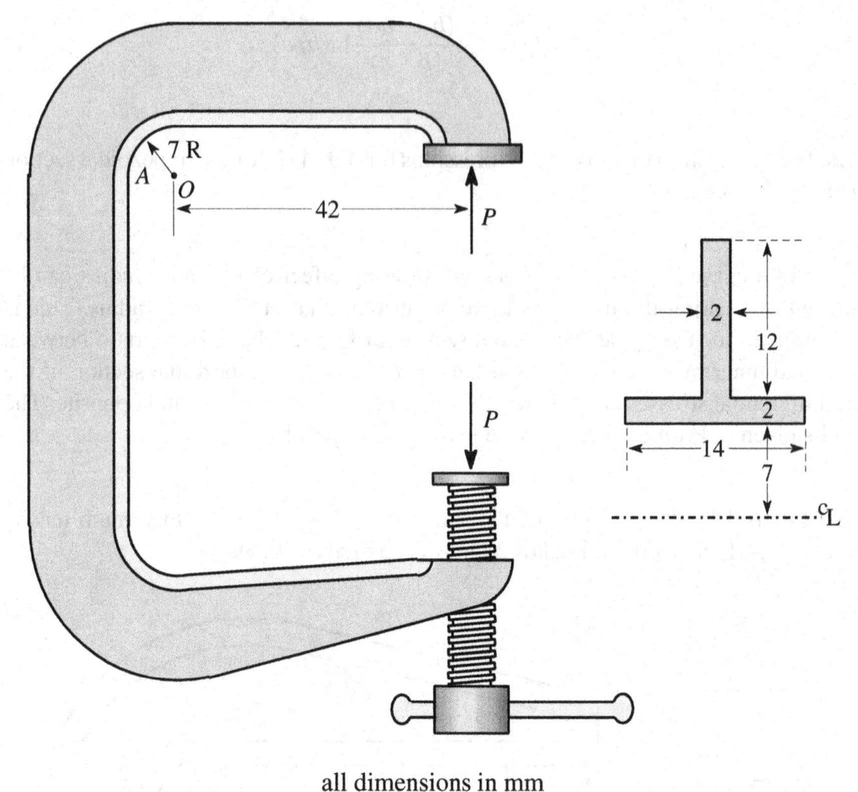


*Figure P11.9*

**11.10.** Figure P11.10 shows a circular ring of mean diameter 350 mm and circular cross section of diameter 100 mm, loaded by equal and opposite 50 kN forces. Find the maximum stress in the ring.

**Method**: This indeterminate problem is similar to that of Example 3.16, except that we need to use equation (11.16) to replace the integrand in (3.110).

Use Castigliano's second theorem to resolve the indeterminacy and hence find the maximum bending moment. Then use the results of §11.1 to determine the maximum tensile stress.

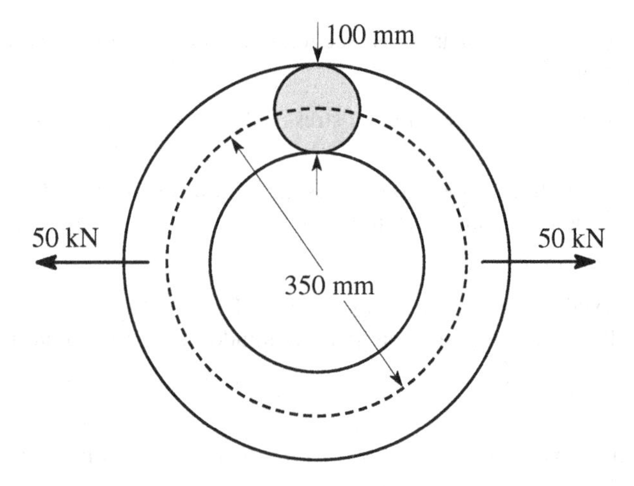


*Figure P11.10*

**11.11.** The steel U-beam of Figure P11.11 has uniform 2 inch square cross section and is loaded by two forces $P$, as shown.

Use equations (11.16, 3.24) to determine the total strain energy in the beam as a function of $P$ and hence use the method of §3.3 to determine the relative displacement of the points $A, B$, when $P = 4000$ lbs ($E_{\text{steel}} = 30 \times 10^6$ psi).

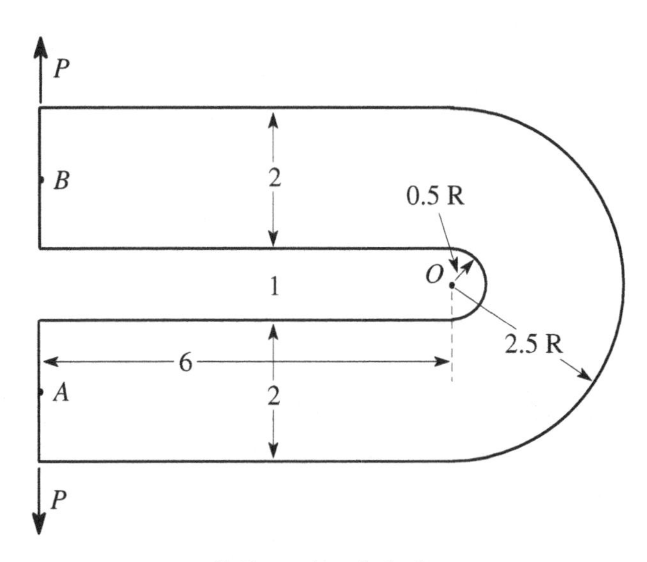


*Figure P11.11*

**11.12.** A curved beam of square cross section $a \times a$ and inner radius $na$ is loaded by a bending moment $M$.

Find the maximum circumferential stress $\sigma_{\theta\theta}^{\max}$ and the maximum radial stress $\sigma_{rr}^{\max}$.

Sketch the variation of these quantities as functions of $n$. Sketch also the ratio $\sigma_{rr}^{\max}/\sigma_{\theta\theta}^{\max}$ and comment on the results.

**11.13.** The curved section of the beam in Figure P11.4 is loaded by a bending moment of 200 lb ft. Find the location and magnitude of the maximum *radial* stress $\sigma_{rr}^{\max}$.

**11.14.** Find the maximum radial stress $\sigma_{rr}^{\max}$ for the U-beam of Figure P11.11 loaded by two forces $P = 4000$ lbs.

# 12

# Elastic Stability

In many design contexts, it is desirable to use components that have the minimum weight and cost compatible with acceptable values of strength and stiffness. The methods of the preceding chapters will often predict that the optimum design according to this criterion is a thin-walled structure. For example, we saw in §6.5 that closed sections are more effective at transmitting torques and that the strength and stiffness of such sections increases with the area enclosed by the section. It follows that the theoretical optimum torsion member for a given mass of material is a thin-walled cylindrical tube and the strength and stiffness theoretically increase without limit as the radius of the tube increases and the wall thickness correspondingly decreases [see equations (6.39, 6.40)].

Thin-walled structures can indeed be quite surprisingly strong and lightweight — a result that is best appreciated by constructing simple structures out of paper and testing them to destruction.

### Experiment

Take a single sheet of $8\frac{1}{2}\times 11$ (or A4) paper, roll it up into a cylindrical tube with a small overlap on the longer edge and tape up the join with Scotch tape. Now stand the tube vertically on a flat table and carefully rest this book on the top of the tube. Then add another book of similar weight.

Probably you will be able to support at least two books, with a total weight exceeding 5 lbs, on this single sheet of paper.[1]

However, add a bit more weight and the complete structure will collapse. The paper will not tear and the joint will stay intact, but the walls will fold in a complex way and no longer support the load. This collapse mode is known as *elastic instability* or *buckling* and it places limits on the thinness of components.

[1] Somehow, it seems even more surprising that with (for example) 10 sheets of paper suitably placed, you can support 50 lbs! Try asking your non-engineering friends to guess how many sheets of paper (and if you want to be exact, inches of tape) are needed to support a 5 lb weight, 11 inches above the table and then do a demonstration.

J.R. Barber, *Intermediate Mechanics of Materials*, Solid Mechanics and Its Applications 175, 2nd ed., DOI 10.1007/978-94-007-0295-0_12, 

Buckling generally only occurs when the component is loaded in compression. This again is easily demonstrated with a sheet of paper. If you grasp two opposite edges and apply a tensile force as shown in Figure 12.1 *(a)*, there will be no instability and in fact the paper is quite strong. It requires considerable strength to fracture it in tension. However, if you try to load it in compression as in Figure 12.1 *(b)*, it will immediately bend out of plane and support hardly any load.

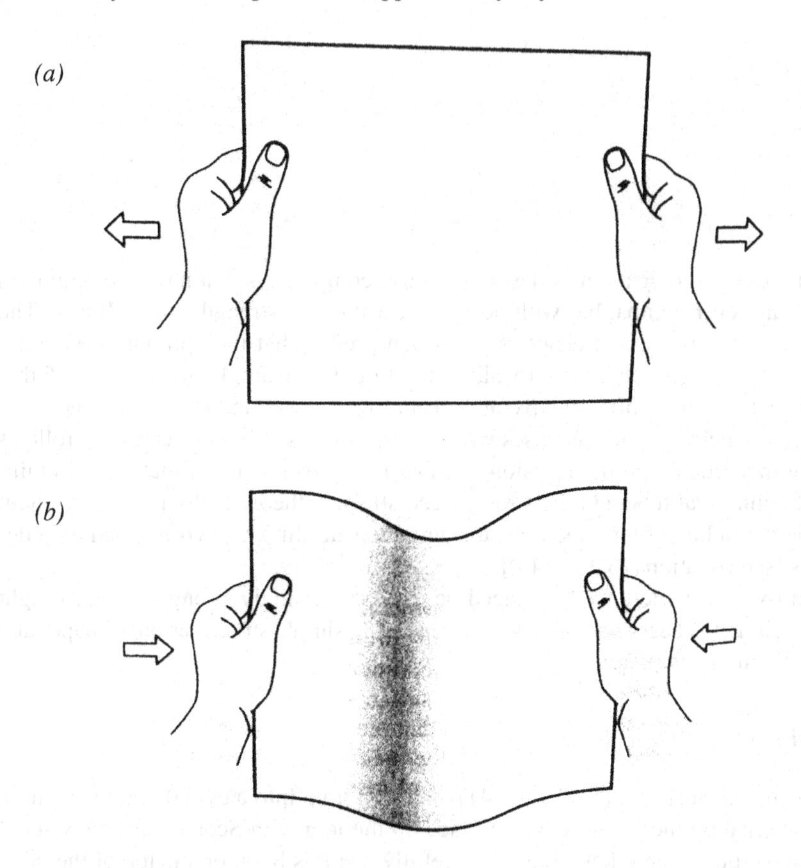


*Figure 12.1: A piece of paper loaded (a) in tension and (b) in compression*

## 12.1 Uniform beam in compression

A simple system that exhibits this behaviour is the uniform beam of Figure 12.2 *(a)* loaded by a compressive axial force. A straight beam loaded purely in compression is sometimes referred to as a *column*. We suppose that the beam is simply-supported at its ends, $z=0,L$, and that the force $P$ is sufficient to cause the beam to deform into the buckled shape shown in Figure 12.2 *(b)*, in which the lateral deflection is described by an as yet unknown function $u(z)$.

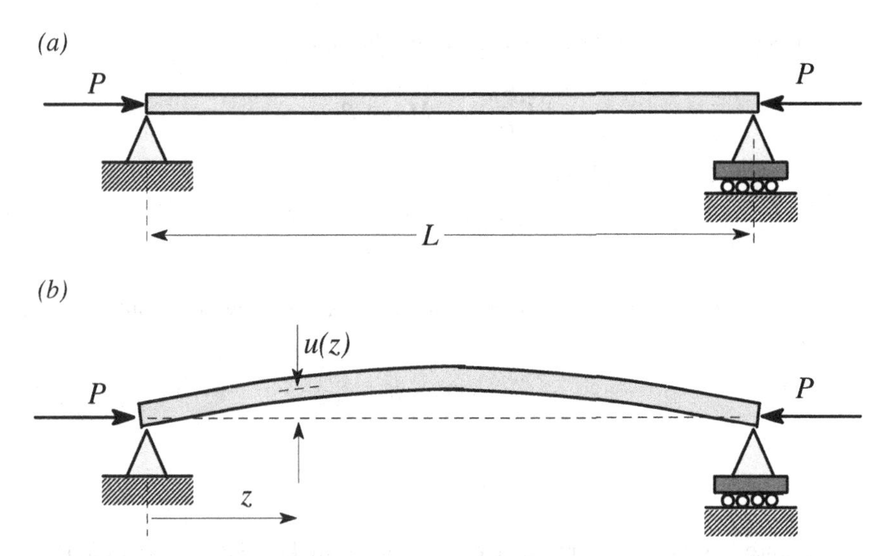


*Figure 12.2: (a) Simply-supported beam loaded in compression; (b) deformed shape*

In general, we anticipate reaction forces $R_1, R_2$ at the simple supports, as shown in the free-body diagram of Figure 12.3 *(a)*, but by taking moments about the ends, $A, B$, respectively, it can be shown that these reactions must both be zero.

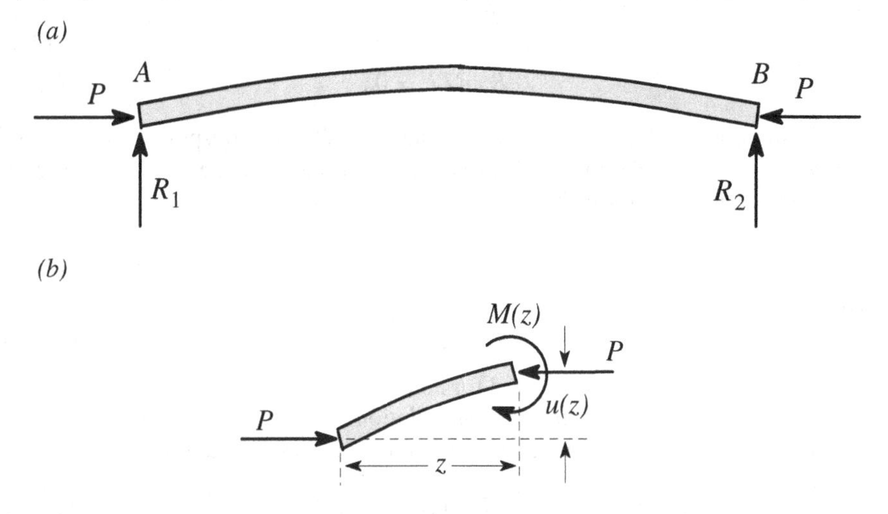


*Figure 12.3: Free-body diagram of (a) the whole beam and (b) a segment of length z at one end*

Figure 12.3 *(b)* shows a free-body diagram for a segment of the beam of length $z$ and by taking moments about one end of this segment, we conclude that the bending moment $M$ is given by

$$M = Pu\ . \tag{12.1}$$

It follows that the displacement $u$ must satisfy the equation

$$EI\frac{d^2u}{dz^2} = -M = -Pu \,, \tag{12.2}$$

from equation (1.17), and hence

$$\frac{d^2u}{dz^2} + \frac{Pu}{EI} = 0 \,. \tag{12.3}$$

This is a homogeneous ordinary differential equation whose solution can be shown by substitution to be

$$u = A\cos\lambda z + B\sin\lambda z \,, \tag{12.4}$$

where

$$\lambda = \sqrt{\frac{P}{EI}} \tag{12.5}$$

and $A, B$ are arbitrary constants to be determined from the boundary conditions

$$u = 0 \;\; ; \;\; z = 0 \tag{12.6}$$

$$u = 0 \;\; ; \;\; z = L \,. \tag{12.7}$$

Substituting (12.4) into (12.6, 12.7), we obtain

$$A = 0 \tag{12.8}$$

$$B\sin\lambda L = 0 \,. \tag{12.9}$$

Clearly this problem has the trivial solution $A=B=0$, corresponding to the case where $u(z)=0$ for all $z$. In other words, the beam transmits the compressive force whilst remaining undeformed.

However, if

$$\sin\lambda L = 0 \,, \tag{12.10}$$

i.e. if

$$\lambda L = \pi,\; 2\pi,\; 3\pi,\; \ldots, \tag{12.11}$$

the boundary conditions are satisfied for non-zero $B$. Indeed, in this case, $B$ can take *any* value. This defines a *neutrally stable* solution. The beam can adopt any one of a continuous series of equilibrium configurations corresponding to different values of the constant, $B$.

A simple system that exhibits neutral stability is the ball resting on a horizontal surface, as shown in Figure 12.4*(a)*. The ball can be placed at any point on the surface and it will remain there in equilibrium. By contrast, if the surface is concave, as in Figure 12.4*(b)*, there is only one position of equilibrium (at the lowest point) and it is stable, because the ball will roll back if it is displaced from the equilibrium position. If instead the surface is convex, as in Figure 12.4*(c)*, there is again a single equilibrium position, but it is unstable because an arbitrarily small displacement will generate forces tending to accelerate the ball *away* from the equilibrium position.

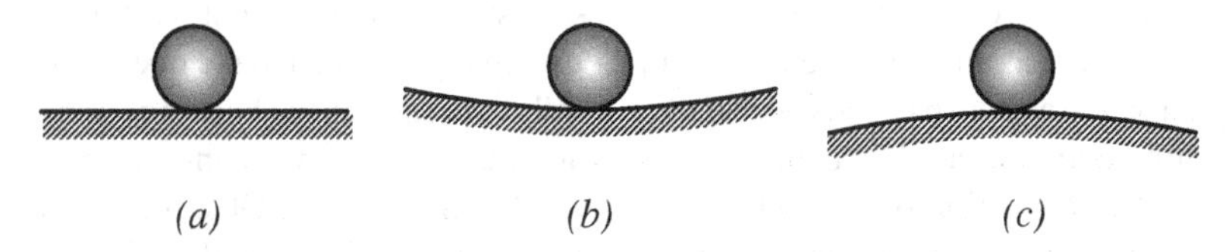


*Figure 12.4: The ball resting on (a) a horizontal surface (neutrally stable); (b) a concave surface (stable) and (c) a convex surface (unstable)*

The neutrally stable system of Figure 12.4*(a)* is intermediate between those of Figures 12.4*(b,c)* and we should therefore not be surprised to find that the neutrally stable configuration of Figure 12.3*(b)* is intermediate between domains of stability and instability. In other words, if $\lambda L<\pi$, the only solution ($A=B=0$) of equations (12.8, 12.9) is stable, whereas if $\lambda L>\pi$ it is unstable. We shall prove this result using an energy argument in §12.7.2 below.

The lowest critical force $P_0$ for neutral stability is defined by

$$\lambda L = \sqrt{\frac{P_0}{EI}}L = \pi \tag{12.12}$$

and hence

$$P_0 = \frac{\pi^2 EI}{L^2} \; . \tag{12.13}$$

If $P>P_0$ the system is unstable, if $P<P_0$ it is stable and if $P=P_0$ it is neutrally stable.

The higher modes corresponding to $\lambda L=2\pi,\ 3\pi$, etc are unattainable unless the system is somehow constrained to inhibit the lowest mode. This can be done by adding a simple support at the mid-point $z=L/2$, since the solution

$$u = B\sin(2\pi z/L) \; , \tag{12.14}$$

corresponding to $\lambda L=2\pi$ is zero at $z=L/2$, whereas that corresponding to $\lambda L=\pi$ is

$$u = B\sin(\pi z/L) \; , \tag{12.15}$$

which is non-zero everywhere except at the end supports. Thus, with a support at the mid-point, the beam will first buckle at the higher force

$$P = \frac{4\pi^2 EI}{L^2} \; , \tag{12.16}$$

corresponding to $\lambda L=2\pi$ and the buckled shape will be of the form shown in Figure 12.5.

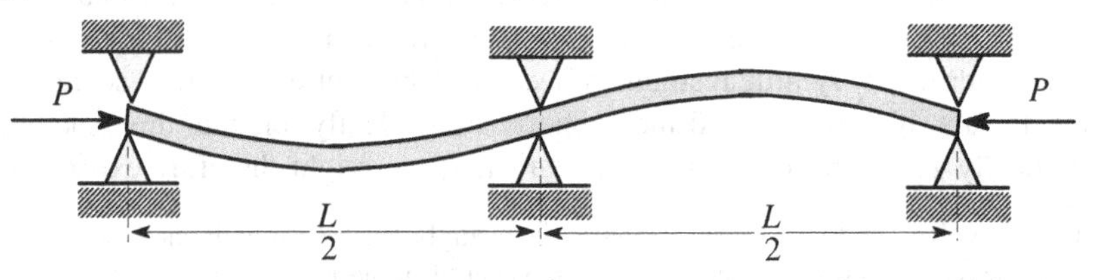


*Figure 12.5: Second buckling mode for the simply supported beam*

The reader is encouraged to confirm these results by experiment. A thin plastic ruler is suitable as an experimental beam. Alternatively, you can make one by cutting a 1 inch × 10 inch strip of fairly substantial cardboard, such as the backing from a pad of writing paper. Load the beam in compression between the two hands, taking care not to restrain rotation of the ends, as shown in Figure 12.6*(a)*. Observe the lowest buckling mode and retain a physical impression of the required buckling force. Now get a friend[2] to restrain the mid-point of the beam by lightly pinching it between two fingers, as in Figure 12.6*(b)*.

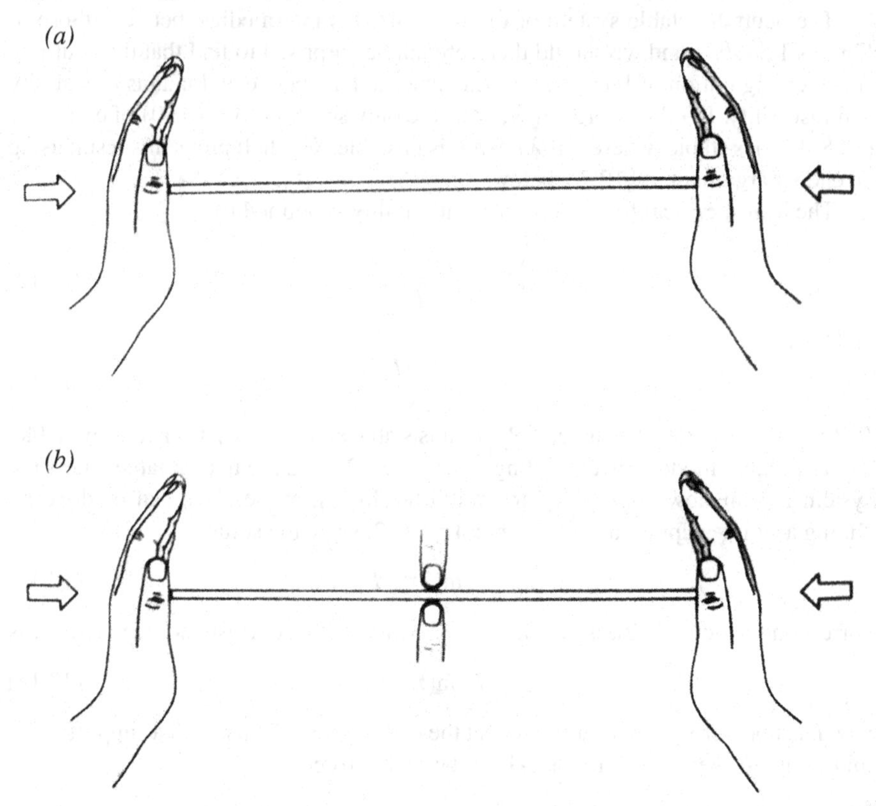


*Figure 12.6: Buckling experiment (a) the simply supported beam; (b) with a central restraint*

You should find that a substantially larger force is now required to buckle the beam and the buckling mode is a complete sine wave, as in Figure 12.5. You should also discover that very little restraint is needed at the centre support to achieve this result. In fact, if the beam and the supports are perfectly aligned, this reaction is theoretically zero. From a design point of view, we might therefore expect to be

[2] If you have no-one to help you, compress the beam by pushing it with one hand against a rigid surface, so as to have a hand free for the central support.

able to increase the load carrying capacity of compression members by providing arbitrarily flimsy intermediate supports. This is not strictly true and we shall return to this issue in §12.5 below. However, it *is* true that design against elastic instability is best achieved by the judicious placing of supports to suppress low order buckling modes.

Equation (12.10) has a denumerably infinite set of solutions $\lambda L = n\pi$ where $n$ is any positive integer. The deflected shape corresponding to these solutions will cross the axis $(n-1)$ times in the length of the beam and the $n$th mode can only be generated if supports are placed at all these nodes.

The problem considered above is a classical *eigenvalue problem*. It defines a problem through a *homogeneous* set of equations [in this case equations (12.3, 12.6, 12.7)], which are homogeneous since the right-hand sides are zero. These equations have only a trivial (null) solution, except for certain *eigenvalues* of the force $P$. At these eigenvalues, the solution has a form $u(z)$ known as the *eigenfunction*, but any linear multiple of this function is a solution, as is shown in this case by the fact that the arbitrary constant $B$ can take any value. It is typical of eigenfunctions that the $n$th eigenfunction crosses zero $(n-1)$ times. Also, eigenfunctions generally form an orthogonal set. In the present case, they are Fourier terms, which are orthogonal to each other.

## 12.2 Effect of initial perturbations

Suppose that the beam of Figure 12.2 *(a)* was not quite straight due to manufacturing errors. We can describe this initial condition by stating that the beam has a deflection $u_0(z)$ even before the compressive force is applied, as shown in Figure 12.7 *(a)*.

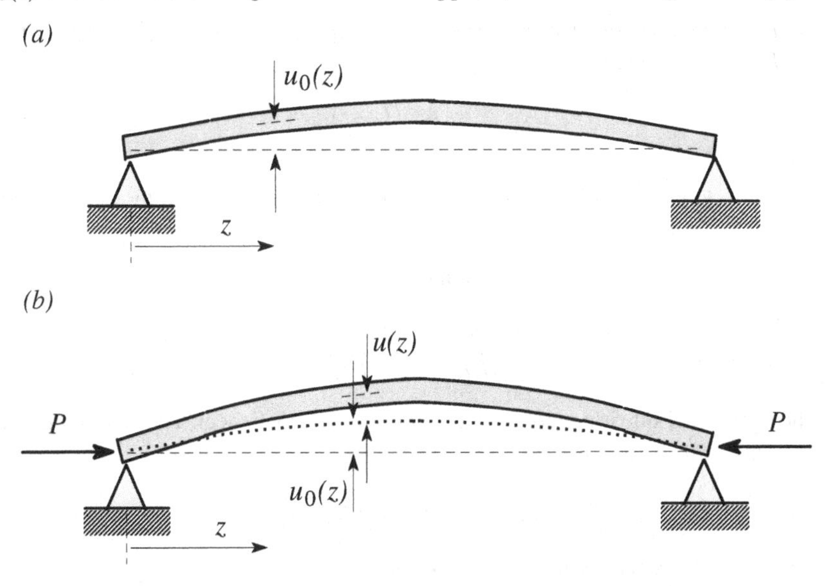


*Figure 12.7: Beam with initial perturbation (a) before loading, (b) after loading*

We now apply a compressive force $P$ causing an *additional* deflection $u(z)$, as shown in Figure 12.7 *(b)*, so that the bending moment at any point is given by

$$M = P(u+u_0) \, . \tag{12.17}$$

The bending moment-curvature relation here involves only the additional deflection, $u(z)$, since no moment was needed to establish the state $u_0(z)$. It follows that $M = -EId^2u/dz^2$ and hence

$$\frac{d^2u}{dz^2} + \frac{Pu}{EI} = -\frac{Pu_0(z)}{EI} \, , \tag{12.18}$$

from (12.17). Thus, the effect of the initial imperfection is to make the governing differential equation inhomogeneous.

The general solution of equation (12.18) can be written as the sum of any particular solution and the general solution of the corresponding homogeneous equation. If $u_0(z)$ is a polynomial, trigonometric or exponential function, a particular solution can be determined by assuming $u(z)$ to have the same general form and then equating coefficients. The homogeneous equation corresponding to (12.18) is (12.3) and hence the general homogeneous solution of (12.18) is given by (12.4). The general solution will contain two arbitrary constants, which are determined from the kinematic boundary conditions.

Since the equation is inhomogeneous, some deflection is produced for all values of the compressive force $P$, but we shall find that it increases without limit as $P$ approaches the critical force $P_0$ of equation (12.13).

**Example 12.1**

*The simply supported beam of Figure 12.7 has length L, flexural rigidity EI and an initial deflection defined by*

$$u_0(z) = \frac{4\varepsilon z(L-z)}{L^2} \, ,$$

*where $\varepsilon \ll L$. Find the deflected shape when the beam is compressed by a force P and sketch the variation of the maximum deflection with P.*

In this case, the differential equation (12.18) takes the form

$$\frac{d^2u}{dz^2} + \frac{Pu}{EI} = -\frac{4P\varepsilon z(L-z)}{EIL^2}$$

and the right-hand side is a second degree polynomial, so we seek a particular integral $u_P(z)$ of the same form, i.e.

$$u_P(z) = C_0 + C_1 z + C_2 z^2 \, ,$$

where $C_0, C_1, C_2$ are as yet unknown constants.

Substituting $u_P(z)$ into the governing equation, we obtain

$$2C_2 + \frac{P}{EI}(C_0 + C_1 z + C_2 z^2) = \frac{4P\varepsilon}{EIL^2}(z^2 - zL) .$$

We now equate coefficients of $z^0, z^1, z^2$, obtaining

$$2C_2 + \frac{PC_0}{EI} = 0 \;\; ; \;\; \frac{PC_1}{EI} = -\frac{4P\varepsilon}{EIL} \;\; ; \;\; \frac{PC_2}{EI} = \frac{4P\varepsilon}{EIL^2} ,$$

with solution

$$C_0 = -\frac{8\varepsilon EI}{PL^2} \;\; ; \;\; C_1 = -\frac{4\varepsilon}{L} \;\; ; \;\; C_2 = \frac{4\varepsilon}{L^2} .$$

The general solution of the governing equation is then obtained by substituting these results into $u_P(z)$ and superposing the general homogeneous solution (12.4), with the result

$$u(z) = \frac{4\varepsilon}{L^2}\left(z^2 - Lz - \frac{2EI}{P}\right) + A\cos\lambda z + B\sin\lambda z .$$

The simply supported boundary conditions $u(0) = u(L) = 0$ then give the two equations

$$-\frac{8\varepsilon EI}{PL^2} + A = 0$$

$$\frac{4\varepsilon}{L^2}\left(L^2 - L^2 - \frac{2EI}{P}\right) + A\cos\lambda L + B\sin\lambda L = 0$$

for the unknown constants $A, B$, with solution

$$A = \frac{8\varepsilon EI}{PL^2} \;\; ; \;\; B = \frac{8\varepsilon EI}{PL^2}\frac{(1-\cos\lambda L)}{\sin\lambda L}$$

and hence

$$u(z) = \frac{4\varepsilon}{L^2}\left(z^2 - Lz - \frac{2EI}{P}\right) + \frac{8\varepsilon EI}{PL^2}\left[\cos\lambda z + \frac{(1-\cos\lambda L)\sin\lambda z}{\sin\lambda L}\right] .$$

The deflected shape of the beam due to the compressive force $P$ is therefore defined by

$$u(z) + u_0(z) = \frac{8\varepsilon EI}{PL^2}\left[\frac{(1-\cos\lambda L)\sin\lambda z}{\sin\lambda L} - (1-\cos\lambda z)\right] .$$

In interpreting this expression, it is important to recognize that the applied force $P$ appears explicitly in the term $8\varepsilon EI/PL^2$, but also implicitly in the terms $\sin\lambda z, \cos\lambda L$, etc., since $\lambda = \sqrt{P/EI}$, from (12.5).

The maximum displacement occurs at the mid-point $z = L/2$, and after some algebra can be written

$$u_{\max} = u\left(\frac{L}{2}\right) + u_0\left(\frac{L}{2}\right) = \frac{8\varepsilon EI}{PL^2}\left[\sec\left(\frac{\lambda L}{2}\right) - 1\right] = \frac{8\varepsilon EI}{PL^2}\left(\sec\sqrt{\frac{PL^2}{4EI}} - 1\right) .$$

This expression becomes unbounded as

$$\frac{\lambda L}{2} \rightarrow \frac{\pi}{2} \quad \text{and hence} \quad P \rightarrow \frac{\pi^2 EI}{L^2} = P_0 ,$$

which is the same critical force determined for the homogeneous problem in equation (12.13).

The maximum deflection is plotted as a function of the ratio $P/P_0$ in Figure 12.8. The deflection is unbounded at $P \rightarrow P_0$, but substantial deflections are produced at significantly lower compressive forces. No manufacturing process is perfect, so it is prudent to avoid column designs in which the maximum axial force exceeds 70% of the theoretical critical force.

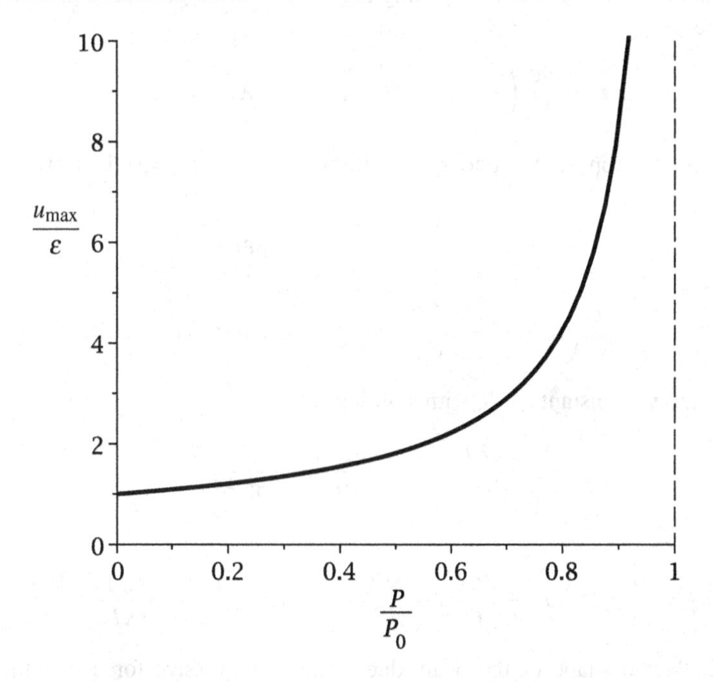


*Figure 12.8: Central deflection as a function of compressive force*

### 12.2.1 Eigenfunction expansions

In Example 12.1, the constant $B \rightarrow \infty$ as $P \rightarrow P_0$ $(\lambda L \rightarrow \pi)$, whilst the constant $A$ remains bounded. As a result, the expression for $u(z)$ is increasingly dominated by the term $B \sin \lambda z$ as the critical force is approached, even though the initial perturbation is not sinusoidal.

This is a general characteristic of eigenvalue problems. As the eigenvalue is approached, the response to any initial perturbation will grow without limit and the shape of the resulting functions will approach that of the corresponding eigenfunction of the homogeneous problem.

This effect can be made more explicit by expanding the initial perturbation as an *eigenfunction expansion*. In the present case, the eigenfunctions are of the sinusoidal form $\sin \lambda z$, with $\lambda L = n\pi$ [see equations (12.4, 12.8, 12.11)] and hence we write

$$u_0(z) = \sum_{n=1}^{\infty} C_n \sin\left(\frac{n\pi z}{L}\right) , \tag{12.19}$$

where $C_n$ is a set of unknown coefficients that can be determined by inverting the Fourier series (12.19).

Substitution of (12.19) into (12.18) will now yield a differential equation whose solution can be written in the form

$$u(z) = \sum_{n=1}^{\infty} A_n \sin\left(\frac{n\pi z}{L}\right) . \tag{12.20}$$

The new coefficients $A_n$ can be found in terms of the $C_n$ by equating coefficients, with the result

$$A_n = C_n \Bigg/ \left(\frac{n^2\pi^2 EI}{L^2 P} - 1\right) . \tag{12.21}$$

Now clearly the $A_n$ will be of the same order as the $C_n$ except when the denominator of (12.21) is small compared with unity, which happens for the term $A_1$ when $P$ is close to $P_0$. Thus, the deflection of the beam [equation (12.20)] will be dominated by the first eigenfunction when the first critical force is approached, whatever form is taken by the initial perturbation.

This technique of expressing the 'forcing term' (in this case the initial perturbation) as an eigenfunction expansion is mathematically analogous to the expansion of the response of a dynamic system in terms of normal modes. It depends on the fact that the eigenfunctions are orthogonal and complete — i.e. that any arbitrary initial beam shape can be expanded in such a series.

## 12.3 Effect of lateral load (beam-columns)

Another class of inhomogeneous stability problems arises if an initially straight beam is subjected to both axial and lateral loading. Beams loaded in this way are sometimes referred to as *beam-columns*, since they combine features of traditional beam bending with axial loading as a column.

The lateral load will produce some lateral deflection even in the absence of the axial force, but the latter exaggerates the deflection, causing it to go unbounded as the critical axial force is approached. The axial force also exaggerates the bending moment due to lateral loading and this effect can be important even when the axial force is significantly sub-critical. The deformed shape approaches more closely to the theoretical eigenfunction for purely axial loading as the compressive axial force is increased.

The analysis of beam-columns proceeds exactly as in the cases considered in previous sections. We sketch a free-body diagram of the beam in an as yet unknown deformed configuration and determine the bending moment as a function of $z$ from equilibrium considerations, taking care to include the moment due to the axial force. We then substitute the moment into the bending equation, solve the resulting ordinary differential equation and determine the two arbitrary constants from the kinematic boundary conditions.

**Example 12.2**

*Figure 12.9 (a) shows a uniform cantilever beam of flexural rigidity $EI$ and length $L$, loaded by an axial force $P$ and a lateral force $F$ at the free end. Find the maximum bending moment in the beam and sketch its variation with $P$ for a given value of $F$.*

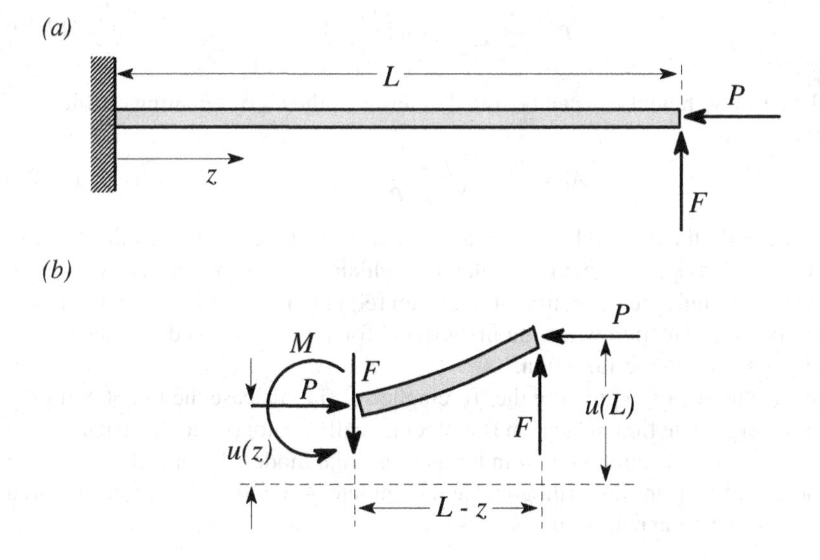


*Figure 12.9: (a) Cantilever loaded by an axial force and a lateral force; (b) free-body diagram for a beam segment*

Figure 12.9 *(b)* shows the free body diagram for a segment of the beam of length $z$ at the loaded end. It is clear that the axial force and the shear force at the cut must be $P$ and $F$, respectively, and taking moments about the cut we have

$$M = -P[u_L - u(z)] - F(L-z) ,$$

where $u_L = u(L)$ is the displacement at the end $z = L$. Notice in particular that the moment arm for the force $P$ [the perpendicular distance between the two equal and opposite forces $P$ in Figure 12.9 *(b)*] is $[u_L - u(z)]$, rather than simply $u(z)$ as in Example 12.1. This is because the end is unsupported and therefore has a non-zero deflection $u_L$. Also, it is convenient to treat $u_L$ as a separate unknown at this stage, only imposing the condition $u_L = u(L)$ when the boundary conditions are being applied.

The governing equation is

$$EI\frac{d^2u}{dz^2} = -M = P\left[u_L - u(z)\right] + F(L-z)$$

and hence

$$\frac{d^2u}{dz^2} + \frac{Pu}{EI} = \frac{Pu_L + FL - Fz}{EI} .$$

Notice how this equation has the same form as the inhomogeneous equation (12.18), only the right-hand side being different. In general, for *any* stability problem for straight beams, the left-hand side of the governing equation [i.e. the terms containg the unknown displacement function $u(z)$] will be identical to that in (12.18) and hence the corresponding homogeneous equation will always be (12.3). This provides a useful check on the sign convention. If you end up with $(d^2u/dz^2 - P/EI)$ as the left-hand side, you have almost certainly[3] made a sign error and would do well to check carefully before proceeding.

The right-hand side of the governing equation is a first order polynomial in $z$ and elementary calculations show that a suitable particular solution is

$$u_P(z) = u_L + \frac{FL}{P} - \frac{Fz}{P} .$$

The general solution is obtained by superposing the homogeneous solution (12.4) and is

$$u(z) = u_L + \frac{FL}{P} - \frac{Fz}{P} + A\cos\lambda z + B\sin\lambda z ,$$

where

$$\lambda = \sqrt{\frac{P}{EI}} .$$

In this case, there are three unknowns $A, B, u_L$, which can be determined from the two kinematic boundary conditions

$$\frac{du}{dz} = u = 0 \ ; \quad z = 0$$

and the definition

$$u = u_L \ ; \quad z = L .$$

Substituting for $u(z)$, we obtain the three simultaneous algebraic equations

$$-\frac{F}{P} + B\lambda = 0$$
$$u_L + \frac{FL}{P} + A = 0$$
$$u_L + A\cos\lambda L + B\sin\lambda L = u_L ,$$

with solution

[3] The exception is if the axial force is *tensile* instead of compressive. See for example, Problem 12.8.

$$A = -\frac{F}{P\lambda}\tan\lambda L \;\; ; \;\; B = \frac{F}{P\lambda} \;\; ; \;\; u_L = \frac{F}{P\lambda}\tan\lambda L - \frac{FL}{P} \, .$$

Substituting these results into the expression for $u(z)$, we obtain[4]

$$u(z) = \frac{F\tan\lambda L}{P\lambda}(1-\cos\lambda z) - \frac{F}{P\lambda}(\lambda z - \sin\lambda z) \, .$$

The first term in this expression increases without limit as $\lambda L \rightarrow \pi/2$, corresponding to the critical force

$$P_0 = \frac{\pi^2 EI}{4L^2} \, .$$

The maximum bending moment ocurs at $z=0$ and is

$$|M_{\max}| = Pu_L + FL = \frac{F}{\lambda}\tan\lambda L - FL + FL = F\sqrt{\frac{EI}{P}}\tan\sqrt{\frac{PL^2}{EI}} \, .$$

This is shown as a function of the dimensionless axial force $P/P_0$ in Figure 12.10. Notice that the axial force causes a significant increase in the maximum bending moment for values as low as $P=0.2P_0$.

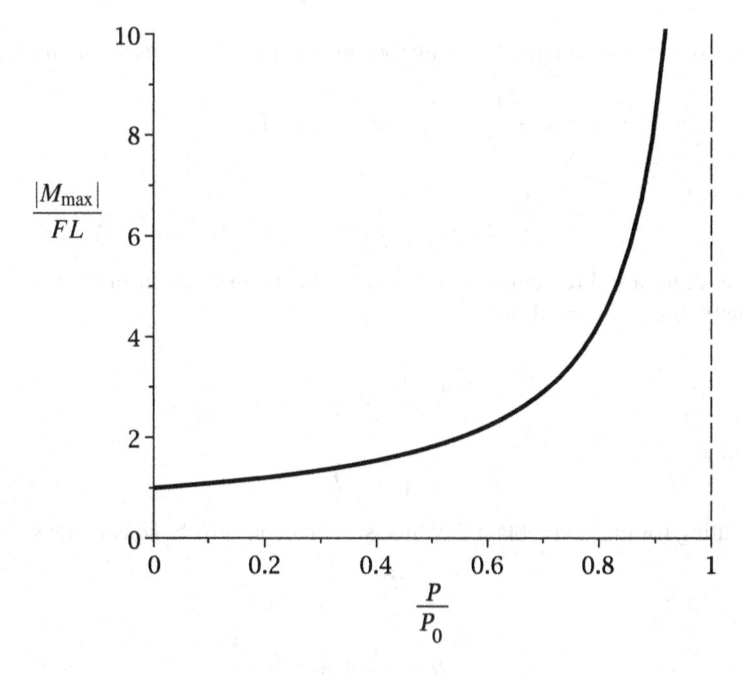


*Figure 12.10: Maximum bending moment as a function of axial compressive force*

[4] The perceptive reader may be puzzled by the fact that the expression for $u(z)$ contains $P$ in the denominator and hence appears to be unbounded as $P \rightarrow 0$. However, in this limit, $\lambda$ also tends to zero and by replacing the trigonometric functions by the first few terms of their power series expansions, it can be shown that the solution does indeed tend to the correct expression for a cantilever beam with an end load $F$, when $P \rightarrow 0$.

## 12.4 Indeterminate problems

Indeterminate problems for axially loaded beams are treated in the same way as classical indeterminate beam bending problems. The redundant reactions are carried through the analysis as additional unknowns until the kinematic boundary conditions are imposed, at which point it will be found that there are just sufficient equations to determine the unknowns.

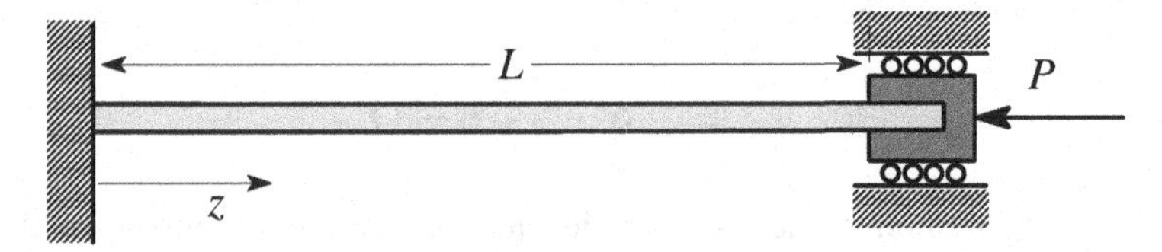


*Figure 12.11: An indeterminate stability problem*

Consider, for example, the beam of Figure 12.11, which is built in at $z=0$ and constrained so as to have zero slope and lateral deflection at $z=L$. The free-body diagram of the buckled beam is shown in Figure 12.12*(a)*, from which we note that there are two unknown reactions $R_1, R_2$ and two unknown restraining moments $M_1, M_2$. Equilibrium conditions provide only two equations for these four unknowns, so the problem is indeterminate.

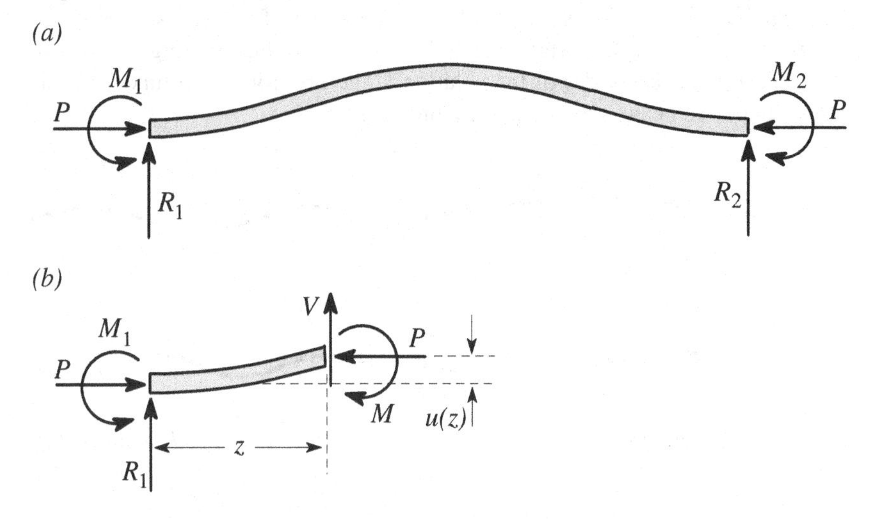


*Figure 12.12: Free-body diagram of (a) the complete beam and (b) a beam segment*

The free-body diagram of a beam segment is shown in Figure 12.12*(b)*, from which we conclude that

$$M = M_1 + Pu - R_1 z \, .$$

Substitution in the bending equation then yields the governing equation

$$\frac{d^2u}{dz^2} + \frac{Pu}{EI} = -\frac{M_1}{EI} + \frac{R_1z}{EI} .$$

The general solution of this ordinary differential equation is

$$u = -\frac{M_1}{P} + \frac{R_1z}{P} + A\cos\lambda z + B\sin\lambda z$$

and the four kinematic boundary conditions

$$u = \frac{du}{dz} = 0 \;\; ; \;\; z = 0 \text{ and } L$$

define a set of four homogeneous algebraic equations for the four unknowns $M_1, R_1$, $A, B$. A non-trivial solution requires that the determinant of the coefficients of these equations be zero, leading to an eigenvalue equation for the critical force $P_0$. The completion of this example is left as an excercise for the reader.

## 12.5 Suppressing low-order modes

We remarked in §12.1 that the buckling force can be increased by the judicious placement of intermediate supports. The best location for these supports can be determined by solving the original stability problem and examining the lowest order eigenfunctions or *modes* of the solution.[5] The first four eigenfunctions for the simply-supported beam are shown in Figure 12.13.

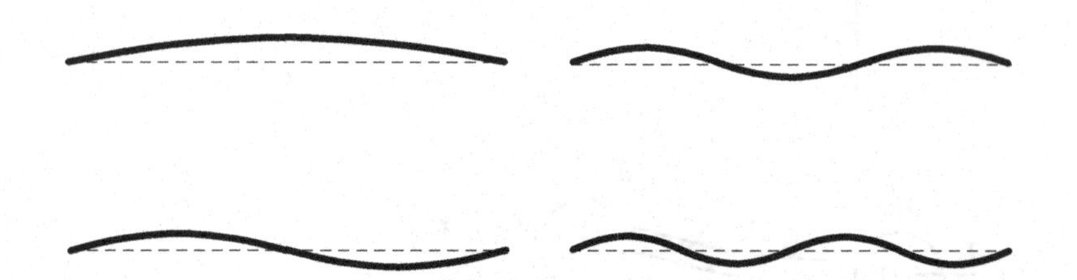

*Figure 12.13: First four eigenfunctions for the simply supported beam of Figure 12.2(a)*

Notice how the displacement for the first eigenfunction has the same sign throughout the beam, whereas for the higher order eigenfunctions there are points, known as *nodes*, where the displacement changes sign. In general, the $n$th eigenfunction has $(n-1)$ nodes, not counting the end points.

If additional supports are placed at each of the $(n-1)$ nodes of the $n$th eigenfunction, the first $(n-1)$ buckling modes will be suppressed and buckling will first occur

[5] i.e. the deformed shapes corresponding to the first few buckling forces.

at the force corresponding to the $n$th eigenvalue. Furthermore, no reactions will be induced at these supports, even when buckling occurs. By contrast, if $(n-1)$ supports are placed at any other locations, then (i) the buckling force will be lower than the $n$th eigenvalue and (ii) reactions will be induced at the supports once buckling occurs.

For these reasons, it is clear that the optimum location for the supports is at the nodes of the first acceptable eigenfunction. Since the additional supports do not carry any load, they do not need to be particularly strong. However, their *stiffness* is important, if they are to perform the function of suppressing low order modes. In general, there is a certain minimum stiffness required, beyond which there is no further advantage in increasing the support stiffness. These results and the method of solution are best illustrated by example.

**Example 12.3**

*Figure 12.14 shows a simply-supported beam of length L and flexural rigidity EI, loaded by an axial force P. It is proposed to increase the buckling force by providing an additional support at the mid-point, which can be modelled as a spring of stiffness k. Find the buckling force as a function of k and hence determine the minimum stiffness for the support if the first buckling mode is to be suppressed.*

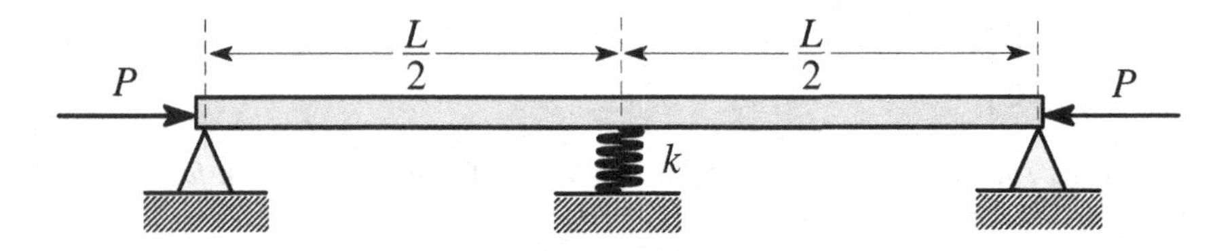


*Figure 12.14: Simply supported beam with an elastic central support*

The system is symmetric about the mid-point and in such cases the eigenfunctions are always either symmetric or antisymmetric — in other words, in any given mode, the displacements of symmetric points are always either equal or equal and opposite. Reference to Figure 12.13 shows that the eigenfunctions of the simply supported beam are alternately symmetric and antisymmetric.

The mid-point must be a node for any antisymmetric mode, so these modes are unaffected by the addition of a central support. In the present case, the lowest antisymmetric mode is that of Figure 12.5, corresponding to the critical force

$$P_1 = \frac{4\pi^2 EI}{L^2} ,$$

from equation (12.16).

The symmetric modes are affected by the flexible support, but the symmetry permits some simplification in the analysis. Figure 12.15*(a)* shows a symmetric mode and the corresponding free-body diagram is shown in Figure 12.15*(b)*.

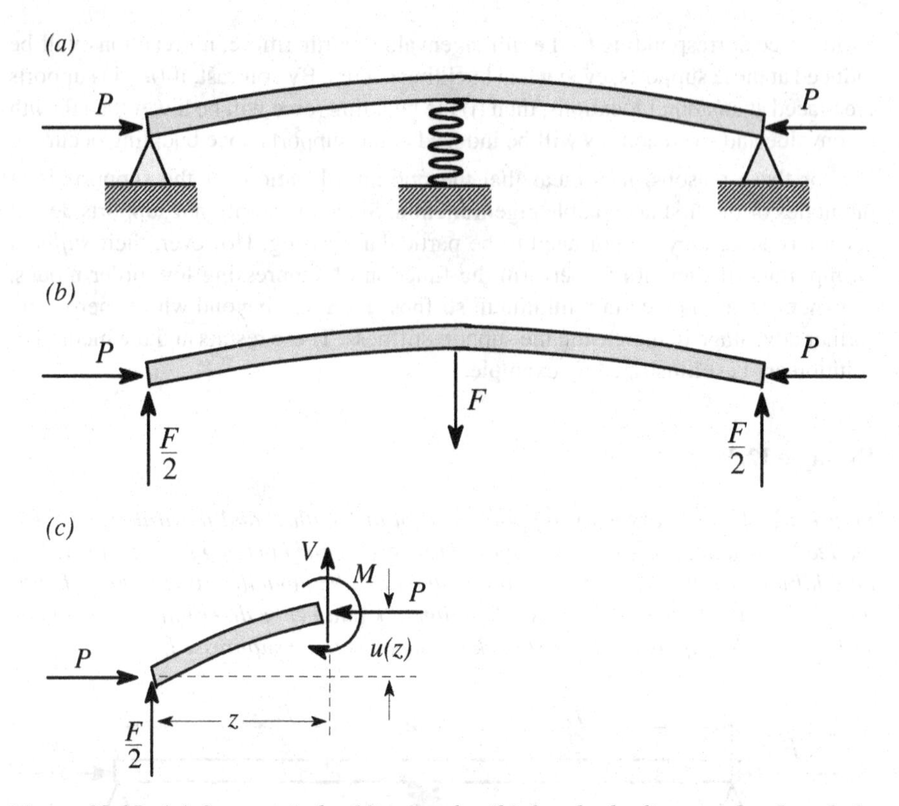


*Figure 12.15: (a) Symmetric buckling mode; (b) free-body diagram for the whole beam and (c) for a beam segment*

The spring exerts an unknown resisting force $F$ and by symmetry and equilibrium, the two reaction forces must both be equal to $F/2$. Also by symmetry, the slope of the beam $du/dz$ must be zero at the mid-point $z=L/2$. These results permit us to solve the problem by considering only the left half of the beam.

The free-body diagram of a beam segment is shown in Figure 12.15 *(c)*, from which we find

$$M = Pu - \frac{Fz}{2} \quad ; \quad 0 < z < \frac{L}{2}$$

and hence

$$\frac{d^2u}{dz^2} + \frac{Pu}{EI} = \frac{Fz}{2EI} \quad ; \quad 0 < z < \frac{L}{2} \, .$$

The general solution of this equation is

$$u = \frac{Fz}{2P} + A\cos\lambda z + B\sin\lambda z$$

and the unknowns $A, B, F$ are determined from the two kinematic boundary conditions

$$u = 0 \;\; ; \;\; z = 0$$
$$\frac{du}{dz} = 0 \;\; ; \;\; z = \frac{L}{2}$$

and the constitutive relation for the spring

$$F = ku\left(\frac{L}{2}\right) .$$

These conditions yield the simultaneous equations

$$A = 0$$
$$\frac{F}{2P} - A\lambda \sin\frac{\lambda L}{2} + B\lambda \cos\frac{\lambda L}{2} = 0$$
$$F\left(\frac{L}{4P} - \frac{1}{k}\right) + A\cos\frac{\lambda L}{2} + B\sin\frac{\lambda L}{2} = 0 ,$$

which have a non-trivial solution if and only if the determinant of the coefficient matrix is zero — i.e.

$$\frac{1}{2P}\sin\frac{\lambda L}{2} = \lambda\left(\frac{L}{4P} - \frac{1}{k}\right)\cos\frac{\lambda L}{2}$$

and hence

$$\tan\sqrt{\frac{PL^2}{4EI}} = \sqrt{\frac{PL^2}{4EI}}\left(1 - \frac{4P}{kL}\right) .$$

The buckling force, $P_0$ is plotted as a function of the dimensionless support stiffness $kL^3/EI$ in Figure 12.16.

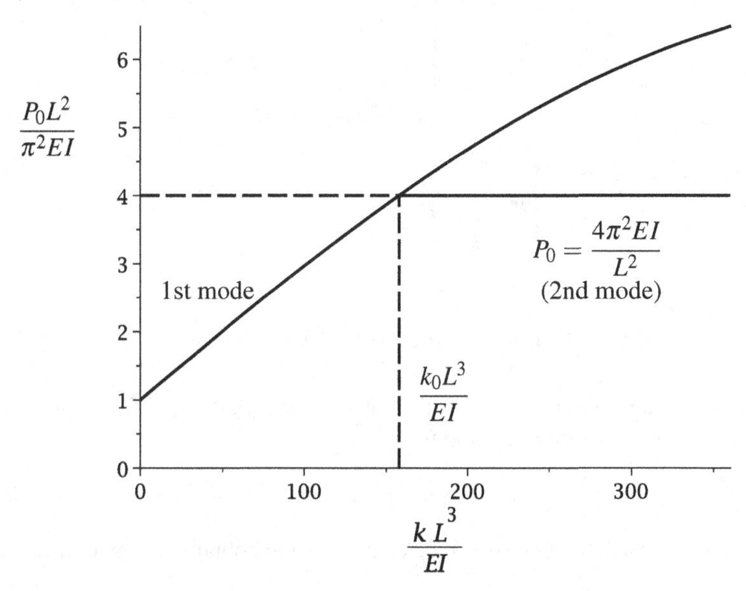


*Figure 12.16: Buckling force as a function of support stiffness*

As we should expect, the first symmetric buckling force is given by the elementary result $P_0 = \pi^2 EI/L^2$ for $k=0$ (no central support) and it increases monotonically with $k$, reaching the buckling force $P_1$ for the first *antisymmetric* mode when

$$k = k_0 = \frac{16\pi^2 EI}{L^3} .$$

Use of a support for which $k > k_0$ would have no effect on the buckling force, since buckling would then occur in the antisymmetric mode at $P = P_1$.

## 12.6 Beams on elastic foundations

The buckling behaviour is qualitatively changed if a beam is supported on an elastic foundation. Figure 12.17 shows a small segment of such a beam [compare with Figure 7.3 *(b)*] and we conclude from equilibrium arguments that

$$\frac{dV}{dz} = w(z) + ku(z) \; ; \quad \frac{dM}{dz} = P\frac{du}{dz} + V . \tag{12.22}$$

Eliminating $V, M$ between these equations and the beam bending equation $M = -EId^2u/dz^2$, we obtain the governing equation

$$\frac{d^4u}{dz^4} + \frac{P}{EI}\frac{d^2u}{dz^2} + \frac{ku}{EI} = -\frac{w(z)}{EI} . \tag{12.23}$$

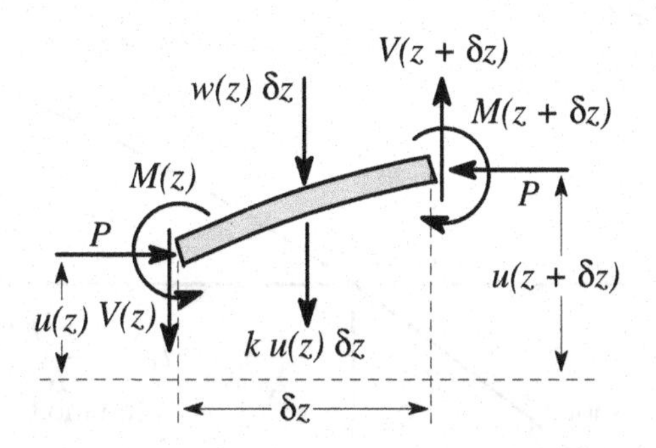


*Figure 12.17: Equilibrium of a beam segment*

If there is no lateral load $[w(z) = 0]$, the solution of equation (12.23) is obtained as in §7.2 by assuming a solution in the form

$$u(z) = Ae^{bz} , \tag{12.24}$$

where $A, b$ are constants. Substituting into (12.23), we then obtain the algebraic equation

$$b^4 + \frac{Pb^2}{EI} + \frac{ku}{EI} = 0 \tag{12.25}$$

and hence

$$b^2 = \frac{1}{2}\left(-\frac{P}{EI} \pm i\sqrt{\frac{4k}{EI} - \left(\frac{P}{EI}\right)^2}\right), \tag{12.26}$$

which can be written in the form

$$b^2 = \rho e^{\pm i\theta}, \tag{12.27}$$

where

$$\rho^2 = \frac{1}{4}\left[\left(\frac{P}{EI}\right)^2 + \frac{4k}{EI} - \left(\frac{P}{EI}\right)^2\right] = \frac{k}{EI} \tag{12.28}$$

and

$$\cos\theta = -\frac{P}{\rho EI} = -\frac{P}{\sqrt{4kEI}}. \tag{12.29}$$

After some algebra, the four solutions for $b$ can be written

$$b = \pm\beta_1 \pm \beta_2 i, \tag{12.30}$$

where

$$\beta_1 = \sqrt[4]{\frac{k}{4EI}}\sqrt{1 - \frac{P}{\sqrt{4kEI}}} \; ; \; \beta_2 = \sqrt[4]{\frac{k}{4EI}}\sqrt{1 + \frac{P}{\sqrt{4kEI}}}. \tag{12.31}$$

The general homogeneous solution of (12.23) is therefore

$$u_H(z) = B_1 e^{\beta_1 z}\cos(\beta_2 z) + B_2 e^{\beta_1 z}\sin(\beta_2 z) + B_3 e^{-\beta_1 z}\cos(\beta_2 z) + B_4 e^{-\beta_1 z}\sin(\beta_2 z). \tag{12.32}$$

Comparison with Chapter 7 shows that this reduces to equation (7.11) in the limiting case where $P = 0$. Two of the functions in (12.32) grow with $z$ and two decay, leading to a state in which the deformation is localized near the ends of the beam, as in §7.2.1, if the beam is sufficiently long. However, $\beta_1 \to 0$ as $P \to P_0$, where

$$P_0 = \sqrt{4kEI}. \tag{12.33}$$

Thus, as $P$ approaches $P_0$, the decay rate of the end disturbance gets slower until at $P = P_0$ there is no decay and end effects penetrate all along the beam. Notice that $P_0$ is also the value of $P$ at which the algebraic equation (12.25) has two identical roots.

For an infinite beam, there are no end conditions and at $P = P_0$ there exists a non-trivial solution of the homogeneous equation of the form

$$u(z) = A\cos(\beta_0 z) + B\sin(\beta_0 z), \tag{12.34}$$

where $A, B$ are arbitrary constants and

$$\beta_0 = \sqrt[4]{\frac{k}{EI}}. \tag{12.35}$$

Thus, for a sufficiently long beam, $P_0$ is the first buckling force and the buckling mode is of sinusoidal form with wavenumber $\beta_0$. This contrasts with the unsupported beam, where the wavenumber and hence the buckling force both depend on the length $L$ [see equations (12.11, 12.13)].

The buckling force for a finite beam will also be $P_0$ if the length is an integer multiple of the half-wavelength $\pi/\beta$ and the ends are simply supported. In other cases, the buckling force will generally be increased by the end constraints, but this increase will be very small if $\beta L \gg 1$. For shorter beams, the solution proceeds as in §12.1, but using (12.32) in place of (12.4).

### 12.6.1 Axisymmetric buckling of cylindrical shells

We saw in Chapter 9 that the axisymmetric bending of cylindrical shells is governed by the same equation as that for a beam on an elastic foundation. The buckling behaviour of these two systems is also closely related. We first need to modify the equilibrium equation (9.16) to include the moment due to the axial force, obtaining

$$V = \frac{dM_z}{dz} - \frac{P}{2\pi a}\frac{du_r}{dz} , \tag{12.36}$$

where $a$ is the mean radius of the shell and

$$P = -2\pi a t \sigma_1 \tag{12.37}$$

is the total axial compressive force[6] applied to the cylinder.

Substituting this modified equation into (9.18) and rearranging terms, we obtain a differential equation whose homogeneous form is

$$\frac{d^4 u_r}{dz^4} + \frac{P}{2\pi a D}\frac{d^2 u_r}{dz^2} + \frac{Etu_r}{Da^2} = 0 . \tag{12.38}$$

Comparing this equation with (12.23) and remembering that the critical force corresponds to the condition where the algebraic equation (12.25) has two identical roots, we conclude that the buckling force $P_0$ for a long cylinder will be given by

$$\left(\frac{P_0}{2\pi a D}\right)^2 = \frac{4Et}{Da^2} \tag{12.39}$$

and hence

$$P_0 = 4\pi\sqrt{EtD} = \frac{2\pi E t^2}{\sqrt{3(1-\nu^2)}} , \tag{12.40}$$

using equation (9.14) for the stiffness $D$.

The corresponding critical axial membrane stress is

[6] Remember that the moment and shear force in equation (9.16) are *per unit circumference*. The axial force per unit circumference is $P/2\pi a$.

$$\sigma_0 = -\frac{P_0}{2\pi at} = -\frac{Et}{a\sqrt{3(1-\nu^2)}}\,, \tag{12.41}$$

and this will generally exceed the yield stress unless the shell is very thin $t/a \ll 1$. Also, a cylindrical shell loaded in compression will generally buckle first in a non-axisymmetric mode (recall the experiment at the beginning of this chapter) unless the axial force is applied in a way which constrains the ends of the cylinder to be perpendicular to the axis.

**Experiment**

Take a short thin-walled tube with plane ends and compress it slowly and carefully in a parallel jaw vice. If you align it carefully, you should be able to produce an axisymmetric buckling mode. However, significant deformations will be produced at the ends before the theoretical buckling force is reached and these will lead to plastic deformation before the rest of the tube buckles. This leads to a form of deformation as shown in Figure 12.18. The tube should then concertina up from the end as the vice is closed. This experiment works best if the tube wall is not too thin, otherwise non-axisymmetric modes tend to dominate the buckling behaviour. However, it is also interesting to try compressing an aluminium beverage can, which has very thin walls and also some quite complex end constraint due to the can end design.

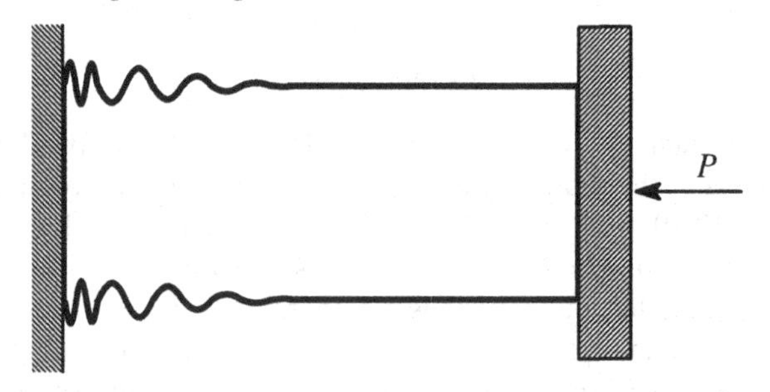


*Figure 12.18: Buckling of a cylindrical shell*

### 12.6.2 Whirling of shafts

A closely related instability occurs in rotating shafts when the speed is sufficiently high. Suppose that for some reason the shaft is deflected from the axis by a displacement $u(z)$ as shown in Figure 12.19.

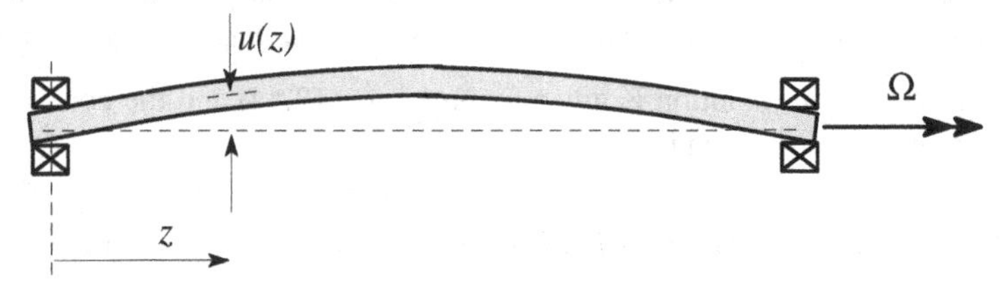


*Figure 12.19: Deflected shape of the rotating shaft*

As a result of the rotation, there will be an acceleration $\Omega^2 u$, where $\Omega$ is the rotational speed. This will result in an equivalent outward inertia force $w(z)=-m\Omega^2 u$, where $m$ is the mass of the shaft per unit length. Notice that the force is proportional to the local displacement, as in the case of a beam on an elastic foundation, but this time the direction of the force is such as to exaggerate the deformation, which has an inherently destabilizing effect.

Using the equations

$$\frac{dV}{dz}=w(z) \;\; ; \;\; \frac{dM}{dz}=V \;\; ; \;\; M=-EI\frac{d^2u}{dz^2}\,, \tag{12.42}$$

we obtain

$$EI\frac{d^4u}{dz^4}=-\frac{d^2M}{dz^2}=-\frac{dV}{dz}=m\Omega^2 u$$

and hence

$$\frac{d^4u}{dz^4}-\frac{m\Omega^2 u}{EI}=0\,. \tag{12.43}$$

This equation has the general solution

$$u(z)=B_1e^{\gamma z}+B_2e^{-\gamma z}+B_3\cos(\gamma z)+B_4\sin(\gamma z)\,, \tag{12.44}$$

where

$$\gamma=\sqrt[4]{\frac{m\Omega^2}{EI}}\,. \tag{12.45}$$

The fours constants $B_1=B_2=B_3=B_4$ are determined from two boundary conditions at the ends of the shaft. For example, if the beam is simply supported at the end as in Figure 12.19, we have $u=0$ and $M=0$. For a built in end, $u=0$ and $u'=0$, whereas for a free end, $M=0$ and $V=0$. These conditions are all homogeneous and the resulting simultaneous equations will therefore have only the trivial solution $B_1=B_2=B_3=B_4=0$ except when the determinant of coefficients is zero. This condition defines the *critical speed* $\Omega_0$, also known as the *whirling speed*. However, if the shaft has some initial imperfection as in §12.2, the governing equation (12.43) will be modified to include an inhomogeneous term and deflection will be produced at all speeds. This can be quite significant even below the critical speed.

**Example 12.4**

*A solid circular shaft of diameter D and length L is simply supported at its ends. Find the critical rotational speed if the material has Young's modulus E and density $\rho$.*

The homogeneous solution is given by equation (12.44) and the corresponding bending moment distribution is

$$M(z)=-EI\frac{d^2u}{dz^2}=EI\gamma^2\left[-B_1e^{\gamma z}-B_2e^{-\gamma z}+B_3\cos(\gamma z)+B_4\sin(\gamma z)\right]\,.$$

Imposing the boundary conditions

$$u = 0,\ M = 0 \quad ; \quad z = 0, L\,,$$

we obtain the four simultaneous homogeneous equations

$$\begin{aligned}
B_1 \quad + \quad B_2 \quad + \quad B_3 \qquad\qquad\qquad\qquad &= 0 \\
-B_1 \quad - \quad B_2 \quad + \quad B_3 \qquad\qquad\qquad\qquad &= 0 \\
B_1 e^{\gamma L} + B_2 e^{-\gamma L} + B_3 \cos(\gamma L) + B_4 \sin(\gamma L) &= 0 \\
-B_1 e^{\gamma L} - B_2 e^{-\gamma L} + B_3 \cos(\gamma L) + B_4 \sin(\gamma L) &= 0\,.
\end{aligned}$$

Elementary algebraic operations show that $B_1 = B_2 = B_3 = 0$ for all $\Omega$, but $B_4$ can take an arbitrary value if $\sin(\gamma L) = 0$ and hence $\gamma L = n\pi$, where $n$ is an integer. Thus, the critical speeds are defined by

$$\gamma = \sqrt[4]{\frac{m\Omega^2}{EI}} = \frac{n\pi}{L}$$

and hence

$$\frac{m\Omega_0^2}{EI} = \frac{n^4\pi^4}{L^4}\,.$$

For the solid circular shaft,

$$m = \frac{\pi D^2 \rho}{4} \quad ; \quad I = \frac{\pi D^4}{64}\,,$$

giving

$$\Omega_0^2 = \frac{n^4\pi^4}{L^4}\frac{E\pi D^4}{64}\frac{4}{\pi D^2\rho} = \frac{n^4\pi^4 E D^2}{16\rho L^4}$$

and hence

$$\Omega_0 = \frac{n^2\pi^2 D}{4L^2}\sqrt{\frac{E}{\rho}}\,.$$

Generally only the first critical speed is of interest, for which $n = 1$ and

$$\Omega_0 = \frac{\pi^2 D}{4L^2}\sqrt{\frac{E}{\rho}}\,. \qquad (12.46)$$

Notice however that as in §12.5, the first critical speed can be suppressed by providing a sufficiently stiff central bearing.

### Design considerations

For typical machine components such as gearshafts and motors, the operating speed will be significantly lower than the critical speed and hence whirling instabilities will not be a deciding factor in the design. For example, for a steel shaft ($E = 210$ GPa, $\rho = 7700$ kg/m$^3$) of diameter 20 mm and length 500 mm, we have

$$\Omega_0 = \frac{\pi^2 \times 0.02}{4 \times 0.5^2} \sqrt{\frac{210 \times 10^9}{7700}} = 1030 \text{ rad/s } = 9836 \text{ rpm},$$

which is significantly higher than any likely operating speed.

Notice however that $L^2$ appears in the denominator of equation (12.46), so whirling instabilities may be important in long transmission shafts. They are also critical in the design of very large machines, such as turbogenerators, which are therefore generally designed to operate between the first and second critical speeds. Whirling instabilities differ from the other elastic instabilities considered in this chapter in that the system is stable at speeds between the critical speeds as well as below the first critical speed. They share this characteristic with other linear vibration problems. However, it is important not to operate too close to a critical speed, since inevitable manufacturing errors would then lead to unacceptably large displacements and stresses. Also, operating conditions have to be chosen carefully to ensure that the process of accelerating through the critical speed range during start-up does not take long enough for the instability to develop to dangerous proportions.

### Gears and pulleys

The whirling speed will be significantly reduced if the shaft carries a gear or some other massive body at some point distant from the bearings. A typical assembly is shown in Figure 12.20, where a gear wheel of mass $M_0$ is attached to the shaft at the mid-point.

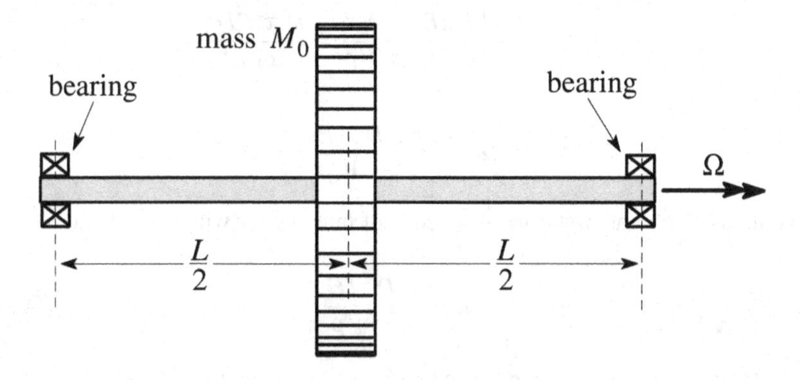


*Figure 12.20: Shaft with a massive wheel supported at the mid-point*

The first whirling mode for this system is shown in Figure 12.21, from which it is clear that the wheel contributes an additional outward force $M_0\Omega^2\delta$ at the mid-point, where

$$\delta = u\left(\frac{L}{2}\right)$$

is the displacement of the mass.

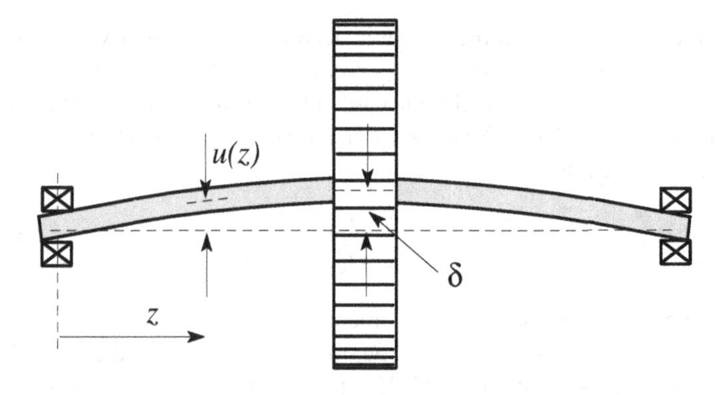


*Figure 12.21: Deflected shape for the shaft of Figure 12.20*

In many cases, the distributed inertia force due to the shaft mass can be neglected relative to that due to the wheel, leading to a very simple estimate of the critical speed. The shaft then acts merely as a spring relating the force at the wheel to its displacement. Elementary beam deflection calculations show that the central force $F$ in Figure 12.22 produces a central displacement

$$\delta = \frac{FL^3}{48EI} \; .$$

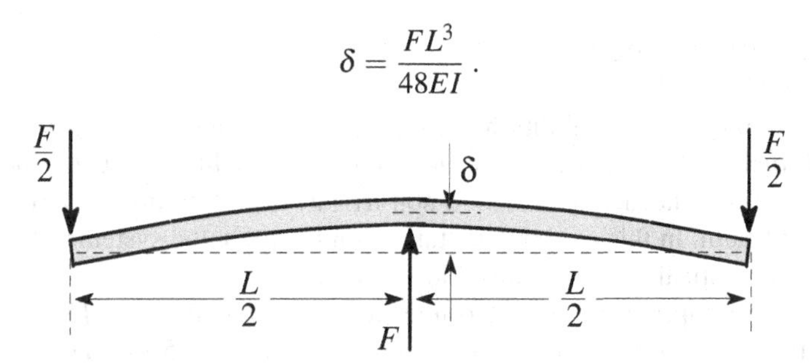


*Figure 12.22: Loading of a simply supported beam by a central force*

Substituting $F = M_0\Omega^2\delta$, we then obtain

$$\delta = \frac{M_0\Omega^2 L^3 \delta}{48EI}$$

and hence the critical speed is

$$\Omega_0 = \sqrt{\frac{48EI}{M_0 L^3}} \; . \tag{12.47}$$

This simple approximation demonstrates that the critical speed can be increased by increasing the stiffness of the supporting shaft and this is most easily done by locating the bearings as close as possible to any massive rotating bodies.

The estimate (12.47) is unfortunately non-conservative, since a more exact calculation taking into account the distributed mass of the shaft will give a *lower* critical speed. However, a quick check on the probable accuracy of the approximation can be made by comparing the critical speed ($\Omega_M$) predicted by (12.47) and for the shaft alone ($\Omega_S$) from (12.46). If (as is usually the case) $\Omega_S \gg \Omega_M$, the true critical speed will not be much less than $\Omega_M$. In fact, a reasonable engineering guess at the combined effect is $\Omega_0$, defined by

$$\frac{1}{\Omega_0^2} = \frac{1}{\Omega_M^2} + \frac{1}{\Omega_S^2} , \tag{12.48}$$

though this is not based on any scientific argument!

A rigorous but more lengthy way of accounting for the combined effect of the mass of the shaft and the wheel in Figure 12.20 is to note that the system is symmetric and hence

$$\frac{du}{dz} = 0 \; ; \; V = \frac{M_0 \Omega^2 \delta}{2} \quad ; \quad z = \frac{L}{2} . \tag{12.49}$$

Applying these boundary conditions and $u=0$ ; $M=0$ at $z=0$, $u=\delta$ at $z=L/2$ in (12.44), and eliminating $B_1, B_2, B_3, B_4, \delta$ from the resulting equations, yields an algebraic equation for the critical speed $\Omega_0$.

## 12.7 Energy methods

Our discussion of elastic stability has so far been focussed on the *neutral stability* or *equilibrium* method, in which we determine the condition (usually the critical compressive force) needed to sustain a non-trivial state of deformation in neutrally stable equilibrium. In this method, we take it on trust that the system will be stable below the first critical force and unstable above it.

A more direct approach is to determine the total potential energy $\Pi$ as a function of the deformed configuration. We have already seen in §3.5 that $\Pi$ is stationary at an equilibrium configuration, but it is also clear that the system will be stable if and only if $\Pi$ is a local minimum, since we can then argue that motion away from equilibrium will be possible only if an external source of energy is provided.

A simple system illustrating this procedure is shown in Figure 12.23*(a)*. The rigid bar $AB$ is pinned at $B$ and loaded by a vertical force $P$ at $A$, which is also restrained by a horizontal spring of stiffness $k$.

The deformed configuration is shown in Figure 12.23*(b)* and we conclude that the potential energy of the force is

$$\Omega = -PL(1 - \cos\theta) ,$$

whereas the spring extension is $L\sin\theta$, giving a strain energy

$$U = \frac{1}{2} kL^2 \sin^2\theta .$$

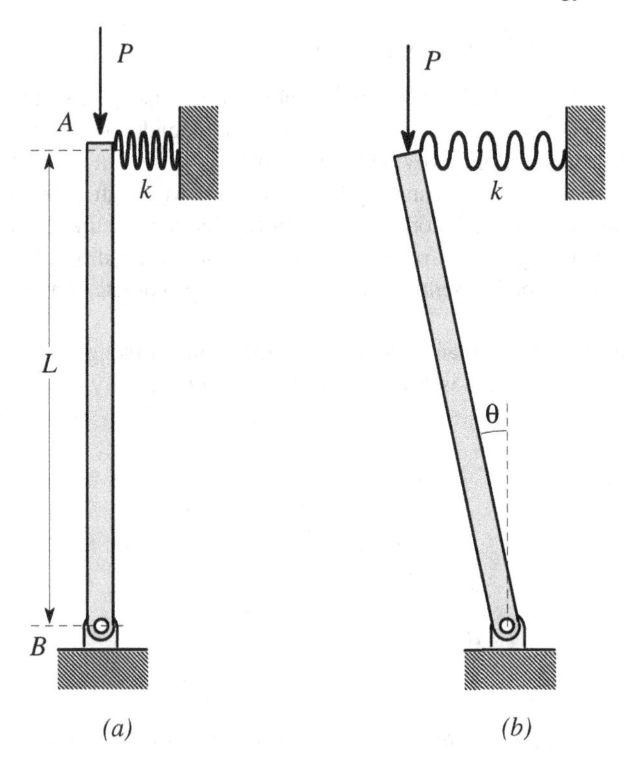


*Figure 12.23: (a) A pinned rigid bar supported by a spring; (b) the deformed configuration*

The total potential energy is therefore

$$\Pi = U + \Omega = \frac{1}{2}kL^2\sin^2\theta - PL(1-\cos\theta) \tag{12.50}$$

and equilibrium configurations are defined by the condition

$$\frac{d\Pi}{d\theta} = kL^2\sin\theta\cos\theta - PL\sin\theta = 0\,. \tag{12.51}$$

The stability of these states can be examined by evaluating the second derivative

$$\frac{d^2\Pi}{d\theta^2} = kL^2(\cos^2\theta - \sin^2\theta) - PL\cos\theta\,. \tag{12.52}$$

An equilibrium state is stable if $d^2\Pi/d\theta^2 > 0$ and unstable if $d^2\Pi/d\theta^2 < 0$.

It is clear from Figure 12.23 *(a)* that the bar is in equilibrium when it is vertical and this is confirmed by the fact that $\theta = 0$ satisfies equation (12.51). To determine whether this state is stable, we set $\theta = 0$ in (12.52), obtaining

$$\frac{d^2\Pi}{d\theta^2} = kL^2 - PL\,, \tag{12.53}$$

which is positive, indicating a stable state, for $P < kL$ and negative (unstable) for $P > kL$. Thus, $P_0 = kL$ is the critical force for the system of Figure 12.23 *(a)*.

### 12.7.1 Energy methods in beam problems

The example of Figure 12.23 is simple because the bar is rigid, so the system has only one degree of freedom $\theta$. The method can be extended to systems with more than one degree of freedom, provided that number is finite, but for continuous systems such as elastic beams, it can only be applied in a Rayleigh-Ritz sense. In other words, we approximate the deformed shape of the beam by a suitable function with a finite number of degrees of freedom and determine the condition for stability by demanding that the total potential energy for this approximate shape be at a local minimum.

We first have to develop an expression for the change in length of a beam due to lateral deflection, since this will enter into the potential energy of the compressive force $P$.

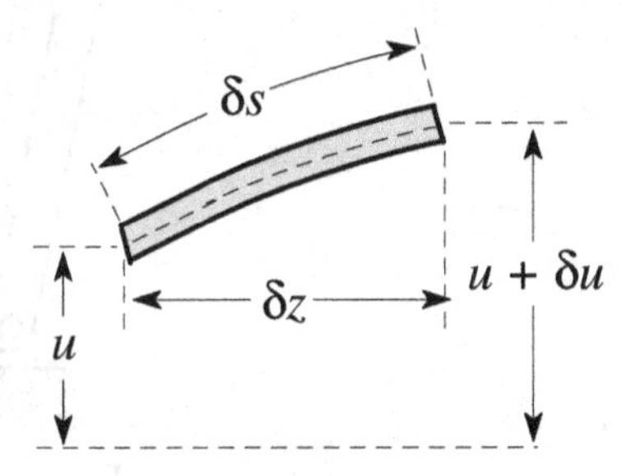


*Figure 12.24: Geometry of beam deflection for a small element*

Consider the segment $\delta s$ of the deformed beam shown in Figure 12.24. From the geometry of this figure we have

$$\delta s = \sqrt{\delta z^2 + \delta u^2} = \delta z \sqrt{1 + \left(\frac{\delta u}{\delta z}\right)^2}$$
$$\approx \delta z \left[1 + \frac{1}{2}\left(\frac{\delta u}{\delta z}\right)^2\right] ,$$

since in the limit $\delta u/\delta z$ is the slope of the beam, which is assumed to be small. Thus,

$$\delta z \approx \delta s - \frac{1}{2}\left(\frac{\delta u}{\delta z}\right)^2 \delta z$$

and the reduction in the distance between the ends of the beam associated with the deformation of this beam segment is

$$\delta s - \delta z = \frac{1}{2}\left(\frac{\delta u}{\delta z}\right)^2 \delta z \,. \tag{12.54}$$

The total change in length of the beam is obtained by taking the limit of equation (12.54) as $\delta s \rightarrow 0$ and then summing the contribution of all infinitesimal beam segments by integration. For a beam of length $L$, we obtain

$$\delta L = \int_0^L \frac{1}{2}\left(\frac{du}{dz}\right)^2 dz\,, \tag{12.55}$$

in terms of which, the potential energy of the compressive force can be written

$$\Omega = -P\delta L\,. \tag{12.56}$$

### 12.7.2 The uniform beam in compression

We first use the energy approach to re-examine the problem of Figure 12.2*(a)*, in which a uniform beam of length $L$ is subjected to a compressive force $P$. In this case, we know from the preceding analysis that the beam will deform into a sinusoidal shape, so we assume the Rayleigh-Ritz form

$$u(z) = C\sin\left(\frac{\pi z}{L}\right)\,. \tag{12.57}$$

The strain energy is then

$$U = \frac{1}{2}EI\int_0^L \left(\frac{d^2u}{dz^2}\right)^2 dz = \frac{EIC^2\pi^4}{2L^4}\int_0^L \sin^2\left(\frac{\pi z}{L}\right) dz\,. \tag{12.58}$$

This integral can be evaluated using the change of variable $x = \pi z/L$, with the result

$$U = \frac{\pi^4 EIC^2}{4L^3}\,. \tag{12.59}$$

Substituting the displacement $u(z)$ from (12.57) into (12.55, 12.56), we obtain

$$\Omega = -\frac{P\pi^2C^2}{2L^2}\int_0^L \cos^2\left(\frac{\pi z}{L}\right) dz = -\frac{P\pi^2C^2}{4L} \tag{12.60}$$

and hence the total potential energy is

$$\Pi = \Omega + U = \frac{\pi^4 EIC^2}{4L^3} - \frac{P\pi^2C^2}{4L} = \frac{\pi^2C^2}{4L}\left(\frac{\pi^2EI}{L^2} - P\right)\,. \tag{12.61}$$

Equilibrium configurations are defined by the condition $d\Pi/dC = 0$, which here yields

$$\frac{\pi C}{2L}\left(\frac{\pi^2EI}{L^2} - P\right) = 0\,. \tag{12.62}$$

This condition is satisfied for all $C$ if $P = P_0 = \pi^2EI/L^2$, but it is only satisfied by $C = 0$ for any other value of $P$. In other words, the trivial undeformed state is the only equilibrium configuration except at the critical force $P = P_0$, where any function of the form (12.57) defines an equilibrium state.

This much was of course established by the equilibrium analysis of §12.1, but we can now make a more definitive statement about stability by evaluating

$$\frac{d^2\Pi}{dC^2} = \frac{\pi}{2L}\left(\frac{\pi^2 EI}{L^2} - P\right) = \frac{\pi}{2L}(P_0 - P) .$$

This expression is positive for $P < P_0$, showing that $\Pi$ is a minimum and hence that the equilibrium state ($C = 0$) is stable. However, if $P > P_0$, we find $d^2\Pi/dC^2 < 0$, $\Pi$ is at a maximum and the equilibrium state is unstable.

**Other approximating functions**

Equation (12.62) predicts a critical force

$$P_0 = \frac{\pi^2 EI}{L^2} ,$$

agreeing with the equilibrium argument of §12.1. In general, we shall find that the Rayleigh-Ritz solution is exact if we are lucky enough to guess exactly the right function to define the deformed shape. However, usually we don't know the exact shape and must use a function that is more or less physically plausible. In this case, the method always *overestimates* the critical force. Unfortunately, this means that the method is non-conservative (i.e. it is not 'fail safe'), but nonetheless it is extremely useful as a way of getting an idea of the magnitude of the critical force,[7] since the equilibrium method often leads to intractable equations, except for very simple geometries.

To examine the effect of using an inexact expression for the deformed shape, suppose we choose to approximate the deformation of Figure 12.2 *(b)* in the parabolic form

$$u = Cz(L - z) , \tag{12.63}$$

which satisfies the kinematic boundary conditions of the problem, as required in all Rayleigh-Ritz applications. We then find

$$U = \frac{1}{2}EI\int_0^L \left(\frac{d^2u}{dz^2}\right)^2 dz = \frac{1}{2}EI\int_0^L (-2C)^2 dz = 2EIC^2L \tag{12.64}$$

$$\Omega = -\frac{P}{2}\int_0^L \left(\frac{du}{dz}\right)^2 dz = -\frac{1}{2}P\int_0^L C^2(L-2z)^2 dz = -\frac{PC^2L^3}{6} \tag{12.65}$$

$$\Pi = U + \Omega = 2EIC^2L - \frac{PC^2L^3}{6} = \frac{C^2L^3}{6}\left(\frac{12EI}{L^2} - P\right) , \tag{12.66}$$

and comparison with the previous example shows that the critical force is

$$P_0 = \frac{12EI}{L^2} , \tag{12.67}$$

which overestimates the exact value (12.62) by 22%.

[7] Often an order of magnitude estimate is sufficient, since we recall from §§12.2, 12.3 that it is important to design compression members to be loaded well below the critical force.

### A better approximation

We could get nearer to the correct value as in §3.6.2 by choosing an approximate shape that gives zero curvature (and hence zero bending moment) at the simple supports $z=0,L$. However, we can do almost as well without improving the deformation guess, by using the equation

$$U=\frac{1}{2EI}\int_0^L M^2dz \tag{12.68}$$

for the strain energy in place of (12.64) and writing

$$M=Pu=PCz(L-z)\ , \tag{12.69}$$

which also has the effect of forcing the contribution of the bending energy near the ends to zero.

Substituting (12.69) into (12.68), we find

$$U=\frac{1}{2EI}\int_0^L P^2C^2z^2(L-z)^2dz=\frac{P^2C^2L^5}{60EI}$$

and hence

$$\Pi=U+\Omega=\frac{P^2C^2L^5}{60EI}-\frac{PC^2L^3}{6}=\frac{PC^2L^3}{6}\left(\frac{L^2}{10EI}-P\right)\ ,$$

corresponding to a critical force

$$P=\frac{10EI}{L^2}\ . \tag{12.70}$$

This overestimates the exact value by only 1.4% and is a considerable improvement on that obtained in equation (12.67).

### 12.7.3 Inhomogeneous problems

The Rayleigh-Ritz method can also be used for inhomogeneous problems, such as those discussed in §§12.2, 12.3. In this case, the expression for total potential energy will contain both linear and quadratic terms in the degree of freedom $C$. The condition $d\Pi/dC=0$ will then yield a non-trival solution for $C$ at all forces, which goes unbounded as $P\rightarrow P_0$.

### Example 12.5

*The simply supported beam of Figure 12.25 is loaded by a compressive force $P$ and a lateral force $F$ applied at the mid-point. Use the Rayleigh-Ritz method to estimate the bending moment at the mid-point.*

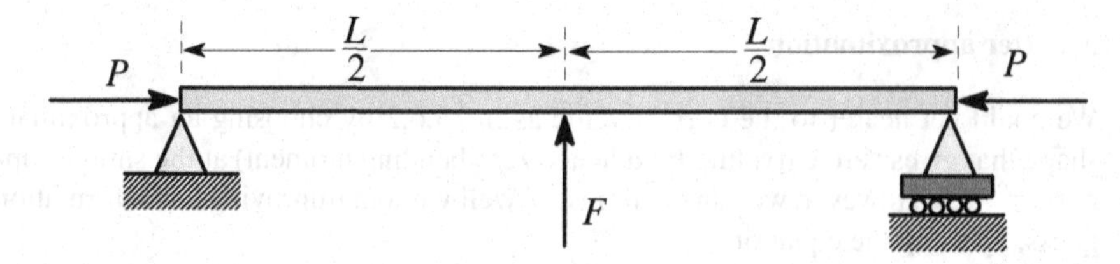


*Figure 12.25*

We approximate the displacement in the parabolic form of equation (12.63), in which case

$$u\left(\frac{L}{2}\right) = \frac{CL^2}{4} .$$

The potential energy $\Omega$ of the applied forces contains two terms, one due to the axial force $P$ and given by (12.65), and the other equal to the work done against the lateral force $F$. Thus

$$\Omega = -\frac{PC^2L^3}{6} - Fu\left(\frac{L}{2}\right) = -\frac{PC^2L^3}{6} - \frac{FCL^2}{4} .$$

The strain energy $U$ is still given by (12.64), so the total potential energy is

$$\Pi = U + \Omega = 2EIC^2L - \frac{PC^2L^3}{6} - \frac{FCL^2}{4} .$$

The equilibrium condition then yields

$$\frac{d\Pi}{dC} = 4EICL - \frac{PCL^3}{3} - \frac{FL^2}{4} ,$$

with solution

$$C = \frac{FL^2}{4\left(4EIL - \frac{PL^3}{3}\right)} = \frac{FL}{16EI\left(1 - \frac{P}{P_0}\right)} ,$$

where $P_0$ is given by (12.67).

It follows that

$$u\left(\frac{L}{2}\right) = \frac{FL^3}{64EI\left(1 - \frac{P}{P_0}\right)}$$

and the bending moment at $z = L/2$ is

$$M\left(\frac{L}{2}\right) = \frac{FL}{4} + Pu\left(\frac{L}{2}\right) = \frac{FL}{4} + \frac{FPL^3}{64EI\left(1 - \frac{P}{P_0}\right)} ,$$

which can also be written

$$M\left(\frac{L}{2}\right) = \frac{FL}{4}\left[1 + \frac{3\frac{P}{P_0}}{4\left(1 - \frac{P}{P_0}\right)}\right] .$$

## 12.8 Quick estimates for the buckling force

We have already seen in §§12.2, 12.3 that initial imperfections or lateral loads can cause substantial deflections of axially loaded members below the theoretical critical force.[8] It is therefore important to design compression members to operate well below the critical force. A good working rule is to stay below half the critical force wherever possible and if circumstances make it advantageous to try to exceed this limit, take great care over the accuracy both of manufacture and loading of the beam.

One incidental advantage of this necessity to design well below the critical force is that it is not necessary to know the theoretical critical force to a very high degree of accuracy. If we are going to apply a safety factor of 2, anything better than 20% accuracy in a calculation is so much wasted effort. The Rayleigh-Ritz method will usually give this kind of accuracy with an elementary function with a single degree of freedom, but care must be taken to ensure that the assumed deformation is similar in shape to that which is likely to occur. If in doubt, perform two or more calculations with different trial functions and use the one which gives the lowest critical force.

In beam problems, an even quicker estimate can be made if the deformed shape can be realistically estimated. The shape of the buckled section is dominated by the sinusoidal functions of equation (12.4) and the buckling force is uniquely related to the wavelength of the resulting sinusoid. Most buckled configurations involve less than a complete wave, so it is convenient to define the length of a quarter wave as $L_{1/4}$. It then follows that the buckling force is

$$P_0^* = \frac{\pi^2 EI}{4L_{1/4}^2} \; . \tag{12.71}$$

Thus, if the expected deformed shape includes $N$ quarter waves in a beam of length, $L$, we have $L_{1/4}=L/N$ and

$$P_0^* = \frac{N^2\pi^2 EI}{4L^2} \; . \tag{12.72}$$

This formula is clearly exact for the modes of the simply-supported beam of Figure 12.2*(a)*, but it also gives a reasonable estimate for more complicated problems. For example, Figure 12.26*(a)* shows a beam that is built in at one end and simply supported at the other, loaded by an axial force $P$. The anticipated deformed shape is shown in Figure 12.26*(b)* and this contains approximately three quarter waves, as indicated by the division of the beam into segments by the dotted lines. Equation (12.72) then predicts that the buckling force will be

$$P_0^* = \frac{9\pi^2 EI}{4L^2} \approx \frac{22EI}{L^2} \; .$$

[8] For this reason, it is quite difficult to get an accurate measurement of the critical force experimentally.

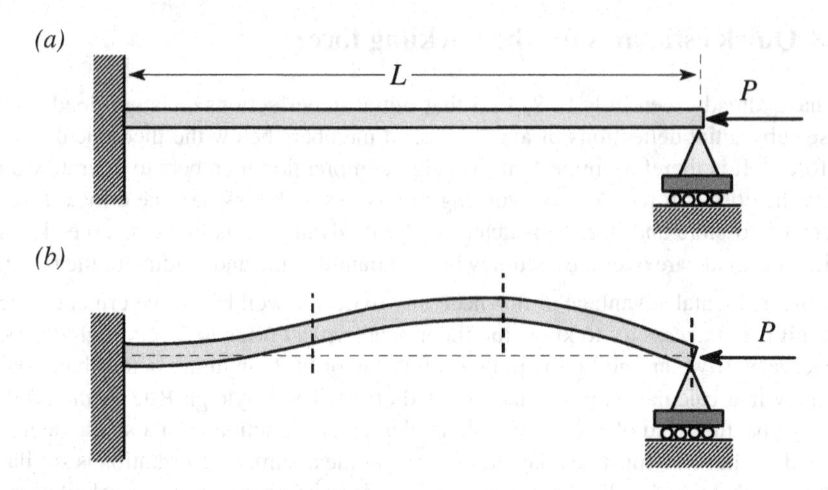


*Figure 12.26: (a) A beam built in at one end and simply supported at the other; (b) the expected deformed shape*

This problem can be solved by the equilibrium method. After a fairly extensive calculation, we obtain the exact value

$$P_0 = \frac{20.1EI}{L^2} ,$$

so the one line estimate of equation (12.72) is really quite good. The reader is advised to check this prediction even in cases where a more accurate result is required, since it provides a useful check against casual calculation errors.

## 12.9 Summary

A perfectly straight beam will suffer no lateral deflection under purely axial loading until a critical compressive force is reached, at which the system is unstable and the beam buckles. A related instability known as whirling occurs in rotating shafts when the rotational speed reaches a critical value.

If the beam is slightly curved or if there are also lateral loads, the lateral deflections are increased by a compressive force and increase without limit as the critical condition is approached. In these cases, significantly enhanced displacements and stresses can be produced well below the critical force, so it is necessary to use a good factor of safety against buckling failure.

Buckling failures are particularly critical in thin-walled members, which in other respects are extremely efficient load bearing structures. Buckling is therefore often a limiting consideration in optimal design.

Two basic methods are available for determining the critical force. In the neutral equilibrium method, differential equations are derived defining the conditions which must be satisfied if the system is to be capable of remaining in equilibrium in a 'non-trivial' state of lateral deformation. Solution of these equations and the imposition of boundary conditions leads to an algebraic equation for the critical condition. Beam problems can generally be solved in this way, but more complex two and three dimensional problems are seldom analytically tractable.

An alternative approach is to use the Rayleigh-Ritz method in which a functional form is assumed for the deformed shape and the stability criterion is based on the requirement that the total potential energy should be a local minimum. This method is very versatile, but it depends crucially on our ability to predict and represent the probable buckled shape of the structure.

## Further reading

J.M.T. Thompson and G.W. Hunt (1973), *A General Theory of Elastic Stability*, John Wiley, London.

S.P. Timoshenko and J.M. Gere (1961), *Theory of Elastic Stability*, McGraw-Hill, New York, 2nd edn.

## Problems

### Section 12.1

**12.1*.** Figure P12.1 shows a beam of flexural rigidity $EI$ and length $L$ which is simply supported at the points $A$ ($z = L/4$) and $B$ ($z = 3L/4$). It is compressed by two forces $P$ whose lines of action are always parallel to $AB$. Find the critical force $P_0$ and sketch the shapes of the first two buckling modes.

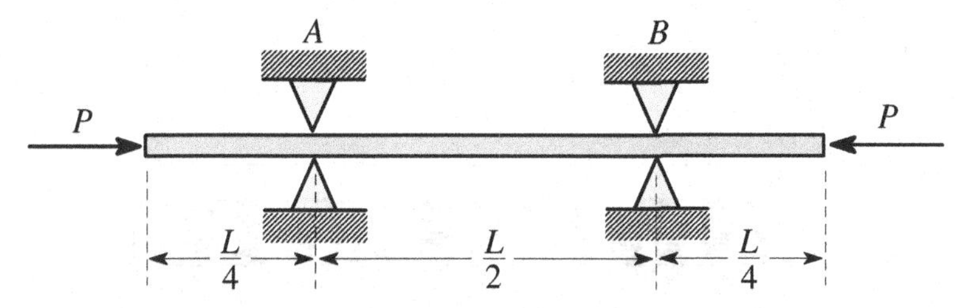


*Figure P12.1*

**12.2.** The composite beam $AB$ of Figure P12.2 consists of three segments. The segments $AC$ and $DB$ are both rigid and of length $a$, whilst the segment $CD$ is of flexural rigidity $EI$ and length $(L-2a)$. The beam is simply supported at $A, B$ and compressed by a force $P$. Find the equation determining the critical value of $P$ for instability and show that your result reduces to equation (12.10) when $a = 0$. Solve the equation for the case where $a = L/4$.

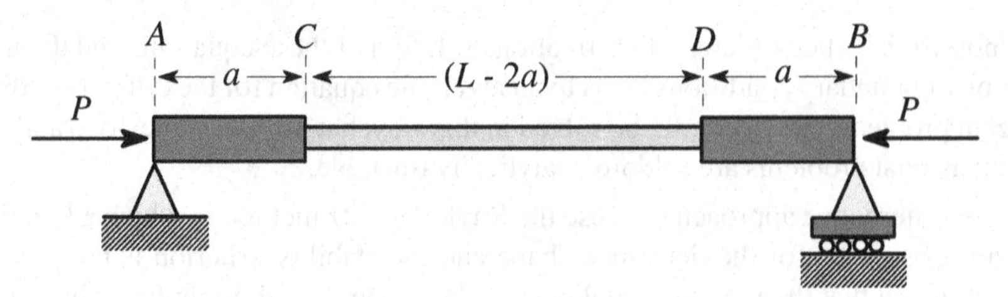


*Figure P12.2*

**12.3.** Find the critical force for the beam of Figure P12.2 for the case where the segment $CD$ is rigid and the segments $AC$, $DB$ have flexural rigidity $EI$.

**12.4.** Figure P12.4*(a)* shows a *rigid* bar $AB$ of length $L$, which is pinned at $B$ and compressed by a horizontal force $P$ at $A$. The side of the bar is bonded to a block of rubber which can be modelled as an elastic foundation of modulus $k$, as shown in Figure P12.4*(b)*. In other words, the rubber exerts a vertical force $w(z)$ per unit length given by

$$w(z) = ku(z) ,$$

where $k$ is a constant and $u(z)$ is the local vertical displacement.

Find the value of $P$ at which the system becomes unstable.

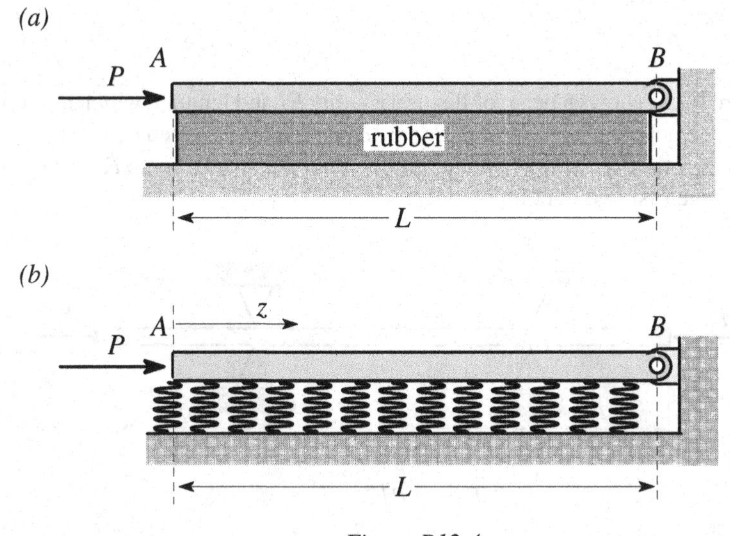


*Figure P12.4*

**Sections 12.2, 12.3**

**12.5.** The cantilever beam of Figure P12.5 of length $L$ and flexural rigidity $EI$ has an initial deflection defined by the equation

$$u_0(z) = \frac{\varepsilon z^2}{L^2} ,$$

where $\varepsilon \ll L$. Find the maximum bending moment in the beam when it is subjected to an axial force

$$P = \frac{\pi^2 EI}{8L^2} .$$

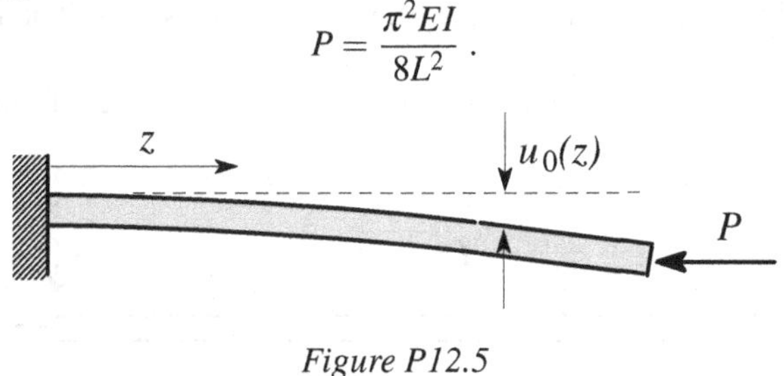


*Figure P12.5*

**12.6.** A uniform simply supported beam of flexural rigidity $EI$ and length $L$ is subjected to a lateral uniformly distributed load $w_0$ per unit length and an axial force $P$ as shown in Figure P12.6. Find the maximum bending moment if $P$ is equal to 70% of the buckling force. What is the ratio between this moment and the value that would be obtained if $P$ were zero?

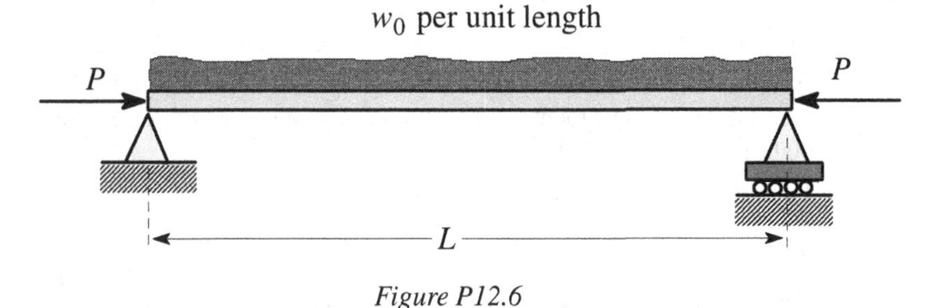


*Figure P12.6*

**12.7.** The uniform beam in Figure P12.7 has a circular cross section of diameter $d$ and length $L$. It is subjected to an axial compressive force $P$ which is lower than the first critical force for the beam, but whose line of action is a distance $d/4$ from the centreline of the beam.

Find the maximum bending moment in the beam.

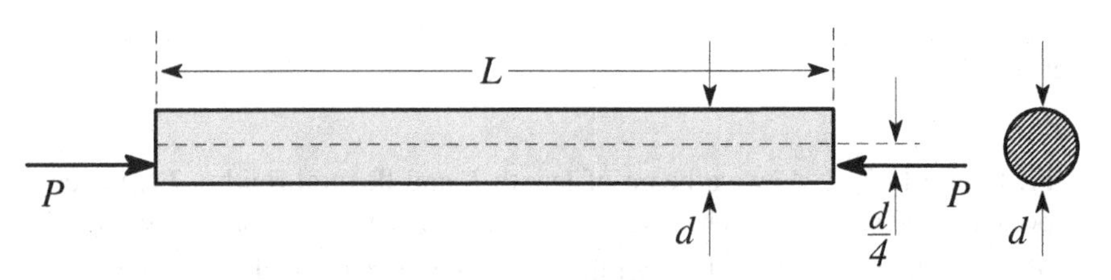


*Figure P12.7*

**12.8.*** Figure P12.8 shows a cantilever beam of length $L$ loaded by a lateral force $F$ and a *tensile* axial force $P$. Show that the bending moment at the support is reduced by the application of the axial force and find this moment as a function of $F$ and $P$.

If the beam is a circular cylinder of diameter, $d$, are there any circumstances in which the maximum tensile stress in the beam is *reduced* as a result of the application of $P$.

**Note**: The homogeneous solution for this problem is different from (12.4).

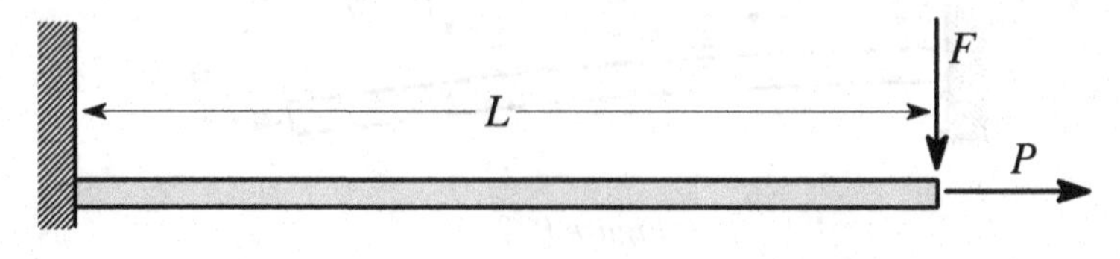


*Figure P12.8*

**Section 12.4**

**12.9.** Find the critical force for the indeterminate problem of Figure 12.11.

**12.10.** The free end of a cantilever beam of length $L$ and flexural rigidity $EI$ is restrained from lateral displacement by a frictionless support, which offers no resistance to rotation (see Figure P12.10).

Find the critical force $P_0$ at which the beam will buckle.

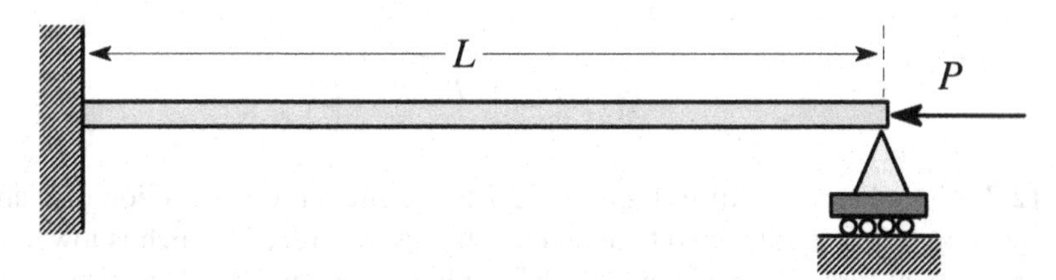


*Figure P12.10*

**Section 12.5**

**12.11.** Figure P12.11 shows a beam of length $L$ and flexural rigidity $EI$, which is supported on two springs, each of stiffness $k$, and loaded by a compressive force $P$.

Treating the beam as rigid, determine the value of $P$ at which the system is unstable. Hence find the critical value $k_0$, beyond which increasing $k$ does not increase the buckling force.

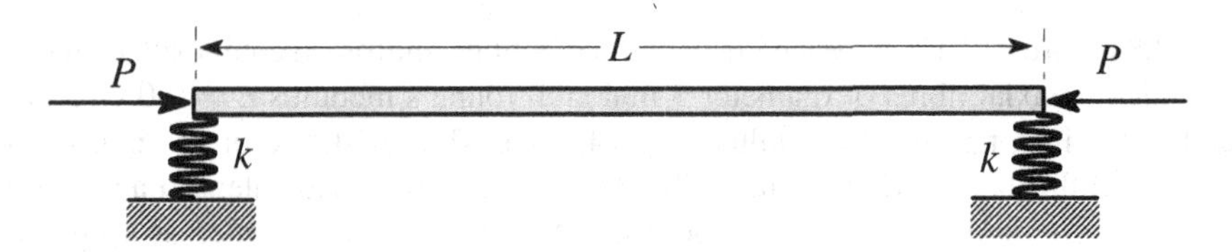


*Figure P12.11*

**12.12.** A beam of flexural rigidity $EI$ and length $L$ is compressed by an axial force $P$. The two ends of the beam are restrained from lateral motion ($u=0$ at $z=0,L$), but rotation (end slope) is resisted by torsion springs which exert moments $M_1, M_2$ respectively, given by

$$M_i = k\theta_i \quad ; \quad i = 1,2 \; .$$

Find the first buckling force $P_0$ for the system. Show that it is a monotonically increasing function of the spring stiffness $k$ and find the value of $k$ for which

$$P_0 = \frac{2\pi^2 EI}{L^2}$$

— i.e. twice the value that would be obtained if $k$ were zero.

## Section 12.6

**12.13.** The mounting assembly of Problem 7.7 is subjected to an axial force $P$ whose line of action passes through the centroid of the steel strip. Find the critical force at which buckling occurs and the wavelength of the resulting deformed configuration.

**12.14.** A long steel beam of second moment of area $I = 5\times 10^6$ mm$^4$ is supported on an elastic foundation of modulus $k = 10$ MPa and loaded by an axial force $P$. Find the buckling force $P_0$ and the wavelength of the resulting deformed configuration ($E_{\mathrm{steel}} = 210$ GPa).

**12.15.** An infinite beam of flexural rigidity $EI$ is supported on an elastic foundation of modulus $k$ and loaded by a concentrated moment $M_0$ at $z=0$. The beam also transmits an axial compressive force $P$. Find the rotation $\theta(0)$ due to the moment as a function of $P$ and hence find the axial force needed to increase this rotation by a factor of two.

**12.16.** A beverage can of diameter 2.5 inch and wall thickness $2\times 10^{-3}$ inch is made of aluminium alloy for which Young's modulus is $E = 11\times 10^6$ psi and $\nu = 0.25$. Find the axial compressive force at which it will buckle in an axisymmetric mode and the wavelength of the resulting deformed configuration. What is the mean compressive stress in the can just before buckling and how does it compare with typical yield strengths for aluminium alloys?

**12.17.*** Figure P12.17 shows an idealization of a fibre-reinforced composite material in which uniaxial fibres of diameter 1 mm and Young's modulus $E = 180$ GPa are embedded in a matrix of modulus $E = 1.4$ MPa. The resistance of the matrix to the lateral displacement of a single fibre is estimated to be equivalent to an elastic foundation of modulus 0.5 MPa. If there are $25 \times 10^3$ fibres per m$^2$ of cross-sectional area, estimate the uniaxial compressive stress at which the composite will fail by fibre buckling when loaded along the fibre axis.

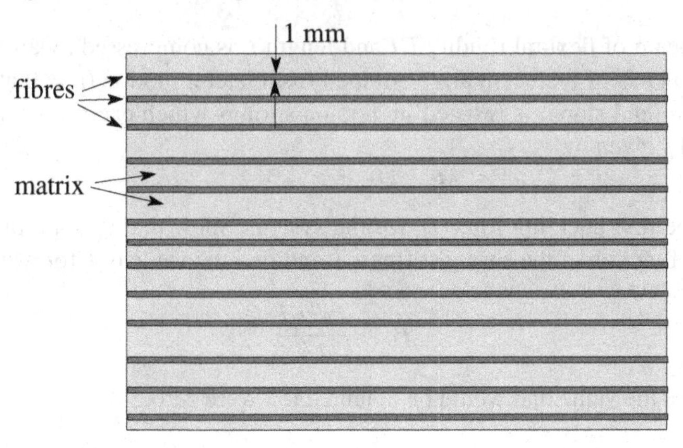


*Figure P12.17*

**12.18.** A solid cylindrical shaft of diameter 10 mm and length 1 m is built in at one end, the other end being unsupported. Find the first critical speed of the shaft, if the material is steel for which $E = 210$ GPa, $\rho = 7700$ kg/m$^3$.

**12.19.** The transmission shaft for a rear wheel drive car consists of a hollow steel tube of mean diameter 30 mm, wall thickness 2 mm and length 1.8 m. Determine the critical rotational speed of the shaft, treating it as simply supported at the ends. ($E_{\text{steel}} = 210$ GPa, $\rho_{\text{steel}} = 7700$ kg/m$^3$.)

**12.20.** A turbo-alternator for a power generation plant can be idealized as a solid steel cylinder of diameter 300 mm and length 14 m. It is supported in three bearings, one at each end and one at the mid-point, all of which can be treated as simple supports. Find the first critical rotational speed ($E_{\text{steel}} = 210$ GPa, $\rho_{\text{steel}} = 7700$ kg/m$^3$).

**12.21.** A solid wheel of diameter 200 mm and thickness 30 mm is supported at the mid-point of a shaft of diameter 20 mm and length 600 mm. The shaft is simply supported at its ends. Find the critical rotational speed if all the components are made of steel ($E = 210$ GPa, $\rho = 7700$ kg/m$^3$).

**12.22.** A student swings a 2 lb weight (mass 1/16 slug) around her head at the end of a spring of stiffness 3 lb/in and length 10 in. What is the theoretical maximum rotational speed she could achieve?

**12.23.*** A thin-walled elastic tube of Young's modulus $E$, wall thickness $t$, radius $a$ and length $L$ is simply supported at its ends. An inviscid fluid of density $\rho$ flows through the tube at velocity $V$. Find the value of $V$ for which the tube will buckle. Neglect gravity.

**Hint**: Treat the tube as a beam of appropriate flexural rigidity. Lateral forces and hence bending moments are generated by the acceleration of the fluid when the tube is in a deformed configuration.

**12.24.*** Find the first whirling speed for the shaft of Figure 12.20, taking into account the mass of both the shaft and the wheel and using the boundary conditions (12.49). Plot a graph of $\Omega_0$ as a function of the dimensionless ratio $M/mL$ (the ratio of the mass of the wheel to that of the shaft) and compare it with the approximate expression (12.48). How good is the approximation?

## Section 12.7

**12.25.** Figure P12.25 shows a rigid bar supported by a torsion spring at one end and loaded by an axial force $P$. Find the critical force $P_0$, if the spring exerts a moment $M = k\theta$, where $\theta$ is the rotation at the support.

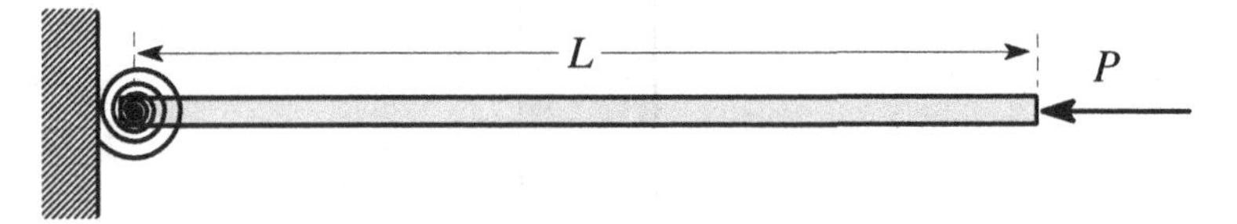


*Figure P12.25*

**12.26.** The uniform cantilever shown in Figure P12.26, of flexural rigidity $EI$ and length $L$, is loaded by an axial force $P$ and its free end is restrained by a spring of stiffness $k$.

Use the Rayleigh-Ritz method to estimate the critical force, assuming the beam deforms into a parabolic shape.

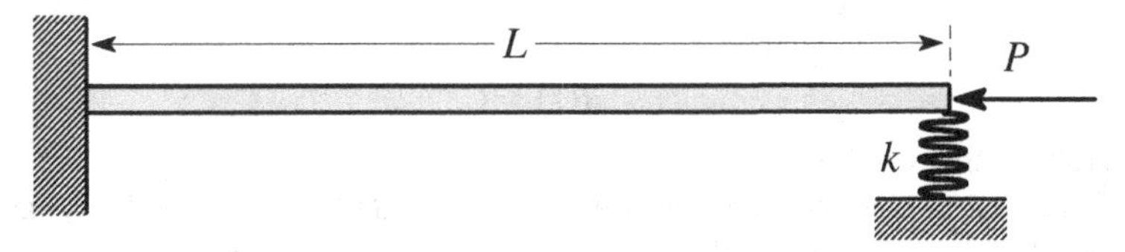


*Figure P12.26*

**12.27.** A uniform beam of flexural rigidity $EI$ and length $L$ is simply supported at its ends and also at a point a distance $b$ from one end. Use the Rayleigh-Ritz method and the trial function $u = Cx(x-b)(x-L)$ to estimate the buckling force for the beam. Use your results to sketch a graph of $P$ as a function of $b$ in the range $0 < b < L$.

**12.28.** Figure P12.28 shows a solid cylindrical tower of height $h$ and diameter $d$, made from an elastic material of density $\rho$ and Young's modulus $E$. Use the Rayleigh-Ritz method to estimate the maximum value of $h$ if the tower is not to buckle under its own weight.

Use a one degree of freedom approximation to the buckled shape. Be particularly careful in determining the expression for the potential energy of the distributed force due to self-weight.

What is the critical height for a steel tower of diameter 10mm? ($E_{\rm steel} = 210$ GPa, $\rho_{\rm steel} = 7700$ kg/m$^3$).

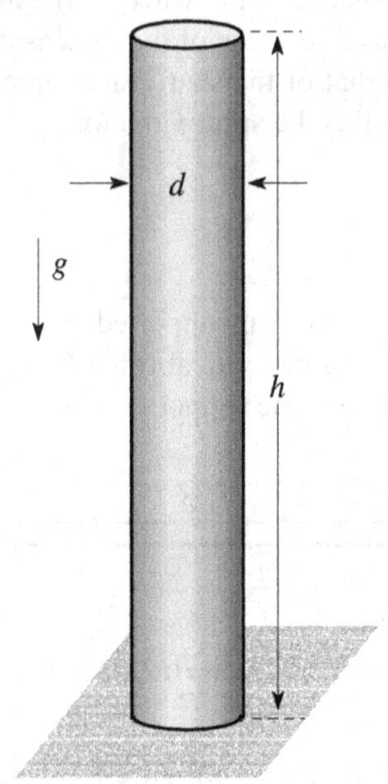


*Figure P12.28*

**12.29.** In an attempt to make the tower of Problem 12.28 higher without using additional material, it is proposed to make the diameter $d$ vary with distance $z$ from the top according to the equation

$$d(z) = Cz^n$$

where $C, n$ are constants.

The total volume of material available is constant and equal to $V$. What value of $n$ should be used to achieve the maximum height $h$ without buckling?

**12.30.** A beam of length $L$ is built in at $z=0$ and loaded by an axial force $P$ at $z=L$ as shown in Figure P12.30. The beam is made of a material with Young's modulus $E$ and has a rectangular cross section of sides $a, b$, where

$$a = c \ ; \ b = c\left(\frac{3}{2} - \frac{z}{L}\right)$$

and $c$ is a constant. Thus, the dimension $a$ is constant along the beam, but $b$ varies linearly along the beam.

Use the Rayleigh-Ritz method to estimate the critical value of $P$ at which the beam will buckle. Use a simple parabolic approximation to the buckled shape.

Notice that the weak axis of bending is parallel to side $a$ near the free end, but parallel to side $b$ near the built-in end. About which axis does your analysis predict that buckling will occur?

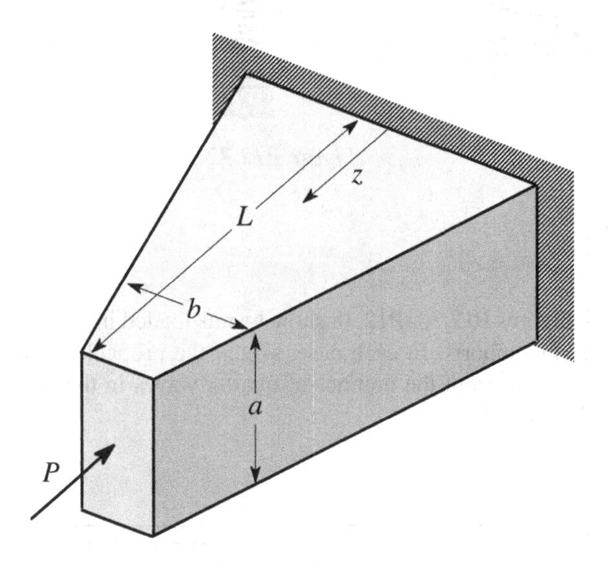


*Figure P12.30*

**12.31.** Figure P12.31 shows a beam of flexural rigidity $EI$ and length $L$, which is built in at one end and simply supported at the other. It is subjected to a uniformly distributed transverse load of $w_0$ per unit length and a compressive axial force $P$.

Use the Rayleigh-Ritz method to estimate the maximum lateral displacement and sketch the result as a function of $P$. Use a one degree of freedom polynomial approximation to the deformed shape.

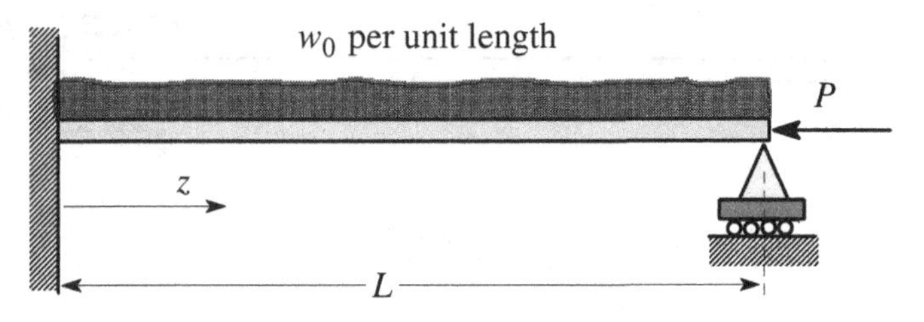


*Figure P12.31*

**12.32.** Figure P12.32 shows a mechanism consisting of two identical rigid bars, each of length $L$, loaded by a force $P$. A spring of stiffness $k$ is attached to the pivot point and is unstretched when the bars are both horizontal. Find the critical force at which the horizontal configuration becomes unstable.

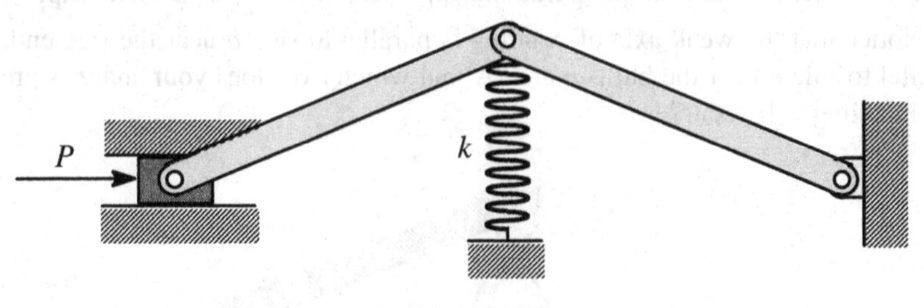


*Figure P12.32*

## Section 12.8

**12.33–12.38.** Figures P12.33–P12.38 show beams loaded in compression with a variety of different supports. In each case, sketch the probable buckled shape at the first buckling force, count the number of quarter waves in the length $L$, and hence estimate the buckling force.

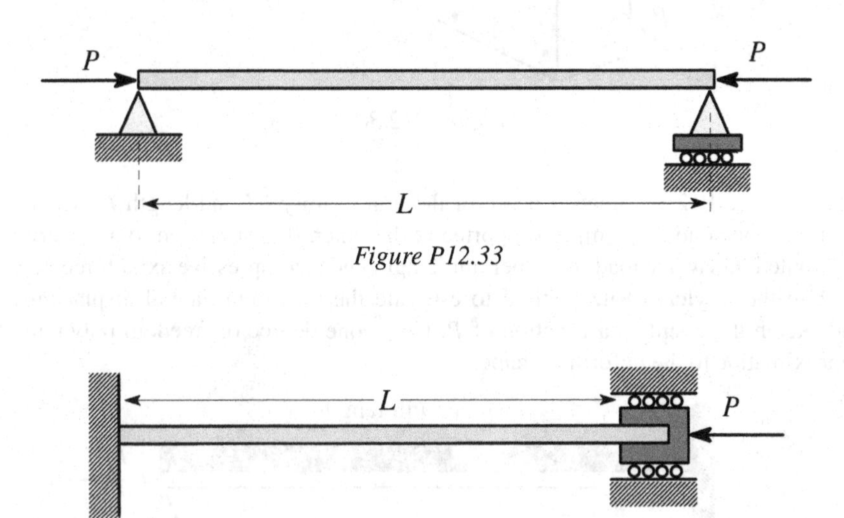


*Figure P12.33*

*Figure P12.34*

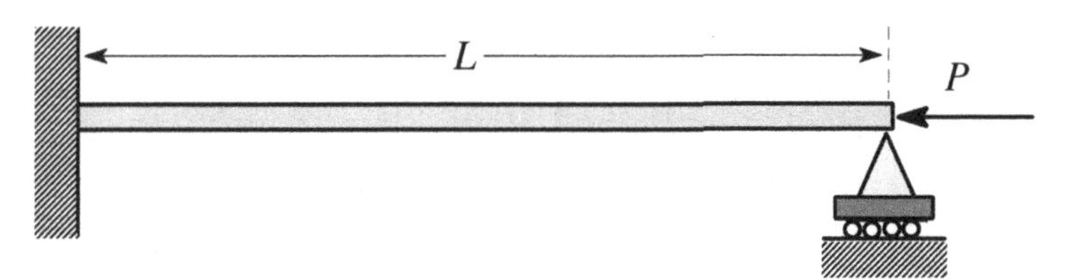


*Figure P12.35*

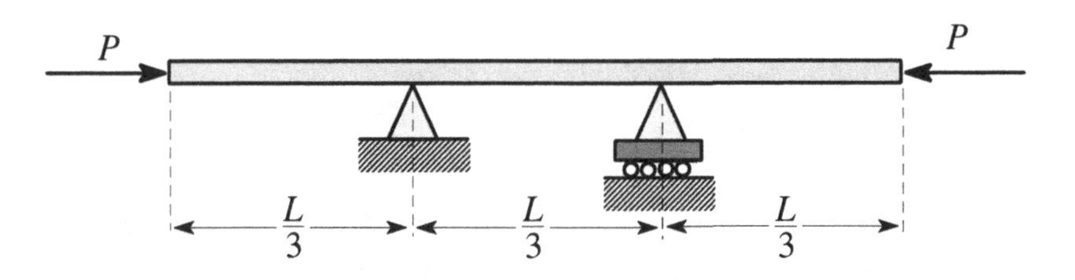


*Figure P12.36*

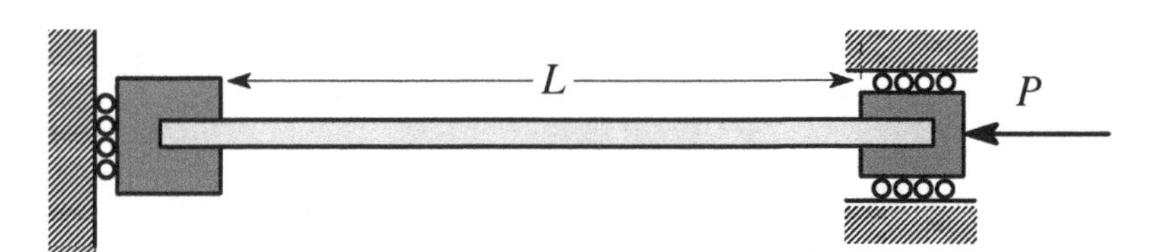


*Figure P12.37*

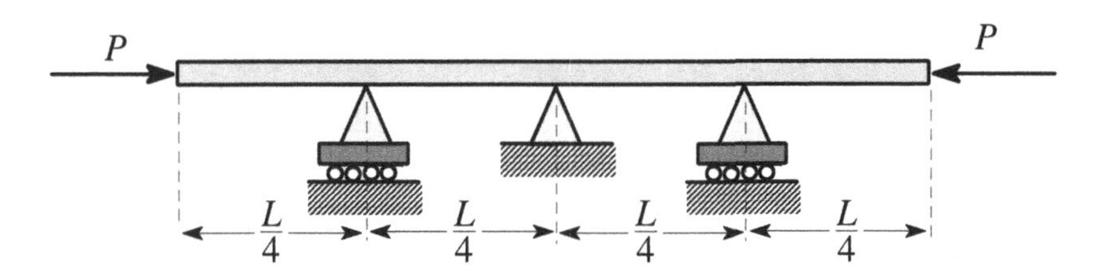


*Figure P12.38*

# A

# The Finite Element Method

Most of this book has been concerned with exact analytical methods for solving problems in mechanics of materials. Real engineering applications seldom involve geometries or loading conditions exactly equivalent to those analyzed, so the use of these methods usually involves some approximation. Nonetheless, the advantages to design offered by a general (symbolic) analytical solution makes them very useful for estimation, even when the idealization involved is somewhat forced.

Numerical methods are appropriate when no plausible idealization of the real problem can be analyzed, or when we require results of greater accuracy than the idealization is expected to produce. By far the most versatile and widely used numerical method is the finite element method and every engineer arguably needs to have at least some acquaintance with it to be considered 'scientifically literate'. There is no room in a book of this length to develop the method to the level where the reader could either write his/her own finite element code or use a commercial code. Fortunately, most commercial codes are these days sufficiently user-friendly that one can learn to use them with minimal introduction from the manual, supplemented by the program help menu. All we seek to do here is to explain the fundamental reasoning underlying the method, introduce some of the terminology, and generally to remove the 'mystique' that can be a barrier to those not familiar with the method.

In the finite element method, the body under consideration is divided into a number of small elements of simple shape, within each of which the stress and displacement fields are represented by simple approximations, usually low order polynomial functions of position. For example, in the simplest case, the stress in each element will be taken as uniform.

Historically, the method was first developed as an extension of the 'stiffness matrix method' of structural analysis, discussed in §3.9. Each element is treated as a simple structural component, whose elastic stiffness matrix can be determined by elementary methods. The problem is thereby reduced to the loading of a set of interconnected elastic elements. The global stiffness matrix for the problem is assembled from that for the individual elements as discussed in §3.9.2.

Early researchers used more or less *ad hoc* methods for determining the element stiffness matrices, but the finite element method was placed on a more rigorous foot-

J.R. Barber, *Intermediate Mechanics of Materials*, Solid Mechanics and Its Applications 175,
2nd ed., DOI 10.1007/978-94-007-0295-0, © Springer Science+Business Media B.V. 2011

ing by the adoption of the principle of stationary potential energy for this purpose. In this formulation, the finite element method is exactly equivalent to the use of the Rayleigh-Ritz method of §3.6 in combination with a piecewise polynomial approximation function for the deformation of the structure.

More recently, 'mathematical' formulations of the finite element method have been developed in which the unknown constants defining the piecewise polynomial approximation (the degrees of freedom) are chosen so as to satisfy the governing equations (e.g. the equilibrium equation) in a 'least squares fit' sense. We already remarked in §3.5 that the principle of stationary potential energy is an alternative statement of the equations of equilibrium, so it should come as no surprise that this formulation leads to the same final matrix equation. However, it has certain advantages, notably that it is readily extended to applications outside mechanics of materials (e.g. heat conduction, electrical conduction or fluid mechanics) and it also provides additional insight into the mathematics underlying the method.

In this brief introduction to the finite element method, we shall illustrate and compare these alternative formulations in the context of the simplest problem in mechanics of materials — the axial loading of an elastic bar — and then extend the argument to the bending of beams. However, we first need to discuss some aspects of approximations in general.

## A.1 Approximation

All numerical methods require that the functions describing physical quantities such as stress and displacement be approximated, typically by the sum of a series of simpler functions with initially unknown linear multipliers. For example, we may choose to approximate the function $f(x)$ by the series

$$f(x) \approx f^*(x) \equiv \sum_{i=1}^{N} C_i v_i(x) , \tag{A.1}$$

where $C_i$ are the multiplying constants and the $v_i(x)$ are known as *shape functions*, which must be linearly independent if the approximation is to be well-defined.

### A.1.1 The 'best' approximation

If we use $f^*(x)$ in place of $f(x)$, it is clear that the governing equations [e.g. the beam bending equation (1.17) or the equation of motion (10.18)] and/or the boundary conditions will not be exactly satisfied and we have to choose the constants $C_i$ so as to achieve in some sense the 'best' approximation. There is no unique definition of 'best' in this context and different criteria might be appropriate in different applications.

In the Rayleigh-Ritz method (§3.6), we made this choice so as to achieve the minimum total potential energy $\Pi$. This has the advantage of reflecting features of the physics of the system, but purely mathematical criteria can be adopted. One such approach is the *collocation method*, in which the constants $C_i$ are chosen so as to make the approximation exact at a set of $N$ *collocation points* $x_j$. In other words, the $C_i$ are determined from the $N$ simultaneous linear algebraic equations

$$f^*(x_j) = \sum_{i=1}^{N} C_i v_i(x_j) = f(x_j) \quad ; \quad j = (1,N) \,. \tag{A.2}$$

The disadvantage of this method is that the approximation may be very inaccurate between the collocation points. This is particularly true if power series terms are used for the shape functions, $v_i(x)$, in which case the resulting approximating function may snake about the exact line, as shown in Figure A.1.

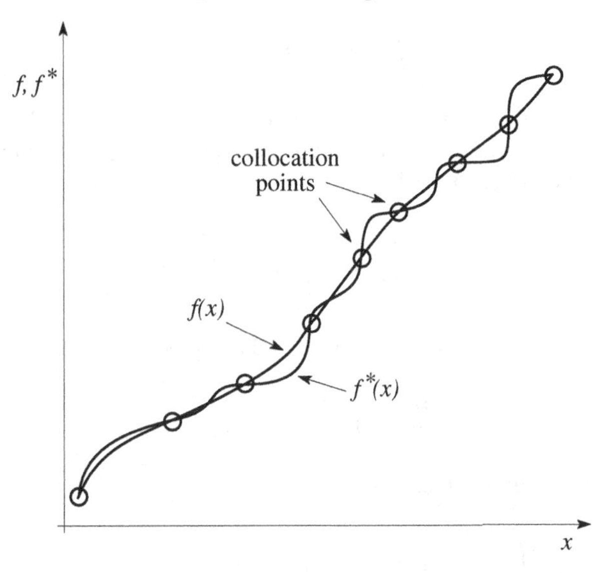


*Figure A.1: Pathological behaviour that can result from high order power series approximations using the collocation method*

A more stable and reliable method is to choose the constants $C_i$ so as to satisfy the $N$ linear equations

$$\int_a^b [f(x) - f^*(x)]\, w_j(x) dx = 0 \quad ; \quad j = (1,N) \,, \tag{A.3}$$

where the approximation is required in the range $a<x<b$ and the $w_j(x)$ are a set of $N$ linearly-independent *weight functions*.

### A.1.2 Choice of weight functions

Many readers will be familiar with the procedure for finding the best straight line through a set of experimental points $(x_k, y_k)$, $k=(1,M)$ by using a 'least squares fit' — i.e. by minimizing the sum of the squares of the error.

This implies minimizing the function

$$E \equiv \sum_{k=1}^{M} [y_k - y^*(x_k)]^2 , \quad (A.4)$$

where

$$y^*(x) = Ax + B \quad (A.5)$$

is the equation of the approximating straight line.

This idea can be extended to the approximation of a continuous function $f(x)$ by minimizing the error defined in the integral form

$$E \equiv \int_a^b [f(x) - f^*(x)]^2 \, dx . \quad (A.6)$$

This is essentially equivalent to extending the points $x_k$ in equation (A.4) to include all real points in the range $a<x<b$. If we represent the approximating function $f^*(x)$ in the form of equation (A.1), we have

$$E = \int_a^b \left( f(x) - \sum_{i=1}^{N} C_i v_i(x) \right)^2 dx \quad (A.7)$$

and we can minimize $E$ with respect to each of the degrees of freedom $C_i$ by enforcing the condition

$$\frac{\partial E}{\partial C_i} = 0 \; ; \; i = (1, N) . \quad (A.8)$$

This leads to the system of equations

$$-2 \int_a^b \left( f(x) - \sum_{i=1}^{N} C_i v_i(x) \right) v_j(x) dx = 0 \; ; \; j = (1, N) \quad (A.9)$$

and hence

$$\int_a^b [f(x) - f^*(x)] \, v_j(x) dx = 0 \; ; \; j = (1, N) . \quad (A.10)$$

Comparison with equation (A.3) shows that this is equivalent to the use of the same functions $v_i(x)$ as both shape functions and weight functions. This is known as the Galerkin formulation. The problem of fitting a best straight line to a continuous function corresponds to the special case $N=2$, $v_1(x)=1$, $v_2(x)=x$.

The set of simultaneous equations (A.9) can be rewritten in the form

$$\sum_{i=1}^{N} C_i \int_a^b v_i(x) v_j(x) dx = \int_a^b f(x) v_j(x) dx \; ; \; j = (1, N) ,$$

or in matrix form

$$\boldsymbol{KC} = \boldsymbol{F} , \quad (A.11)$$

where

$$K_{ji} = \int_a^b v_i(x)v_j(x)dx \tag{A.12}$$

$$F_j = \int_a^b f(x)v_j(x)dx\,. \tag{A.13}$$

Notice that the coefficient matrix $\boldsymbol{K}$ is simply the integral over the domain of the product of the shape and weight functions.

### A.1.3 Piecewise approximations

Everything we have said so far could be applied to any form of approximating function, including for example power series or Fourier series. The distinguishing feature of the finite element method is that a piecewise approximation is used. In other words, the range of the function is divided into a number of *elements* within each of which a fairly simple approximate form is used — usually a low order power series.

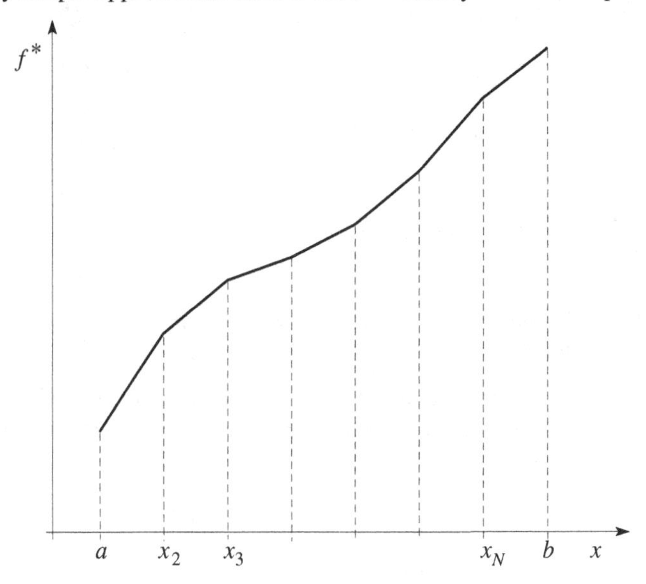


*Figure A.2: A piecewise linear approximating function*

The simplest approximation of this kind is the piecewise linear approximation illustrated in Figure A.2. Here, the range $a<x<b$ has been divided into $N$ elements, $a\equiv x_1 <x<x_2$, $x_2<x<x_3$, ..., $x_N<x<x_{N+1}\equiv b$, in each of which the function $f(x)$ is approximated by a straight line. The $(N+1)$ points $x_1,x_2,...,x_{N+1}$ are known as *nodes* and the corresponding values

$$f_i^* \equiv f^*(x_i)\,, \quad i=(1,N+1) \tag{A.14}$$

of the piecewise linear approximating function $f^*(x)$ are *nodal values*. The function $f^*(x)$ is completely defined if all the nodal values are known.

*(a)*

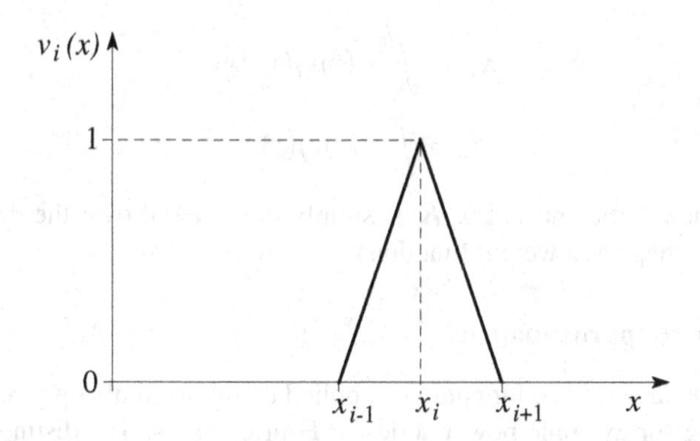


*(b)*

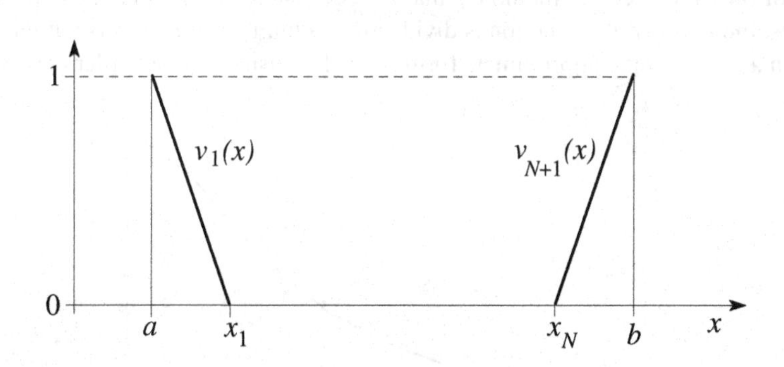


*Figure A.3: Shape functions for the piecewise linear approximation, (a) interior elements, (b) terminal elements*

The piecewise linear function of Figure A.2 can be expressed in the form of equation (A.1) by using shape functions of the form shown in Figure A.3 *(a)* and defined by

$$\begin{aligned} v_i(x) &= 0 && ; \; x < x_{i-1} \text{ and } x > x_{i+1} \\ &= \frac{x - x_{i-1}}{x_i - x_{i-1}} && ; \; x_{i-1} < x < x_i \\ &= \frac{x_{i+1} - x}{x_{i+1} - x_i} && ; \; x_i < x < x_{i+1} \,. \end{aligned} \tag{A.15}$$

If the elements are all of equal length

$$\Delta = \frac{b-a}{N} \,,$$

these equations simplify to

$$
\begin{aligned}
v_i(x) &= 0 && ; \quad x < x_{i-1} \text{ and } x > x_{i+1} \\
&= \frac{x - x_{i-1}}{\Delta} && ; \quad x_{i-1} < x < x_i \\
&= \frac{x_{i+1} - x}{\Delta} && ; \quad x_i < x < x_{i+1} \, .
\end{aligned}
\tag{A.16}
$$

For the terminal nodes, $i=(1,N+1)$, these functions take the form illustrated in Figure A.3 *(b)*.

In particular, we have

$$
\begin{aligned}
v_i(x_j) &= 1 \; ; \quad i = j \\
&= 0 \; ; \quad i \neq j
\end{aligned}
\tag{A.17}
$$

and hence

$$
f_i^* = C_i \, , \tag{A.18}
$$

from equations (A.1, A.14, A.17). In other words, the coefficients $C_i$ are also the nodal values $f_i^*$.

The piecewise linear shape functions (A.16) are zero everywhere except in the two elements adjacent to node $i$ and hence most of the components $K_{ji}$ of the coefficient matrix (A.12) will be zero. Non-zero values are obtained only when $|j-i|=0$ or 1 and in these cases the only contributions to the integral (A.12) come from at most two elements. Substituting (A.16) into (A.12) and performing the resulting elementary integrations, we find

$$
\begin{aligned}
K_{ji} &= \frac{2\Delta}{3} && ; \quad i = j \\
&= \frac{\Delta}{6} && ; \quad |i-j| = 1 \\
&= 0 && ; \quad |i-j| > 1 \, ,
\end{aligned}
\tag{A.19}
$$

except for the end nodes, $i=1$ and $i=N+1$, where the functions on Figure A.3 *(b)* must be used and we have

$$
\begin{aligned}
K_{ji} &= \frac{\Delta}{3} && ; \quad i = j = 1 \text{ or } N+1 \\
&= \frac{\Delta}{6} && ; \quad |i-j| = 1 \\
&= 0 && ; \quad |i-j| > 1 \, .
\end{aligned}
\tag{A.20}
$$

For example, if there are 7 elements and 8 nodes, the matrix $\boldsymbol{K}$ will be

$$
\boldsymbol{K} = \frac{\Delta}{6}
\begin{bmatrix}
2 & 1 & 0 & 0 & 0 & 0 & 0 & 0 \\
1 & 4 & 1 & 0 & 0 & 0 & 0 & 0 \\
0 & 1 & 4 & 1 & 0 & 0 & 0 & 0 \\
0 & 0 & 1 & 4 & 1 & 0 & 0 & 0 \\
0 & 0 & 0 & 1 & 4 & 1 & 0 & 0 \\
0 & 0 & 0 & 0 & 1 & 4 & 1 & 0 \\
0 & 0 & 0 & 0 & 0 & 1 & 4 & 1 \\
0 & 0 & 0 & 0 & 0 & 0 & 1 & 2
\end{bmatrix} .
\tag{A.21}
$$

Notice that the matrix is symmetric [as is clear from equation (A.12)] and it is also *banded* — i.e. non-zero values occur only on or near the diagonal. The reader will recall from §3.9 that the stiffness matrix for an elastic system is also symmetric and banded. Indeed, when the finite element method is applied to problems in linear elasticity, the resulting matrix $\boldsymbol{K}$ will be identical to the stiffness matrix of the system of elements, each regarded as elementary elastic components (generalized springs). For this reason, the matrix $\boldsymbol{K}$ is generally known as the *stiffness matrix* and this terminology is often retained even in cases where the finite element method is applied to other physical processes, such as the conduction of heat.

**Example A.1**

*Find a piecewise linear approximation to the exponential function* $e^x$ *in the range* $0<x<1$, *using two equal elements and three nodes,* $0, 0.5, 1$. *Plot a graph comparing the function and its approximation. Plot also the piecewise linear function obtained using the collocation method with the same three points and comment on the comparison.*

In this case, we have $\Delta = 0.5$ and the coefficient matrix is

$$\boldsymbol{K} = \frac{1}{12}\begin{bmatrix} 2 & 1 & 0 \\ 1 & 4 & 1 \\ 0 & 1 & 2 \end{bmatrix} .$$

For $F_1$, we have

$$F_1 = \int_0^1 e^x v_1(x)dx = \int_0^{0.5} e^x(1-2x)dx = \left[3e^x - 2xe^x\right]_0^{0.5} = 2e^{0.5} - 3$$
$$= 0.2974$$

Similar calculations for $F_2, F_3$ yield

$$F_2 = 2e - 4e^{0.5} + 2 = 0.8417$$
$$F_3 = 2e^{0.5} - e = 0.5792$$

Substituting into (A.11) and solving the resulting equations, we obtain

$$C_1 = f_1^* = f^*(0) = 0.9779$$
$$C_2 = f_2^* = f^*(0.5) = 1.6135$$
$$C_3 = f_3^* = f^*(1) = 2.6682 .$$

The resulting piecewise linear approximation is compared with the exponential function in Figure A.4. If the collocation method is used instead, we obtain the dotted curve in this Figure. It is clear that the collocation approximation is less accurate in this case, since it always lies on one side of the exact curve, except at the collocation points.

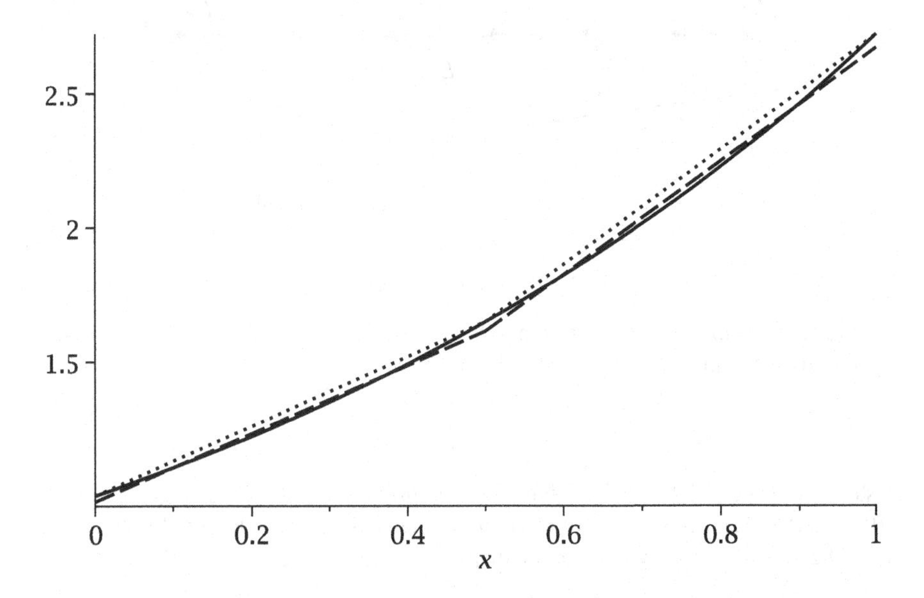


*Figure A.4: The exact exponential curve (———), the piecewise linear approximation (— — —), and a piecewise linear approximation using the collocation method (······).*

## A.2 Axial loading

The simplest application of the finite element method in mechanics of materials concerns the determination of the displacement of a bar subjected to axial loading. We shall use this example to illustrate the arguments underlying the method and also to explore the relation between the structural mechanics and Rayleigh-Ritz formulations.

### A.2.1 The structural mechanics approach

Figure A.5 shows an elastic bar of length $L$, supported at $x=0$ and loaded by a distributed load $p(x)$ per unit length and an end load $F_0$ at $x=L$. In the most general case, the cross-sectional area $A$ and Young's modulus $E$ can vary along the bar and hence be functions of $x$, but for the moment we shall restrict attention to the simpler case where $E,A$ are independent of $x$.

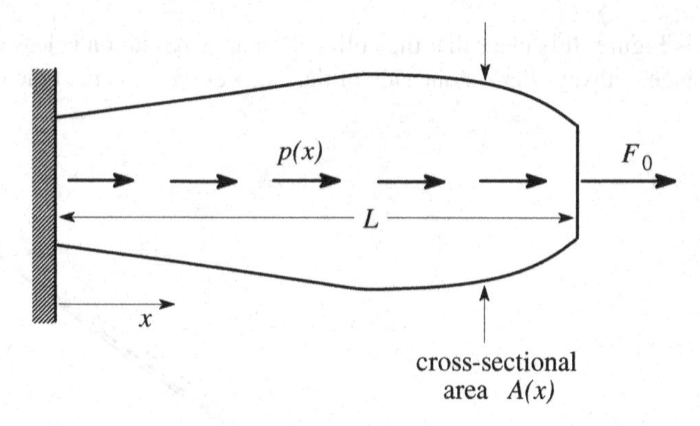


Figure A.5: Axial loading of the bar

To obtain an approximate solution of the problem, we consider the bar to be made up of a set of $N$ equal elements, each of length

$$\Delta = \frac{L}{N} ,$$

as shown in Figure A.6 *(a)*. The elements are numbered $i=(1,N)$, starting from the left end. As in §A.1.3, the deformation of the bar will be characterized by the set of *nodal displacements* $u_i$, which are the displacements of the points $x_i = iL/N$.

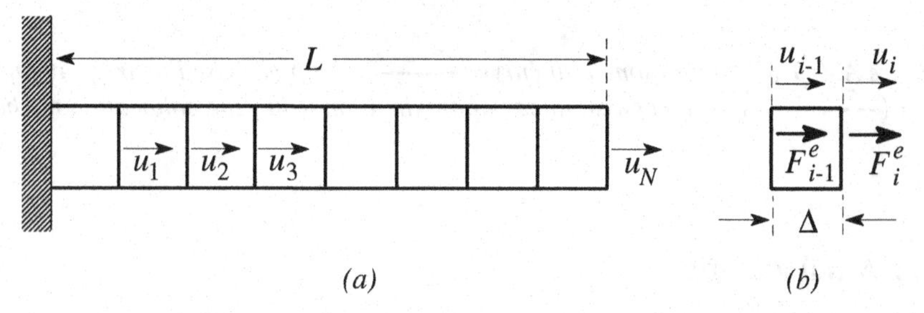


*Figure A.6: (a) Subdivision of the bar into N equal elements, (b) Nodal forces and displacements for the i-th element*

The first stage in the analysis is to determine the relation between the nodal forces on an individual element and the displacements of its ends. The *i*-th element is shown in Figure A.6 *(b)*. The displacements of its ends are $u_{i-1}, u_i$, and the corresponding nodal forces are $F^e_{i-1}, F^e_i$, respectively. The element is a uniform bar of length $\Delta$, cross-sectional area $A$ and Young's modulus $E$, so it behaves as a linear spring of stiffness

$$k = \frac{EA}{\Delta} .$$

The extension of the element is

$$\delta_i = u_i - u_{i-1}$$

and hence the element nodal forces and displacements are related by the equation

$$F_i^e = -F_{i-1}^e = \frac{EA}{\Delta}(u_i - u_{i-1}) \,.$$

This equation can be written in the matrix form

$$\frac{EA}{\Delta}\begin{bmatrix} 1 & -1 \\ -1 & 1 \end{bmatrix}\begin{Bmatrix} u_{i-1} \\ u_i \end{Bmatrix} = \begin{Bmatrix} F_{i-1}^e \\ F_i^e \end{Bmatrix} , \tag{A.22}$$

where the matrix

$$\boldsymbol{K_e} = \frac{EA}{\Delta}\begin{bmatrix} 1 & -1 \\ -1 & 1 \end{bmatrix} \tag{A.23}$$

is known as the *element stiffness matrix.*

### A.2.2 Assembly of the global stiffness matrix

The next stage is to use the properties of the individual elements to assemble the stiffness matrix $\boldsymbol{K}$, for the whole structure, defined such that

$$\boldsymbol{F} = \boldsymbol{K}\boldsymbol{u} \,,$$

where $\boldsymbol{F}$ is the vector of nodal forces *for the structure.* The matrix $\boldsymbol{K}$ is known as the *global stiffness matrix.*

Consider the special case in which the components of the nodal displacement vector $\boldsymbol{u}$ are given by $u_i = \delta_{ik}$. In other words, the $k$-th node is given a unit displacement ($u_k = 1$) and the other nodal displacements are zero ($u_i = 0, i \neq k$). The corresponding nodal forces are then given by

$$F_j = \sum_{i=1}^{N} K_{ji}\delta_{ik} = K_{jk}$$

and hence the $k$-th column of the global stiffness matrix $\boldsymbol{K}$ is equal to the nodal force vector $\boldsymbol{F}$ required to produce the displacement field $u_i = \delta_{ik}$. This in turn is the sum of the force vectors needed to deform the elements adjacent to node $k$ and is obtained by summing the corresponding components of the element stiffness matrices (A.23) at their appropriate points in the global matrix.

Proceeding element by element, we obtain

$$\frac{EA}{\Delta}\begin{bmatrix} 1 & 0 & 0 & 0 \ldots \ldots \\ 0 & 0 & 0 & 0 \ldots \ldots \\ 0 & 0 & 0 & 0 \ldots \ldots \\ 0 & 0 & 0 & 0 \ldots \ldots \\ \ldots & \ldots & \ldots & \ldots \ldots \ldots \\ \ldots & \ldots & \ldots & \ldots \ldots \ldots \end{bmatrix} + \frac{EA}{\Delta}\begin{bmatrix} 1 & -1 & 0 & 0 \ldots \ldots \\ -1 & 1 & 0 & 0 \ldots \ldots \\ 0 & 0 & 0 & 0 \ldots \ldots \\ 0 & 0 & 0 & 0 \ldots \ldots \\ \ldots & \ldots & \ldots & \ldots \ldots \ldots \\ \ldots & \ldots & \ldots & \ldots \ldots \ldots \end{bmatrix}$$

$$+\frac{EA}{\Delta}\begin{bmatrix} 0 & 0 & 0 & 0 \ldots \ldots \\ 0 & 1 & -1 & 0 \ldots \ldots \\ 0 & -1 & 1 & 0 \ldots \ldots \\ 0 & 0 & 0 & 0 \ldots \ldots \\ \ldots & \ldots & \ldots & \ldots \ldots \ldots \\ \ldots & \ldots & \ldots & \ldots \ldots \ldots \end{bmatrix} + \frac{EA}{\Delta}\begin{bmatrix} 0 & 0 & 0 & 0 \ldots \ldots \\ 0 & 0 & 0 & 0 \ldots \ldots \\ 0 & 0 & 1 & -1 \ldots \ldots \\ 0 & 0 & -1 & 1 \ldots \ldots \\ \ldots & \ldots & \ldots & \ldots \ldots \ldots \\ \ldots & \ldots & \ldots & \ldots \ldots \ldots \end{bmatrix} + \ldots , \tag{A.24}$$

leading, for example, to

$$K = \frac{EA}{\Delta}\begin{bmatrix} 2 & -1 & 0 & 0 & 0 & 0 & 0 \\ -1 & 2 & -1 & 0 & 0 & 0 & 0 \\ 0 & -1 & 2 & -1 & 0 & 0 & 0 \\ 0 & 0 & -1 & 2 & -1 & 0 & 0 \\ 0 & 0 & 0 & -1 & 2 & -1 & 0 \\ 0 & 0 & 0 & 0 & -1 & 2 & -1 \\ 0 & 0 & 0 & 0 & 0 & -1 & 1 \end{bmatrix}, \tag{A.25}$$

for the case where $N=7$. Notice how the resulting matrix is banded and is very similar in form to that of equation (A.21).

An alternative way of obtaining (A.25) is to find the force vector corresponding to $u_i=\delta_{ik}$ directly. Clearly only the two adjacent elements, $k, k-1$ will be deformed and by superposition, using equation (A.22), we conclude that the required non-zero forces are

$$F_{k-1} = -\frac{EA}{\Delta} \;\; ; \;\; F_k = \frac{2EA}{\Delta} \;\; ; \;\; F_{k+1} = -\frac{EA}{\Delta} \; ,$$

agreeing with the $k$-th column of (A.25).

### A.2.3 The nodal forces

In order to complete the solution of the problem of Figure A.5, we need to determine the nodal forces corresponding to the distributed loading $p(x)$. In the simple case where the load is uniformly distributed $[p(x)=p_0]$, it is reasonable to 'share out' the load $p_0\Delta$ acting on the $i$-th element equally between the two adjacent nodes $x_{i-1}, x_i$. This leads to the set of nodal forces

$$F_i = p_0\Delta \;\; ; \;\; i \neq N \tag{A.26}$$

$$F_N = \frac{1}{2}p_0\Delta + F_0 \; . \tag{A.27}$$

Notice that the end node $(i=N)$ receives a contribution only from the element on its left, but also experiences the end load $F_0$.

We now have enough information to set up the equation system

$$\boldsymbol{K}\boldsymbol{u} = \boldsymbol{F} \tag{A.28}$$

for the nodal displacements $u_i$.

#### Example A.2

*A uniform bar of cross-sectional area A and length L is loaded by a uniform force $p_0$ per unit length. The end $x=0$ is fixed and the end $x=L$ is unloaded. Estimate the displacement at the free end using the finite element method with four equal length elements. The material has Young's modulus E.*

For this problem, $\Delta = L/4$ and the stiffness matrix is obtained by analogy with (A.25) as

$$\boldsymbol{K} = \frac{4EA}{L}\begin{bmatrix} 2 & -1 & 0 & 0 \\ -1 & 2 & -1 & 0 \\ 0 & -1 & 2 & -1 \\ 0 & 0 & -1 & 1 \end{bmatrix} .$$

The nodal forces are

$$\boldsymbol{F} = \frac{p_0 L}{4}\left\{1,\ 1,\ 1,\ \frac{1}{2}\right\}^T ,$$

from (A.26, A.27). Notice that for the end node, $j=4$, there is no additional term from $F_0$ in (A.27), since the end of the bar is unloaded.

Substituting into (A.28) and solving for $\boldsymbol{u}$, we obtain

$$\boldsymbol{u} = \frac{p_0 L^2}{16EA}\{3.5,\ 6,\ 7.5,\ 8\}^T$$

and hence the end displacement is

$$u_x^*(L) = u_4 = \frac{p_0 L^2}{2EA} .$$

### A.2.4 The Rayleigh-Ritz approach

The structural mechanics arguments of §A.2.1 has the advantage of being conceptually simple. The structure is essentially represented by a system of springs and the properties of these springs are determined using the elementary concepts of Hooke's law, as applied to the axial loading of a uniform bar. Notice however that we had to make a decision as to how to distribute the continuous axial loading $p(x)$ between the nodes and we developed the theory only for the simple case where the cross-sectional area $A$ and Young's modulus $E$ are indpendent of $x$.

For more general cases, the structural mechanics formulation can appear rather arbitrary and to avoid this, the stationary potential energy principle is used to develop the corresponding vectors and matrices. This is equivalent to the use of the Rayleigh-Ritz method in combination with a suitable discrete approximation function, such as the piecewise linear function of Figure A.2.

Following equation (A.1), we define the approximation

$$u^*(x) = \sum_{i=1}^{N} u_i v_i(x) , \tag{A.29}$$

where $v_i(x)$ is the piecewise linear shape function of equation (A.16). In the $i$-th element ($x_{i-1}<x<x_i$), this gives

$$u^*(x) = u_{i-1}\left(\frac{x_i - x}{\Delta}\right) + u_i\left(\frac{x - x_{i-1}}{\Delta}\right) \tag{A.30}$$

and hence the longitudinal strain is

$$e_{xx} = \frac{\partial u_x^*}{\partial x} = \left( \frac{u_i - u_{i-1}}{\Delta} \right) .$$

The strain energy density (2.32) can be written in terms of strain using Hooke's law, giving

$$U_0 = \frac{1}{2} E e_{xx}^2 = \frac{E(u_i - u_{i-1})^2}{2\Delta^2} .$$

The total strain energy in the $i$-th element is therefore given by equation (3.32) as

$$U_e = \iiint_e U_0 dV = \int_{x_{i-1}}^{x_i} \frac{E(x)(u_i - u_{i-1})^2}{2\Delta^2} A(x) dx$$

$$= \frac{(u_i - u_{i-1})^2}{2\Delta^2} \int_{x_{i-1}}^{x_i} E(x) A(x) dx .$$

If there are element nodal forces $F_{i-1}^e, F_i^e$ at $x_{i-1}, x_i$, the corresponding potential energy will be

$$\Omega_e = -F_{i-1}^e u_{i-1} - F_i^e u_i \tag{A.31}$$

and the total potential energy is

$$\Pi_e = U_e + \Omega_e = \frac{(u_i - u_{i-1})^2}{2\Delta^2} \int_{x_{i-1}}^{x_i} E(x) A(x) dx - F_{i-1}^e u_{i-1} - F_i^e u_i .$$

The principle of stationary potential energy then gives

$$\frac{\partial \Pi_e}{\partial u_i} = 0 \; ; \quad \frac{\partial \Pi_e}{\partial u_{i-1}} = 0$$

and hence

$$k_i \begin{bmatrix} 1 & -1 \\ -1 & 1 \end{bmatrix} \begin{Bmatrix} u_{i-1} \\ u_i \end{Bmatrix} = \begin{Bmatrix} F_{i-1}^e \\ F_i^e \end{Bmatrix} , \tag{A.32}$$

where

$$k_i = \frac{1}{\Delta^2} \int_{x_{i-1}}^{x_i} E(x) A(x) dx . \tag{A.33}$$

Notice that equation (A.32) is similar in form to (A.22), with an element stiffness matrix

$$\boldsymbol{K}_e = k_i \begin{bmatrix} 1 & -1 \\ -1 & 1 \end{bmatrix} . \tag{A.34}$$

In the special case where $E(x), A(x)$ are independent of $x$, this reduces to (A.23), as we should expect. Notice however that the Rayleigh-Ritz method automatically selects an appropriate approximation for the element stiffness in the case where $E$ and $A$ are not constant.

If instead of element nodal forces $F_{i-1}^e, F_i^e$ we have a possibly variable distributed load $p(x)$ per unit length, the potential energy of the load will be

$$\begin{aligned}\Omega_e &= -\int_{x_{i-1}}^{x_i} p(x)u^*(x)dx \\ &= -\frac{u_{i-1}}{\Delta}\int_{x_{i-1}}^{x_i} p(x)(x_i - x)dx - \frac{u_i}{\Delta}\int_{x_{i-1}}^{x_i} p(x)(x - x_{x-1})dx\,,\end{aligned}$$

from equation (A.30). Comparing this result with equation (A.31), we see that the distributed load is equivalent to a pair of element nodal forces equal to

$$F_{i-1}^e = \frac{1}{\Delta}\int_{x_{i-1}}^{x_i} p(x)(x_i - x)dx \tag{A.35}$$

$$F_i^e = \frac{1}{\Delta}\int_{x_{i-1}}^{x_i} p(x)(x - x_{x-1})dx\,. \tag{A.36}$$

In the special case where $p(x) = p_0$ is independent of $x$, equations (A.35, A.36) give $F_{i-1}^e = F_i^e = p_0\Delta/2$, as assumed in §A.2.3. For the more general case, they define an algorithm for sharing out the distributed load $p(x)$ between adjacent nodes according to the lever rule — e.g., a force $F$ at the point $x$ in the range $x_{j-1} < x < x_j$ contributes $F(x - x_{i-1})/\Delta$ to $F_i^e$ and $F(x_i - x)/\Delta$ to $F_{i-1}^e$.

The global stiffness matrix can now be assembled by superposing the element stiffness matrices (A.34), following the pattern of (A.23). The resulting matrix $\boldsymbol{K}$ is defined by

$$\left.\begin{aligned} K_{ji} &= \frac{(EA)_j^- + (EA)_j^+}{\Delta} && ; \quad i = j \neq N \\ &= \frac{(EA)_j^-}{\Delta} && ; \quad i = j = N \\ &= -\frac{(EA)_j^-}{\Delta} && ; \quad i = j-1 \\ &= -\frac{(EA)_j^+}{\Delta} && ; \quad i = j+1 \\ &= 0 && ; \quad |i - j| > 1\,, \end{aligned}\right\} \tag{A.37}$$

where

$$(EA)_j^- = \frac{1}{\Delta}\int_{x_{j-1}}^{x_j} E(x)A(x)dx \;\; ; \;\; (EA)_j^+ = \frac{1}{\Delta}\int_{x_j}^{x_{j+1}} E(x)A(x)dx \tag{A.38}$$

are the mean values of $E(x)A(x)$ in the element to the left and right respectively of node $j$.

Also, each nodal force for the complete system is the sum of the corresponding elemental nodal forces from the two adjacent elements, given by equations (A.35, A.36). We therefore have

$$F_j = \frac{1}{\Delta}\int_{x_{j-1}}^{x_j} p(x)(x - x_{j-1})dx + \frac{1}{\Delta}\int_{x_j}^{x_{j+1}} p(x)(x_{j+1} - x)dx\,, \tag{A.39}$$

for $j \neq N$ and

$$F_N = \frac{1}{\Delta}\int_{x_{N-1}}^{x_N} p(x)(x - x_{N-1})dx + F_0\,. \tag{A.40}$$

**Example A.3**

*Figure A.7 shows a vertical bar of length L supported at the top, $x=0$, and loaded only by its own weight. The cross-sectional area of the bar varies according to the equation*

$$A(x) = A_0\left(1-\frac{x}{2L}\right),$$

*where x is the vertical distance from the support. Estimate the displacement at the free end using the finite element method with four equal length elements. The material has density $\rho$ and Young's modulus E.*

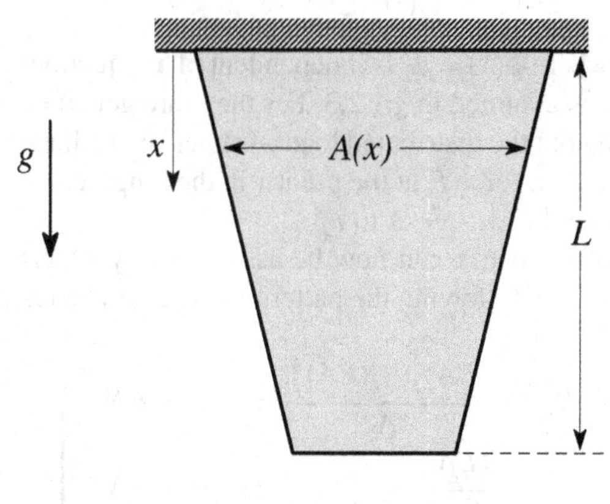


*Figure A.7*

For this problem, the element length is $\Delta = L/4$. We first evaluate the four integrals

$$\frac{1}{\Delta}\int_0^{x_1} E(x)A(x)dx = \frac{4EA_0}{L}\int_0^{L/4}\left(1-\frac{x}{2L}\right)dx = \frac{15EA_0}{16}$$

$$\frac{1}{\Delta}\int_{x_1}^{x_2} E(x)A(x)dx = \frac{4EA_0}{L}\int_{L/4}^{L/2}\left(1-\frac{x}{2L}\right)dx = \frac{13EA_0}{16}$$

$$\frac{1}{\Delta}\int_{x_2}^{x_3} E(x)A(x)dx = \frac{4EA_0}{L}\int_{L/2}^{3L/4}\left(1-\frac{x}{2L}\right)dx = \frac{11EA_0}{16}$$

$$\frac{1}{\Delta}\int_{x_3}^{x_4} E(x)A(x)dx = \frac{4EA_0}{L}\int_{3L/4}^{L}\left(1-\frac{x}{2L}\right)dx = \frac{9EA_0}{16}.$$

Substituting these results into equations (A.38, A.37), we obtain the global stiffness matrix as

$$\boldsymbol{K} = \frac{EA_0}{4L}\begin{bmatrix} 28 & -13 & 0 & 0 \\ -13 & 24 & -11 & 0 \\ 0 & -11 & 20 & -9 \\ 0 & 0 & -9 & 9 \end{bmatrix}.$$

The nodal forces are given by equations (A.39, A.40) as

$$F_1 = \frac{1}{\Delta}\int_0^{L/4} p(x)xdx + \frac{1}{\Delta}\int_{L/4}^{L/2} p(x)\left(\frac{L}{2}-x\right)dx$$

$$F_2 = \frac{1}{\Delta}\int_{L/4}^{L/2} p(x)\left(x-\frac{L}{4}\right)dx + \frac{1}{\Delta}\int_{L/2}^{3L/4} p(x)\left(\frac{3L}{4}-x\right)dx$$

$$F_3 = \frac{1}{\Delta}\int_{L/2}^{3L/4} p(x)\left(x-\frac{L}{2}\right)dx + \frac{1}{\Delta}\int_{3L/4}^{L} p(x)\left(L-x\right)dx$$

$$F_4 = \frac{1}{\Delta}\int_{3L/4}^{L} p(x)\left(x-\frac{3L}{4}\right)dx\,,$$

where the distributed load $p(x)$ is due to the weight of the body and is

$$p(x) = \rho g A(x) = \rho g A_0\left(1-\frac{x}{2L}\right)$$

per unit length. Substituting this result into the above expressions and evaluating the integrals, we obtain

$$\boldsymbol{F} = \rho g A_0 L\left\{\frac{7}{32},\frac{3}{16},\frac{5}{32},\frac{13}{192}\right\}^T .$$

The equation system

$$\boldsymbol{K}\boldsymbol{u} = \boldsymbol{F}$$

can then be solved to give

$$\boldsymbol{u} = \frac{\rho g L^2}{E}\{0.168,\, 0.295,\, 0.376,\, 0.406\}^T$$

and in particular, the end deflection is

$$u(L) = \frac{0.406\rho g L^2}{E} .$$

This problem can in fact be solved exactly,[1] the resulting end deflection being

$$u(L) = \frac{[3-2\ln(2)]\rho g L^2}{4E} \approx \frac{0.403\rho g L^2}{E} .$$

Thus, the four element approximation gives better than 1% accuracy for the end deflection.

[1] To obtain this solution, use equilibrium arguments to determine the axial force and hence the stress as a function of $x$. Find the strain using Hooke's law and substitute the resulting expression into equation (1.3). Integration with respect to $x$ and susbtitution of the boundary condition $u(0)=0$ then yields the expression for $u(x)$.

### A.2.5 Direct evaluation of the matrix equation

In the preceding examples, we first determined the properties of the individual elements and then assembled the system of simultaneous equations, using the pattern of equation (A.24). With the Rayleigh-Ritz formulation, it is not necessary to go through this preliminary step. The complete system of equations can be generated by applying the Rayleigh-Ritz method directly to the approximation of equation (A.29).

The longitudinal strain is given for all $x$ by

$$e_{xx} = \frac{\partial u_x^*}{\partial x} = \sum_{i=1}^{N} u_i v_i'(x) , \tag{A.41}$$

from equation (A.29) and hence the strain energy density is

$$U_0 = \frac{1}{2}E(x)e_{xx}^2 = \frac{1}{2}E(x)\left[\sum_{i=1}^{N} u_i v_i'(x)\right]^2 = \frac{1}{2}E(x)\sum_{i=1}^{N}\sum_{j=1}^{N} u_i u_j v_i'(x)v_j'(x) . \tag{A.42}$$

Notice that to expand the square of the summation, we have to use a different dummy index $i, j$ in each sum. Integrating over the volume of the bar, we then obtain the total strain energy

$$U = \frac{1}{2}\sum_{i=1}^{N}\sum_{j=1}^{N} u_i u_j \int_0^L E(x)v_i'(x)v_j'(x)A(x)dx . \tag{A.43}$$

The potential energy of the loads in Figure A.5 is

$$\Omega = -\int_0^L p(x)u^*(x)dx - F_0 u^*(L) = -\sum_{j=1}^{N} u_j \int_0^L p(x)v_j(x)dx - F_0 u_N \tag{A.44}$$

and hence the total potential energy is

$$\begin{aligned}\Pi = U + \Omega = \frac{1}{2}\sum_{i=1}^{N}\sum_{j=1}^{N} u_i u_j \int_0^L E(x)A(x)v_i'(x)v_j'(x)dx \\ -\sum_{j=1}^{N} u_j \int_0^L p(x)v_j(x)dx - F_0 u_N .\end{aligned} \tag{A.45}$$

The principle of stationary potential energy requires that

$$\frac{\partial \Pi}{\partial u_j} = 0$$

and hence

$$\sum_{i=1}^{N} u_i \int_0^L E(x)A(x)v_i'(x)v_j'(x)dx - \int_0^L p(x)v_j(x)dx - F_0 \delta_{jN} = 0 . \tag{A.46}$$

This equation can be expressed in the matrix form $\boldsymbol{Ku}=\boldsymbol{F}$, where

$$K_{ji} = \int_0^L E(x)A(x)v_i'(x)v_j'(x)dx \tag{A.47}$$

$$F_j = \int_0^L p(x)v_j(x)dx + F_0\delta_{jN} \; . \tag{A.48}$$

Differentiating (A.16), we have

$$\begin{aligned} v_i'(x) &= 0 \quad ; \quad x < x_{i-1} \text{ and } x > x_{i+1} \\ &= \frac{1}{\Delta} \quad ; \quad x_{i-1} < x < x_i \\ &= -\frac{1}{\Delta} \quad ; \quad x_i < x < x_{i+1} \; . \end{aligned} \tag{A.49}$$

Substitution in (A.47) and evaluation of the integrals then leads directly to equations (A.37, A.38).

Also, substitution of (A.16) into (A.48) and evaluation of the resulting integral gives

$$F_j = \frac{1}{\Delta}\int_{x_{j-1}}^{x_j} p(x)(x-x_{j-1})dx + \frac{1}{\Delta}\int_{x_j}^{x_{j+1}} p(x)(x_{j+1}-x)dx \; , \tag{A.50}$$

for $j \neq N$ and

$$F_N = \frac{1}{\Delta}\int_{x_{N-1}}^{x_N} p(x)(x-x_{N-1})dx + F_0 \; , \tag{A.51}$$

agreeing with (A.39, A.40).

This method of developing the solution is in some respects conceptually more straightforward than the assembly from elemental values described in §A.2.2. However, there are practical reasons for preferring the assembly method in more complex problems. For example, in a large problem different types of elements and hence shape functions might be used in different parts of the structure. Efficient solvers for the resulting equations also depend on the matrix being assembled piece by piece, as discussed in §A.6.1 below.

## A.3 Solution of differential equations

In §§A.2.4, A.2.5, we developed the finite element method by applying the principle of stationary potential energy to a piecewise linear approximation to the displacement. This is still the most widely used method for problems in mechanics of materials, but an alternative method is to apply the 'least squares fit' arguments of §A.1 directly to the governing differential equations of equilibrium. The principle of stationary potential energy is mathematically equivalent to a variational statement of the equations of equilibrium, so the two methods lead to the same final equations, but one advantage of the least squares method is that it can be used to obtain an approximate solution of any problem governed by a differential equation.

To apply the methods of §A.1.3 to differential equations, two minor modifications are required:-

1. We need to make allowance for appropriate boundary conditions.
2. It is usually necessary to perform one or more integrations by parts to retain appropriate continuity in the approximating function.

We shall illustrate this procedure in the context of the axial loading problem of Figure A.5 above.

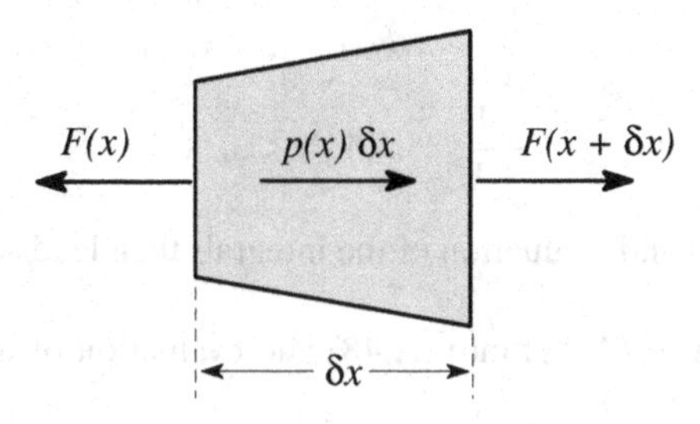


*Figure A.8: Equilibrium of an infinitesimal bar segment*

We first need to develop the differential equation of equilibrium for the bar. Figure A.8 shows the forces acting on an infinitesimal segment $\delta x$ of the bar. The axial force $F$ is assumed to be a function of $x$. Equilibrium of the beam segment then requires that $F(x+\delta x)+p(x)\delta x-F(x)=0$ and hence

$$\frac{F(x+\delta x)-F(x)}{\delta x}+p(x)=0\,.$$

Taking the limit as $\delta x\rightarrow 0$, this gives the differential equation

$$\frac{dF}{dx}+p(x)=0\,. \tag{A.52}$$

The corresponding stress is

$$\sigma_{xx}=\frac{F(x)}{A(x)}=E(x)e_{xx}=E(x)\frac{du_x}{dx} \tag{A.53}$$

and substitution of (A.53) into (A.52) gives

$$\frac{d}{dx}\left[E(x)A(x)\frac{du_x}{dx}\right]+p(x)=0\,. \tag{A.54}$$

This is a second-order ordinary differential equation requiring two boundary conditions, which in this case are $u_x(0)=0$ and $F(L)=F_0$. The latter can be expressed in terms of $u_x$ using (A.53), with the result

$$u_x(0)=0\quad ;\quad E(L)A(L)u_x'(L)=F_0\,. \tag{A.55}$$

Equations (A.54, A.55) define the continuum formulation of the problem.

A piecewise linear approximation satisfying the first of (A.55) in a collocation sense is

$$u_x^*(x) = \sum_{i=1}^{N} u_i v_i(x) \; , \tag{A.56}$$

where the shape functions $v_i(x)$ are given by (A.16).

A set of conditions for the unknowns $u_i$ is then obtained by enforcing

$$\int_0^L \left\{ \frac{d}{dx} \left[ E(x)A(x)u_x^{*\prime}(x) \right] + p(x) \right\} v_j(x)dx = 0 \;\; ; \;\; j = (1,N) \; , \tag{A.57}$$

where as in §A.1 we use the $v_i(x)$ for both shape and weight functions.

The obvious next step is to substitute (A.56) into (A.57) to obtain $N$ simultaneous equations for the $u_i$, but a difficulty is encountered in that $u^{*\prime\prime}(x)$ is ill-defined for the piecewise linear function (A.16). To overcome this difficulty, we apply partial integration to the first term in (A.57) obtaining

$$E(x)A(x)u_x^{*\prime}(x)v_j(x)\big|_0^L - \int_0^L E(x)A(x)u_x^{*\prime}(x)v_j'(x)dx$$
$$+ \int_0^L p(x)v_j(x)dx = 0 \;\; ; \;\; j = (1,N) \; .$$

Substituting for $u_x^*(x)$ from (A.56) and for $u_x^{*\prime}(L)$ from the second of the boundary conditions (A.55), we then have

$$\boldsymbol{Ku} = \boldsymbol{F} \; , \tag{A.58}$$

where

$$K_{ji} = \int_0^L E(x)A(x)v_i'(x)v_j'(x)dx \tag{A.59}$$

$$F_j = \int_0^L p(x)v_j(x)dx + F_0 \delta jN \; , \tag{A.60}$$

agreeing with the results (A.47, A.48) obtained using the Rayleigh-Ritz method in §A.2.5.

## A.4 Finite element solutions for the bending of beams

Piecewise linear approximations cannot be used for the displacement of beams in bending, since they involve discontinuities in slope at the nodes. To avoid this difficulty, we use a discretization in which the nodal displacements $u_i$ and slopes $\theta_i$ are treated as independent variables with separate shape functions of higher order polynomial form.

For the nodal displacements $u_i$, we require shape functions $v_i^u(z)$ satisfying the conditions

$$\begin{aligned} v_i^u(z_j) &= 1 \ ; \ j=i \\ &= 0 \ ; \ j \neq i \\ \theta_i^u(z_j) \equiv (v_i^u)'(z_j) &= 0 \ ; \ \text{all } j \, . \end{aligned} \tag{A.61}$$

This corresponds to the situation where all the nodes except $j=i$ are restrained against displacement and rotation, whilst node $j=i$ is subjected to unit displacement without rotation, as shown in Figure A.9 *(a)*.

*(a)*

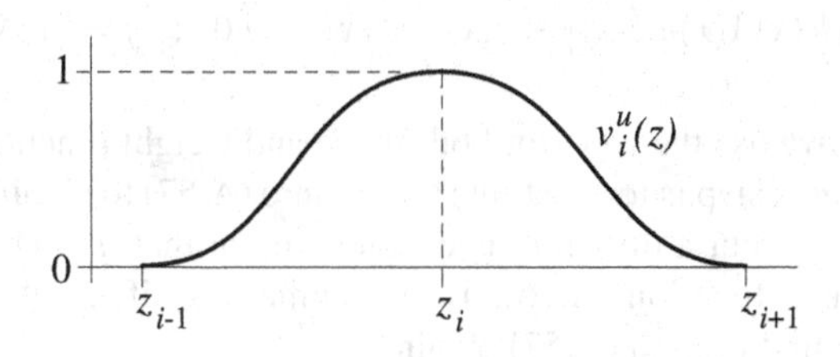

*(b)*

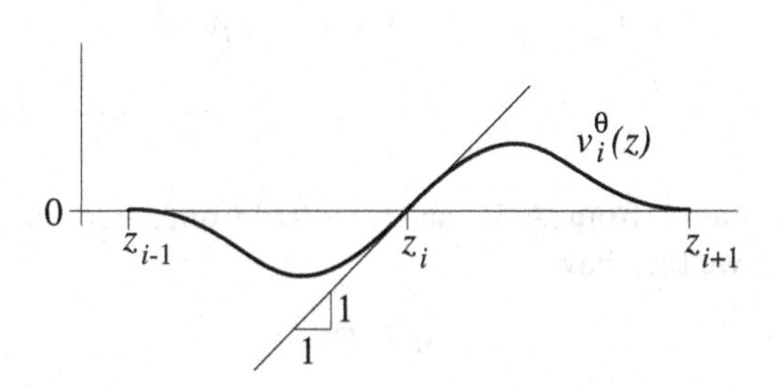


*Figure A.9: Shape functions for (a) nodal displacements $u_i$ and (b) nodal rotations $\theta_i$*

For the nodal slopes $\theta_i$, we require shape functions $v_i^\theta(z)$ satisfying

$$\begin{aligned} \theta_i^\theta(z_j) \equiv (v_i^\theta)'(z_j) &= 1 \ ; \ j=i \\ &= 0 \ ; \ j \neq i \\ v_i^\theta(z_j) &= 0 \ ; \ \text{all } j \, , \end{aligned} \tag{A.62}$$

which correspond to the deformation of Figure A.9 *(b)*, where node $i$ is rotated without displacement and all other nodes are restrained against both displacement and rotation.

The lowest order piecewise polynomial functions satisfying these conditions are

$$\begin{aligned} v_i^u(z) &= \frac{3(z-z_{i-1})^2}{\Delta^2} - \frac{2(z-z_{i-1})^3}{\Delta^3} \ ; \ z_{i-1} < z < z_i \\ &= \frac{3(z_{i+1}-z)^2}{\Delta^2} - \frac{2(z_{i+1}-z)^3}{\Delta^3} \ ; \ z_i < z < z_{i+1} \\ &= 0 \ ; \ z < z_{i-1} \text{ and } z > z_{i+1} \end{aligned} \tag{A.63}$$

$$
\begin{aligned}
v_i^\theta(z) &= -\frac{(z-z_{i-1})^2}{\Delta} + \frac{(z-z_{i-1})^3}{\Delta^2} \quad ; \quad z_{i-1} < z < z_i \\
&= \frac{(z_{i+1}-z)^2}{\Delta} - \frac{(z_{i+1}-z)^3}{\Delta^2} \quad ; \quad z_i < z < z_{i+1} \\
&= 0 \;\; ; \;\; z < z_{i-1} \text{ and } z > z_{i+1} \, .
\end{aligned} \tag{A.64}
$$

The reader can easily verify by substitution that these expressions[2] satisfy the conditions (A.61, A.62).

If there are $N$ elements and $N+1$ nodes, the discrete approximation to the beam displacement can then be written

$$
u^*(z) = \sum_{i=1}^{N+1} u_i v_i^u(z) + \sum_{i=1}^{N+1} \theta_i v_i^\theta(z) \, . \tag{A.65}
$$

Consider the $i$-th element of length $\Delta$ between nodes $i$ and $i+1$, in which the displacement is approximated as

$$
u_e^*(z) = u_i v_i^u(z) + \theta_i v_i^\theta(z) + u_{i+1} v_{i+1}^u(z) + \theta_{i+1} v_{i+1}^\theta(z) \, . \tag{A.66}
$$

The element strain energy is

$$
U_e = \frac{1}{2} \int_{z_i}^{z_{i+1}} EI \left[ \frac{d^2 u_e^*(z)}{dz^2} \right]^2 dz \tag{A.67}
$$

and if the flexural rigidity $EI$ is independent of $z$, we can substitute for the shape functions $v_i^u(z), v_i^\theta(z), v_{i+1}^u(z), v_{i+1}^\theta(z)$ from (A.63, A.64) and perform the integrations, obtaining

$$
\begin{aligned}
U_e = \frac{EI}{2\Delta^3} & (12u_i^2 + 4\Delta^2\theta_i^2 + 12u_{i+1}^2 + 4\Delta^2\theta_{i+1}^2 + 12\Delta u_i\theta_i - 24u_iu_{i+1} \\
& + 12\Delta u_i\theta_{i+1} - 12\Delta\theta_i u_{i+1} + 4\Delta^2\theta_i\theta_{i+1} - 12\Delta u_{i+1}\theta_{i+1}) \, .
\end{aligned} \tag{A.68}
$$

The four columns of the $4\times4$ element stiffness matrix are now obtained by differentiating $U_e$ with respect to the four nodal variables $u_i, \theta_i, u_{i+1}, \theta_{i+1}$. We obtain

$$
\frac{\partial U_e}{\partial u_i} = \frac{EI}{\Delta^3}(12u_i + 6\Delta\theta_i - 12u_{i+1} + 6\Delta\theta_{i+1}) \tag{A.69}
$$

$$
\frac{\partial U_e}{\partial \theta_i} = \frac{EI}{\Delta^3}(6\Delta u_i + 4\Delta^2\theta_i - 6\Delta u_{i+1} + 2\Delta^2\theta_{i+1}) \tag{A.70}
$$

$$
\frac{\partial U_e}{\partial u_{i+1}} = \frac{EI}{\Delta^3}(-12u_i - 6\Delta\theta_i + 12u_{i+1} - 6\Delta\theta_{i+1}) \tag{A.71}
$$

$$
\frac{\partial U_e}{\partial \theta_{i+1}} = \frac{EI}{\Delta^3}(6\Delta u_i + 2\Delta^2\theta_i - 6\Delta u_{i+1} + 4\Delta^2\theta_{i+1}) \tag{A.72}
$$

---

[2] More generally, the shape functions of equations (A.63, A.64) define a set of cubic interpolation functions that can be used in the sense of §A.1 to define a discrete approximation to any function that preserves continuity in both value and derivative at the nodes. This is known as a *cubic spline* approximation.

and hence the element stiffness matrix is

$$\boldsymbol{K}_e = \frac{EI}{\Delta^3}\begin{bmatrix} 12 & 6\Delta & -12 & 6\Delta \\ 6\Delta & 4\Delta^2 & -6\Delta & 2\Delta^2 \\ -12 & -6\Delta & 12 & -6\Delta \\ 6\Delta & 2\Delta^2 & -6\Delta & 4\Delta^2 \end{bmatrix} . \tag{A.73}$$

The global stiffness matrix can be assembled by superposition, as in §A.2.2. For example, for a beam with 4 elements and 5 nodes, $i=(1,5)$, we obtain

$$\boldsymbol{K} = \frac{EI}{\Delta^3}\begin{bmatrix} 12 & 6\Delta & -12 & 6\Delta & 0 & 0 & 0 & 0 & 0 & 0 \\ 6\Delta & 4\Delta^2 & -6\Delta & 2\Delta^2 & 0 & 0 & 0 & 0 & 0 & 0 \\ -12 & -6\Delta & 24 & 0 & -12 & 6\Delta & 0 & 0 & 0 & 0 \\ 6\Delta & 2\Delta^2 & 0 & 8\Delta^2 & -6\Delta & 2\Delta^2 & 0 & 0 & 0 & 0 \\ 0 & 0 & -12 & -6\Delta & 24 & 0 & -12 & 6\Delta & 0 & 0 \\ 0 & 0 & 6\Delta & 2\Delta^2 & 0 & 8\Delta^2 & -6\Delta & 2\Delta^2 & 0 & 0 \\ 0 & 0 & 0 & 0 & -12 & -6\Delta & 24 & 0 & -12 & 6\Delta \\ 0 & 0 & 0 & 0 & 6\Delta & 2\Delta^2 & 0 & 8\Delta^2 & -6\Delta & 2\Delta^2 \\ 0 & 0 & 0 & 0 & 0 & 0 & -12 & -6\Delta & 12 & -6\Delta \\ 0 & 0 & 0 & 0 & 0 & 0 & 6\Delta & 2\Delta^2 & -6\Delta & 4\Delta^2 \end{bmatrix} \tag{A.74}$$

In this matrix, the internal rectangles indicate the way in which the $4\times4$ element matrices are overlaid along the diagonal to construct the global stiffness matrix.

The stiffness matrix of equation (A.74) contains terms for all the nodal displacements that are possible with the given number of nodes, but in practice some of these degrees of freedom must be constrained if the beam is to be kinematically supported. If the beam is simply supported at node $i$, the corresponding nodal displacement $u_i=0$ and ceases to be an unknown, whilst the corresponding nodal force $F_i$ becomes an unknown reaction. We therefore remove $u_i, F_i$ from the equation system by deleting the row and column corresponding to $u_i$ from the stiffness matrix. In the same way, the row and column corresponding to $\theta_i$ will be deleted if node $i$ is prevented from rotating, as at a built in support.

**Example A.4**

*A beam of length $2L$ is built in at $z=0$, simply supported at $z=L$ and loaded by a force $F$ at the end $z=2L$, as shown in Figure A.10. Using a discretization with two equal elements each of length $L$, estimate the deflection under the force.*

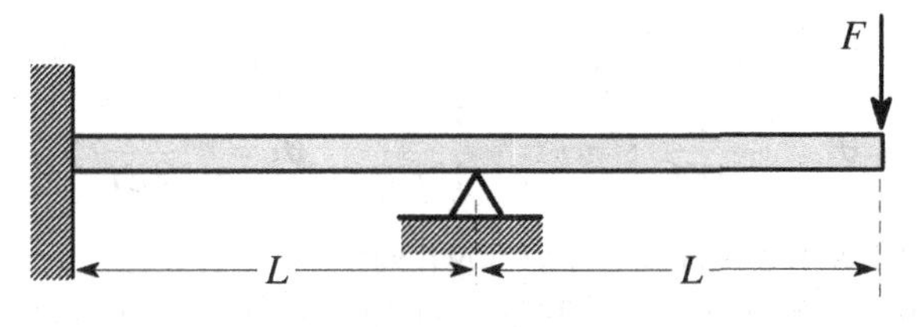


*Figure A.10*

Figure A.11 shows the discretization, with 2 elements of length $L$ and three nodes, for which the full stiffness matrix is

$$\boldsymbol{K} = \frac{EI}{L^3}\begin{bmatrix} 12 & 6L & -12 & 6L & 0 & 0 \\ 6L & 4L^2 & -6L & 2L^2 & 0 & 0 \\ -12 & -6L & 24 & 0 & -12 & 6L \\ 6L & 2L^2 & 0 & 8L^2 & -6L & 2L^2 \\ 0 & 0 & -12 & -6L & 12 & -6L \\ 0 & 0 & 6L & 2L^2 & -6L & 4L^2 \end{bmatrix}, \tag{A.75}$$

by analogy with (A.74).

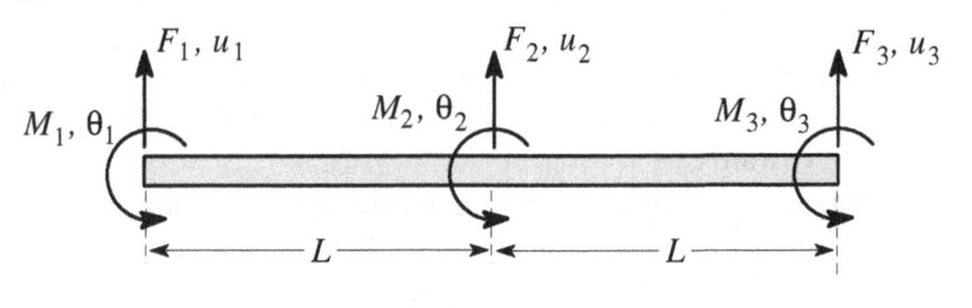


*Figure A.11*

However, the boundary conditions at the supports require that $u_1 = u_2 = \theta_1 = 0$ and we therefore delete the first three rows and columns from $\boldsymbol{K}$, leaving

$$\boldsymbol{K} = \frac{EI}{L^3}\begin{bmatrix} 8L^2 & -6L & 2L^2 \\ -6L & 12 & -6L \\ 2L^2 & -6L & 4L^2 \end{bmatrix}. \tag{A.76}$$

The remaining nodal forces and moments (i.e. the external forces on the beam) are $M_2 = 0, F_3 = -F, M_3 = 0$ and hence

$$\frac{EI}{L^3}\begin{bmatrix} 8L^2 & -6L & 2L^2 \\ -6L & 12 & -6L \\ 2L^2 & -6L & 4L^2 \end{bmatrix}\begin{Bmatrix} \theta_2 \\ u_3 \\ \theta_3 \end{Bmatrix} = \begin{Bmatrix} 0 \\ -F \\ 0 \end{Bmatrix}, \tag{A.77}$$

with solution

$$\theta_2 = -\frac{FL^2}{4EI} \;\; ; \;\; u_3 = -\frac{7FL^3}{12EI} \;\; ; \;\; \theta_3 = -\frac{3FL^2}{4EI} \, . \tag{A.78}$$

Thus, the end deflection is $7FL^3/12EI$ downwards.

The deleted columns of the stiffness matrix could be reinstated to determine the unknown reactions. For example, the reaction at the simple support is

$$F_2 = \frac{EI}{L^3}([0]\theta_2 - 12u_3 + 6L\theta_3) = \frac{5F}{2} \, ,$$

after substituting for $\theta_2, u_3, \theta_3$.

### A.4.1 Nodal forces and moments

In Example A.4, the only loads were applied at the nodes, so the load vector was immediately known. If there is a distributed load, $w(z)$ per unit length acting on the beam, it must be shared out between adjacent nodes and this can be done using the principle of stationary potential energy, as in §A.2.4.

The potential energy of the distributed load is

$$\Omega = \int_0^L u^*(z)w(z)dz \, , \tag{A.79}$$

where $L$ is the length of the beam, $u^*(z)$ is the discrete approximation to the displacement in the entire beam defined by equation (A.65) and the distributed load is assumed to act downwards.

Differentiating with respect to the nodal displacements, we have

$$\frac{\partial \Omega}{\partial u_i} = \int_0^L v_i^u(z)w(z)dz \tag{A.80}$$

$$\frac{\partial \Omega}{\partial \theta_i} = \int_0^L v_i^\theta(z)w(z)dz \tag{A.81}$$

and we conclude that the distributed load contributes

$$F_i = -\int_0^L v_i^u(z)w(z)dz \tag{A.82}$$

$$M_i = -\int_0^L v_i^\theta(z)w(z)dz \tag{A.83}$$

to the nodal forces and moments at node $i$.

If there is a uniformly distributed load $w_0$ per unit length in the element $z_i < z < z_{i+1}$ and no load elsewhere, substitution into equations (A.82, A.83) shows that the only non-zero nodal forces and moments occur at the ends of the loaded element and are

$$F_i = F_{i+1} = -\frac{w_0\Delta}{2} \;\; ; \;\; -M_i = M_{i+1} = \frac{w_0\Delta^2}{12} \, . \tag{A.84}$$

As we might expect, the total load $w_0\Delta$ is shared equally between the adjacent nodes, but less obviously, equal and opposite nodal moments are generated as well. The distributed loading and the equivalent nodal loading for this case are illustrated in Figure A.12 *(a,b)* respectively.

*(a)*

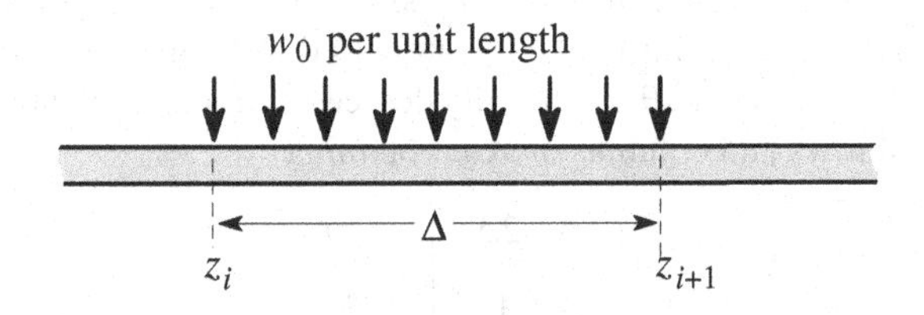


*(b)*

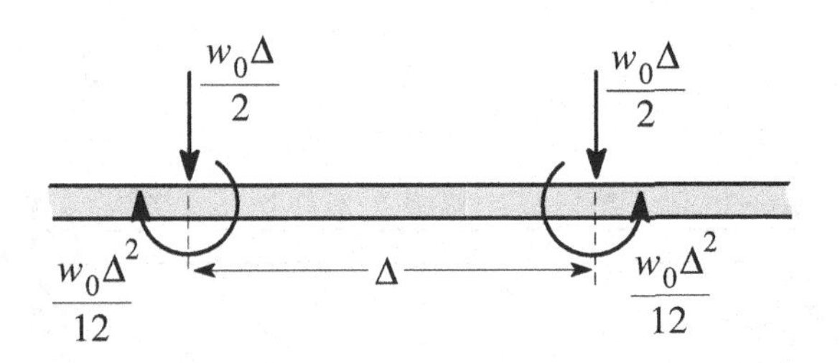


*Figure A.12: Uniform loading of a single element: (a) continuous loading; (b) equivalent nodal forces and moments*

If the whole beam is subjected to a uniformly distributed load, the equivalent nodal forces and moments of equations (A.84) will be superposed to yield the discrete loading of Figure A.13. Notice that each node receives half of the load on each of the adjacent elements, whereas the moments from adjacent elements are equal and opposite and hence cancel each other out except at the two terminal nodes.

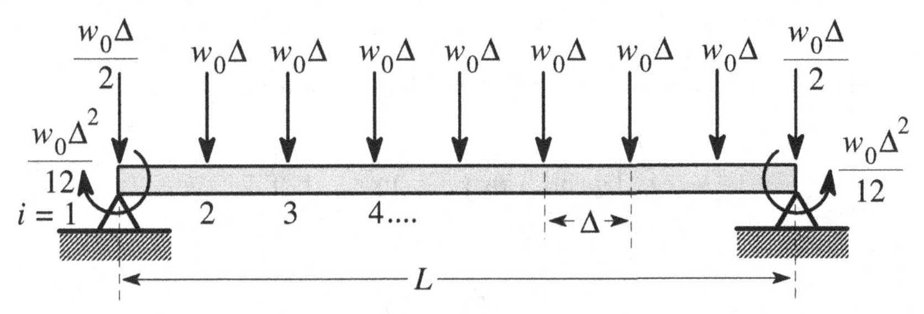


*Figure A.13: Nodal forces and moments equivalent to uniform loading of the whole beam*

The existence of these terminal moments is perhaps best explained by reference to the case where the beam is modelled by a single element of length, $L$, with two terminal nodes, as in the following example.

**Example A.5**

*A beam of length L is simply supported at its ends and subjected to a uniformly distributed load $w_0$ per unit length. Find an approximate solution for the deflection, representing the beam as a single element of length L.*

Since there is only one element, the stiffness matrix is the same as $\boldsymbol{K}_e$ of equation (A.73). Furthermore, the beam is simply supported at both nodes, so $u_1 = u_2 = 0$, leaving the nodal rotations $\theta_1, \theta_2$ as the only degrees of freedom. We therefore delete the first and third rows and columns in $\boldsymbol{K}_e$, obtaining

$$\boldsymbol{K} = \frac{EI}{\Delta^3}\begin{bmatrix} 4\Delta^2 & 2\Delta^2 \\ 2\Delta^2 & 4\Delta^2 \end{bmatrix} = \frac{EI}{L}\begin{bmatrix} 4 & 2 \\ 2 & 4 \end{bmatrix}, \tag{A.85}$$

since here the element length $\Delta = L$

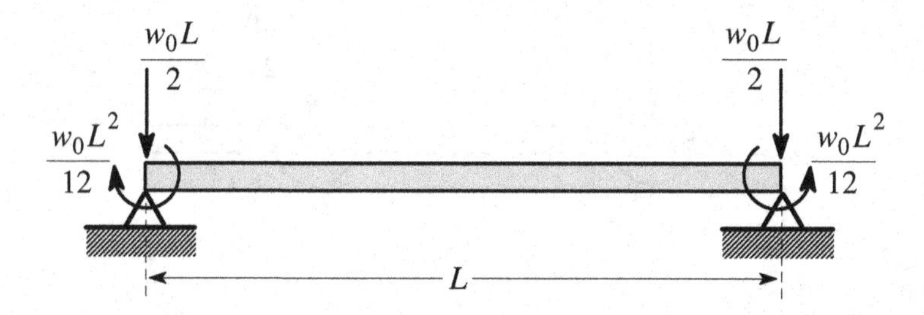


*Figure A.14: Nodal forces and moments for the uniformly loaded beam modelled by a single element of length L*

The nodal forces and moments corresponding to the distributed load are given by equation (A.84) and are illustrated in Figure A.14. The nodal forces act directly above the supports and hence produce no beam deflection, but the nodal moments

$$-M_1 = M_2 = \frac{w_0 L^2}{12},$$

do produce deformation, which can be found from the equations

$$\frac{EI}{L}\begin{bmatrix} 4 & 2 \\ 2 & 4 \end{bmatrix}\begin{Bmatrix} \theta_1 \\ \theta_2 \end{Bmatrix} = \frac{w_0 L^2}{12}\begin{Bmatrix} -1 \\ 1 \end{Bmatrix}, \tag{A.86}$$

with solution

$$-\theta_1 = \theta_2 = \frac{w_0 L^3}{24EI}.$$

With only two rotational degrees of freedom, equation (A.65) reduces to

$$u^*(z) = \theta_1 v_1^\theta(z) + \theta_2 v_2^\theta(z) = \frac{w_0 L^3}{24EI}\left[-\frac{(L-z)^2}{L} + \frac{(L-z)^3}{L^2} - \frac{z^2}{L} + \frac{z^3}{L^2}\right]$$

$$= -\frac{w_0 L^2 z(L-z)}{24EI}. \tag{A.87}$$

This is precisely the same result that would be obtained to this problem by applying the Rayleigh-Ritz method of §3.6, using a third degree polynomial approximation to the deformed shape (see Problem A.19).

## A.5 Two and three-dimensional problems

The scope of this book has been restricted to systems that can be analyzed using ordinary differential equations and hence it is inappropriate to give any extensive discussion of the application of the finite element method in two and three dimensions, which in the continuum formulation leads to partial differential equations. However, the broad use of the method in engineering is largely due to the fact that it can be extended to such problems with scarcely any modification or added complexity.

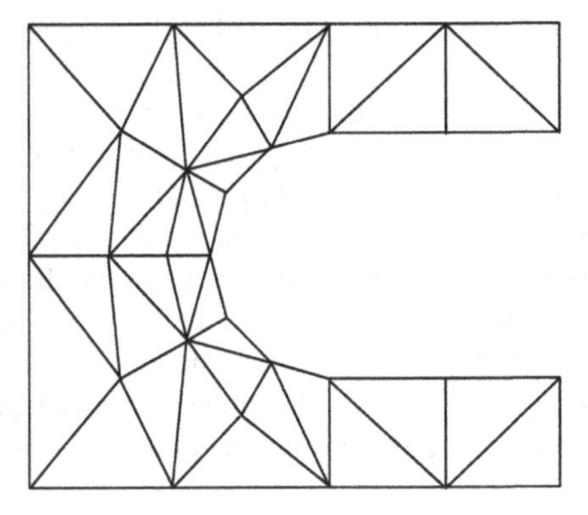

*Figure A.15: Discretization of a two-dimensional component using triangular elements*

Conceptually, the simplest approach is to use a piecewise linear approximation function, which leads naturally to a discretization with triangular elements in two dimensions or tetrahedral elements in three dimensions. Figure A.15 shows a two-dimensional component and a possible discretization using triangular elements. Notice how larger numbers of smaller elements are placed in the re-entrant corner. This gives a better approximation to the curved boundary, but more importantly, it provides additional degrees of freedom to approximate the rapidly varying stress and displacement fields due to the stress concentration.

Each node in two dimensions will have two degrees of freedom $u_x^i, u_y^i$, which are conveniently combined as the vector

$$\boldsymbol{u}_i = \begin{Bmatrix} u_x^i \\ u_y^i \end{Bmatrix} .$$

Inside each triangular element, the displacement field is described by the linear function

$$\boldsymbol{u} = \boldsymbol{A}x + \boldsymbol{B}y + \boldsymbol{C} \tag{A.88}$$

and the conditions

$$\boldsymbol{u}_i = \boldsymbol{A}x_i + \boldsymbol{B}y_i + \boldsymbol{C} \tag{A.89}$$

applied at the three corners of the triangle are necessary and sufficient to determine the three vector constants $\boldsymbol{A},\boldsymbol{B},\boldsymbol{C}$ and hence define the complete piecewise linear field in terms of the nodal displacement vectors $\boldsymbol{u}_i$, which therefore constitute the degrees of freedom in the formulation.

The two components of the vector shape functions $\boldsymbol{v}_i(x,y)$ corresponding to this discretization are those piecewise linear functions that satisfy equations (A.88, A.89) with

$$\boldsymbol{u}_i = \begin{Bmatrix} 1 \\ 0 \end{Bmatrix} \text{ and } \begin{Bmatrix} 0 \\ 1 \end{Bmatrix},$$

respectively.

Notice the similarities between this formulation and the beam bending formulation of §A.4, where the discretization also involves two degrees of freedom $\{u_i, \theta_i\}$ at each node.

The algebraic equations defining the $\boldsymbol{u}_i$ can be obtained as before by developing expressions for the strain energy $U$ and the potential energy of the applied forces $\Omega$, and then applying the principle of stationary potential energy. Alternatively we can multiply the governing partial differential equation by the piecewise linear weight functions $\boldsymbol{v}_j(x,y)$ and integrate over the domain in which the unknown function is defined — here the two dimensional domain occupied by the body, so this will lead to a double integral. Generally it is necessary to integrate by parts to preserve appropriate continuity conditions and, in two and three dimensions, this requires the use of the divergence theorem, but further discussion of this topic is beyond the scope of this book.

## A.6 Computational considerations

The finite element method is widely used to solve problems for engineering systems of considerable complexity and it is common for the resulting discrete model to involve very large numbers of elements and hence of unknowns in the resulting equations. Various techniques are avaliable for reducing the computational time required to solve these equations, most of which take advantage of the fact that the stiffness matrix (i.e. the coefficient matrix for the corresponding system of equations) is banded.

When the matrix is banded, it follows that each unknown (nodal value) appears in only a few equations. It is therefore possible to eliminate any given variable from the system with a modest number of operations. Sequential use of this technique leads to a solution that is much more efficient than matrix inversion or other techniques that would typically be needed when the coefficient matrix is full (i.e. not banded).

**Example A.6**

*Solve the system of 5 equations*

$$\begin{bmatrix} 2 & 1 & 0 & 0 & 0 \\ 1 & 4 & 1 & 0 & 0 \\ 0 & 1 & 4 & 1 & 0 \\ 0 & 0 & 1 & 4 & 1 \\ 0 & 0 & 0 & 1 & 2 \end{bmatrix} \begin{Bmatrix} C_1 \\ C_2 \\ C_3 \\ C_4 \\ C_5 \end{Bmatrix} = \begin{Bmatrix} 4 \\ 2 \\ 3 \\ 4 \\ 2 \end{Bmatrix} . \tag{A.90}$$

The equations can be expanded as

$$2C_1 + C_2 = 4 \tag{A.91}$$

$$C_1 + 4C_2 + C_3 = 2 \tag{A.92}$$

$$C_2 + 4C_3 + C_4 = 3 \tag{A.93}$$

$$C_3 + 4C_4 + C_5 = 4 \tag{A.94}$$

$$C_4 + 2C_5 = 2 . \tag{A.95}$$

The first unknown, $C_1$ appears only in the first two equations, so it can be eliminated by multiplying (A.92) by 2 and subtracting (A.91) from it, with the result

$$7C_2 + 2C_3 = 0 . \tag{A.96}$$

We next eliminate $C_2$ by multiplying (A.93) by 7 and subtracting (A.96) from it, with the result

$$26C_3 + 7C_4 = 21 . \tag{A.97}$$

Continued use of the same technique yields

$$98C_4 + 26C_5 = 83 \tag{A.98}$$

and finally

$$170C_5 = 113 , \tag{A.99}$$

or $C_5 = 0.6765$. Equation (A.98) then gives

$$C_4 = \frac{83 - 26 \times 0.6765}{98} = 0.6675$$

and the remaining unknowns are recovered one by one from equations (A.97, A.96, A.91) respectively as $C_3 = 0.6280$, $C_2 = -0.1794$, $C_1 = 2.0897$.

Anyone who has attempted to solve a system of five simultaneous equations by hand will realize that this procedure, which only works when the matrix is banded, is a great deal quicker than other methods. For a system of $N$ simultaneous equations, it involves only $N$ matrix row operations to find the last unknown, followed by another $N$ operations to recover the full set of unknowns.

This is a particularly simple example where the bandwidth of the matrix is only three — i.e. there are only three terms in each of the algebraic equations. In more complex problems, the band will be wider. For example it is of width 6 for the beam bending formulation of equation (A.74). However, the computational effort involved in this elimination procedure is still very much less than that required for direct inversion of the global stiffness matrix.

### A.6.1 Data storage considerations

When the system being analyzed is very complex, the amount of numerical data to be transmitted and stored is very large and this in itself requires special consideration. For example, the stiffness matrix contains a large number of components, but most of these components are zeros and considerable savings can be achieved by identifying the bandwidth of the matrix and storing only the non-zero values. We can also take advantage of the symmetry of the matrix ($K_{ji}=K_{ij}$) to store only half of the off-diagonal components.

In the solution procedure for banded matrices used in Example A.6, only a few components of the stiffness matrix are needed at each stage and this permits some added efficiency in data handling. Even more efficiency is obtained by only assembling parts of the stiffness matrix as they are needed in the solution procedure. In this way, only a small part of the large stiffness matrix is stored in the computer at any given stage. The resulting algorithm is known as a *front solver*[3] and it makes it possible to solve problems involving millions of degrees of freedom.

## A.7 Use of the finite element method in design

The finite element method is typically used at a later stage in the design process when the basic geometry of the system is largely finalized and we need to establish more accurate values for the stresses and displacements.[4] It is extremely widely used in industry and the chances are very high that you will be called upon to use it at some time in your career as an engineer. However, commercial finite element codes are very flexible and user friendly and they cover the vast majority of design needs. It is therefore almost never necessary to write a program yourself. The primary purpose of this appendix is to explain to the reader in a simple context some of the fundamental processes involved in these codes and to introduce some of the terminology used.

The manuals and other supporting materials for the best known codes contain numerous examples and it is often possible to modify one of these to cover the problem under consideration, merely by redefining the number of nodes and their location. The best advice here is to start formulating the problem after the bare minimum of introductory reading, since the output from the program will often teach you how to correct errors or improve the model. Make sure you can reproduce the example solution on your computer system before you make the modifications.

The error messages you will get in early attempts can be obscure and frustrating. If you know someone in the company who has used the code before, their advice

[3] B.M. Irons (1970), A frontal solution program for finite element analyses, *International Journal for Numerical Methods in Engineering,* Vol.2, pp.5–32.

[4] It is rather paradoxical that we use an avowedly approximate method when we want more accuracy. The reason of course is that by using a finer discretization, the accuracy of the finite element method can be improved as much as we need. By contrast, the 'exact' methods discussed elsewhere in this book make an implicit approximation at the beginning when the component is idealized, for example as a beam or a shell.

will be invaluable. Alternatively, if you cut and paste an obscure error message into Google or another search engine, often it will direct you to an explanation of the problem and a way of fixing it.

Once the program appears to be working satisfactorily, there is a strong temptation to breathe a sigh of relief and accept the results as accurate. This is very risky, particularly since most commercial codes have sophisticated postprocessors that can make any results look very impressive and plausible. If the design is important enough for you to perform a finite element analysis, it is also important enough to make every effort to be sure that there are no errors in your formulation.[5] There are three tests you might perform here:-

(i) Compare the finite element results with the analytical solution of the simplest idealized problem you can think of that is somewhat like the real problem. The results should certainly be of the same order of magnitude and there is reason to be suspicious if they differ by more than a factor of 2.
(ii) Try to find a system of loads and/or a small modification of the geometry that would cause the complex system to have a simple stress field such as uniform stress or a linearly varying bending stress. Make sure the finite element program confirms this result.
(iii) Test the convergence of the program by making several runs with different degrees of mesh refinement, particularly in critical areas of high stress.

## A.8 Summary

In this chapter, the basic principles of the finite element method were introduced and applied to problems in one dimension. In mechanics of materials problems, the method is equivalent to the use of the Rayleigh-Ritz method with a piecewise polynomial approximation to the displacement. The approximation is defined through a finite set of nodal values that constitute the degrees of freedom for the problem.

An alternative mathematical formulation can be obtained from the governing equilibrium equation by choosing the nodal values to minimize the integral of the square of the error over the domain. This leads to the Galerkin approximation in which the same functions are used as approximating (shape) functions and weight functions. The advantage of this method is that it can be applied to any problem governed by a differential equation.

The finite element method was applied to problems of axial loading and the bending of beams. The extension of the method to problems in two and three dimensions was also briefly discussed.

---

[5] This argument cuts both ways. If you do not feel high accuracy and reliability are crucial, it is probably not worth doing the finite element analysis, unless of course you simply want to convince your boss of your thoroughness!

## Further reading

K.J. Bathe (1982), *Finite Element Procedures in Engineering Analysis*, Prentice-Hall, Englewood Cliffs, NJ.
Y.K. Cheung and M.F. Yeo (1979), *A Practical Introduction to Finite Element Method*, Pittman, London.
T.J.R. Hughes (1987), *The Finite Element Method*, Prentice Hall, Englewood Cliffs, NJ, Chapters 1,2.
H. Kardestuncer and D.H. Norrie, eds. (1987), *Finite Element Handbook*, McGraw-Hill, New York.
Y.W. Kwon and H. Bang (1996), *The Finite Element Method using MATLAB*, CRC Press, Boca Raton, LA.
G. Strang and G.J. Fix (1973), *An Analysis of the Finite Element Method*, Prentice-Hall, Englewood Cliffs, NJ.
O.C. Zienkiewicz (1977), *The Finite Element Method*, McGraw-Hill, New York.

## Problems

### Section A.1

**A.1.** Find a piecewise linear approximation to the parabola $y = x^2$ in the range $0 < x < 1$, using two equal elements and three nodes, $0, 0.5, 1$. Plot a graph comparing the function and its approximation.

**A.2.** Find a piecewise linear approximation to the function $\sin(\pi x/2)$ in the range $0 < x < 1$, using two equal elements and three nodes, $0, 0.5, 1$. Plot a graph comparing the function and its approximation.

**A.3.** Find the best straight line fit to the exponential function $e^x$ in the range $0 < x < 1$. Plot a graph comparing the function and its approximation.

**A.4*.** Find the straight line fit to the exponential function $e^x$ in $0 < x < 1$ that minimizes the *percentage error*,

$$\frac{f(x) - f^*(x)}{f(x)}$$

rather than the absolute error, $f(x) - f^*(x)$.

### Section A.2

**A.5.** A uniform bar of length $L$, cross-sectional area $A$ and Young's modulus $E$, is supported at $x = 0$ and subjected to a distributed load $p(x) = p_0 x/L$ per unit length.

Develop a finite element solution for the problem, using two equal length elements and hence estimate the displacement at the free end $x = L$.

**A.6.** The bar of Figure PA.6 is 100 mm long and has a cross-sectional area $A$ that increases linearly from 100 mm$^2$ at one end to 200 mm$^2$ at the other. The bar is subjected to a tensile axial force $F_0 = 300$ kN.

Use the stiffness matrix of equation (A.37) with four equal length elements to estimate the increase in length of the bar, if Young's modulus for the material is 80 GPa.

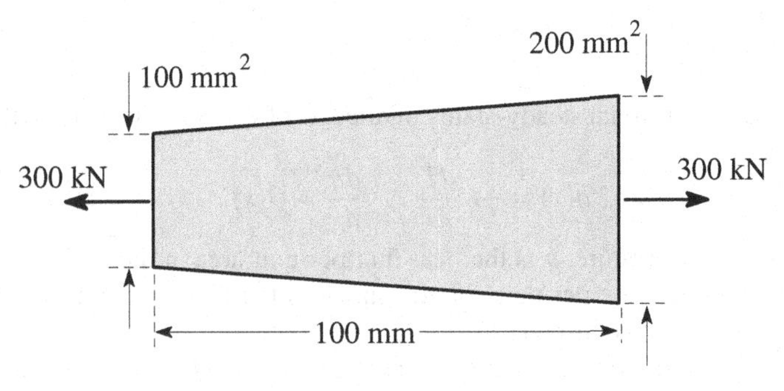


*Figure PA.6*

**A.7.** The composite bar of Figure PA.7 is supported at $A$ and $B$ and loaded only by its own weight. The upper segment is made of steel ($\rho g = 75$ kN/m$^3$, $E = 210$ GPa) and has cross-sectional area $A = 100$ mm$^2$, whilst the lower segment is of aluminium alloy ($\rho g = 27$ kN/m$^3$, $E = 80$ GPa) and cross-sectional area $A = 200$ mm$^2$.

Use the finite element method with two equal length elements in each segment to estimate the displacement at $C$.

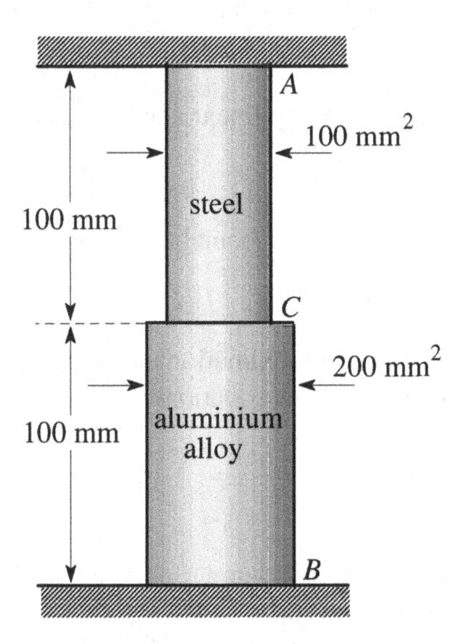


*Figure PA.7*

**A.8.** A uniform bar $0<x<L$ of length $L$, cross-sectional area $A$ and Young's modulus $E$ is supported at both ends and subjected to a distributed load $p(x)=p_0x/L$ per unit length.

Develop a finite element solution for the problem, using four equal length elements, and hence estimate the support reaction at $x=L$.

## Section A.3

**A.9.** The one-dimensional steady-state conduction of heat is governed by the equations

$$q(x) = -K\frac{dT}{dx} \quad ; \quad \frac{dq}{dx} = Q(x) \; ,$$

where $T$ is the temperature, $q$ is the heat flux per unit area in the $x$-direction, $Q(x)$ is the heat generated rate per unit volume and $K$ is the thermal conductivity of the material (here assumed constant).

In the body $0<x<a$ there is uniform heat generation $Q(x)=Q_0$ and the boundaries $x=0,a$ are both maintained at zero temperature, $T(0)=T(a)=0$. Using a piecewise linear approximation to the temperature with two elements and three equally-spaced nodes, estimate the temperature at the mid-point $x=a/2$.

**A.10.** Use the shape functions (A.16) to develop an appropriate piecewise linear approximation for the displacement $u$ governed by equation (12.3) with boundary conditions (12.6, 12.7), based on the subdivision of the length $L$ into four equal elements.

Use the method of §A.3 to obtain a set of homogeneous equations for the unknown constants and hence estimate the buckling load $P_0$. Compare your result with the exact value (12.13).

**A.11.** It is proposed to use a piecewise linear approximation for the ordinary differential equation

$$f''(x) = g(x) \quad ; \quad 0<x<L \; ,$$

where $g(x)$ is a known function and the boundary conditions are

$$f(0) = 0 \quad ; \quad f'(L) = 0 \; .$$

Define an appropriate piecewise approximation using $N$ degrees of freedom and develop the resulting set of simultaneous equations.

## Section A.4

**A.12.** Figure PA.12 shows a beam of flexural rigidity $EI$ and length $L$, built-in at $z=0$ and loaded by a force $F$ at the free end.

Develop a one element solution to this problem, noting that the displacement $u$ and the rotation $\theta$ at the free end will be the only degrees of freedom.

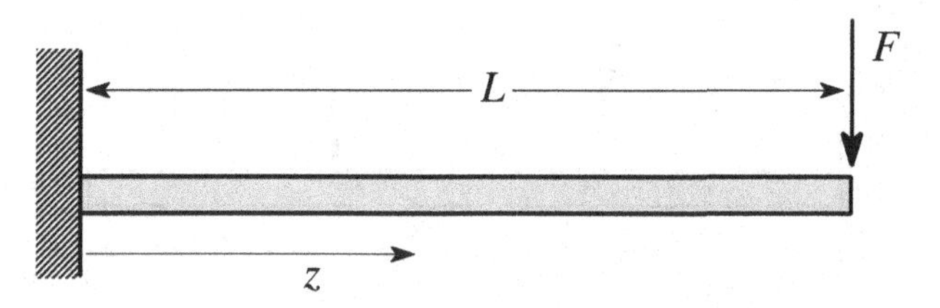


*Figure PA.12*

**A.13.** The beam of Figure PA.13 is simply supported at the points $z=0, 2L$ and loaded by equal forces $F$ at $z=L, 3L$. The flexural rigidity of the beam is $EI$. Estimate the displacement under each of the loads, using a discretization with three equal elements, each of length $L$. What is the location and magnitude of the maximum displacement?

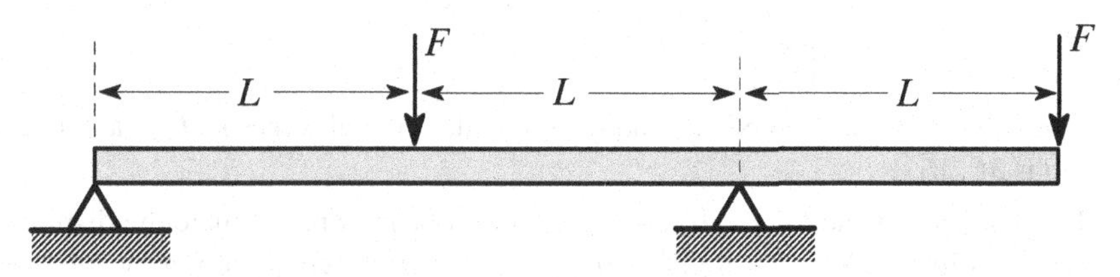


*Figure PA.13*

**A.14.** Figure PA.14 *(a)* shows one element of length $\Delta$ of a beam of flexural rigidity $EI$, loaded by a concentrated force a distance $a$ from one end.

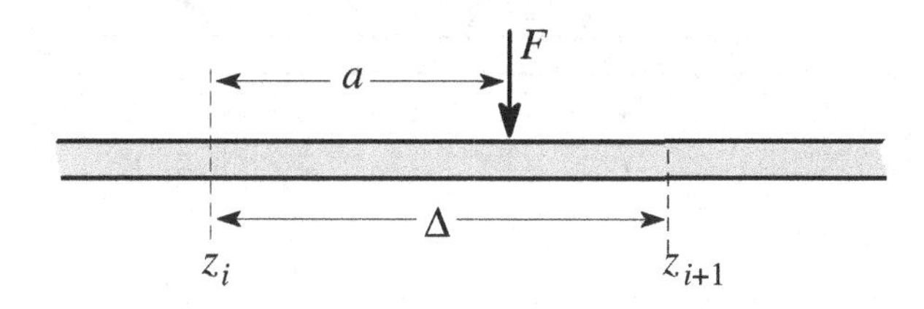


*Figure PA.14(a)*

Use equations (A.82, A.83) to find the equivalent nodal forces $F_i, F_{i+1}$ and nodal moments $M_i, M_{i+1}$.

Use your results and the stiffness matrix in (A.86) to approximate the displacement $u(z)$ of the beam of Figure PA.14 *(b)*, based on a representation of the beam as a single element of length $L$.

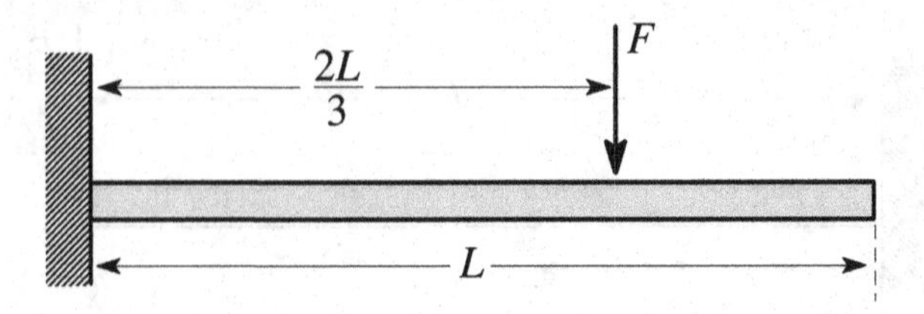


*Figure PA.14(b)*

**A.15.** Figure PA.15 *(a)* shows one element of length $\Delta$ in a beam discretized by the finite element method. The only loading is the linearly varying load

$$w(z) = \frac{w_0(z - z_i)}{\Delta} \quad ; \quad z_i < z < z_{i+1}$$

per unit length, where $\Delta = z_{i+1} - z_i$.

Use equations (A.82, A.83) to find the equivalent nodal forces $F_i, F_{i+1}$ and nodal moments $M_i, M_{i+1}$.

Use your results and the stiffness matrix in (A.86) to approximate the displacement $u(z)$ of the beam of Figure PA.15 *(b)*, based on a representation of the beam as a single element of length $L$. The flexural rigidity of the beam is $EI$.

*(a)*

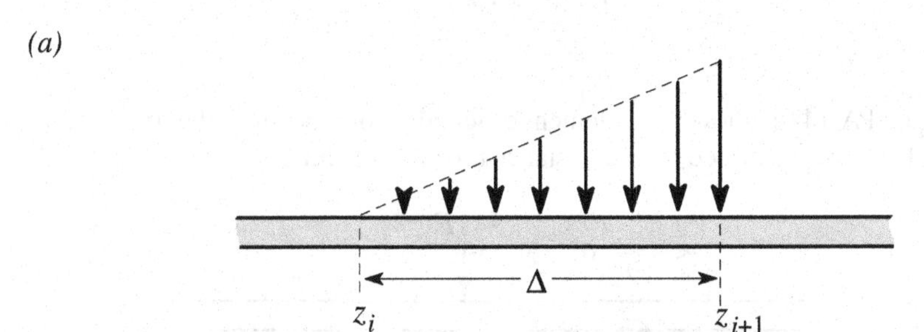


*(b)*

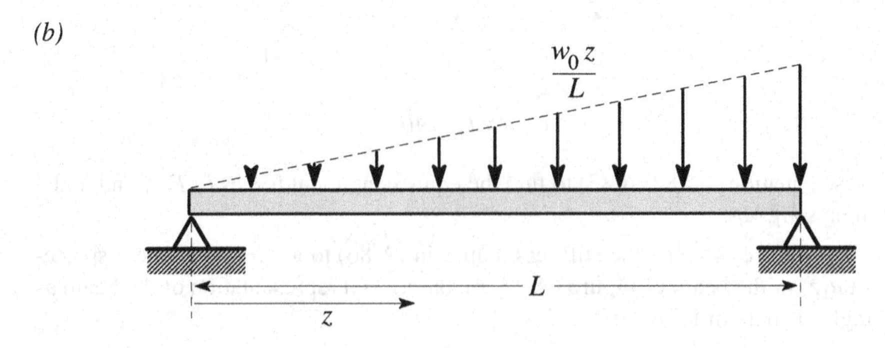


*Figure PA.15*

**A.16.** Solve the problem of Figure A.13, using two elements with two terminal nodes and one internal node at the mid-point.

Find the displacement predicted at the mid-point and compare it with the exact value $5w_0L^4/384EI$.

If you are familiar with matrix operations and have access to Maple, MathCad, Mathematica or Matlab etc., re-solve the problem with 4,6,8 elements, etc and plot the percentage error in the predicted central displacement as a function of the number of elements used.

How many elements are needed to achieve 0.1% accuracy?

**A.17.** The beam of Figure PA.17 has flexural rigidity $EI$, is built in at both ends $z = 0, L$, and is subjected to a uniformly distributed load $w_0$ per unit length.

Estimate the central displacement, using two elements of length $L/2$.

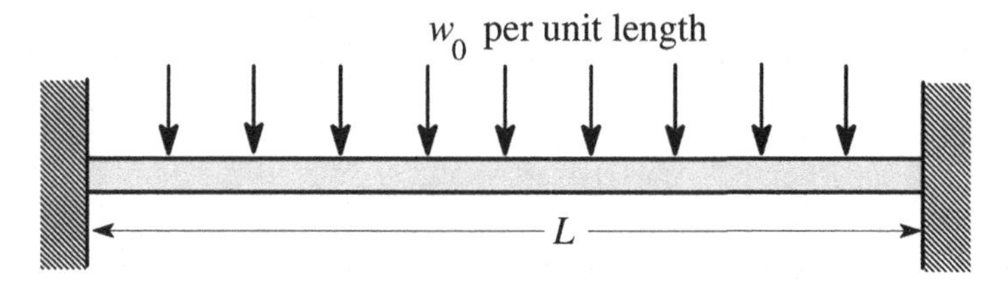


*Figure PA.17*

**A.18*.** A beam of length $2L$ is simply supported at the ends and at the mid-point. It is subjected to a uniformly distributed load $w_0$ per unit length. Estimate the reaction at the central support, using a discretization with two equal elements, each of length $L$.

**A.19.** Use the Rayleigh-Ritz method to obtain an approximate solution for the deflection of a beam of flexural rigidity $EI$ and length $L$, simply supported at its ends and subjected to a uniformly distributed load $w_0$ per unit length. Use a one degree of freedom quadratic approximation to the deformed shape.

Verify that the predicted displacement is identical to that obtained in equation (A.87).

# B

# Properties of Areas

In §4.3 we found that the coordinates $\bar{x}, \bar{y}$ of the centroid of a plane area $A$ can be determined from equations (4.28, 4.29) — i.e.

$$A\bar{x} = \iint_A x' dA \;\; ; \;\; A\bar{y} = \iint_A y' dA \,, \tag{B.1}$$

whilst the second moments of area about the general set of axes $O'x'y'$ are defined by the integrals

$$I'_x = \iint_A y'^2 dA \;\; ; \;\; I'_y = \iint_A x'^2 dA \;\; ; \;\; I'_{xy} = \iint_A x'y' dA \,. \tag{B.2}$$

In many practical cases, the area can be decomposed into simpler shapes for which the appropriate quantities are tabulated and the required results can then be obtained by superposition, as demonstrated in §4.3. Otherwise, the first and second moments must be obtained directly from the definitions by performing the integrations.

The biggest challenge in this procedure is to establish the appropriate limits on the double integral. Once this has been done, the double integral operator can be applied to each of the quantities in equations (B.1,B.2) and the resulting integrals are generally straightforward.

One of the two integrations can often be performed by inspection. For example, if we wish to determine $I'_x$ for the section of Figure B.1(a), we can reduce it immediately to a single integral by dividing the section into strips parallel to the $x'$-axis as shown. Thus,

$$I'_x = \iint_A y'^2 dA = \int_a^b w(y') y'^2 dy' \,, \tag{B.3}$$

where $w(y')$ is the width of the section (the length of the strip) as a function of $y'$. In effect, all we have done here is to perform the integration with respect to $x'$ in our heads, since the integrand $(y'^2)$ does not depend on $x'$.

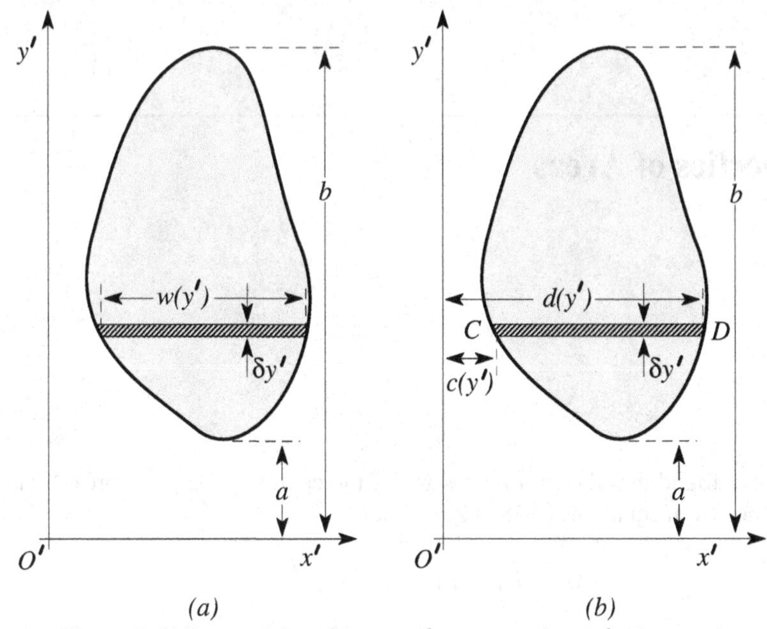


*Figure B.1: Determining the second moment of area by integration*

However, if we have to calculate all the first and second moments, it is safer to make both stages of integration explicit. For example, in rectangular coordinates the element of area $dA$ is $dx'dy'$ and if we perform the $x'$ integration first, we have

$$\iint_A dA = \int_a^b \int_{c(y')}^{d(y')} dx'dy' \; . \tag{B.4}$$

In this expression, notice how the limits $c,d$ on the inner integral are functions of the *outer* variable $y'$, whereas the limits $a,b$ on the outer integral are constants. This is illustrated in Figure B.1(b). The inner integral is performed over the strip $CD$ and its limits depend on which strip (which value of $y'$) is being considered, whereas the outer integral sums the contributions of the separate strips.

Usually we require the centroidal second moments, but the coordinates of the centroid may not be known initially. We therefore choose any convenient origin for the integration and then use the parallel axis theorem of §4.3.2 to deduce the centroidal values.

**Example B.1**

*Determine the location of the centroid and the centroidal second moments $I_x, I_y, I_{xy}$ for the quarter circle of radius a shown in Figure B.2.*

The equation of the circular boundary is

$$x'^2 + y'^2 = a^2$$

and hence the right-hand boundary of the area at height $y'$ is

$$x' = \sqrt{a^2 - y'^2} \; .$$

A suitable realization of the double integral is therefore

$$\iint_A dA = \int_0^a \int_0^{\sqrt{a^2-y'^2}} dx' dy' \; .$$

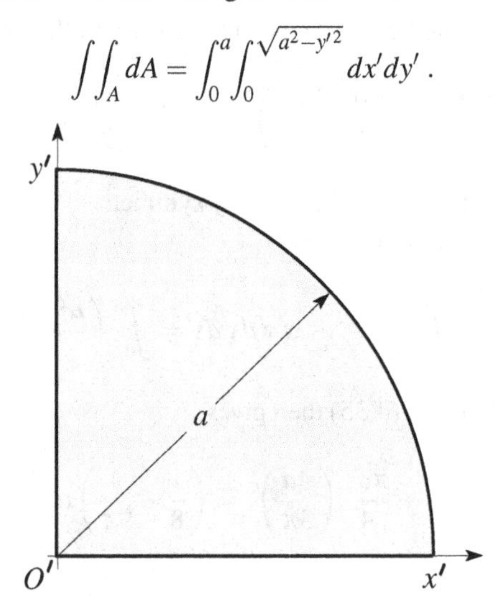


*Figure B.2*

Using this expression, we have

$$A = \int_0^a \int_0^{\sqrt{a^2-y'^2}} dx' dy' = \int_0^a \sqrt{a^2 - y'^2} dy' = \frac{\pi a^2}{4} \; ,$$

which of course we could have written down as the area of a quarter circle of radius $a$.

Applying the same operator to the definition of $A\bar{x}$, we have

$$A\bar{x} = \iint_A x' dA = \int_0^a \int_0^{\sqrt{a^2-y'^2}} x' dx' dy' = \int_0^a \left( \frac{a^2 - y'^2}{2} \right) dy' = \frac{a^3}{3}$$

and hence

$$\bar{x} = \frac{A\bar{x}}{A} = \frac{a^3}{3} \Big/ \frac{\pi a^2}{4} = \frac{4a}{3\pi} \; .$$

A similar procedure could be used to determine the coordinate $\bar{y}$, but the symmetry of the section shows that this must also be

$$\bar{y} = \frac{4a}{3\pi} \; .$$

Since we now know the location of the centroid, it would be possible to transfer the origin there and calculate the centroidal second moments of area directly. However, It is more efficient to calculate the second moments about the axes through $O'$ and then use the parallel axis theorem to transfer the values to the centroidal axes.

We have

$$I_x' = \int\!\!\int_A y'^2 dA = \int_0^a \int_0^{\sqrt{a^2-y'^2}} y'^2 dx' dy' = \int_0^a y'^2 \sqrt{a^2-y'^2} dy' = \frac{\pi a^4}{16} .$$

The centroidal value $I_x$ can then be deduced from the parallel axis theorem (4.33) as

$$I_x = I_x' - A\bar{y}^2 = \frac{\pi a^4}{16} - \frac{\pi a^2}{4}\left(\frac{4a}{3\pi}\right)^2 = \left(\frac{\pi}{16} - \frac{4}{9\pi}\right) a^4 = 0.05488 a^4 .$$

The same results are obtained for $I_y'$ and $I_y$, by symmetry.

For the product inertia, we have

$$I_{xy}' = \int\!\!\int_A y'^2 dA = \int_0^a \int_0^{\sqrt{a^2-y'^2}} x'y' dx' dy' = \int_0^a \left(\frac{a^2-y'^2}{2}\right) y' dy' = \frac{a^4}{8}$$

and the parallel axis theorem (4.35) then gives

$$I_{xy} = I_{xy}' - A\bar{x}\bar{y} = \frac{a^4}{8} - \frac{\pi a^2}{4}\left(\frac{4a}{3\pi}\right)^2 = \left(\frac{1}{8} - \frac{4}{9\pi}\right) a^4 = -0.01647 a^4 .$$

# C

# Stress Concentration Factors

These figures are plotted from the approximate expressions given by W.D. Pilkey (1994), *Formulas for Stress, Strain and Structural Matrices*, John Wiley, New York and R.J. Roark and W.C. Young (1975), *Formulas for Stress and Strain*, McGraw-Hill, New York, 5th edn. A wider range of charts for stress intensity factors can be found in R.E. Peterson (1974), *Stress Concentration Factors*, John Wiley, New York and in numerous books on mechanical design. The reader should be warned that these various sources often disagree signficantly as to numerical values, particularly at small values of the fillet radius in Figures C.3 – C.7.

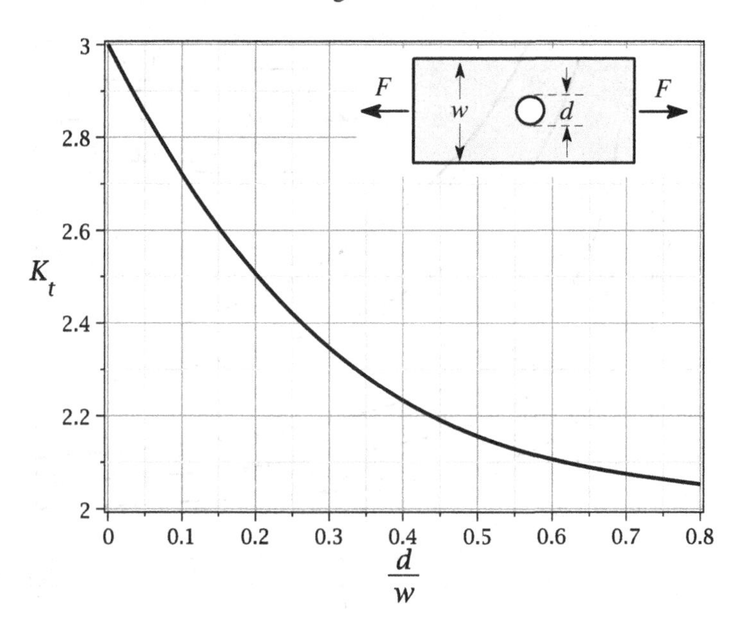


*Figure C.1: Rectangular bar with a transverse hole in tension or compression;* $\sigma_{\rm nom} = F/A$*, where* $A = (w-d)h$ *is the reduced cross-sectional area at the hole and* $h$ *is the plate thickness.*

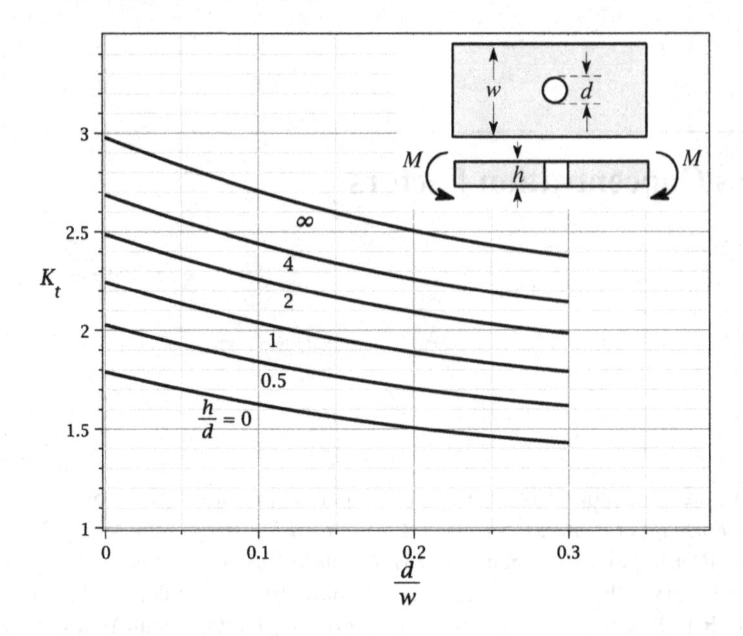


*Figure C.2: Rectangular bar with a hole in transverse bending;* $\sigma_{\text{nom}} = Mh/2I$*, where* $I = (w-d)h^3/12$ *is based on the reduced cross section at the hole and h is the plate thickness.*

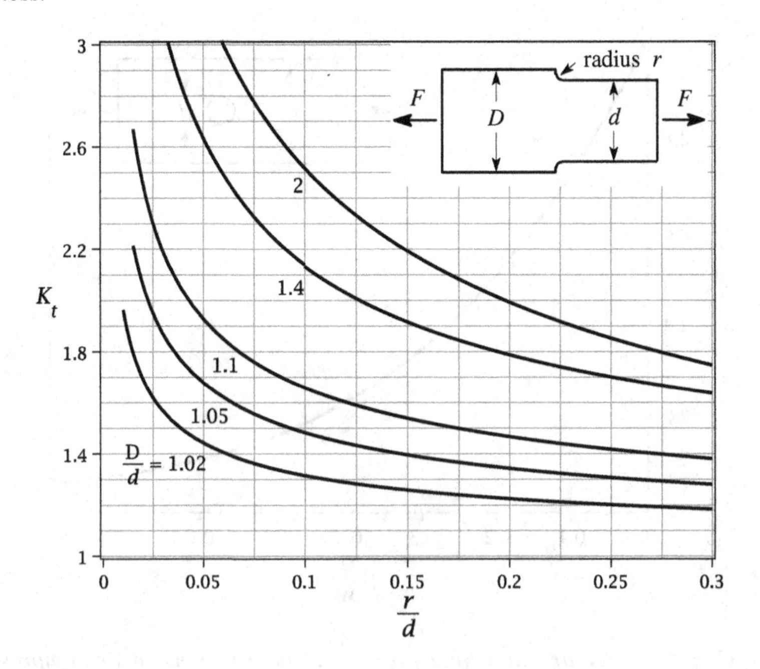


*Figure C.3: Rectangular bar with a change in section through a fillet radius r, loaded in tension or compression;* $\sigma_{\text{nom}} = F/dh$*, where h is the plate thickness.*

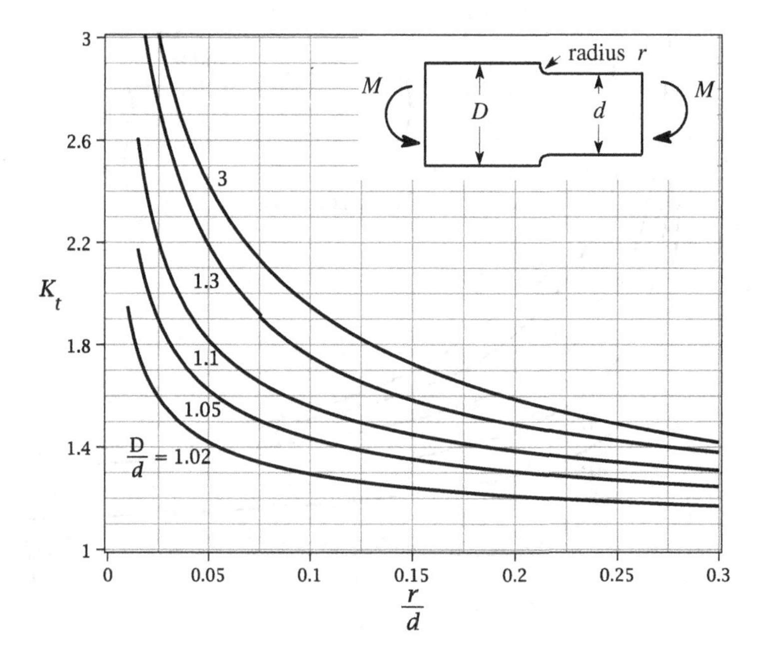


*Figure C.4: Rectangular bar with a change in section through a fillet radius r, loaded in bending;* $\sigma_{\mathrm{nom}} = 6M/hd^2$*, where h is the plate thickness.*

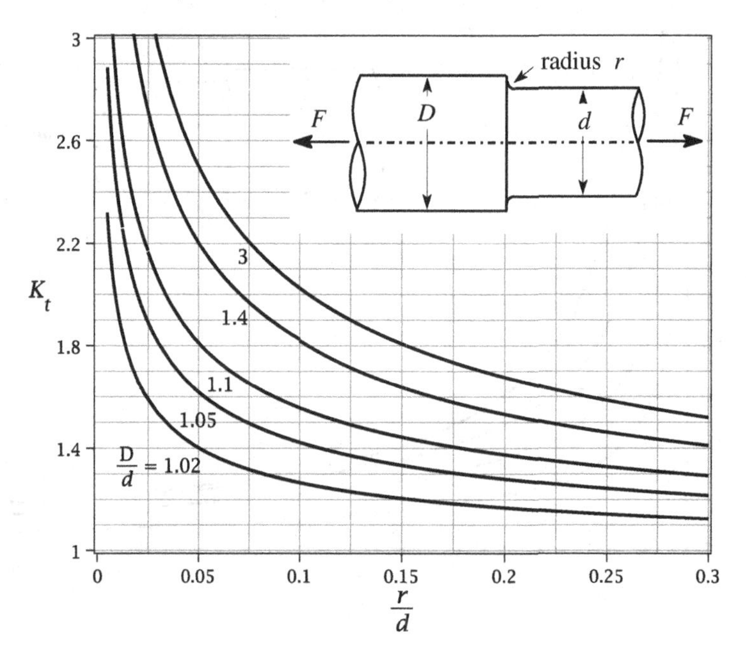


*Figure C.5: Cylindrical bar with a change in section through a fillet radius r, loaded in tension or compression;* $\sigma_{\mathrm{nom}} = 4F/\pi d^2$*.*

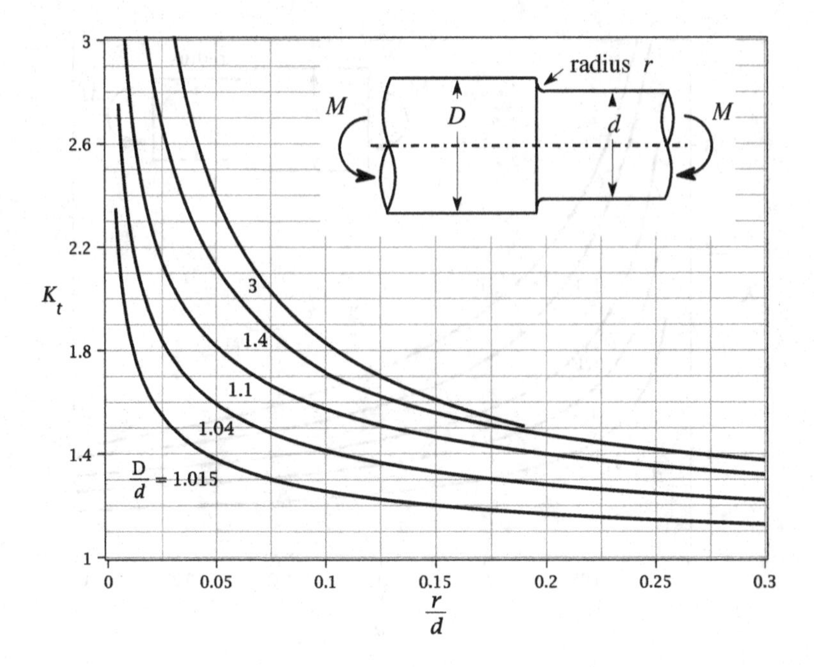


*Figure C.6: Cylindrical bar with a change in section through a fillet radius r, loaded in bending;* $\sigma_{nom} = 32M/\pi d^3$.

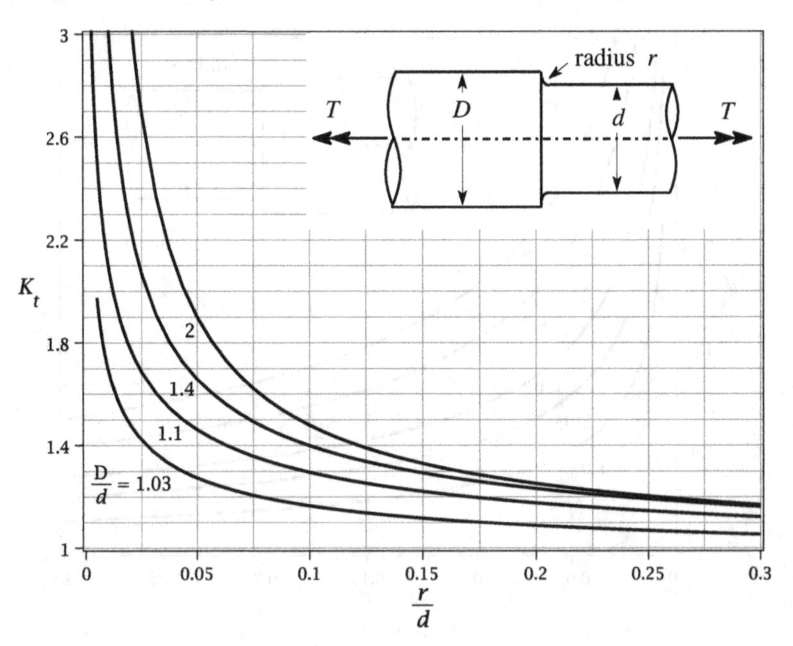


*Figure C.7: Cylindrical bar with a change in section through a fillet radius r, loaded in torsion;* $\sigma_{nom} = 16T/\pi d^3$.

# D

# Answers to Even Numbered Problems

## Chapter 2

**2.2**: $\tau_{max} = 20.06$ ksi, $\theta_s = 2.15^o$ clockwise.
**2.4**: $\sigma_1 = 114$ MPa, $\sigma_2 = -14$ MPa, $\theta_1 = 19.3^o$ clockwise.
**2.6**: $\tau_{max} = 2.5$ ksi.
**2.8**: $\sigma_1 = 303$ MPa, $\sigma_2 = -8$ MPa, $\sigma_3 = 145$ MPa, $\{l,m,n\} = \{-0.692, 0.715, 0.095\}$.
**2.10**: $\sigma_1 = 3$ ksi, $\sigma_2 = -6$ ksi, $\sigma_3 = 3$ ksi, $\{l,m,n\} = \{1/\sqrt{3}, 1/\sqrt{3}, 1/\sqrt{3}\}$.
**2.12**: $U_0 = [\sigma_1^2 + \sigma_2^2\sigma_3^2 - 2\nu(\sigma_1\sigma_2 + \sigma_2\sigma_3 + \sigma_3\sigma_1)]/2E$.
**2.14**: $\sigma_{xx} + \sigma_{yy} + \sigma_{zz} < 0$, $\sigma_{xx}\sigma_{yy} + \sigma_{yy}\sigma_{zz} + \sigma_{zz}\sigma_{xx} - \sigma_{xy}^2 - \sigma_{yz}^2 - \sigma_{zx}^2 > 0$, $\sigma_{xx}\sigma_{yy}\sigma_{zz} - \sigma_{xx}\sigma_{yz}^2 - \sigma_{yy}\sigma_{zx}^2 - \sigma_{zz}\sigma_{xy}^2 + 2\sigma_{xy}\sigma_{yz}\sigma_{zx} = 0$.
**2.16**: $\sigma_1 = 330.4$ MPa, $\sigma_2 = 89.6$ MPa, $\tau_{max} = 130.2$ MPa.
**2.18**: $\sigma_1 = 338$ MPa, $\sigma_2 = 71$ MPa, $\sigma_3 = 96$ MPa, $\tau_{max} = 133.5$ MPa.
**2.20**: $\tau = 193$ MPa (Tresca) and 223 MPa (Von Mises).
**2.24**: $\tau_E = \sqrt{(\sigma_{xx}^2 + \sigma_{yy}^2 + \sigma_{zz}^2 - \sigma_{xx}\sigma_{yy} - \sigma_{yy}\sigma_{zz} - \sigma_{zz}\sigma_{xx} + 3\sigma_{xy}^2 + 3\sigma_{yz}^2 + 3\sigma_{zx}^2)/3}$
**2.26**: $S.F. = 1.31$.
**2.28**: $M_Y = 5$ kNm (Tresca) and 4.33 kNm (Von Mises).
**2.30**: $451^o$.
**2.34**: 35 kN.
**2.36**: $F = \pi D^3 S_t/8L$.
**2.38**: $S = S_c[\sin(2\theta) - \mu - \mu\cos(2\theta)]/2$, maximum when $\theta = \frac{1}{2}\tan^{-1}(-1/\mu)$.
**2.40**: 17.7 mm.
**2.42**: $\mu = 0.42$.
**2.44**: $525 \sim 675$ MPa.
**2.46**: 0.53.
**2.48**: $b = 33$, $S_0 = 405$ MPa.
**2.50**: $\sigma_{max} = 137$ MPa.
**2.52**: $S.F. = 1.29$.
**2.54**: A machined shaft is sufficient.
**2.56**: $S.F. = 2.0$.
**2.58**: Probability of failure = 0.2 percent at $A, C$.
**2.60**: $S.F. = 3.1$.

# Chapter 3

**3.2**: $U = 16T^2L/\pi d^4G$.
**3.4**: $k = k_1k_2/(k_1+k_2)$.
**3.6**: $a^3[\sigma_{xx}^2+\sigma_{yy}^2+\sigma_{zz}^2-2\nu(\sigma_{xx}\sigma_{yy}+\sigma_{yy}\sigma_{zz}+\sigma_{zz}\sigma_{xx})]/2E$.
**3.8**: $F[4+3\cos(2\alpha)+\sqrt{3}\sin(2\alpha)]/k$, $u_{\max}/u_{\min}=7$.
**3.10**: $u = 2.517FL^3/Ec^4$.
**3.12**: $u = 4.108FL/EA$.
**3.14**: $\theta = 0$ is stable if $W < 2kL$.
**3.18**: $F = 2ka(1-2\sin\theta)\cos\theta \bigg/ \sin\theta\left(1+\frac{a\cos\theta}{\sqrt{b^2-a^2\sin^2\theta}}\right)$.
**3.20**: $u(L) = FL^3/84EI$.
**3.22**: $C_i = 2L^4w_0(-1)^{i+1}/\pi^4EIi^4$.
**3.24**: $u(L/4) = -2.6\times10^{-4}FL^3/EI$.
**3.26**: $u_{\max} = M_0L^2/9\sqrt{3}EI$.
**3.28**: $\Delta_{AC} = 0.5FR^3/\pi EI$.
**3.30**: See 3.18 above.
**3.32**: $u_B = M_0/kL$.
**3.34**: The block rises by $4\nu F_2/\pi DE$.
**3.36**:

$$\boldsymbol{K} = k\begin{bmatrix} 1 & -1 & 0 & 0 \\ -1 & 3 & -2 & 0 \\ 0 & -2 & 5 & -3 \\ 0 & 0 & -3 & 7 \end{bmatrix} \;;\; \boldsymbol{C} = \frac{1}{k}\begin{bmatrix} \frac{25}{12} & \frac{13}{12} & \frac{7}{12} & \frac{1}{4} \\ \frac{13}{12} & \frac{13}{12} & \frac{7}{12} & \frac{1}{4} \\ \frac{7}{12} & \frac{7}{12} & \frac{7}{12} & \frac{1}{4} \\ \frac{1}{4} & \frac{1}{4} & \frac{1}{4} & \frac{1}{4} \end{bmatrix}.$$

**3.38**:

$$\boldsymbol{K} = \begin{bmatrix} 44630\text{ N/mm} & 11656\text{ N/mm} & 875\text{ N} \\ 11656\text{ N/mm} & 11670\text{ N/mm} & -3350\text{ N} \\ 875\text{ N} & -3350\text{ N} & 1408246\text{ N mm} \end{bmatrix}.$$

**3.40**:

$$\begin{bmatrix} F \\ M_1 \\ M_2 \end{bmatrix} = \begin{bmatrix} \left(\frac{12EI_1}{L_1^3}+\frac{12EI_2}{L_2^3}\right) & \frac{6EI_2}{L_2^2} & \frac{6EI_1}{L_1^2} \\ \frac{6EI_2}{L_2^2} & \left(\frac{GK_1}{L_1}+\frac{4EI_2}{L_2}\right) & 0 \\ \frac{6EI_1}{L_1^2} & 0 & \left(\frac{GK_2}{L_2}+\frac{4EI_1}{L_1}\right) \end{bmatrix} \begin{bmatrix} u \\ \theta_1 \\ \theta_2 \end{bmatrix}.$$

**3.42**: $\alpha = 22.5^\circ$ or $-67.5^\circ$.
**3.44**: $W_{\max} = 4.4$ lb.
**3.46**: $u_A = 7FL^3/2EI$.
**3.48**: $\Delta_{BD} = 0.1366FR^3/EI$.
**3.50**: $u_C = M_0L^2/2EI$.
**3.52**: $R = 3F/2$, $u = 0.0169FL^3/EI$.
**3.54**:

$$u_B = \frac{\pi FR^3}{4EI} + \frac{FR^3}{GK}\left(\frac{3\pi}{4}-2\right).$$

**3.56**: $0.11^\circ$ (horizontal) and $0.04^\circ$ (vertical).
**3.58**: $0.37\times10^{-6}$ radians.

## Chapter 4

**4.2**: $\sigma_{max} = 154$ MPa.
**4.4**: Inside a diamond shaped region with corners at $(0, \pm b/6), (\pm a/6, 0)$.
**4.6**: $\alpha = -47.5^o$, $\sigma_{max} = 58.3$ MPa.
**4.8**: $\sigma_{max} = 70$ MPa.
**4.10**: $u_y^{max} = -3.98$ mm, $u_x^{max} = 4.11$ mm.
**4.12**: $u_x = 0.739$ in.
**4.14**: $I_x = 560,000\ \text{mm}^4$, $I_y = 290,000\ \text{mm}^4$, $I_{xy} = 300,000\ \text{mm}^4$.
**4.16**: $I_x = 22.6 \times 10^6\ \text{mm}^4$, $I_y = 3.84 \times 10^6\ \text{mm}^4$, $I_{xy} = -5.14 \times 10^6\ \text{mm}^4$.
**4.18**: $I_x = I_y = 54.0a^4,\ \ I_{xy} = 1.01a^4$.
**4.20**: $I_x = 171,000\ \text{mm}^4,\ \ I_y = 180,271\ \text{mm}^4,\ \ I_{xy} = -65,250\ \text{mm}^4$.
**4.22**: $I_x = 23a^3t/16,\ \ I_y = 37a^3t/48,\ \ I_{xy} = 7a^3t/16$.
**4.24**: $I_1 = 315 \times 10^6\ \text{mm}^4$, $I_2 = 101 \times 10^6\ \text{mm}^4$, $\theta_1 = 58.3^o$ clockwise from $x$.
**4.26**: $I_1 = 754,000\ \text{mm}^4$, $I_2 = 96,000\ \text{mm}^4$, $\theta_1 = 32.9^o$ clockwise from $x$.
**4.28**: $I_1 = 35.04a^3t,\ \ I_2 = 4.30a^3t,\ \ \theta_1 = 32.8^o$ anticlockwise from $x$.
**4.30**: 76.2 MPa, –67.8 MPa.
**4.32**: $0.6M_0/a^2t$.
**4.42**: $I_1/I_2 < 11.5$.

## Chapter 5

**5.2**: $0.198\sigma_0 bh^2$.
**5.4**: $M_Y = 15\pi S_Y a^3/8$, $M_P = 28S_Y a^3/3$, $f = 1.584$.
**5.6**: $M = 9.86$ kNm, $R = 15$ m.
**5.8**: $M = 398$ Nm.
**5.10**: $M_P = 934$ Nm.
**5.12**: $0.828S_Y a^2 t$.
**5.14**: $M_P = 10.6$ kNm, $M = 9.7$ kNm for second plastic zone.
**5.16**: $M = 0.934$ kNm.
**5.18**: $M_P = 1.66$ kNm, $\theta = 0.0466^o$.
**5.20**: $\theta = (1 - 4\cot^2\alpha)/2(1 + 4\cot\alpha)$.
**5.22**: For $-4 < \tan\alpha < 4$, $\tan\alpha = -\cot\theta + \sqrt{\cot^2\theta + 48}$,
$M_P = \sqrt{6400 - (2300/9)\tan^2\alpha + (25/9)\tan^4\alpha}$.
**5.24**: $\sigma_{zz} = S_Y(\text{sgn}(y) - 8y/3\pi a)$.
**5.26**: $R_u = 1573$ in.
**5.28**:
$$\begin{aligned}\sigma_{zz} &= -200 - 4.883y\ ; && -73.71 < y < 0\\ &= -200\ ; && 0 < y < 16.29\\ &= 200 - 4.883y\ ; && 16.29 < y < 36.29\end{aligned}$$
**5.30**: $F_P = M_P L/a(L - a)$.
**5.32**: $d = 0.419L$, $w_0 = 22.8M_P/L^2$.
**5.34**: $F_P = 2M_P/L$.
**5.36**: $F = \pi D^3 S_t/8L$.
**5.38**: $S = S_c(\sin(2\theta) - \mu - \mu\cos(2\theta))/2$, maximum when $\theta = \frac{1}{2}\tan^{-1}(-1/\mu)$.

**5.40**: 17.7 mm.
**5.42**: $\mu = 0.42$.
**5.44**: 525 ~ 675 MPa.
**5.46**: 0.53.

## Chapter 6

**6.2**: 2.88 ksi.
**6.4**: $\tau(y) = 4V_y(a^2 - y^2)/5a^3t$ (inclined segment),
$\tau(y) = V_y(7a^2 - 4y^2)/10a^3t$ (vertical segment).
**6.6**: $\tau(\phi) = 3V_y \cos\phi / 4at_0$.
**6.8**: 3.52 bolts per foot.
**6.10**: 25 ksi.
**6.12**: $c = 5a/8$.
**6.14**: $c = 3\pi a/8$.
**6.16**: $c = 0.533a$.
**6.18**: $c = 2.03a$.
**6.20**: $\tau_{max} = V_y(4-\varepsilon)/4\pi a t_0(1-\varepsilon)$.
**6.22**: $\tau_A = 4V_y/\pi a t$, $d = 2a$.
**6.24**: $\tau = V_y(0.059 + 0.259\cos\theta)/at$ (curved segment)
$\tau = V_y(0.06477x^2 - 0.2652a^2)/a^3t$ (straight segments, $x$ meas. from corner).
**6.26**: $\tau = 2T(1+r)^2/p^2rt$, $K = p^3r^2t/4(1+r)^4$.
**6.28**: $6.4a^3t$.
**6.30**: $\tau_{max} = T/2\pi a^2t_0(1-\varepsilon)$, $\theta/L = T/2\pi G a^3 t_0\sqrt{1-\varepsilon}$
**6.32**: 67.5 MPa.
**6.34**: $V_y(2 - 2\sqrt{1-\varepsilon^2} + \varepsilon^2)/4\pi G a^2 t_0 \varepsilon\sqrt{1-\varepsilon^2}$.
**6.36**: $\theta/L = 0$.
**6.38**: $c/a = -(2 - 2\sqrt{1-\varepsilon^2} + \varepsilon^2)/2\varepsilon$.
**6.40**: The shear centre is at $O$.
**6.42**: $\tau_{max} = 0.13T/a^2t$, $K = 6.4a^3t$.
**6.44**: $K = 0.444at_0^3$.
**6.46**: 45.1°/m.
**6.48**: $\tau_{max} = 45.7$ MPa, $\theta/L = 15.2$°/m.
**6.50**: $\tau_{max} = 74$ MPa, $\theta/L = 17.7$°/m.
**6.52**: $\theta = 10.3°$.

## Chapter 7

**7.2**: $u(0) = -6.17$ mm, $\theta(0) = 0.88°$.
**7.4**: $F = -3564$ N, $|M|_{max} = 1.02$ kNm.
**7.6**: $u(0) = 2.58$ mm, $\sigma_{max} = 103$ MPa.
**7.8**: $u_{max}$ 26% less, $M_{max}$ 10% less
**7.10**: $\sigma_{max} = 42.9$ MPa, $|F_s|_{max} = 1.367$ N.
**7.12**: $u(z) = -w_0\sin(az)/(a^4EI + k)$, $M(z) = w_0a^2EI\sin(az)/(a^4EI + k)$.

**7.14**:

$$\begin{aligned} u(z) &= -\frac{F_0}{4Lk}[2 - f_1(\beta(z+L)) + f_1(\beta(L-z))] \;\; ; \;\; -L < z < L \\ &= -\frac{F_0}{4Lk}[f_1(\beta(z-L)) - f_1(\beta(z+L))] \;\; ; \;\; z > L \\ M(z) &= -\frac{F_0}{8\beta^2 L}[f_2(\beta(z+L)) + f_2(\beta(L-z))] \;\; ; \;\; -L < z < L \\ &= -\frac{F_0}{8\beta^2 L}[f_2(\beta(z+L)) - f_2(\beta(z-L))] \;\; ; \;\; z > L\,. \end{aligned}$$

**7.16**: $M(z) = w_0 f_3(|\beta z|)/8\beta^3$.
**7.18**: $M_{max} = 0.14 w_0/\beta^2$ at $\beta' z = -0.699$, where $\beta$ is calculated for the segment $z > 0$ and $\beta'$ for $z < 0$.
**7.20**: Yes.
**7.22**: –647 Nm at the center, 1303 Nm at the supports, 0.166 mm central deflection.
**7.24**: $u(0) = -0.0103 w_0/k$, $M(L/2) = 0.0833 w_0/\beta^2$.

## Chapter 8

**8.2**: A selection of answers:
$CC$, $R_1 = 300$ mm, $R_2 = 1025$ mm, $\sigma_1 = 51$ MPa, $\sigma_2 = -72$ MPa.
$DD$, $R_1 = \infty$, $R_2 = 924$ mm, $\sigma_1 = 46$ MPa, $\sigma_2 = 92$ MPa.
**8.4**: $\tau_{max} = 20.4$ ksi.
**8.6**:

$$\begin{aligned} \sigma_1 &= -\frac{\rho g R^2}{3t\sin^2\phi}\left(\frac{4}{27} - \cos^2\phi + \cos^3\phi\right)\,. \\ \sigma_2 &= \frac{\rho g R^2}{3t\sin^2\phi}\left(-\frac{50}{27} + 3\cos\phi + \cos^2\phi - 2\cos^3\phi\right)\,. \end{aligned}$$

**8.8**:

$$\sigma_1 = \frac{pa(2b - a\sin\phi)}{2t(b - a\sin\phi)} \;\; ; \;\; \sigma_2 = \frac{pa}{2t}\,.$$

**8.10**: $\sigma_1^{max} = -p_0 a/2t$, $\sigma_2^{max} = -p_0 a/t$, $|\sigma_1 - \sigma_2|_{max} = p_0 a/t$.
**8.12**: $t_{min} = 244$ mm.
**8.14**:

$$\begin{aligned} \sigma_1 &= -\frac{\rho g a(5 - 3\cos\phi)}{2(4 - 3\cos\phi)(1 + \cos\phi)} \\ \sigma_2 &= -\rho g a\cos\phi + \frac{\rho g a(5 - 3\cos\phi)}{2(4 - 3\cos\phi)(1 + \cos\phi)}\,. \end{aligned}$$

**8.16**: Include only internal pressure. $\sigma_{max} = 425$ kPa.
**8.18**: $u_r = -\rho_s g h^2 \tan\alpha[(2 - \nu)\tan^2\alpha - \nu]/2E$.
**8.20**: $u_r = 0.405$ mm at $A_1$ and $0.468$ mm at $A_2$.
**8.22**: At the equator, $u_r = \rho\Omega^2 a^3/E$.

## Chapter 9

**9.2**: 19.6 ksi.
**9.4**: $\sigma_{max} = 215$ MPa.
**9.6**: $\alpha > F_0\sqrt{3(1-\nu^2)}/\pi f E t^2$.
**9.8**: $R = 0.288a/\sqrt{(1-\nu^2)}$ (independent of $p, t_0$).
**9.10**: $M_z(0) = p/16\beta^3 b$.
**9.12**: $F_0 = p_0 a \tan\alpha/2$, $A = (18t/\beta) - (10at\tan\alpha/(2-\nu))$.
**9.14**: $M_0 = E\theta I_\eta/a^2$.
**9.16**: $\sigma_{max} = 22.86$ MPa.

## Chapter 10

**10.2**: 632 MPa.
**10.4**: $-1875$ psi; $0.86\times10^{-3}$ in.
**10.6**: $S.F. = 7.9$.
**10.8**: $p_Y = (5-8\nu)S_Y/4(1-2\nu)$.
**10.12**: $p = E\delta(n^2-1)/2n^2a$, $\sigma_{max} = E\delta(n^2+1)/2n^2a$.
**10.14**: 9900 psi.
**10.16**: $0.090S_Y$.
**10.18**: $p_Y = 2S_Y(\sqrt{(b/a)}-1)$ ; $\sigma_{rr} = -2S_Y(\sqrt{(b/r)}-1)$ ; $\sigma_{\theta\theta} = S_Y(2-\sqrt{(b/r)})$.
**10.20**: 11,543 rpm.
**10.22**: $c_{max} = 3.66$ in.
**10.24**: $p(r) = p_0 + \rho\Omega^2(r^2 - a^2/2)$.
**10.26**:

$$\sigma_{rr} = -S_Y\ln(r/a) + \frac{S_Y\ln(n)}{(1-1/n^2)}\left(1-\frac{a^2}{r^2}\right)$$

$$\sigma_{\theta\theta} = -S_Y\ln(r/a) + \frac{S_Y\ln(n)}{(1-1/n^2)}\left(1+\frac{a^2}{r^2}\right) .$$

$$n_{max} = 2.218 .$$

## Chapter 11

**11.2**: 229 Nm.
**11.4**: 19.4 ksi.
**11.8**: 6.35 ksi.
**11.10**: 36.1 MPa (compressive), 27.0 MPa (tensile).
**11.12**:

$$\sigma_{\theta\theta}^{max} = \frac{M}{a^3}\frac{(1-n\ln(1+1/n))}{\left[\left(n+\frac{1}{2}\right)\ln(1+1/n)-1\right]}$$

$$\sigma_{rr}^{max} = \frac{M(\exp[n\ln(1+1/n)-1]-1)}{a^3\left[\left(n+\frac{1}{2}\right)\ln(1+1/n)-1\right]} .$$

**11.14**: 10.7 ksi.

## Chapter 12

**12.2**: $\lambda L = 3.44$.
**12.4**: $kL^3/3$.
**12.6**: $M_{max} = 0.428 w_0 L^2$.
**12.8**: $M_{max} = (F/\lambda)\tanh(\lambda L)$.
**12.10**: $20.2EI/L^2$.
**12.12**: $k = 3.382EI/L$.
**12.14**: $P_0 = 6.48 \times 10^6$ N, $l_0 = 3.58$ m,
**12.16**: $P_0 = 165$ lbs, $l_0 = 0.17$ in.
**12.18**: 438 rpm.
**12.20**: 753 rpm.
**12.22**: 66 rpm.
**12.24**: The approximation is almost indistinguishable from the exact result.
**12.26**:

$$P_0 = \frac{3EI}{L^2} + \frac{3kL}{4}.$$

**12.28**: 5.93 m.
**12.30**:

$$P_0 = \frac{Ec^4}{4L^2}.$$

**12.32**: $P_0 = kL/2$.
**12.34**: $4\pi^2 EI/L^2$.
**12.36**: $\pi^2 EI/L^2$.
**12.38**: $\pi^2 EI/L^2$.

## APPENDIX A

**A.2**:

$$\{C_1, C_2, C_3\} = \left\{-\frac{1}{24}, \frac{5}{24}, \frac{23}{24}\right\}.$$

**A.4**: $C_1 = 0.91941$, $C_2 = 2.49937$.
**A.6**: 2.592 mm.
**A.8**: $-p_0 L/3$.
**A.10**: $P_0 = 16.47EI/L^2$.
**A.12**: $u_2 = -FL^3/3EI$, $\theta_2 = -FL^2/2EI$.
**A.14**:

$$F_i = -\frac{F(\Delta - a)^2(\Delta - 2a)}{\Delta^3};\ F_{i+1} = -\frac{Fa^2(3\Delta - 2a)}{\Delta^3};$$

$$M_i = -\frac{Fa(\Delta - a)^2}{\Delta^2};\ M_{i+1} = \frac{F(\Delta - a)a^2}{\Delta^2};u^*(z) = -\frac{Fz^2(27L - 10z)}{81EI}.$$

**A.16**: $u_2 = -w_0 L^4/128EI$.
**A.18**: $5w_0 L/4$.

# Index